D0145845

Experimental Biochemistry

Experimental Biochemistry

Robert L. Switzer

UNIVERSITY OF ILLINOIS

AT URBANA-CHAMPAIGN

Liam F. Garrity

PIERCE CHEMICAL COMPANY

W. H. Freeman and Company

New York

Publisher: Michelle Russel Julet
Cover Designer: Brian Switzer
Production Coordinator: Susan Wein
Manufacturing: Hamilton Printing Company
Text Design and Composition: York Production Services
Executive Marketing Manager: John Britch

Library of Congress Cataloging-in-Publication Data

Switzer, Robert L.
Experimental biochemistry / Robert L. Switzer and Liam F. Garrity.
—3rd ed.
p. cm.
Includes bibliographical references and index.
ISBN 0-7167-3300-5
1. Biochemistry—Laboratory manuals. I. Garrity, Liam F.
II. Title.
QP519.S95 1999 99-12268
572—dc21 CIP

Printed in the United States of America

First printing 1999

W. H. Freeman and Company
41 Madison Avenue, New York, NY 10010
Houndmills, Basingstoke RG21 6XS, England

Contents

Preface

"Progress in science is made by observation and experiment. This is particularly evident in biochemistry, where the remarkable advances of the past two decades have been made possible largely by the introduction of new methods and techniques." These words, which began the Preface of the Second Edition of *Experimental Biochemistry* (1977), by John M. Clark, Jr., and Robert L. Switzer, are equally appropriate as we complete the Third Edition.

Goals of the Third Edition. Although the foundations laid by twentieth century biochemists have proven to be quite solid, the more than twenty years since the previous edition was published have seen so many advances in both the fundamental understanding and the technology of biochemical research that extensive and continuous revision of the content of general biochemistry laboratory instruction is imperative. This edition has grown out of our efforts, and the efforts of our colleagues at the University of Illinois at Urbana-Champaign, to introduce useful new experimental techniques, tools, and materials into our own biochemistry laboratory instruction. The new and revised experimental exercises in this edition were developed in the Department of Biochemistry at the University of Illinois and have been extensively tested in typical instructional settings in our biochemistry laboratory course for advanced undergraduates and beginning graduate students. We are confident that the new edition can be used in a wide variety of settings for instruction in the most important techniques of modern biochemistry. Students without previous biochemical laboratory experience can develop mastery of the theory and practice necessary for advanced laboratory work or for their first experiences in biochemical research. Such education in biochemical techniques is even more important today than in the past. Developments in modern chemistry and biology have shown that biochemistry must today be regarded as a "root science." As science moves toward more integrated forms of investigation, virtually every student of biology and chemistry requires substantial formal training in modern biochemical laboratory techniques to be adequately prepared for a scientific career in the twenty-first century. It is our hope that *Experimental Biochemistry* Third Edition will provide a valuable tool for educators seeking to meet that goal.

Features of the Third Edition. As was true of previous editions of *Experimental Biochemistry*, this book attempts to provide a broad range of hands-on experiences with the most important and commonly used techniques of contemporary biochemistry. We have included exposure to all the major classes of biological molecules and topics of current biochemical and molecular biological research: proteins and enzymes, nucleic acids, lipids and membranes, carbohydrates, as well as metabolic systems, regulation, and the newly emerging field of biochemical information science. In no area of research have advances since 1977 been more extensive and revolutionary than in molecular biology. The section on nucleic acids and molecular biology has therefore been extensively revised. This book cannot be regarded as a comprehensive laboratory text for a molecular biology or molecular genetics course, however. This specialty is well served by a number of other texts.

Traditional Strengths Were Retained. In preparing this revision, we have attempted to strike a balance between experiments using entirely new techniques and materials and revisions of more classical experiments from previous editions of *Experimental Biochemistry* that we find valuable for teaching and strengthening the fundamental principles of the discipline. Thus, experiments such as those in Section I, Experiments 5, 6, 7, 8, 12, 14, 19, and 20 present experimental exercises in "classical" biochemistry, which we believe still form an important part of a thorough training. These experiments have all been significantly revised to incorporate improvements in methods or materials, where applicable, and to improve clarity of presentation. Other experiments, which we felt were less useful or were outdated, have been omitted from the Third Edition.

Extensive New Material Has Been Incorporated. Many sections and experiments are entirely new in the Third Edition. These include Experiments 10, 15, and 16, an entirely new section on Immunochemical Techniques, which includes Experiments 17 and 18, Experiments 21, 22, 23, and 24 in the Nucleic Acids and Molecular Biology section, as well as Section VI on biochemical information science. All of these new experiments have been performed many times in our own laboratory classes. We are confident that they will provide reliable and stimulating educational experiences in all instructional settings.

Use of Radioisotopes in Experimental Biochemistry. Some reviewers have expressed concerns about the use of radioisotopes in some of the experiments in this text. Although it is true that alternative techniques are being developed that lessen the dependence of biochemical research on the use of radioisotopes, we firmly believe that radioisotopes will continue to be important in the foreseeable future, and that an adequately trained biochemist must be aware of these techniques and the power of radioisotopic techniques in biochemical research. The experiments in this text are designed to minimize the levels and exposure of students and staff to radioisotopes. They are completely safe for students and present no contamination hazards if the students are adequately supervised. Handling of significant amounts of radioisotopes is restricted to the teaching staff in the preparation of the experiments. In the same vein, we have attempted to identify clearly and forcefully in the text all instances where good laboratory practice requires special precautions and safety considerations.

Flexible Organization and Uses of the Contents. This book is intended as a text and laboratory manual for a one-semester laboratory course, but it is not realistic for the students in such a course to complete all 25 experiments. Rather, we expect that instructors will choose those portions of the text that suit their schedules, interests, and laboratory facilities. Some further thoughts about optimal and individualized uses of *Experimental Biochemistry* in the teaching laboratory are presented in the section entitled "To the Instructor" in the introductory chapter to Section I. Each of the six Sections is preceded by an extensive Introductory Chapter that reviews relevant background and fundamental biochemistry of a particular class of biological molecules, with a strong emphasis on experimental methodology. It has been our experience with previous Editions of *Experimental Biochemistry* that many instructors and students find these introductory chapters valuable even when they are performing lab exercises other than those in the text. Also, many students continue to use *Experimental Biochemistry* as a reference text long after completing their formal lab course. We certainly hope that the new edition will be even more valuable in these auxiliary uses.

Acknowledgments. The preparation of this laboratory manual was not solely the product of the labor of the two authors. We owe a great debt of gratitude to many colleagues and teaching assistants at the University of Illinois, who over the past twenty years have contributed ideas, techniques, and experience to the repeated revisions of our laboratory exercises. First, we acknowledge our former colleague, Professor John M. Clark, Jr., who retired from the University of Illinois in 1990; his many important contributions as author, and co-author, respectively, of the First and Second Editions of *Experimental Biochemistry* continue to be found in many forms throughout the Third Edition. In addition, we gratefully acknowledge Professor David Kranz and Dr. Michael Saulmon for their major contributions to Section IV (Immunochemical Techniques) and Section VI (Information Science), respectively. Dr. Charles Matz contributed numerous improvements and refinements during the period when he served as an instructor in Illinois' biochemistry laboratory course. We thank Professor Peter Imrey for preparing the section on statistics in the Introductory Chapter to Section I, Professor JoAnn Wise for her contributions to experiments in molecular biology, Dr. Kenneth Harlow for his assistance in developing the application of HPLC to amino acid analysis, Professor Evan Kantrowitz of Boston College for contributions to the modernization of Experiment 9, Dr. Charles Matz and Daryl Meling for assistance in development of Experiment 24, Christopher Wall for experimental testing of revised form of Experiment 9, and Kim Hines of the Pierce Chemical Co. for her expertise and contributions of reagents that were important in developing the chemiluminescent assay of Western blots. We are grateful to Professor Peter Orlean and Professor Michael Glaser for critical readings of the Introductory Chapter to Section III (Biomolecules and Biological Systems). Finally, we express our appreciation to our colleagues on other campuses who reviewed portions of our text prior to publication: Frank N. Chang, Temple University; Harry L. Price, University of Central Florida; Mike Panigot, Arkansas State University; Theodore Chase Jr., Cook College, Rutgers University; Larry Byers, Tulane University; Larry G. Ferren, Olivet Nazarene University; James H. Hageman, New Mexico State University; Nat Ramani, SUNY at Westbury; Joe Provost, Moorhead State University; and Robert Lindquist, San Francisco State University.

The cover design for the third edition of *Experimental Biochemistry* is by Brian Switzer; Meta Design, London, UK.

About the Authors

Robert L. Switzer is Professor of Biochemistry at the University of Illinois at Urbana-Champaign, where he has been on the faculty since 1968. He holds a Ph.D. in Biochemistry from the University of California at Berkeley and was a postdoctoral fellow in the Laboratory of Biochemistry of the National Heart, Lung, and Blood Institute, National Institutes of Health, Bethesda, Maryland, from 1966 to 1968. His research centers on the regulation of enzyme activity and the control of gene expression by novel mechanisms in bacteria. He was a Guggenheim Fellow in 1975. Professor Switzer has had a career-long interest in biochemistry laboratory instruction and has twice received the School of Chemical Sciences Excellence in Teaching Award. He was a co-author with John M. Clark, Jr., of the Second Edition of *Experimental Biochemistry*.

Liam F. Garrity received his B.S. degree in Biochemistry from the University of Wisconsin-Madison in 1991 and his Ph.D. in Biochemistry from the University of Illinois at Urbana-Champaign in 1996. His graduate work focused on the regulation of kinase activity and the phosphoryl group transfer cascade governing chemotaxis in *Bacillus subtilis*. From 1996 to 1998 he served as a Visiting Lecturer in the Department of Biochemistry at the University of Illinois at Urbana-Champaign, where he supervised the biochemistry laboratory and lecture course for undergraduates and beginning graduate students. He received the Excellence in Teaching Award from the School of Chemical Sciences for his instruction in the general chemistry program while serving as a graduate student. Dr. Garrity is now with Pierce Chemical Company in Rockford, Illinois, which manufactures products for protein chemistry, immunochemistry, and molecular biology.

SECTION I

Basic Techniques of Experimental Biochemistry

SECTION I

Basic Techniques of Experimental Biochemistry

Introduction

Biochemistry is the chemistry of biological systems. Biological systems are complex, potentially involving a variety of organisms, tissues, cell types, subcellular organelles, and specific types of molecules. Consequently, biochemists must separate and simplify these systems to define and interpret the biochemical process under study. For example, biochemical studies on tissue slices or whole organisms are followed by studies on cellular systems. Populations of cells are disrupted, separated, and their subcellular organelles are studied. Biological molecules are studied in terms of their specific mechanisms of action. By dividing the system under study and elucidating the action of its component parts, it is possible to then define the function of a particular biological molecule or system with respect to the cell, tissue, and/or organism as a whole.

Biochemical approaches to the simplification and understanding of biological systems require two types of background. First, biochemists must be thoroughly skilled in the basic principles and techniques of chemistry, such as stoichiometry, photometry, organic chemistry, oxidation and reduction, chromatography, and kinetics. Second, biochemists must be familiar with the theories and principles of a wide variety of biological and physical disciplines often used in biochemical studies, such as genetics, radioisotope tracing, bacteriology, and electronics. This need reflects the biochemists' ready acceptance and use of theories and techniques from allied areas and disciplines.

It is not possible or appropriate for this book to summarize the many disciplines and principles used in biochemistry. However, students will find that a review of the basic principles and units used in the quantitative aspects of experimental biochemistry is quite useful. This section is intended to provide such a review. In addition, it is valuable for students to understand the methods often used in experimental biochemistry. Experiments 1 to 4 of this section deal specifically with these techniques: spectrophotometry, chromatography, radioisotope tracing, and electrophoresis. Finally, it is imperative that students understand the intricacies of data analysis. The final part of this Introduction discusses the principles underlying the basics of statistical analysis that are critical to the ability to determine the precision or error associated with quantitative data obtained in biochemical experiments.

Requirements for a Student of Experimental Biochemistry

This course is aimed at developing your interest in and understanding of modern biochemical and molecular biological experimentation. This goal necessitates a careful emphasis on the experimental design, necessary controls, and successful completion of a wide variety of experiments. This goal will require additional efforts if you are to benefit fully from *Experimental Biochemistry.* First, you should familiarize yourself with general background material concerning each experiment. Three elements have been incorporated into the text to aid you in this effort: (1) Each experiment is preceded by a short introduction designed to aid you in understanding the various theories and techniques underlying the exercises. (2) The experiments are divided into sections that deal with a specific class of biological molecules. The introduction preceding each section will serve as a review to provide you with enough information to understand the experiments. This material is intended to reinforce and supplement the knowledge you have gained from biochemistry lecture courses and textbooks of general biochemistry, which you should review as needed. (3) Each experiment is followed by a set of exercises and related references that will allow you to further develop your interest and understanding of a particular method, technique, or topic.

Second, you must keep in mind that the ability to complete the experiments within allocated times requires you to be familiar with the protocol of the experiment *before the start of the laboratory session.* Each of the experiments contains a detailed, class tested, step-by-step protocol that will enable you to perform, analyze, and interpret the experiments on your own. Your success will depend on your ability to organize and understand the experimental procedures, making efficient use of your time.

Third, efficient use of *Experimental Biochemistry* requires that you perform and interpret many calculations *during the course of the laboratory sessions.* Specifically, laboratory work for introductory biochemistry, unlike many introductory laboratory courses, frequently requires you to use the results of one assay to prepare and perform additional assays. Thus, you will have to understand fully what you are doing at each step and why you are doing it.

Finally, it is *imperative* that you maintain a complete research notebook containing all your data, calculations, graphs, tables, results, and conclusions. Your notebook should be so clear and complete that *anyone* can quickly understand what was done and what results were obtained. Your instructor may provide additional specific instructions for your laboratory reports; the following suggestions may be helpful:

1. Use a large, bound notebook, preferably one with gridded pages. Such notebooks permit the direct construction of data tables and allow you to attach records of primary accessory data, such as computer-derived graphs, chromatograms, dried SDS-PAGE gels, and photographs.
2. Never record your data on separate sheets of paper. Rather, record all your data directly in your notebook. You may consider using one side of the notebook for raw data and calculations and the other side for results and interpretation.
3. All graphs and tables must be clearly and unambiguously labeled. Be particularly careful to specify units on the ordinate (y-axis) and abscissa (x-axis) of every graph.
4. The laboratory report for all experiments should include:

   a. a brief statement of purpose (why you are doing the experiment and what you wish to determine);
   b. a brief account of the theory and design of the experiment, including a summary or flow chart of the principal manipulative steps;
   c. the raw data;
   d. all calculations (if analysis requires a single, repetitive calculation, a sample calculation for one of a series is acceptable);
   e. results;
   f. conclusions and interpretations (the information that you can derive from the results of the experiment).

As stated earlier, all the experiments in this textbook have been class tested by hundreds of students. The experiments, therefore, show a high rate of success. Still, there may be times when your experimental results are not particularly useful, or when they yield unexpected results that require an explanation. If this is the case for a particular experiment, discuss in the results section of your laboratory report what may have gone wrong. Did you make an improper dilution of one of the reagents? Did you accidentally omit one of the experimental steps? In the conclusion section, discuss what you may have expected to see if the experiment had been successful. Your knowledge of the theory underlying the techniques, along with your understanding of the experimental protocol, should be sufficient to allow you to determine what type of data you may have obtained under ideal conditions. By doing this, you are likely to turn what appears to be a failed experiment into a valuable learning opportunity. It is never sufficient to say, "the experiment did not work." Attempt to understand why a particular experiment may not have worked as expected.

Laboratory Safety

Experimental Biochemistry employs the use of potentially hazardous reagents. Strong acids, strong bases, volatile compounds, flammable compounds, mutagenic compounds, corrosive compounds, radioisotopes, electricity, and sharp objects are the tools of the biochemist. Like any other tool, these are hazardous only when handled improperly. At the beginning of each experimental protocol, we draw your attention to potential hazards that may be associated with a particular reagent you are about to use.

Safety goggles/glasses must be worn in the laboratory at all times. The main purpose of eye protection is to prevent chemical damage to the eye. Laboratory eye protection should also be shatterproof to protect against debris that would be produced from broken glass in the event of an accident. Although you may feel confident that you will not be the cause of such an accident, it is impossible to ensure that your laboratory partner or neighboring groups will not have accidents.

It is advised that you wear latex or vinyl exam gloves at all times in the laboratory. Even if a particular experiment does not require the use of hazardous chemicals, one can never be sure that those from a previous experiment have been properly disposed of. If volatile compounds are used, they should be stored under a fume hood at all times. If possible, students should work with these materials under the fume hood as well. The large amounts of materials that are often required for a laboratory group may soon fill the room with unpleasant and potentially hazardous vapors. This is particularily important if the reagent vapors are flammable (see Experiment 6) or radioactive (see Experiment 12).

Laboratory coats may be worn if desired. It is a good idea to wear them when working with radioisotopes, since very small quantities of a radioactive solution can carry a significant amount of activity. It is also a good idea to wash your hands thoroughly with soap before leaving the laboratory to ensure that you do not take any chemicals outside the laboratory. When working with radioisotopes such as ^{32}P, it is necessary to check your hands and shoes with a Geiger counter before leaving the laboratory.

The laboratory will be equipped with safety showers, eyewash stations, emergency exits, sharps containers, and fire extinguishers. Take the time to become familiar with the location of all of these safety components. All "sharps" (razor blades, Pasteur pipettes, broken glass, etc.) should be placed in the labeled "sharps" containers. Your laboratory supervisor will instruct you on the proper use and disposal of all hazardous reagents. If you do become injured or have any questions about your health risk during the course of the experiment, *immediately* notify the laboratory instructor. Most laboratory supervisors have had training in dealing with fires and exposure to different chemicals. Have fun with the experiments, be safe, and always leave a clean laboratory workbench for the beginning of the next laboratory session.

Units of Biochemistry

Biochemistry employs a decade system of units based on the metric system. Thus, biochemists use units such as the mole or the liter and various subdivisions that differ by three orders of magnitude (Table I-1). With knowledge of the molecular weight of a particular molecule and equation I-1, a given mass of a molecule can be converted to units of moles:

(I-1)

$$\text{Number of moles} = \frac{\text{number of grams of molecule}}{\text{molecular weight of molecule}}$$

As indicated in Table I-1, grams may be converted to milligrams and moles can be converted to millimoles simply by multiplying each of the appropriate values by 10^3. For example, 0.025 mol of a molecule is equal to 25 mmol:

$$0.025 \text{ mol} \times \frac{10^3 \text{ mmol}}{1 \text{ mol}} = 25 \text{ mmol}$$

Volume and mole values define the *concentration* terms of molar (M), millimolar (mM), and micromolar (μM) as shown in equation I-2:

Table I-1 Basic Units Used in Biochemistry

Mole Units	Liter Units
1 mole	1 liter
1 millimole (mmol) = 10^{-3} moles	1 milliliter (ml) = 10^{-3} liter
1 micromole (μmol) = 10^{-6} moles	1 microliter (μl) = 10^{-6} liter
1 nanomole (nmol) = 10^{-9} moles	1 nanoliter (nl) = 10^{-9} liter
1 picomole (pmol) = 10^{-12} moles	
Gram Units	**Equivalent Units**
1 gram	1 equivalent (Eq)
1 milligram (mg) = 10^{-3} g	1 milliequivalent (mEq) = 10^{-3} Eq
1 microgram (μg) = 10^{-6} g	1 microequivalent (μEq) = 10^{-6} Eq
1 nanogram (ng) = 10^{-9} g	
1 picogram (pg) = 10^{-12} g	
1 femtogram (fg) = 10^{-15} g	

(I-2)

$$\text{Concentration (molar)} = \frac{\text{number of moles}}{\text{volume (in liters)}}$$

$$\text{Concentration (millimolar)} = \frac{\text{number of millimoles}}{\text{volume (in liters)}} = \frac{\text{number of micromoles}}{\text{volume (in milliliters)}}$$

$$\text{Concentration (micromolar)} = \frac{\text{number of micromoles}}{\text{volume (in liters)}} = \frac{\text{number of nanomoles}}{\text{volume (in milliliters)}}$$

Similarly, volume and equivalent values define the *concentration* term of normality (N) commonly used in the expression for acid (H^+) or base (OH^-) strength, as indicated by equation I-3:

(I-3)

$$\text{Concentration (normal)} = \frac{\text{number of equivalents}}{\text{volume (in liters)}} = \frac{\text{number of milliequivalents}}{\text{volume (in milliliters)}}$$

Because these units involve basic metric principles, one can make use of the metric interconversions of mass (grams), fluid volumes (liters or milliliters), and spatial volumes (cubic centimeters, cc). Specifically, under most laboratory conditions, 1 ml of water or dilute aqueous solution weighs approximately 1 g and occupies 1 cc of volume (1 ml = 1.000027 cc).

These simple interrelationships of moles, weights, volumes, and so forth are often covered in introductory or freshman level college chemistry textbooks. Yet, practical experience reveals that these basic concepts are a major source of difficulty for many students in their initial exposure to experimental biochemistry. Therefore, we strongly suggest that students thoroughly review these concepts before conducting the experiments described in this textbook.

Analysis and Interpretation of Experimental Data

In nearly all of the experiments outlined in this textbook, you will be asked to collect, analyze, and interpret experimental data. Whether you are determining the concentration of a molecule in an unknown solution, the activity of an enzyme, the absorbance of a solution at a particular wavelength, or the activity of a particular isotope in a biological sample, the exercise will require you to perform a quantitative measurement and calculate a specific value. There are several questions that frequently arise during the analysis of experimental data: How do you determine the level of precision of a set of measurements? How many data values or trials of an experiment must you perform before a measurement can be deemed precise? If you have a single value in a data set that is not in agreement with other members of the set, how do you determine whether it is statistically acceptable to ignore the aberrant value? In the subsections below, each of these issues is addressed.

Accuracy, Precision, and Bias of a Quantitative Measurement

When interpreting laboratory data, it is important to recognize that individual measurements, such as the concentration of a biological molecule observed in an assay, are never entirely accurate. For instance, the serum cholesterol measured by a medical laboratory from a blood sample is not the exact average serum cholesterol in the patient's blood at the time the sample was drawn. There are a number of reasons for this, the most obvious being that cholesterol may not have been quite uniformly mixed throughout the bloodstream. The patient's blood and the sample drawn from it are never totally homogeneous, the reagents used in the test are never totally pure or totally stable if repeatedly used, and the calibration of the autoanalyzer is never exactly correct or totally stable. Even such small deviations from the ideal execution of the assay may sometimes noticeably affect the results, and additional undetected errors in execution sometimes produce substantial errors. For these reasons, carrying out the same experiment more than once, or even repeatedly assaying the same sample, is bound to produce somewhat different numerical results each time.

Now, although the quantity being measured is a property of the particular sample under study, the degree of expected fluctuation from one measurement to another depends most fundamentally on the measurement process itself—that is, how the assay is conducted—rather than on the particular sample. Since, depending on the circumstances, the amount of fluctuation among attempts to measure the same quantity may be trivial or crucially important, we now consider briefly some basic concepts that help the biochemist deal with variability among measurements.

In performing an assay, the biochemist aims for *accuracy*. An assay method is accurate when the chance is high that its result will be quite close to the true value being measured. Since individual assay results invariably fluctuate, an accurate assay method must be (1) highly *precise* (equivalently, reproducible), having little variability when repeated, and (2) nearly *unbiased*, meaning that almost all of the time the average result from a large number of repeated assays of the same sample must be very close to correct. Conversely, an assay method can be poor because it is imprecise, biased, or both. For instance, a highly reproducible assay based on a very poorly calibrated instrument may yield almost the same, but grossly incorrect result, every time it is applied to a given sample. Another assay may be unbiased but also never close to correct, because its frequently large overestimates are balanced out by equally large and frequent underestimates.

The above concepts become more precise when expressed mathematically. Let the Greek letter μ represent the true characteristic of a sample that we are trying to measure, and suppose the n observations x_i, $i = 1, \ldots, n$, represent the results of entirely separate executions of an assay procedure. Then $(x_i - \mu)$ is the error of the ith assay. If we square these errors and take their arithmetic mean, we obtain the *mean squared error* $= (1/n)\Sigma(x_i - \mu)^2$ of the group of n repetitions. If we could repeat the assay an extremely large number of times, so that n is very large, the average result $\bar{x} = (1/n)\Sigma x_i$ would eventually stabilize at a limiting value $\bar{X}$, the "long-run" average value of the assay for the given sample. The mean squared error would similarly stabilize at a value, MSE, that can be used as an index of the assay's inherent accuracy. In principle, a perfect assay method has MSE = 0 for all samples, but no such assay exists. The bias of the assay is $(\bar{X} - \mu)$ and its square, $(\bar{X} - \mu)^2$, is a component of the MSE. The difference MSE $-$ Bias^2 = MSE $-$ $(\bar{X} - \mu)^2 = \sigma^2$ is a measure of inherent variability of the assay, known as its *variance*. While we can never determine the variance of an assay exactly, because that would require performing the assay an impossibly large number of times, we can estimate it by

$$s^2 = \frac{1}{n-1} \sum (x_i - \bar{x})^2$$

known technically as the *sample variance*. By taking its square root,

$$s = \sqrt{\frac{1}{n-1}\sum (x_i - \overline{x})^2}$$

we obtain a measure of variability in the same units as the individual assay results. This measure of fluctuation among repeated assays is known as the *sample standard deviation*, abbreviated SD. The mean, $\overline{x}$, and standard deviation, SD, together represent a compact summary of a group of repeated assays. For example, if the absorbance values of three identically prepared solutions at a particular wavelength are determined to be 0.50, 0.44, and 0.32, then their mean value is $\overline{x} = 0.42$ and their SD is

$$s = \sqrt{\frac{0.08^2 + 0.02^2 + 0.10^2}{2}} = \sqrt{0.0084} = 0.09$$

It is common to report such results as $\overline{x} \pm$ SD, for example, 0.42 ± 0.09, although this notation can lead to some unfortunate confusion, as we shall see below.

Precision of a Replicated Assay

Intuitively, it seems that one way to improve the accuracy of an assay is to do it more than once and take the average of the repetitions as your result. In principle, random overestimation and underestimation by the individual results will cancel one another out in the average, leaving a more accurate result than can be obtained from only one measurement. This is true if the assay has been properly calibrated, so that its bias is very low. In that case, the accuracy of the assay depends almost entirely on its precision, which is represented by the SD, with a precise assay having a small SD. Using the SD, we may determine how precision improves with each additional repetition. Suppose we replace the procedure of a single assay A with standard deviation s by a new procedure, which involves n repetitions of A, with only the average result reported. Call this "improved" assay A_n. Thus, each single result of the assay A_n is an average of n results of A. Though it is beyond the scope of this book to demonstrate, it may be shown that the standard deviation of the assay A_n is just $s/\sqrt{n}$. Thus, the improvement in precision obtained by replicating an assay n times is a reduction in the SD by a factor of $[1 - (1/\sqrt{n})]$, meaning that two repetitions yield about a 29% reduction in SD, three repetitions a 42% reduction, and four repetitions a 50% reduction. However, the reduction obtained by A_{n+1} relative to that achieved by A_n is $[1 - (\sqrt{n}/\sqrt{n+1})]$, or 18%, for three versus two repetitions, 13% for four versus three, and 11% for five versus four. Thus, the relative benefit of each additional repetition declines with n, and the comparison of the benefit to the cost of an additional repetition generally becomes less favorable to additional repetitions as n increases. Nevertheless, in principle any desired level of precision may be achieved by using enough repetitions, although the extra repetitions will not change the bias of the assay.

The quantity s^2/n, which represents the variability among arithmetic means (i.e., simple averages) of n repetitions, is known as the *standard error of the mean*, and is abbreviated either as SEM or SE. Results of assays involving n repetitions, such as the absorbance results reported earlier as $\overline{x} \pm \text{SD} = 0.42 \pm 0.09$ are also frequently reported as $\overline{x} \pm$ SE, which in this case is $0.42 \pm (0.09/\sqrt{3}) = 0.42 \pm 0.05$. Confusion can arise if either this $\overline{x} \pm$ SE or the $\overline{x} \pm$ SD format is used without specific indication of whether it is SD or SE that follows the reported mean. The latter is preferable whenever the purpose is to represent the precision of the assay's summary result rather than the variability of the individual measurements that have contributed to it. This is almost always the case with chemical analyses, and we recommend the $\overline{x} \pm$ SE notation, supplemented by the number of replicates, n, for general use. Thus, in the example, we would report an absorbance of 0.42 ± 0.05 (SE), from three replications.

When faced with a choice between two assays you may compare either their SDs, or their standard errors for any fixed n, to determine which assay is most useful. The more precise assay is generally preferred if biases are similar, assuming costs and other practical considerations are comparable as well. In such circumstances, an assay with SD = $s = 0.09$ is much preferable to one with $s = 0.36$; similarly, a triplicate assay procedure with SE = 0.02 is greatly preferable to another triplicate procedure with SE = 0.07.

For an unbiased assay the SE may also often be used to form a range, centered on the reported assay result, that has a known probability of containing the true value μ for the sample being studied. This range, known as a confidence interval, summarizes the information that the assay provides about the sample in a manner that incorporates the underlying fuzziness of our knowledge due to random variability in the assay process. For instance, $\bar{x} \pm 4.31 \times$ SE gives a 95% confidence interval for the true value estimated by a triplicate assay, and $\bar{x} \pm 3.19 \times$ SE gives a 95% confidence interval for an assay in quadruplicate. For the triplicate absorbance data, we have $0.42 \pm 4.31 \times 0.05 = 0.42 \pm 0.22$, or 0.20 to 0.64. In the long-run, 19 of 20 (95%) of such ranges obtained from triplicate assays using the given method will include the true absorbance of the sample, though we cannot say exactly where within the interval that true value lies. However, for 1 in 20 assays (5%), the true absorbance will be outside the interval. For a higher confidence such intervals may be widened, while intervals obtained using a smaller multiplier will exclude the true concentration more than 5% of the times they are employed. The appropriate multiplier of the SEM depends on both n and the desired degree of confidence, here 95%. When the SD of an assay is known to good accuracy (e.g. reported by an instrument manufacturer based on considerable data on the performance of the instrument over the range of likely values) confidence intervals may be constructed using reported rather than observed variability. These intervals require smaller multipliers, and hence will tend to be narrower, than those based only on the observed replications for an individual assay. The choice of multiplier is beyond the scope of this book.

A consequence of the above ideas is a rule of thumb that to obtain adequate precision with some assurance that gross error has been avoided, at reasonable cost, you will often be well served to perform three or four trials of a single experiment. The experiments outlined in this textbook rarely call for such replication, due to time and financial constraints of the educational process. *Remember, however, that a properly designed experiment should be performed multiple times, and that the data should be presented with a well-defined statistical analysis to allow the reader to ascertain the precision of the experiment.*

Outlying Data Values

Suppose that you perform an identical assay four times and obtain the following values: 0.47, 0.53, 1.53, and 0.45. Within this small data set, the value of 1.53 stands out as apparently different. If all four of these values are included in the reported assay result, we report 0.75 ± 0.26 (SE) from four replicates. This measurement does not appear particularly precise, since the error associated with the measurement is about 35% of its value. By ignoring the apparently aberrant value of 1.53, however, you may report 0.48 ± 0.02. It is tempting to assume that the 1.53 must have resulted from a mistake, and that the latter result is likely more accurate and far more precise than the former. Unfortunately, in many such instances this is fallacious, and deleting the apparently outlying number results in a less accurate assay result with misrepresentation of its precision. The problem results from our intuitive inclination to regard the 1.53 as probably wrong because it is so deviant from the other three values, when an alternative plausible explanation is that any of the other three values is somewhat, but not unusually, lower than might be expected. When three or four numbers are obtained from a situation partially governed by chance, it is not at all unusual for one of them to depart further from the others than intuition might suggest. When such numbers are deleted, the resulting report may be very misleading.

Yet there are circumstances in which single gross errors contaminate otherwise valid assays, and it is desirable to be able to delete such errors and use the remaining repetitions of the assay. How can we tell when to delete such outlying values, and when it is necessary to retain them and accept the possibility that our assay is less precise than we had hoped?

The first and best approach should be to examine the assay procedure that was used to try to find an explanation (e.g., a technical error or even a data recording error) for the outlying value. If a substantial technical error is found, then the outlying value may be discarded, and if possible the assay should be run again to replace it. If no such explanation is found, however, we still may wish to discard grossly aberrant values rather than repeat the entire set of n repetitions of the assay. One approach is to assume a particular mathematical model that describes, in terms of probability, the amounts by which repetitions of the same assay vary from one another, and then to discard an outlying value only when it seems quite discrepant from what that model predicts. The model commonly used is the gaussian (or "normal") probability distribution, which is a reasonable approximation of the behavior of random fluctuations for many assays (though by no means all). This model suggests the following procedure:

1. Calculate the mean and SD for all observations other than the outlying value.
2. Subtract the calculated mean from the outlying value, and divide by the calculated SD. For the data above, this yields (1.530 − 0.483)/0.024 = 43.6.
3. Discard the outlying value if the absolute value of this number is too large. In other words, the outlying value is discarded when it is too far from the mean of the other values, in units of their SD.

But how many "SD units" is too far? Specifically, is the ratio of 43.6 large enough to discard the outlying datum? To answer this question, we must decide how much good data we are willing to throw away, each time biasing our result and exaggerating its precision, in order to protect ourselves against error resulting from including genuinely erroneous values. Suppose you feel that genuine gross errors are not unusual, and thus would be willing to throw away 1 in 20 good results in order to obtain such protection. Then a calculation involving probabilities indicates that ratios larger than 6.2 justify discarding the outlier. However, if you believe that genuine gross errors are rarer, and are willing to throw out only 1% of good results in order to protect against it, the required multiple is 14.1. In either case, the value of 1.53 in the above example would be discarded. The observed ratio of 43.6 is slightly below the criterion ratio of 44.7 that should be used if you are willing to discard only 1 in 1000 good data points. Although that policy may well be too conservative, a biochemist adhering to it would retain the value of 1.53, and report the result of 0.75 ± 0.26 (SE) from four replications.

The numerical criteria given above (for deciding when the calculated ratio is large enough to justify discarding a datum) are specifically for quadruplicate assays such as the example. Their counterparts for triplicate assays are higher, respectively, 25.5, 127.3, and 1273.3. Therefore, ratios that justify discarding data points from a quadruplicate assay may well indicate they should be retained in a triplicate assay. *Note that many a practicing biochemist uses numbers far lower than any of these, without appreciating that a substantial fraction of good results is being discarded. Unless true large errors are common, this frequently yields assays that have lower precision and accuracy than the full data would provide, with far lower precision and accuracy than is claimed for them.*

Presentation of Experimental Data

There are several options available in the presentation of experimental data. Undoubtedly the two most popular formats are tables and graphs, since both allow a great deal of material to be presented in a concise manner. Considerations for each of these formats are presented below.

Data Tables

Remember that a data table should be designed to present a set of data as clearly and concisely as possible. All columns and rows within the table should be clearly defined with respect to the identity and units associated with each value. In addition, the table should be titled to allow the reader to determine quickly what features of the table are relevant to the study or experiment. Table I-2 presents examples of both a poorly designed and a properly designed data table. Note that the poorly designed table has no title, is very redundant, is cluttered, and is presented in a manner that does not allow the reader to easily see differences between (to compare) trials of the same experiment. The properly designed data table, in contrast, features the experimental values and quickly draws the reader's attention to differences between the values.

Graphs

As with data tables, graphs are a good tool for presenting a large amount of experimental data in a concise manner. It is probably a better format to use if you wish to draw the reader's attention quickly

Table I-2 Examples of a Poorly Designed and a Well-Designed Data Table

A poorly designed data table:

Absorbance values for Reaction #1 at 360, 420, and 540 nm	Absorbance values for Reaction #2 at 360, 420, and 540 nm	Absorbance values for Reaction #3 at 360, 420, and 540 nm
0.876 (360 nm)	0.885 (360 nm)	0.823 (360 nm)
0.253 (420 nm)	0.250 (420 nm)	0.244 (420 nm)
0.164 (540 nm)	0.163 (540 nm)	0.157 (540 nm)

A properly designed data table:

Table 1: Absorbance values for three trials of the experiment at 360, 420, and 540 nm.

Trial #	A_{360}	A_{420}	A_{540}
1	0.876	0.253	0.164
2	0.885	0.250	0.163
3	0.823	0.244	0.157

to differences in experimental data. The two most commonly used types of graphs in experimental biochemistry are the *bar graph* and the *line graph*. Figure I-1 presents two simple sets of enzyme kinetic data in the bar graph and the line graph format. In general, line graphs are preferable to bar graphs for the presentation of data in which the *x*-axis variables are numbers along a continuum, such as pH, time, wavelength, etc. Line graphs should always be designed with the *controlled variable* as the abscissa (*x*-axis) and the *experimentally observed variable* as the ordinate (*y*-axis). For example, if you determined the activity of an enzyme at several different pH values, you should plot units of enzyme activity on the *y*-axis versus pH on the *x*-axis. As with the data table, the axes of any graph should be clearly labeled with respect to identity and units. Bar graphs are more frequently used when the controlled variable is not numerical, as in the example shown in Figure I-1.

Whenever possible, line and bar graphs should be constructed with the use of computer graphing programs (CricketGraph, Excel, Lotus, etc.). Aside from the fact that they produce graphs that are uniform and visually attractive, they have the added capability of fitting non-ideal data to a "best-fit" line or polynomial equation. This operation of fitting a set of data to a best-fit line becomes extremely important when you are attempting to determine the activity of an enzyme solution or trying to determine a specific value for an unknown sample with the use of a standard curve. Although it is possible to determine a best-fit line through a set of non-ideal data manually with the use of the least squares formula, the process can be quite time consuming and more prone to human error than the computer-based method. Experiment with different software graphing programs until you find one with which you are comfortable.

To the Instructor

The third edition of *Experimental Biochemistry* is an introductory biochemistry laboratory textbook targeted to juniors, seniors, and first-year graduate students in biochemistry, chemistry, microbiology, biology, and/or all related disciplines. As such, we assume that students using this book will: (1) have completed an introductory lecture and laboratory course in organic chemistry; (2) be thoroughly familiar with the mathematics of introductory chemistry; and (3) be enrolled in, or have completed, an introductory biochemistry lecture course. Education in biology, bacteriology, and/or physical chemistry, although helpful, is not essential for this course.

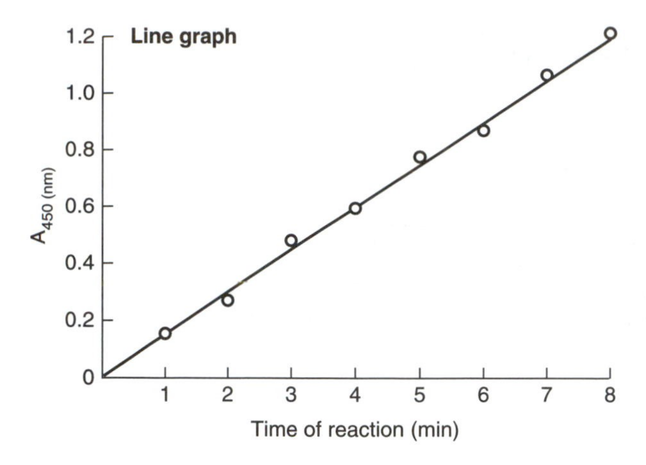

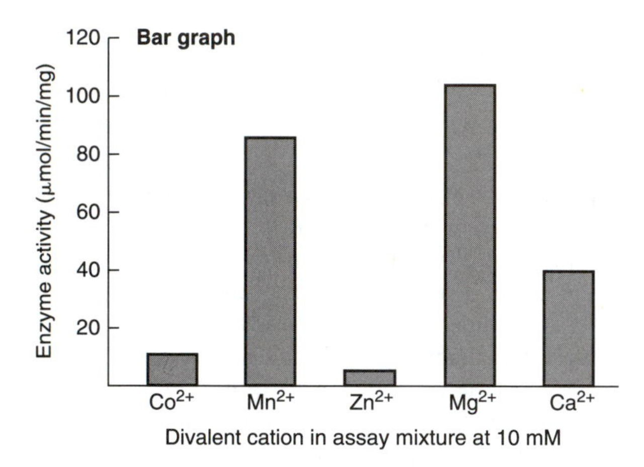

Figure I-1 Examples of line graph and bar graph format for presentation of experimental data.

The introductory sections and experiments in the third edition of *Experimental Biochemistry* introduce many of the basic techniques of modern biochemical research. Thus, the material in this book will provide students with the theoretical and quantitative background to understand and conduct meaningful research in biochemistry.

This goal may be achieved without undertaking all of the experiments in this book, or even all of the protocols within any particular experiment. We are intentionally repetitive in an attempt to introduce a wide variety of principles, procedures, tissue sources, and logistic requirements. *We know from experience that the majority of students in an introductory biochemistry laboratory course have difficulty in translating a very generalized set of instructions into a detailed protocol that will allow them to conduct the experiment and obtain reasonable results.* Students who are forced to do this waste a great deal of time designing and carrying out experiments that are unsuccessful. Such students may become frustrated, lose interest in the material, and gain little practical experience from the course. To avoid this situation, this edition of *Experimental Biochemistry* includes very detailed, step-by-step protocols that will guide the students through each of the 25 experiments. This by no means reduces the experiments to a "cookbook" format; there are many portions throughout the book that require students to demonstrate the depth of their knowledge on each individual topic. This same format is also valuable to instructors, limiting the time needed to modify and test the experiments to fit their individual needs and requirements.

We urge instructors to tailor any of the individual experiments in *Experimental Biochemistry* to fit the structural and time constraints of their course. We at the University of Illinois schedule a 14-week, 1-semester course that meets 3 times per week for a 4-hour period. In this time, we can typically perform 12 to 14 of the individual experiments in their entirety. The daily-schedule format of the protocols for each experiment reflect what we can perform with a class of up to 50 students in a reasonable period of time. If you are limited by time, you may choose to perform fewer experiments or selected sections from a number of individual experiments.

The majority of the daily experiments outlined in *Experimental Biochemistry* can be completed by a large group of students in 2 to 4 hours with relatively limited assistance by the instructor. This is especially true if the bulk reagents are prepared by the teaching assistants and instructors prior to the beginning of the laboratory period. As with the second edition, the third edition of *Experimental Biochemistry* includes a detailed list of reagents required to perform the experiments, as well as a detailed Appendix describing how to prepare each reagent and the quantities needed for students to perform the experiment.

We have made an effort to design experiments that require a minimum of equipment and, therefore, afford maximum flexibility. Modern biochemistry, however, makes use of increasingly more sophisticated procedures and equipment. We would be deficient if we avoided contact with sophisticated modern techniques such as scintillation counting, high-performance liquid chromatography (HPLC), sodium dodecyl sulfate–polyacrylamide gel electrophoresis (SDS-PAGE), agarose gel electrophoresis, and the like. This book must therefore represent compromises between teaching requirements and budgetary realities. Where appropriate, we offer optional methods for particular applications that yield similar experimental results.

We urge instructors to add slight modifications to the experiments to fit their needs. If you are able to find a more inexpensive source for a reagent, we do not discourage you from using it. If you think that you can perform a similar experiment with a less expensive apparatus or instrument, we encourage you to try it. The sources for the commercially available reagents that we suggest for each experiment are included simply to aid the instructor in obtaining the materials. We have found, however, that the quality control procedures performed on commercially available reagents are more reliable, especially in the case of complex reagents.

One final point is worth specific attention. We strongly encourage that the experiment be performed by the instructor and/or teaching assistants in its entirety prior to the beginning of the class experiment. Further, we strongly recommend that this "pre-lab" be performed using the exact same bulk reagents that the class will be using. This avoids many of the potential problems of variations in tissue samples, lot-to-lot variation among commercially available reagents, and variation in reagent composition resulting from scaling up the preparation of solutions. If you find that you must make slight variations of the protocol for a particular trial of an experiment due to the performance of a particular set of reagents (volumes, incubation times, etc.), make the change and bring it to the class's attention at the beginning of the experiment. The protocols provided for each experiment are the "ideal" conditions that we have established from years of class testing.

We are confident that you will enjoy this textbook and find it to be very valuable in your efforts to design and/or update your biochemistry laboratory course.

REFERENCES

Garret, R. H., and Grisham, C. M. (1995). *Biochemistry.* Orlando, FL: Saunders.

Lehninger, A. L., Nelson, D. L., and Cox, M. M. (1993). *Principles of Biochemistry*, 2nd ed. New York: Worth.

Lockhart, R. S. (1998). *Introduction to Statistics and Data Analysis.* New York: Freeman.

Mathews, C. K., and van Holde, K. E. (1990). *Biochemistry.* Redwood City, CA: Benjamin/Cummings.

Moore, D. S. (1998). *The Basic Practice of Statistics.* New York: Freeman.

Stryer, L. (1995). *Biochemistry*, 4th ed. New York: Freeman.

Voet, D., and Voet, J. G. (1990). *Biochemistry.* New York: Wiley.

Wilson, K., and Walker, J. (1995). *Principles and Techniques of Practical Biochemistry*, 4th ed. New York: Cambridge University Press.

Photometry

Theory

Light can be classified according to its wavelength. Light in the short wavelengths of 200 to 400 nm is referred to as ultraviolet (UV). Light in the longer wavelengths of 700 to 900 nm is referred to as near infrared (near-IR). Visible light falls between the wavelengths of 400 and 700 nm. Within this range of wavelengths is found all of the colors visible to the human eye. For example, light at 400 to 500 nm will appear blue, while light at 600 to 700 nm will appear red. Any solution that contains a compound that absorbs light in the range of 400 to 700 nm will appear colored to the eye. The solution is colored because specific wavelengths of light are removed (absorbed) as they pass through the solution. The only light that the eye will perceive, then, are the wavelengths of light that are transmitted (not absorbed).

This principle is illustrated in the absorption spectrum for riboflavin (Fig. 1-1). An absorption spectrum is a plot representing the absorbance of a solution at a number of wavelengths. Why does a solution of riboflavin appear yellow to the eye? As shown in Figure 1-1, riboflavin absorbs light at 450 nm, which is in the blue region of visible light. Because of this, red and yellow light will be transmitted through the solution and detected by the eye. Figure 1-1 also shows that riboflavin absorbs light strongly at 260 and 370 nm. Although this will not influence the apparent color of the solution (since these wavelengths lie outside the range of visible light), these absorption events in the UV range can be detected with a spectrophotometer.

Spectrophotometry can be used to identify and quantitate specific compounds in both pure and impure solutions. Spectrophotometry is based on two physical principles: Lambert's law and Beer's law. Lambert's law states that the proportion of light absorbed by a solution is independent of the intensity of light incident upon it. In addition, it states that each unit layer of a solution will absorb an equal fraction of the light passing through it. For example, if the intensity of light incident upon a solution is 1.00 and each unit layer absorbs 10% of the light passing through it, then the light transmitted through the solution will be diminished by 10% per unit layer (1.00, 0.90, 0.81, 0.73, 0.66, . . .).

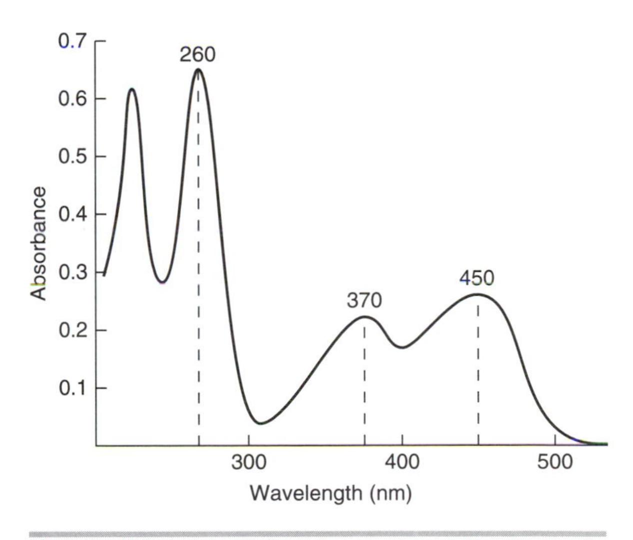

Figure 1-1 Absorption spectrum of riboflavin (22 μM in 0.1 M sodium phosphate, pH 7.06, in 1-cm light path).

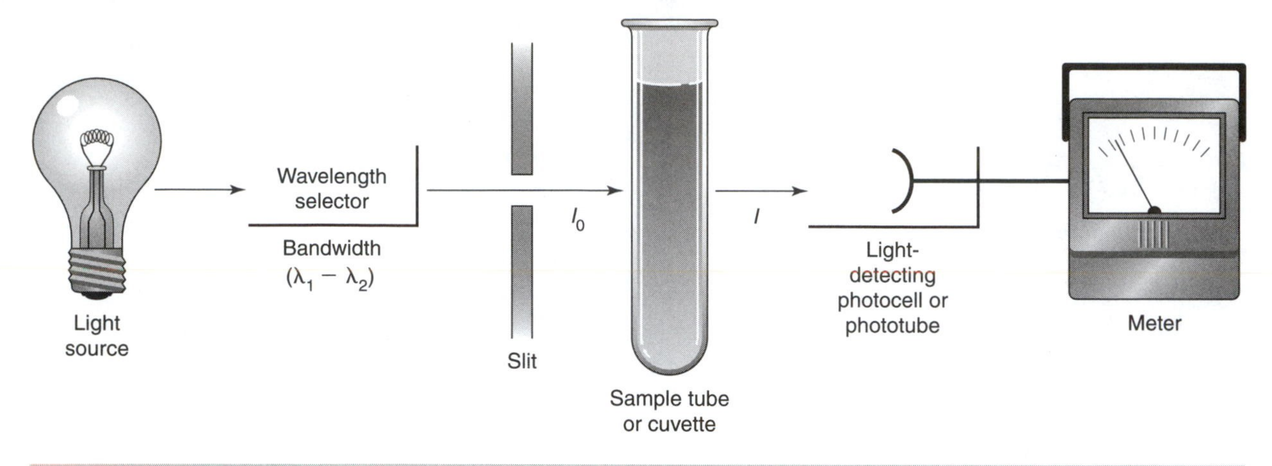

Figure 1-2 Operational diagram of a photometer or spectrophotometer.

Lambert's law gives rise to the following equation (see Fig. 1-2):

(1-1)

$$I = I_0 \cdot e^{-\alpha l}$$

where I_0 = intensity of the incident light; I = intensity of the transmitted light; l = length of light path (in centimeters); and α = absorption coefficient of the medium. Converting Equation 1-1 to the logarithmic form, we have:

(1-2)

$$\ln I_0/I = \alpha l$$

Using logarithms to the base 10, the absorption coefficient (α) is converted to a proportionality constant (K):

(1-3)

$$\alpha = 2.303\mathrm{K}$$

Thus,

(1-4)

$$\log_{10} I_0/I = Kl$$

The $\log_{10} I_0/I$ is termed "absorbance" (A) or "optical density" (OD). "Absorbance" is currently the preferred term. It is the absorbance of a solution at a particular wavelength that is of use in the discipline of spectrophotometry. As is apparent from these equations, absorbance is a unitless value.

Beer's law recognizes the relationship between the absorbance of a solution and the concentration of light absorbing compound(s) in that solution. It states that light absorption is proportional to the number of molecules per unit volume of light-absorbing compound through which the light passes. Beer's law is based on the observation that a solution at concentration c and length l absorbs twice as much light as a solution at concentration $c/2$ and length l. In other words, as the concentration of the light-absorbing compound doubles, so too will the amount of light absorbed over a given pathlength of light.

Because of the concentration factors α and K, Beer's law and Lambert's law can easily be combined to create Equation 1-5. Beer's law states that the proportionality constant K is related to the concentration of the absorbing solute.

(1-5)

$$\varepsilon c = K$$

where ε is the extinction coefficient and c is the concentration of the light-absorbing compound. This equation leads to:

(1-6)

$$I = I_0 \cdot 10^{-\varepsilon c l}$$

This equation can then be rearranged to:

(1-7)

$$\log_{10} I_0/I = A = \varepsilon c l$$

The extinction coefficient ε is constant at a given wavelength for a given solute that absorbs light. As discussed later, the extinction coefficient may depend strongly on both the ionic strength and the pH of the solution in which the solute resides.

What is the extinction coefficient? This term allows you to relate the concentration of the light-absorbing compound and the pathlength of incident light to the absorbance of a solution. The extinction coefficient for a compound in solution is dependent on both the solvent used and the wavelength of light at which the absorbance is measured. Clearly, riboflavin absorbs more light at 260 nm than at 450 nm. This fact will be reflected in the extinction coefficients for riboflavin at these two wavelengths (the extinction coefficient will be higher at 260 nm than at 450 nm).

Since absorbance values are unitless, the extinction coefficient is most often expressed in units of inverse concentration times inverse pathlength (i.e., M^{-1} cm^{-1} or mM^{-1} cm^{-1}), since $\varepsilon = A/cl$. The molar extinction coefficient for compound X is equal to the absorbance of a 1 M solution of that compound in a particular solvent at a particular wavelength in a pathlength of 1 cm. The millimolar extinction coefficient for compound X is equal to the absorbance of a 1 mM solution of that compound in a particular solvent at a particular wavelength in a pathlength of 1 cm. Extinction coefficients are also commonly expressed in units of $(mg/ml)^{-1}$ cm^{-1}. The greater the extinction coefficient under a particular condition (solvent and wavelength), the greater the amount of light absorbed. This idea becomes important when considering the sensitivity associated with quantifying a compound by measuring the absorbance of a solution at a particular wavelength. For example, since riboflavin has a greater extinction coefficient at 260 nm than at 450 nm, an absorbance reading taken at 260 nm would be much more useful in attempting to determine the concentration of a dilute solution of this compound.

Qualitative Spectrophotometric Assays

Because many compounds of biological interest absorb light in the UV, visible, and/or near-IR wavelengths, spectrophotometry can be a very useful technique in identifying unknown compounds. When given a sample of a pure compound, an absorption spectrum can be generated by measuring the absorbance of the compound in solution at a variety of wavelengths. Modern spectrophotometers are often equipped with an automatic scanning capability that allows the investigator to produce a full absorption spectrum easily and rapidly. By comparing the absorption spectrum of an unknown compound with a number of spectra produced from solutions containing known compounds, insight can be gained into the identity of the compound(s) in the unknown solution.

Often, it is not necessary to produce a full absorption spectrum to identify a compound, particularly if the molecule absorbs light at unique or unusual wavelengths. For example, the nitrogenous bases that comprise nucleic acids are known to absorb strongly at 260 nm. The aromatic rings on tryptophan and tyrosine (amino acids that are found in proteins) are known to absorb strongly at 280 nm. Many different iron-containing (heme) proteins or cytochromes show a distinct absorbance maxima in the visible range from 500 to 600 nm, as well as at about 400 nm. The profiles of these absorption spectra may even provide information about the oxidation state of the heme group in the protein (whether it is in the oxidized or reduced form).

You must keep in mind that all absorption spectra and extinction coefficients are presented in the context of a defined pH and salt concentration (ionic strength). As demonstrated in Figure 1-3, changing one of these conditions may have a dramatic effect on the position and height of one or more peaks within the spectrum. This is particularly true if the compound has ionizable groups within or adjacent to its chromophoric (light-absorbing) center. Recall that the ability to absorb light is dependent on the arrangement and energy state of the electrons surrounding the atoms that make up the molecule. If the local environment of the chromophoric center is altered by changes in pH, ionic strength, or solvent composition, this could potentially have a great effect on the molecule's ability to interact with photons of specific wavelengths, altering the profile of the absorption spectrum.

Quantitative Spectrophotometric Assays

As stated earlier, spectrophotometry can also be used to quantitate the amount of a compound in

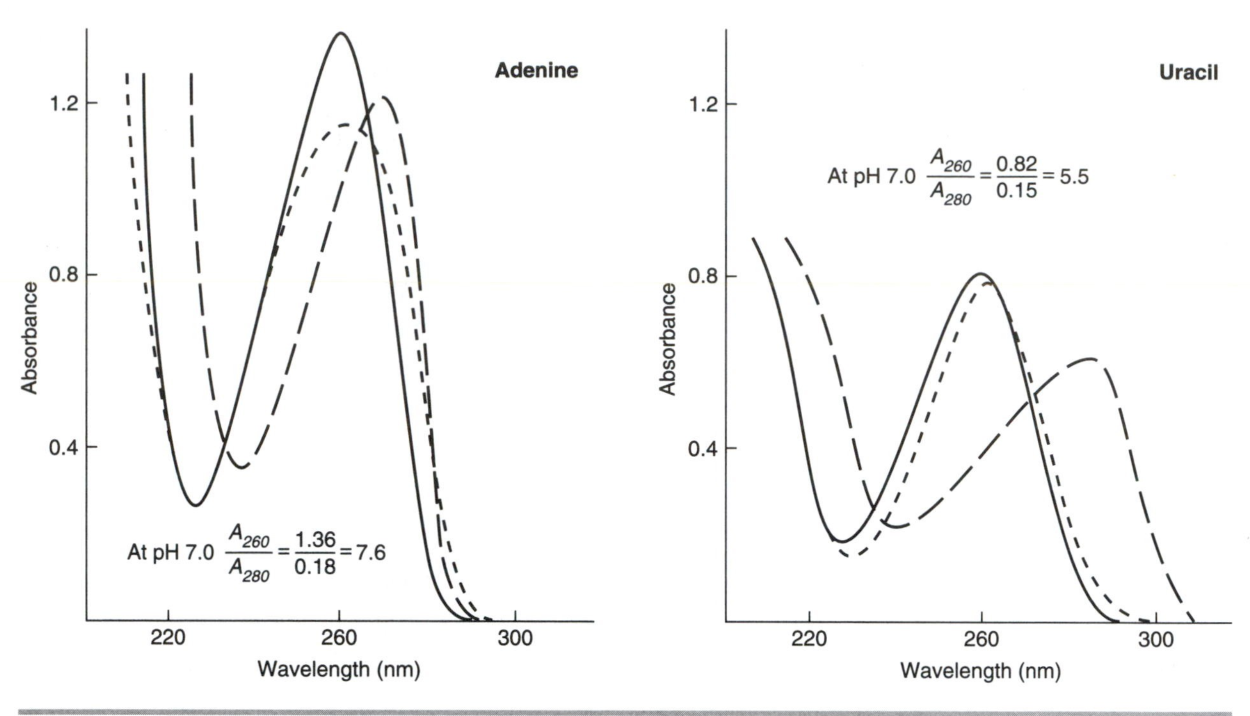

Figure 1-3 Absorption spectra of 0.1 mM adenine and uracil in a 1 cm cuvette (- - - - -) = in 6 N HCl. (———) = pH 7.0. (— — —) = pH 13.0.

solution. The Beer–Lambert law ($A = \varepsilon cl$) allows you to relate the absorbance of a solution to the concentration of a particular solute in that solution using the light pathlength and the extinction coefficient for that solute. The extinction coefficient for hundreds of molecules at specified conditions have been published. The Beer–Lambert law therefore allows you to predict the absorbance that will result from a solution containing a known concentration of a solute. Similarly, you can determine the concentration of a molecule in solution, provided the extinction coefficient is known. If an extinction coefficient is not available, a value can be generated for a particular condition and wavelength, provided you know the concentration of the compound and the pathlength of light through the photocell or cuvette. Beer's law predicts that absorbance is directly proportional to the concentration of the compound. Though this is true for most molecules, it is a good practice to measure the absorbance of several concentrations of the compound when determining the extinction coefficient. Deviations from Beer's law can occur if the chemical nature of the solute changes with concentration. For instance, adenosine is known to form dimers at high concentrations. Adenosine dimers absorb less light than adenosine monomers. Because of this, the extinction coefficient for an adenosine solution at high concentration will be less than that of a dilute adenosine solution at a given wavelength.

Is it possible to determine the concentration of a single compound in solution with other compounds? This is possible if the molecule of interest absorbs light at a wavelength where the other molecules present in solution do not absorb light. Suppose that you have a mixture of compounds, X, Y, and Z, as well as extinction coefficients for each at a variety of wavelengths. Compound X absorbs light at 230, 280, and 460 nm, while Y and Z absorb light at only 230 and 280 nm. An absorbance reading at 460 nm will be sufficient to allow you to calculate the concentration of compound X in the impure solution. Because compound X is the only molecule that absorbs light at this wavelength, you can attribute all of the absorbance at 460 nm to compound X and accurately quantify this molecule in the presence of Y and Z. The same is not true for absorbance readings taken at 230 or 280 nm, since all three molecules contribute to the absorbance of the solution at these wavelengths.

Suppose you knew the concentrations of Y and Z in the mixture of X, Y, and Z above. In this case, it would be possible to calculate the concentration of compound X with an absorbance reading measured at 230 or 280 nm. If you know the extinction coefficients for Y and Z at these wavelengths, you could use these and the concentrations of Y and Z to subtract the absorbance contributions from these two molecules. The remainder of the absorbance at 230 or 280 nm, along with the extinction coefficients for compound X at these wavelengths, could then be used to calculate the concentration of X in the mixture. Note that this type of analysis is possible only if you know that X, Y, and Z are the only compounds present in the mixture that absorb light at 230 nm or 280 nm.

The principles of quantitative spectrophotometric assays that we have discussed involve direct photometric measurements of light-absorbing compounds. It is not always the case, however, that compounds of biological interest absorb light at a unique wavelength. If a molecule does not absorb light, it can often undergo a reaction with other molecules to produce a compound that does absorb light (Equation 1-8).

(1-8)

Colorless compound to be assayed + Excess of color-forming reagents ⟶ Color proportional to amount of colorless compound

You can quantify the amount of the colored compound (and, therefore, the amount of the colorless compound) if the extinction coefficient and the reaction stoichiometry are known. It is also possible to quantify the amount of colorless compound in an unknown by preparing a "standard curve." Here, the color-forming reagents react with a known series of increasing concentrations of the colorless compound, and the absorbance at a defined wavelength is measured against a "blank" containing only the color-forming reagents (no colorless compound) (Fig. 1-4). If the same procedure is performed on an unknown (test) solution of the colorless compound, the absorbance can be used to determine the concentration of the compound in solution by comparing where the absorbance reading lies in relation to the concentrations of the compound used to produce the standard curve.

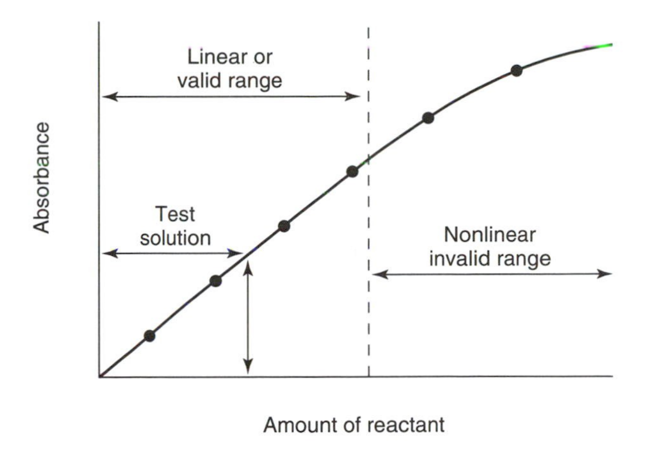

Figure 1-4 Standard curve for a color-forming quantitative reaction. The absorbance of the test solution can be used with the standard curve to determine the concentration of a compound in the test solution.

Two concepts are very important to keep in mind when designing a colorimetric assay in which the color forms as the result of a chemical reaction: *the color-forming reagents must be present in excess, and the data used to produce the standard curve must show a linear relationship between absorbance and concentration of the compound (changes in absorbance must be directly proportional to changes in concentration of the compound under study).* If the color-forming reagents are not in excess, the amount of colored product formed may be limited by these reagents rather than by the amount of colorless compound being analyzed. In other words, the reaction system will become saturated at high concentrations of the colorless compound. The result of this is that the change in absorbance may no longer be directly proportional to the amount of the colorless compound present, but by the amount of the color-forming reagents present. *These nonlinear data at high concentrations of the compound are the greatest source of student error in quantitative spectrophotome-*

try. If the absorbance of the unknown solution is found to lie outside the linear range of the standard curve, the experiment must be performed again on a dilution of the original unknown solution. The effect of the dilution can then be taken into account when determining the concentration of the compound in the original solution. *It is not acceptable to dilute the completed reaction that formed a colored solution with water or other reagent to force the absorbance value into the linear range of the standard curve.* If each of the completed colored solutions used to generate the standard curve were diluted to the same extent, the unknown solution would once again be found to have an absorbance value in the nonlinear portion of the standard curve.

Construction and Properties of Spectrophotometers

Photometers, colorimeters, and spectrophotometers employ the basic components indicated in Figure 1-2. Each of these components can have a marked effect on the efficiency of any colorimetric assay. Accordingly, it is essential to consider each of the components in turn.

Light Source. The light source must be capable of emitting a steady amount of light of sufficient energy in the range of wavelengths required for the analysis of the sample. Most spectrophotometers employ a constant voltage-regulated tungsten lamp for spectral analyses in the range of 340 to 900 nm. More sophisticated spectrophotometers, which also have the capability for analyses in the UV range, employ an additional constant voltage-stabilized H_2-deuterium lamp that emits light in the range of 200 to 360 nm.

Wavelength Selector. Spectrophotometry requires assay of absorbance at defined wavelengths. Usually, it is not possible to have light representative of a single wavelength. Therefore, when we speak of monochromatic light of, say, 500 nm, we usually mean a source that has its maximum emission at this wavelength, with progressively less energy at longer and shorter wavelengths. Thus, a totally correct description of this light must also specify a spectral bandwidth (Fig. 1-2), which is the range of wavelengths represented. For example, 95% of the energy from a 500-nm light source falls between 495 and 505 nm. From any light source, less intensity is obtained with light of greater purity. On the other hand, the greater the spectral purity of monochromatic light, the greater the sensitivity and resolution of the measurements.

Spectrophotometers generate the desired wavelength of light by use of a wavelength selector. The simplest wavelength selectors are one or more absorption filters that screen out light above and below specific wavelengths. Many of the older colorimeters employ such filters because of their simplicity and low cost. However, these filters generally have broad transmission ranges, and the resolution of absorption spectra is low (i.e., peaks are "smeared" or obscured). Moreover, measurements can be made only at wavelengths for which filters are available. Most modern spectrophotometers overcome these deficiencies through use of monochrometers containing a prism or diffraction grating. Such monochrometers generate relatively pure light at any wavelength over a wide range.

A wavelength in the region of maximum absorption is usually employed in assays of the concentration of colored compounds. This is not absolutely necessary, however, since Beer's law may be expected to hold true at all wavelengths at which there is appreciable absorption. Thus, when an interfering substance absorbs at the wavelength of maximal absorption of the substance being measured, another wavelength at which the compound of interest absorbs light may be utilized.

Slit. The intensity of light emitted through any filter or monochrometer may be too intense or too weak for a light-sensing device to record. It is therefore necessary to be able to adjust the intensity of the incident light (I_0) by placing a pair of baffles in the light path to form a slit. Simple colorimeters often have a fixed slit, but more sophisticated spectrophotometers usually have a variable-slit-width adjustment mechanism.

Sample Tubes or Cuvettes. It follows from Lambert's law that if absorbance is to be used to measure concentration, the length of the light path traversed by light in the sample container (i.e., a sample tube if it is test-tube shaped, or sample cuvette if rectilinear in shape) must be the same as that in the blank. Thus, the sample tube or cuvette must have the same internal thickness as the blank. This is generally true for rectilinear cuvettes, but it cannot be assumed that all sample tubes are identical in this respect. You should always check the absorbance characteristics of sample tubes and obtain a standardized set for spectrophotometric measurements.

Light-Detecting Phototubes or Photocells. Spectrophotometers use either a photovoltaic cell or a vacuum phototube to detect the transmitted light (I) in a light system. Photovoltaic cells contain a photosensitve semiconductor (e.g., selenium) sandwiched between a transparent metal film and an iron plate. Light falling on the surface of the cell causes a flow of electrons from the selenium to the iron, thereby generating an electromotive force. If the circuit is closed through an ammeter, the current induced is proportional, within a certain range, to the intensity of light transmitted on the selenium (Fig. 1-5). The cell is limited by low sensitivity (it does not detect light of very low intensity) and is insensitive to wavelengths shorter than 270 nm and longer than 700 nm.

Vacuum phototubes have two electrodes that have a maintained potential difference. The cathode (negative electrode) consists of a plate coated with a photosensitive metal. Radiation incident upon the photosensitive cathode causes an emission of electrons (the photoelectric effect), which are collected at the anode (the positive electrode). The resulting photocurrent can be readily amplified and measured (Fig. 1-5).

Because of differences among photosensitive cathodes, certain phototubes are more sensitive in certain regions of the light spectrum. Others are more sensitive when used in combination with preliminary filters that screen out specific regions of the spectrum. Accordingly, certain spectrophotometers require an additional filter or a special red-sensitive phototube when they operate in the red or near-IR ranges of the light spectrum.

Photomultiplier tubes are a variation of the conventional phototube. Such tubes have several intermediate electrodes, known as dynodes, in addition to the primary photocathode and anode. Electrons emitted from the cathode strike the first of these dynodes, whereupon by secondary emission, several secondary electrons are emitted. Each of these, in turn, strikes the second dynode. The process is repeated until the original photocurrent is amplified by as much as 100,000 times. Such photomultipliers are very sensitive light detectors.

Supplies and Reagents

Colorimeter or spectrophotometer
Sample tubes and cuvettes (or colorimeter tubes for the colorimeter)
5×10^{-5} M riboflavin and 5×10^{-5} M adenosine or any other ribonucleoside
1 mg/ml solution of lysozyme
Solution of lysozyme at unknown concentration
1% (wt/vol) $CuSO_4 \cdot 5H_2O$
2% (wt/vol) sodium tartrate
2% (wt/vol) Na_2CO_3 in 0.1 N NaOH
Folin–Ciocalteau reagent (2 N)

Protocol

Examination of an Absorption Spectrum

This exercise can be performed easily and rapidly using a Beckman DU-64 spectrophotometer or any spectrophotometer having a "scanning wavelength" capacity. Place a 1-ml sample of 5×10^{-5} M riboflavin in a 1-cm pathlength quartz cuvette. Measure the absorbance of the sample at wavelengths ranging from 220 to 620 nm. If your spectrophotometer is equipped with an RS232 outlet, the absorbances at these wavelengths can be relayed to a printer to obtain the full absorption spectrum. If the spectrophotometer that you are using is not automatically able to scan a range of wavelengths, the same procedure can be done manually by increasing the wavelength at 10-nm intervals over this range (220 to 620 nm). *Note that each change of wavelength will require you to adjust the slit width or sensitivity control for the blank to a new 100% transmission (zero absorbance) setting before you measure the absorbance of the riboflavin solution at that wavelength.* In the range of wavelengths at which the absorbance reading is high, measure the absorbance at 5-nm intervals to identify the wavelength of maximum absorbance. Attempt to produce an absorption spec-

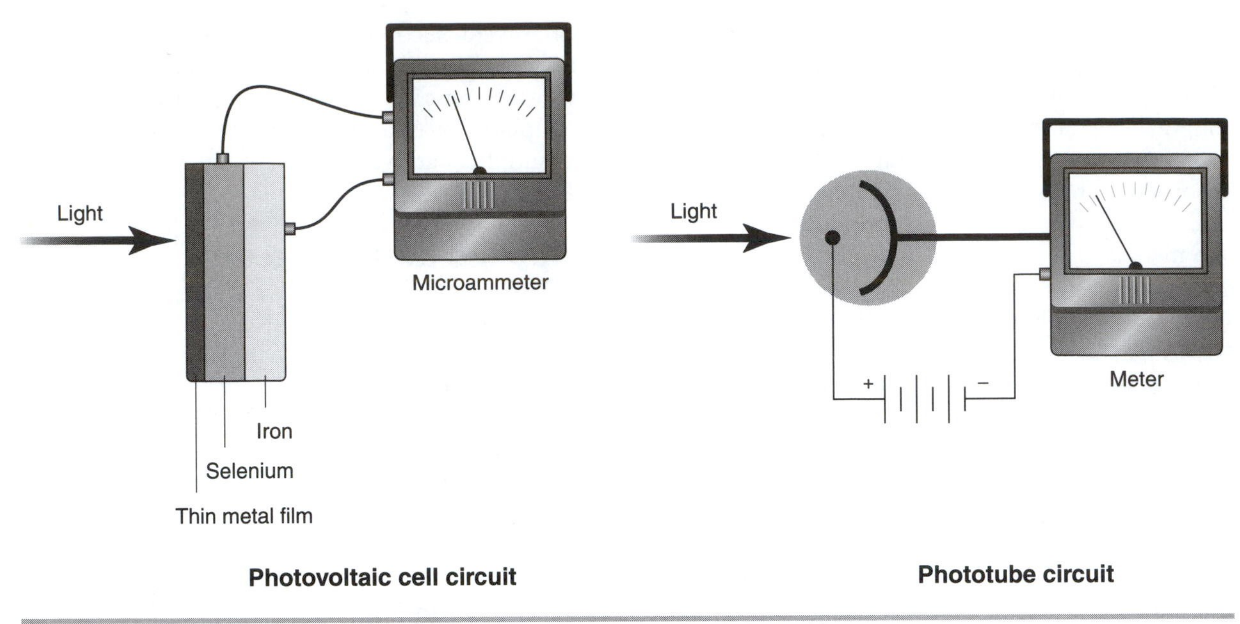

Figure 1-5 Schematic representation of photovoltaic cell and phototube.

trum by plotting absorbance (vertical axis) versus wavelength (horizontal axis) as shown in Figure 1-1. If time permits, perform the same exercise using a 5×10^{-5} M solution of adenosine. For this ribonucleoside, it is suggested that you measure the absorbance of the sample over a range of wavelengths from 220 to 320 nm.

Once you have obtained an absorption spectrum for riboflavin and/or adenosine under a given set of conditions, calculate a molar extinction coefficient (ε) for riboflavin at 450 nm and for adenosine at 260 nm using the Beer–Lambert equation ($A = \varepsilon cl$). Remember that the calculation of the extinction coefficient requires that you know the exact pathlength of light through the cuvette and the exact concentration of the molecule of interest in the solution. Most modern spectrophotometers are designed for a rectilinear, 1-cm-pathlength cuvette. Other photometers, however, are designed for a 13×100-mm colorimeter tube (pathlength of 1.3 cm).

If time permits, dilute the riboflavin and adenosine samples twofold and measure the absorbances of these samples at the same wavelengths used to calculate the molar extinction coefficients (450 nm for riboflavin and 260 nm for adenosine). If Beer's law holds true for these compounds, each of these assays should yield the same molar extinction coefficient values as those determined using the 5×10^{-5} M samples. Is this true for riboflavin and adenosine?

Photometric Assay of Color Developed by a Chemical Reaction

WARNING

Safety Precaution: **Avoid skin contact with the Folin–Ciocalteau phenol reagent. Wear gloves and safety goggles at all times.**

The Folin–Ciocalteau Assay of Protein Concentration. The Folin–Ciocalteau assay is one of the most sensitive and most commonly used assays to determine protein concentration (sensitive to about 10 μg/ml protein). This procedure employs two color-forming reactions to assay protein concentration photometrically. In the first reaction (a biuret reaction), compounds with two or more peptide bonds form a dark blue-purple color in the presence of alkaline copper salts. In the second reaction, tryptophan and tyrosine side chains react with the Folin solution to produce cuprous ions. This reaction is most efficient under basic conditions (pH 10.0–10.5). The cuprous ions produced in this reaction act to reduce components in the Folin reagent (sodium tungstate, molybdate, and phosphate), yielding an intense blue-green solution that absorbs light at 500 nm. Since the combined levels of tryptophan and tyrosine are generally constant in soluble proteins, the blue-green color of the Folin–Ciocalteau reaction is proportional to protein concentration in a complex mixture of proteins. *Since the number of aromatic residues can vary from protein to protein, one must keep in mind that this assay may have to be standardized when measuring the concentration of a purified protein.* In such cases, it may be best to use a protein standard that has the same relative proportion of aromatic residues as the protein under study.

This assay to determine the concentration of protein in a sample should not be used if the sample contains significant concentrations of detergents (SDS, Tween, Triton, etc.), sulfhydryl-containing reducing agents (dithiothreitol), copper-chelating reagents, carbohydrates (glucosamine and N-acetylglucosamine), glycerol (above 10% [vol/vol]), Tris (above 10 mM), tricine, and K^+ (above 100 mM). All of the compounds listed above have been reported to interfere with this assay. If a sample does contain these compounds, you may use acetone or trichloroacetic acid (see Day 6 protocol for Experiment 8) in precipitating the proteins from the sample and resuspending them in a more suitable buffer. *For a more detailed discussion of the different types of quantitative protein assays available, refer to the Section II introduction.*

1. Prepare a series of tubes containing the following:

Tube #	Volume of Water (ml)	Volume of 1 mg/ml Lysozyme (ml)	Volume of Unknown Lysozyme Solution (ml)
1	1.20	—	—
2	1.15	0.05	—
3	1.10	0.10	—
4	1.00	0.20	—
5	0.90	0.30	—
6	—	—	1.20
7	0.70	—	0.50
8	1.10	—	0.10
9	1.15	—	0.05
10	1.19	—	0.01

2. Separately prepare 100 ml of fresh alkaline copper reagent by mixing together in order:

   1 ml of 1% $CuSO_4 \cdot 5H_2O$
   1 ml of 2% sodium tartrate
   98 ml of 2% Na_2CO_3 in 0.1 N NaOH

3. Add 6 ml of this alkaline copper reagent to each of the 10 tubes and *mix immediately after the addition to avoid precipitation.*
4. Incubate the tubes at room temperature for 10 min. Add (*and immediately mix in*) 0.3 ml of Folin–Ciocalteau reagent to each of the 10 tubes.
5. Incubate for 30 min at room temperature and read the absorbance of all 10 tubes at 500 nm. Use tube 1 (blank) to zero your spectrophotometer.
6. Prepare a standard curve of your results by plotting A_{500} versus micrograms of protein for tubes 2 to 5. Determine whether A_{500} is proportional to the mass of protein from 0 to 300 μg. (Does the plot yield a straight line through the data points?)
7. Determine which of tubes 6 to 10 give absorbance readings that lie in the linear portion of the standard curve. What mass of protein (in micrograms) is indicated by these positions on the standard curve?
8. Based on the volume of the unknown protein solution present in tubes 6 to 10, calculate the concentration of lysozyme in the unknown solution (in milligrams per milliliter).
9. If you have more than two data points for the unknown solution that lie in the linear portion of the standard curve, it is possible to calculate several independent values for the protein concentration in the unknown solution. These values can then be averaged and presented along with a standard deviation from the mean.
10. If you do not have more than two data points for the unknown solution that lie in the linear portion of the standard curve, the assay can be repeated again, in triplicate, using a volume of the unknown solution known to give a value in the linear portion of the standard curve.
11. It is also important to realize that there is some error associated with the data points used to produce the standard curve. If possible, it would be best to perform the assays used to produce the standard curve in triplicate so that

each point could be reported with a standard deviation. Most computer graphing programs allow one to place "error bars" on the standard curve at each data point and generate a best-fit curve through these points. After this best-fit curve is calculated, the absorbance values for the unknown solutions can be introduced into the equation that defines the curve and used to calculate a value for the protein concentration.

There are three points about the Folin–Ciocalteau assay that deserve additional comment. First, the greatest source of error in this assay stems from inadequate mixing of reagents. Be sure to mix all of the components immediately after each reagent is added. Second, you may find that this assay is slightly nonlinear. Despite this fact, the assay is valid up to about 300 μg of protein. Finally, the color produced by this assay is largely formed during the first few minutes after mixing. If you find that the color in your unknown tubes (6–10) is more intense than that in tube 5 (the greatest mass of protein used in making the standard curve) after 2 or 3 minutes, it would be wise to immediately dilute your unknown protein solution by half and prepare tubes 6 to 10 again using this newly made solution. In doing so, you will avoid waiting 30 minutes for experimental data that you cannot use to obtain information by comparing to your standard curve.

You must also realize that there are some practical limitations associated with making spectrophotometric measurements. In general, you should not trust absorbance measurements greater than 1.0 and less than 0.1. If $A = 1.0$, then only 10% of the incident light has been transmitted. If $A = 2.0$, then only 1% of the incident light has been transmitted. Since most spectrophotometers do allow a small amount of ambient light or scattered light from the incident light source to reach the phototube and be recorded as transmitted light, absorbance readings above 1.0 are generally not accurate. The most accurate absorbance readings are those that lie in the range of 0.1 to 0.5.

Exercises

1. You are given 500 ml of a solution at pH 7.0 containing an unknown concentration of adenosine-5′-triphosphate (ATP, $\varepsilon = 14{,}300$ M^{-1} cm^{-1} at 260 nm and pH 7.0). You place 4 ml of this solution in a 1-cm-wide, 1-cm-long, 4-cm-high quartz cuvette and determine that the A_{260} of the solution is 0.143. What is the molarity of the ATP solution? How many moles of ATP are in 500 ml of this solution? How many micromoles of ATP are in the cuvette?
2. Given the data shown in Figure 1-3, determine the molar extinction coefficient for uracil at pH 7.0 and 260 nm.
3. Most pure proteins have extinction coefficients at 280 nm in the range of 0.5 to 3.0 $(mg/ml)^{-1}$ cm^{-1}. Explain why this value may differ from one protein to another.
4. A solution of bovine serum albumin (BSA) was prepared by mixing 0.5 ml of a stock solution with 9.5 ml of water. Given that the extinction coefficient for BSA at 280 nm is 0.667 $(mg/ml)^{-1}$ cm^{-1} and the concentration of BSA in the stock solution was 2.0 mg/ml, predict what absorbance would result from this solution at 280 nm. You may assume that a 1-cm-pathlength cuvette was used.

REFERENCES

Beaven, G. H., Holiday, E. R., and Johnson, E. A. (1955). *Optical Properties of Nucleic Acids and Their Components* (The Nucleic Acids, Vol. 1). New York: Academic Press.

Cresswell, C. J. (1972). *Spectral Analysis of Organic Compounds: An Introductory Programmed Text*, 2nd ed. Minneapolis: Burgess.

Jaffe, H. H. (1962). *Theory and Applications of Ultraviolet Spectroscopy*. New York: Wiley.

Lehninger, A. L., Nelson, D. L., and Cox, M. M. (1993). Amino Acids and Peptides. In: *Principles of Biochemistry*, 2nd ed. New York: Worth, pp. 111–133.

Lowry, O. H., Rosebrough, N. J., Farr, A. L., and Randall, R. J. (1951). Protein Measurements with the Folin Phenol Reagent. *J Biol Chem* **193**:265.

Schenk, G. H. (1973). *Absorption of Light and Ultraviolet Radiation, Fluorescence and Phosphorescence Emission*. Boston: Allyn and Bacon.

Stryer, L. (1995). Exploring Proteins. In: *Biochemistry*, 4th ed. New York: Freeman, pp. 45–74.

Chromatography

Theory

Biological molecules can be separated from one another by exploiting differences in their size, charge, or affinity for a particular ligand or solid support. Chromatography is the laboratory technique that allows separation of molecules based on their differential migration through a porous medium. Although there are many different types of chromatography, the principle behind the separation of the molecules is the same: a mixture of compounds will have different affinities for the stationary phase (solid support or matrix) on which it is adsorbed and the mobile phase (buffer or solvent) passing through the stationary phase.

General Theory

Modern preparative and analytical chromatography is most often performed in a column format. Here, the porous matrix or solid support is enclosed in a durable cylinder or column saturated with aqueous buffer or organic solvent (Fig. 2-1). The column is loaded with a solution containing a mixture of compounds at the top of the column by allowing it to flow into the porous medium. After the compounds have entered the solid matrix, they can be differentially *eluted* from the column either by continuous buffer flow or by changing the nature of the mobile phase passing through the porous matrix. This process of eluting compounds from the column is termed "development." As the different compounds emerge from the column, the eluted solution can be separated into multiple "fractions" or "cuts" that can be analyzed for the presence of a molecule of interest.

Partition Coefficient and Relative Mobility. As stated earlier, molecules adsorbed on a solid support (stationary phase) will partition between it and a mobile phase passing through the stationary phase. To predict the behavior of different molecules during chromatography, one must be able to define the affinity of a compound for the stationary and/or mobile phase. The term used to describe the affinity of a compound for the stationary phase is the *partition coefficient* (α). It is defined as the fraction of the compound that is adsorbed on the stationary phase at any given point in time. The partition coefficient can have a value between 0 and 1. The greater the value of α, the greater the affinity of the compound for a particular stationary phase. For example, a molecule with $\alpha = 0.4$ will be 40% adsorbed on the stationary phase at any given point in time. It has less affinity for the stationary phase than another molecule with $\alpha = 0.7$ (70% adsorbed at any given point in time). Mathematically, the partition coefficient can be expressed as:

$$\alpha = \frac{\text{molecules adsorbed on stationary phase}}{\text{molecules in stationary and mobile phase}}$$

The affinity of a molecule for the mobile phase is described in terms of relative mobility (R_f). This term describes the rate of migration of the molecule relative to the rate of migration of the mobile phase passing through the solid support. Mathematically, R_f is equal to $1 - \alpha$. For instance, a molecule with $\alpha = 0.4$ will be found in the mobile phase

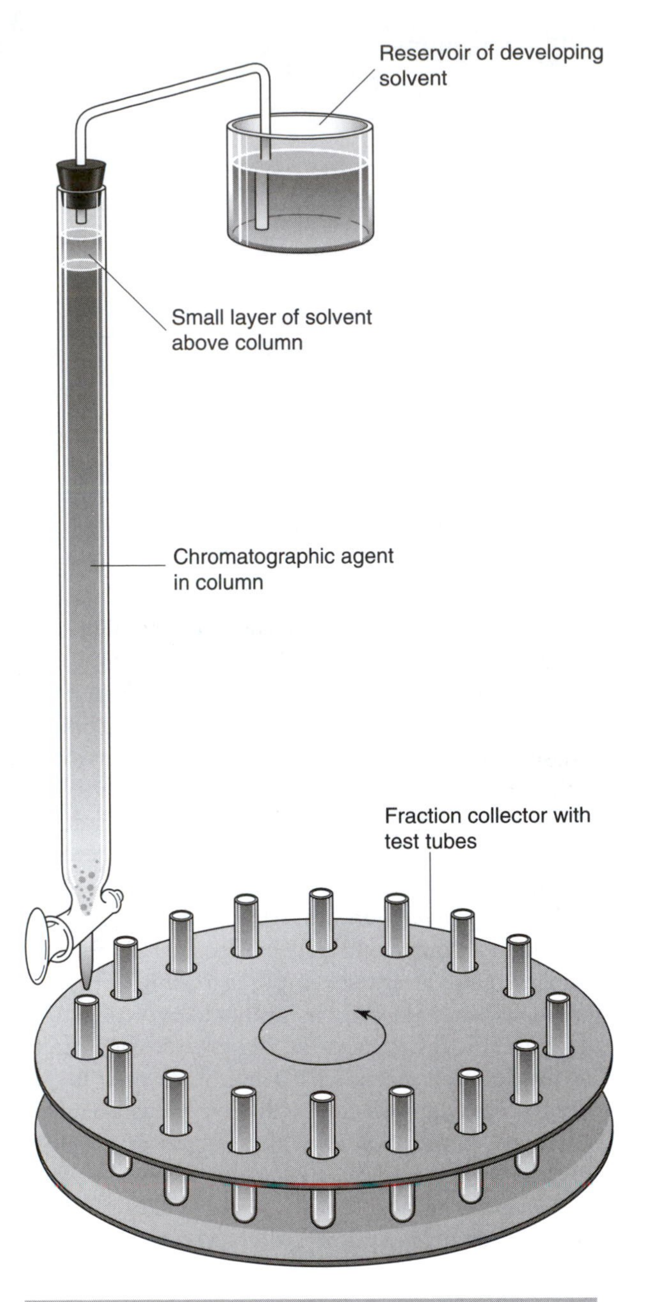

Figure 2-1 Typical column chromatography system.

60% of the time (1 − 0.4), whereas a molecule with $\alpha = 0.7$ will be found in the mobile phase 30% of the time. A comparison of R_f values indicates that the molecule with $\alpha = 0.4$ will migrate through the column twice as fast as a molecule with $\alpha = 0.7$. The following formula defines the mobility of an adsorbed molecule on the stationary phase relative to the flow rate of the mobile phase:

$$F_x = F_s \times R_f$$

where F_x is the flow rate of the compound with a given relative mobility R_f, and F_s is the flow rate of the mobile phase.

Chromatographic "Plates." In theory, a chromatographic "plate" can be described as the largest area of the column where two molecules with different partition coefficients will have the opportunity to display different rates of migration. In practice, this will often be determined by the dimensions (volume) of the column relative to the volume of sample that is loaded. The larger the difference of α between two compounds, the fewer chromatographic plates they will be required to pass through before achieving separation. Whereas two molecules with α values of 0.2 and 0.8 may require passage through only 5 chromatographic plates to achieve separation, two molecules with α values of 0.8 and 0.7 may require passage through as many as 50 plates to achieve separation.

Zone Spreading. From the concept of chromatographic plates just described, it would appear that you could separate any two molecules with slightly different partition coefficients provided that the column contained a sufficient number of plates. Due to a phenomenon termed "zone spreading," this is not always the case. As the number of plate transfers increases, so too does the volume in which a particular molecule is found (the molecule is diluted as it passes through the column). A molecule with $\alpha = 0.5$ may elute from a column in a sharp peak after 20 plate transfers and a very broad peak after 50 to 100 plate transfers. Since zone spreading will occur independently on two different molecules with similar partition coefficients, you may find that the elution peaks of the two will overlap after a large number of plate transfers. In general, zone spreading increases with increasing partition coefficient: a molecule with $\alpha = 0.2$ will show less zone spreading after 100 plate transfers than a molecule with $\alpha = 0.7$. Zone spreading is the consequence of simple diffusion: A sample of molecules introduced to the top of a column will move from

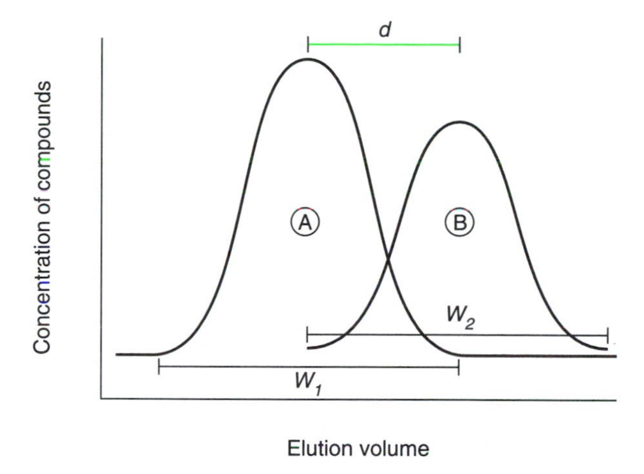

Figure 2-2 Resolution of two compounds (A and B) emerging from a chromatography column.

areas of high concentration to areas of low concentration as the mobile phase continuously passes through the stationary phase.

Resolution. The goal of any chromatographic step is to maximize resolution or to minimize zone spreading. Resolution is a function of the position of the maximum elution peak height and the elution peak width (Fig. 2-2). The greater the resolution between two elution peaks, the greater the degree of separation between the two molecules. Mathematically, resolution (R) is two times the distance between two elution peak maxima (*d*) divided by the sum of the widths of the two elution peaks ($W_1 + W_2$):

$$R = \frac{2d}{(W_1 + W_2)}$$

For any given chromatographic step, you must achieve a compromise between the number of plate transfers and the resolution. In other words, you must provide enough plate transfers to allow separation while preventing excess zone spreading.

Isocratic versus Gradient Elution. The principles of chromatography just described hold true for isocratic elution schemes, in which the nature of the mobile phase remains constant throughout the development of the column. Many chromatographic columns, however, are eluted with some sort of mobile phase gradient. Here, the nature of the mobile phase changes continuously throughout development with respect to ionic strength, pH, ligand concentration, or organic:aqueous solvent ratio. As one or more of these parameters changes, so too does the partition coefficient of one or more molecules adsorbed on the stationary phase. Although this greatly complicates the calculation of α, it can greatly enhance the ability of a given sized column to separate a number of different compounds. By continuously changing the partition coefficients of molecules passing through the stationary phase, you can effectively increase the number of plate transfers occurring over a given length of the column. A 20-ml column that provides 10 plate transfers utilizing an isocratic elution scheme may provide 100 to 200 plate transfers when used in conjunction with an elution gradient.

Remember that one problem encountered with increasing the number of plate transfers is an increase in zone spreading, decreasing the resolution of two elution peaks containing compounds with similar partition coefficients. Gradient elution schemes often work to minimize zone spreading. Recall that molecules with lower α values show less zone spreading after a given number of plate transfers than molecules with large α values. If a molecule that has a relatively large α value under loading conditions can be eluted with a gradient that continually decreases its α value as it passes through the column, zone spreading can be minimized, and resolution between neighboring elution peaks can be maintained. In general, mobile phase gradients are most often used to optimize a specific chromatographic procedure by maximizing both the number of plate transfers and the resolution of the elution peaks.

Types of Chromatography

Gel Filtration or Size Exclusion. Gel filtration is perhaps the easiest type of chromatography to understand conceptually (Fig. 2-3). It relies on the ability of a given size molecule to enter the uniformly sized pores of a solid support or matrix. Imagine that you have a solution containing two molecules: one of 8000 Da and one of 50,000 Da. If this solution is passed through a porous matrix

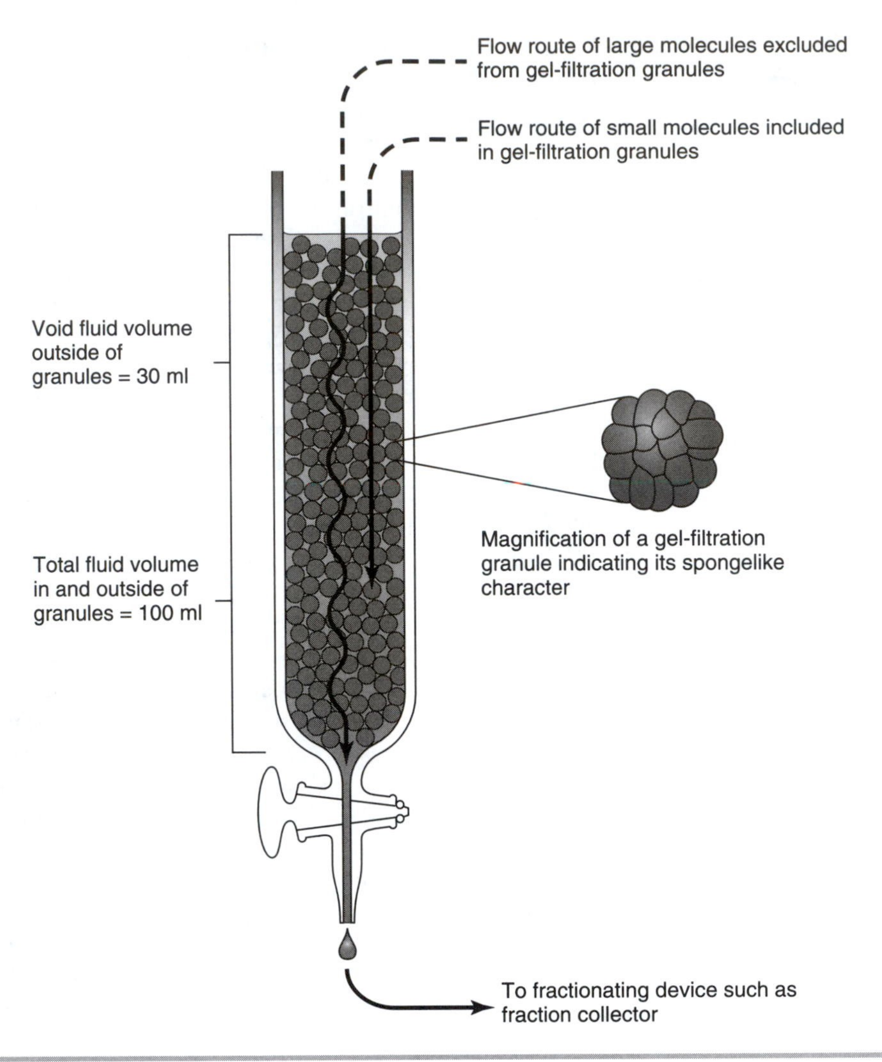

Figure 2-3 Diagram of the gel-filtration process.

capable of excluding molecules larger than 10,000 Da, then the two molecules will be separated from one another as they pass through the column. While the 8000-Da molecule will be able to enter and occupy the pores within the matrix (total volume of the column), the 50,000-Da molecule will be excluded from the mobile phase occupying the pores and be forced to occupy only the volume outside of the matrix (void volume of the column). Since the 8000-Da molecule can occupy a larger volume of the column than the 50,000-Da molecule, the migration of the smaller molecule through the column will be retarded relative to the rate of migration of the mobile phase through the column. Therefore, the 8000-Da molecule will elute later in the development than the 50,000-Da molecule, which will elute in the void volume of the column.

In the last example, one molecule was excluded from the pores of the matrix while the other was able to occupy the volume between the pores. Gel-filtration chromatography can also separate molecules that are only differentially excluded from the pores of the matrix. Suppose that you pass a solution containing molecules of 10,000 Da and 45,000 Da through a porous matrix capable of excluding molecules larger than 60,000 Da. Although both molecules will be able to occupy the volume between the pores of the matrix, the mobility of the

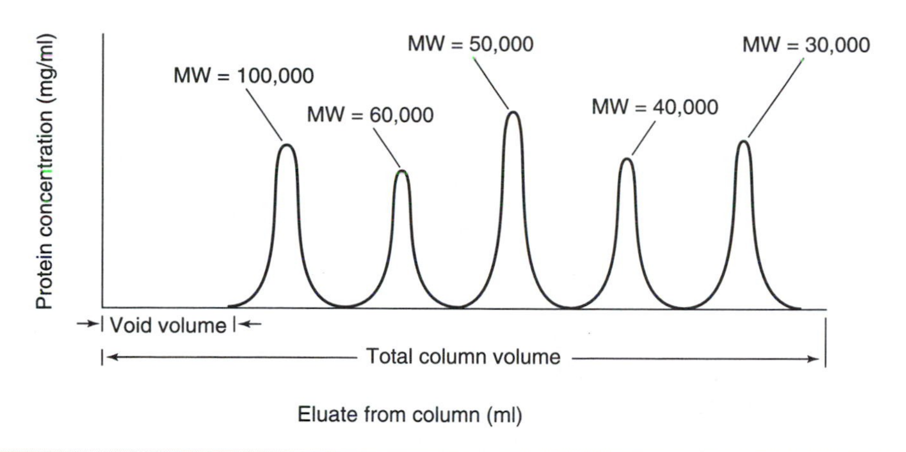

Figure 2-4 Gel filtration of excluded, partially included, and fully included proteins. (Partially included, spherical proteins generally elute in an order directly proportional to the log of their molecular weights. Gel filtration of a protein of unknown molecular weight along with proteins of known molecular weights can therefore yield the molecular weight of the "unknown" protein.)

10,000-Da molecule will be retarded more than that of the 45,000-Da molecule relative to the migration or flow rate of the mobile phase. As a result, the 45,000-Da molecule would elute from the column before the 10,000-Da molecule. The volume at which both molecules elute would be somewhere between the void volume and the total volume of the column (Fig. 2-4).

Gel filtration is usually performed with an isocratic mobile-phase gradient. It is most often used to separate molecules that differ significantly in size. It can also be used as a method to exchange the buffer in which a molecule of interest resides. Suppose that you have purified a 50,000-Da protein in a solution containing 50 mM phosphate buffer at pH 8.0 containing 0.5 M NaCl. The NaCl could easily be removed from the protein solution by passing it through an appropriate gel-filtration column with a small pore size using 50-mM phosphate buffer at pH 8.0 as the mobile phase. The pore size of the matrix should be small enough to fully exclude the protein but still large enough to fully include the ions that you wish to remove.

Ion-Exchange Chromatography. Ion exchange is perhaps the most classic and popular type of chromatography used in biochemistry. It relies on the differential electrostatic affinities of molecules carrying a surface charge for an inert, charged stationary phase or solid support. There are two types of ion-exchange chromatography, determined by the charges present both on the stationary phase and on the molecules that will be adsorbed to it. *Cation exchange* refers to a situation in which the stationary phase carries a negative charge. It will have affinity for molecules in solution that carry a net positive surface charge (i.e., cations). The opposite is true for *anion exchange:* the stationary phase carries a positive charge that will have affinity for molecules that carry a net negative surface charge (i.e., anions).

Anion and cation exchange resins can be classified as being either strong or weak. Strong ion exchangers maintain a net charge at extremes of pH while weak ion exchangers are subject to titration at these extremes (see Table 2-1). The most commonly used anion-exchange resin is diethylaminoethyl (DEAE) cellulose. It is a weak ion-exchange resin that carries a positive charge below pH 8.5 (pKa ~ 10.0). The most common cation exchange resin in use today is carboxymethyl (CM) cellulose. It is also a weak ion exchanger that carries a negative charge above pH 4.5 (pKa ~ 4.0). Since most enzymes display optimal activity and stability between pH 4.0 and 9.0, these two ion-exchange resins will suffice for most protein purification schemes. A variety of ion-exchange resins,

Table 2-1 Commercially Available Chromatography Adsorbents

Gel-Filtration Chromatography

Adsorbent	MW Fractionation Range (Da)*	Matrix Composition	Supplier
Superdex	1×10^4 to 6×10^5	Dextran/agarose	Pharmacia
Superose	1×10^3 to 5×10^6	Agarose	Pharmacia
Sephacryl	1×10^3 to 8×10^6	Acrylamide	Pharmacia
Bio-Gel P	1×10^2 to 1×10^6	Acrylamide	BIORAD
Sepharose	1×10^4 to 4×10^7	Agarose	Pharmacia
Sephadex	7×10^2 to 2.5×10^5	Dextran	Pharmacia
Bio-Beads (SX)	600 to 1400	Styrene divinylbenzene	BIORAD
(for organic compounds and other hydrophobic molecules)			

*Fractionation range depends on the particle size of the gel.

Ion-Exchange Chromatography

Adsorbent	Functional Group	Application	Supplier
Mono Q	$-CH_2N^+(CH_3)_3$	Strong anion	Pharmacia
Q Sepharose	$-N^+(CH_3)_3$	Strong anion	Pharmacia
AG 1	$-CH_2N^+(CH_3)_3$	Strong anion	BIORAD
AG 2	$CH_2N^+(CH_3)_2C_2H_4OH$	Strong anion	BIORAD
DEAE-Sepharose	$-CH_2N^+H(CH_3)_2$	Weak anion	Pharmacia
AG 4-X4	$-CH_2N^+H(CH_3)_2$	Weak anion	BIORAD
Mono S	$-CH_2SO_3^-$	Strong cation	Pharmacia
AG 50W	$-SO_3^-$	Strong cation	BIORAD
SP Sepharose	$-SO_3^-$	Strong cation	Pharmacia
CM-Sepharose	$-CH_2COO^-$	Weak cation	Pharmacia
Bio-Rex 70	$-COO^-$	Weak cation	BIORAD

Hydrophobic Interaction Chromatography

Adsorbent	Functional Group	Supplier
Methyl HIC	$-CH_3$	BIORAD
T-butyl HIC	$-CH(CH_3)_3$	BIORAD
Phenyl Sepharose	benzene ring	Pharmacia
Butyl Sepharose	$-CH_2-CH_2-CH_2-CH_3$	Pharmacia
Octyl Sepharose	$-CH_2-(CH_2)_6-CH_3$	Pharmacia

each highly resolving, sturdy, and able to maintain a high flow rate under pressure are now commercially available (see Table 2-1).

How are molecules adsorbed on ion-exchange columns eluted? Two options are available: changing the ionic strength or the pH of the mobile phase (Fig. 2-5). Unlike with gel-filtration columns, ion-exchange columns are most often developed using a mobile-phase gradient that continuously alters one of these two parameters. A molecule containing one or more ionizable groups will be charged (positive or negative) over a given pH range. Suppose that a molecule of interest contains a carboxyl group (COO^-, COOH) with a pKa value of 4.2. Above pH 4.2, this group will take on a negative charge (COO^- form). In order for a molecule con-

Table 2-1 Continued

Affinity Chromatography

Adsorbent	Functional Group	Application	Supplier
Arginine-Sepharose	L-Arginine	Serine proteases	Pharmacia
Affi-Gel Blue	Cibacron Blue	Albumin, interferons,	BIORAD
Blue Sepharose		NAD^+-requiring enzymes	Pharmacia
Red Sepharose	Procion Red	$NADP^+$-requiring enzymes	Pharmacia
		Carboxy-peptidase G	
Calmodulin Sepharose	Calmodulin	Calmodulin-binding proteins, neurotransmitters	Pharmacia
Glutathione Sepharose	Glutathione	*S*-transferases	Pharmacia
Affi-Gel Heparin	Heparin	DNA-binding proteins,	BIORAD
Heparin Sepharose		growth factors, steroid receptors	Pharmacia
		Lipases, lipoproteins	
Lysine Sepharose	Lysine	Ribosomal RNA	Pharmacia
Affi-Gel Protein A	Protein A	Immunoglobulin G	BIORAD
Protein A Superose			Pharmacia
Protein G Superose			Pharmacia
Con A Sepharose			Pharmacia
Lentil Lectin Sepharose	Lectins	Membrane proteins, glycoproteins, polysaccharides	Pharmacia
Wheat Germ Lectin Sepharose		T lymphocytes	Pharmacia
Poly A Sepharose	Polyadenylic acid	mRNA-binding proteins	Pharmacia
		Viral RNA, RNA polymerase	
Poly U Sepharose	Polyuridylic acid	mRNA, reverse transcriptase, interferons	Pharmacia
Poly T Sepharose	Polydeoxythymidylic acid	mRNA	Pharmacia
DNA Agarose	Calf thymus DNA	DNA/RNA polymerase, polynucleotide kinase	Pharmacia
		Endonucleases and exonucleases	
5′AMP Sepharose	5′AMP	NAD^+-requiring enzymes, kinases	Pharmacia
Affi-Gel Polymixin	Polymixin	Endotoxins	BIORAD
Affi-Gel 501	Sulfhydryls	Proteins with sulfhydryl groups	BIORAD
Affi-Gel 601	Boronate	Sugars, nucleotides, glycopeptides	BIORAD

taining this group to adsorb to a DEAE column, you must ensure that the pH of the mobile phase is well above 4.2 (say, pH 6.0). If the pH of the mobile phase is then lowered to 3.0, the molecule would no longer be negatively charged (the pH is below the pKa value for this group) and would no longer have affinity for this anion-exchange resin. This principle underlies the use of a pH gradient in the elution of molecules from an ion-exchange column: the molecule is adsorbed to the column at

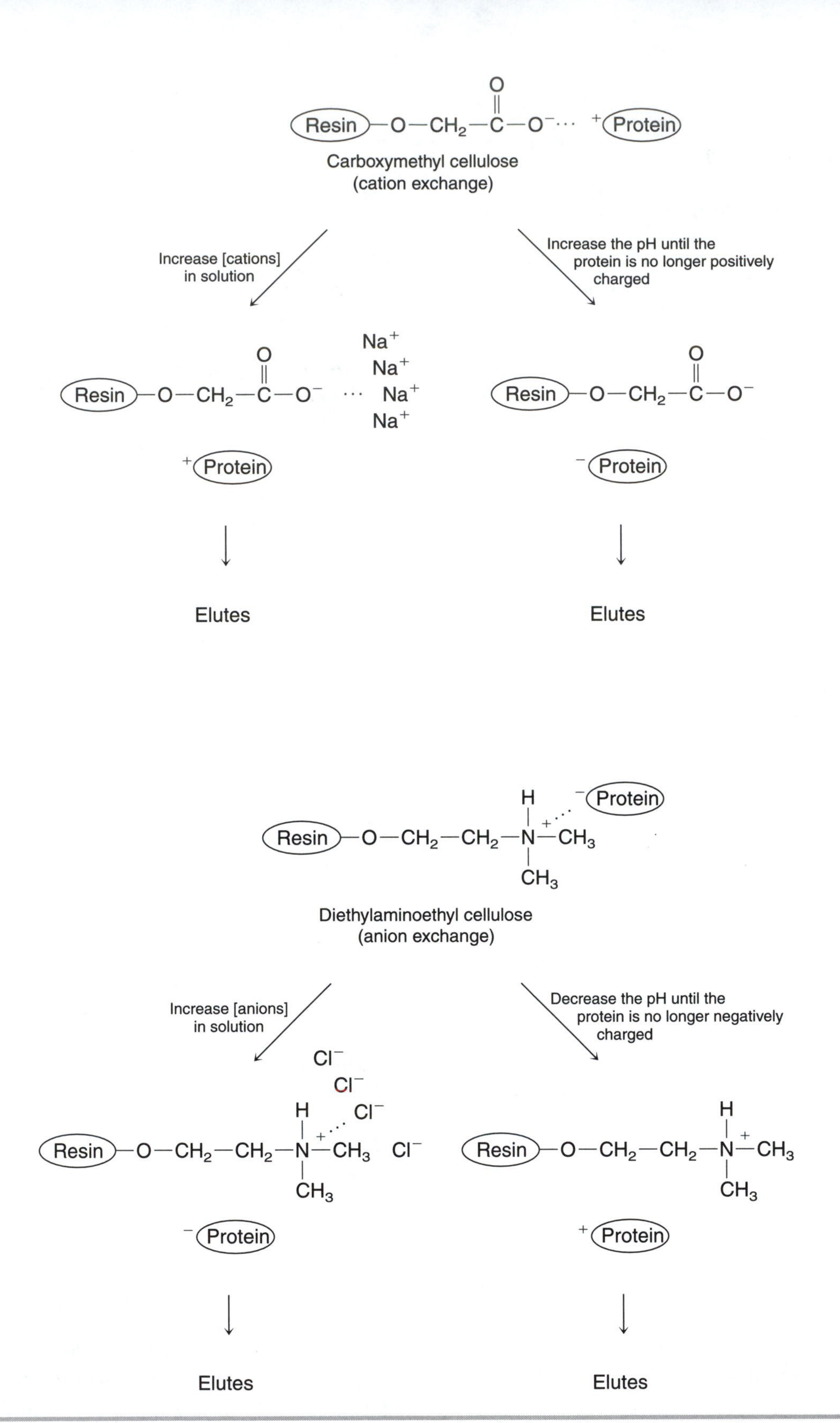

Figure 2-5 Elution of proteins from ion-exchange resins by altering pH or ionic strength.

a pH that gives it a charge opposite that of the resin, and is eluted by a change in pH that causes it to lose its charge or affinity for the resin. Anion-exchange columns are usually developed with a decreasing pH gradient, while cation-exchange columns are usually developed with an increasing pH gradient.

Although elution with a pH gradient appears theoretically feasible, a number of practical aspects complicate its use in the laboratory. First, as the pH of the mobile phase is altered, the charges on both the resin and the adsorbed molecules are subject to change. This is particularly true if you are using a weak ion-exchange resin near the pKa of the substituent (charged) group. Second, the adsorbed molecules contribute significantly to the overall pH of the mobile phase, particularly in the microenvironment of the resin where the exchange process is taking place. Finally, the buffering power contributed by the adsorbed molecules as they elute from the column can turn what appear to be small changes in the pH dictated by the introduced gradient into large changes in pH that cause the elution of a large number of different molecules from the column. For example, the decrease in the pH of 0.5 that is introduced by a gradient may translate into a pH decrease of 1.5 in the microenvironment of the resin because of the high concentration of titratable molecules trapped within it as compared with the concentration of buffering species in the bulk mobile phase.

Because of the practical aspects just described for pH gradients, ionic strength (salt) gradients are most often used for the elution of molecules from ion-exchange columns. As the concentration of counter-ions in the mobile phase increases, they begin to compete electrostatically with adsorbed molecules bound to the charged resin. Suppose that you have a positively charged molecule bound to CM-cellulose resin. As the concentration of positively charged ions in the mobile phase increases (through an increasing gradient of KCl or NaCl), the Na^+ and K^+ will compete with the positively charged molecules for binding to the resin. Molecules with lower α values will elute at a lower salt concentration than molecules with larger α values (in this case, the α value will be dictated by the magnitude of the positive charge on the molecule). Since the concentration of salt in the mobile phase is continually increasing, the top of the band of molecules moving down the column will always experience a slightly higher salt concentration (lower α value) than the portion of the band toward the bottom of the column. This, in effect, compacts the bands or zones of eluting molecules, minimizing zone spreading and maximizing resolution. By using gradient elution schemes in ion-exchange chromatography, one can offset the deleterious effects (mainly, zone spreading) that accompany chromatographic procedures that require a large number of plate transfers to achieve separation.

Choosing an appropriate buffer for ion-exchange chromatography is an important consideration. In general, there is only one rule to keep in mind: *choose a buffer that will provide the maximum buffering power at a particular pH while contributing as little as possible to the overall ionic strength of the mobile phase.* Recall that a buffering species in solution will have its greatest buffering capacity at or very near its pKa value. Therefore, a buffer with a pKa of 6.5 would be a poor choice for an ion-exchange column equilibrated and loaded at pH 8.0. As a general rule, it is suggested that the ionic strength of the mobile phase be limited to 5 to 10 mM with a buffer that has a pKa 0 to 0.5 pH units away from the operating pH of the column. Remember that the charged groups on the ion-exchange resin will electrostatically attract any molecule or ion carrying an opposite charge. Therefore, it is preferable to use a buffer whose charged form is of the same sign as that on the resin. For example, Tris is a buffer commonly used in conjunction with anion-exchange resins. Tris can either exist in the protonated form (positively charged) or the unprotonated (neutral) form (Cl^- is the counter-ion for Tris in the protonated form). Since the anion exchange column is positively charged, the concentration of negatively charged species in the mobile phase should be minimized. Since Tris can exist only in the positively charged or neutral form, the concentration of Cl^- counter-ion is the only species contributing to the effective ionic strength of the mobile phase. The same is not true for a mobile phase buffered by, say, phosphate or acetate ions. Here, at least one form of the buffer carries a charge opposite that of the anion-exchange resin, potentially affecting the ability of negatively charged molecules in the sample to adsorb to the resin (a decrease in the capacity of the column). Any buffering ion that carries a negative charge would be better to use in conjunction with a cation-exchange column, where the resin also carries a negative charge. A list of buffers recommended for use in anion- and cation-exchange chromatography can be found in Table 2-2.

Hydrophobic Interaction Chromatography (HIC). As many proteins fold into their native, three-dimensional structure, a number of hydrophobic regions or "patches" become exposed on their surfaces. The more of these hydrophobic regions present on the surface, the less water soluble the protein is likely to be. This difference in hydrophobicity between proteins can be exploited as a means of purification by a technique termed "hydrophobic interaction chromatography." If a stationary phase contains a large number of aliphatic (multicarbon) chains on it, hydrophobic interactions may allow proteins with different hydrophobic surface character to adsorb to it with variable strength. The principles that underlie this technique are quite similar to those forces at work during the process of "salting out" of proteins under high ionic strength conditions (see Section II, Experiment 8, "Purification of Glutamate–Oxaloacetate Transaminase from Pig Heart").

In general, the conditions of the mobile phase that promote the binding of proteins to hydrophobic adsorbents are exactly opposite those that promote binding of proteins to cation-exchange resins: low pH and high ionic strength. Whereas high ionic strength serves to weaken protein-adsorbent electrostatic interactions in ion-exchange chromatography, high ionic strength serves to promote hydrophobic interactions. High ionic strength is a much more critical parameter than low pH, especially since many proteins are subject to denaturation under mildly acidic conditions. The more hydrophobic the protein of interest, the less hydrophobic the aliphatic chain on the adsorbent is required for binding. While a very hydrophobic protein may bind a C_4 column under conditions of high ionic strength, a slightly hydrophobic protein may require a C_8 or C_{10} resin (more hydrophobic) for adsorption.

Elution of proteins from hydrophobic adsorbents can be achieved by altering the mobile phase in a manner that disrupts the hydrophobic interactions between the molecule and the adsorbent. Since high ionic strength conditions promote such interactions, elution can often be induced by simply lowering the ionic strength of the mobile phase. In other cases, it may be necessary to include a component in the mobile phase that actively disrupts the hydrophobic interactions between the protein and the resin. Organic solvents, non-ionic (nondenaturing) detergents such as Triton X-100, and polyethylene glycol (PEG) are commonly used. Finally, some hydrophobic interactions can also be weakened by increasing the pH of the mobile phase, provided that it is not detrimental to the stability of the protein of interest.

Affinity Chromatography. Affinity chromatography is perhaps the most rapidly advancing type of chromatography in the field of biochemistry. Considering that the goal of any chromatographic step is to achieve the highest yield and the highest degree of purity possible, affinity chromatography presents itself as the most promising candidate for this purpose. If one considers a biological molecule such as an enzyme, what distinguishes one enzyme from the hundreds that are found in the cell? The answer is in the ligand(s) (substrates, ions, cofactors, peptides, etc.) for which the enzyme has affinity. Whereas the other types of chromatography discussed thus far separate biological molecules based on the more general characteristics of size, surface charge, and hydrophobicity, affinity chromatography attempts to separate individual molecules by exploiting characteristics specific to them.

Substrate Affinity Chromatography. A number of adsorbents designed to isolate specific classes of biological molecules are currently in use in the field of biochemistry. DNA-binding proteins are often purified with adsorbents containing heparin (a sulfonated polysaccharide with high negative charge) or calf-thymus DNA. Oligosaccharides and glycoproteins can be purified with lectin columns. For example, concanavalin A (a lectin) is known to have strong affinity for both mannose and glucose. Resins derivatized with 5′AMP, NAD^+, or $NADP^+$ are often used in place of dye-ligand adsorbents (see below) as an alternative in the purification of ki-

Table 2-2 Buffers for Use in Ion-Exchange Chromatography

For Anion Exchange

Buffer	pKa
Histidine	6.0
bis-[2-Hydroxyethyl]iminotris[hydroxymethyl]methane (BIS-TRIS)	6.5
1,3-bis[Tris(Hydroxymethyl)methylamine]propane (BIS-TRIS PROPANE)	6.8
Imidazole	7.0
Triethanolamine (TEA)	7.8
Tris[Hydroxymethyl]aminomethane (TRIZMA)	8.1
N-Tris[Hydroxymethyl]methylglycine (TRICINE)	8.1
N,N-bis[2-Hydroxyethyl]glycine (BICINE)	8.3
Diethanolamine	8.9

For Cation Exchange

Buffer	pKa
Lactic acid	3.9
Acetic acid	4.8
2-[*N*-morpholino]ethanesulfonic acid (MES)	6.1
N-[2-Acetamido]-2-iminodiacetic acid (ADA)	6.6
2-[(2-Amino-2-oxoethyl)amino]ethanesulfonic acid (ACES)	6.8
Piperazine-*N,N'*-bis[2-ethanesulfonic acid] (PIPES)	6.8
3-[*N*-morpholino]-2-hydroxypropanesulfonic acid (MOPSO)	6.9
N,N-bis[2-Hydroxyethyl]-2-aminoethanesulfonic acid (BES)	7.1
3-[*N*-Morpholino]propanesulfonic acid (MOPS)	7.2
N-Tris[Hydroxymethyl]methyl-2-aminoethanesulfonic acid (TES)	7.4
N-[2-Hydroxyethyl]piperazine-N'-[2-ethanesulfonic acid] (HEPES)	7.5
3-[*N,N*-bis(2-Hydroxyethyl)amino]-2-hydroxypropanesulfonic acid (DIPSO)	7.6
3-[*N*-Tris(Hydroxymethyl)methylamino]-2-hydroxypropanesulfonic acid (TAPSO)	7.6
N-[2-Hydroxyethyl]piperazine-N'-[2-hydroxypropanesulfonic acid] (HEPPSO)	7.8
Piperazine-*N,N'*-bis[2-hydroxypropanesulfonic acid] (POPSO)	7.8
N-[2-Hydroxyethyl]piperazine-N'-[3-propanesulfonic acid] (EPPS)	8.0
N-Tris[Hydroxymethyl]methyl-3-aminopropanesulfonic acid (TAPS)	8.4
3-[(1,1-Diethyl-2-hydroxyethyl)amino]-2-hydroxypropanesulfonic acid (AMPSO)	9.0
2-[*N*-Cyclohexylamino]ethanesulfonic acid (CHES)	9.3
3-[Cyclohexylamino]-2-hydroxy-1-propanesulfonic acid (CAPSO)	9.6
3-[Cyclohexylamino]-1-propanesulfonic acid (CAPS)	10.4

nases and dehydrogenases. Poly dT columns are often used to purify messenger RNA from eukaryotic cells that often end in a poly A tail (A will form hydrogen bonds with T in DNA). In all instances, the molecules of interest are eluted by the addition of excess free ligand, change in ionic strength, change in pH, addition of low concentrations of chaotropic salts or denaturing agents, addition of organic solvents, or addition of some type of detergent (see below).

In order to perform affinity chromatography, you must have a means of covalently attaching the desired substrate molecule to the resin. A number of different techniques can be used to chemically

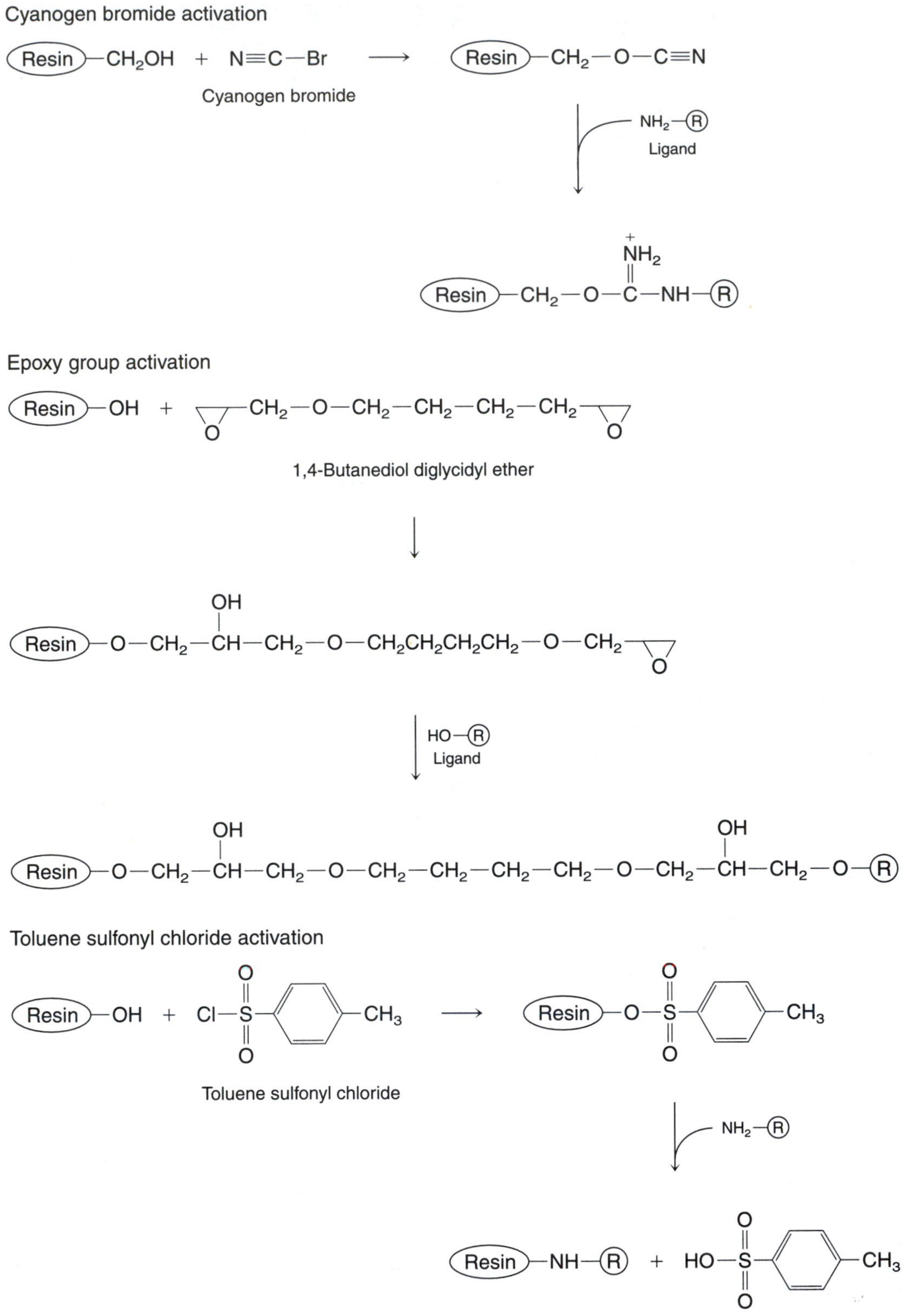

Cyanogen bromide activation

$$\text{Resin}-CH_2OH + N{\equiv}C-Br \longrightarrow \text{Resin}-CH_2-O-C{\equiv}N$$

Cyanogen bromide

NH_2-(R) Ligand

$$\text{Resin}-CH_2-O-\overset{\overset{+}{NH_2}}{\overset{\|}{C}}-NH-(R)$$

Epoxy group activation

$$\text{Resin}-OH + \text{(epoxide)}-CH_2-O-CH_2-CH_2-CH_2-CH_2-\text{(epoxide)}$$

1,4-Butanediol diglycidyl ether

$$\text{Resin}-O-CH_2-\overset{OH}{\overset{|}{CH}}-CH_2-O-CH_2CH_2CH_2CH_2-O-CH_2-\text{(epoxide)}$$

HO—(R) Ligand

$$\text{Resin}-O-CH_2-\overset{OH}{\overset{|}{CH}}-CH_2-O-CH_2-CH_2-CH_2-CH_2-O-CH_2-\overset{OH}{\overset{|}{CH}}-CH_2-O-(R)$$

Toluene sulfonyl chloride activation

$$\text{Resin}-OH + Cl-SO_2-C_6H_4-CH_3 \longrightarrow \text{Resin}-O-SO_2-C_6H_4-CH_3$$

Toluene sulfonyl chloride

NH_2-(R)

$$\text{Resin}-NH-(R) + HO-SO_2-C_6H_4-CH_3$$

Figure 2-6 Activation of carbohydrate-based resins for covalent attachment of affinity ligands (R).

"activate" carbohydrate-based resins for derivatization with a ligand of interest (Fig. 2-6). Most of these chemical reactions take place on the hydroxyl groups of the resin and create intermediates that are subject to nucleophilic attack by a free amino group on the desired ligand.

Dye-Ligand Chromatography. Of all the types of chromatography presented in this chapter, the principles governing the technique of dye-ligand chromatography are perhaps the least well understood. Purely by accident in the 1960s, a blue dye known as Cibacron Blue was found to have strong affinity for two enzymes found in yeast: pyruvate kinase and phosphofructokinase. As shown in Fig. 2-7, Cibacron Blue consists of a series of individual and fused aromatic rings containing several conjugated double bonds common to molecules that absorb light in the visible range of wavelengths. Ring structures similar to this are present in purine nucleotides. This molecule also contains three sulfonic acid (SO_3^-) groups that may be mistaken for phosphate (PO_4^-) groups by an enzyme. Based on these two pieces of evidence, it is currently believed that Cibacron Blue acts as a substrate analog of ADP-ribose, explaining why the molecule has been found to have affinity for many different NAD^+-requiring enzymes (dehydrogenases) and kinases (commonly bind purine nucleotides).

Unfortunately, the specificity of this blue dye for the types of enzymes just described is not as great as originally thought. Indeed, Cibacron Blue has been proven to have great affinity for a number of seemingly unrelated proteins, such as γ-interferon and blood serum albumin. The nonspecific interactions between the protein and the dye may be due to cation exchange effects (due to the SO_3^- groups) or hydrophobic and/or hydrogen-bonding effects between the proteins and the aromatic rings on the dye.

A number of methods have proven successful in eluting proteins from Cibacron Blue. Most often, the greatest purification is achieved through some sort of affinity elution scheme. Here, a protein of interest bound at its nucleotide binding site to the resin is eluted from it through the addition of its true substrate or a high concentration of a substrate analog to the mobile phase. This free ligand will compete with the "fixed" groups on the resin for binding to the protein, lowering its α value. Substrates or ligands commonly used to elute proteins from Cibacron Blue columns include: *S*-adenosylmethionine (SAM), ATP, GTP, NAD^+, NADH, ADP, AMP, and cAMP. If a suitable ligand cannot be identified that will cause the protein of interest to elute from the column, elution can be achieved using the same mobile-phase gradients commonly used with cation-exchange columns: increasing pH or ionic strength. Two commercially available dye-ligand chromatography adsorbents are shown in Table 2-1.

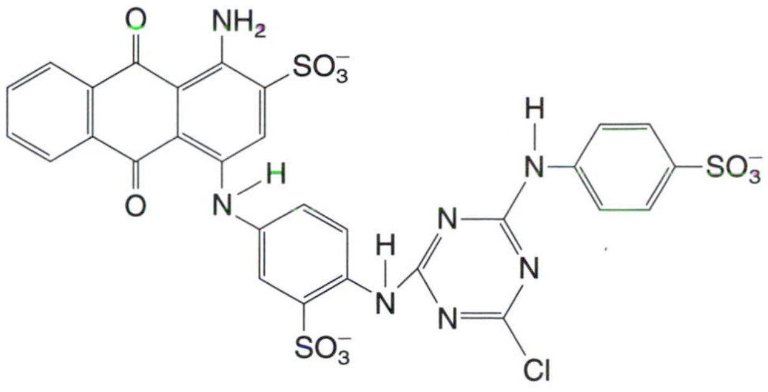

Figure 2-7 Structure of Cibacron Blue F3GA.

Immunoaffinity Chromatography. As will be described in Section IV, an antibody is capable of selectively recognizing and binding a single protein or other antigen in the presence of thousands of other proteins and other biological molecules. Naturally, then, immunoaffinity chromatography offers perhaps the greatest potential in attempting to purify a protein in a single step. A purified antibody (polyclonal or monoclonal) is covalently bound to a stationary-phase resin through which a solution containing a protein of interest is passed. As the column is washed with an appropriate buffer, all of the proteins not recognized by the antibody will elute from the column. Ideally, only the protein of interest will remain adsorbed to the resin. Unlike the other types of chromatography discussed thus far, the challenge in immunoaffinity chromatography is in trying to determine the conditions of the mobile phase that will lower the α value of the protein, causing it to elute from the column.

Antigen–antibody interactions rank as some of the strongest found among biological molecules, with dissociation constants (K_Ds) as high as 10^{-12} M.

Typical dissociation constants for these interactions range from about 10^{-8} to 10^{-10} M. Because the binding between an antibody and its antigen is so strong, eluting the protein from the column without denaturing it or the column-bound antibodies is not a trivial matter. Part of the problem is due to the variety of chemical forces known to play a role in antigen–antibody complexes, such as hydrophobic interactions, electrostatic interactions, and hydrogen bond formation. A mobile-phase gradient designed to alter the pH drastically may elute the protein, but it may also denature it and the antibodies derivatized to the column (a very expensive and time-consuming column to prepare). A mobile-phase gradient of increasing ionic strength may disrupt electrostatic interactions between the protein and the antibody but increase the affinity of the two caused by hydrophobic interactions.

Although the use of immunoaffinity chromatography may seem problematic, there are several precautions that can be taken to increase the odds of successful, nondenaturing elution. First and foremost, monoclonal antibodies work best for immunoaffinity chromatography. Remember that a monoclonal antibody will recognize a specific region or epitope on a protein with a defined affinity (K_D). This situation is much more predictable and subject to control than that using polyclonal antibodies, which contain many different antibodies that recognize a number of different epitopes with different affinities. During the screening process for monoclonal antibodies, you could select for a clone producing an antibody with an affinity for the protein that is great enough to allow for purification but not so great as to present problems during the elution step (K_D on the order of 10^{-6} to 10^{-7}). Second, you may need to experiment with different mobile phases to try to determine which forces predominate in the specific antigen–antibody complex under study (hydrophobic, electrostatic, or hydrogen bonding). Reagents and conditions commonly used to elute proteins from immunoaffinity columns include introduction of a decreasing pH gradient, low concentrations of denaturing agents such as urea or guanidine hydrochloride, low concentrations of chaotropic (denaturing) salts such as lithium bromide and potassium thiocyanate, and low concentrations of non-ionic (nondenaturing) detergents such as Triton X-100.

Immobilized Metal Affinity Chromatography (IMAC). Immobilized metal affinity chromatography is growing in popularity as a means of purifying proteins that contain histidine residues on their surface (Fig. 2-8). A resin derivatized with a metal ion-chelating agent is first equilibrated with a buffer containing a metal ion. Salts containing a divalent cation are most often used for this (Zn^{2+}, Fe^{3+}, Ni^{2+}, Cu^{2+}). A protein solution is then passed through the column and washed extensively with buffer to remove unbound proteins. Proteins that bind the column through their exposed histidine residues can be eluted by one of three methods: (1) lowering the pH of the mobile phase, causing the histidine residue to become positively charged and lose its affinity for the metal ion, (2) addition of a metal-chelating agent (i.e., EDTA) stronger than that on the IMAC resin with which the metal ion is complexed, or (3) increasing the concentration of a compound in the mobile phase that will compete with the protein for metal binding (i.e., imidazole or histidine). Though sound in theory, certain practical problems prevent widespread use of this type of chromatography. First, since the resin carries a strong negative charge due to the presence of the metal chelator, it is possible that cation exchange effects may occur if all the sites containing the chelator are not complexed (saturated) with the metal ion in the mobile phase. Second, the long spacer arms commonly used in IMAC resins often bind a number of proteins nonspecifically through hydrophobic interactions. This latter problem is compounded by the fact that IMAC columns are often developed under conditions of high ionic strength in an attempt to minimize the potential ion-exchange effects. Despite these problems, IMAC has proven to be a powerful tool in the purification of a number of different proteins.

Partition Chromatography. Unlike the other types of chromatography discussed thus far, this type is usually used for analytical (rather than preparative) applications. In addition, this type of chromatography is often performed in a thin-layer (rather than a column) format. The stationary phase in partition chromatography is usually a glass plate (rigid) or polyester sheet (flexible) coated with a very thin layer of the desired adsorbent. For most applications, the adsorbent is a cellulose, silica, polyamine, or aluminum oxide–based matrix. In

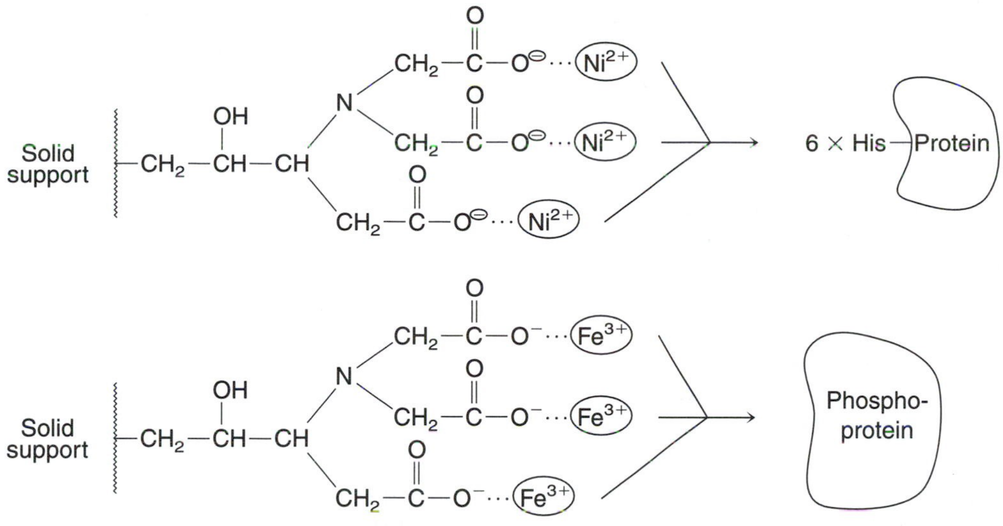

Figure 2-8 Immobilized metal affinity chromatography resin (nitrilotriacetic acid).

some cases, these types of adsorbents can be derivatized with DEAE, polyethyleneimine, acetyl, or C_{18} groups.

In normal-phase partition chromatography, the stationary phase is polar and the mobile phase used in development is nonpolar. In reverse-phase partition chromatography, the stationary phase is nonpolar and the mobile phase used in development is polar. In thin-layer partition chromatography, the samples are spotted in a straight line across the bottom of the solid support (at the origin). Care must be taken to avoid widespread diffusion of these samples (small aliquots of the total applied sample are spotted and allowed to dry). After all the applied samples have dried, the plate is placed, origin side down, in a chamber containing a small amount of the mobile-phase solvent on the bottom. It is essential that the origin lies above the level of the mobile-phase solvent in the chamber (see Fig. 6-8) and that the chamber is saturated with the vapors of the mobile phase before development is begun.

The mobile phases used in the development of thin-layer chromatography plates are usually combinations of organic and aqueous solvents. Ethanol, t-amyl alcohol, acetonitrile, and methanol are commonly used organic solvents. Acetic acid is the most commonly used aqueous solvent. In normal-phase partition chromatography, the polar stationary phase is developed with relatively nonpolar mobile phase. Compounds will display different rates of migration through the stationary phase depending on their polarity. More-polar compounds will have more affinity for the polar stationary phase than for the nonpolar mobile phase. As a result, the more-polar compounds will travel less distance across the plate than the nonpolar compounds, which have a higher affinity for the nonpolar mobile phase. The opposite is true for reverse-phase partition chromatography: the more polar the compound, the greater its affinity for the mobile phase, the further that compound will travel across the plate in a given period of time. The behavior of a compound in thin-layer partition chromatography is usually described in terms of relative mobility (R_f):

$$R_f = \frac{\text{Distance traveled by sample (cm)}}{\text{Distance traveled by solvent front (cm)}}$$

The larger the value of R_f, the more affinity the compound has for a particular mobile phase solvent.

The major advantages of thin-layer partition chromatography are cost and versatility. Compared to column chromatography and HPLC, thin-layer chromatography is very inexpensive to perform. Experimentation with different adsorbents and mobile-phase compositions will allow the separation of a wide variety of biological molecules. Thin-layer chromatography has been used successfully in analysis of proteins, hormones, amino acids, amino sugars, carbohydrates, nucleotides, antibiotics, fatty acids, pesticides, and a number of other drugs and metabolic intermediates.

A variation of this same technique uses paper as the stationary phase. Since all commercially available chromatography paper is composed of a polysaccharide matrix saturated with water or some other polar solvent, paper chromatography is most often carried out in the normal-phase format using a nonpolar mobile phase for development. It is even possible to perform reverse-phase paper chromatography by saturating the paper with some nonpolar solvent (such as liquid paraffin) and developing it with a more-polar solvent. The drawback to this technique, however, is that this type of adsorbent is not commercially available and must be used immediately after it is prepared. The techniques of thin-layer and paper chromatography are described in greater detail in Experiment 6.

One final variation of partition chromatography exploits the affinity of different molecules in the gas state for a solid adsorbent. This technique is termed "gas chromatography," and it works particularly well for the separation of nonpolar compounds. In this technique, a liquid sample is vaporized and delivered to a thin-diameter, heated coil coated with a solid matrix or viscous liquid (grease) possessing a very high boiling point. The vaporized sample is delivered to the coil using an inert gas, such as nitrogen, helium, or argon. Volatile molecules that have affinity for the solid or liquid matrix will be retarded as the mobile-phase gas carries the sample through the coil. A detector linked to the system at the end of the column will allow you to identify different compounds as they elute from the coil.

High-Performance (Pressure) Liquid Chromatography. Successful chromatography relies on the ability to maximize the number of chromatographic plates transfers while maintaining good resolution (minimizing zone spreading). Therefore, it would be desirable to minimize the volume of the column and still provide a reasonable number of chromatographic plates to allow separation of the molecule of interest. As stated earlier, a chromatographic plate can be described as the largest area of the column where two molecules with different partition coefficients will have an opportunity to display different rates of migration. The number of chromatographic plates in a given column volume, then, will be related to the surface area of the adsorbent (stationary phase) with which the compounds passing through the column can interact. A 5-ml column containing a stationary phase matrix with a 200-μM pore size will provide less effective surface area for the compounds to interact with than a 5-ml column containing a stationary-phase matrix with a 5-μM pore size. In other words, the 200-μM-pore-size stationary phase will provide fewer chromatographic plates per unit volume than the 5-μM-pore-size stationary phase.

Although stationary phases with small pore sizes provide more chromatographic plates per unit volume than those with larger pore sizes, smaller-pore-size matrices provide more resistance to the flow of the mobile phase passing through it. As a result, high-pressure mobile-phase delivery systems (pumps) have been developed that are capable of driving the mobile phase through small-pore matrices at pressures exceeding 7000 psi. Since the pressure on the stationary phase increases with the flow rate of the mobile phase, the flow rate that a particular HPLC column can accommodate will depend on the strength or rigidity of the matrix. In general, the pressure on the stationary phase increases as the column diameter decreases and column length increases. Therefore, a 20-mm-by-100-cm column will experience less pressure at a flow rate of 1 ml/min than will a 5-mm-by-100-cm column. HPLC columns of different dimensions and different matrix composition are now commercially available that can sustain flow rates as high as 100 ml/min or as low as 0.5 ml/min.

Another feature of modern HPLC systems that makes them desirable for both analytical and preparative applications is the complex mobile-phase gradients that they are capable of producing. Many systems come equipped with a pump integrator or controller (computer) that allow a number of different mobile-phase solvents to be simultaneously mixed and delivered to the stationary phase. Since this process is automated, complex gradients used for a particular application are quite reproducible.

As with most technologies, HPLC is not without some disadvantages. First, and perhaps foremost, is cost. Modern HPLC systems, along with the specialized tubing, fittings, and columns needed to operate them, are quite expensive to acquire and maintain. Second, extreme care must be taken to ensure that both the sample and the mobile-phase solvents are free of any particulates that could clog the column, increase the internal pressure, and crush the stationary phase. Third, the mobile-phase solvents must be degassed to prevent the formation of air bubbles in the system. If this occurs, the integrity and reproducibility of the applied gradient will be compromised. Finally, the smaller dimensions of most HPLC columns as compared with more traditional gravity-driven columns limits the size of the sample that can be applied. If HPLC columns are "overloaded," resolution will decrease and the number of available chromatographic plates per unit volume will not be realized.

Chromatography systems that operate with pressurized mobile-phase gradients are undoubtedly the wave of the future. Solid supports capable of withstanding the pressures associated with HPLC have already been developed for many of the types of chromatography discussed in this section (normal- and reverse-phase partition, ion exchange, gel filtration, etc.). In recent years, the trend in technology has been toward the development of high-capacity, large-pore-size solid supports that can operate at high flow rates (low pressure) and still maintain good resolution. Systems that operate under this principle are commonly referred to as fast protein liquid chromatography (FPLC) systems or low-pressure liquid chromatography (LPLC) systems.

There is no doubt that the technique of chromatography (particularly affinity-based) will continue to make large advances as our understanding of the molecular mechanisms governing protein folding and enzyme–substrate interactions increases. Thousands of proteins and other molecules of biological interest have been purified using combinations of the different types of chromatography described in this chapter. Although no type of chromatography is without problems, each type can provide a great deal of purification when used at the correct time under optimal conditions as determined through thoughtful and carefully performed experiments.

Supplies and Reagents

5- to 8-mm (inside diameter) glass tubing, or equivalent chromatography columns
CM-Sephadex (G-50) in 0.1 *M* potassium acetate, pH 6.0
CM-Sephadex (G-50) in 1.0 *M* potassium acetate, pH 6.0
0.1 *M* Potassium acetate, pH 6.0
1.0 *M* Potassium acetate, pH 6.0
Glass wool
Mixture of blue dextran (1 mg/ml), cytochrome *c* (2 mg/ml) and DNP-glycine (1 mg/ml)
Tape
Pasteur pipette with dropper bulb

Protocol

Demonstration of Chromatographic Separation on CM-Sephadex (G-50)

This simple exercise demonstrates the chromatographic separation of the components of a mixture by a procedure that relies on two of the principles discussed above, namely ion-exchange chromatography and size-exclusion chromatography. The components to be separated (see Table 2-3) differ greatly from one another both in their molecular

Table 2-3 Compounds in Colored Solution Applied to CM-Sephadex

Name	Color	Molecular Weight	Ionic Character
Blue dextran	Blue	>500,000	Nonionic
Cytochrome *c*	Red	12,400	Ionic protein, pI = 10.7 (i.e., net cation below pH 10.7)
DNP-glycine	Yellow	241	Anion above pH 3.0

weight (the basis of their separation by size-exclusion chromatography on Sephadex G-50) and in their net charge at pH 6.0 (exploited for their separation by ion exchange on CM-Sephadex). Each component also has a distinctive chromophore, so that their separation is easily followed by simply observing the color of the components as they separate.

1. Heat a 14- to 16-in. piece of 5- to 8-mm inside diameter glass tubing and pull out a constriction in the center of the tube. Cut the tube at the constriction and mount each of the 2 sections as shown in Figure 2-9.
2. Using a small-diameter glass rod or a Pasteur pipette, push a small piece of glass wool into the throat of each column, being careful not to push the glass wool beyond the beginning of the constriction of the column. Next, label two strips of tape, one "CM-Sephadex in 0.1 M potassium acetate, pH 6.0" and the other "CM-Sephadex in 1.0 M potassium acetate pH 6.0." Attach the two tapes to the separate columns (see Fig. 2-9).
3. Using a Pasteur pipette equipped with a dropper bulb, begin to add well-mixed slurries of CM-Sephadex in potassium acetate solutions to the appropriate columns. Allow the columns to drip while adding increasing increments of the slurried CM-Sephadex until the settled resin bed reaches a height of 4–5 in. *Do not allow the columns to run dry;* that is, do not allow the top fluid surface to penetrate into the settled resin bed. Instead, add small aliquots of 0.1 M potassium acetate buffer, pH 6.0, or 1.0 M potassium acetate buffer, pH 6.0, to the appropriate columns to keep the fluid surfaces above the resin bed.
4. Once a 4- to 5-in. defined resin bed has accumulated in each column, pipette off most of the top fluid with a Pasteur pipette, and then allow the columns to drip until the fluid surface just reaches the resin bed.
5. Immediately add 0.2 ml of the colored solution (containing each of the components in Table 2-3) so that it forms a thin band on top of the resin. (*NOTE:* Do not disturb the top of the resin.) Allow the colored solution to penetrate the columns until the fluid surface again meets the resin bed, then gently add a few drops of the appropriate potassium acetate buffer, pH 6.0, as designated by the taped label on each column. Allow this buffer to penetrate into the column, and then slowly fill up the rest of the column with the potassium acetate buffer of the concentration that matches the column label. Replenish these upper fluid volumes as necessary while observing the course of the elution.

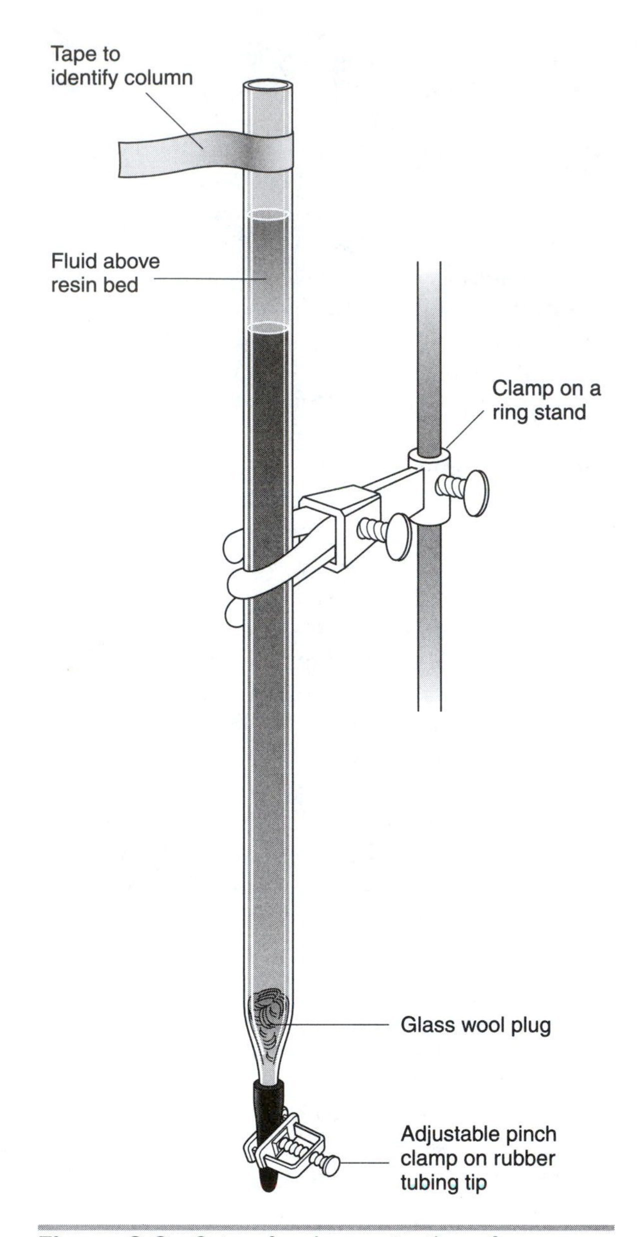

Figure 2-9 Setup for demonstration of properties of CM-Sephadex.

6. Prepare a description of your results that includes an explanation for the degree of migration of each colored component in both columns. Once the first colored reagent begins to exit from the low-salt column (i.e., the column containing 0.1 M potassium acetate buffer, pH 6.0), replace this elution fluid with the high-salt buffer (i.e., 1.0 M potassium acetate, pH 6.0), and observe and explain the results. Which colored component is blue dextran, which is cytochrome c, and which is DNP-glycine? Explain the order in which they elute in terms of their physical properties and the nature of the CM-Sephadex G-50 column. Finally, predict and justify the results that would have occurred if the same experiment had been run with a CM-Sephadex column equilibrated in and eluted with 0.1 M HCl.
7. After the relative movements of all the colored components have been observed under the various elution conditions specified, remove the columns from their clamps, take them to a central container for used CM-Sephadex, and empty the used gel-filtration ion exchanger into the container by shaking or blowing out the resin. (Discard the glass wool plug; do not leave it in the CM-Sephadex resin. The resin will be "regenerated" by appropriate washings and used again.)

Exercises

1. What is the chemical nature of the chromophore of each of the three materials separated on CM-Sephadex in this experiment; that is, what is the light absorbing component that gives it a visible color? Describe the nature of the linkage of the chromophore to the rest of the molecule.
2. You have an aqueous solution containing:

   alanine (a monoamino, monocarboylic amino acid)
   fructose (a non-ionic monosaccharide)
   glycogen (a non-ionic large polysaccharide)
   ribose-5-phosphate (an anionic monosaccharide phosphate)
   tRNA (a polyanionic nucleic acid, MW ~ 30,000)

   Assuming you have a distinctive assay for each of these compounds, what procedures would you use to obtain gram quantities of each of these compounds free of each of the other compounds?

REFERENCES

Block, R. J., Durrum, E. L., and Zweig, G. (1955) *A Manual of Paper Chromatography and Paper Electrophoresis.* New York: Academic Press.

Brenner, M., and Neiderwieser, A. (1967). Thin-Layer Chromatography (TLC) of Amino Acids. Enzyme Structure. *Methods Enzymol* **11**:39.

Fischer, L. (1969). *An Introduction to Gel Chromatography.* New York: American Elsevier.

Jakoby, W. B. (1971). Enzyme Purification and Related Techniques. *Methods Enzymol* **22**.

Lehninger, A. L., Nelson, D. L., and Cox, M. M. (1993). An Introduction to Proteins. In: *Principles of Biochemistry,* 2nd ed. New York: Worth.

Robyt, J. F., and White, B. J. (1990). Chromatographic Techniques. In: *Biochemical Techniques: Theory and Practice.* Prospect Heights, IL: Waveland Press.

Scopes, R. K. (1994). *Protein Purification: Principles and Practice.* New York: Springer-Verlag.

Wilson, K., and Walker, J. (1995). Chromatographic Techniques. In: *Principles and Techniques of Practical Biochemistry,* 4th ed. Hatfield, U.K.: Cambridge University Press.

EXPERIMENT 3

Radioisotope Techniques

Theory

Nuclear Transformations

An atom can be conceptualized as a central nucleus, composed of protons and neutrons, surrounded by a specific cloud of electrons, whose energy levels are defined by the laws of quantum mechanics. Protons carry a mass of 1 and a charge of +1, neutrons carry a mass of 1 but lack a charge, and electrons have a relatively insignificant mass but carry a charge of −1. In a neutral atom the number of electrons equals the number of protons in the nucleus. Ions are atoms from which one or more electrons have been added or removed, creating a species with a net charge. The atomic number (i.e., the number of protons in the nucleus) and the corresponding cloud of electrons largely determine the chemical properties of an atom. Neutrons present in any nucleus therefore increase the mass of the atom in question, but do not influence the charge or atomic number of the atom. Atoms with a constant nuclear charge (constant number of protons) but with varied masses (varied number of neutrons) are called different "isotopes" of an atom. Different isotopes of an atom react alike, having very nearly the same chemical properties.

Chemists and physicists use specific symbols to depict these mass and charge relationships. For any atom X, superscript values denote the nuclear mass of the atom (number of protons plus neutrons), whereas subscript values denote the atom's atomic number (number of protons). Thus ${}^{12}_{6}C$, ${}^{13}_{6}C$, and ${}^{14}_{6}C$ are three different isotopic forms of the element carbon. In this convention, the letters define the element and atomic number in question. The subscript value is often omitted so that ${}^{12}C$, ${}^{13}C$, and ${}^{14}C$ depict the three isotopes of carbon.

Certain combinations of nuclear protons and neutrons for an atom are stable and result in multiple stable isotopes of an atom (e.g., ${}^{12}C$ and ${}^{13}C$). In contrast, other combinations of protons and neutrons are unstable, which results in spontaneous nuclear transformations of nuclear mass and charge (e.g., ${}^{14}C$). There are a variety of nuclear transformation mechanisms by which unstable isotopes achieve a stable state. Many of these mechanisms result in emission of specific particles. Such unstable, particle-emitting isotopes are called "radioisotopes," or radioactive forms of an element. Fortunately for biochemists, the emitted particles can be detected readily. Thus, introduction of a radioactive isotope into a specific biologically important population of molecules allows tracing of the passage or conversion of that radiolabeled molecular species through biochemical reactions or biological processes. It is this ability to follow the rate and extent of flow of specific molecules through biological processes—radiotracer technology—that makes the use of radioisotopes important to biochemists.

Any radioisotope decays by one or, at most, only a few nuclear transformation mechanisms. The characteristics of the emitted particles are sufficiently different from one another that the procedures for detection of the different particles must also be different. Therefore, a fundamental knowledge of radioisotope techniques in biochemistry begins with a study of the particles produced during different mechanisms of nuclear transformation. Because the great majority of biochemical applica-

Table 3-1 Decay Properties of Some Biochemically Important Isotopes

Isotopic Tracer or Label	Reaction	Average Energy (MeV) of β^- Particle	Maximum Energy (MeV) of β^- Particle (E_{max})	Half-Life
^3H	$^3_1H \longrightarrow {}^3_2He + \beta^-$	0.0055	0.015	12.3 years
^{14}C	$^{14}_6C \longrightarrow {}^{14}_7N + \beta^-$	0.05	0.15	5500 years
^{32}P	$^{32}_{15}P \longrightarrow {}^{32}_{16}S + \beta^-$	0.70	1.71	14.3 days
^{35}S	$^{35}_{16}S \longrightarrow {}^{35}_{17}Cl + \beta^-$	0.0492	0.167	87.1 days
^{33}P	$^{33}_{15}P \longrightarrow {}^{33}_{16}S + \beta^-$		0.25	25 days

tions involve only two types of radioactive decay, namely, β^- emission and γ emission, only these two kinds of nuclear transformation and emitted particles will be discussed in detail.

β^- Particle Emission

The process of β^- particle emission occurs in the biologically important radioisotopes indicated in Table 3-1. β particles are negatively charged particles of negligible mass (i.e., electronlike particles). β particles are emitted along with antineutrinos during the conversion of neutrons to protons.

$$\text{Neutron} \longrightarrow \text{proton}^+ + \beta^- + \text{antineutrino}$$

Because the newly formed proton remains in the nucleus after discharge of the β^- particle, the product atom has an atomic number increased by one; for example, carbon is transformed to nitrogen and phosphorus is transformed to sulfur (see Table 3-1). Antineutrinos have no charge and essentially no mass, and they are therefore hard to detect. Because of this, antineutrinos are of little significance to biochemists. β^- Particles are readily detected, however, because of their charge and the energy with which they are expelled from the decaying nucleus. The total energy available from a neutron-to-proton transformation is shared randomly by the β^- particle and the antineutrino. Thus, the β^- particle may be emitted at a variety of different energies (a spectrum) up to the maximum energy available in the transformation. The maximum energies and the associated β^- particle energy profiles for a given β^--emitting radioisotope are therefore distinct (Fig. 3-1). These differences in β^- emission energies play an important role in radioisotope detection and differentiation in so-called double-label experiments, in which more than one radioactive isotope is assayed in a reaction or process (see Experiment 11).

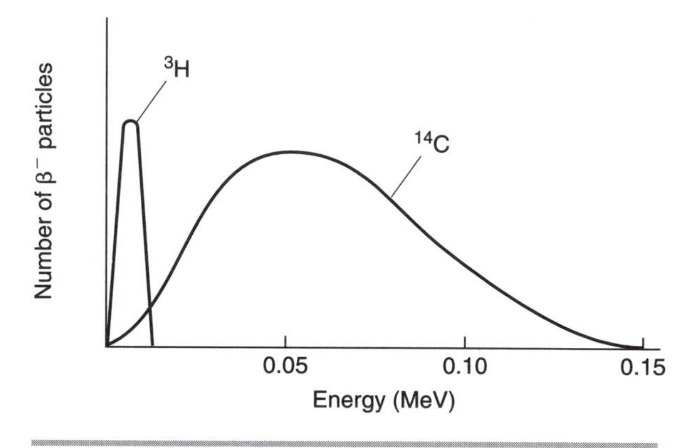

Figure 3-1 β^- energy profiles of ^3H and ^{14}C.

γ Particle Emission

Many particle emission processes in radioisotopes of higher nuclear masses follow a more complex, multistep decay process that results in both particle emission (e.g., β^- particle emission) and emission of high energy photons called γ particles or γ rays. Figure 3-2 depicts a typical scheme for the origin of γ rays during a nuclear decay event.

γ Ray decay systems are different from β^- decay events in three ways. First, these systems may produce more than one γ ray. Second, the resultant γ rays possess distinct energies rather than a spec-

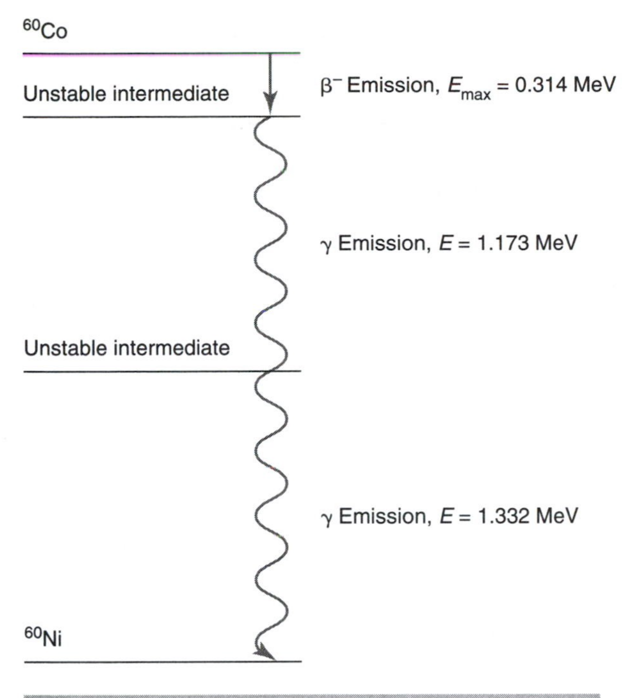

Figure 3-2 Predominant decay scheme for ^{60}Co.

trum of energies. Third, a γ ray decay system may originate from unstable isotopes that initially emit particles other than β^- particles (e.g., α particles). γ Ray emission usually occurs soon after the emission of the primary particle (β^- in Fig. 3-2). The emission of the initiating primary α or β^- particle is often hard to detect, whereas fairly efficient systems exist to detect γ rays. γ Ray spectroscopy is therefore the best way to detect γ ray emitting radioisotopes, such as ^{125}I, ^{59}Fe, or ^{60}Co.

Other Particles

A number of other nuclear transformation mechanisms are found in nature. α Particle decay, positron emission, and electron capture events are examples of such nuclear transformations. Techniques have been developed for measuring such decay events. However, the radioisotopic atoms that yield such decay products are used relatively infrequently in biochemistry and biology. If your research requires analysis of such atoms, you should consult a practical textbook in modern physics or radioisotopic techniques.

Nuclear Decay Rates

All nuclear transformations proceed spontaneously at rates that are not altered by ordinary chemical or physical processes. For any population of unstable atoms, the rate of nuclear transformation or radioactive decay is first order; that is, proportional to the number, N, of decomposing nuclei present:

(3-1)

$$\frac{dN}{dt} = \lambda N$$

where t is time and λ is a disintegration constant characteristic for a particular isotope.

At $t = 0$, N_0 atoms are present. At other times, N atoms are present. Integration of Equation 3-1 within the limits of $t = 0$ and t is:

(3-2)

$$\int_{N_0}^{N} \frac{dN}{N} = -\lambda \int_0^t dt$$

which yields:

(3-3)

$$N = N_0 e^{-\lambda t}$$

or in natural logarithms:

(3-4)

$$\ln \frac{N}{N_0} = -\lambda t$$

Scientists frequently express Equation 3-4 in terms of half-life ($t_{1/2}$), which is the time required for one half of the population of unstable atoms to decay. In other words, after one half-life, the rate of particles emitted has decreased to one half the rate of the starting population of unstable atoms. Substitution in Equation 3-4 yields Equation 3-5:

(3-5)

$$\ln \frac{1/2\ N_0}{N_0} = \ln (1/2) = -\lambda t_{1/2} \quad \text{or} \quad t_{1/2} = \frac{0.693}{\lambda}$$

This equation, in combination with half-lives (see Table 3-1) and Equation 3-4, allows calculation of the quantity of unstable atoms that will be present

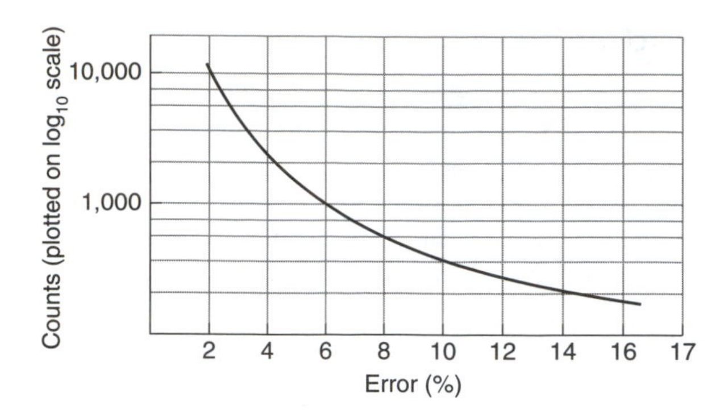

Figure 3-3 Counts versus percent error relationship.

at any time t in a population of atoms containing a known quantity of unstable atoms or known radioactive decay rate.

Statistics and Units of Radioactive Decay

Radioactive particle emission is a random process; that is, the timing of emission of individual particles over time is random. You must therefore detect a large number of particles or counts to define statistically significant count rates (e.g., counts per minute, cpm). Figure 3-3 illustrates the relationship between total recorded counts and the statistical accuracy of the observed count rate. In general, all radioactive samples should be counted to a value within the ±5% error limit of other experimental errors (e.g., pipetting error) within an experiment. In some cases, the design of the experiment may justify accumulating counts per minute to a higher statistical accuracy, such as ±1 or 2%. These statistically significant data of counts per minute can then be converted to data of disintegrations per minute with knowledge of the counting efficiency of the system (see below).

The basic unit of radioactive particle emission is the curie, Ci. Originally, the curie was defined as the number of disintegrations per second emitted from a gram of impure radium (3.7×10^{10}/s or 2.2×10^{12}/min). This unit and its subdivisions the millicurie, mCi (2.2×10^9 dpm), the microcurie, μCi (2.2×10^6 dpm), are the fundamental units to define radioactive particle emission rates. Radioactive compounds display different rates of particle emission depending on the number of unstable, potentially radioactive atoms present and the disintegration constant of the radioisotope. These differences in radioactivity among compounds are expressed as radioactive specific activity; that is, particle emission rates per amount of compound. The most convenient units of radioactive specific activity are curies and moles and their respective subdivisions (e.g., curies per mole, millicuries per millimole, microcuries per mole), but certain instances may require radioactive specific activity units such as millicuries per milligram or counts per minute per gram.

You may also encounter the SI unit of radioactivity, the Becquerel (Bq), which is defined as 1 disintegration per second (dps). In this system 1 curie equals 3.7×10^{10} Bq. For interconversion, it is useful to note that 1 μCi = 3.7×10^4 Bq and 10^6 Bq = 27.027 μCi.

Measurement of Radioactive Decay

Biochemists most frequently make practical use of radioisotopic compounds that emit β^- and γ particles. A number of different techniques have been developed for detecting the rates and energies with which these particles are released during radioactive decay. We will discuss in detail two of the methods most commonly used in biochemical research. These are liquid and solid scintillation counting and radioautography using photographic and phosphor storage techniques.

Scintillation Counting. Scintillation counting is used to detect both β^- and γ particles. Differences exist in the detection process for these different particles, but in both cases, emitted particles cause a series of brief light flashes that are detected by a photocell and recorded as counts for quantitative interpretation. Solid scintillation counting is preferred for the measurement of γ particles. The dense array of atoms in the detection crystal, which is usually sodium iodide, optimizes the chances of absorption of the γ particle, which can be described as a highly energetic x-ray. On the other hand, liquid scintillation counting is preferred for the detection of β particles, which are of relatively low energy and must be intimately mixed with the molecules used for detection.

Liquid Scintillation Counting of β^- Emissions. Liquid scintillation counting is the most efficient technique to detect β^- particle emission. The light-emitting process that is the basis of scintillation counting occurs as a result of a radioactive sample being placed in a "scintillation cocktail" containing an excitable solvent and one or more fluorescent compounds, called "fluors." Emitted β^- particles contact solvent molecules (S) and transfer some of their energy to these molecules. This yields an excited solvent molecule (S*) and a β^- particle containing less energy (E):

(3-6)

$$\beta^- + S \longrightarrow S^* + (\beta^- - E)$$

The energy required to excite a solvent molecule in this reaction is often small compared to the energy of a given β^- particle. Therefore, the residual β^- particle (β^- − E in Equation 3-6) may excite many solvent molecules before its energy is depleted. The eventual number of excited solvent molecules is therefore proportional to the energy of the β^- particle in question.

Excited solvent molecules transfer their energy to other solvent molecules, and eventually to a primary fluor molecule (F_1^*), which then emits a photon ($h\nu_1$) and decays to its ground state (Equations 3-7 and 3-8). A widely used fluor is 2,5-diphenyloxazole (PPO).

(3-7)

$$S^* + F_1 \longrightarrow F_1^* + S$$

(3-8)

$$F_1^* \longrightarrow F_1 + h\nu_1$$

Some scintillation cocktails may employ a secondary fluor that absorbs photons from the primary fluor decay. The excited secondary fluor then reemits new photons at a wavelength more favorable for detection of experimental materials in the scintillation cocktail or more favorable to the phototubes of the particular scintillation counter. However, most modern scintillation counters no longer require the use of a secondary fluor.

The photons emitted by the fluor are detected by a phototube-photomultiplier system. The lifetime of the β^- particle passage, solvent excitation, and eventual photon emission from fluors is short

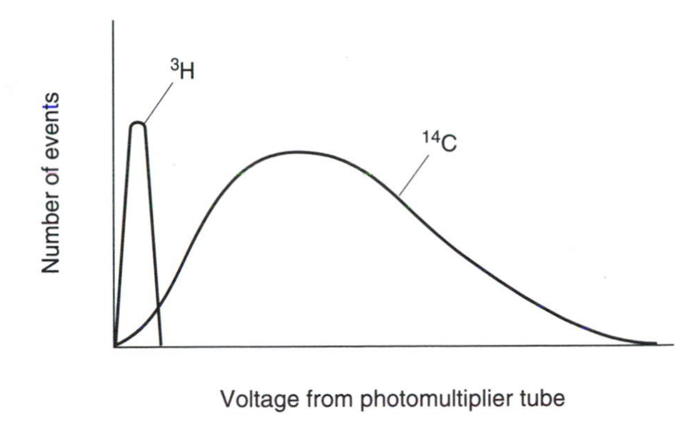

Figure 3-4 Photomultiplier-tube responses from ^{3}H and ^{14}C decay events in a scintillation cocktail.

compared with the response time of the photomultiplier tube system. Therefore, *all the events from a single β^- emission are scored by the photomultiplier tube system as a single "pulse," the energy (voltage) of which is proportional to the number of photons developed by the β^- particle emission event.* Summation of a series of such pulses from a given isotope recreates, in electrical terms (voltage), a curve proportional to the energy profile of an isotope (Fig. 3-4; compare this to Fig. 3-1). The individual voltage pulses that generate the curves of Figure 3-4 (i.e., the areas under the curves) are passed on by the circuitry of the scintillation counter to a scaler system. This system records the pulses as individual counts, and accumulation of them over a period of time leads to a counts per minute (cpm) value.

This is the basic liquid scintillation counting process. However, a working knowledge of scintillation counting requires familiarity with additional features of modern scintillation counters. Consider a modular diagram of a typical scintillation counter (Fig. 3-5). (Frequent design changes and design differences among manufacturers make it difficult to define a typical scintillation counter. Figure 3-5 is a generalization.)

The overall energy of the photons from a single β^- emission is often quite small. The photomultiplier tube used to detect such low-energy events must therefore be very sensitive, but this sensitivity also causes it to detect "photomultiplier noise," or spurious counts unrelated to actual β^- particle emissions. This noise is greatly reduced by

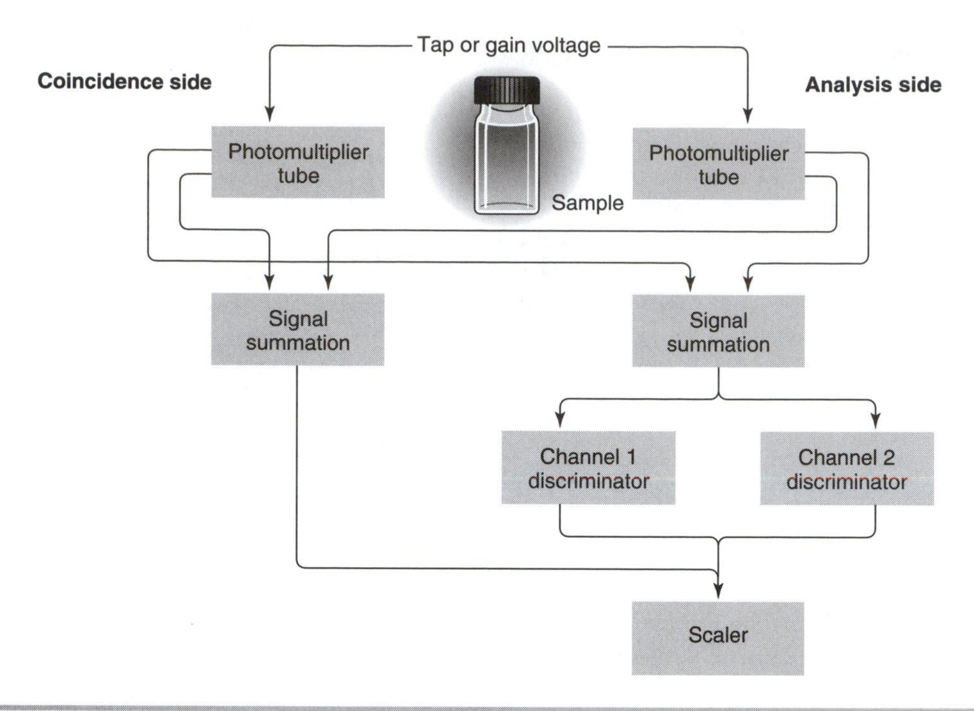

Figure 3-5 Unit diagram of a typical scintillation counter.

use of a coincident circuit that employs two photomultiplier tubes. With such coincidence systems, both photomultiplier tubes must simultaneously detect a light flash to yield a count in the scaler. Thus, most noise events, which occur in a single phototube independently of the other phototube, are screened out. However, the coincident circuitry is also potentially detrimental in that it can exclude detection of low-energy β^- emissions. For example, certain low-energy β^- emissions may not generate enough excited solvent molecules to allow photon detection by both photomultiplier tubes (e.g., only one photon may be formed). This limitation prevents coincident circuit scintillation counters from achieving 100% efficiency (i.e., detection of all of the β^- particle decay events), and is the major limitation in the detection of low-energy β^- particles such as those from ^{3}H.

As seen in Figure 3-5, both the analysis side and the coincidence side of the scintillation counter contain discriminators, which are particularly useful in experiments that employ two or more β^- emitting isotopes. Discriminators contain electronic "gates" that limit the portions of an energy profile observed by each channel. This is best illustrated by the example of a sample containing both ^{3}H and ^{14}C (Fig. 3-6). You can adjust the discriminator in Channel 1 to detect only events that produce the energies or voltages between A and B, and adjust the Channel 2 discriminator to detect only events between B and C. Such settings dictate that Channel 2 will detect ^{14}C independently of any quantity of ^{3}H present in the sample. Channel 1 will detect ^{3}H, but will also detect a significant number of counts from low-energy ^{14}C decay events. Counting a ^{14}C standard (i.e., a ^{14}C sample with a known quantity of radioactivity) in both Channel 1 and Channel 2 allows you to determine the percentage of the total ^{14}C counts detected in Channels 1 and 2. This allows you to correct for the "overlap" of ^{14}C counts within the ^{3}H channel (Channel 1). Thus, independent channels and appropriate discriminator settings provide a useful technique for double-label experiments (see Experiment 11).

The "gain" voltage applied to the photomultipliers of the phototube–photomultipliers defines the degree of amplification these units provide to

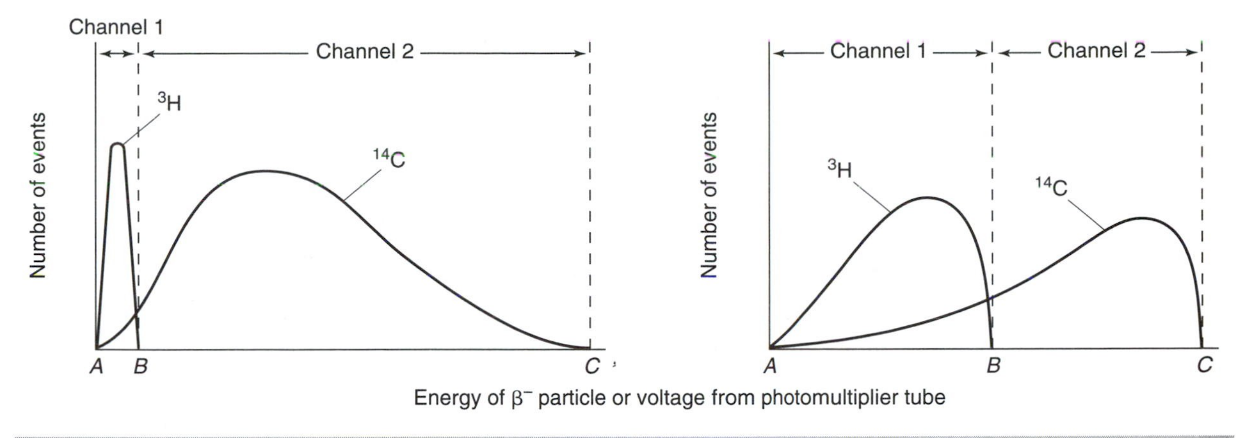

Figure 3-6 Discriminator settings for simultaneous detection of ^{3}H and ^{14}C in linear amplification (e.g., Packard Inst.) at left, and logarithmic amplification (e.g., Beckman), at right.

the voltage pulses from the phototubes. This has a direct bearing on the desired discriminator settings of the channels. Consider the two gain settings of Figure 3-7. The solid lines in Figure 3-7 represent an optimal gain voltage on the photomultipliers that cause the energy profiles to fall within the preset channels. In contrast, use of a gain setting that is too high (dotted lines) yields profiles where some ^{3}H decay events appear as ^{14}C events, while some high-energy ^{14}C events go unrecorded (i.e., are above the ^{14}C gate). It is important to remember that no single gate setting is optimal for all counting conditions; the nature of the scintillation fluid used will determine the photomultiplier gain to be used.

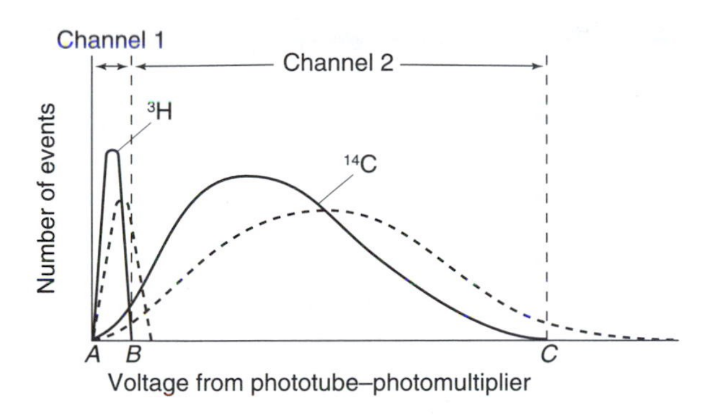

Figure 3-7 Effect of gain voltage on energy profiles of ^{3}H and ^{14}C.

As this example illustrates, gain voltage and discriminator or channel settings are interrelated. Different makes of scintillation counters vary in the control they allow the operator. Many current models have fixed (preset) channels and variable gain voltage, which is adjusted to bring the observed counts into the preset channels. Others have highly flexible channels and sophisticated amplification of the energy profiles. Most manufacturers also offer models with varying numbers of independent channels and varying counting sequences on these channels. A clear understanding of the interaction of amplification of energy profiles and control of discriminators will suffice for all these differences in makes and models of scintillation counters.

Measurement of radioactive decay can also be affected by various components present in, or added to, the scintillation cocktail. These components can cause quenching; that is, they can decrease the efficiency of the scintillation process. Scintillation counting provides data in counts per minute (cpm). Quenching dictates that the counts per minute detected is less than the actual decay rate, or disintegrations per minute (dpm). Almost every sample encountered experimentally is quenched to some degree; for example, O_2 picked up by the scintillation fluid from contact with air serves as a quencher. Therefore, researchers frequently count an additional sample containing a standard of known decay rate (dpm) along with the experimental sample vials. Comparison of the observed counts per minute with the known disintegrations per minute allows you to determine the counting efficiency of a given scintillation system.

(3-9)

$$\%\ \text{Counting efficiency} = \frac{\text{cpm}}{\text{dpm}} \times 100\%$$

Quenching can occur in three ways. Some of these are quite simple, as in *color quenching*, in which the color of the sample absorbs photons emitted from the fluor before they are detected by the photomultiplier tube. Another type of quenching is *point quenching*, in which the sample is not solubilized in the scintillation fluid, and β^- particles are absorbed at their origin before they contact the solvent. Color quenching can usually be overcome by refining the sample before adding the scintillation fluid; point quenching can frequently be eliminated by adding specific solubilizers or detergents. The most important and subtle quenching is *chemical quenching*. This phenomenon occurs when various compounds present in the scintillation vial decrease the counting efficiency by interacting with excited solvent or fluor molecules so as to dissipate the energy without yielding photons. Chemical quenching usually cannot be eliminated, so methods have been developed to quantify and correct for it.

The methods often used to analyze chemical quenching make use of the characteristic impact that chemical quenching has on the apparent energy profile of radioisotopes that emit β^- particles (Fig. 3-8). Some radioisotopes are more prone to chemical quenching than others (see Fig. 3-8). In general, the higher the average energy of the emitted β^- particles, the less chemical quenching is observed. Thus, high-energy β^- emitters (such as ^{32}P) are not sensitive to chemical quenching. Also, chemical quenching agents vary in their quenching efficiency per gram or per mole of quenching agent. Thus, the exact degree of quenching is potentially highly variable, yet the overall effect of chemical quenching is a reduced counting efficiency.

If we set up arbitrary channels, such as $A \rightarrow B$, $B \rightarrow C$, and $A \rightarrow C$ on the ^{14}C energy profiles of Figure 3-8, we can measure chemical quenching for this radioisotope. An examination of the ratio of

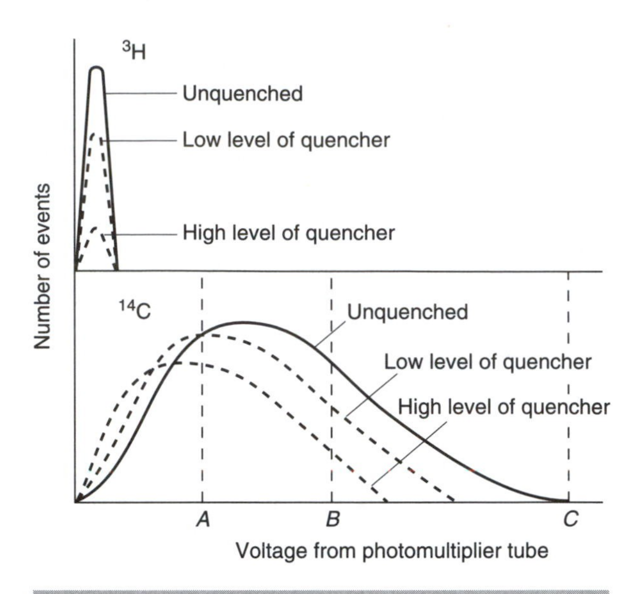

Figure 3-8 Effect of quenching on observed β^- energy profiles.

counts in the $A \rightarrow B$ window to the counts in the $A \rightarrow C$ window shows that increased quenching increases the $A \rightarrow B/A \rightarrow C$ counts ratio. You can construct a plot that relates decreases in counting efficiency (that is, degree of quenching) to the $A \rightarrow B/A \rightarrow C$ counts ratio. Such a plot (Fig. 3-9) forms the basis for the *channels ratio method* of quench analysis. These plots are obtained from determinations of channels ratio values on a series of samples containing a known quantity of a known isotope and increasing amounts of a chemical quencher.

The channels ratio method makes use of existing counts within the sample vial. This method is suitable when large numbers of counts are present, but it becomes very time consuming with samples containing few counts, because a long time is required to accumulate sufficient counts for statistical accuracy. Most modern scintillation counters therefore employ an automatic external standardization system of quench analysis to avoid the time required for the internal channels ratio method. This method utilizes a specially selected external γ radiation source carried in a lead-shielded chamber that is buried in the instrument. Before the regular counting of the sample, the external standard is

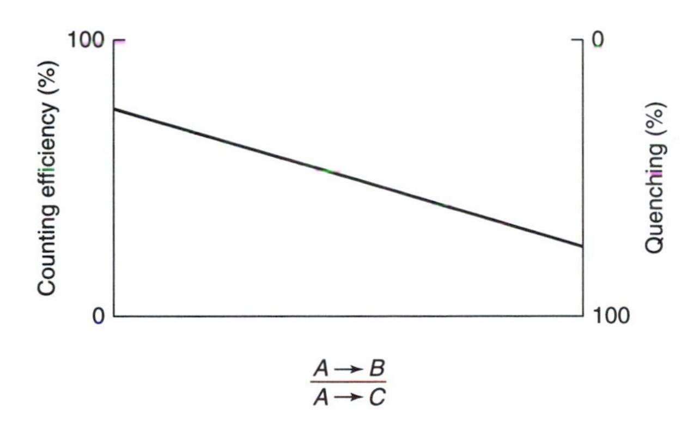

Figure 3-9 Channels ratio quench correction plot for ^{14}C. Quench correction data may curve slightly up or down depending on the make and model of counter used.

automatically placed next to the sample vial to be analyzed for quenching (Fig. 3-10). γ Rays from the external standard penetrate the sample vial and collide with orbital electrons of atoms in molecules within the scintillation cocktail. Because of the choice of γ emitter selected, these collisions result in a spectrum of Compton electrons that possess an energy spectrum analogous to, but of higher energy than a β^- spectrum. These electrons then interact with the solvent and fluors to produce a collection of light flashes of varied intensity that are similar to, but higher in energy than that produced from β^- particle interaction with the scintillation cocktail. Chemical quenching agents will quench spectra that are induced by γ rays in a manner identical to chemical quenching in β^- particle analysis. Therefore, analysis of quenching in this induced spectrum in the high-energy range above the β^- particle range provides an accurate analysis of the quenching of a β^- spectrum that would occur in the absence of the internal standard. (The β^- particle-emitting sample within the sample vial does not influence the external standard system because the energy of the β^- particles do not fall in the high-energy Compton range.)

Various scintillation counters analyze the quenching of the induced spectrum of the external standard differently. Some machines relate decreases in the count rate of a selected portion of the induced spectrum with the quenching in any β^--emitting sample spectrum. Others utilize a channels ratio analysis of selected channels within the external or induced spectrum. In all instances, the scintillation counter generates an external standard number that reflects the method of quenching analysis employed on the external or induced spectrum. This external standard number is directly related to the counting efficiency or quenching observed in a β^- emission system, as indicated in the quench curve of Figure 3-11. Use of the channels ratio method or the automatic external standardization method of quench analysis allows the conversion of counts per minute data to true disintegrations per minute data. Such disintegrations per minute data are necessary for conversion of counting data to curies, the basic units of radioactivity.

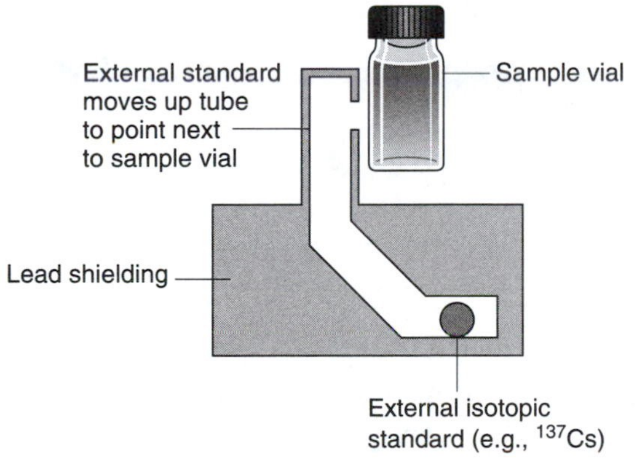

External standard moves up tube to point next to sample vial
Sample vial
Lead shielding
External isotopic standard (e.g., ^{137}Cs)

Figure 3-10 Example of external standard mechanics.

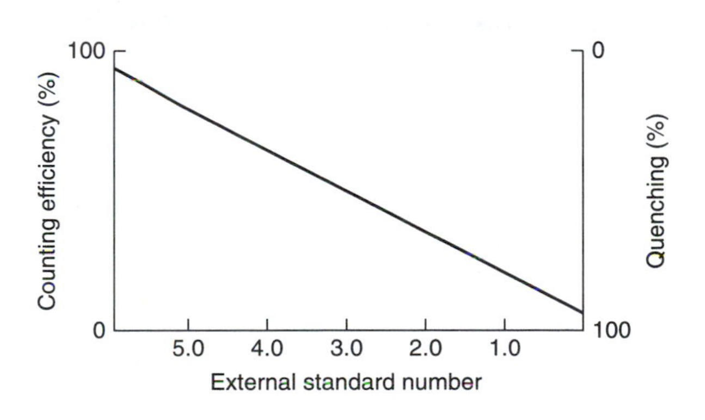

Figure 3-11 Typical quench curve.

Preparation of Samples for Liquid Scintillation Counting. Quenching in its various forms constitutes the major concern during sample preparation for liquid scintillation counting. The following sections discuss some of the problems commonly encountered and also present some remedies to overcome these problems.

1. Color quenching can only be eliminated by removal of the colors. This requires refining of the sample before counting. If the compound to be counted is colored, little can be done except to minimize and equalize the quantities of colored material added to the sample vials.
2. Incomplete solubility and associated point quenching constitute a major problem in scintillation counting. Efficient scintillation counting requires that the sample be fully soluble in the excitable organic solvents of the scintillation fluid. However, biological systems, which are usually aqueous systems or assays, frequently contain water or hydrophilic molecules that will not dissolve in standard toluene-based scintillation cocktails.

   Several approaches have been used to circumvent these problems. First, scintillation cocktails have been developed that will accept more water and associated hydrophilic compounds. Bray's solution (4 g PPO, 0.2 g POPOP, 60 g naphthalene, 20 ml ethylene glycol, 100 ml methanol, and dioxane up to 1 liter), Kinard solution (xylene, *p*-dioxane, ethanol (5:5:3) containing 0.5% PPO, 0.005% α-NPO, and 6% naphthalene), and ethanol systems (e.g., 3 parts ethanol, 4 parts 0.8% PPO, 0.01% POPOP in toluene) are notable examples. In these examples, PPO denotes 2,5-diphenyloxazole, POPOP denotes 1,4-bis-2-(5-phenyloxazolyl)-benzene, and α-NPO denotes 2-(1-naphthyl)-5-phenyloxazole. (POPOP and α-NPO are secondary fluors and can usually be omitted with modern scintillation counters.)

   A second approach has been developed, in part because of difficulties encountered in disposing of scintillation fluids containing naphthalene, which is insoluble in water. In this approach, solubilizer or detergents are added to the standard toluene-based scintillation cocktails. Hyamine hydroxide, NCS, Soluene-100, Aquasol, and the Bio-solve series are sold commercially for this purpose.

   Third, difficult samples of materials containing ^{3}H or ^{14}C (e.g., tissue slices, etc.) can be converted to $^{3}H_2O$ or $^{14}CO_2$ in commercially available "sample oxidizers." These units burn samples under controlled conditions and then separate the $^{3}H_2O$ and the $^{14}CO_2$ for counting in an acceptable scintillation cocktail. Sample oxidation also eliminates color-quenching problems.

   A fourth solution is the "internal standard" method. In this procedure, a sample is counted that is known to have solubility or point-quenching problems (e.g., aqueous samples, insoluble materials embedded in paper or cellulose derivative filters, etc.). A known amount of isotope (labeled) is added to the sample (i.e., to the fluid or dried paper) and the counting process is repeated. Evaluation of the counting efficiency of the added isotope (the internal standard) allows determination of the counting efficiency of the original sample. (*NOTE:* Alternatively, the internal standard method may employ addition of internal standard to one pair of duplicate samples).
3. Chemical quenching is always a problem in liquid scintillation counting. Certain compounds (e.g., chlorinated hydrocarbons) are notorious chemical quenchers. Dissolved O_2 from the air, water, even the solubilizers mentioned before, chemically quench scintillation counting somewhat. Thus, it is unlikely that chemical quenching can be avoided completely. The best approach for minimizing chemical quenching is to design experimental protocols that add the simplest or most refined assay components to the final scintillation cocktail. Further, when chemical quenchers are added (including water, buffers, etc.) to scintillation vials for counting, the same amounts should be added to all sample vials to equalize quenching errors.

Solid Scintillation Counting of β^- Emissions. To avoid some of the problems associated with liquid scintillation counting, a method has been developed for measuring β^- radioactivity by scintillation of solid samples. Liquid samples containing the radioactive material to be counted are placed in small sample cups coated on the bottom with a solid matrix containing yttrium, which acts as the scintillant. The sample is dried and counted in the same instrument as used for liquid scintillation counting (with some changes in the counting "window" and coincidence circuit settings). The volume of liquid that can be analyzed is limited to 200 μl, and the radioactive material cannot be volatile. Point quenching can still present a problem for samples that contain substantial amounts of solid matter. However, other kinds of quenching are minimized, and use of expensive and toxic liquid scintillation fluid, which is also difficult to dispose of properly, is avoided. Counting efficiencies depend on the counting equipment used, but can be nearly as high as with liquid scintillation counting under optimal conditions.

Solid Scintillation Counting of γ Emissions. Scintillation counting is also the most efficient method for detection of γ rays. γ Rays are high-energy photons that lack both charge and mass. As such, they are highly penetrating and require dense materials to be absorbed or recorded efficiently. The soluble scintillation cocktails employed with β^- particle assays in scintillation counters are not sufficiently dense to absorb γ-ray emissions effectively. Scintillation counting of γ rays is usually accomplished by placing the γ ray source in a solid NaI crystal scintillation cell (Fig. 3-12). The brief light flashes emitted by the NaI crystal scintillator are measured as counts by phototube–photomultipliers. They are

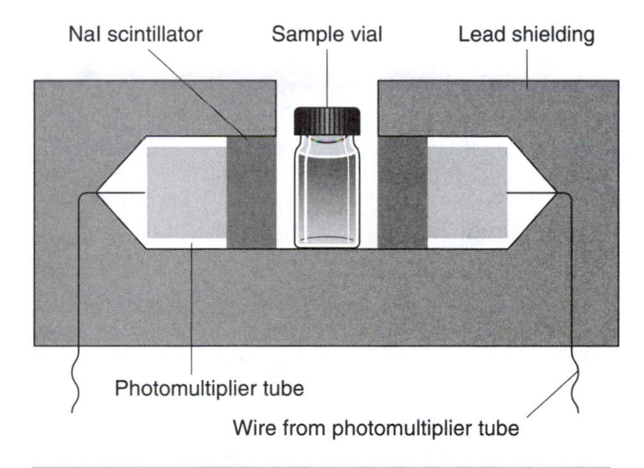

Figure 3-12 NaI crystal scintillation cell.

amplified and processed by the conventional scintillation counter circuitry and then recorded by the scaler element of the unit. Thus, a liquid scintillation counter can be converted to a γ-ray counter by substitution of an NaI crystal scintillator counting cell into the conventional circuitry.

In contrast to β^- particles, γ rays are emitted with distinct or single energy values. Because of this property, the identity of unknown γ-emitting radioisotopes can be determined through analysis of the energies of emitted γ particles. More important, the progress of biological reactions involving various components labeled with different γ-ray-emitting radioisotopes can be followed. Thus, double- and triple-label experiments are possible with γ-emitting radioisotopes if those that emit γ rays of rather different energies are selected, and a γ ray scintillation counter equipped with an energy channel analyzer is used.

Cerenkov Counting. β^- particles of very high energy (above 0.5 MeV) pass through aqueous media with a velocity greater than the speed of light in water. This causes the emission of light, which is known as the Cerenkov effect. The amount of light emitted is proportional to the number of radioactive nuclei that emit high energy β^- particles, and it can be readily detected in the same scintillation counters as are used for liquid scintillation counting. Because Cerenkov counting does not require any special solvents or fluors, it is not subject to chemical quenching and is an extremely convenient and inexpensive means of determining the radioactivity of samples of suitable isotopes. Of the isotopes commonly used in biochemical research, only ^{32}P emits β^- particles of sufficiently high energy to be counted by the Cerenkov method. Not all of the β^- particles emitted by ^{32}P have energies above 0.5 MeV, of course, so this isotope can be detected with only about 40% efficiency. This provides little practical limitation to the use of Cerenkov counting for ^{32}P, however. Other isotopes occasionally used in biological research with β^- particles of sufficient energy for Cerenkov counting include ^{22}Na, ^{36}Cl, and ^{42}K.

Autoradiography. In biochemical research, it is frequently desired to determine the location of one or more radioactive components after analysis by chromatographic or electrophoretic separations. For example, an enzyme reaction may be assayed by including a radioactive substrate and separating the reactants and products after a certain reaction time by paper or thin-layer chromatography. To determine the rate of the enzyme-catalyzed reaction, it is necessary to separate the radioactive product from the radioactive substrate and to determine the amount of radioactivity in the product. Examples of other situations that require the identification of the location of radioactive materials after electrophoretic or chromatographic analysis include the separation of radioactive oligonucleotides in DNA sequencing, separation of protein–DNA or protein–RNA complexes from unbound radioactive DNA or RNA in electrophoretic gel mobility shift analysis, and analysis of the protein content of whole cells by radiolabeling and two-dimensional gel separation of the very complex mixtures of radioactive proteins found in such cells. Autoradiography offers a very convenient solution to such problems.

Photographic Autoradiography. The most commonly used form of autoradiography involves exposure of a paper chromatogram or electrophoretic gel to a sheet of photographic film, such as x-ray film. Radioactive emissions are able to activate silver halide granules in photographic film for photoreduction in much the same way as visible light or x-rays do. Thus, after a period of exposure of the film to a sheet of paper or gel containing radioactive zones, the film can be developed to reveal a pattern of dark spots showing the location and shape of the radioactive zones. The location of the spots can then be compared to the locations of standard compounds and used for identification. The most common and most reliable use of autoradiography is qualitative analysis. However, the radioactive zones can be cut out of the original gel or paper and their radioactivity can be quantitated by liquid scintillation counting (although care must be taken to eliminate or correct for point quenching of the radioactivity by the paper or gel matrix in which the radioactive material is embedded).

The efficiency of various radioactive isotopes in producing autoradiograms is related to the energy of their decay. Thus, isotopes that emit high-energy β^- particles cause more darkening of x-ray film per decay event than weak β^- emitters. Very weak β^- emitters, such as ^3H, may require very long exposure to the film because many of the radioactive particles are too weak to escape the chromatographic or electrophoretic medium (e.g., paper or gel), or the wrapping needed to protect the film from sticking to the gel. ^3H can be detected in gels by fluorography, where the gel is dehydrated with dimethylsulfoxide, saturated with a fluor, and rehydrated. When the weak β^- particles excite the fluor, light is given off that is more effective at darkening the photographic film than the original β^- particle. Weak β^- emitters are infrequently used for autoradiography of chromatograms or electropherograms, but they are valuable for autoradiography of very thin tissue samples, in which you wish to determine the localization of the isotope in a biological sample. In this latter application, a very-high-energy β^- emitter, such as ^{32}P, is less suitable for localization of radioactivity with high resolution because the high energy decay leaves too long of a trail of darkened silver particles on the film.

The darkness of the spots on photographic film after exposure to a radioactive substance will be a function of several variables. The radioactivity of the material (i.e., the disintegrations per minute present in the original spot) and the time of exposure to the photographic film are obviously important, because the more radioactive decay events occur during the exposure of the film, the more grains of silver halide will be reduced to elemental silver. In a crude semiquantitative sense, then, you can conclude that the darker the spots on the autoradiogram, the more radioactivity was present in the position of the spot on the original chromatogram or electropherogram. Eventually, however, all of the silver halide near the radioactive material will be reduced, and the spot cannot become darker on longer exposure or exposure to higher amounts of radioactivity. In other words, the film may become saturated in a particular zone that contains a high level of radioactivity. Under very carefully controlled conditions, and within the limits of avoiding overexposure, the intensity (or darkness) of spots on autoradiograms can be used as a quantitative measure of the amount of radioactive material on the original chromatogram or electropherogram. This technique requires the use of a densitometer for quantitative determination of the darkness of the spots and careful standardization of the densitometer response using samples of known radioactivity separated by the same method.

Autoradiography Using Storage Phosphor Technology. A powerful new method for detection of radioactive materials on chromatograms or electropherograms that also provides a convenient means for quantitative determination of the amount of radioactive material in each location has been provided by the use of storage phosphor techniques. Instead of using a photographic film to detect the radioactive decay events on a chromatogram or electropherogram, this technique uses a screen coated with a thin layer of an inorganic phosphor in which a chemical change is induced by interaction with a strong β^- or γ particle. The changed phosphor is then detected in a second stage in which the screen is scanned with a laser beam, which induces the local emission of light only from the altered phosphor molecules. The emitted light is collected by a high-resolution photocell system, so that its position and intensity can be stored in digital form, for later analysis.

An example of storage phosphor technology is that used by the PhosphorImager (Molecular Dynamics). This device uses imaging screens that are coated with a finely crystalline layer of BaFBr:Eu^{2+}. A radioactive particle from a nearby decay event causes oxidation of the phosphor to BaFBr:Eu^{3+}. This change is stable, so that the screen eventually acquires a chemical image of the adjacent radioactive sheet in which the position and intensity of the radioactive zones is recorded on the phosphor screen. The chemical image is read by placing the screen in a device in which it is scanned with a fine beam of laser light (633 nm), which is absorbed only by the activated phosphor. The absorbed light allows reduction and phosphorescent decay of the BaFBr:Eu^{2+} back to BaFBr:Eu^{3+} with the release of a photon at 390 nm.

$$\text{BaFBr:Eu}^{2+} + \beta^- \longrightarrow \text{BaFBr:Eu}^{3+} \xrightarrow{h\nu\text{ (laser)}} \text{BaFBr:Eu}^{2+*} \longrightarrow \text{BaFBr:Eu}^{2+} + h\nu$$

Thus, the pattern of radioactive decay events is transformed into a pattern of localized light emissions, which is collected digitally and can be viewed on a computer screen, printed onto ordinary paper, and, most importantly, analyzed quantitatively. The amount of light released from the phosphor screen is proportional to the number of decay events detected over a rather broad range. The technology is useful for all of the β^--emitting isotopes commonly used in biochemical research except ^{3}H. Unlike x-ray film, which has a dynamic range of about 10, some phosphor screens have a dynamic range on the order of 10^5.

Safety Precautions for Work with Radioisotopes

Use of radioisotopes in the laboratory generates potential biological hazards for investigators, and for all of society during subsequent radioisotope removal and disposal. Accordingly, those who work with radioisotopes should follow these safety precautions during all applications that employ radioisotopes.

1. Do not ingest radioisotopes. Specifically, never pipette radioactive solutions by mouth. Instead, use a Propipette bulb or a micropipetter to withdraw and dispense radioactive solutions. Use an appropriate ventilated hood when working with volatile radioactive compounds.
2. Avoid contact of radioisotopes with your skin; wear disposable gloves and long-sleeved washable clothing when possible. Most biochemically important radioisotopes are metabolites that are readily incorporated into the body on ingestion or penetration.
3. Avoid close contact with high-energy radioisotopes. The ^{3}H and ^{14}C systems, and even the ^{32}P systems presented in this textbook, do not represent excessive radiation hazards. Yet, all experiments with radioisotopes present some radiation hazard. Those who plan experiments with compounds that emit γ-ray and high-energy β^- particles should recognize this and use appropriate Plexiglas and lead shielding.
4. Avoid contamination of the laboratory with radioisotopes. Work over disposable surfaces (e.g., blotter paper) and remove and clean all "spills" in a thorough manner (consult the laboratory supervisor). Further, do not leave radioisotope solutions in the laboratory; they may support bacterial or mold growth and eventual production of ^3H_2O or $^{14}CO_2$, which are volatile.
5. Wash all lightly contaminated equipment (glassware, pipettes, hose tubing, etc.) thoroughly to avoid future contamination of yourself or your experiments.
6. Place all highly contaminated solutions and all disposable items (planchets, used filters, etc.) in

radioactive waste containers for prompt removal and disposal as specified by law.

7. Survey your hands and all work areas with a Geiger-Müller meter after each use of isotopes that emit γ rays or high-energy β^- particles. After each use of low-energy β^- isotopes, such as ^{3}H, sample your hands and work areas with a damp cotton swab and place the swab in a vial containing an appropriate scintillation fluid to screen for possible contamination. Survey the bottoms of your shoes as well, to prevent "tracking" of radioisotopes around the laboratory.
8. If significant levels of high-energy β^- emitters or γ emitters are used, you may be required to wear a radiation badge to monitor your cumulative exposure. Persons who regularly work with ^{125}I should have their throats surveyed routinely to monitor possible accumulation of radioiodine in the thyroid gland.

Supplies and Reagents

Marking pen
^{3}H toluene, ^{14}C toluene unknowns
^{14}C toluene standard of known disintegrations per minute per milliliter
Volumetric dispensing bottles
Scintillation fluid (0.5% PPO in toluene)
Chloroform
Quenched ^{14}C unknown
Scintillation counter

Protocol

Analysis of a Labeled Unknown Containing Both ^{3}H and ^{14}C by Scintillation Counting

This experiment tests your ability to determine the counts per minute of both ^{3}H and ^{14}C in a mixture that contains unknown amounts of ^{3}H and ^{14}C toluene. The unknown and a ^{14}C toluene standard employed in the experiment are both equally quenched (by dissolved O_2, etc.), so quenching differences do not influence the final result. The experiment can therefore employ measurement of counting efficiencies of the unknown and the standard in separate ^{3}H and ^{14}C above ^{3}H channels, followed by correction for the overlap of ^{14}C counts in the ^{3}H channel.

1. Using a volumetric dispensing bottle, tip 5 or 10 ml of scintillation fluid into each of two scintillation vials, depending on the size of scintillation vials to be used. Label the bottles "A" and "B" at points on the bottle above the fluid level or on the cap. In the same fashion, label the bottles with your name for later identification.
2. Add the following:

   Bottle A: 0.1 ml of unknown radioactive toluene solution (unknown quantities of ^{3}H and ^{14}C)
   Bottle B: 0.1 ml of standard ^{14}C toluene solution

3. Cap each bottle, invert several times to mix, and count both bottles for 1 min (no automatic external standardization) in both the ^{3}H channel (channel 1) and the ^{14}C-over-^{3}H channel (channel 2), which have been designated by your instructor.
4. Using the following equations, calculate the total number of counts per minute of both ^{3}H and ^{14}C in your unknown solution:

   A. Determine the percent of the total ^{14}C cpm detected in channel 2 using the following equation:

$$\%\,^{14}\text{C cpm in channel 2} = \frac{^{14}\text{C cpm in channel 2 for bottle B}}{\text{total }^{14}\text{C cpm in bottle B}}$$

   B. Determine the total number of ^{14}C cpm in the unknown sample using the following equation:

$$\text{Total }^{14}\text{C cpm in unknown} = \frac{\text{channel 2 cpm for bottle A}}{\%\,^{14}\text{C cpm in channel 2}}$$

   C. Determine the percent of the total ^{14}C cpm detected in channel 1 using the following equation:

$$\%\,^{14}\text{C cpm in channel 1} = \frac{^{14}\text{C cpm in channel 1 for bottle B}}{\text{total }^{14}\text{C cpm in bottle B}}$$

D. Determine the ^{14}C cpm detected in Channel 1 for the unknown using the following equation:

$$^{14}C \text{ cpm in channel 1 for unknown} = \frac{\text{channel 1 cpm for bottle A}}{\% \ ^{14}C \text{ cpm in channel 1}}$$

E. Determine the total ^{3}H cpm in the unknown using the following equation:

^{3}H cpm in unknown = (total cpm in channel 1 for bottle A) − (^{14}C cpm in channel 1 for unknown)

Determination of a Quench Curve and Analysis of a Quenched ^{14}C Sample

This experiment demonstrates two methods for analysis of quenching within samples, namely, the channels ratio method and the automatic external standardization method. Either or both of the methods may be demonstrated with the quench series of bottles described in the following protocol, depending on the capabilities of your particular scintillation counter.

1. Using a volumetric dispensing bottle, tip 5 or 10 ml of scintillation fluid into each of five scintillation vials. Label the vials 1 to 5 on the outside, above the level of the fluid in the vial or on the cap. Label each vial with your name for later identification.
2. Add to the vials as indicated below:

Sample	^{14}C toluene standard of known dpm/ml	Chloroform	Quenched ^{14}C unknown
1	1.0 ml	—	—
2	1.0 ml	0.1 ml	—
3	1.0 ml	0.2 ml	—
4	1.0 ml	0.3 ml	—
5	—	—	1.0 ml

If the channels ratio method is used, proceed with Steps 3 to 8. If the automatic external standardization method is used, proceed with Steps 9 to 13.

3. Determine three different but specific gain and channels settings, each of which will detect all or a portion of the ^{14}C counts in the samples described above. This may have already been done for you by the laboratory instructor (see Fig. 3-8).
4. Using these three prescribed channel settings, count each sample for 1 min in each of the channels.
5. Using the data obtained from samples 1 to 4, the known number of ^{14}C disintegrations per minute added to each of these samples, and considering the counts per minute data obtained from the widest window as a reflection of overall counting efficiency (the total possible number of counts per minute that can be detected in each sample), construct a channels ratio quench correction curve similar to that shown in Figure 3-9.
6. What do you notice about the relationship between the chloroform concentration in the sample and the counting efficiency for ^{14}C?
7. Based on the $A \rightarrow B/A \rightarrow C$ ratio of sample 5, calculate the number of disintegrations per minute and Ci of ^{14}C present in the unknown quenched ^{14}C sample using the following formula:

$$^{14}C \text{ dpm} = \frac{\text{cpm } ^{14}C \text{ in } A \rightarrow C \text{ channel}}{\text{counting efficiency}}$$

8. What is the maximal ^{14}C counts per minute value that you would expect to find for this sample in the absence of any of the chloroform-quenching agent?

Proceed with Steps 9 to 13 only if you are using the automatic external standardization method.

9. Count each of the samples in the prescribed ^{14}C-above-^{3}H channel for 1 min in the automatic external standardization mode. The machine reports an "external standard number" (often termed an "H" number for many machines) for each of the samples.
10. Using the data obtained from samples 1 to 4, the known number of ^{14}C disintegrations per minute added to each of these samples, and considering the counts per minute data obtained from sample 1 (unquenched sample) as a reflection of overall counting efficiency (the total possible number of counts per minute that can be detected in each sample), construct a

channels ratio quench correction curve similar to that shown in Figure 3-11.

11. What do you notice about the relationship between the chloroform concentration in the sample and the counting efficiency for ^{14}C? What effect does increased quenching have on the external standard number reported by your instrument?
12. Based on the external standard number reported for sample 5, calculate the number of disintegrations per minute and curies of ^{14}C present in the unknown quenched ^{14}C sample using the following formula:

$$^{14}\text{C dpm} = \frac{\text{cpm } ^{14}\text{C in prescribed window}}{\text{counting efficiency}}$$

13. What is the maximal ^{14}C counts per minute value that you would expect to find for this sample in the absence of any of the chloroform-quenching agent?

Exercises

1. You obtain a 1.0-ml aqueous sample containing 100 μl of uniformly labeled ^{14}C-glycine (specific activity = 10 mCi/mmol). What size aliquot of the total sample should you remove, and how should you subsequently dilute this aliquot with ^{12}C-glycine, to obtain 10 ml of 0.01 M ^{14}C-glycine containing 2.2×10^5 dpm/ml?
2. Most liquid scintillation counters that count β particles utilize PPO (2,5-diphenyloxazole) as a fluor. Excited PPO emits a light flash in the blue region of the visible spectrum. Recognizing this fact, would 1 nCi of yellow-colored DNP-^{14}C-alanine count more or less efficiently that 1 nCi of colorless ^{14}C-alanine when both are counted in identical PPO scintillation cocktails? Explain your answer.
3. During the course of using automatic external standardization to prepare a ^{14}C quench curve, you determine that you have mistakenly counted your series of quenched samples in a wide ^{14}C channel instead of a ^{14}C-above-^{3}H channel. Will your data generate a quench curve with a greater or lesser slope than the curve you would obtain from counting the same quenched samples in a narrow ^{14}C-above-^{3}H channel? Explain your answer.
4. Is a quench curve determined for ^{14}C applicable for analysis of quenching of ^{3}H compounds? Why or why not?

REFERENCES

Hawkins, E. F., and Steiner, R. (1994). *Scintillation Supplies and Sample Preparation Guide.* Fullerton, CA: Beckman Instuments.

Horrocks, D. L., and Peng, C. (1971). *Organic Scintillation and Liquid Scintillation Counting.* New York: Academic Press.

Kobayashi, Y., and Maudsley, D. (1969). Practical Aspects of Liquid Scintillation Counting. *Methods Biochem Anal* **17**:55.

Johnston, R. F., Pickett, S. C., and Barker, D. L. (1990). Autoradiography Using Storage Phosphor Technology. *Electrophoresis* **11**:355.

Steiner, R. (1996) *Advanced Technology Guide for LS 6500 Series Scintillation Counters.* Fullerton, CA: Beckman Instruments.

Wang, C. H., and Willis, D. L. (1965). *Radiotracer Methodology in Biological Science.* Englewood Cliffs, NJ: Prentice-Hall.

Wilson, K., and Walker, J. (1994). *Principles and Techniques of Practical Biochemistry*, 4th ed. Chapter 5. Cambridge, UK: Cambridge University Press.

Electrophoresis

Theory

Basic Principles

Electrophoresis is the process of migration of charged molecules through solutions in an applied electric field. Electrophoresis is often classified according to the presence or absence of a solid supporting medium or matrix through which the charged molecules move in the electrophoretic system. Solution electrophoresis systems employ aqueous buffers in the absence of a solid support medium. Such systems can suffer from sample mixing due to diffusion of the charged molecules, with resultant loss of resolution during sample application, separation, and removal steps. Thus, solution electrophoresis systems must employ some means of stabilizing the aqueous solutions in the electrophoresis cell. For example, soluble-gradient electrophoresis systems use varying densities of a non-ionic solute (e.g., sucrose or glycerol) to minimize diffusional mixing of the materials being separated during electrophoresis (Fig. 4-1). Even with these refinements, solution electrophoresis systems have only limited application, usually when preparative scale electrophoretic separation is required.

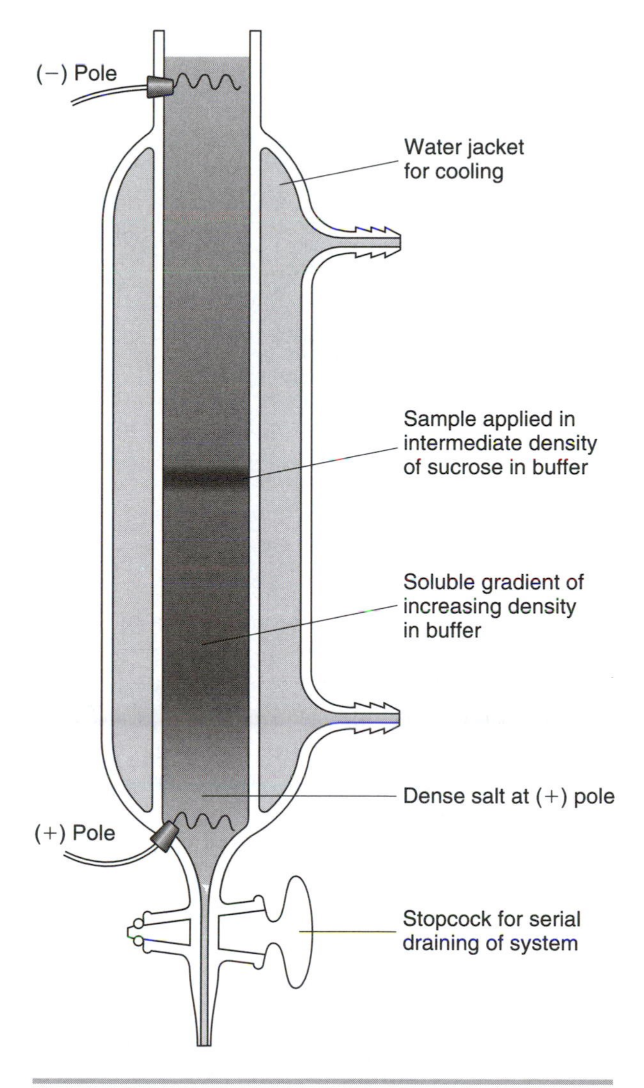

Figure 4-1 A solution electrophoresis system.

Most practical applications of electrophoresis in biochemistry employ some form of zonal electrophoresis, in which the aqueous ionic solution is carried in a solid support and samples are applied as spots or bands of material. Paper electrophoresis, cellulose acetate strip and cellulose nitrate strip, and gel electrophoresis are all examples of zonal

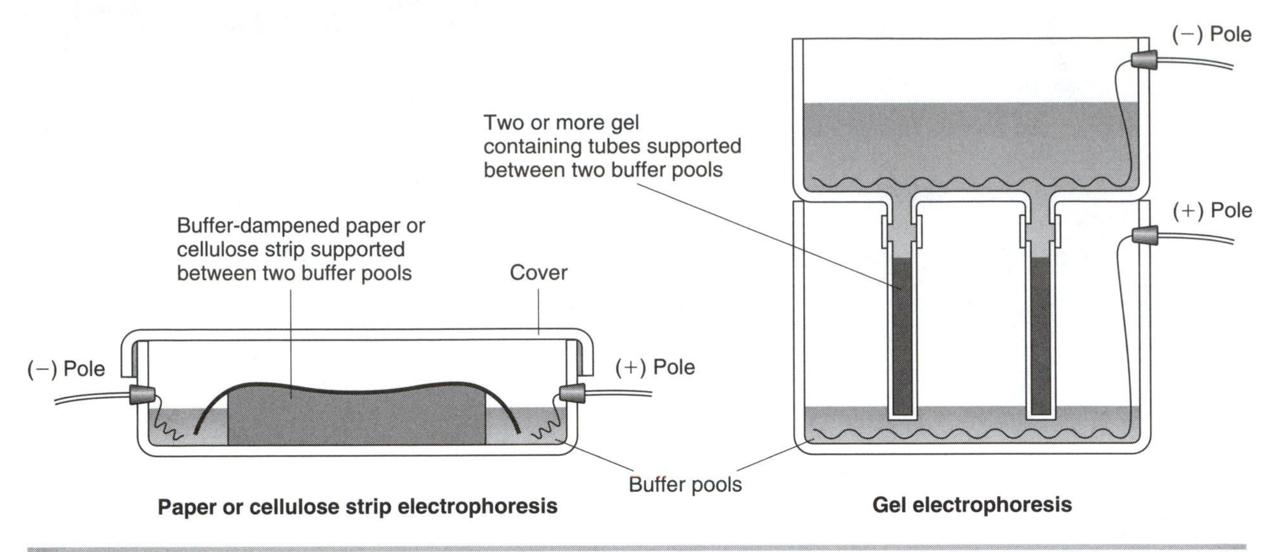

Figure 4-2 Two zonal electrophoresis systems.

electrophoresis systems (Fig. 4-2). Such systems are typically employed for analytical, rather than preparative scale, separations.

All types of electrophoresis are governed by the single set of general principles illustrated by Equation 4-1:

(4-1)

$$\text{Mobility of a molecule} = \frac{\text{(applied voltage)(net charge on the molecule)}}{\text{(friction of the molecule)}}$$

The mobility, or rate of migration, of a molecule increases with increased applied voltage and increased net charge on the molecule. Conversely, the mobility of a molecule decreases with increased molecular friction, or resistance to flow through the viscous medium, caused by molecular size and shape. Total actual movement of the molecules increases with increased time, since mobility is defined as the rate of migration.

An understanding of the relationships described in Equation 4-1 is essential for many practical aspects of experimental biochemistry. Most electrophoretic systems employ an equal and constant voltage on all of the cross-sectional areas of the paper strips, gels, or solutions employed in the electrophoretic separation. These electric fields are best defined in terms of volts per linear centimeter. However, Ohm's law ($V = IR$) dictates that voltage (V) is a function of current (I) and resistance (R). The nature of the electrophoresis apparatus and buffer composition dictates the resistance in the system. Therefore, current (e.g., mA) is often used to define the voltage requirements of an electrophoretic separation. The resistance of the system is important because it will determine the amount of heat generated during electrophoresis. Since electrophoretic mobility is also a function of temperature, heating of the separation matrix must be controlled. If significant heating occurs during electrophoresis, it will be necessary to provide some means of cooling the apparatus so as to maintain a constant temperature. The "smiling" pattern often seen on slab gel electrophoresis (see below) is the result of nonuniform heating of the gel.

If the voltage or current applied to an electrophoresis system is constant throughout an electrophoretic separation, the mobilities of the molecules being resolved will reflect the other terms of Equation 4-1, namely, the net charge and frictional characteristics of the molecules in the sample. Consider the paper electrophoretic separation of glutamic acid, glutamine, asparagine methyl ester, and glycinamide at pH 6.0. The charges and molecular weights of these compounds at pH 6.0 are indicated in Figure 4-3. As seen, the equal-sized amino acids and the asparagine methyl ester separate strictly as a function of their net charges at pH 6.0, whereas

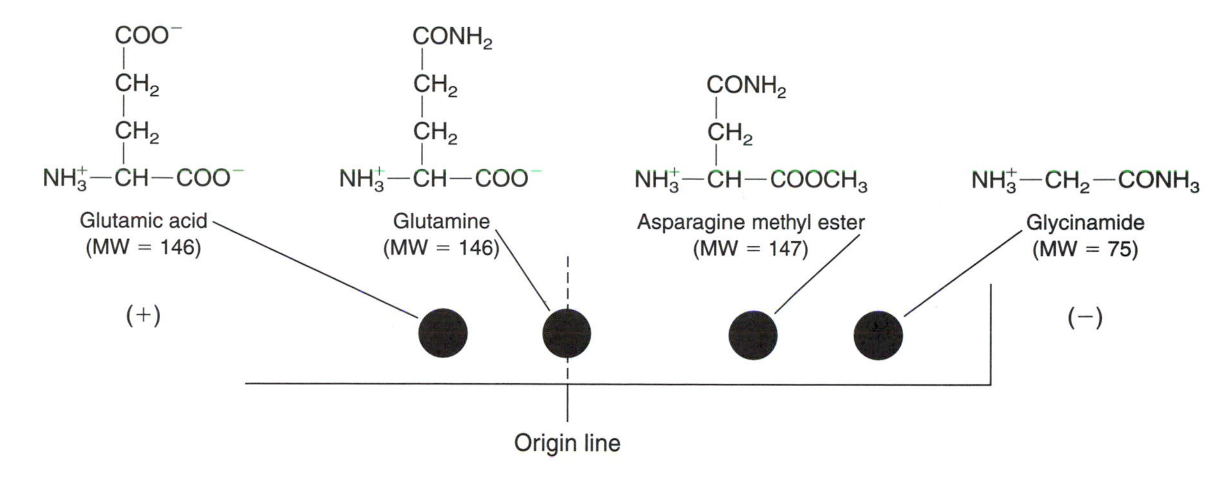

Figure 4-3 Paper electrophoretic separation of glutamic acid, glutamine, asparagine, methyl ester, and glycinamide at pH 6.0. The glycinamide, with a charge of +1 and a molecular weight of 75, may not necessarily migrate twice the distance migrated by asparagine methyl ester, with the same charge and ~2 times larger size. Specifically, the character of the buffer can influence the expression of the electrophoretic frictional contribution of small molecules. Higher ionic strength decreases the frictional contribution.

the glycinamide (which has the same net charge as asparagine methyl ester but has only half its size) migrates proportionally further toward the (−) pole of the system.

Generally, the principles illustrated in Figure 4-3 can be used to predict the relative mobility of small ionic molecules. These principles are particularly useful for predicting the potential for electrophoretic separation of small molecules containing weakly acidic or weakly basic groups that carry average partial charges in the pH ranges associated with their titration, as, for example, during the electrophoretic separation of nucleotides or amino acids.

The electrophoretic separation of larger macromolecules follows the general principles of Equation 4-1, but other factors influence the resolution of macromolecules. The friction experienced by molecules during electrophoretic migration reflects both molecular size and molecular shape. If the electrophoresis is carried out in a medium that offers significant barriers to the movement of macromolecules through it (as is the case with polyacrylamide and agarose, two very commonly used systems), *molecular size may prove to be the most important determinant of mobility*. If the charge to mass ratio on the macromolecules being separated is approximately equal (as in several of the cases discussed below), molecular size becomes the sole determinant of electrophoretic mobility. These conditions are exploited for the determination of the molecular weight of protein subunits by electrophoresis in polyacrylamide gels containing sodium dodecyl sulfate (SDS-PAGE), and in the electrophoretic separation of oligonucleotide "ladders" during DNA sequencing. Molecular shape is not very significant in small molecules, in which bonds are free to rotate, so size alone defines their friction. However, macromolecules often have defined shapes with specific axial ratios (i.e., length to width ratios). As a result, both size and shape influence migration. Molecules with high axial ratios demonstrate lower electrophoretic mobility than more spherical molecules that have equal weight and equal charge. In addition, macromolecules may deviate from the electrophoretic principles of Equation 4-1 because of interaction with ions or because of charge-dependent intermolecular associations.

Specific Forms of Electrophoresis Commonly Used in Biochemistry

Paper Electrophoresis. Paper electrophoresis is a commonly used electrophoretic method for analysis and resolution of small molecules. This method is not used to resolve macromolecules (e.g., proteins) because the adsorption and surface tension associated with paper electrophoresis usually alter or denature the macromolecules, causing poor resolution.

Two different methods are used routinely to apply samples to the electrophoretic paper. In the dry application procedure, a sample of solutes dissolved in distilled water, or a volatile buffer, is applied as a small spot or thin stripe on a penciled "origin line" on the paper. Appropriate standards of known compounds are applied at other locations on the origin line. If you anticipate electrophoretic migration toward both poles of the system, the origin line should be in the center of the paper. If you anticipate migration in only one direction, the origin line should be near one end of the paper. After the solvent containing the samples has evaporated, the paper is dampened with the electrophoresis buffer, either by uniform spraying or by dipping and blotting the ends of the paper so that wetting of the paper from both ends meets at the origin line simultaneously. In the wet application procedure, samples dissolved as concentrated solutions in distilled water are applied to paper predampened with electrophoresis buffer. The dry application procedure has the advantage of allowing small initial sample spots and better resolution of similarly mobile compounds. However, this method is awkward because the dipping or spraying requires considerable skill to avoid spreading the applied samples. The wet application procedure is simpler to perform, but usually yields larger spots and poorer resolution because of sample diffusion.

After the sample is applied to the dampened piece of paper, the paper is placed in the electrophoresis chamber so that both ends are in contact with reservoirs of the electrophoresis buffer at the electrodes (Fig. 4-4). If the origin line is not in the center of the paper, the paper must be positioned to allow maximum migration toward the correct electrode. After the chamber is covered or closed to protect against electric shock, an electric field is applied to the system.

Application of the electric field and the resultant resistance to current flow in the buffered paper generates heat. *This is the greatest source of difficulty with paper electrophoresis.* Heat dries the paper, which in turn leads to more resistance to current flow, which causes greater resistance, and so forth. Even if paper drying is prevented, heating will change the current flow and resistance properties of the system, which will distort the migration of molecules.

Because of these difficulties, modern paper electrophoresis systems have been designed to compensate for potential heating problems. Most low-voltage systems are portable and can be operated in cold rooms or refrigerated chambers. In contrast, most high-voltage systems either employ a cooled flat-bed system to dissipate the heat or operate in a cooled bath of inert and nonpolar solvent (e.g., Varsol, a petroleum distillate). This solvent absorbs the heat generated by the system without mixing with the water, buffers, or samples on the paper (Fig. 4-4). After electrophoretic resolution of the samples on the paper for the time and voltage required for optimal separation, the current is turned off, the paper is removed and dried, and the presence and location of the molecules of interest are determined.

Capillary Electrophoresis. A new method of analytical electrophoresis that is rapidly finding increasing application in biochemical research is capillary electrophoresis. As the name suggests, the material to be analyzed and the electrophoresis medium (a conducting liquid, usually aqueous) are placed in a long, fine-bore capillary tube, typically 50 to 100 cm long and 25 to 100 μm inside diameter. A very small sample (in the nanoliter range) is placed at one end of the capillary and subjected to electrophoresis under fields up to 20 to 30 kV. The analytes are separated by the principles illustrated in Equation 4-1 and detected as they emerge from the other end of the capillary by any of the methods commonly used in high-performance liquid chromatography (see Experiment 2). Capillary electrophoresis offers the advantages of extremely high resolution, speed, and high sensitivity for the analysis of extremely small samples, but is obviously not useful as a preparative method. It has proven especially useful in the separation of DNA molecules that differ in size by as little as only a single nucleotide. Because of its high resolution, capillary electrophoresis is the basis of separation of polynucleotides in some of the newer designs of DNA sequencers. Capillary electrophoresis can also be adapted to the separation of uncharged molecules by including charged micelles of a detergent (such

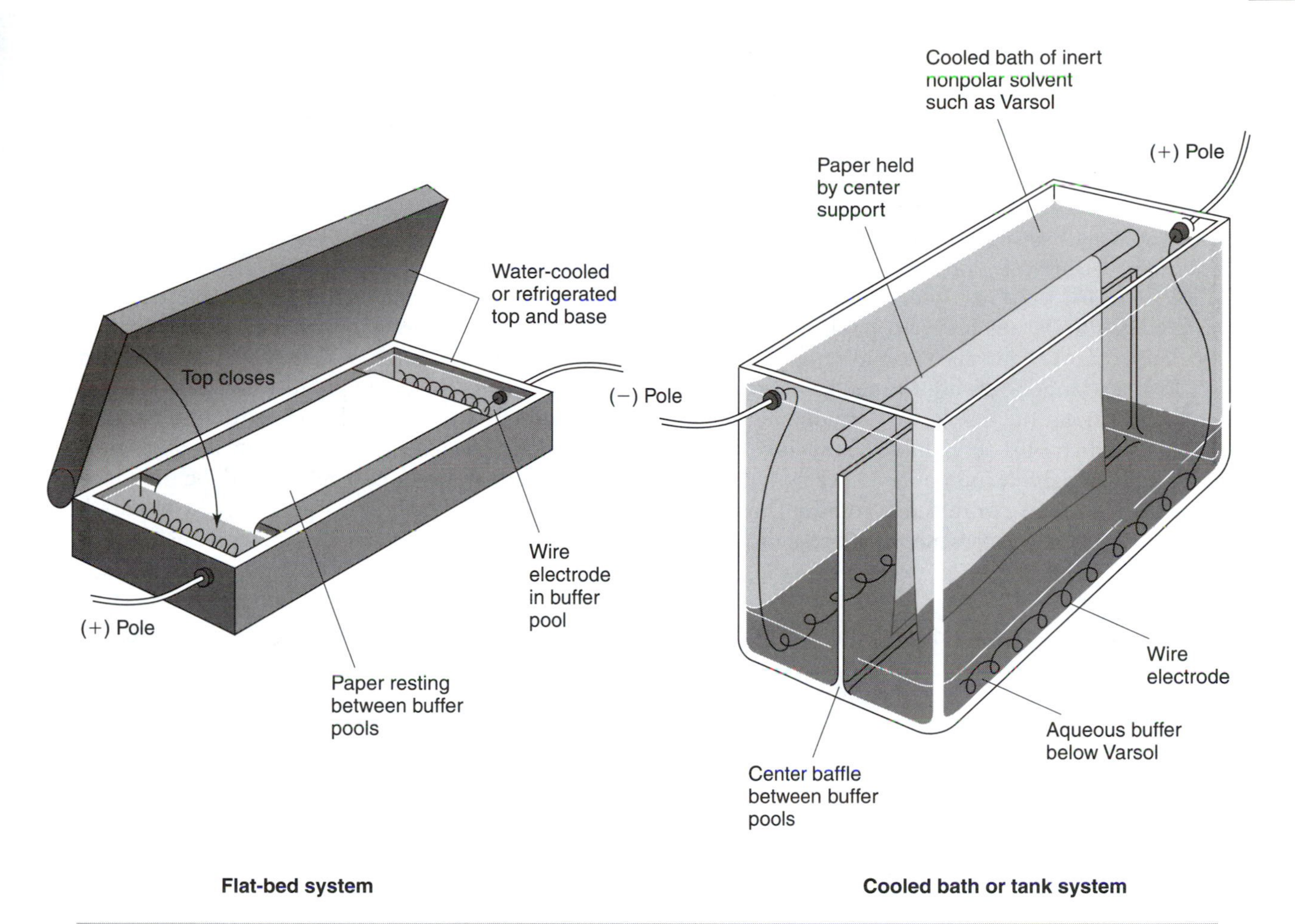

Figure 4-4 Two paper electrophoresis systems.

as SDS) in the aqueous electrophoresis medium. If a mixture of solute molecules that partitition between the aqueous medium and the hydrophobic interior of the micelles is introduced into such a system, they can be separated by electrophoresis. Capillary electrophoresis is a highly adaptable method, and the range of its applications and optimal methodology are still being explored.

Gel Electrophoresis. In gel electrophoresis, molecules are separated in aqueous buffers supported within a polymeric gel matrix. Gel electrophoresis systems have several distinct advantages. First, they can accommodate larger samples than most paper electrophoresis systems, and so can be used for preparative scale electrophoresis of macromolecules. Second, the character of the gel matrix can be altered at will to fit a particular application. This is possible because the gel enhances the friction that governs the electrophoretic mobility (see Equation 4-1). Low concentrations of matrix material or a low degree of cross-linking of the monomers in polymerized gel systems allow them to be used largely as a stabilizing or anticonvection device with relatively low frictional resistance to the migration of macromolecules. Alternatively, higher concentrations of matrix material or a higher degree of cross-linking of monomers are used to generate greater friction, which results in molecular sieving. Molecular sieving is a situation in which viscosity and pore size largely define electrophoretic mobility and migration of solutes. As a result, the migration of macromolecules in the system will be substantially determined by molecular weight.

Many gel-like agents are used in electrophoretic systems. Agarose (a polygalactose polymer) gels have proven quite successful, particularly when applied to very large macromolecules such as nucleic acids, lipoproteins, and others. Polyacrylamide gels are among the most useful and most versatile in gel electrophoretic separations because they readily resolve a wide array of proteins and nucleic acids (see Tables 4-1 and 4-2 for instructions on how to prepare SDS-PAGE gels and polyacrylamide gels for low molecular weight nucleic acid samples).

Polyacrylamide gels are formed as the result of polymerization of acrylamide (monomer) and *N,N′*-methylene-bis-acrylamide (cross-linker) (Fig. 4-5). The acrylamide monomer and cross-linker are stable by themselves or mixed in solution, but polymerize readily in the presence of a free-radical generating system. Biochemists use either chemical or photochemical free-radical sources to induce the polymerization process. In the chemical method (the most commonly used method), the free radical initiator, ammonium persulfate (APS), is added along with a *N,N,N′,N′*-tetramethylethylenediamine (TEMED) catalyst. These two components, in the presence of the monomer, cross-linker, and appropriate buffer, generate the free radicals needed to induce polymerization. In the photochemical method (less widely used), ammonium persulfate is replaced by a photosensitive compound (e.g., riboflavin) that will generate free radicals when irradiated with UV light. Many modifications can be made to produce a gel that will be useful for a particular application. If larger pores are required, you could decrease the amount of monomer and/or

Table 4-1 Recipe for Polyacrylamide Gels with Various Percent Acrylamide Monomer for Use with SDS-PAGE

	% Acrylamide in Resolving Gel				
Component (ml)	**7.5**	**10**	**12**	**15**	**20**
Distilled water	9.6	7.9	6.6	4.6	2.7
30% acrylamide solution	5.0	6.7	8.0	10.0	11.9
1.5 M Tris chloride (pH 8.8)	5.0	5.0	5.0	5.0	5.0
10% (wt/vol) SDS	0.2	0.2	0.2	0.2	0.2
TEMED	0.008	0.008	0.008	0.008	0.008
10% (wt/vol) ammonium persulfate	0.2	0.2	0.2	0.2	0.2

Prepare the ammonium persulfate fresh and add last to induce the polymerization process (polymerization of the resolving gel will take approximately 30 min).

4% Acrylamide Stacking Gel for SDS-PAGE

Component	**Volume (ml)**
Distilled water	2.7
30% acrylamide	0.67
1.0 M Tris chloride (pH 6.8)	0.5
10% (wt/vol) SDS	0.04
TEMED	0.004
10% (wt/vol) ammonium persulfate	0.04

Prepare the ammonium persulfate fresh and add last to induce the polymerization process (polymerization of the resolving gel will take approximately 30 min).

30% Acrylamide Solution

Dissolve 29.2 g of acrylamide and 0.8 g of *N,N′*-methylene-bis-acrylamide in 100 ml of distilled water. Filter through a 0.45-μM-pore-size membrane (to remove undissolved particulate matter) and store in a dark bottle at 4°C.

Table 4-2 Recipe for a 15% Polyacrylamide Gel for Use in Separating Nucleic Acids Smaller Than 500 Bases in Length

Component	Volume or Mass
Urea	15 g
40% acrylamide solution	11.3 ml
10× TBE buffer	3.0 ml
Distilled water	4.45 ml
TEMED	30 ml
10% (wt/vol) ammonium persulfate	200 ml

Prepare the ammonium persulfate fresh and add last to induce the polymerization process. This recipe will be enough to cast 3 to 4 gels, depending on the size of the plates that you use. The recipe for a 0.5× TBE solution is shown in Table 4-5. A 5× concentrated stock can be prepared and stored at room temperature for a short time by multiplying all of the masses and volumes of the Tris base, boric acid, and EDTA components by 5. After several days at room temperature, a precipitate may begin to form in the 5× TBE stock. When this occurs, a fresh solution should be made. This 5× working stock should be diluted 1 : 10 with distilled water to produce a 0.5× running buffer for use with separation of low-molecular-weight nucleic acids by polyacrylamide gel electrophoresis.

40% Acrylamide Solution

Dissolve 76.0 g of acrylamide and 4.0 g of *N,N′*-methylene-bis-acrylamide in 200 ml of distilled water. Filter through a 0.45-μM-pore-size membrane (to remove undissolved particulate matter) and store in a dark bottle at 4°C.

cross-linker in the polymerization solution. If smaller pores are required, one may increase the concentration of monomer and/or cross-linker.

The pore size required for a particular electrophoretic separation will depend on the difference in size of the compounds that you wish to resolve. For instance, if you wish to resolve two small proteins of 8,000 Da and 6,000 Da, you will require a small-pore-size gel for this application (~15–20% acrylamide). This same percent acrylamide gel would not permit the resolution of two larger proteins of say 150,000 Da and 130,000 Da. A larger-pore-size gel would be required for this application (~7.5–10% acrylamide). A list of components required to produce polyacrylamide gels of varying percent acrylamide is shown in Table 4-1. A list of the effective separation range of proteins on acrylamide gels made with various percent acrylamide is shown in Table 4-3.

Most important, any methods used to produce polyacrylamide gels must be followed exactly each time they are cast, since reproducible electrophoretic separations require uniform gel-forming conditions. Current literature contains many electrophoresis procedures and applications for polyacrylamide gels. Most of these employ the polyacrylamide gel in some sort of "slab" format. One of the most commonly used of these procedures is discussed below.

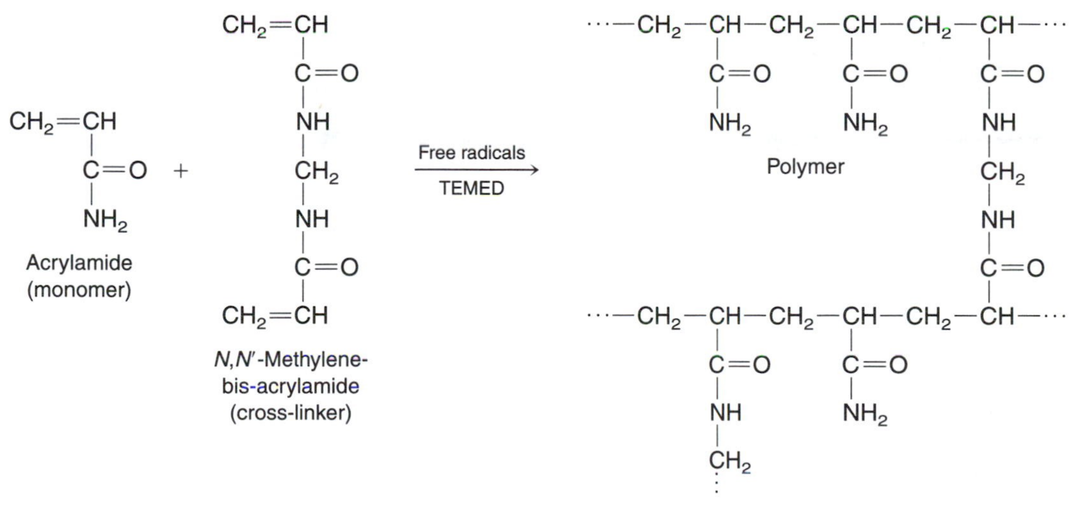

Figure 4-5 Formation of polyacrylamide gels.

Table 4-3 The Effective Separation Range of Polyacrylamide Gels of Various Percent Acrylamide Monomer for Use With SDS-PAGE

% Acrylamide in Resolving Gel	Effective Separation Range (Da)
7.5	45,000–200,000
10	20,000–200,000
12	14,000–70,000
15	5,000–70,000
20	5,000–45,000

Sodium Dodecyl Sulfate–Polyacrylamide Gel Electrophoresis (SDS-PAGE). Sodium dodecyl sulfate (SDS) gel electrophoresis systems are used to determine the number and size of protein chains or protein subunit chains in a protein preparation. Initially, the protein preparation is treated with an excess of soluble thiol (usually 2-mercaptoethanol) and SDS. Under these conditions, the thiol reduces all disulfide bonds (–S–S–) present within and/or between peptide units, while the SDS (an ionic or denaturing detergent) binds to all regions of the proteins and disrupts most noncovalent intermolecular and intramolecular protein interactions. These two components result in total denaturation of the proteins in the sample, yielding unfolded, highly anionic (negatively charged) polypeptide chains (Fig. 4-6).

The anionic polypeptide chains are then resolved electrophoretically within a polyacrylamide gel saturated with SDS and the appropriate current-carrying buffer. The excess SDS is included in the gel to maintain the denatured state of the proteins during the electrophoretic separation. The SDS-coated polypeptides (carrying approximately one SDS molecule per two amino acids) creates a situation in which the charge-to-mass ratio of all of the proteins in the sample is approximately the same. At this point, the intrinsic charge on the individual polypeptide chains (that is, in the absence of SDS) is insignificant as compared with the negative charge imposed on them by the presence of the SDS. The friction experienced by the population of molecules as they migrate through the polyacrylamide matrix is now the major factor influencing differences in their mobility. In addition, the fric-

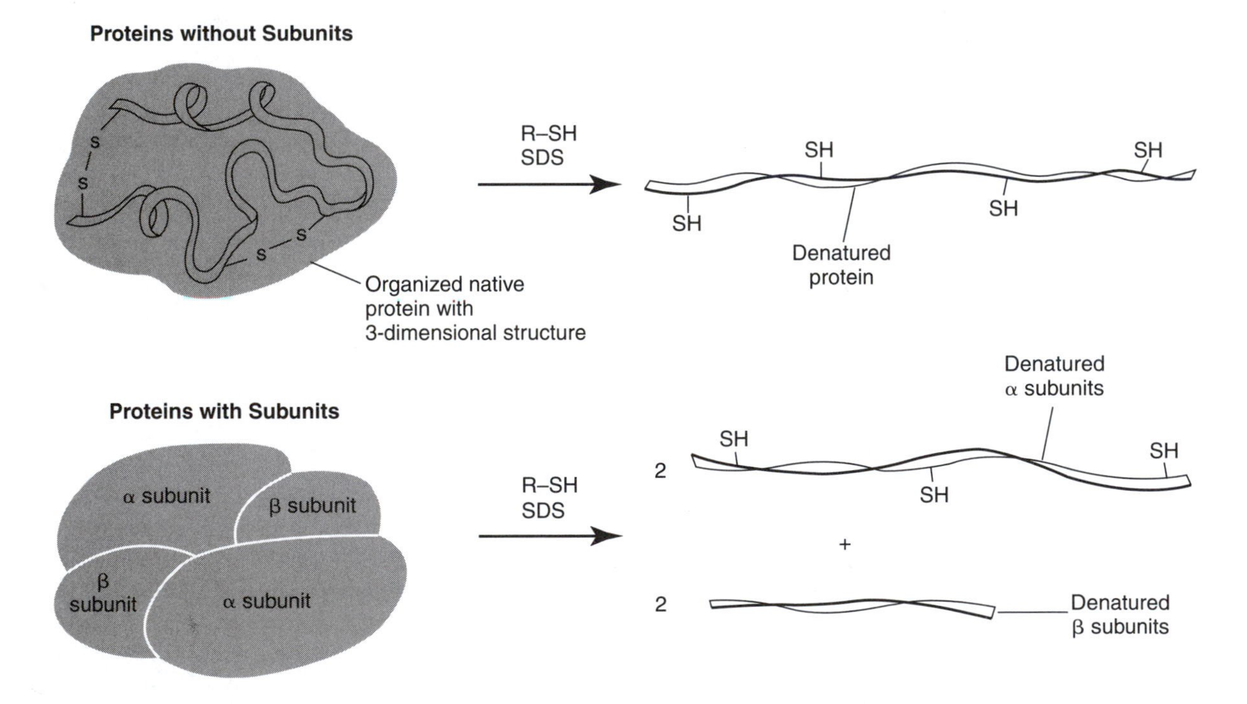

Figure 4-6 Disruption of proteins with excess thiol and SDS.

tion experienced by the molecules during the separation is governed by the pore size of the polyacrylamide matrix: *larger polypeptides will experience greater friction when passing through a gel of defined pore size (will migrate more slowly) than smaller polypeptides (which will migrate more rapidly).* In summary, SDS-PAGE allows the separation of proteins on the basis of size.

The principles underlying the most commonly used form of SDS-PAGE are best illustrated by a description of the step-by-step progress of a protein migrating in the gel. As seen in Figure 4-7 and Table 4-1, SDS-PAGE employs two buffer/polyacrylamide gel compositions in a single slab. These are referred to as the "stacking gel" and the "running (or resolving) gel." Protein samples are first introduced into wells cast within the stacking gel after they are mixed with a viscous sample buffer (30% glycerol) containing SDS and thiols. After the samples have been loaded, voltage is applied to the system (current is carried through the gel, ~3–4 V/cm^2) between two separated pools of glycine buffer, pH 8.3 (Table 4-4).

Remember that ion (current) flow must follow the principles dictated in Equation 4-1, namely, greater charge increases mobility, whereas greater size leads to greater friction and decreased mobility. Glycine carries an average charge of about −0.1 per molecule at pH 8.3 (running buffer) and almost no net negative charge at pH 6.8 (stacking gel pH). In contrast, the SDS-coated proteins in the system carry a high negative charge that is essentially independent of the pH of the system. At the low pH in the stacking gel, glycine anions lose negative charge and display decreased mobility in the system. In contrast, the chloride ions contained in the stacking gel migrate ahead of the proteins because of their small size (low friction) and full negative charge. Since the negatively charged proteins in the system have a larger frictional factor, they migrate into the stacking gel at a rate that is slower than that of the chloride population, but faster than that of the virtually uncharged glycine anion population. The resultant scale of anion mobilities in the low percentage acrylamide stacking gel ($Cl^- >$ proteins$^- >$ glycine$^-$) causes the proteins to accumulate ahead of the advancing glycine front (Fig. 4-7*b*) and eventually stack into concentrated, narrow bands at the interface between the stacking gel and the running gel (Fig. 4-7*c*). Meanwhile, the chloride

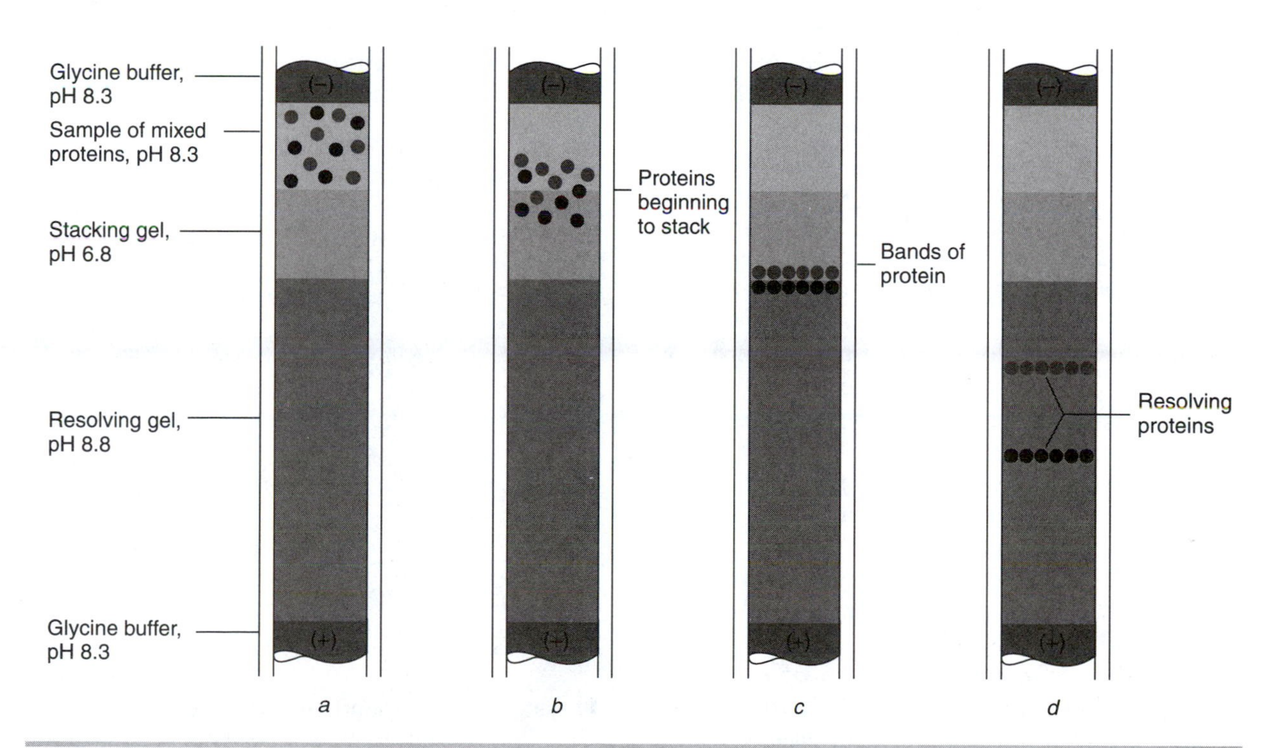

Figure 4-7 Migration of proteins through stacking and resolving gels during SDS-PAGE.

Table 4-4 Recipes for Buffers Used in Agarose Gel Electrophoresis of Nucleic Acids and SDS-PAGE

Tris/Borate/EDTA (TBE) Buffer (0.5×, working solution)
Dissolve 5.4 g of Tris base and 2.75 g of boric acid in 700 ml of distilled water. Add 2 ml of 0.5 M EDTA, pH 8.0. Bring the final volume of the solution to 1 liter with distilled water. *NOTE:* A 5× concentrated stock can be prepared and stored at room temperature for a short time by multiplying all of the masses and volumes of the Tris base, boric acid, and EDTA components by 10. This 5× concentrated stock will be diluted 1:10 with distilled water to give a 0.5× working solution.
Tris/Acetate/EDTA (TAE) Buffer (1×, working solution)
Dissolve 4.84 g of Tris base in 700 ml of distilled water. Add 1.14 ml of glacial acetic acid and 2 ml of 0.5 M EDTA, pH 8.0. Bring the final volume of the solution to 1 liter with distilled water. *NOTE:* A 50× concentrated stock can be prepared and stored at room temperature for long term by multiplying all of the masses and volumes of the Tris base, glacial acetic acid, and EDTA components by 50. This 50× concentrated stock will be diluted 1:50 with distilled water to give a 1× working solution.
Electrophoresis Buffer for SDS-PAGE (1X working solution)
Dissolve 3.02 g of Tris base and 18.8 g of glycine in 700 ml of water. Add 10 ml of 10% (wt/vol) SDS and adjust the pH of the solution to 8.3 with HCl. Bring the final volume of the solution to 1 liter with distilled water. *NOTE:* A 5× concentrated stock can be prepared by multiplying the masses and volumes of all of the components by five. This 5× concentrated stock will be diluted 1:5 with distilled water to give a 1× working solution described above.

ions in the stacking gel readily migrate into the running gel.

As the advancing front of anions enters the running gel, the high concentration of pH 8.8 buffer and the high concentration of acrylamide (decreased pore size) in the gel triggers two events. First, the high pH buffer imparts a greater negative charge to the glycine anions, whose migration was retarded in the pH 6.8 buffer in the stacking gel. Because of their small size and increased anionic character, the glycine ions quickly overtake the proteins as they migrate through the system. Second, the reduced pore size of the gel imparts a significant frictional component to the mobility of each individual protein present in the sample. The equal charge-to-mass ratio imparted on the proteins by the SDS present in the system (see above) now dictates that all of the proteins in the system will migrate through the gel on the basis of size.

As shown in Figure 4-8, the relative mobility of each of the anionic polypeptide chains is a function of the logarithm of its molecular weight. If a set of polypeptides of known molecular weight are included with the sample during the electrophoretic separation, their relative mobilities may be determined and plotted as log molecular weight versus relative mobility. The standard curve produced from this analysis of the polypeptides of known molecular weights can then be used, along with the relative mobilities of the unknown polypeptides in the sample, to estimate their molecular weight (see Fig. 4-8).

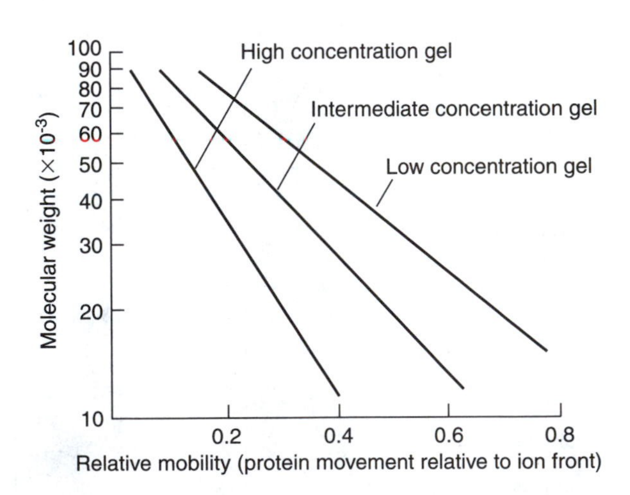

Figure 4-8 Relation between molecular weight and relative mobility of proteins on SDS gels.

How do you visualize the proteins that have been separated following SDS-PAGE? Two visualization methods are most often used. The choice between them depends largely on the sensitivity of detection that is required for your application. The first method involves saturating the gel with a solution of acetic acid, methanol, and water containing Coomassie Brilliant Blue R-250 dye. As the methanol and acetic acid in the solution work to "fix" the proteins within the gel matrix, Coomassie Brilliant Blue binds to the proteins in the gel. The interaction between the Coomassie dye and proteins has been shown to be primarily through arginine residues, although weak interactions with tryptophan, tyrosine, phenylalanine, histidine, and lysine are also involved. After the gel is "destained" with the same aqueous acetic acid/methanol solution without the dye (to remove the dye from portions of the gel that do not contain protein), the proteins on the gel are visible as dark blue "bands" on the polyacrylamide gel. When conducted properly, this method of detection is sufficiently sensitive to detect a protein band containing as little as 0.1 to 0.5 μg of a polypeptide. An alternative staining method, silver staining, is sensitive to about 10 ng of protein contained in a single band on the acrylamide gel. Silver staining begins by saturating the gel with a solution of silver nitrate. Next, a reducing agent is added to cause the reduction of Ag^+ ions to metallic silver (Ag), which precipitate on the proteins in the gel and cause the appearance of protein bands that are black in color. Silver staining is technically more difficult, and therefore is used only when extreme sensitivity is required.

Isoelectric Focusing. Isoelectric focusing is an electrophoretic technique that separates macromolecules on the basis of their isoelectric points (pI, pH values at which they carry no net charge). As with SDS-PAGE, this process can be carried out in a "slab" format. A pH gradient is established in the polyacrylamide gel with the aid of ampholytes, which are small (~5000 Da) polymers containing random distributions of weakly acidic and weakly basic functional groups (e.g., carboxyls, imidazoles, amines, etc.). A polyacrylamide gel containing these ampholytes is connected to an electrophoresis apparatus that contains dilute acid solution (H^+) in the anode chamber and a dilute base (OH^-) solution in the cathode chamber (Fig. 4-9).

When voltage is applied to the system, the current flow will be due largely to migration of the charged ampholyte species present in the gel. The ampholytes migrate toward either pole in a manner consistent with their charge distributions; ampholytes with lower isoelectric points (e.g., those that contain more carboxyl groups and have a net negative charge) migrate toward the anode, while ampholytes with higher isoelectric points (e.g., those that contain more amine groups and have a net positive charge) migrate toward the cathode. Eventually, each ampholyte in the system will reach a position in the gel that has a pH equal to its isoelectric point. When this occurs, these ampholytes carry no net charge and will no longer migrate in the gel. The net effect is that, after a sufficient period of electrophoresis, the population of ampholytes will act as local "buffers" to establish a stable pH gradient in the polyacrylamide gel. This pH gradient forms the basis for the separation of macromolecules that are present in the system (or are subsequently introduced into the system).

Proteins present in the system act like ampholytes in that they migrate as a function of their net charge. As the ampholytes establish a pH gradient in the gel, the proteins in the system also migrate toward their respective poles until they reach a pH in the gel at which they too carry no net negative charge (the pI of the protein). At this point, the attractive forces applied to the protein by the anode and cathode are equal, and the protein no longer migrates in the system. In effect, the different proteins in the sample are "focused" to particular portions of the gel where the pH is equal to their pI values.

How do you determine when the electrophoretic separation is complete? As stated above, the separation is complete when the ampholytes and proteins reach a pH value in the gel equal to their pI and no longer migrate. Once the proteins and ampholytes no longer migrate, the current (I) in the system decreases dramatically. Because isoelectric focusing is performed in the presence of fairly low concentrations of ampholytes (2–4%), high voltage potential (150 V/cm^2) can be applied to the system without the generation of excess amounts of heat caused from high currents.

Since this method is designed to separate proteins on the basis of isoelectric point, one must take care not to impose a significant "friction factor" to

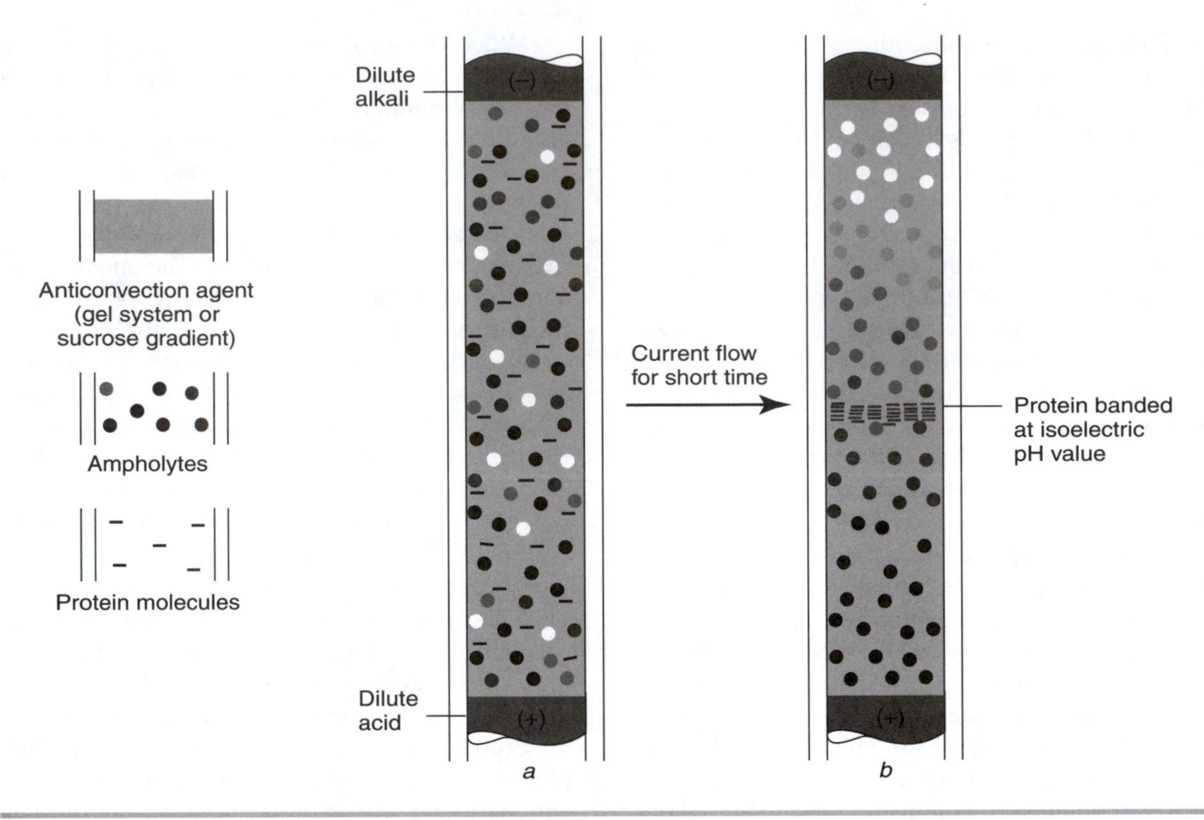

Figure 4-9 Steps of an isoelectric focusing procedure.

the macromolecules being separated. In other words, the pore size of the gel should be large enough that all of the macromolecules in the system can migrate freely to their appropriate isoelectric points. If the pore size of the matrix limits the rate of migration of the proteins with higher molecular weights, the separation may turn out to be influenced more, or as much as, by size as by isoelectric point. If polyacrylamide gels are used for this purpose, the concentration of acrylamide in the gel should not exceed 8%. If desired, you can perform isoelectric focusing using protocols that employ other "slab" matrix materials, such as agarose, or in a liquid matrix format that uses a sucrose density gradient as the anticonvection agent. You should also be aware that ampholytes are commercially available in a wide array of pH ranges that may be required for separation of a molecule of interest (e.g., pH 3–5, pH 2–10, pH 7–9, etc.). In the past, isoelectric focusing has been performed in the "slab" format and employed largely as an analytical technique. More recently, systems have been developed that will allow isoelectric focusing to be carried out in a preparative (large-scale, nondenaturing) format. Although the principles of the separation are the same as those described above, preparative isoelectric focusing is carried out using a liquid anticonvection agent rather than a gel slab.

Agarose Gel Electrophoresis. Agarose gel electrophoresis is the principal technique used to determine the size of high-molecular-weight nucleic acids (DNA and RNA). Agarose is a long polymer of galactose and 3,6-anhydrogalactose linked via α $(1 \rightarrow 4)$ glycosidic bonds. This material is readily isolated from seaweed. Agarose polymers may contain up to 100 monomeric units, with an average molecular weight of around 10,000 Da. Agarose gels are cast by dissolving the white agarose powder in an aqueous buffer containing EDTA and either Tris-acetate or Tris-borate as the buffering species (TAE or TBE buffer, respectively, see Table 4-4). When the sample is heated to just below boiling, the agarose powder dissolves in the buffer to form a clear solution. As the solution slowly cools to room temperature, hydrogen bonding within and between the polygalactose units in the solution will cause the formation of a rigid gel with a relatively uniform pore size. The induction of this polymerization event involves no chemical reaction, unlike the polymerization process described above for polyacrylamide gels.

As with polyacrylamide gels, the pore size of the gel can be controlled by the percentage of the agarose dissolved in the solution. A high percent agarose gel (say, 3% wt/wt) will have a smaller pore size than a lower (0.8% wt/wt) agarose gel. The percent of agarose to be cast in the gel will be determined by the size of the various molecules to be resolved during electrophoresis; the smaller the molecular weight of the molecules to be resolved, the higher percent agarose (smaller pore size) the gel should contain. A list of the effective separation ranges with agarose gels of various percent agarose is shown in Table 4-5.

Once the solution is heated to dissolve the agarose, the solution is cooled momentarily and poured into a slab mold fitted at one end with a Teflon or plastic comb. After the solution polymerizes, the comb is removed to create wells into which the desired DNA or RNA samples will be applied. The gel is then transferred to an electrophoresis chamber and is completely covered with the same TAE or TBE buffer that was used to cast the gel. Next, the nucleic acid sample is mixed with a viscous buffer (30% glycerol) containing one or more tracking dyes that will be used to monitor the progress of the electrophoresis. Bromophenol blue dye will migrate at the same rate as a DNA molecule of about 500 base pairs, while xylene cyanole dye will migrate at the same rate as a DNA molecule of about 4000 base pairs.

Table 4-5 The Effective Separation Range of Agarose Gels of Various Composition for Separation of Nucleic Acids

% Agarose (wt/vol)	Effective Separation Range (base pairs)
0.8	700–9000
1.0	500–7000
1.2	400–5000
1.5	200–3000
2.0	100–300

The nucleic acid samples are loaded into the wells of the gel, along with a sample of DNA fragments of known molecular weight (number of base pairs) in one of the wells. These standards will be used following the electrophoretic separation to aid in the determination of the size of the nucleic acid samples present in the unknown sample (see below). Remember that the DNA is negatively charged. Because of this, the cathode (negative electrode) should be connected to the side of the apparatus nearest the wells in the gel, while the anode (positive electrode) is connected to the opposite side of the apparatus. A voltage of about 4 to 6 V/cm is applied to the system (~50–70 mA of current), and the electrophoresis is continued until the bromophenol blue dye front reaches the end of the gel. You will notice that agarose gel electrophoresis, unlike SDS-PAGE, does not employ the use of a stacking gel. Since the nucleic acids in the sample have a much greater frictional component in the gel than they do in the buffer contained in the wells, the nucleic acids focus very quickly at the buffer–gel interface before entering the matrix.

How do you visualize the nucleic acids following the electrophoretic separation? Ethidium bromide is a fluorescent dye that has the ability to intercalate between the stacked bases of nucleic acid duplexes (i.e., the double helix of DNA). After electrophoresis, the gel is placed in a solution of TBE or TAE buffer containing ethidium bromide, which diffuses into the gel and associates with the nucleic acids. Following a short destaining period in buffer without ethidium bromide (to remove it from the areas on the gel that do not contain nucleic acids), the gel is briefly exposed to ultraviolet (UV) light (256–300 nm), revealing the nucleic acid fragments as orange or pink fluorescent bands in the gel. The pattern of fluorescent bands is recorded by photographing the gel in a dark chamber while the gel is exposed to UV light from below. A filter is included to screen out UV light, so that most of the light that exposes the film is from the fluorescence of the nucleic acid bands that contain tightly bound ethidium bromide.

The method of determining the molecular weight of an unknown nucleic acid sample is exactly the same as that described for the determination of protein molecular weight in SDS-PAGE.

The log molecular weight of the nucleic acid samples of known sizes are plotted against their relative mobilities to produce a standard curve. From the relative mobility of an unknown nucleic acid fragment on the same gel, the molecular weight and number of base pairs that the fragment contains can be readily determined. (A single base has an average molecular weight of approximately 320 Da, while a single base pair has an approximate average molecular weight of 640 Da).

In the experiment that follows, you will have an opportunity to gain experience with some of the foregoing principles by using SDS-PAGE to determine the number of polypeptides and its (their) molecular weight(s) in an unknown protein sample.

Supplies and Reagents

1.5 M Tris chloride, pH 8.8
1.0 M Tris chloride, pH 6.8
30% acrylamide (292 g/liter acrylamide, 8 g/liter *N,N'*-methylenebisacrylamide) [*CAUTION, ACRYLAMIDE IS TOXIC!*]
10% (wt/vol) SDS in water
TEMED (*N,N,N',N'*-tetraethylethylenediamine)
10% (wt/vol) ammonium persulfate in water—freshly prepared
Running buffer (25 mM Tris, 250 mM glycine, 0.1% (wt/vol) SDS, pH to 8.3 with HCl)
4X Sample buffer:

- 0.25 M Tris chloride, pH 7.0
- 30% (vol/vol) glycerol
- 10% (vol/vol) 2-mercaptoethanol
- 8% (wt/vol) SDS
- 0.001% (wt/vol) bromophenol blue

Coomassie Blue staining solution (40% methanol, 10% glacial acetic acid, 0.25% (wt/vol) Coomassie Brilliant Blue R-250 in water)
Destaining solution (40% methanol, 10% glacial acetic acid in water)
Power supply and electrophoresis apparatus
Protein molecular weight standards (suggested standards: phosphorylase [97,400 Da], bovine serum albumin [66,200 Da], ovalbumin [45,000 Da], carbonic anhydrase [31,000 Da], soybean trypsin inhibitor [21,500 Da], and lysozyme [14,400 Da] at 1 mg/ml each in a single mixture)
Unknown protein solutions (approximately 1 mg/ml)

Protocol

NOTE: The quantities of materials to be used and details of preparation of the PAGE gels for electrophoresis will depend on the number of students and the characteristics of the electrophoresis apparatus to be used. A typical apparatus for SDS-PAGE in vertical slab gels is shown in Figure 4-10; the instructions that follow assume that you are using such a system. Your instructor will provide detailed instructions and will demonstrate the techniques to be used in preparing the gels.

1. Prepare the running gel. (The volume to be prepared will depend on the size of the electrophoresis apparatus you will use and the number of gels to be prepared; consult your instructor. The procedure described here is for 20 ml of a 10% monomer running gel.) Mix the following, in order, in a clean 100-ml beaker: 7.9 ml distilled H_2O, 6.7 ml 30% acrylamide solution, 5.0 ml 1.5 M TrisHCl, pH 8.8, 200 μl 10% SDS, 8 μl TEMED, and 200 μl freshly dissolved 10% ammonium persulfate. The ammonium persulfate should be added last, as it initiates polymerization.
2. Pour the freshly mixed running gel solution into the spaces between parallel glass or plastic gel-forming sheets of the electrophoresis apparatus, leaving suficient space at the top for the stacking gel to be added later. (This is usually about 1 cm from the top, but the exact distance depends on the apparatus being used; consult your instructor.) Create a flat surface above the running gel solution by gently adding a water layer on top of the more dense gel solution, and allow the gel to polymerize for at least 30 min.
3. Pour off the water from above the polymerized running gel. This should leave a very sharp flat surface at the top of the gel.
4. Prepare the stacking gel. (The procedure described here is for 4 ml of a 4% monomer stacking gel; again, consult your instructor to see if you should prepare a different amount.) Mix the following, in order, in a clean test tube: 2.7 ml distilled H_2O, 0.67 ml 30% acrylamide solution, 0.5 ml 1.0 M TrisHCl, pH 6.8, 40 μl 10% SDS, 4 μl TEMED, and, added last, 40 μl freshly dissolved 10% ammonium persulfate.

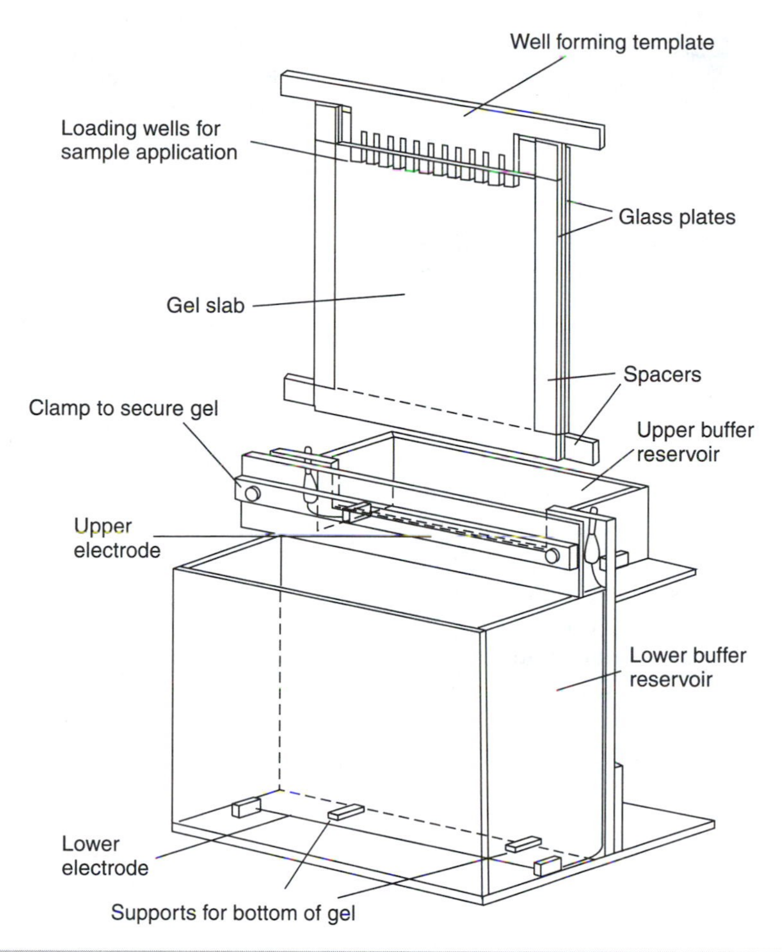

Figure 4-10 A typical apparatus for vertical slab SDS-PAGE. (From Wilson, K., and Walker, J. (1994) *Principles and Techniques of Practical Biochemistry,* 4th ed. Cambridge, UK. Cambridge University Press, Fig. 9-1. Reprinted with permission.)

Mix the solution gently and proceed immediately to Steps 5 and 6.

5. Pour the stacking gel into the space above the running gel.
6. Immediately insert the comb to create sample wells in the stacking gel before it polymerizes. Be sure that the comb is clean and free of foreign material and that no air bubbles form around the comb. If bubbles do form, tilt the gel slightly and tap the glass near the area of the bubble to dislodge it and allow it to rise to the top. Be sure that you fill the stacking gel slightly over the top of the small plate, as the stacking gel will shrink slightly during polymerization. Allow the gel to polymerize for at least 30 min at room temperature.
7. Remove the bottom spacer from between the plates, and place the polymerized gel in the plates in which it was formed into the electrophoresis apparatus.
8. Add running buffer to the chambers at the bottom and top of the gel. Gently remove the comb from between the plates. Be sure that no bubbles interfere with uniform contact of the liquid in the running buffer with the gel at the bottom or at the top. The wells in the stacking portion of the gel should be uniformly filled with buffer with no bubbles or gel debris in them. If bubbles need to be removed, use a syringe and buffer to gently force them out. Also, be sure that there is no leaking between the two buffer chambers. Remember that the current

will take the path of least resistance, and it will not pass through the gel unless the two buffer chambers are separated.

9. Prepare the standard and unknown protein samples for analysis by mixing 10 μl of each with 5 μl of 4X sample buffer in microcentrifuge tubes, mixing gently, and heating at 100°C for 5 min in a boiling water bath. This step is to ensure that all of the proteins in the sample are completely denatured. Remove the samples from the water bath and allow them to cool to room temperature.
10. Load the samples onto the gel with a syringe, taking care to place the needle so that the dense blue solution settles gently in a layer at the bottom of the sample wells, displacing the running buffer upward as it is added. Be careful not to tear or distort the soft stacking gel as you add the samples. Several students can share a gel; for example, if a 10-well gel is used, eight students can analyze their unknowns and two samples of standards can be analyzed, one in a center well and one in a well at the edge of the gel.
11. Connect the electrodes of the apparatus to the power source. The anode (+ pole) must be at the bottom of the apparatus. (Can you explain why?)
12. Apply a constant 200-V field to the apparatus, and continue the electrophoresis until the bromophenol blue tracking dye reaches the very bottom of the gel. Bromophenol blue is a small anionic dye, which will migrate faster than any protein during electrophoresis.
13. Turn off the power supply, and disconnect the electrodes from it.
14. Gently remove the gel from the apparatus. Using a spatula or razor blade, separate the two plates by gently prying up one corner of a single plate. The gel should adhere to one of the plates. Using the same spatula or razor blade, separate the soft stacking gel from the more rigid running gel. Discard the stacking gel in the specified container (polyacrylamide is toxic and should be properly disposed of).
15. Gently transfer the running gel to a pan filled with approximately 50 ml of Coomassie Blue staining solution and allow it to soak in the solution for 1 hour.
16. Pour off the staining solution (save it, because it can be reused many times), and replace it with destaining solution for 1 hour, replacing it with fresh destaining solution every 15 min. Destaining proceeds more rapidly if the gel is gently agitated in the solution continuously during destaining. Destaining is complete when the proteins appear as blue bands on a transparent gel background.
17. The gel can now be photographed or placed between two sheets of cellophane (using water to make the gel easier to slide around, then gently squeezing out the excess water, but leaving no bubbles) and dried under a vacuum on a commercial gel drier to keep a permanent record of the electropherogram.

Data Analysis

1. For each protein band in the lanes in which standards and the unknown sample were analyzed, measure the distance from the bottom of the well (in mm) to the center of the band. Also measure the distance from the bottom of the well to center of the dye front (in millimeters) in the same lane.
2. Calculate the R_f value for each protein band according to:

$$R_f = \frac{\text{distance migrated by the protein band (mm)}}{\text{distance migrated by the dye front (mm)}}$$

3. Use the R_f values for the protein standards to prepare a standard curve that plots the log of the molecular weight of each standard protein on the ordinate versus the R_f value for that protein on the abscissa. Do all of the values fall on a straight line? Can you predict what the curve would look like for proteins that are much larger and much smaller than the standards you used?
4. From the R_f value(s) of the protein band(s) in your unknown sample, determine the molecular weights of these proteins by interpolation on your standard curve. How many bands did you observe in your unknown? What was/were the molecular weight(s)? If there was more than one band in your unknown, what can you say from the intensity of Coomassie staining of the bands about the relative abundance of the proteins in the sample?

Exercises

1. You wish to resolve the compounds in Figure 4-11 by paper electrophoresis in a system that uses a pH 7.0 buffer. At pH 7.0, will all of these compounds migrate in one direction, so that you can place the origin line at one end of the paper? If so, would you place the origin line near the anode (+ pole) or cathode (− pole)? What is the predicted relative order of electrophoretic migration of these compounds at pH 7.0?
2. Will the three compounds in Figure 4-11 be clearly separated from one another by electrophoresis at pH 11? If so, describe the electrode toward which they will migrate and the order of migration. If not, why not?

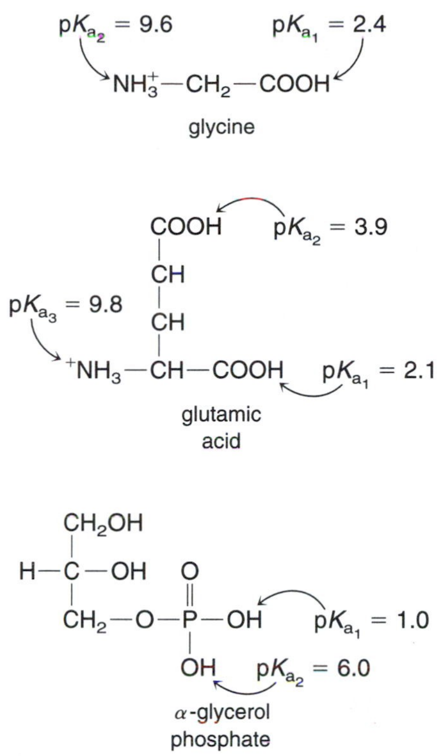

REFERENCES

Dunbar, B. S. (1987). *Two Dimensional Electrophoresis and Immunological Techniques.* New York: Plenum Press.

Gersten, D. M. (1996). *Gel Electrophoresis* (Essential Techniques Series). New York: Wiley.

Kuhr, W. G. (1990). Capillary Electrophoresis. *Anal Chem* **62**:403R–414R.

Laemmli, U. K. (1970). Cleavage of Structural Proteins during the Assembly of the Head of Bacteriophage T4. *Nature* **227**:680.

Sambrook, J., Fritsch, E. F., and Maniatis, T. (1989). *Molecular Cloning: A Laboratory Manual*, 2nd ed. Cold Spring Harbor, NY: Cold Spring Harbor Laboratory Press.

Tal, M., Silberstein, A., and Nusser, E. (1980). Why Does Coomassie Brilliant Blue Interact Differently with Different Proteins? *J Biol Chem* **260**:9976–9980.

Wilson, K., and Walker, J. M. (1994). *Principles and Techniques of Practical Biochemistry*, Chapter 9. Cambridge, UK: Cambridge University Press.

$pK_{a_2} = 9.6$ $pK_{a_1} = 2.4$

$NH_3^+-CH_2-COOH$

glycine

$pK_{a_2} = 3.9$ (COOH); $pK_{a_3} = 9.8$ ($^+NH_3$); $pK_{a_1} = 2.1$ (COOH)

$^+NH_3-CH(COOH)-CH_2-CH_2-COOH$

glutamic acid

$CH_2OH-CH(OH)-CH_2-O-P(=O)(OH)-OH$

$pK_{a_1} = 1.0$ $pK_{a_2} = 6.0$

α-glycerol phosphate

Figure 4-11 Compounds to be separated in the Exercises, questions 1 and 2.

SECTION II

Proteins and Enzymology

SECTION II

Proteins and Enzymology

Introduction

Most of the astonishing variety of structural features and biochemical activities of cells is a consequence of an equally extraordinary variety in the structures of the proteins in those cells. In some cases the relation between the structure and function of a protein is fairly obvious. For example, the major structural proteins of hair and collagen fibrils, keratin and tropocollagen, have high molecular weights and regularly repeated amino acid sequences that endow them with the ability to form long, regularly structured strands. These proteins are quite different in size and amino acid composition from lower-molecular-weight, soluble proteins, such as enzymes and hormones, for which the relationship between protein structure and biochemical function is far less obvious. Another example is provided by integral membrane or transmembrane proteins, which usually contain predictable sequences of amino acids that fold into helices with hydrophobic surfaces that adapt these regions of the protein to stable residence in the hydrophobic interior of the membrane. The variations in chemical structure among other proteins are small or subtle; yet these small differences may be associated with major differences in physiological function. For example, hemoglobin S differs from hemoglobin A by only a single amino acid residue, yet the behavior of these two proteins in red blood cells is very different. Many other cases are known in which replacement of a single amino acid residue by mutation radically alters the biological function of a protein. In fact, the technique of *in vitro* mutagenesis, the process of engineering known changes into the DNA encoding a protein so as to introduce specific amino acid substitutions into it, has become a powerful and very widely used tool for the investigation of structure–function relationships in proteins. Whether major or minor, such differences in structure are the consequences of differences in the total number and chemical nature of amino acids in a protein, the primary sequence of these amino acids, and the manner in which the polypeptide chain is folded in space. To understand the way biological function is conferred on proteins it is necessary to study the chemistry of amino acids and proteins, as well as the methods for elucidating the higher levels of structure.

Amino Acids: Identification and Quantitative Determination

The individual amino acids of a protein can be liberated by hydrolyzing the peptide (amide) bonds (–CO–NH–) that link them. The usual procedure is to dissolve the peptide or protein in 6 N HCl and heat the solution in a sealed, evacuated tube at 100°C for 8 to 72 hr (or for shorter times at higher temperatures). All amide linkages (including the side chain amides of glutamine and asparagine) are cleaved under these conditions. Certain amino acids are entirely (tryptophan) or partially (serine, threonine, tyrosine, cysteine) destroyed, so that special precautions are required for the quantitative determination of these amino acids.

The amino acids in a protein hydrolysate can be conveniently separated for qualitative analysis by paper or thin-layer chromatography or by electrophoresis. Qualitative and quantitative analysis of amino acid mixtures can be achieved with much greater convenience and accuracy using automated ion-exchange chromatography. A common means of detecting amino acids is by determination of their reaction with ninhydrin (triketohydrindene hydrate). Ninhydrin decarboxylates α amino acids producing an intensely colored blue-purple product, CO_2, water, and aldehydes derived from the carbon skeleton of the amino acid (see Fig. 6-9, Experiment 6). Ninhydrin yields a similar blue-purple product on reaction with primary amines or ammonia. The reagent also reacts with imino acids such as proline to yield a yellow product. Under controlled conditions the color development is quantitative. Amino acids also react with fluorescamine or with orthophthalaldehyde and mercaptoethanol to yield fluorescent products that provide a more sensitive means of analyzing amino acid mixtures.

Another widely used method for qualitative and quantitative analysis of amino acid mixtures is high-performance liquid chromatography (HPLC) (see Experiments 2 and 6). The mixture of amino acids is first subjected to reaction with phenylisothiocyanate (PITC) to convert them to the phenylthiocarbamyl–amino acid derivatives, which are then subjected to chromatographic separation. The derivativatization of the amino acids serves two purposes: it attaches a UV-absorbing "tag," which makes their quantitative determination easy, and it converts them to a more hydrophobic form, which is necessary for good separation on the reverse-phase system commonly used with this technique. This method of amino acid analysis will be used in Experiment 6.

Amino Acids: Ionic Properties

All amino acids contain ionizable groups that act as weak acids or bases, giving off or taking on protons when the pH is altered. As is true of all similar ionizations, these ionizations follow the Henderson–Hasselbalch equation:

(II-1)

$$\text{pH} = \text{p}K_a + \log_{10} \frac{[\text{unprotonated form (base)}]}{[\text{protonated form (acid)}]}$$

The pK_a is the negative $\log_{10}$ of the acid dissociation constant of the ionizable group. Examination of this equation leads to a further understanding of pK_a. When the concentration of the unprotonated form equals that of the protonated form, the ratio of their concentrations equals 1, and $\log_{10} 1 = 0$. Hence, *pK_a can be defined as the pH at which the concentrations of unprotonated and protonated forms of a particular ionizable species are equal.* The pK_a also equals the pH at which the ionizable group is at its best buffering capacity; that is, the pH at which the solution resists changes in pH most effectively. The pK_a value of the weak acid or base in a buffering system is therefore an important consideration in choosing a buffer of a given pH. For example, if you wanted to prepare a buffer at pH 7.0, you might choose inorganic phosphate (pK_{a_2} = 6.8). pK_a values are derived with unit activity concentrations (approximately 1.0 M) of components. The pK_a values of many biochemically important ionic compounds shift slightly on dilution to more physiological concentrations. Biochemists therefore use the symbol pK_a' to designate pK_a values in dilute ionic systems.

Consider applying the Henderson–Hasselbalch equation to the titration of glycine with acid and base. Glycine has two ionizable groups: a carboxyl group and an amino group, with pK_a values of 2.4 and 9.6, respectively. In water at pH 6.0, glycine exists as a dipolar ion, or zwitterion, in which the carboxyl group is unprotonated ($-COO^-$) and the amino group is protonated to give the substituted ammonium ion ($-NH_3^+$). (Verify this, using the Henderson–Hasselbalch equation, pH 6.0, and the respective pK_a values.) Addition of acid to the solution lowers the pH rapidly at first and then more slowly as the buffering action of the carboxyl group is exerted (see Fig. II-1). At pH 2.4 the pK_a is reached and half of the acid has been consumed; the carboxyl group is half ionized and effectively buffers the solution. Further addition of acid results in titration of the remainder of the carboxylate ions in the solution (e.g., at pH 1.4 only 1 carboxyl in 11 is ionized). Titration of the protonated α-amino group with base follows a similar curve into the alkaline region. The concentration of the protonated and unprotonated species of each ion at any pH can be calculated from the Henderson–Hasselbalch equation. The intersection between the titration of the carboxyl group and the titration of the amino group defines the pH at which glycine has no net charge, and is called the "isoelectric point" (pI, Fig. II-1). Experiment 5 provides the student with an

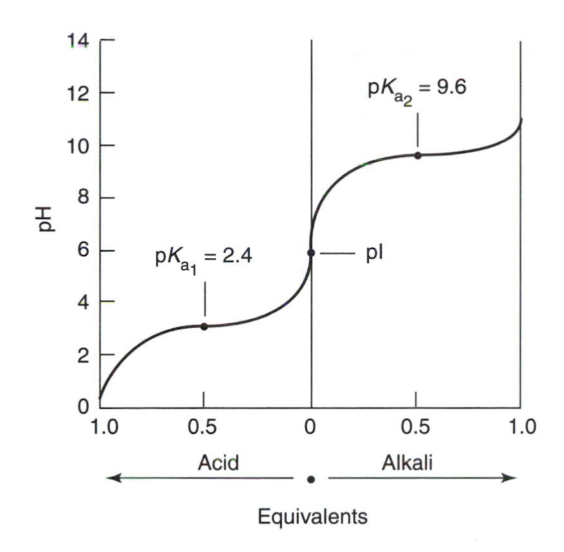

Figure II-1 Titration curve of glycine.

opportunity to develop a further understanding of the acid–base properties of amino acids as these are studied by titration.

Most amino acids contain carboxyl and amino groups having pK_a values similar to those of glycine. In addition to these groups, many amino acids contain other ionizable groups, which introduce other "steps" or pK_a values into their titration curves. It is largely these side chain groups—those not linked in peptide bonds to adjacent amino acids—that account for the ionic properties of proteins. Table II-1 lists the ionizable groups of proteins and amino acids, the nature of their ionizations, and their approximate pK_a values. Notice that the pK_a values for groups on proteins are given as ranges rather than as single values. This is because the ionization of various groups on proteins is often modified by the presence of other groups on the protein. This is evident from a comparison between the typical pK_a values for the α-amino and α-carboxyl groups of a free amino acid, 2.3 and 9.5, respectively, and the corresponding values for these groups when they occur at the ends of a peptide, even a peptide as small as a dipeptide, which are about 3.5 and 8.5. The close proximity of the charged $-COO^-$ and $-NH_3^+$ groups in an amino acid affects the ionization of the other group by electrostatic attraction and repulsion. These effects are largely lost when these groups are farther apart in peptides.

Protein Structure

The elements of protein structure are divided into four classes: primary, secondary, tertiary, and quaternary. *Primary structure* refers to the linear sequence of amino acids linked by amide bonds along protein chains. These polymer chains vary in length from a few amino acid residues (oligopeptides) to molecules containing 2000 or more amino acids. Most proteins are from 100 to 500 amino acid residues in length. One or more of each of 20 natural amino acids may be present in each protein molecule. In some cases the amino acids undergo posttranslational chemical modification, which introduces still more variety into protein structure.

Both secondary and tertiary structure refer to the arrangement of the polypeptide chain in a three-dimensional structure. *Secondary structure* generally describes the arrangement of neighboring amino acids into more or less regular patterns that are held together by hydrogen bonding (H-bonding) between the carbonyl oxygens and the amide hydrogens (C=O···H–N) of the peptide backbone. Several such regular patterns have been discovered by x-ray crystallographic analysis of many proteins. The two patterns of secondary structure that occur most frequently are the α-helix and the β-pleated sheet (or simply β-strands) (Fig. II-2). Certain fibrous proteins may consist almost entirely of one of these secondary structural elements. In contrast, H-bonded secondary structures account only for a portion of the peptide chain folding in globular proteins (such as most enzymes) and tend to occur in shorter segments connected by bends or turns to fold them into a compact structure.

The complete pattern of folding of the polypeptide chain of a protein, whether regular or irregular, is called the "tertiary structure." The tertiary structure of any protein is the sum of many forces and structural elements, many of which are the result of interactions between the side chain groups of amino acids in the protein. Some of these interactions are described below.

Disulfide Bonds. The sulfhydryl groups of the amino acid cysteine often (but not always) form partners in intrachain or interchain disulfide bonds, and they thus serve to fix a protein molecule into a three-dimensional structure. While such bonds are,

Table II-1 Dissociation of Ionizable Groups of Amino Acids and Proteins

Group	Ionization	pK_a Range (on a protein)	pK_a (free amino acid)
Carboxyl (α)	$-COOH \rightleftharpoons H^+ + -COO^-$	2.0–4.0	1.8–2.4
Carboxyl (ω, of glu, asp)	$-COOH \rightleftharpoons H^+ + -COO^-$	3.0–4.7	3.8–4.3
Ammonium (α)	$-NH_3^+ \rightleftharpoons H^+ + -NH_2$	7.5–8.5	8.9–9.7
sec-Ammonium (proline)	$>NH_2^+ \rightleftharpoons H^+ + >NH$	10.2–10.8	10.6
Ammonium (ϵ, lysine)	$-NH_3^+ \rightleftharpoons H^+ + -NH_2$	9.4–10.6	10.5
Phenolic hydroxyl (tyrosine)	$-C_6H_4-OH \rightleftharpoons H^+ + -C_6H_4-O^-$	9.8–10.4	10.1
Imidazolium (histidine)	imidazolium (N$^+$–H, N–H) $\rightleftharpoons H^+ +$ imidazole (N, N–H)	5.6–7.0	6.0
Guanidinium (arginine)	$-NH-C(=NH_2^+)-NH_2 \rightleftharpoons H^+ + -NH-C(=NH)-NH_2$	11.6–12.6	12.5
Sulfhydryl (cysteine)	$-SH \rightleftharpoons H^+ + -S^-$	9.4–10.8	10.4

of course, part of the primary covalent linkages of the protein molecule, it is more useful to think of them as contributing to the chain folding.

Hydrogen Bonds. Hydrogen atoms that are attached to the electronegative atoms oxygen or nitrogen (e.g., tyrosine-OH, glutamine, and asparagine amides) frequently interact by H-bonding with other electronegative atoms (usually oxygen, as in carboxyl oxygen). This noncovalent bonding can stabilize the interactions of amino acid side chains with other side chains or with the carboxyls and amides of the protein backbone.

Ionic Attractions and Repulsions. Attractions and repulsions occur between ionized functional groups.

Hydrophobic and Hydrophilic Interactions. Although proteins are generally soluble in water or dilute salt solutions, they contain many amino acids with aliphatic or aromatic side-chain residues that have low solubility in water, apparently because of the highly organized "icelike" structure that must be formed around them. Hence, such hydrophobic residues are frequently found clustered inside the protein molecule, where there are weak van der Waals associations among them and they are "removed" from energetically unfavorable contact with water molecules. The net energy gained from such "oil drop" arrangements contributes substantially to overall protein folding. Similarly, amino acids with highly polar or charged side chain residues tend to be exposed to the solvent, where their interaction with water molecules is energetically favorable.

Proline Residues. Since proline is an imino acid, unlike the other amino acids, it does not fit into a regularly organized helical or pleated sheet structure. Hence, proline residues are likely to be found in disorganized regions within the polypeptide chain or in "bends" associated with a change in the overall direction of the peptide chain.

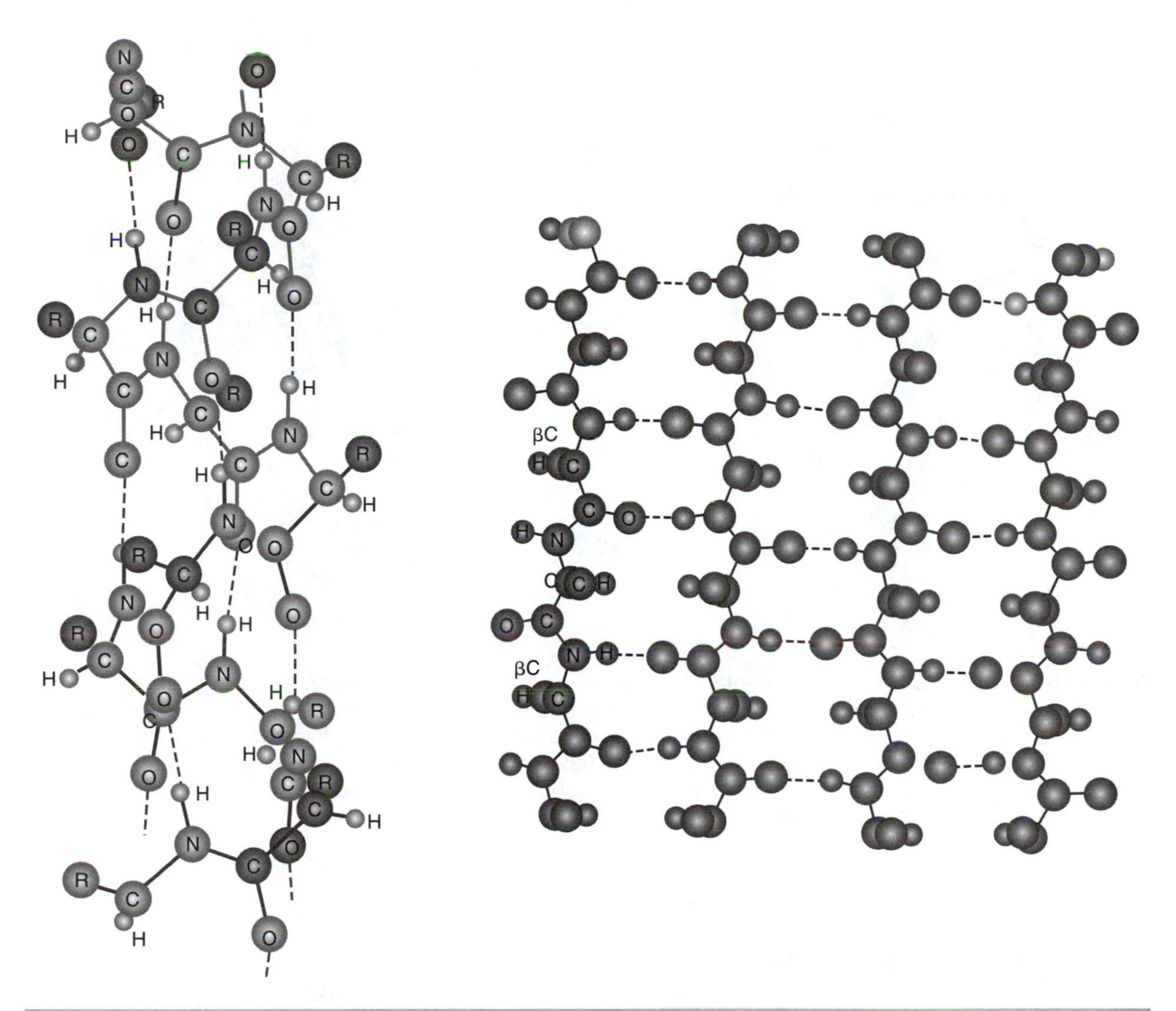

Figure II-2 Major elements of secondary structure of proteins. Left, the α-helix; right, representation of the antiparallel pleated sheet structures for polypeptides. (After Pauling, L., and R. B. Corey (1951). *Proc Natl Acad Sci USA* **37**:729).

Many of these features are evident in the three-dimensional structures of ribonuclease and myoglobin shown in Figure II-3. To understand the common and distinguishing features of the secondary and tertiary structure of various proteins as determined by x-ray crystallographic analysis, the student should study several examples in standard textbooks of biochemistry.

Some proteins can be fully described by primary, secondary, and tertiary structure, but many others, perhaps most, have an additional level of structural organization—quaternary structure. *Quaternary structure* refers to the organization of identical or different polypeptide chains into noncovalently linked aggregates containing a fixed number of polypeptide subunits in a single functional protein. Thus, many proteins have been found to consist of 2, 3, 4, 6, 8, 12, or other numbers of polypeptide chains, usually bound in regular arrangements. Some examples are shown in Figure II-4.

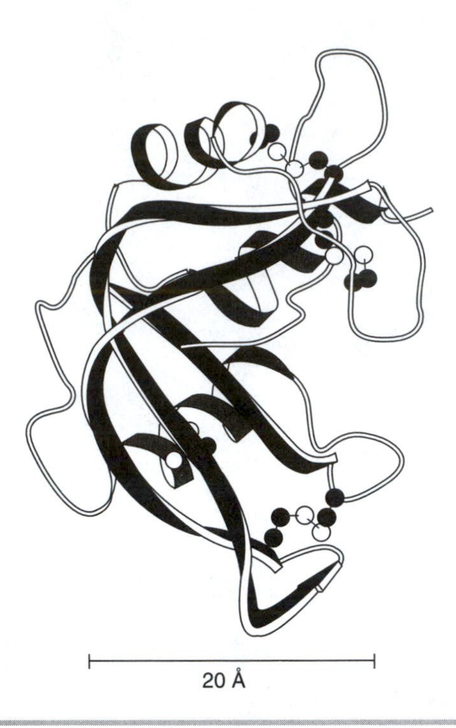

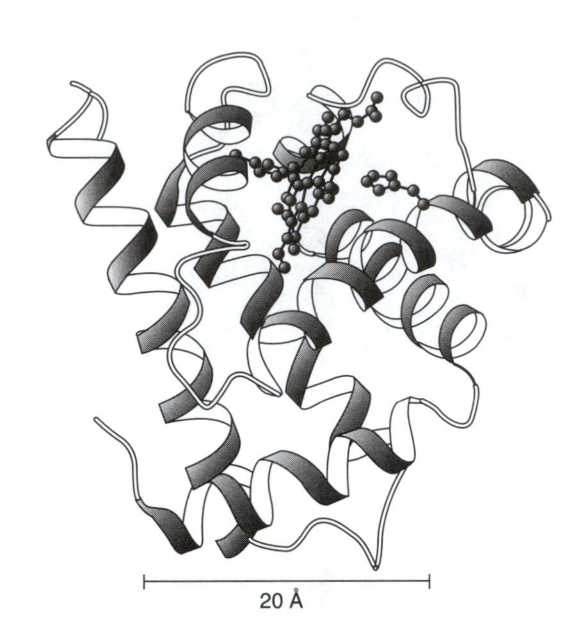

Figure II-3 Ribbon diagrams showing the three-dimensional structures of ribonuclease A (left) and myoglobin (right). Helical segments are presented as coiled ribbons, β-strands as smoothly curving ribbons, and regions of irregular structure as thin tubes. The four disulfide bonds of ribonuclease are shown as ball and stick models as are the atoms of the heme A prosthetic group and the side chains of two histidine residues that ligate to the heme in myoglobin. (From Stryer, L. (1995). *Biochemistry,* 4th ed. New York: Freeman, Figures 2-45 and 2-43. Reprinted with permission.)

Determination of Protein Structure

The primary structure (i.e., the amino acid sequence) of a protein can be determined by stepwise chemical degradation of the purified protein. By far the most powerful and commonly used technique for doing this is the automated Edman degradation. The amino terminal amino acid residue of the polypeptide is reacted with Edman's reagent (phenylisothiocyanate) to form the phenylthiocarbamyl derivative, which is removed without hydrolysis of the other peptide bonds by cyclization in anhydrous acid. The amino acid derivative is converted to the more stable phenylthiohydantoin and identified by HPLC. The process can be repeated many times, removing the amino acids from the amino terminus of the polypeptide one residue at a time and identifying them until the entire sequence is determined. These reactions are shown in Fig. 6-5 (Experiment 6). When long polypeptides are degraded stepwise, the accumulated effects of small imperfections in the chemistry of the process eventually makes it impossible to deduce the sequence. Under favorable conditions, up to 100 residues can be determined, and as little as 5 pmol of protein can be analyzed. In the future, it is likely that mass spectrometry will also provide increasingly powerful means of determining the sequence of polypeptides.

In contemporary biochemistry the amino acid sequence of a protein is much more likely to be determined indirectly from the sequence of the DNA that encodes the protein or, in the case of eukaryotes, from the sequence of cDNA made from mRNA encoding it. This is because DNA sequences can be determined far more rapidly and accurately than protein sequences. Since the genetic code is known, it is a simple matter to deduce the

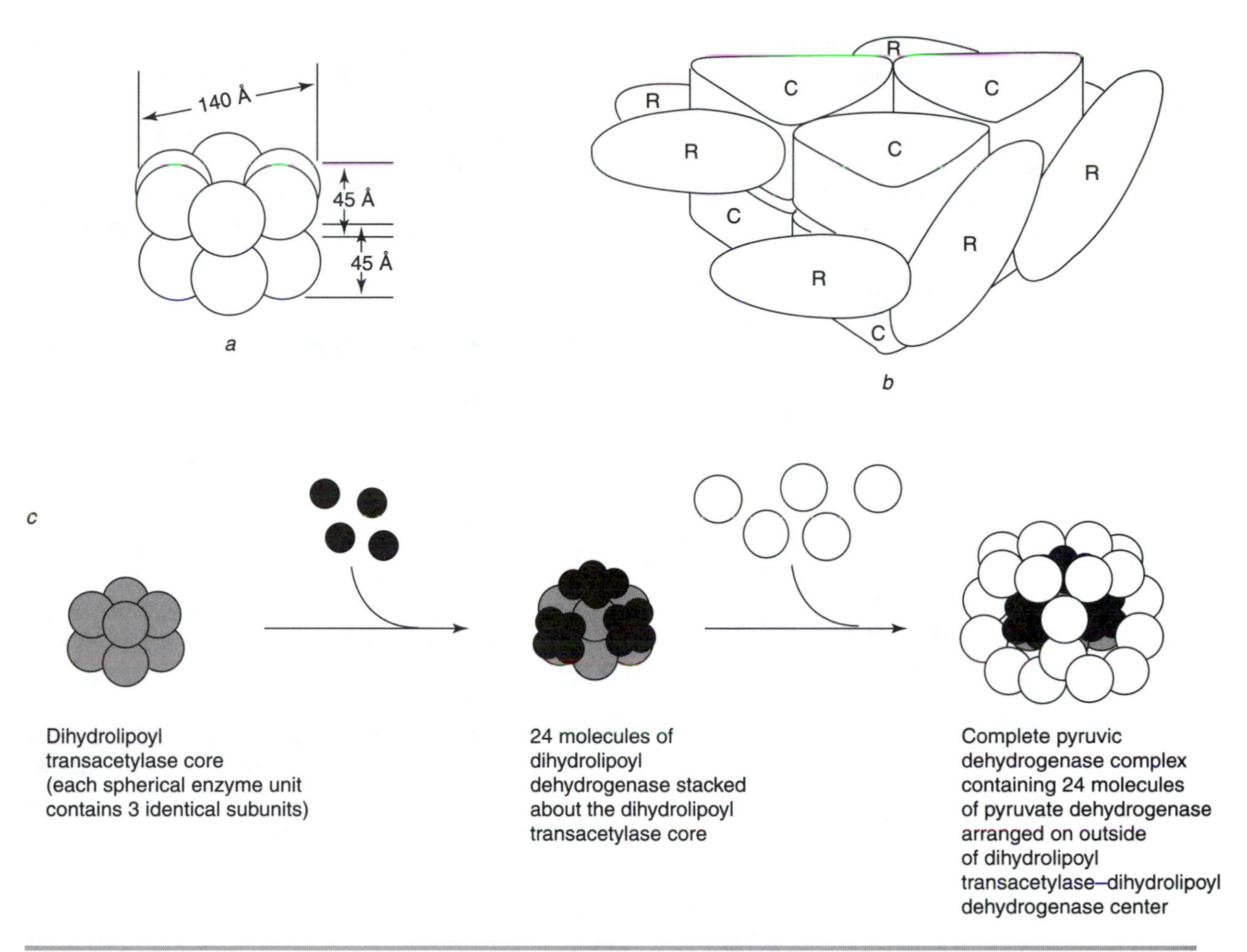

Figure II-4 Examples of the quaternary structure of proteins. (*a*) A drawing of glutamine synthetase of *E. coli* showing the orientation of the 12 identical subunits of the enzyme. (*b*) A drawing of aspartate transcarbamylase of *E. coli* showing the proposed orientation of the 6 catalytic subunits (labeled C, each MW = 33,000), and 6 regulatory subunits (labeled R, each MW = 17,000), of the enzyme. (*c*) A drawing of the pyruvate dehydrogenese complex of *E. coli* showing the proposed orientation of the various proteins in the multienzyme complex.

amino acid sequence of a protein from the DNA sequence of the gene that specifies it. In fact, computer programs have been developed to do this (see Experiment 25).

The determination of the secondary and tertiary structure—that is, the details of the three-dimensional folding of the polypeptide chain of a protein at high resolution—relies on one of two powerful techniques: x-ray diffraction analysis of protein crystals and multidimensional high-field nuclear magnetic resonance (NMR) spectroscopy. Both methods provide very detailed structural information at the level of atomic resolution, and several thousand structures of proteins have now been determined using these techniques. The standard repository for such structural information is the Brookhaven Protein Data Base (http://www.pdb.bnl.gov/). These structures provide a rich source of information about the nature of protein architecture and the relationship between protein structure and biological function. A detailed description of these methods is beyond the scope of this book, but a few general comments should be made. X-ray diffraction requires that you be able to prepare suitable crystals of the pure protein. The method is applicable to both large and small proteins. Structures of very large protein aggregates, e.g., the yeast proteasome, have been determined by x-ray diffraction. NMR spectroscopy does not require that the protein be in crystalline form; analysis is conducted on concentrated solutions of the protein, so the method is especially valuable for proteins that are difficult to crystallize. At present, structure determination by NMR is limited to proteins with molecular weights lower than about 30,000; this limit may be raised in the future by advances in the technology.

At a much lower level of resolution, it is possible to estimate the α-helical and β-strand content of proteins by measuring their circular dichroic spectra. Such determinations give a general idea of the secondary structural features of a protein, but do not determine which segments of the primary structure possess which secondary structures. This can be predicted by computer analysis of the primary sequence, however, since analysis of many protein structures determined from crystallography and the study of synthetic polypeptides has given a good indication of the propensity of a given sequence of amino acids to fold into α-helices or β-strands (see Experiment 25). You should be aware that the predictions of such analyses have been found by subsequent structure determination to be correct only about 70% of the time. Although circular dichroism may be a crude tool for determining secondary structures of proteins, it is quite valuable for studying changes in the structures of proteins, particularly their unfolding and refolding, as occurs during protein denaturation (see below).

Determination of Molecular Weights and Quaternary Structures of Proteins

The quaternary structure of a protein can be largely determined by careful molecular weight determination of the native protein, followed by dissociation of the protein into its constituent polypeptide chains with denaturing agents. If nonidentical subunits are found, they must be separated from each other. Then the molecular weight of the isolated subunits must be determined, usually under denaturing conditions. Usually a small, integral (most commonly an even number) of subunits comprise the native species. That is, the molecular weight of native species is a simple multiple of the molecular weight(s) of the subunits. Hence, techniques for determining the molecular weights of macromolecules are central to determining quaternary structure.

Determination of Subunit Molecular Weight

The determination of the primary structure (i.e., the complete amino acid sequence) of a protein automatically allows the calculation of the molecular weight of the polypeptide from the sum of the residue weights of the amino acids. This is normally done by the same computer programs that facilitate the determination of the amino acid sequence from the sequence of the gene (see Experiment 25). This is probably the most common way of determining subunit molecular weights in contemporary biochemistry.

If small amounts of a purified protein are available, but the amino acid sequence is not known, two methods are generally available for determination of its subunit molecular weight.

Electrophoresis in Sodium Dodecyl Sulfate (SDS). (See also, Experiment 4.) Heating a protein at 100°C in an SDS-containing solution (usually in the presence of thiols to reduce any disulfide bonds) will completely dissociate and denature all polypeptide chains from one another. The polypeptide chains bind a large quantity of SDS (about one SDS molecule per two amino acid residues), thereby gaining a strongly net negative charge. The binding of so much SDS gives all of the proteins approximately the same (negative) charge to mass ratio, regardless of what this ratio might be for the native proteins at a given pH. Thus, the rate at which such SDS-polypeptide complexes migrate during electrophoresis on polyacrylamide gels is in most instances determined by viscous resistance to their flow in the gel (i.e., by their frictional coefficients), rather than by net charge. This phenomenon generates an empirical relation between molecular weight and electrophoretic mobility: the logarithm of the molecular weight is proportional to the relative mobility. Standards of known molecular weight are used to calibrate the results. Obviously, only molecular weights of constituent polypeptide chains, not native aggregates, can be determined with this method. Proteins that bind abnormal amounts of SDS, proteins with very unusual amino acid compositions, glycoproteins, and intrinsic membrane proteins often fail to migrate as expected on SDS–polyacrylamide gel electrophoresis.

Mass Spectrometry. Advances in the technology of mass spectrometry have made it possible to determine the molecular weights of polypeptides of the sizes commonly found in proteins. This has resulted primarily from the development of novel techniques for obtaining molecular ions of these large, nonvolatile molecules. Two such techniques are electrospray ionization (ESI) and matrix-assisted laser desorption ionization (MALDI). When these methods for obtaining molecular ions of proteins are combined with improvements in technology for resolution and mass determination of such ions, it has become feasible to determine molecular weights of protein subunits to within 0.01% of their value, which is more precise than any method other than determination of the complete primary structure. In fact, mass spectrometry is now frequently used to check for errors in the determination of the sequence of proteins from DNA sequencing.

Determination of Native Molecular Weight

Gel-Filtration or Size-Exclusion Chromatography. Protein molecules will diffuse into the pores of appropriately chosen gel-filtration beads (e.g., Sephadex or Bio-Gel). The molecules partition between the solvent outside and inside of the gel beads according to their ease of diffusion, which is related to their molecular weight (see Experiment 2). For most globular proteins of generally similar shapes, a linear relationship between the logarithm of the molecular weight and the elution volume will be observed, within certain molecular weight limits. Hence, comparison of the elution volume of a protein to the elution volume of standards of known molecular weights on the same gel column will provide a reasonably good estimate of the molecular weight of the protein. The method is unreliable, however, if the protein in question has a shape that is significantly different from that of the standards used or if the protein subunits participate in a rapid equilibrium between states of aggregation containing different numbers of subunits.

Direct Determination of Quaternary Structure. None of the above methods is capable of revealing the precise geometric arrangements of subunits in a native protein. Two methods have proven useful for this purpose: x-ray crystallographic analysis and electron microscopy of the protein molecules using negative staining. The examples of quaternary structure shown in Figure II-4 were determined by these techniques.

Protein Denaturation

Proteins in their natural form are called "native proteins." Any significant change in the structure of a protein from its native state is called "denaturation." Denaturation results from changes in the secondary, tertiary, and quaternary structures of proteins. Generally, denaturation is to be avoided, because you wish to study a protein in as close to its native state as possible. Occasionally, however, you may wish to denature proteins deliberately. For example, proteins are often quickly denatured to stop an enzyme reaction when determining the rates of enzyme reactions. Proteins are also often deliberately denatured so that the detailed nature of the unfolding and refolding of their polypeptide chains can be studied. Denaturation of proteins usually leads to greatly decreased solubility, so it can be a useful way to separate them from other classes of biological molecules during purification. The following are a few agents and techniques commonly used to denature proteins.

Chaotropic Agents. Compounds such as urea and guanidine hydrochloride usually cause denaturation when present in high concentrations (4 to 8 M). The chemical basis of this disruption of protein structure is not completely understood. Urea and guanidine are thought to form competing hydrogen bonds with the amino acid residues of the peptide chain, thereby disrupting the internal hydrogen bonding that stabilizes the native structure.

These agents also alter the solvent properties of water so that hydrophobic interactions in the protein are weakened. They probably act by a combination of these effects. Urea- or guanidine-induced denaturation is frequently partially or completely reversed when the concentration of the competitive hydrogen bonding agent in the protein solution is lowered by dialysis or dilution.

Detergents. Sodium dodecyl sulfate (SDS) and other ionic detergents are extremely effective protein denaturants. Such agents usually have a highly polar end and a hydrophobic end. SDS has been shown to bind tightly to polypeptides, probably by hydrophobic interactions, and to convert them into rodlike structures that have a strong negative charge. Associations between polypeptide chains (quaternary structure) are also disrupted by detergents.

Heat. Heat denatures most dissolved proteins when the temperature reaches higher than about 50°C. Harsh heat treatment alters the secondary and tertiary structures of proteins. Heat denaturation usually results in eventual protein precipitation as a result of destruction of the secondary structure and formation of random aggregates. Some proteins have been found to be heat-stable, especially when ligands are bound to them (i.e., many enzymes are protected against heat by their substrates). This property can be exploited during purification (see Experiment 8).

Acid or Alkali. Proteins are amphoteric polyelectrolytes; thus, changes in pH would be expected to affect the existence of salt bridges, which reinforce the tertiary structure of proteins. Further, when localized areas of a protein acquire a large net positive or negative charge, the ionizable groups repel each other and thus place the molecule under conformational strain. For example, denaturation of the proteolytic enzyme, pepsin, occurs at alkaline pH values. This may be attributed to internal strain arising from mutual repulsions of the ionized carboxyls of aspartic and glutamic acids. Moderate concentrations of certain acids (for example, trichloroacetic, phosphotungstic, or perchloric acid) usually produce complete denaturation and precipitation. Proteins commonly are treated with these acids to stop enzymatic reactions and to concentrate protein samples prior to SDS-PAGE analysis (see Experiment 8).

Oxidation or Reduction of Sulfhydryl Groups. Many, but not all, proteins are sensitive to alterations in the oxidation–reduction potential of their environment. The effect is caused in part by oxidation of sulfhydryl groups or reduction of disulfide bonds. Not all proteins are equally sensitive to such alterations, but when they are, it is critical to be aware of their sensitivity. The purification or assay of some proteins can be accomplished only by providing reducing conditions (reduced glutathione, free cysteine, dithiothreitol, or mercaptoethanol) in all buffer solutions.

Enzymatic Action. Proteinases (also called proteases) present in crude or unpurified protein preparations often catalyze the breakdown of proteins by hydrolyzing the peptide bonds. Since these enzymes act more slowly at low temperatures, crude protein solutions are frequently kept cold (0 to 2°C) during the early stages of purification. In some cases it is also necessary to add inhibitors of proteinases, such as *p*-methylphenylsulfonyl fluoride (PMSF).

Protein Purification

There is no single or simple way to purify all proteins. Procedures and conditions used in the purification of one protein may result in the denaturation of another. Further, slight chemical modifications in a protein may greatly alter its structure and thus affect its behavior during purification. Nevertheless, there are certain fundamental principles of protein purification on which most fractionation procedures are based. In all cases, the experimenter takes advantage of small differences in the physical and/or chemical properties of the many proteins in a crude mixture to bring about separation of the proteins from one another. Since the precise nature of these differences is not usually known in advance, the development of a purification procedure necessarily involves considerable trial and error. A procedure leading to a homogeneous protein from a natural source usually requires the sequential application of a number of the following techniques.

Overexpression of the Recombinant Protein. Advances in molecular biology have provided an extremely powerful means of purifying proteins by overexpressing the genes encoding them from recombinant plasmids in suitable host cells grown in culture. Instead of having to isolate a protein from the cell or tissue in which it normally resides, often as a tiny fraction of the total protein, genomic DNA or cDNA specifying the protein is spliced into appropriately designed plasmids, called "overexpression vectors," which can be induced to direct the synthesis of very large amounts of the protein (see Section V). It is not unusual for such recombinant proteins to be made at levels from 10 to 50% of the total cell protein on induction. This greatly simplifies the task of purifying the protein: instead of having to devise a procedure for a 1000- to 100,000-fold enrichment of the protein, as is typical for isolation of a protein from its natural source, purification of a recombinant protein by 2- to 10-fold yields essentially pure protein.

Although the use of recombinant genes for protein production is often very successful, problems can arise. Some genes or cDNAs are not expressed well in the commonly used vector/host systems. Occasionally, the overproduced proteins do not fold properly and aggregate into insoluble inclusion bodies in which the recombinant protein is denatured. Posttranslational modifications, such as phosphorylation or glycosylation, that occur in the native cells and are important for the function of the protein may not occur in the recombinant expression system.

Heat Denaturation. Because all proteins are not equally stable when heated in aqueous solution, fractionation can often be achieved by controlled heating. As noted above, the presence of a substrate, a product, or an inhibitor often stabilizes a protein against heat denaturation.

Protein–Nucleic Acid Complexes. During protein purification it occasionally becomes necessary to remove basic proteins or contaminating nucleic acids. The controlled addition of sparingly soluble material of opposite charge (e.g., addition of a basic protein such as protamine for nucleic acid removal) results in the coprecipitation of the added and unwanted material (e.g., as a protamine–nucleic acid complex).

Solubility Properties. The solubility of most proteins in aqueous solutions can be attributed to the hydrophilic interaction between the polar molecules of water and the ionized groups of the protein molecules. Reagents that change the dielectric constant or the ionic strength of an aqueous solution therefore would be expected to influence the solubility of proteins. The addition of ethanol, acetone, or certain salts (such as ammonium sulfate), usually at low temperatures to avoid denaturation, frequently results in precipitation of proteins. Such agents are thought to act by effectively removing water molecules that normally solvate the surface of proteins and favor their solubility. As the water molecules are removed, nonspecific protein–protein interactions become dominant over protein–water interactions and protein molecules aggregate and form insoluble precipitates. Of these precipitating agents, fractionation with ammonium sulfate is by far the most commonly used. Use of organic solvents is less generally useful, because many proteins are denatured by them. Step-by-step application of such techniques to heterogeneous protein solutions often results in fractionation (purification) because of the differing degrees of solubility among the proteins in the solution.

The precipitations noted above usually result when the attractive forces among protein molecules exceed those between the protein molecules and water. Proteins are polyelectrolytes with different numbers and types of ionizable groups; the net charge on any given protein molecule and hence its attraction to another protein molecule are functions of the pH and ionic strength of the medium. Therefore, any treatment designed to separate proteins by making use of relative solubility must take into consideration the pH of the medium. Often, alteration of only pH under a set of conditions is sufficient to effect an "isoelectric" precipitation. Choosing the correct conditions of pH, ionic strength, or dielectric constant to carry out a fractionation is somewhat arbitrary and is usually determined empirically.

Adsorptive Properties. Under certain conditions of pH and low ionic strength, certain proteins will be adsorbed by various substances. Calcium phosphate gel and alumina Cγ gel, for example, are frequently used to adsorb specific proteins from heterogeneous mixtures. The adsorbed proteins can frequently be released from the insoluble support either by altering the pH or increasing the ionic strength. Thus, purification can be obtained by removing either the extraneous unwanted proteins or the desired protein with the gel.

Chromatographic Separation. Chromatographic separation of protein mixtures has become one of the most effective and widely used means of purifying individual proteins. A more detailed discussion of the principles of chromatography and their application to the separation of protein mixtures is presented in Experiment 2 and is also illustrated in Experiments 8 and 10.

Proteins, as polyelectrolytes, can be separated by ion-exchange chromatography in much the same way as smaller molecules. Because of the size, ease of denaturation, and complex charge distribution of protein molecules, certain special techniques are required. The most generally useful ion-exchange materials for proteins are derivatives of cellulose, dextran, agarose, and polyacrylamide (e.g., Sephadex, Sepharose, BioGel, etc.). The most commonly used anion exchangers are substituted with diethylaminoethyl (DEAE) or quaternary ammonium ethyl (QAE) groups; useful cation exchangers are substituted with carboxymethyl (CM), phosphoryl (P), or sulfoethyl (SE) groups. Excellent separation of protein mixtures can often be brought about by step-by-step or gradient elution of the sample from a column of exchange material in which the ionic strength of the eluting buffer is increased, the pH altered, or both.

Chromatographic separation of proteins on the basis of molecular size (or, more precisely, ease of diffusion) is readily accomplished by size-exclusion chromatography on dextran, agarose, or polyacrylamide beads.

Affinity Chromatography. Affinity chromatography is a very useful new technique that takes advantage of the natural affinity of a protein such as an enzyme for specific molecules, for instance, the substrate or a very effective inhibitor of that enzyme. The molecule that binds to the protein is covalently attached to an insoluble support, such as agarose. This insoluble material can then be used to adsorb the appropriate protein selectively from other proteins that do not bind to the molecule attached to the support material. Some remarkably effective purifications have been accomplished by use of this useful technique. A much more detailed discussion of the technique of affinity chromatography is presented in Experiment 2. An alternative approach to isolation of proteins by affinity chromatography is to use genetic engineering to add sequences coding for extra amino acids or even an entire protein to the gene encoding the protein you wish to isolate; the extra polypeptide is chosen to provide a chemical "tag" that binds tightly and specifically to an affinity adsorbent. The desired protein with its covalently attached tag can then be easily isolated from all the other proteins in the cells in which the recombinant protein was produced. The protein is then released from the adsorbent by dissociating the tag from the adsorbent. This approach is illustrated in Experiment 10.

Electrophoretic Separation. The net charge on a particular protein varies with the pH of the medium in which it is dissolved. Accordingly, application of an electric field to a buffered, heterogeneous protein solution often results in their differential migration in free solution or in heterogeneous systems with an inert supporting material, such as a polyacrylamide gel. Because of the difficulty of "scaling up" electrophoretic procedures, they are more commonly used for analytical applications than for the preparative scale purification of proteins. The use of electrophoresis is discussed in more detail in Experiment 4.

Criteria of Protein Purity. There are no tests for protein purity, only for impurities. Crystallinity, the absence of contaminating enzymes, and homogeneous behavior in centrifugal fields (ultracentrifugation), in electric fields (electrophoresis), in ion-exchange chromatography, and in immunological reactions may all be consistent with purity, but they do not prove it. We attempt to prove inhomogeneity with as many tests as are available. Practically, if a protein preparation behaves as a single entity in careful studies using these criteria, it may be assumed to have a high degree of purity.

Quantitative Determination of Proteins

In general, there is no completely satisfactory single method to determine the concentration of protein in any given sample. The choice of the method depends on the nature of the protein, the nature of the other components in the protein sample, and the desired speed, accuracy, and sensitivity of assay. Several of the methods commonly used for protein determination are discussed in the sections that follow.

Biuret Test. Compounds containing two or more peptide bonds (e.g., proteins) take on a characteristic purple color when treated with dilute copper sulfate in alkaline solution. The name of the test comes from the compound biuret, which gives a typically positive reaction. The color is apparently caused by the coordination complex of the copper atom and four nitrogen atoms, two from each of two peptide chains (Fig. II-5). The biuret test is fairly reproducible for any protein, but it requires relatively large amounts of protein (1 to 20 mg) for color formation. Because of its low sensitivity, the biuret assay is no longer widely used.

Folin–Ciocalteu (Lowry) Assay. The quantitative Folin–Ciocalteu assay (also often called the "Lowry assay") can be applied to dried material as well as to solutions. In addition, the method is sensitive; samples containing as little as 5 μg of protein can be analyzed readily. The color formed by the Folin–Ciocalteu reagent is thought to be caused by the reaction of protein with the alkaline copper in the reagent (as in the biuret test) and the reduction of the Cu^{2+} (cupric) ions in the reagent to Cu^{1+} (cuprous) by the tyrosine and tryptophan residues of proteins. The cuprous ions in turn reduce the phosphomolybdate–phosphotungstate salts in the reagent to form an intensely blue complex. Because the content of these two amino acids varies substantially within proteins, the color yield per milligram of protein is not constant. It may differ substantially from that of a protein standard. Nevertheless, the method is very useful for following changes in protein content, as, for example, during purification of a protein. This assay is used in most of the experiments in this book, because it is convenient and inexpensive.

Bicinchoninic Acid (BCA) Assay. The bicinchoninic acid assay for proteins is based on the same reactions as the Folin–Ciocalteau assay. Proteins are again reacted with alkaline cupric ions to form the biuret complex, and these ions are reduced to cuprous ions by the aromatic amino acids in the proteins. In this case, however, the Cu^{1+} ions form a complex with bicinchoninic acid (Fig. II-6), which has an intense absorbance maximum at 562 nm. This assay shows the same variation from protein to protein as the Folin–Ciocalteau assay, but is more convenient experimentally and can be made somewhat more sensitive.

Dye-Binding (Bradford) Assay. The binding of proteins to Coomassie Brilliant Blue 250 causes a shift in the absorbance maximum of the dye from 465 nm to an intense band at 595 nm. Determination of the increase in absorbance at 595 nm as a function of protein added provides a sensitive assay

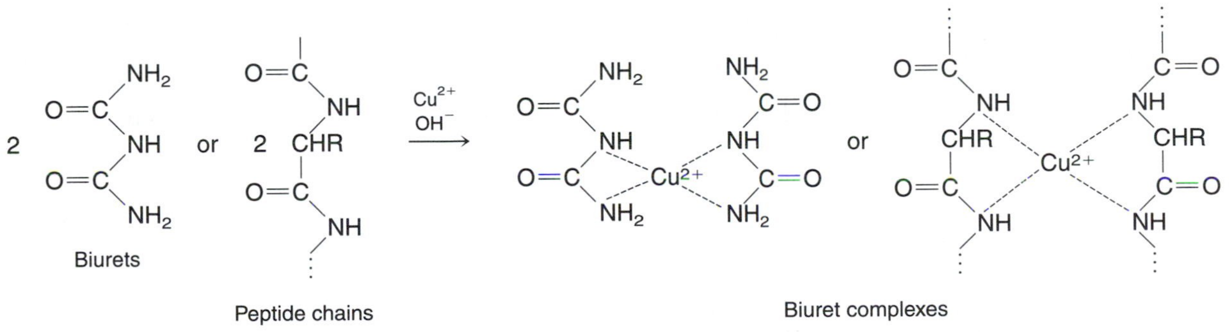

Figure II-5 The biuret reaction.

Protein + Cu^{2+} --→ Protein–Cu^{2+} complex (biuret reaction)

Protein–Cu^{2+} complex --→ Cu^{1+} + Uncharacterized protein oxidation products

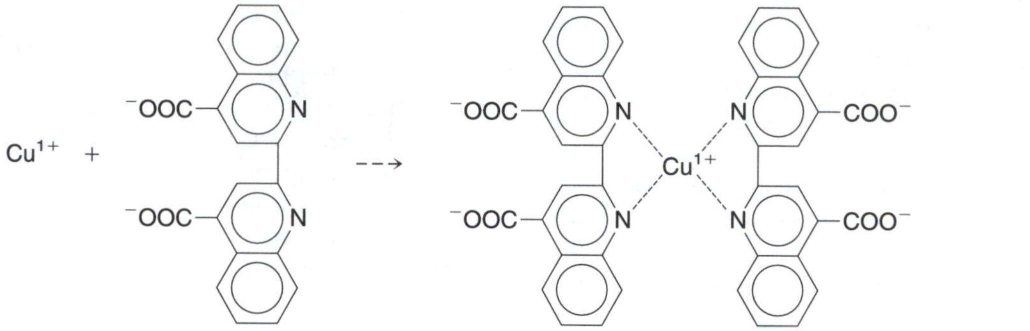

Cu^{1+} + BCA --→ BCA–Cu^{1+} complex

Figure II-6 Chemistry of the bicinchoninic acid protein assay.

of protein that is quite constant from protein to protein and relatively free from interference from other cellular components, commonly used salts, etc. This assay uses commercially available reagents and is about four times as sensitive as the Folin–Ciocolteu assay.

Spectrophotometric Assay. The tyrosine and tryptophan residues of proteins exhibit an ultraviolet absorbance at approximately 275 nm and 280 nm, respectively. Because the combined content of these amino acids is roughly constant in many proteins, the concentration of protein (in the absence of other ultraviolet-absorbing materials) is generally proportional to the absorbance at 280 nm. (Note that because individual proteins can vary substantially in their content of tryptophan and tyrosine, the absorption spectra of pure proteins and their absorbance at 280 nm may vary widely.) Many protein solutions, especially solutions containing mixtures of many proteins, at 1 mg of protein per milliliter exhibit an absorbance at 280 nm of about 0.8 when viewed through a 1-cm light path. Thus, the concentration of most proteins can rapidly be assayed by measurement of the absorbance at 280 nm, if they are free of other materials that absorb light at 280 nm. Such assays are advantageous because they are rapid and allow full recovery of the protein for subsequent analyses.

Unfortunately, other compounds present in natural materials also exhibit absorbance at 280 nm. Specifically, nucleic acids, which have maximal absorbance at 260 nm, exhibit extensive absorbance at 280 nm. Corrections can be made for nucleic acid absorbance at 280 nm if the ratio of absorbance at 280/260 nm is known. Pure proteins have a 280 nm/260 nm ratio of approximately 1.75 (variations are due to differences in number and type of aromatic amino acids present), whereas nucleic acids have a 280 nm/260 nm ratio of 0.5. Intermediate ratios corresponding to various mixtures of protein and nucleic acid can be used to compute the concentration of each.

Dry Weight. Most of the above methods are suitable for comparisons, such as following the increasing specific activity of an enzyme during purification. However, for careful chemical work with a pure protein, in which the mass must be known accurately, it is generally necessary to measure directly the dry weight of a salt-free sample of the protein. This is done after drying a sample of protein, which has been extensively dialyzed against distilled water or a volatile buffer (e.g., ammonium carbonate) to constant weight in a vacuum at 50 to 100°C over a drying agent. This mass can then be related to one of the more convenient assays, such as the absorbance of the protein at 280 nm.

Enzymology

Theory of Enzyme Action

Enzymes are proteins that catalyze biochemical reactions. *As is true of other catalysts, enzymes influence the rate at which equilibrium is obtained, but do not affect the overall equilibrium of the reaction.* The reaction is accelerated by providing a reaction route having a lower free energy of activation for the transition of substrate to products than the uncatalyzed process. These two fundamental points are illustrated by the highly simplified free energy profiles of hypothetical enzyme-catalyzed and non–enzyme-catalyzed reactions depicted in Figure II-7. The free energy levels of substrates and products are the same in these two systems, and therefore the net free energy change in the overall reaction, ΔF (also called ΔG in some textbooks); is identical for both processes. The equilibrium constant K_{eq} is directly related to $\Delta F°$ as in Equation II-2:

(II-2)

$$-\Delta F° = RT \ln K_{eq}$$

where $\Delta F°$ is ΔF at standard states (one molar) of reactants and products, R = gas constant, and T = absolute temperature. It follows that the equilibrium constants for both enzymatic and nonenzymatic processes are identical as long as much less than stoichiometric amounts of catalyst are used.

The rate of a process is determined by the free energy levels of the rate-limiting transition state in the reaction pathway (the higher the free energy barriers the slower the rate). This relationship is defined by the absolute reaction rate theory as in Equation II-3.

(II-3)

$$k_{vel} = \frac{k_B T}{h} e^{-\Delta F^{\ddagger}/RT}$$

where k_{vel} = reaction velocity constant, k_B = Boltzmann's constant, h = Planck's constant, R = gas constant, and T = absolute temperature. The free energy differences between intermediates and transition states are indicated by $\Delta F^{\ddagger}$; the superscript ‡ emphasizes that the thermodynamic parameter refers to transition state complexes determining the rate of the reaction.

Kinetic Analysis of Enzyme Activity

Unlike most uncatalyzed reactions, the first observed rate or initial velocity (v_0) of enzyme-catalyzed reactions increases with increasing sub-

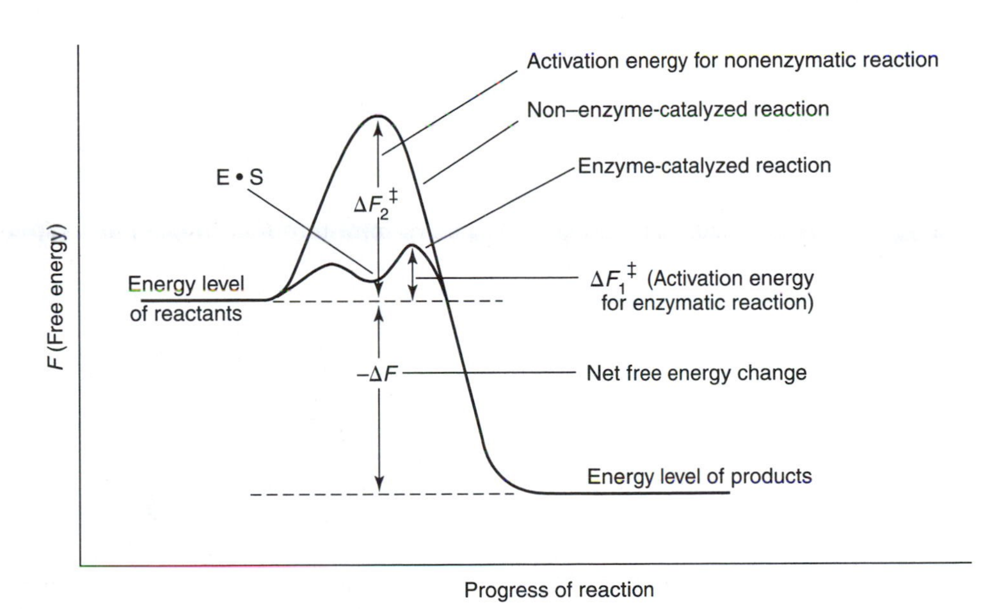

Figure II-7 Energy profiles of enzymatic and nonenzymatic catalysis of a reaction.

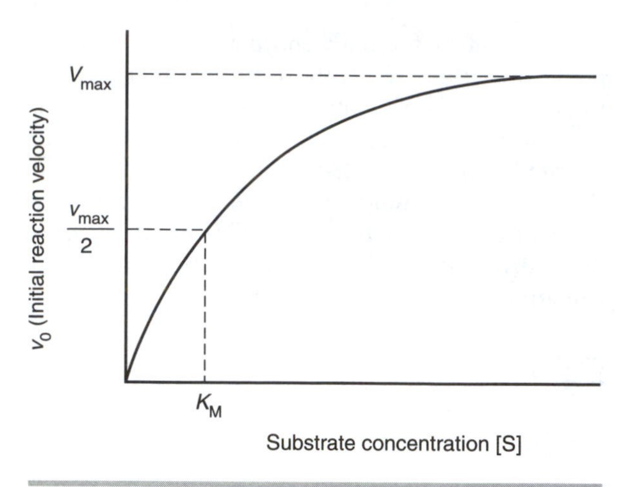

Figure II-8 The relationship of substrate concentration and initial velocity in enzymatic reactions.

strate concentration only until a substrate level is reached beyond which further additions of substrate do not increase the initial rate. This phenomenon and the maximum initial velocity (V_{max}) achieved at substrate saturation, are illustrated in Figure II-8. These observations, as they apply to enzymatic reactions that employ a single substrate or the hydrolysis of a single substrate, have been explained by the following postulated reaction scheme (Equation II-4), where E is enzyme, S is substrate, and P is product:

(II-4)

$$E + S \underset{k_2}{\overset{k_1}{\rightleftharpoons}} ES \underset{k_4}{\overset{k_3}{\rightleftharpoons}} E + P$$

Thus, the reaction proceeds by means of an obligate intermediate enzyme–substrate complex (ES). When all of the enzyme is in the ES state (i.e., the system is saturated with substrate), the observed rate of reaction is maximal and the reaction is at V_{max}.

Michaelis and Menten, and later Briggs and Haldane, used the scheme shown in Equation II-4 to derive a mathematical expression that describes the relation between initial velocity and substrate concentration. (Consult a biochemistry textbook for the step-by-step derivation of this relationship, because it is important to be aware of the assumptions that were used.) The Michaelis–Menten equation (Equation II-5) is:

(II-5)

$$v_0 = \frac{V_{max}\,[S]}{K_M + [S]}$$

In this expression, V_{max} and v_0 are velocities as defined earlier, [S] is the substrate concentration, and K_M is a new constant, the Michaelis constant, equal to $k_2 + k_3/k_1$ where the k values are the specific rate constants of Equation II-4.

If we rearrange the Michaelis–Menten equation to Equation II-6,

(II-6)

$$K_M = [S]\,(V_{max}/v_0 - 1)$$

we can readily see that K_M is numerically equal to a substrate concentration; K_M therefore has the dimensions of moles per liter (or millimolar or micromolar, depending on the units used for the substrate concentration). *The substrate concentration that yields a velocity, v_0, equal to half the maximum velocity, V_{max}, is equivalent to K_M* (Fig. II-8). The value of the K_M reflects the stability of the enzyme–substrate interaction and is of great practical value. K_M is not, however, the true dissociation constant of the enzyme–substrate complex. This dissociation constant is commonly termed K_S. Only when $k_2 \gg k_3$ does

(II-7)

$$K_M = K_S = k_2/k_1 = \frac{[E]\,[S]}{[ES]}$$

The reciprocal of K_S is the affinity constant of enzyme for the substrate. That is, the greater the affinity of the enzyme for the substrate, the smaller the K_S. *It is not correct to consider K_S and K_M to be equivalent without additional evidence, but this assumption is a common error.*

Although K_M may be determined using data similar to that presented in Figure II-8, it is more accurate and convenient to use one of the linear forms of the Michaelis–Menten equation to determine K_M. Lineweaver and Burk first pointed out that equation II-8 can be obtained by inversion of the Michaelis–Menten equation:

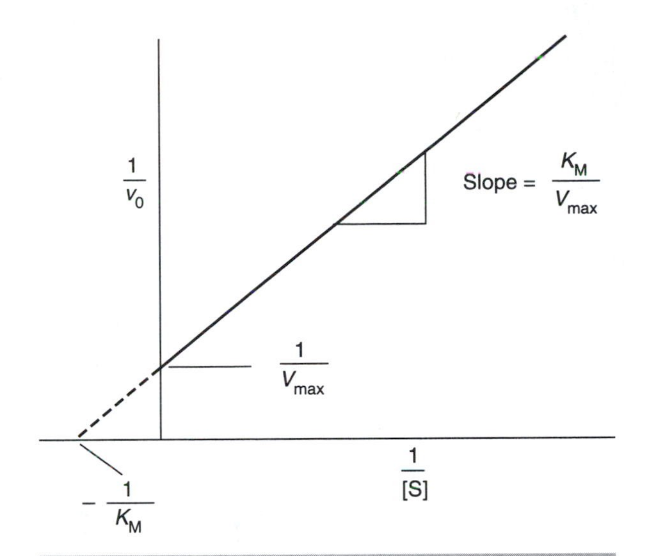

Figure II-9 Lineweaver–Burk plot.

(II-8)

$$\frac{1}{v_0} = \frac{K_M}{V_{max}} \frac{1}{[S]} + \frac{1}{V_{max}}$$

Velocities at different substrate concentrations may be plotted as reciprocals, as shown in Figure II-9. Values for K_M and V_{max} are obtained by extrapolation of the line, as indicated. Plots derived from other formulations of the Michaelis–Menten equation may also be used to obtain the same constants (Equations II-9 and II-10).

(II-9)

$$\frac{[S]}{v_0} = \frac{1}{V_{max}}[S] + \frac{K_M}{V_{max}} \text{ (for a Hanes–Woolf plot)}$$

and

(II-10)

$$v_0 = -K_M \frac{v_0}{[S]} + V_{max} \text{ (for an Eadie–Hofstee plot)}$$

Plots derived from these equations (e.g., v_0 versus v_0/[S]) are often preferred because they spread out points that are close together on Lineweaver–Burk plots. In current research, it is preferable to use computer programs, such as KinetAsyst (IntelliKinetics, State College, PA), KINSIM, or numerous others, to fit kinetic data for the Michaelis–Menten equation, or for the other equations shown below because these programs provide a precise estimate of the statistical reliability of the fit of the data and calculate standard error values for the kinetic constants. Graphic representations are still conventionally used to display the data, but computer-based statistical analysis is essential for accurate analysis.

Most enzymes react with two or more substrates. For this reason, the Michaelis–Menten equation is inadequate for a full kinetic analysis of these enzyme reactions. Nonetheless, the same general approach can be used to derive appropriate equations for two or more substrates. For example, most enzymes that react with two substrates, A and B, are found to obey one of two equations if initial velocity measurements are made as a function of the concentration of both A and B (with product concentrations equal to zero). These are

(II-11)

$$v_0 = \frac{V_{max}[A][B]}{K_{ia}K_b + K_b[A] + K_a[B] + [A][B]}$$

for "sequential" mechanisms, which require intermediate formation of an EAB complex (K_{ia} is a product inhibition constant) or

(II-12)

$$v_0 = \frac{V_{max}[A][B]}{K_b[A] + K_a[B] + [A][B]}$$

for "ping-pong" mechanisms in which a product is released from the enzyme before substrate B binds. K_a and K_b are Michaelis constants for the substrates A and B, respectively, in both equations. It is readily shown by converting these equations to their double reciprocal form, analogous to the Lineweaver–Burk equation, that they can be distinguished by experiments in which v_0 is measured as a function of one substrate at several fixed levels of the other substrate. For sequential mechanisms, both the intercepts at $1/[S] = 0$ and the slopes of a double reciprocal plot (e.g., $1/v_0$ versus 1/[A]) will be changed if the concentration of B is changed. That is, a family of intersecting lines will be obtained. For ping-pong mechanisms only the intercepts are altered by changing the concentration of

B; that is the double reciprocal plots for a ping-pong kinetic mechanism are parallel lines—the slopes are not changed by altering the concentration of the second substrate.

V_{max} is the velocity of an enzyme-catalyzed reaction when the enzyme is saturated with all of its substrates and is equal to the product of the rate constant for the rate-limiting step of the reaction at substrate saturation (k_{cat}) times the total enzyme concentration, E_T, expressed as molar concentration of enzyme active sites. For the very simple enzyme reaction involving only one substrate described by Equation II-4, $k_{cat} = k_3$. However, more realistic enzyme reactions involving two or more substrates, such as described by Equations II-11 and II-12, require several elementary rate constants to describe their mechanisms. It is not usually possible to determine by steady-state kinetic analysis which elementary rate constant corresponds to k_{cat}. Nonetheless, it is common to calculate k_{cat} values for enzymes by dividing the experimentally determined V_{max}, expressed in units of moles per liter of product formed per minute (or second), by the molar concentration of the enzyme active sites at which the maximal velocity was determined. The units of k_{cat} are reciprocal time (min^{-1} or sec^{-1}) and the reciprocal of k_{cat} is the time required for one enzyme-catalyzed reaction to occur. k_{cat} is also sometimes called the "turnover number" of the enzyme.

Kinetics of Inhibition of Enzymes

Many substances interact with enzymes to lower their activity; that is, to inhibit them. Valuable information about the mechanism of action of the inhibitor can frequently be obtained through a kinetic analysis of its effects. To illustrate, let us consider a case of *competitive inhibition*, in which an inhibitor molecule, I, combines only with the free enzyme, E, but cannot combine with the enzyme to which the substrate is attached, ES. Such a competitive inhibitor often has a chemical structure similar to the substrate, but is not acted on by the enzyme. For example, malonate ($^{-}OOCCH_2COO^{-}$) is a competitive inhibitor of succinate ($^{-}OOCCH_2CH_2COO^{-}$) dehydrogenase. If we use the same approach that was used in deriving the Michaelis–Menten equation together with the additional equilibrium that defines a new constant, an inhibitor constant, K_I,

(II-13)

$$K_I = \frac{[E]\,[I]}{[EI]}$$

we will obtain

(II-14)

$$v_0 = \frac{V_{max}\,[S]}{K_M(1 + [I]\,/\,K_I) + [S]}$$

As before, it is most useful to convert this equation to its reciprocal form:

(II-15)

$$\frac{1}{v_0} = \frac{K_M}{V_{max}}(1 + [I]\,/\,K_I)\frac{1}{[S]} + \frac{1}{V_{max}}$$

Measurement of v_0 at various substrate levels and known inhibitor concentration yields data in the form shown in Figure II-10. The various kinetic constants may be obtained as shown in the figure. A convenient way to obtain K_I is to plot the values of the slopes of lines such as in Figure II-10 at several inhibitor concentrations. This replot should be a straight line with an intercept on the x-axis equal to $-K_I$. A statistically reliable determination of inhibition constants, however, should always involve

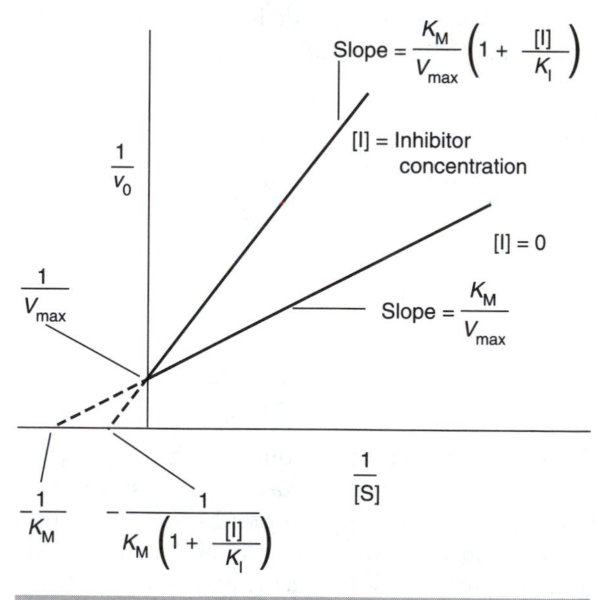

Figure II-10 Lineweaver–Burk plot: competitive inhibition.

fitting the data to the appropriate equation using a computer program.

Biochemists observe other kinds of enzyme inhibition. *Noncompetitive inhibition* consists of cases in which an inhibitor combines with either the E or the ES form of the enzyme. This requires definition of two new inhibitor constants:

(II-16)

$$K_{Ie} = \frac{[E]\,[I]}{[EI]} \quad \text{and} \quad K_{Is} = \frac{[ES]\,[I]}{[ESI]}$$

Derivation of a rate expression in a manner similar to that used to derive the Michaelis–Menten equation, and subsequent inversion of this rate expression yields Equation II-17.

(II-7)

$$\frac{1}{v_0} = \frac{K_M}{V_{max}}(1 + [I]\,/\,K_{Is})\frac{1}{[S]} + \frac{1}{V_{max}}(1 + [I]\,/\,K_{Ie})$$

In some instances, $K_{Ie} = K_{Is}$; that is, the extent of combination of inhibitor with the E and ES forms of the enzyme are identical. The left panel of Figure II-11 illustrates a plot of such data. In other cases of noncompetitive inhibition, K_{Ie} is different (usually larger) than K_{Is} as illustrated in the right panel of Figure II-11.

In *uncompetitive inhibition*, the inhibitor combines only with the ES form of the enzyme. This pattern of inhibition is not seen frequently except in studies of inhibition by the products of enzyme reactions. If you define an inhibitor constant

(II-18)

$$K_I = \frac{[ES][I]}{[ESI]}$$

and derive a rate expression for uncompetitive inhibition, its inverted form is

(II-19)

$$\frac{1}{v_0} = \frac{K_M}{V_{max}}\frac{1}{[S]} + \frac{1}{V_{max}}(1 + [I]\,/\,K_I)$$

as illustrated in Figure II-12. Note that these three kinetic patterns of inhibition can be distinguished most easily by obtaining initial velocity data as a function of substrate and inhibitor concentration and plotting the results in double-reciprocal form (Figures II-10 to II-12). The pattern will differ in the effects of the inhibitor on the slopes and intercepts of these plots (see Table II-2).

It is important to recognize that the three patterns of inhibition just described are observed in many different situations (e.g., product inhibition)

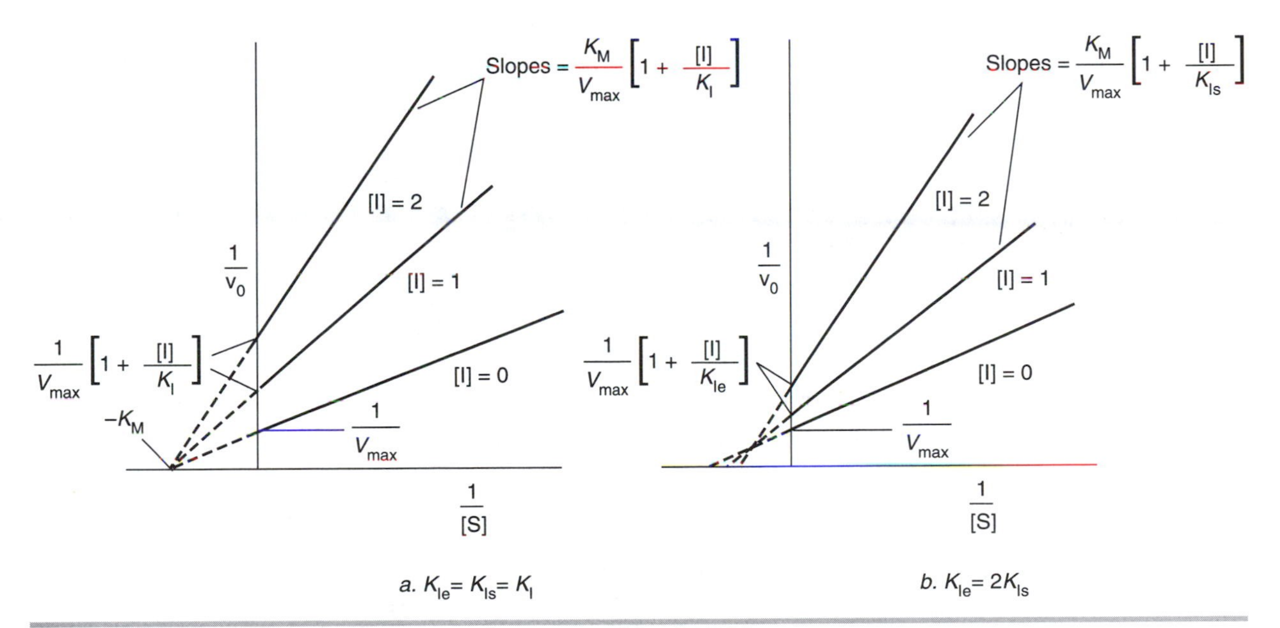

Figure II-11 Double reciprocal plots of noncompetitive inhibition.

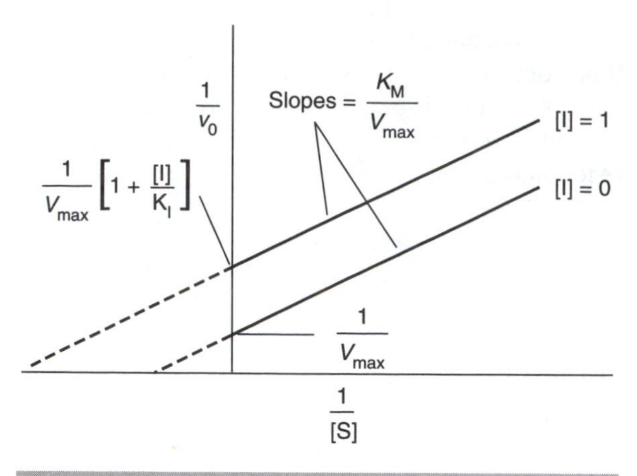

Figure II-12 Double reciprocal plot of uncompetitive inhibition.

in addition to the somewhat oversimplified cases described above. In all cases the type of inhibition can be identified and analyzed by the equations shown, but the physical meaning of the inhibition constants and the exact mechanism of inhibition may not be the same as given in our simple examples.

Other Factors Affecting Enzyme Activity

A variety of factors other than substrate and enzyme concentration affects the activity of enzymes. Some of the most important of these are discussed very briefly below and are studied in the laboratory in Experiment 7.

Temperature. In general, the dependence of enzyme activity on temperature will be described by the curve illustrated in Figure II-13. At the lower temperatures, the temperature dependence of V_{max} will be described by the Arrhenius equation (Equation II-20).

(II-20)

$$\log_{10} V_{max} = -\frac{E_a}{2.3\,R}\frac{1}{T} + \text{constant}$$

Thus, a plot of $\log_{10} V_{max}$ versus the reciprocal of the absolute temperature T is linear with the slope equal to $-E_a/2.3\,R$, where E_a is an empirical quantity called the Arrhenius activation energy and R is the gas constant, 1.98 cal/deg-mol. At higher temperatures the maximal velocity generally is much lower than predicted; this deviation is the result of denaturation and inactivation of the enzyme at elevated temperatures.

pH. Most enzymes show a strong dependence of activity on the pH of the medium. For this reason it is essential to conduct enzyme assays in buffered solutions. The pH dependence of enzyme reactions is the consequence of changing degrees of ionization of groups in the enzyme, in the substrate, or in both. In some cases it is possible to make edu-

Table II-2 Inhibition Patterns Can Be Distinguished by the Effects of Inhibitors on Double Reciprocal Plots

	Effect of Increasing [I] upon:	
Type of Inhibition	**Slope**	**Intercept** $\left(\text{at } \frac{1}{v_0} = 0\right)$
Competitive	increases	no change
Noncompetitive	increases	increases
Uncompetitive	no change	increases

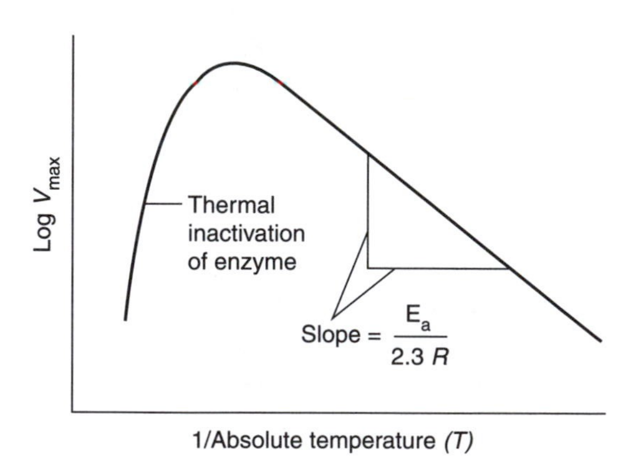

Figure II-13 Relation of enzyme activity to temperature. This is a typical Arrhenius plot.

cated guesses as to functional groups in enzymes that are involved in enzyme activity by determining pK_a values associated with changes in enzyme activity and comparing them to values for known protein groups (Table II-1).

Cofactors. Many enzymes require relatively loosely bound organic cofactors (coenzymes) or metal ions for activity. A biochemist must be alert for indications of such a requirement during purification of an enzyme, because it may be necessary to add such cofactors to the assay solution to obtain enzyme activity. The interactions of coenzyme with enzyme can be analyzed by kinetic and spectral techniques. Similar techniques are used to evaluate interactions with metal ions.

Allosteric Effectors. Many enzymes are subject to metabolic regulation through interaction with metabolites that often act at allosteric sites, which are distinct from the active site. The kinetic behavior of such enzymes is often more complex than the behavior we have discussed above, and such complex kinetics may serve as an indication that you are dealing with an allosteric enzyme. Further discussion of this subject is found in Experiments 9 and 15.

Enzyme Assay

During the isolation of an enzyme from any source, you must determine the amount of enzyme present in the preparations. You usually accomplish this indirectly, by measuring the catalytic activity of the enzyme under some fixed conditions of pH, temperature, and added ions. Examination of another formulation of the Michaelis–Menten equation (Equation II-21) shows us that this is straightforward.

(II-21)

$$v_0 = \frac{V_{max}\,[S_0]}{K_M + [S_0]} = \frac{k_3\,[E_T]\,[S_0]}{K_M + [S_0]}$$

(In this formulation, S_0 equals the initial substrate concentration and E_T equals the total enzyme, E + ES). Because k_3 and K_M are fundamental constants of the system and S_0 is known from the assay conditions, the initial velocity v_0 is directly proportional to $[E_T]$. Careful measurement of v_0 values, therefore, allows an estimation of the enzyme concentration. Formally, v_0 is expressed as the initial rate of change in concentration of substrate or product per unit time (e.g., millimolar per minute), but if you always use the same reaction volume for the assay, it is satisfactory to express v_0 simply as the amount of substrate used or product formed per unit of time (e.g., micromole per minute). This convention is used in Experiments 7, 8, and 9.

The simplest procedure for obtaining v_0 is to use such high concentrations of substrate that a negligible loss occurs during the reaction: [S] always ≅ $[S_0]$. In such cases the velocity after 5 min, or even 30 min, may be the same as the initial velocity. Eventually, the substrate will become limiting or inhibitory concentrations of products will be formed. If a low [S] is used (i.e., [S] ≅ K_M), the velocity will continually decrease as substrate concentration decreases, and it may be difficult to estimate v_0.

An accurate determination of v_0 requires that an appropriate amount of enzyme be used. You can use so little enzyme that the velocity cannot be accurately determined (see Assay 1, Figure II-14). Similarly, use of too much enzyme makes it difficult to determine an extremely high velocity in the time required to make the assay measurements (see Assays 3 and 4, Figure II-14).

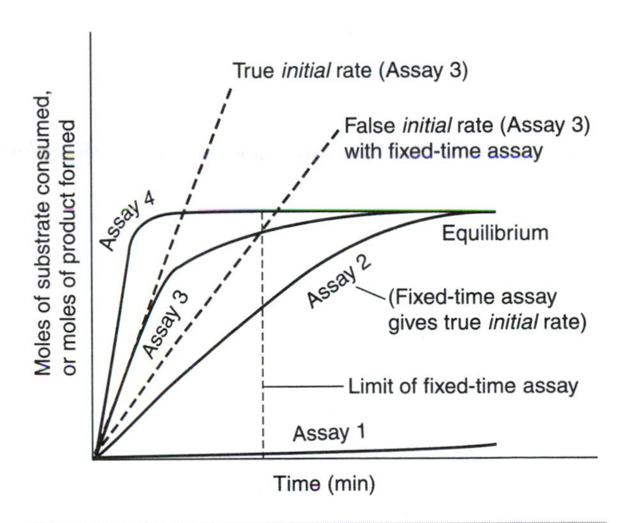

Figure II-14 Time course of enzyme-catalyzed reactions.

A so-called kinetic assay, in which the reaction rate is followed continuously, is advantageous because it is possible to observe directly the linearity or nonlinearity of the response with respect to time. Many enzyme assays, however, are based on a single measurement at a defined time, a so-called fixed-time assay. It is usually not possible to predict the appropriate amount of enzyme in either kinetic or fixed-time assays to obtain an optimum velocity like that of Assay 2 in Figure II-14. This may be empirically determined by a dilution experiment in two stages. At first, constant volumes of serial 10-fold dilutions of enzyme are assayed to find the range of dilution in which the calculated activity is maximal and constant (see Figure II-15).

Such "range finding" may frequently straddle the optimal enzyme concentration for assay. Thus, a second series of enzyme dilutions is assayed to pinpoint the enzyme concentration over a narrower range. Usually four to five dilutions over a 10- to 20-fold concentration range are adequate.

The prime test of the validity of v_0 assays is a graphical test to establish that the rates observed are a linear function of enzyme concentration, as illustrated in Figure II-16. This test should be applied to all assays in which reaction rates are used to measure enzyme concentrations. This procedure is especially important when fixed-time assays are used (see Figure II-14).

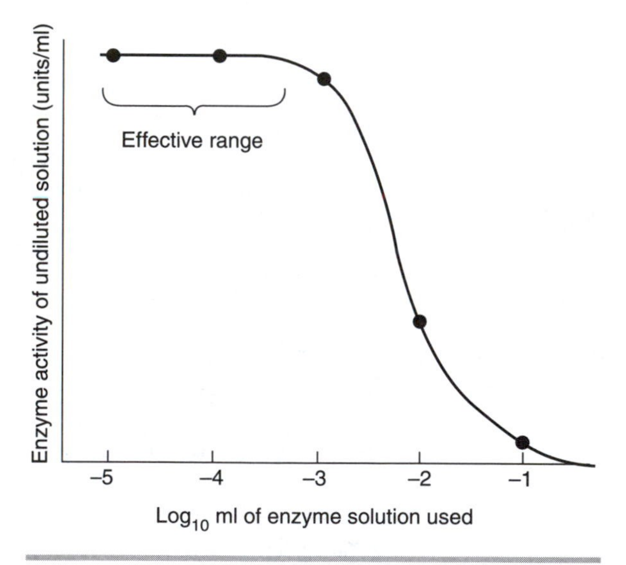

Figure II-15 Effect of enzyme dilution on apparent assay.

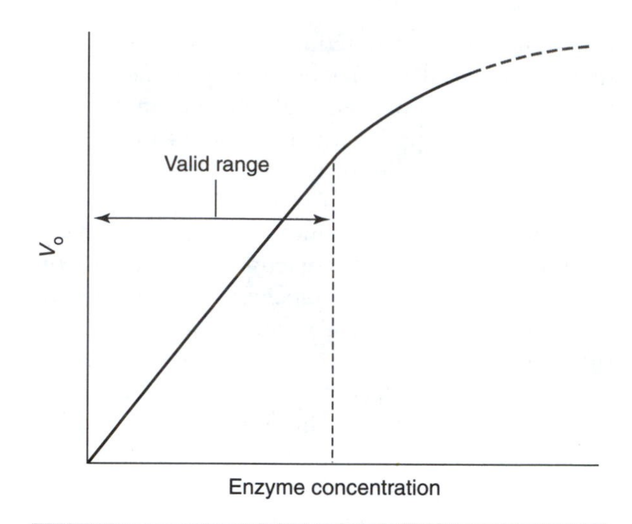

Figure II-16 Relationship of enzyme activity to enzyme concentration.

Assay Procedure during Enzyme Purification

Because enzyme (protein) purification involves the selective removal of other proteins, it is necessary to assess the amount of enzymatic activity relative to the amount of protein present. A measure of enzymatic activity per milligram of protein is usually employed to indicate the degree of purity of the enzyme in the various fractions obtained during purification.

This quantity, the specific activity, is calculated from a measure of enzymatic activity or *units*, usually expressed in terms of micromole of product formed per minute, and a protein determination, as in Equation II-22.

(II-22)

Specific activity = units per milligram of protein

= micromole of substrate consumed per minute per milligram of protein

= micromole of product formed per minute per milligram of protein

The efficiency of an enzyme purification is also assessed by determining how much of the enzyme activity is recovered from each purification step. This requires that you calculate the total activity in each fraction.

Total activity = (specific activity) ×
(total mg protein in preparation)

or

(II-23)

Total activity = (activity per milliliter) ×
(volume of the preparation in milliliter)

Once you know these values for any preparation of an enzyme, you can proceed with the purification of the enzyme by applying one or more of the techniques discussed elsewhere in this section and in Experiments 2, 8, and 10. It is essential to calculate specific activity, total activity, and percent yield for every fraction obtained during a purification.

(II-24)

$$\text{Percent yield} = \frac{\text{total activity of a given fraction}}{\text{total activity of the starting material}} \times 100\%$$

The usefulness of a particular step may then be evaluated with reference to the increase in the specific activity of the enzyme and percent yield in the fractions of greatest enrichment. An increase in specific activity indicates a purification. An ideal fractionation would provide complete enrichment (pure enzyme) in 100% yield; in practice, few procedures are so selective, and various, less ideal fractionation steps are combined. Of course, if enrichment is great, a lower yield may be allowed in a particular step. Table II-3 shows a convenient and usual way to present the results obtained during a purification procedure. Because protein concentration and enzyme activity are usually determined as milligram per milliliter and units per milliliter, respectively, it is also always necessary to record the volume of each enzyme fraction. Experiment 8 provides the student with experience in the application of these concepts to the purification of an enzyme.

REFERENCES

Ackers, G. K. (1970). Analytical Gel Chromatography of Proteins. *Adv Protein Chem* **24**:343.

Andrews, P. (1965). The Gel-Filtration Behavior of Proteins Related to their Molecular Weights over a Wide Range. *Biochem J* **96**:595

Banks, J. F., Jr., and Whitehouse, C. M. (1996). Electrospray Ionization Mass Spectrometry. *Methods Enzymol* **270**:486.

Beavis, R. C., and Chait, B. T. (1996). Matrix-Assisted Laser Desorption Ionization Mass-Spectrometry of Proteins. *Methods Enzymol* **270**:519.

Benson, J. R., and Hare, P. E. (1975). o-Phthalaldehyde: Fluorogenic Detection of Primary Amines in the Picomole Range. *Proc Natl Acad Sci* **72**:619.

Bradford, M. M. (1976). A Rapid and Sensitive Method for the Quantitation of Microgram Quantities of Protein Utilizing the Principle of Protein-Dye Binding. *Anal Biochem* **72**:248.

Carter, C. W., and Sweet, R. M. (1997). Macromolecular Crystallography, Parts A and B. *Methods Enzymol* **276** and **277**.

Table II-3 Typical Summary of an Enzyme Purification Procedure

Purification Step	Volume (ml)	Activity (μmol/min/ml)	Specific Activity (μmol/min/mg protein)	Fold Purification	Total Activity (μmol/min)	Yield (%)
Crude extract	50	20	0.1	(1.0)	1000	(100)
First $(NH_4)_2SO_4$ precipitate	20	40	0.7	7.0	800	80
DEAE-Sephadex chromatography	15	30	20	200	450	45
Affinity chromatography	3	100	450	4500	300	30

Dixon, M., and Webb, E. C. (1964). *Enzymes*, 2nd ed. New York: Academic Press.

Gordon, J. A., and Jencks, W. P. (1963). The Relationship of Structure to the Effectiveness of Denaturing Agents for Proteins. *Biochemistry* **2**:47.

Groll, M., L. Ditzel, Löwe, J., Stock, D., Bochtler, M., Bartunik, H. D., and Huber, R. (1997). Structure of 20S Proteasome from Yeast at 2.4 Å Resolution. *Nature* **386**: 463.

Haschemeyer, R. H. (1970). Electron Microscopy of Enzymes. *Adv Enzymol* **33**:71.

Hirs, C. H. W., Stein, W. H., and Moore, S. (1954). The Amino Acid Composition of Ribonuclease. *J Biol Chem* **211**:941.

Hunkapillar, M. W., Hewick, R. M., Dreyer, W. J., and Hood, L. E. (1983). High-Sensitivity Sequencing with a Gas-Phase Sequenator. *Methods Enzymol* **91**:399.

Layne, E. (1957). Spectrophotometric and Turbidometric Methods for Measuring Proteins. *Methods Enzymol* **3**:447.

Lowry, O. H., Rosebrough, N. J., Farr, A. L., and Randall, R. J. (1951). Protein Measurement with the Folin Phenol Reagent. *J Biol Chem* **193**:265.

Moore, S., and Stein, W. H. (1963). Chromatographic Determination of Amino Acids by Use of Automatic Recording Equipment. *Methods Enzymol* **6**:819.

Reynolds, J. A., and Tanford. C. (1970). The Gross Conformation of Protein-Sodium Dodecyl Sulfate Complexes. *J Biol Chem* **245**:5161.

Scopes, R. K. (1994). *Protein Purification: Principles and Practice.* New York: Springer-Verlag.

Segel, I. H. (1975). *Enzyme Kinetics.* New York: Wiley.

Smith, P. K., Krohn, R. I., Hermanson, G. T., Mallia, A. K., Gartner, F. H., Provenzano, M. D., Fujimoto, E. K., Goeke, N. M., Olson, B. J., and Klenk, D. C. (1985). Measurement of Protein Using Bicinchoninic Acid. *Anal Biochem* **150**:76.

Woody, R. W. (1995). Circular Dichroism. *Methods Enzymol* **246**:34.

Acid–Base Properties of Amino Acids

Theory

Most biologically important molecules, large and small, are water-soluble and ionic. Their ionic character derives from the association and dissociation of protons to and from weakly acidic or basic groups in their structure. It is difficult to overestimate the importance of these ionizing groups. They confer water solubility on the molecules, which is important because water is the fundamental intracellular medium. They are also crucial for the formation of native three-dimensional structure, for molecular recognition, and for biological function, as for example, in the catalytic activity of enzymes. The degree of protonation of acidic and basic groups, and hence the charge on such groups, is a function of the pH of the surrounding medium. This explains why pH is so important in biochemical systems.

In this experiment we shall examine the effects of pH on the ionization (protonation) of amino acids, but it is important to remember that the same principles and quantitative relationships can be used to describe the ionization behavior of all biological molecules that contain weak acids or bases. As discussed in the Introduction to Section II, all amino acids contain at least one acidic (α-carboxyl) group and one basic (α-amino) functional group. Some amino acids contain other side chain groups that ionize readily: amino, carboxyl, *p*-hydroxyphenyl, sulfhydryl, guanidinium, and imidazole groups. Because the α-amino and α-carboxyl groups of amino acids all participate in peptide linkages (except for those at the amino and carboxyl termini of the polypeptide), the ionic character of polypeptides and proteins is largely due to the additional ionizable groups on the side chains.

This experiment is a study of the reaction of typical amino acids with hydrogen ions. As an acid or base is added to a solution of an amino acid, a change in the pH of the solution is observed. After allowances for the dilution of the acid or base during titration, the results can be predicted by use of the general Henderson–Hasselbalch equation. The Henderson–Hasselbalch equation was presented in the Introduction to Section II (Equation II-1), and its application to the analysis of acid and base titration of a typical amino acid was described. Be sure that you understand this discussion well before proceeding with this experiment. Test your understanding further by predicting the titration curves of other amino acids with ionizable side chain groups, such as histidine, glutamic acid, and lysine.

Amino acids (and peptides) are characterized not only by their acid dissociation constants (pK_as), but also by an isoelectric point (pI), which is the pH at which the amino acid or peptide has no net charge. For a simple amino acid with only an α-carboxyl and an α-amino group the pI is determined by a balance between the tendency of the protonated amino group to lose a proton and the unprotonated carboxyl group to bind a proton. It can be readily shown that this pH is

(5-1)

$$\mathrm{pI} = \frac{\mathrm{p}K_{a1} + \mathrm{p}K_{a2}}{2}$$

However, for an amino acid or peptide with three or more ionizable groups, you must avoid the trap of thinking that pI is the average of pK_a values. Rather, it is always determined by the balance of only two pK_a values:

(5-2)

$$pI = \frac{pK_n + pK_{n+1}}{2}$$

pK_n and pK_{n+1} are the two pK_a values that describe the ionization of the species with a net zero charge; that is *the first ionization that adds a proton to the neutral species and gives it a net charge of +1 and the first ionization that removes a proton from the neutral species and gives it a net charge of −1.* Test your understanding of this concept by calculating the pI values for aspartic acid, histidine, and lysine.

In this experiment you will determine pK_a and pI values for an unknown amino acid by titrating it with acid and base. You will also determine the molecular weight of the amino acid by determining the number of equivalents of standardized acid or base that were required to titrate one functional group—an α-carboxyl group or an α-amino group. From the known mass of the amino acid titrated and the equivalents required to titrate one group, the molecular weight of the amino acid can be readily calculated.

Note that the amino acid may be given to you in one of several ionic states. For example, it could be given to you as a zwitterionic (isoelectric) form, or as a hydrochloride salt or a sodium salt. See Figure 5-1 for an illustration of this concept as applied to the amino acid glutamic acid. This amino acid could be isolated as the hydrochloride, the zwitterionic form, the monosodium salt, or the disodium salt. The initial pH of a solution of glutamic acid gives you information about its ionic form. For example, the zwitterionic form would yield a solution with a pH of about 3, whereas the monosodium salt would give a solution with a pH near 7. The titration curves for glutamic acid would be the same for all forms, except that the groups titrated by acid or base depend on the initial state. For example, titration of the hydrochloride would reveal three groups titrated by base and none with acid. The monosodium salt would require one equivalent of base and two equivalents of acid to be fully titrated.

Your instructor may also give you unknowns that are monofunctional weak acids or bases, but that are not amino acids. Careful examination of your titration curves should enable you to distinguish among these possibilities. If you receive an amino acid as an unknown, you should be able to identify it from its molecular weight and pK_a values. Table 5-1 lists the pK_a values and molecular weights for the common amino acids.

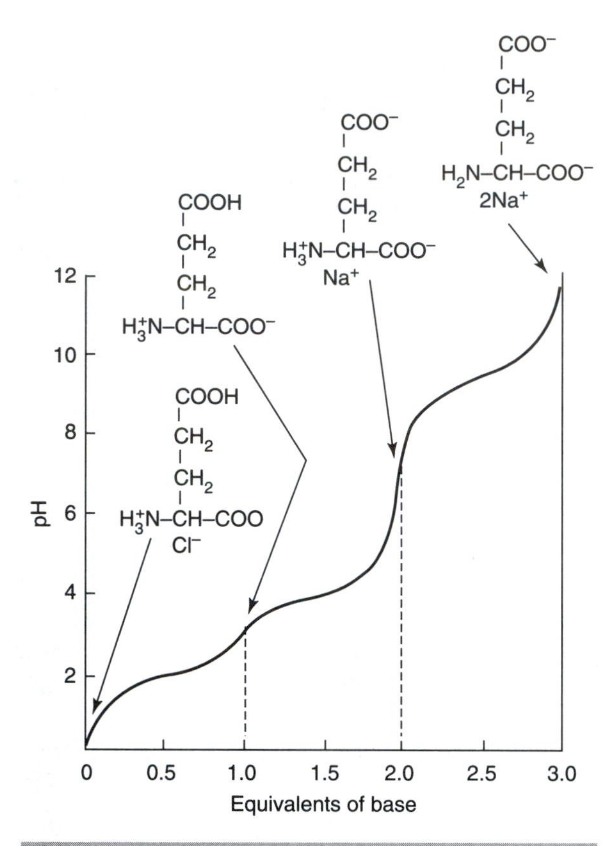

Figure 5-1 Titration of glutamic acid hydrochloride with base, showing possible ionic states of the amino acid as a function of pH.

Supplies and Reagents

Unknown (amino acid, monofunctional weak acid or weak base) in powder form

2 N NaOH
2 N H_2SO_4
pH meter
Magnetic stirring motor and bar
10-ml burette
Burette holder
Buffers for standardization of pH meter

Table 5-1 Molecular Weights and pK_a Values for Common Amino Acids

Amino Acid	MW (g/mol)	pK_a of α-Carboxyl Group	pK_a of α-Amino Group	pK_a of Side Chain Group
Glycine	75	2.35	9.78	
Alanine	89	2.35	9.87	
Valine	117	2.29	9.74	
Leucine	131	2.33	9.74	
Isoleucine	131	2.32	9.76	
Methionine	149	2.13	9.28	
Proline	115	2.95	10.65	
Phenylalanine	165	2.16	9.18	
Tryptophan	204	2.43	9.44	
Serine	105	2.19	9.21	
Threonine	119	2.09	9.10	
Asparagine	133	2.10	8.84	
Glutamine	146	2.17	9.13	
Tyrosine	181	2.20	9.11	10.13
Cysteine	121	1.92	10.78	8.33
Lysine	147	2.16	9.18	10.79
Arginine	175	1.82	8.99	12.48
Histidine	155	1.80	9.20	6.00
Aspartic acid	132	1.99	9.90	3.90
Glutamic acid	146	2.10	9.47	4.07
Chloride	35.5			
Sodium	23			

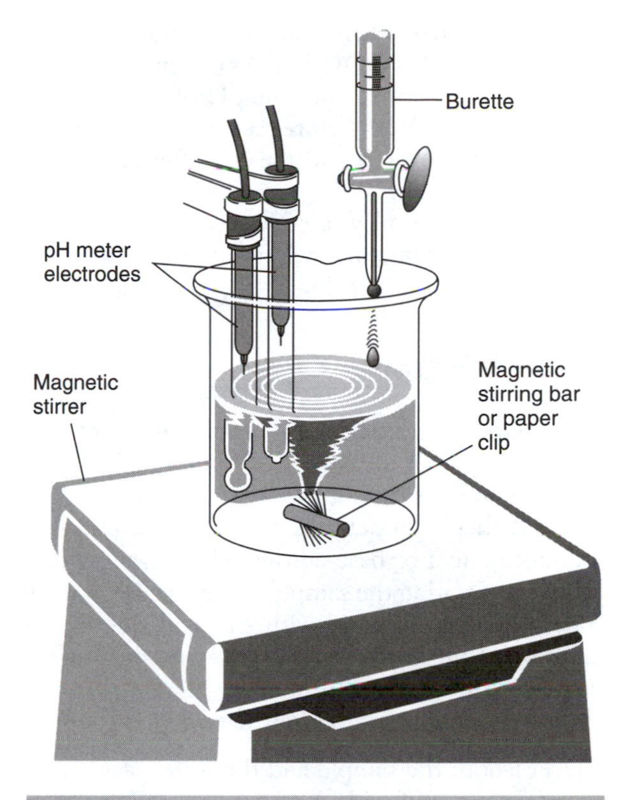

Figure 5-2 Setup for titration.

Protocol

1. Using a sample of standard buffer, familiarize yourself with the operation of the pH meter and standardize the instrument. For titrations with acid, standardize the pH meter with the pH 4.0 standard buffer; for titrations with base use the pH 9.0 or 10.0 standard buffer. *Be sure to rinse the electrodes with deionized water after standardization.*
2. After recording your unknown number, weigh about 400 mg of the unknown, recording the mass to the nearest milligram, and dissolve it in 20 ml of deionized H_2O. If the sample does not dissolve readily, gently heat the solution to 50 to 60°C and allow it to dissolve. Allow the solution to cool to room temperature before beginning the titration.
3. Set up the electrodes of the pH meter, a magnetic stirring bar (a small metal paper clip can be substituted for a stirring bar) and stirring motor, and a burette as shown in Figure 5-2, so that the sample can be titrated with continuous stirring. Take extreme care to ensure that the electrodes are not damaged by the stirring bar or by rough treatment.
4. Titrate the dissolved unknown sample with 2 N H_2SO_4, recording the burette readings (and from these, the volume of H_2SO_4 added) and the pH at frequent intervals throughout the titration until the solution reaches pH 1.0.
5. Titrate another 20-ml H_2O sample (a water blank containing no unknown) with 2 N H_2SO_4 1 drop at a time for the first 10 drops, 2 drops at a time for the next 10 drops, and finally 4 drops at a time until the solution reaches pH 1.0. Record the pH and volume of acid delivered after each addition.
6. Dissolve another sample of the unknown (again approximately 400 mg, weighed to the nearest

milligram) in 20 ml of H_2O and titrate this sample with 2 N NaOH in the same manner as for the acid titration until the pH of the solutions reaches 12.0. As before, record the volume of NaOH delivered and the pH after each addition.

7. Titrate a 20 ml water blank with 2 N NaOH to pH 12.0 in a manner similar to the first water blank (see step 5).

Data Analysis

To determine the true titration curve of any substance, you must measure how much acid or base is consumed in titrating the solvent (water) to each pH and then subtract this amount from the total amount of acid or base consumed in reaching that pH when titrating the sample (amino acid + water). The following example with the acid side of the titration of an amino acid illustrates the method for correcting for such acid or base dilution that you should use in correcting your data.

1. For both the sample and the water blank, plot the volume of acid added versus the pH reached (see the example in Fig. 5-3).
2. From the graph (or from the original data), prepare a table like Table 5-2, listing the amount of acid required to attain each pH, determining a value for as many points as needed to define the titration curve. Then, subtract the volume of acid required to bring the water blank to any pH from the volume of acid required to bring the sample to the same pH. This difference represents the amount of acid consumed in the titration of the sample only.
3. Using the data from your table, plot the pH versus the number of equivalents of acid needed to titrate the amino acid sample to any pH (see the example in Fig. 5-4).

Table 5-2 Amounts of Acid Required to Titrate Sample and Water Blank in Solution

	Volume (ml) of Acid (2 N HCl)		
pH	Water plus Sample	Water Blank	Difference
3.5	0.103	0.003	0.100
3.0	0.335	0.020	0.315
2.5	0.667	0.032	0.635
2.2	1.063	0.063	1.000
2.0	1.425	0.200	1.225

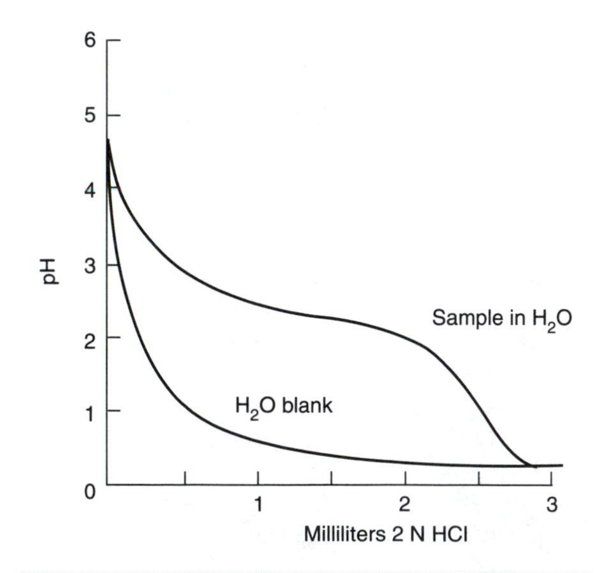

Figure 5-3 Uncorrected titration curves of water blank and sample.

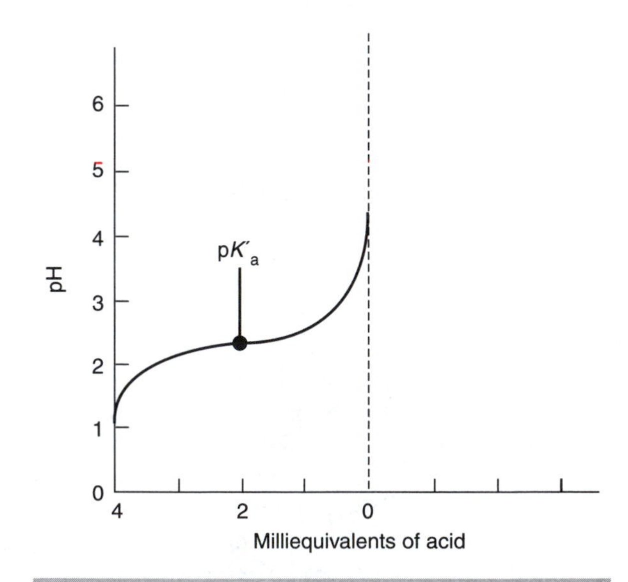

Figure 5-4 Corrected titration curve of sample.

4. Use the same method to correct for titration of the water blank with NaOH. Prepare a complete, corrected NaOH titration curve for the unknown that you titrated.
5. Combine the corrected acid and base titration curves into a single titration curve for the unknown sample for the entire pH range from 1 to 12. If your unknown was a simple amino acid with no ionizable side chains, this figure will look like Figure II-1 in the Introduction to Section II. The number of equivalents of acid or base consumed in passing through one full inflection of a curve (in the zone of a single ionization, as in Fig. 5-4) represents the quantity of acid required to titrate one ionizable group in the quantity of amino acid you have assayed. Using this relationship and noting the number of full inflections of your curve, calculate the molecular weight for your unknown. The end point of the titration should be recognizable as the point at which the pH rises (or falls) sharply with addition of titrant. This point is usually most accurately determined from the titration of the amino acid with base.
6. Compare the molecular weights obtained from the titration of each ionizable group. How well do they agree?
7. Using the Henderson–Hasselbalch equation, calculate the pK_a value for each ionizable group titrated.
8. What is the identity of your unknown? Justify your conclusion by comparing the observed molecular weight and pK_a values to those for all amino acids that might have been your unknown (Table 5-1). If your unknown was not an amino acid, what can you suggest about its chemical nature? How could you confirm your identification of your unknown?

Exercises

1. How do the true titration curves of aspartic acid and lysine differ from that of glycine? Explain your answer.
2. Is the true titration curve of leucyl valine the same as that of glycine? Explain your answer.
3. The quantity 432 mg of a monoamino monocarboxylic amino acid is observed to consume 7.39 mEq of acid and base in the titration range pH 0.8 to 12.0. What is the name of the amino acid?
4. Give an example of a polypeptide that is not a polyelectrolyte.
5. Calculate the isoelectric point for the following polyfunctional amino acids: (a) histidine, (b) lysine, and (c) aspartic acid. (See Table 5-1.)
6. A useful method to assay many hydrolytic enzymes is to follow enzymatic activity by determining production or uptake of hydrogen ions during the reaction. For example, consider a protease that catalyzes cleavage of internal peptide bonds in a protein:

$$\text{–C(=O)–NH–CH(R}_1\text{)–C(=O)–NH–CH(R}_2\text{)–C(=O)–NH–} \xrightarrow[\text{H}_2\text{O}]{\text{protease}} \text{–C(=O)–NH–CH(R}_1\text{)–COOH} + \text{H}_2\text{N–CH(R}_2\text{)–C(=O)–NH–}$$

   a. Assume the enzyme is fully active at pH 8.8, that the pK_a of an α-carboxyl group in a peptide is 3.5, and that the pK_a of an α-amino group in a peptide is 8.5. How will the action of this protease on a protein affect the pH of an unbuffered reaction mixture (starting at pH 8.8)? Explain your answer.
   b. The protease reaction is run in a pH-stat, a device that maintains a constant pH in a reaction mixture by continuous, automatic addition of acid or base. The instrument has a recorder that plots the volume of acid or base added versus time of reaction. If the reaction in (a) is run in a pH-stat maintained at pH 8.8, calculate the millimoles of titrant that must be added per millimole equivalent of peptide bonds hydrolyzed by the enzyme.

REFERENCES

Christensen, H. N. (1963). *pH and Dissociation.* Philadelphia: Saunders.

Lehninger, A. L., Nelson, D. L., and Cox, M. M. (1993). Amino Acids and Peptides. In: *Principles of Biochemistry*, 2nd ed. New York: Worth.

Stryer, L. (1995). *Biochemistry*, 4th ed. New York: Freeman, pp. 17–23, 42–43.

Sequence Determination of a Dipeptide

Theory

The compound 1-fluoro-2,4-dinitrobenzene (FDNB) reacts with free amino, imidazole, and phenolic groups at neutral to alkaline pH to yield the corresponding, colored dinitrophenyl (DNP) compounds. Thus, FDNB will react with the free, unprotonated α-amino groups on amino acids, as well as with the side chains of lysine, histidine, and tyrosine (Fig. 6-1). Dansyl chloride is another compound that is known to react with the unprotonated, *N*-terminal amino groups of peptides. Derivatization of peptides with this compound yields fluorescent products that provide a very sensitive method of detection of the amino acid derivatives (Fig. 6-2).

The reaction of FDNB with a peptide proceeds by the attack of an uncharged amino group on the electron-deficient carbon atom adjacent to the fluorine atom. The pK_a value for the α-amino group on most peptides is approximately 8.5. For this reason, it is critical for the reaction solution to be buffered at or slightly above pH 9.0. Sodium bicarbonate ($NaHCO_3$), a weak base, is used to buffer the reaction, since OH^- ions (strong base) will react with FDNB to form 2,4-dinitrophenol as a side product (Fig. 6-3). As shown in Figure 6-1, acid (HF) is produced on the reaction of a peptide with FDNB. As a result, additional $NaHCO_3$ may have to be added as the reaction proceeds to maintain an alkaline pH.

If the DNP derivative of a peptide is acid hydrolyzed and extracted with ether at low pH, all nonpolar (uncharged) compounds will be extracted into the ether phase while all polar (charged) compounds will remain in the aqueous phase. At low pH, the *N*-terminal DNP-amino acids will extract into the ether phase because their α-carboxyl groups are protonated (uncharged). In contrast, all amino acids that are not *N*-terminal will remain in the aqueous phase because their α-amino groups are protonated (charged) at low pH (see Fig. 6-4). The only DNP-amino acids that will not enter the ether phase at low pH conditions are DNP-arginine and DNP-cysteic acid, since these side chains continue to carry a charge at low pH. Once the *N*-terminal DNP-amino acid derivatives are separated from those amino acids or DNP-amino acid derivatives that are not *N*-terminal in the peptide, the DNP derivatives and free amino acids may be identified by a variety of chromatographic techniques. These data will yield part of the sequence of the peptide under study.

Although the reaction of peptides with FDNB is useful in identifying the *N*-terminal amino acid, the acid hydrolysis step destroys the remainder of the peptide (hydrolyzes all of the remaining peptide bonds). It would be extremely valuable to have a means of cleaving a derivative of the *N*-terminal amino acid from the peptide without hydrolyzing the other peptide bonds. With such a technique, you could determine the entire sequence of a peptide merely by repeating the derivatization and cleavage cycle, removing and identifying one amino terminal residue at a time. Such a technique is currently available with the use of Edman's reagent, phenylisothiocyanate (PITC). This reagent reacts to form a phenylthiocarbamyl derivative of the free α-amino group of a peptide, which is then cyclized in anhydrous HF to cleave the amino terminal

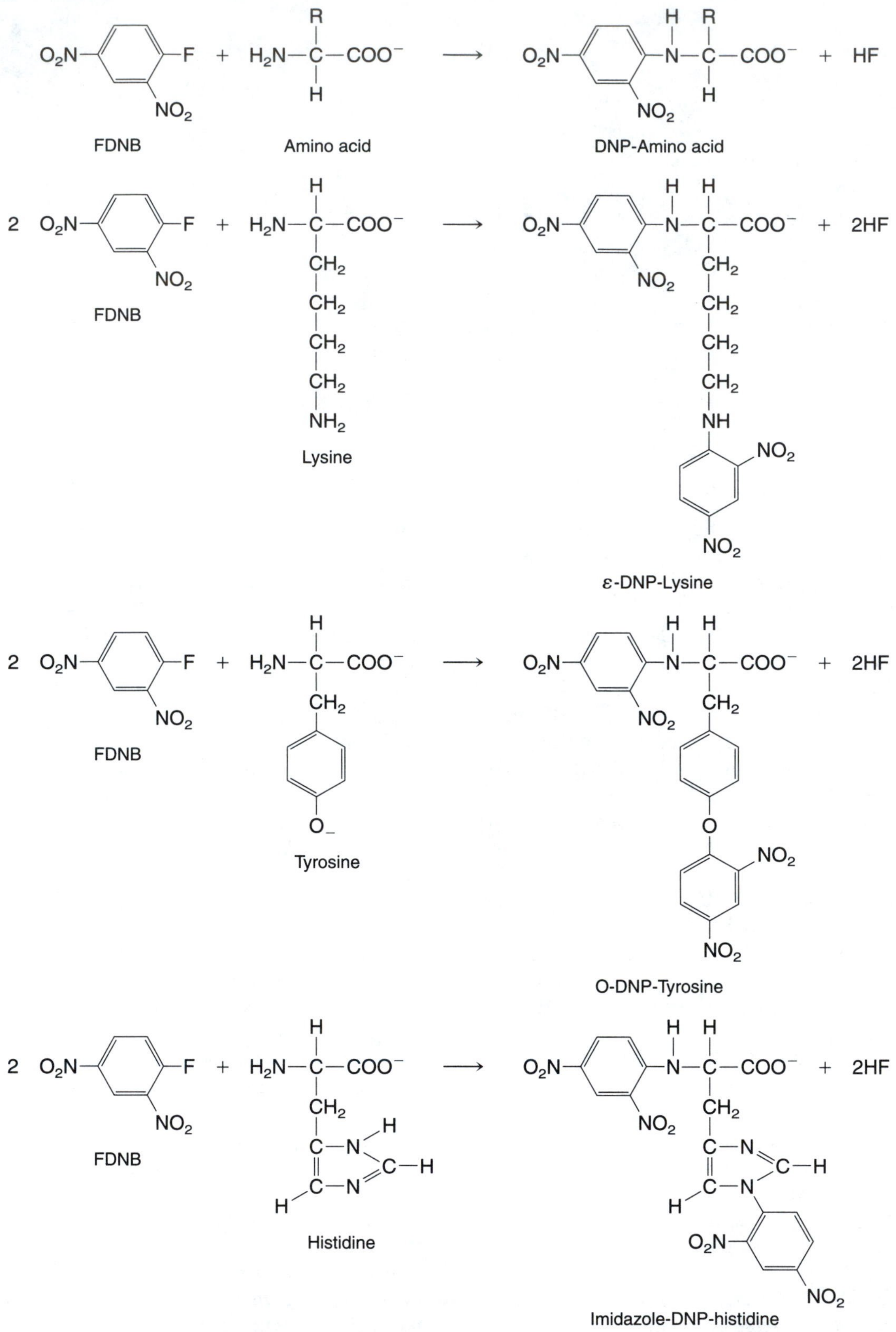

FDNB + Amino acid ⟶ DNP-Amino acid + HF

2 FDNB + Lysine ⟶ ε-DNP-Lysine + 2HF

2 FDNB + Tyrosine ⟶ O-DNP-Tyrosine + 2HF

2 FDNB + Histidine ⟶ Imidazole-DNP-histidine + 2HF

Figure 6-1 Reaction of amino acids with 1-fluoro-2,4-dinitrobenzene (FDNB). DNP = dinitrophenyl; HF = hydrofluoric acid

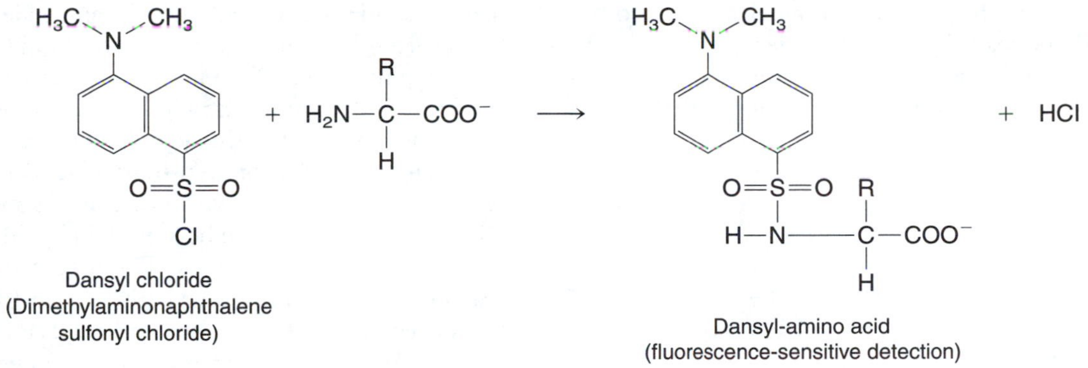

Dansyl chloride
(Dimethylaminonaphthalene sulfonyl chloride)

$+ HCl$

Dansyl-amino acid
(fluorescence-sensitive detection)

Figure 6-2 Reaction of amino acids with dansyl chloride.

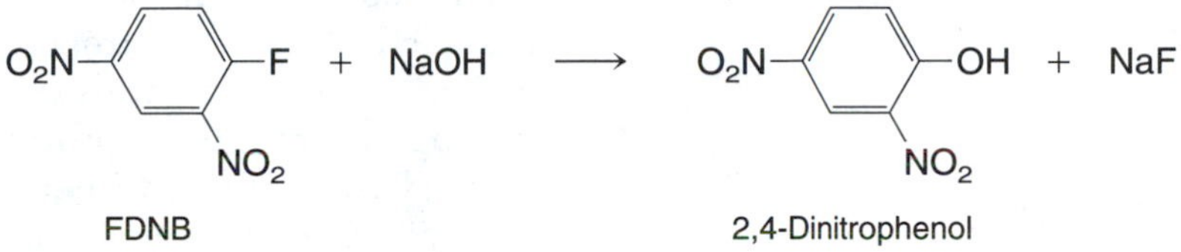

$+ NaOH \longrightarrow$ $+ NaF$

FDNB

2,4-Dinitrophenol

Figure 6-3 Reaction of NaOH with 1-fluoro-2,4-dinitrobenzene (FDNB).

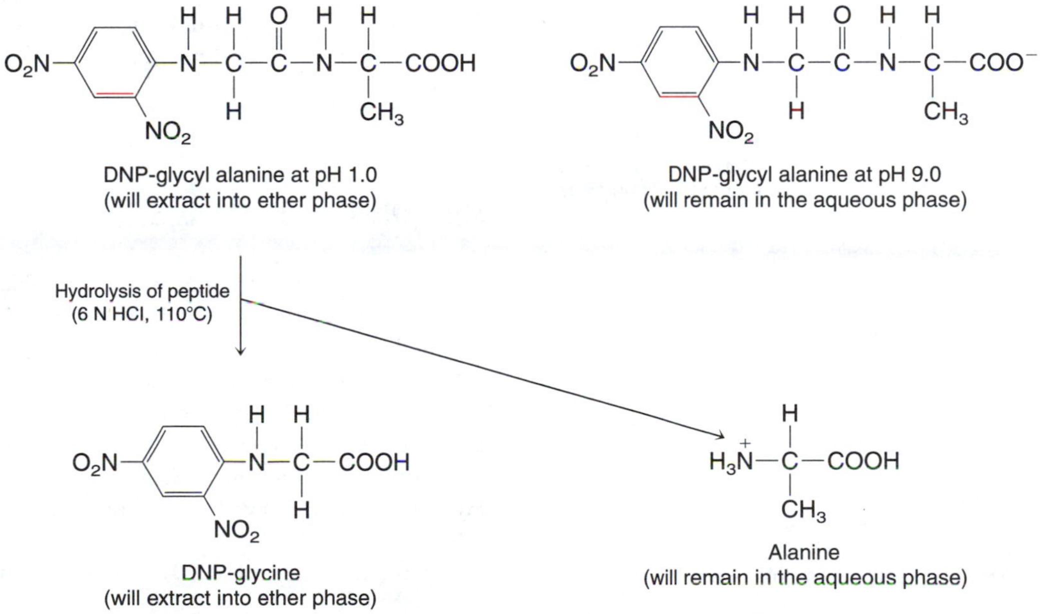

DNP-glycyl alanine at pH 1.0
(will extract into ether phase)

DNP-glycyl alanine at pH 9.0
(will remain in the aqueous phase)

Hydrolysis of peptide
(6 N HCl, 110°C)

DNP-glycine
(will extract into ether phase)

Alanine
(will remain in the aqueous phase)

Figure 6-4 Partitioning of an acid hydrolyzed DNP-dipeptide between the aqueous and ether phases is pH-dependent.

residue from the peptide without cleaving the other peptide bonds (Fig. 6-5). This derivative of the *N*-terminal amino acid (the anilinothiazolinone derivative) can be converted to a stable phenylthiohydantoin (PTH-amino acid), which can be identified by reverse-phase high-performance liquid chromatography (HPLC). This procedure is then repeated to determine the identity of the next *N*-terminal amino acid in the peptide. This technique has been automated for step-by-step sequence determination of peptides and proteins. As many as 60 to 100 amino acids can be identified by this technique under ideal conditions. The first step in the reaction of an amino acid or peptide with PITC is the formation of the phenylthiocarbamyl (PTC) derivative. As in the case of reaction of amino acids with FDNB, this reaction takes place at unprotonated α-amino groups (as well as at the ε-amino group on lysine) under alkaline conditions. In this exercise, you will prepare PTC derivatives of the amino acids released by the hydrolysis of the unknown dipeptide and identify them by reverse-phase HPLC and spectrophotometry. The PTC moiety on the amino acids serves two purposes in this experiment: first, it provides a method of detection (absorbs light at 254 nm). Second, it adds a hydrophobic component to all of the amino acids that will make them much easier to separate by reverse-phase HPLC partition chromatography (see Experiment 2).

In this three-period experiment, milligram amounts of an unknown dipeptide will be examined by the procedures that we have just discussed. In Figure 6-6, a flow diagram of the experiment is provided to aid your understanding of the various steps.

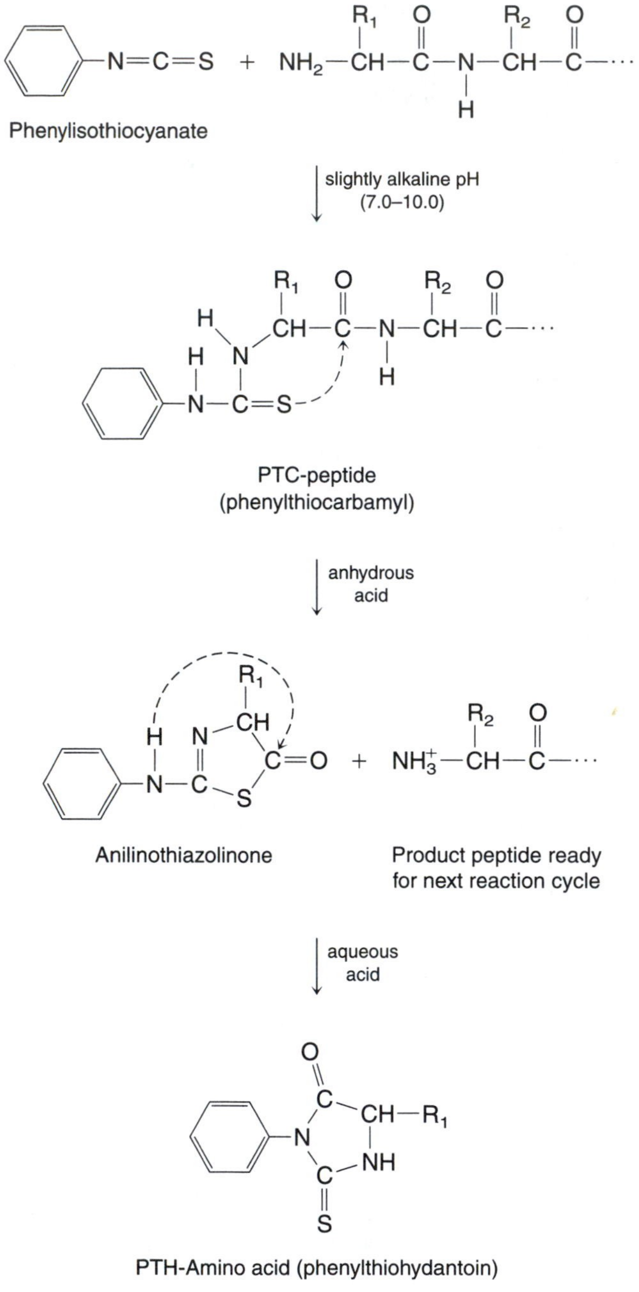

Phenylisothiocyanate

slightly alkaline pH (7.0–10.0)

PTC-peptide (phenylthiocarbamyl)

anhydrous acid

Anilinothiazolinone

Product peptide ready for next reaction cycle

aqueous acid

PTH-Amino acid (phenylthiohydantoin)

Figure 6-5 The steps of the Edman degradation of a polypeptide.

Supplies and Reagents

P-20, P-200, P-1000 Pipetman with disposable tips
Glass test tubes (13 by 100 mm and 16 by 125 mm)
Dipeptides
DNP-amino acid standards (in acetone)
Amino acid standards (in acetone)
Peroxide-free ether
Whatman 3 MM filter paper (8.5 in. by 8.5 in.)
Hydrolysis vial (4 ml) with Teflon-lined screw caps
6 N HCl
FDNB solution (0.25 ml of FDNB to 4.75 ml of absolute ethanol)
4.2% $NaHCO_3$ (4.2 g of $NaHCO_3$ in 100 ml of water)
Pasteur pipettes

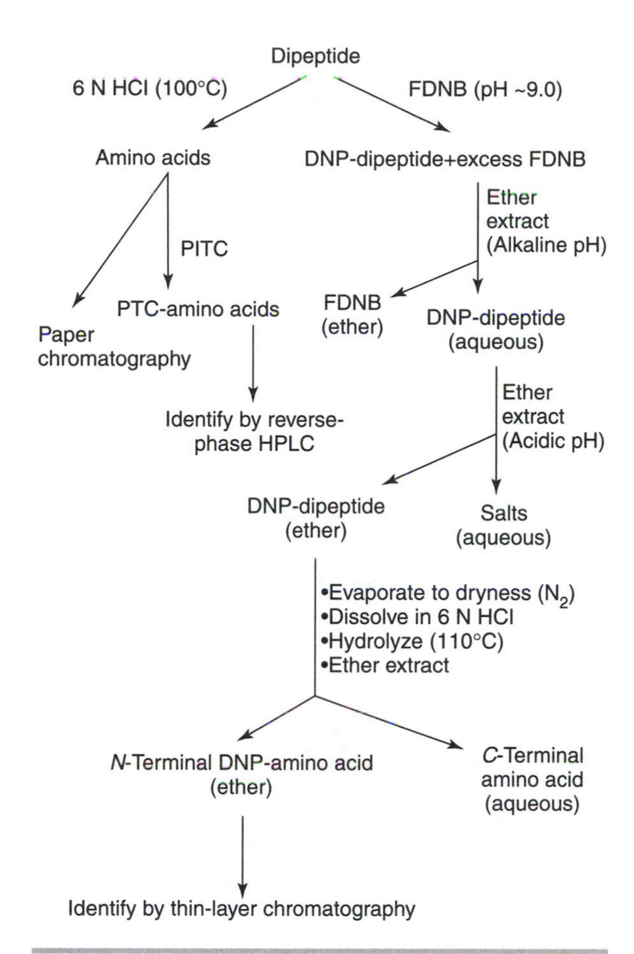

Figure 6-6 Flow diagram for sequence determination of a dipeptide.

Ninhydrin spray solution (Dissolve 1 g of ninhydrin in 500 ml of *n*-butanol. Add 0.5 ml of 2,4-collidine just before use.)
Silica-gel thin-layer chromatography plates (Fisher catalog 05-713-317)
pH paper (1–11)
Pressurized N_2
Heating block
Acetone
Chloroform
t-Amyl alcohol
Glacial acetic acid
n-Butanol
Ethanol
Triethylamine
Phenylisothiocyanate (Sigma catalog P-1034)
PicoTag Eluant A (Waters catalog 88108)
Amino Acid Standard H (Sigma catalog AA-S-18)
HPLC unit
Alltech Adsorbosphere HS C_{18} reverse-phase column (250 × 4.6 mm)
Spectrophotometer
Chart recorder
Vacuum microcentrifuge (Speed-Vac)

Protocol

> **WARNING**
> **Safety Precautions: Avoid getting FDNB on your skin by wearing rubber or latex gloves at all times. Ether is very noxious and very flammable. Keep the ether in the fume hoods at all times and avoid having open flames in the laboratory. FDNB can cause burns to the skin. Wear safety goggles at all times. Ninhydrin vapors are corrosive, so use caution. Perform all ether extractions in glass test tubes, since polypropylene tubes will dissolve in ether.**

Day 1: Acid Hydrolysis of the Peptide and Preparation of the DNP-Derivatized Dipeptide

1. Obtain duplicate samples of the unknown dipeptide, each containing about 2 mg. One sample will be in a microcentrifuge tube and the other will be in a 13-by-100-mm glass test tube. *Record the number of your unknown sample in your notebook.*
2. To the sample in the glass test tube, add the following:

   200 μl distilled water
   50 μl of 4.2% $NaHCO_3$
   400 μl of FDNB reagent

3. Allow the sample to incubate for 1 hr at room temperature, mixing gently every 3 or 4 min. During this hour, check the pH every 10 min by dropping a spot of the reaction solution on pH paper. Maintain the pH at ~9.0 by adding drops of 4.2% $NaHCO_3$ as needed. Proceed with steps 4 to 5 during this 1-hr incubation.
4. To the sample in the microcentrifuge tube, add 200 μl of 6 N HCl and mix gently to dissolve the sample. Using a Pasteur pipette, transfer

the sample to a 4-ml glass vial and seal with a Teflon-lined screw cap.

5. Label the top of the vial (not the sides) with your name and place in a 110°C heating block overnight (~12 hr).
6. To the sample in the glass test tube (after the 1-hr incubation), add 1 ml of distilled water and 0.5 ml of 4.2% $NaHCO_3$. Make sure that the pH of the solution is above 8.0 before continuing.
7. Add an equal volume (~2 ml) of ether and gently swirl to mix for 10 sec. Allow the aqueous and ether phases to separate and remove and discard the ether (upper) phase containing excess (unreacted) FDNB.
8. Repeat Step 7 two more times or until the ether phase no longer shows any yellow color.
9. To the resulting aqueous phase (containing the DNP-dipeptide), add 100 μl of 6 N HCl. Make sure that the pH is ~1.0 by putting a small drop of the reaction on pH paper after mixing. You may notice some bubbles forming in the solution as the H^+ causes the formation of carbon dioxide gas and water from the formation of H_2CO_3.
10. Extract the acidified, aqueous solution with 2 ml of ether as in step 7. After mixing, allow the phases to separate, remove the ether phase (containing the DNP dipeptide) and transfer to a 16-by-125-mm glass test tube.
11. Repeat step 10 two more times. The ether phases from all three extractions can be saved in a single 16-by-125 mm glass test tube.
12. In the fume hood, evaporate the ether containing the DNP-dipeptide by placing the tube in a beaker of warm (not hot!) water and directing a gentle stream of N_2 over the solution (see Fig. 6-7). Continue with this procedure until a dry, yellow solid remains at the bottom of the tube.
13. Dissolve the dried DNP dipeptide (yellow solid) in 0.5 ml of acetone and transfer the sample to a 4-ml glass vial.
14. Evaporate the acetone under N_2 in the fume hood and redissolve the dried sample in 0.5 ml of 6 N HCl.
15. Seal the vial with a Teflon-lined screw cap, label the top (not the sides) with your name, and place it in a 110°C heating block overnight (~12 hr).

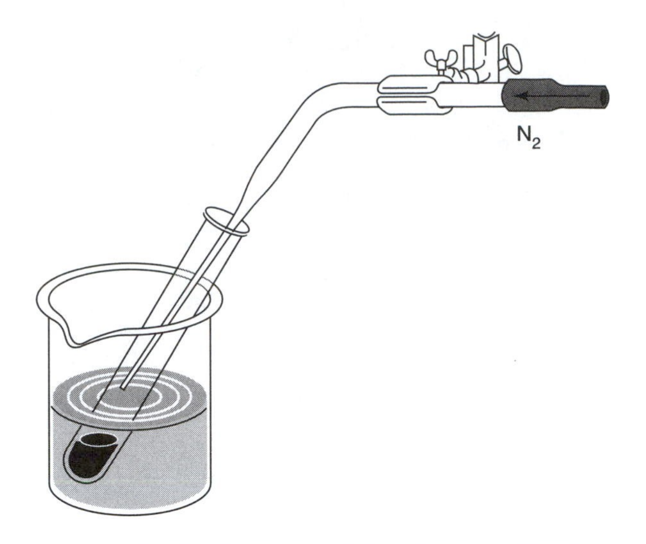

Figure 6-7 Setup for ether extract evaporation.

Day 2: Thin Layer Chromatography of the DNP-Amino Acid and Paper Chromatography of the Acid-Hydrolyzed Dipeptide

Thin-Layer Chromatography of the DNP-Amino Acid

1. Transfer the contents of the 4-ml vial containing the acid-hydrolyzed DNP dipeptide (yellow sample) to a clean 13-by-100-mm glass test tube.
2. Add 2 ml of distilled water and 2 ml of ether. Swirl for 10 sec, allow the phases to separate, and transfer the ether phase to a clean 16-by-125-mm glass test tube. Repeat this ether extraction two more times and combine the ether phases in the same glass test tube. This final ether extraction is done to separate the *N*-terminal DNP-amino acid from the other amino acid (or DNP-amino acid) in the sample. Remember that the *N*-terminal DNP-amino acid carries no charge at low pH (ether phase), while the *C*-terminal amino acid has a charged α-amino group at low pH (aqueous phase).
3. Evaporate the ether extracts (containing your *N*-terminal DNP-amino acid) to dryness under N_2 as before.

4. "Activate" a silica-gel thin-layer chromatography (TLC) plate by heating for 5 min in a 100°C oven. This procedure drives off excess moisture from the plate so that the TLC separation is more reproducible.
5. Make a small scratch on the edge of the TLC plate (short axis) about 2 cm from the bottom. This will indicate the position of the origin where the samples will be spotted (do not mark the plate with pencil and avoid touching the face of the plate with your fingers). Refer to Fig. 6-8 for proper setup of thin-layer chromatography experiment.
6. Carefully spot 5 μl of the DNP-amino acid standards (provided by the instructor) along the origin of the plate. To prevent diffusion of the sample, it is a good idea to spot 1 μl of the sample, allow it to dry, spot another 1-μl aliquot, allow it to dry, and so forth.
7. Dissolve the dried DNP-amino acid (*N*-terminal) in 100 μl of acetone.
8. Using the same technique as described in step 6, carefully spot a 5-μl aliquot of your sample along the same origin as the DNP-amino acid standards. If you did not recover a sufficient amount of the sample in the last ether extraction, you may also want to spot a 10-μl sample next to this. The sample spot should be yellow enough in color that you will be able to identify it once the plate is developed.
9. When all of the spots have dried, place the TLC plate (sample side toward the bottom) in a TLC tank containing 1 cm of the mobile-phase solvent [chloroform:t-amyl alcohol:glacial acetic acid (70:30:3)] (see Fig. 6-8). *Remember to make a map or drawing of the TLC plate in your notebook so that you will remember which spot corresponds to which sample.*
10. Allow the mobile phase to migrate across (up) the plate until the solvent front is about 1 cm from the top of the plate (~2 hr).
11. Remove the plate from the tank and mark the position of the solvent front with a pencil. After the plate has dried, circle the position of all of the spots on the plate with a pencil. Also, draw a line across the bottom of the plate in pencil to mark the position of the origin.
12. For each spot on the plate, calculate a relative mobility value (R_f):

$$R_f = \frac{\text{distance traveled by sample (cm)}}{\text{distance traveled by solvent front (cm)}}$$

If the sample spot showed some diffusion during development, use the center of the spot in determining the R_f value. Also, if the solvent front migrated unevenly across the plate, measure the distance traveled by the sample (cm) relative to the distance traveled by the solvent front (cm) in each lane.

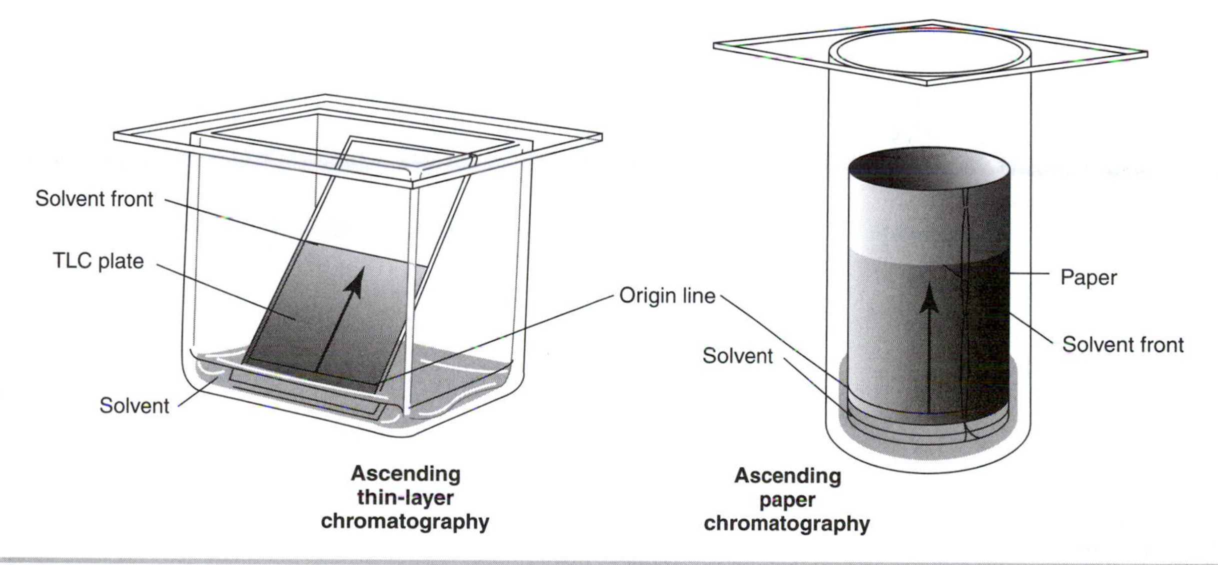

Figure 6-8 Ascending thin-layer and paper chromatography.

13. From the R_f value of your unknown sample compared to the R_f values of the DNP-amino acid standards, what is the identity of the *N*-terminal amino acid in your unknown dipeptide? Discuss any uncertainty that may be involved in your identification, as well as how these uncertainties may be eliminated through modifications in the experiment.

Paper Chromatography of Acid-Hydrolyzed Dipeptide

1. Recover the 4-ml vial containing your acid hydrolyzed dipeptide (colorless sample).
2. Obtain an 8.5 in. by 8.5 in. piece of 3 MM Whatman filter paper from the instructor and draw a straight line across the bottom using a soft lead pencil about 1.5 in. from the edge of the paper. This line will designate the origin where your samples will be spotted.
3. On both edges of the paper perpendicular to the origin, draw two straight lines from the top to the bottom that intersect the origin 1 in. from either edge of the paper. All of the samples will be spotted on the origin between these two lines.
4. Spot 3 μl of each of the amino acid standards along the origin. As with the thin-layer chromatography plate, spot each sample 1 μl at a time. Indicate the identity of each sample spot below the origin using a soft lead pencil.
5. Next to these standards, spot a 1-μl and a 3-μl sample of your unknown acid hydrolyzed dipeptide. *Save the remainder of this sample for use on Day 3.*
6. When all of the spots have dried, roll the paper into the form of a cylinder so that the spotted samples face inward toward the bottom of the paper (see Fig. 6-8). Staple the paper in three places (top, middle, and bottom) *avoiding any overlap of the paper on the edges.*
7. Place the paper (origin side down) in a large jar (10 in. in height with a 5.5-in. diameter) containing 2 cm of mobile-phase solvent on the bottom (butanol:acetic acid:water [60:15:25]).
8. Cap the jar and allow the mobile phase to migrate through (up) the paper until it reaches 1 in. from the top (~4–6 hr). Remove the paper from the jar, mark the position of the solvent front with a pencil, and allow the paper to dry in a fume hood.

Day 3: Analysis of the Paper Chromatogram and Preparation and Analysis of PTC-Amino Acids

Analysis of Paper Chromatogram

1. Spray the paper chromatogram lightly with ninhydrin reagent and heat for 5 min in a 100°C oven. The amino acids will appear as blue or purple spots (except proline, which will appear yellow). Figure 6-9 shows the reaction that amino acids undergo with ninhydrin.

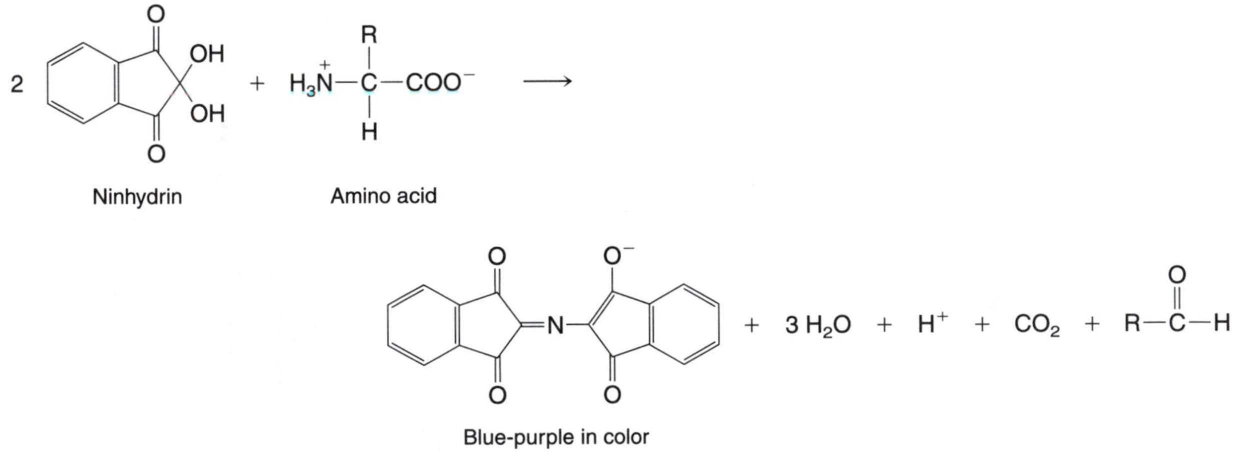

$$2\ \text{Ninhydrin} + H_3\overset{+}{N}-CH(R)-COO^- \longrightarrow \text{(Blue-purple in color)} + 3\,H_2O + H^+ + CO_2 + R-\overset{O}{\overset{\|}{C}}-H$$

Ninhydrin

Amino acid

Blue-purple in color

Figure 6-9 Reaction of amino acids with ninhydrin.

2. Circle the position of all spots with a pencil and note the color, which may give some clue as to the identity of the amino acid.
3. Calculate the R_f values of the amino acids in your unknown sample as well those of the amino acid standards, using the same procedure and equation as that used on Day 2.
4. Based on the R_f values and colors of the spots present in your unknown sample compared to those produced by the amino acid standards, what two amino acids are present in your unknown dipeptide? Can you now determine the sequence of your unknown dipeptide based on these results and those from Day 2? As before, discuss any uncertainty that may be involved in your identification, as well as how these uncertainties may be eliminated through modifications in the experiment.

Preparation of PTC-Amino Acids

1. Transfer 10 μl of your acid-hydrolyzed dipeptide sample (colorless) to a microcentrifuge tube and add 20 μl of distilled water. In addition, pipette 0.25 ml of Amino Acid Standard H into a separate microcentrifuge tube. Perform steps 2 to 6 on both of these samples.
2. Dry the sample in the vacuum microcentrifuge (Speed-Vac) under the high heat setting for 30 min or until all of the HCl has evaporated.
3. Add 20 μl of wash reagent (ethanol:triethylamine:water [2:2:1]), mix, and dry the sample down in the Speed-Vac under the high heat setting. Repeat this wash procedure one more time. The amino acids standards are supplied in a solution of 0.1 N HCl. The washes are performed to remove all traces of acid, which may interfere with the coupling reaction.
4. Add 15 μl of coupling reagent (ethanol:triethylamine:PITC:water [7:1:1:1]), mix, and incubate at room temperature for 20 min.
5. Dry the PTC-amino acids in the Speed-Vac under the low heat setting.
6. Resuspend the sample in 1 ml of PicoTag Eluant A. Centrifuge the sample for 5 min at 12,000 $\times$ *g* to pellet any undissolved material. Transfer the top 0.7 ml of the clarified sample to a fresh microcentrifuge tube. This sample will be analyzed by reverse phase HPLC as described below.

HPLC Analysis of PTC-Amino Acids

1. Fill HPLC Buffer Reservoir A with PicoTag Eluant A and Buffer Reservoir B with 60% (vol/vol) acetonitrile in HPLC grade water.
2. Equilibrate the C_{18} reverse-phase HPLC column with a solvent composition of 98% Buffer A and 2% Buffer B (flow rate at 1 ml/min).
3. Inject 100 μl of the PTC-amino acid sample and apply the following mobile-phase gradient during development (mark the time of injection, or $T = 0$, on the chart recorder):

Time (min)	% Buffer A	% Buffer B
0	98	2
12	90	10
13	80	20
23	60	40
25	0	100
31	0	100

4. Detect the PTC-amino acids as they elute with a spectrophotometer set at 254 nm. When the signal output (10 mV maximum) is relayed to a chart recorder with a speed of 0.5 cm/min, the PTC-amino acid peaks will show good height and good resolution.
5. Inject a 100-μl sample of the 20 PTC-amino acid standards and apply the same mobile-phase gradient. The sequence of the PTC-amino acids eluting from the column developed with this gradient is shown in Table 6-1.
6. Measure the retention times (R_T) of the unknown PTC-amino acids and compare these to those of the PTC-amino acid standards. The retention time (R_T) can be calculated as follows:

$$R_T = \frac{\text{distance from injection point } (T = 0) \text{ to the middle of the elution peak (cm)}}{0.5 \text{ cm/min}}$$

Table 6-1 Retention Times of PTC-Amino Acids on a C_{18} Reverse-Phase HPLC Column at a Flow Rate of 1 ml/min

Amino Acid	R_T (min)
PTC-aspartic acid	7.0
PTC-glutamic acid	7.4
PTC-serine	8.4
PTC-glycine	9.4
PTC-histidine	10.4
PTC-arginine	12.4
PTC-threonine	12.4
PTC-alanine	13.4
PTC-proline	14.0
PTC-tyrosine	22.0
PTC-valine	23.0
PTC-methionine	24.0
PTC-cysteine	25.6
PTC-isoleucine	26.4
PTC-leucine	26.6
PTC-phenylalanine	28.6
PTC-lysine	30.4

7. Based on these results, what two amino acids are present in your unknown dipeptide? As shown in Table 6-1, it may be difficult to resolve some PTC-amino acids (PTC-arginine and PTC-threonine). Still, this mobile-phase gradient will resolve the majority of the PTC-amino acid derivatives. How might the conditions of this experiment be changed to resolve molecules such as PTC-arginine and PTC-threonine?

   NOTE: Several amino acids will be modified or destroyed during the harsh acid hydrolysis step:

   Tryptophan is destroyed.
   Glutamine is converted to glutamate.
   Asparagine is converted to aspartate.
   Cysteine is oxidized to cysteic acid.

   Because of this, we suggest that you perform the experiment with dipeptides that do not contain these amino acids.

   NOTE: In addition to the HPLC mobile-phase system described above, you can replace Pico-Tag Eluant A with 140 mM KOAc, pH 5.3 (sodium acetate is not an acceptable alternative). If this is done, Resevoir B must contain 84% (vol/vol) acetonitrile in HPLC-grade water. This change in mobile-phase composition will require the flow rate to be 2 ml/min (for good resolution), with the following gradient:

Time (min)	% Buffer A	% Buffer B
0	94	6
0.1	93	7
2	91	9
8.5	89	11
9	78	22
11	77	23
15	76	24
17	74	26
18.6	64	36
20.5	64	36
21.5	0	100
22.5	0	100

   The retention times for each of the PTC-amino acids described above (Table 6-1) will not apply to a chromatography experiment done with this mobile-phase system. If this alternative gradient is used, these values will have to be determined independently.

Exercises

1. Draw the chemical structures of PTC-valine and PTH-valine. With what biochemical technique is the formation of PTH-amino acid derivatives associated?
2. The five dipeptides listed below were treated as follows: (a) The dipeptide was reacted exhaustively with FDNB. (b) The reaction was then acidified and extracted with ether. (c) The ether phase from this extraction was dried down, resuspended in 0.5 ml of 6 N HCl, and heated at 110°C for 12 hr. (d) 1 ml of water and 2 ml of ether were then added to the sample. For each of the dipeptides listed below, list the peptide-derived compounds that you would expect to find in the final ether phase and the final aqueous phase following the series of four treatments described above.

   Alanyl-asparagine
   Glutamyl-lysine
   N-acetyl-alanyl-tryptophan
   Valyl-arginine
   Arginyl-valine

3. A peptide (Compound A) was acid hydrolyzed and found to contain equimolar amounts of Arg, Val, Tyr, Glu, Lys, Ala, and Gly.

   a. Exhaustive treatment of Compound A with trypsin yields the following compounds: Arg, Ala-Lys, a peptide (Compound B) containing Glu, Gly, Tyr, Val. Treatment of compound B with chymotrypsin yields Val-Tyr and Glu-Gly.
   b. Short-term treatment of Compound A with carboxypeptidase yields free Gly as the first detectable amino acid.
   c. *N*-terminal analysis of Compound A with FDNB yields DNP-alanine in the ether phase.

   What is the sequence of the dipeptide?

4. With what groups on amino acids can FDNB react? Why could an amino acid that is not *N*-terminal, but still reacts with FDNB on its side chain, fail to be extracted into the ether phase at low pH conditions?

REFERENCES

Fraenkel-Conrat, H., Harris, J. I., and Levy, A. L. (1955). *Recent Developments in Techniques for Terminal and Sequence Studies in Peptides and Proteins.* (Methods of Biochemical Analysis, Vol. 2). New York: Wiley.

Konigsberg, W. (1972). Subtractive Edman Degradation. Enzyme Structure, Part B. *Methods Enzymol* **25**:326.

Laursen, R. A. (1972). Automatic Solid-Phase Edman Degradation. Enzyme Structure, Part B. *Methods Enzymol* **25**:344.

Lehninger, A. L., Nelson, D. L., and Cox, M. M. (1993). An Introduction to Proteins. In: *Principles of Biochemistry,* 2nd ed. New York: Worth.

Schroeder, W.A. (1972). Degradation of Peptides. Enzyme Structure, Part B. *Methods Enzymol* **25**:298.

Stryer, L. (1995). Exploring Proteins. In: *Biochemistry,* 4th ed. New York: Freeman.

Tarr, G. E. (1985). *Microcharacterization of Polypeptides: A Practical Manual* (Manual Edman Sequencing System). Totowa, NJ: Humana Press.

Wilson, K., and Walker, J. (1995). Chromatographic Techniques. In: *Principles and Techniques of Practical Biochemistry,* 4th ed. Hatfield, UK: Cambridge University Press.

EXPERIMENT 7

Study of the Properties of β-Galactosidase

Theory

The enzyme β-galactosidase allows lactose metabolism in *Escherichia coli.* As shown in Figure 7-1, this enzyme catalyzes the hydrolysis of lactose into its monosaccharide units, galactose and glucose. A side product of the β-galactosidase reaction, allolactose, is known to control the expression of the enzyme at the level of transcription.

The gene for β-galactosidase *(lacZ)* resides within a set of genes and regulatory DNA sequences collectively referred to as the *lac operon* (Fig. 7-2). An operon is a set of genes whose expression is coordinately regulated. Often, these genes encode proteins or enzymes related in a particular function or to a particular biological pathway. The *lac* operon consists of a promoter sequence, an operator (regulatory) sequence, and three genes. As mentioned above, *lacZ* encodes β-galactosidase. *LacY* encodes galactoside permease, a transmembrane transport protein that allows lactose to enter the cell. *LacA* encodes thiogalactoside transacetylase, an enzyme whose function is not fully understood. The promoter sequence within the operon is the site where

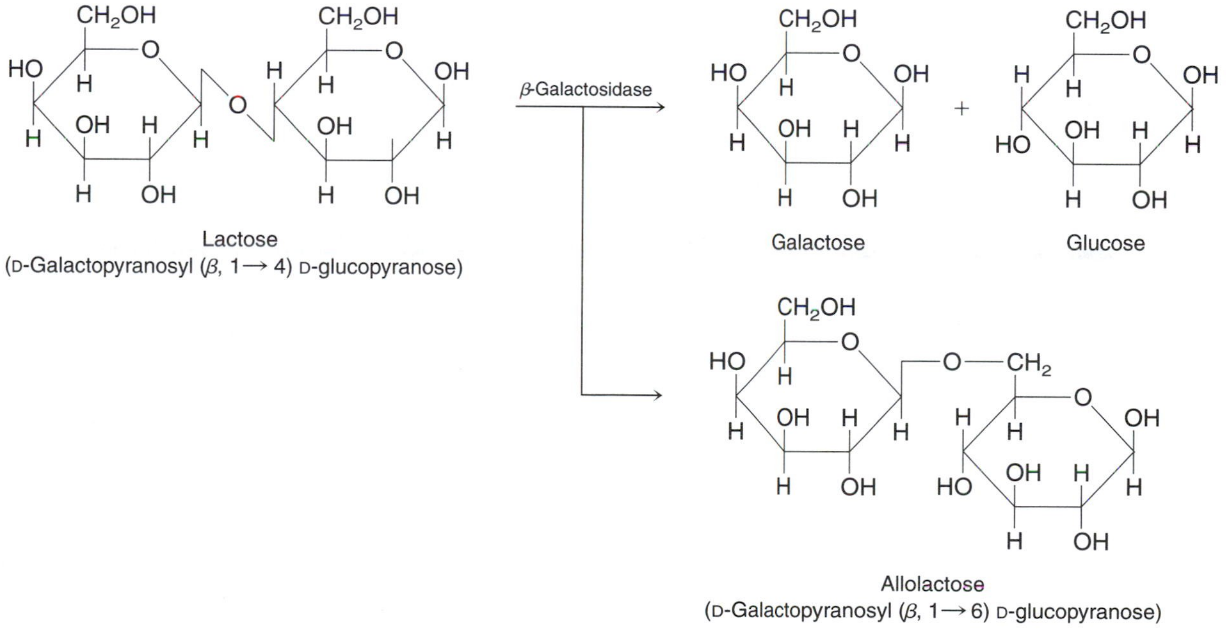

β-Galactosidase

Lactose
(D-Galactopyranosyl (β, 1⟶ 4) D-glucopyranose)

Galactose

Glucose

Allolactose
(D-Galactopyranosyl (β, 1⟶ 6) D-glucopyranose)

Figure 7-1 Reaction catalyzed by β-galactosidase.

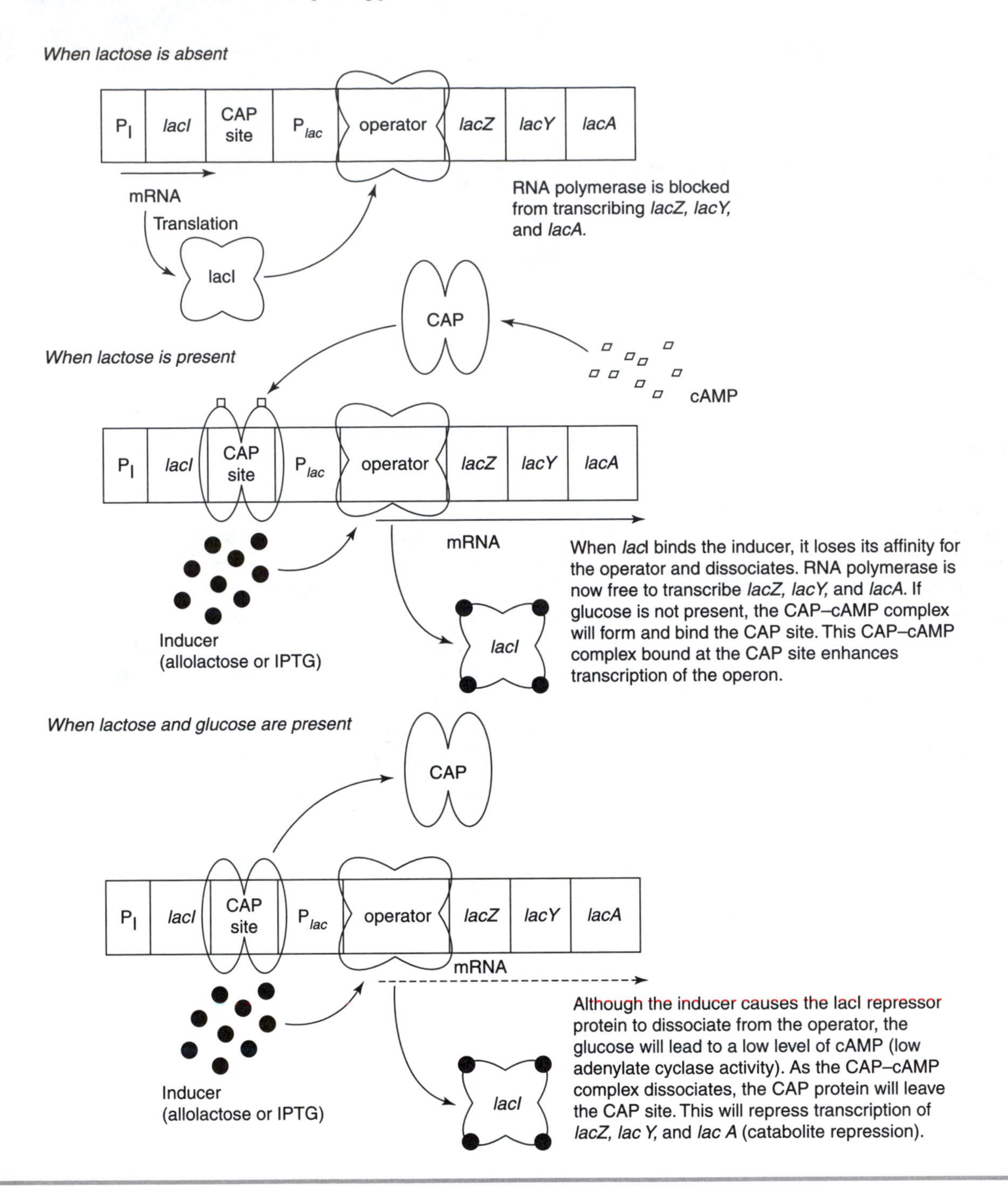

Figure 7-2 Regulation of the expression of the *lac* operon.

RNA polymerase binds the DNA and initiates transcription of *lacZ*, *lacY*, and *lacA*. As we shall see, the operator sequence adjacent to the promoter is an important regulatory sequence that will participate in the control of the expression of the *lac* operon.

Transcription of the *lac* operon can be *induced* (turned on) in the presence of lactose and repressed in the absence of lactose when the components required for its metabolism are not needed. This regulatory system ensures that the cell will not waste valuable energy producing proteins that it does not immediately require. Expression of the *lac* operon is repressed in the presence of the *lac repressor protein.* This regulatory protein, encoded by the *lacI* gene immediately 5′ to the *lac* operon, is a constitutively expressed DNA binding protein that has a high affinity for the *lac* operator sequence. When the *lac* repressor binds the operator sequence, it blocks RNA polymerase from transcribing the *lac* operon. As stated earlier, transcription of the *lac* operon is induced in the presence of lactose. This phenomenon is due to the effect that allolactose (the inducer) has on the affinity of the *lac* repressor for the operator sequence (Fig. 7-2). As the *lac* repressor (*LacI*) binds allolactose, it loses its affinity for the operator sequence. RNA polymerase is then able to transcribe the *lac* operon to produce the proteins required for lactose metabolism. In the presence of lactose, expression of the *lac* operon can increase by up to as much as 1000-fold. Although allolactose is the natural inducer of the *lac* operon, it is not the only compound that will increase its level of expression. Isopropylthiogalactoside (IPTG, Fig. 7-3) is a *gratuitous inducer* of the *lac* operon. Although it is not a substrate for β-galactosidase, it too will complex with the *lac* repressor protein and lower its affinity for the operator sequence.

The regulation of the *lac* operon just described is a form of *negative regulation: LacI* represses expression of the operon in the absence of the inducer. It turns out, however, that transcription of the *lac* operon is also subject to *positive regulation.* Immediately 5′ to the *lac* promoter is another regulatory DNA region called the CAP (catabolite gene activator protein) site. When CAP is bound at this site, transcription of the *lac* operon can be increased 20-fold to 50-fold. By itself, CAP has low affinity for the CAP site in the operon. The affinity of CAP for this site increases dramatically, however, when the protein binds cAMP (Fig. 7-2). Since the *lac* promoter by itself is relatively weak, the cAMP–CAP complex bound at the CAP site is required for significant expression of the *lac* operon. It is apparent, then, that the expression of the *lac* operon is closely tied to the level of intracellular cAMP.

Adenylate cyclase is the enzyme responsible for catalyzing the conversion of ATP to cAMP (Fig. 7-4). This enzyme is inactive in the cell when glucose levels are high. As a result, cAMP levels decrease in the cell when glucose is present. The net effect of this is that the cAMP–CAP complex is not able to form, and transcription of the *lac* operon is decreased. *This will be the case even if lactose is also*

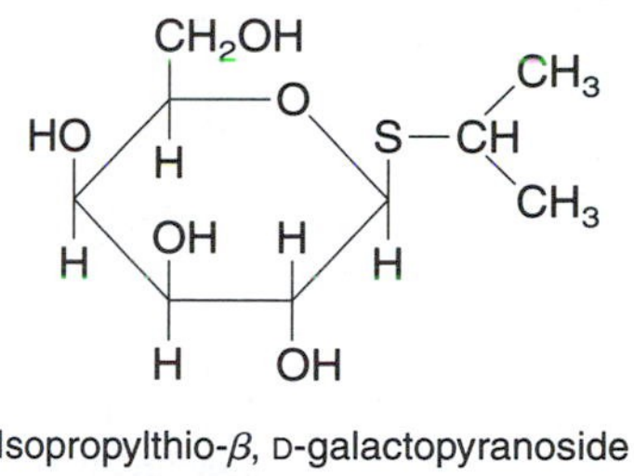

Isopropylthio-β, D-galactopyranoside

Figure 7-3 Isopropylthiogalactoside, a gratuitous inducer of the *lac* operon.

Adenosine triphosphate (ATP) — Adenylate cyclase → Adenosine-3′, 5′-cyclic monophosphate (cAMP) + PP_i

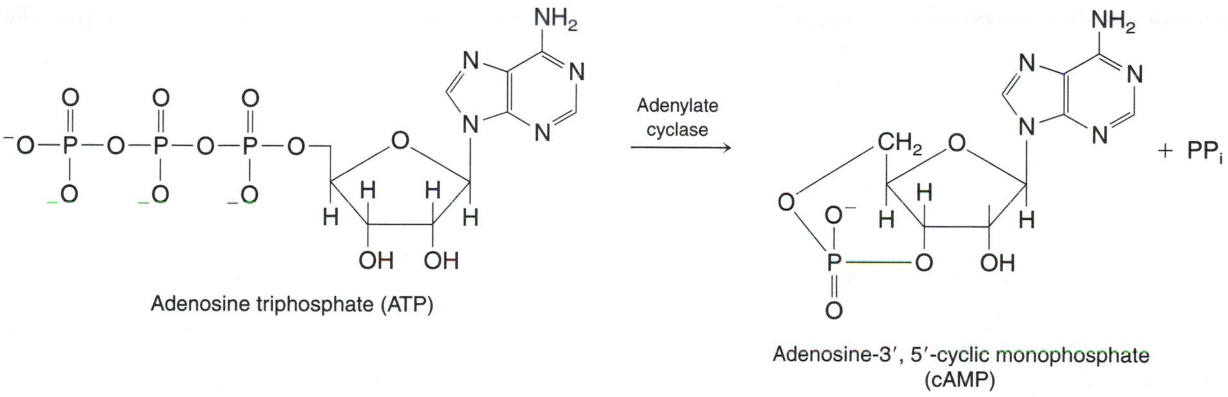

Figure 7-4 Reaction catalyzed by adenylate cyclase. The activity of adenylate cyclase is decreased when glucose is present in the growth medium.

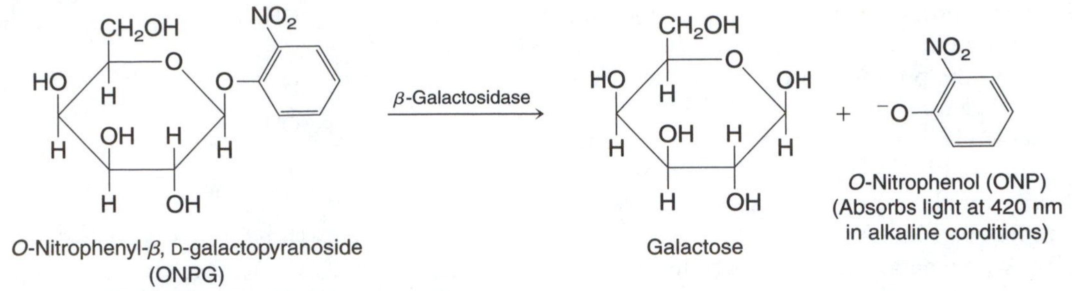

O-Nitrophenyl-β, D-galactopyranoside (ONPG)

β-Galactosidase

Galactose

O-Nitrophenol (ONP) (Absorbs light at 420 nm in alkaline conditions)

Figure 7-5 Colorimetric assay for β-galactosidase activity.

present. The ability of glucose to prevent expression of the *lac* operon even in the presence of its inducer is an example of *catabolite repression.* Since glucose and gluconate are the preferred carbon sources for *E. coli,* these compounds are very effective catabolite repressors of the *lac* operon. Other carbon sources, such as glycerol, acetate, and succinate, are less effective catabolite repressors. Since catabolite repression is due to the effect of high glucose levels on adenylate cyclase activity, the effect of catabolite repression can be overcome (reversed) in vivo through the addition of cAMP to the growth medium.

You will use *o*-nitrophenyl-β,D-galactopyranoside (ONPG), rather than lactose, as the substrate for the β-galactosidase assays performed in this experiment. β-galactosidase will hydrolyze the ONPG substrate to produce two products, one of which absorbs light at 420 nm under alkaline conditions (Fig. 7-5). By monitoring the change in A_{420} of a solution containing β-galactosidase and ONPG, you will be able to *continuously* measure the change in concentration of the product (*o*-nitrophenol, ONP) over time.

In this four-period experiment, you will study and characterize the kinetics of β-galactosidase at varying pH and temperature as well as in the presence of two different inhibitors of the enzyme. In these exercises, you will determine the K_M and V_{max} for the enzyme with respect to its ONPG substrate, as well as the inhibition constant (K_I) for the two inhibitors. In addition, the temperature studies will allow you to calculate the activation energy for the β-galactosidase catalyzed reaction. Finally, you will utilize a number of different *E. coli* strains in an in vivo β-galactosidase assay to demonstrate the intricate regulatory system of the *lac* operon and the concept of catabolite repression. Before you begin the experiment, review the section on enzymology in the Introduction to Section II.

Supplies and Reagents

0.08 M sodium phosphate buffer, pH 7.7
0.08 M sodium phosphate buffer, pH 6.4
0.08 M sodium phosphate buffer, pH 6.8
0.08 M sodium phosphate buffer, pH 7.2
0.08 M sodium phosphate buffer, pH 8.0
β-galactosidase solution (~10 units/milliliter in 0.08 M sodium phosphate buffer, pH 7.7, supplemented with 1 mg/milliliter bovine serum albumin)
o-nitrophenyl-β,D-galactopyranoside solutions (2.5 mM and 10 mM)
13 × 100 mm glass test tubes
16 × 125 mm glass test tubes with caps (sterile)
1.5-ml plastic microcentrifuge tubes
P-20, P-200, P-1000 Pipetmen with disposable tips
ice
methyl-β,D-galactopyranoside (MGP) solution (750 mM)
methyl-β,D-thiogalactoside (MTG) solution (150 mM)
isopropylthiogalactoside (IPTG) solution (200 mM)—filter-sterilized
cAMP solution (100 mM)—filter sterilized
1 M Na_2CO_3
0.1% (wt/vol) sodium dodecyl sulfate (SDS) solution
chloroform
YT broth (1% wt/vol bactotryptone, 0.5% wt/vol yeast extract, 0.5% wt/vol NaCl)
20% (wt/vol) glucose solution—filter sterilized
spectrophotometer to read absorbance at 280 nm, 420 nm, 600 nm, and 550 nm
heating blocks or temperature-controlled water baths
colorimeter tubes and quartz cuvettes

Protocol

Day 1: Determination of the Activity and Specific Activity of the β-Galactosidase Solution

1. Set up the following reactions in six 13-by-100-mm glass test tubes. *Keep the enzyme solution on ice and add it to the tubes below, which are at room temperature.*

Tube	Volume of 0.08 M Sodium Phosphate, pH 7.7 (ml)	Volume of 2.5 mM ONPG (ml)
1	4.0	0.5
2	4.0	0.5
3	4.0	0.5
4	4.0	0.5
5	4.0	0.5
6	4.0	0.5

2. Add 0.5 ml of sodium phosphate buffer, pH 7.7 to tube 1 (blank) and mix *gently* in a Vortex mixer. Blank your spectrophotometer to read zero absorbance at 420 nm against this solution.
3. Add 0.5 ml of undiluted β-galactosidase solution to tube 2. Mix *gently* with a Vortex mixer for 5 sec and place the solution in your spectrophotometer. Exactly 30 sec after the addition of the enzyme, record the A_{420} of the solution in your notebook. Continue to record the A_{420} value at 30-sec intervals until the A_{420} value no longer changes (indicates that the reaction is complete). This final A_{420} value will be used later to determine the activity of β-galactosidase in each reaction (ΔA_{420max}).
4. Prepare a twofold dilution of the enzyme by adding 250 μl of the β-galactosidase stock solution to 250 μl of 0.08 M sodium phosphate buffer, pH 7.7. Add this entire 0.5-ml solution to tube 3. Exactly 30 sec after the addition of the enzyme, record the A_{420} of the solution in your notebook. Continue to record the A_{420} value at 30-sec intervals. *Unlike with tube 2, it is not necessary that this reaction is followed to completion. Rather, you are looking to obtain a set of A_{420} values that are linear with respect to time over about 2 min.*
5. Prepare a fivefold dilution of the enzyme by adding 100 μl of the β-galactosidase stock solution to 400 μl of 0.08 M sodium phosphate buffer, pH 7.7. Add this entire 0.5-ml solution to tube 4. Exactly 30 sec after the addition of the enzyme, record the A_{420} of the solution in your notebook. Continue to record the A_{420} value at 30-sec intervals. *Unlike with tube 2, it is not necessary that this reaction is followed to completion. Rather, you are looking to obtain a set of A_{420} values that are linear with respect to time over about 5 min.*
6. Prepare a 10-fold dilution of the enzyme by adding 50 μl of the β-galactosidase stock solution to 450 μl of 0.08 M sodium phosphate buffer, pH 7.7. Add this entire 0.5-ml solution to tube 5. Exactly 30-sec after the addition of the enzyme, record the A_{420} of the solution in your notebook. Continue to record the A_{420} value at 30-sec intervals. *Unlike with tube 2, it is not necessary that this reaction is followed to completion. Rather, you are looking to obtain a set of A_{420} values that are linear with respect to time over about 5 min.*
7. Prepare a 20-fold dilution of the enzyme by adding 25 μl of the β-galactosidase stock solution to 475 μl of 0.08 M sodium phosphate buffer, pH 7.7. Add this entire 0.5-ml solution to tube 6. Exactly 30 sec after the addition of the enzyme, record the A_{420} of the solution in your notebook. Continue to record the A_{420} value at 30-sec intervals. *Unlike with tube 2, it is not necessary that this reaction is followed to completion. Rather, you are looking to obtain a set of A_{420} values that are linear with respect to time over about 5 min.*
8. Prepare a plot of A_{420} versus time for the data obtained from the reactions in tubes 2 to 6. Plot the data from all five reactions on a single graph for comparison.
9. Draw a "best-fit" curve through each set of data points. For each curve, determine over what time frame the reaction kinetics appear linear (where $\Delta A_{420}/\Delta$time yields a straight line). Calculate the slope of the *linear portion* of each curve.
10. Use the following equation to determine the β-galactosidase activity (micromoles of ONPG hydrolyzed per minute/per milliliter of enzyme) in each reaction:

$$\text{Activity} = \frac{(\Delta A_{420}/\Delta \text{time})(1.25/\Delta A_{420\text{max}})}{V_p}$$

where $1.25/\Delta A_{420max}$ is a conversion factor relating the number of micromoles of ONPG hydrolyzed to the change in absorbance at 420 nm, and Vp is the total volume (in milliliters) of β-galactosidase solution present in each reaction. Taking the volumes of the stock β-galactosidase solution (undiluted) used in each reaction into consideration, what is the activity of β-galactosidase in the stock solution of the enzyme? *NOTE:* You should be able to calculate several activity values for the stock β-galactosidase solution using the reactions in tubes 2 to 6. These values should be averaged and presented along with a deviation from the mean.

11. Obtain two 1.5-ml plastic microcentrifuge tubes. To one, add 0.5 ml of distilled water and 0.5 ml of the stock β-galactosidase solution. To the other, add 0.8 ml of distilled water and 0.2 ml of the stock β-galactosidase solution. Cap each tube and invert several times to mix.
12. Read the absorbance of both solutions in a quartz cuvette at 280 nm using water as a blank to zero your spectrophotometer. Record the A_{280} value of both solutions in your notebook.
13. Based on these values, estimate the total protein concentration (in milligrams per milliliter) of the stock β-galactosidase solution using Beer's law ($A = \varepsilon cl$). Assume an extinction coefficient of 0.8 $(mg/ml)^{-1}$ cm^{-1}. *This extinction coefficient is based on the average value of aromatic amino acids found in most proteins.*
14. Calculate the *specific activity* of the stock β-galactosidase solution (micromoles of ONPG hydrolyzed per minute per milligram of protein) using the following equation:

$$\text{Specific activity} = \frac{\text{activity}}{[\text{protein}]\ (\text{mg/ml})}$$

Day 2: Determination of K_M and V_{max} and the Effect of Inhibitors on β-Galactosidase Activity

Determination of K_M and V_{max}

1. From the data obtained on Day 1, determine a dilution of the stock β-galactosidase solution that showed good linear kinetics over a 5-min period (where $\Delta A_{420}/\Delta$time produced a straight line over 5 min). Use a dilution of the enzyme that gives a change in absorbance at 420 nm of about 0.1 per min. Prepare 9 ml of a dilute β-galactosidase solution by making the appropriate dilution of the β-galactosidase stock solution in sodium phosphate buffer, pH 7.7. This dilute β-galactosidase solution will be used in both of the assays described below. *Keep this solution on ice and add to the tubes below at room temperature when you are ready to perform the assay.*
2. Set up the following reactions in six 13-by-100 mm glass test tubes:

Tube	Volume of Phosphate Buffer, pH 7.7 (ml)	Volume of 2.5 mM ONPG (ml)
1	4.00	0.50
2	4.00	0.50
3	4.20	0.30
4	4.30	0.20
5	4.35	0.15
6	4.40	0.10

3. Add 0.5 ml of sodium phosphate buffer, pH 7.7, to tube 1 (blank) and mix by gently swirling. Blank your spectrophotometer to read zero absorbance at 420 nm against this solution.
4. Add 0.5 ml of the *dilute* β-galactosidase solution to tube 2. Mix gently with a Vortex mixer for 5 sec and place the solution in your spectrophotometer. Exactly 30 sec after the addition of the substrate, record the A_{420} of the solution in your notebook. Continue to record the A_{420} value at 30-sec intervals for 5 min.
5. Follow the exact same procedure as described in step 3 for tubes 3 to 6. Record the A_{420} at 30-sec intervals for at least 5 min for each reaction.
6. Prepare a plot of A_{420} versus time for the data obtained from the reactions in tubes 2 to 6. Plot the data from all five reactions on a single graph for comparison.
7. Calculate the slope of each line. If your enzyme dilution prepared in step 1 was correct, the ΔA_{420} per minute for each reaction should be linear for at least 5 min.

8. Use the following equation to determine the *initial velocity* of each β-galactosidase reaction (micromoles of ONPG hydrolyzed per minute):

$$\text{Initial velocity } (V_0) = (\Delta A_{420}/\Delta\text{time})(1.25/\Delta A_{420\text{max}})$$

Note how the slope of each line (initial velocity) changes with decreasing ONPG concentration.

9. Prepare a plot of $1/V_0$ versus 1/[ONPG] (a Lineweaver–Burk plot) using the data obtained from tubes 2 to 6. Do the data points "fit" to a straight line? From the *x*-intercept of this graph, can you determine the K_M (mM) for β-galactosidase with respect to ONPG? From the *y*-intercept of this graph, can you determine the V_{max} for the β-galactosidase solution (micromoles of ONPG hydrolyzed per minute)?

Effect of Inhibitors on β-Galactosidase Activity

1. Set up the following reactions in twelve 13-by-100-mm glass test tubes:

Tube	Volume of Phosphate Buffer, pH 7.7 (ml)	Volume of MGP (ml)	Volume of MTG (ml)	Volume of 2.5 mM ONPG (ml)
1	3.9	0.1	—	0.50
2	3.9	0.1	—	0.50
3	4.1	0.1	—	0.30
4	4.2	0.1	—	0.20
5	4.25	0.1	—	0.15
6	4.30	0.1	—	0.10
7	3.9	—	0.1	0.50
8	3.9	—	0.1	0.50
9	4.1	—	0.1	0.30
10	4.2	—	0.1	0.20
11	4.25	—	0.1	0.15
12	4.3	—	0.1	0.10

2. Add 0.5 ml of sodium phosphate buffer, pH 7.7, to tube 1 (blank) and mix by *gently* swirling. Blank your spectrophotometer to read zero absorbance at 420 nm against this solution.
3. Add 0.5 ml of the *dilute* β-galactosidase solution to tube 2. Swirl *gently* for 5 sec and place the solution in your spectrophotometer. Exactly 30 sec after the addition of the substrate, record the A_{420} of the solution in your notebook. Continue to record the A_{420} value at 30-sec intervals for 5 min.
4. Follow the exact same procedure as described in step 3 for tubes 3 to 6. Record the A_{420} at 30-sec intervals for at least 5 min for each reaction.
5. Add 0.5 ml of sodium phosphate buffer, pH 7.7, to tube 7 (blank) and mix by *gently* swirling. Zero your spectrophotometer at 420 nm against this solution.
6. Add 0.5 ml of dilute β-galactosidase solution to tube 8. Swirl *gently* for 5 sec and place the solution in your spectrophotometer. Exactly 30 sec after the addition of the substrate, record the A_{420} of the solution in your notebook. Continue to record the A_{420} value at 30-sec intervals for 5 min.
7. Follow the exact same procedure as described in step 6 for tubes 9 to 12. Record the A_{420} at 30-sec intervals for at least 5 min for each reaction.
8. Prepare two plots of A_{420} versus time: one plot should include data obtained from the reactions in tubes 2 to 6, the other should include data obtained from the reactions in tubes 8 to 12. Plot the data from reactions containing the same inhibitor on a single graph for comparison.
9. Use the following equation to determine the initial velocity of each these β-galactosidase reactions (micromoles of ONPG hydrolyzed per minute):

$$\text{Initial velocity } (V_0) = (\Delta A_{420}/\Delta\text{time})(1.25/\Delta A_{420\text{max}})$$

10. Prepare a plot of $1/V_0$ versus 1/[ONPG] (a Lineweaver–Burk plot) using the data obtained from tubes 2 to 6. Compare the *x*-intercept and *y*-intercept of this graph with those obtained in the absence of inhibitor (to determine K_M and V_{max}). Are they the same? What type of inhibition does methyl-β,D-galactopyranoside display toward β-galactosidase? Can you calculate an inhibition constant (K_I) for this inhibitor (remember that the methyl-β,D-galactopyranoside stock solution was at a concentration of 750 mM)?
11. Prepare a plot of $1/V_0$ versus 1/[ONPG] (a Lineweaver–Burk plot) using the data obtained

from tubes 8 to 12. Compare the x-intercept and y-intercept of this graph with those obtained in the absence of inhibitor (to determine K_M and V_{max}). Are they the same? What type of inhibition does methyl-β,D-thiogalactoside display toward β-galactosidase? Can you calculate an inhibition constant (K_I) for this inhibitor (remember that the methyl-β,D-thiogalactoside stock solution was at a concentration of 150 mM)?

12. Based on the K_I values for methyl-β,D-galactopyranoside and methyl-β,D-thiogalactoside, which compound is the more potent inhibitor of β-galactosidase?
13. Based on the type(s) of inhibition displayed by these two inhibitors, can you determine the site on the β-galactosidase enzyme with which these two compounds interact? Explain. Would you expect to be able to reverse the effects of these inhibitors by changing the concentration of ONPG used in these assays? Explain.

Day 3: Effect of pH and Temperature on β-Galactosidase Activity

Effect of pH on β-Galactosidase Activity

1. Prepare two different dilutions of the stock β-galactosidase solution: the first dilution should be the same as that done for the experiments on Day 2 (one that gave a linear change in absorbance at 420 nm per minute of about 0.1). Designate this solution "β-gal 1." The second solution should be twice as dilute as the first solution. Designate this solution "β-gal 2." You will need a total of 1.0 ml *of each* dilute β-galactosidase solution for the assays described below.
2. Obtain 10 ml of sodium phosphate buffers ranging in pH from 6.4 to 8.0:

   Buffer 1 = pH 6.4
   Buffer 2 = pH 6.8
   Buffer 3 = pH 7.2
   Buffer 4 = pH 7.7
   Buffer 5 = pH 8.0

3. Set up the following reactions in thirteen 13-by-100-mm glass test tubes. All volumes are given in milliliters:

Tube	Buffer 1	Buffer 2	Buffer 3	Buffer 4	Buffer 5	β-gal 1	β-gal 2
1	4.0	—	—	—	—	—	—
2	3.9	—	—	—	—	0.1	—
3	3.9	—	—	—	—	—	0.1
4	—	3.9	—	—	—	0.1	—
5	—	3.9	—	—	—	—	0.1
6	—	—	3.9	—	—	0.1	—
7	—	—	3.9	—	—	—	0.1
8	—	—	—	3.9	—	0.1	—
9	—	—	—	3.9	—	—	0.1
10	—	—	—	—	3.9	0.1	—
11	—	—	—	—	3.9	—	0.1
12	—	—	—	—	4.0	—	—
13	—	—	—	3.5	—	—	—

4. To tube 1 (blank), add 0.5 ml of 2.5 mM ONPG and 0.5 ml 1 M Na_2CO_3, mix, and use the solution to zero your spectrophotometer at 420 nm.
5. Add 0.5 ml of 2.5 mM ONPG to tube 2, mix *gently* with a Vortex mixer for 5 sec, and incubate at room temperature for 4 min.
6. After the 4 min incubation, stop the reaction by adding (and mixing in) 0.5 ml of 1 M Na_2CO_3. *The addition of this weak base will raise the pH of the reaction enough to completely stop the reaction.*
7. Perform steps 5 and 6 on tubes 3 to 7. Record the final A_{420} of all these solutions in your notebook.
8. To tube 12 (blank), add 0.5 ml of 2.5 mM ONPG and 0.5 ml of 1 M Na_2CO_3, mix, and use the solution to zero your spectrophotometer at 420 nm.
9. Add 0.5 ml of 2.5 mM ONPG to tube 8, mix gently with a Vortex mixer for 5 sec, and incubate at room temperature for 4 min.
10. After the 4-min incubation, stop the reaction by adding (and mixing in) 0.5 ml of 1 M Na_2CO_3.
11. Perform steps 9 and 10 on tubes 9 to 11. Record the final A_{420} of all these solutions in your notebook.
12. To tube 13, add 0.5 ml of the undiluted β-galactosidase stock solution. Add 0.5 ml of 2.5 mM ONPG, mix gently with a Vortex mixer for 5 sec, and incubate at room temperature for 10 min, or until the $\Delta A_{420}/\Delta$time is equal to zero (indicates that the reaction is complete).

13. Add 0.5 ml of 1 M Na_2CO_3, mix, and record the final A_{420} value of this solution in your notebook. This value ($A_{420max2}$) will be used to calculate the initial velocity of each of these β-galactosidase reactions. *It is necessary to calculate this new conversion factor since these fixed time-point assays are being stopped by increasing the pH. Recall from Experiment 1 that the absorbance of a solute is strongly dependent on the pH and ionic strength of the solution in which it resides. Since the pH at which you are measuring the absorbance of these solutions is not the same as the pH at which you were measuring the absorbance of the solutions on Day 1, a new conversion factor must be calculated.*
14. Calculate the initial velocity of each β-galactosidase reaction using the following equation (micromoles of ONPG hydrolyzed per minute):

$$\text{Initial velocity } (V_0) = (\Delta A_{420}/4 \text{ min})(1.25/\Delta A_{420max2})$$

NOTE: Two dilutions of the enzyme were used at each pH because this is a fixed time-point assay. Rather than continuously measuring the change in absorbance at 420 nm over time, you stopped each reaction at 4 min by increasing the pH. If the enzyme dilution that showed linear kinetics over 4 min at pH 7.7 did not show linear kinetics at a different pH, the solution with a lower concentration of the enzyme (β-gal 2) will produce linear kinetics over 4 min. If β-gal 1 and β-gal 2 both displayed linear kinetics over 4 min at a particular pH, then the initial velocity of the solution containing β-gal 1 will be twice that of the solution containing β-gal 2.

15. Prepare a plot of V_0 versus pH based on the results of this assay. From this plot, can you determine the pH optimum for the reaction catalyzed by β-galactosidase?
16. Based on the pH optimum for β-galactosidase and your knowledge of the pK_a values of the various amino acid side chains, can you hypothesize which amino acids might be present or important in the active site of the enzyme?

Effect of Temperature on β-Galactosidase Activity

1. Prepare the following reactions in nine 13-by-100 mm glass test tubes:

Tube	Volume of Phosphate Buffer, pH 7.7 (ml)	Volume of 2.5 mM ONPG (ml)
1	4.0	0.5
2	3.9	0.5
3	3.9	0.5
4	3.9	0.5
5	3.9	0.5
6	3.9	0.5
7	3.9	0.5
8	3.9	0.5
9	3.9	0.5

2. Place tubes 2 and 6 in a 25°C water bath. Place tubes 3 and 7 in a 30°C water bath. Place tubes 4 and 8 in a 37°C water bath. Place tubes 5 and 9 in a 48°C water bath. Incubate all of these tubes at the indicated temperatures for 10 min.
3. Add and mix 0.1 ml of β-gal 1 to tubes 2 to 5. *Note the exact time that the dilute β-galactosidase solution was added to each of the tubes.* After exactly 4 min, add 0.5 ml of 1 M Na_2CO_3 to each of tubes 2 to 5.
4. Add and mix 0.1 ml of β-gal 2 to tubes 6 to 9. *Note the exact time that the dilute β-galactosidase solution was added to each of the tubes.* After exactly 4 min, add 0.5 ml of 1 M Na_2CO_3 to each of tubes 6 to 9.
5. To tube 1 (blank), add 0.5 ml of 1 M Na_2CO_3, mix, and use the solution to zero your spectrophotometer at 420 nm.
6. Measure and record the final A_{420} values of the solutions in tubes 2 to 9 in your notebook.
7. Calculate the initial velocity of each β-galactosidase reaction using the following equation (micromoles of ONPG hydrolyzed per minute):

$$\text{Initial velocity } (V_0) = (\Delta A_{420}/4 \text{ min})(1.25/\Delta A_{420max\ 2})$$

NOTE: Two dilutions of the enzyme were used at each temperature because this is a *fixed time-point assay*. Rather than continuously measuring the change in absorbance at 420 nm over time, you stopped each reaction at 4 min by increasing the pH. If the enzyme dilution that showed linear kinetics over 4 min at a particular temperature did not show linear kinetics at a different temperature, the solution with a lower concentration of the enzyme (β-gal 2) will produce linear kinetics over 4 min. If β-gal 1 and β-gal 2 both displayed linear kinetics over 4 min at a particular temperature, then the initial velocity of the solution containing β-gal 1 will be twice that of the solution containing β-gal 2.

8. Construct a plot of log V_0 versus 1/absolute temperature (°K). Determine the slope of the line through the *linear portion* of this curve. From this slope, determine the Arrhenius activation energy (E_a) for the reaction catalyzed by β-galactosidase using the following equation:

$$\text{Slope} = -\frac{E_a}{2.3R}$$

where R is the gas constant (1.98 cal/deg.-mol).

9. What do you notice about the activity of the enzyme at higher temperatures? Explain what you think is happening to the enzyme at higher temperatures.

Day 4: Catabolite Repression and Regulation of the *lac* Operon in Vivo

We suggest that steps 1 through 3 be performed by the instructor prior to the beginning of the experiment.

1. The night before the experiment, grow 5-ml overnight cultures of each of the following *E. coli* strains in YT broth with shaking at 37°C.

   EM1: *E. coli* K-12 ($lacI^+,cya^+,crp^+,thi^-$)
   EM1327: *E. coli* K-12 ($lacI^+,cya^+,\Delta crp,thi^-$)
   EM1328: *E. coli* K-12 ($lacI^+,\Delta cya,crp^+,thi^-$)
   EM1329: *E. coli* K-12 ($\Delta lacI,cya^+,crp^+,thi^-$)

   Recall that the *lacI*, *crp*, and *cya* genes encode the *LacI* repressor protein, the catabolite gene activator protein, and adenylate cyclase, respectively.

2. The next morning (~3–4 hr before the experiment), dilute the cultures 1:10 to 1:200 in 50 ml of fresh YT broth. *We have found that these strains display different growth rates. It is very important that you perform growth trials beforehand to determine the correct dilution for each strain to ensure that the cultures will grow to the same optical density prior to the beginning of the experiment.*

3. Grow these cultures at 37°C with shaking to late log phase (A_{600} ~1.0). This will take approximately 3 to 4 hr.

4. Set up the following inoculations in 12 *sterile* (16 × 125 mm) glass test tubes:

Tube	YT Broth (ml)	200 mM IPTG (ml)	100 mM cAMP (ml)	20% Glucose (ml)
1	5.0	—	—	0.05
2	5.0	0.05	—	0.05
3	5.0	0.05	0.05	0.05
4	5.0	—	—	0.05
5	5.0	0.05	—	0.05
6	5.0	0.05	0.05	0.05
7	5.0	—	—	0.05
8	5.0	0.05	—	0.05
9	5.0	0.05	0.05	0.05
10	5.0	—	—	0.05
11	5.0	0.05	—	0.05
12	5.0	0.05	0.05	0.05

5. Using sterile technique (consult the instructor), add 100 μl of late-log-phase strain EM1 culture to tubes 1 to 3. Add 100 μl of late-log-phase strain EM1327 culture to tubes 4 to 6. Add 100 μl of late-log-phase strain EM1328 culture to tubes 7 to 9. Add 100 μl of late-log-phase strain EM1329 culture to tubes 10 to 12.

6. Grow all of the cultures with shaking at 37°C for 2.5 to 3.0 hr.

7. Set up 12 16 × 125 mm glass test tubes (one for each culture) containing the following:

   3.75 ml of sodium phosphate buffer, pH 7.7
   50 μl of chloroform
   50 μl of 0.1% (wt/vol) sodium dodecyl sulfate (SDS) solution

   Number the tubes 1 to 12. The reaction for each numbered culture will be done in the corresponding numbered reaction tube.

8. Add 0.25 ml of each culture to the corresponding reaction tube. Mix *vigorously* in a Vortex mixer for 5 sec. The chloroform and SDS will disrupt the integrity of (solubilize) the cell membrane, allowing the ONPG substrate to come into contact with the β-galactosidase present in the cell.

9. Add 0.5 ml of 10 mM ONPG to each reaction tube. Mix and incubate at room temperature

for 15 to 20 min or until a yellow color begins to appear *(be sure that some yellow color begins to appear in at least some of the reaction tubes. If it does not appear after 20 min, continue the reaction until it does. If you see yellow color appear after 2 or 3 min, decrease the incubation time to 5 min). It is important that you note the exact time that the ONPG is added to each tube.*

10. Stop each of the 12 reaction tubes by adding and mixing in 0.5 ml of 1 M Na_2CO_3. *It is important that you note the exact time that the Na_2CO_3 is added to each tube to stop the reaction.*
11. Prepare a blank solution in a large glass test tube containing the following:

    3.75 ml of sodium phosphate buffer, pH 7.7
    50 μl of chloroform
    50 μl of 0.1% (wt/vol) SDS solution
    0.25 ml of YT broth
    0.5 ml of 10 mM ONPG
    0.5 ml of 1M Na_2CO_3

    Use this solution to zero your spectrophotometer at both 550 nm and 420 nm in the absorbance determinations described below.
12. Measure the A_{420} and A_{550} of the reaction solutions in each of the 12 reaction tubes. Record these values in your notebook.
13. Measure the A_{600} of each of the 12 *cultures (not the reaction tubes)*. For the absorbance measurements taken at 600 nm, the spectrophotometer should be blanked to read zero absorbance of a solution containing only YT broth.
14. Calculate the activity of β-galactosidase (nanomoles of ONPG hydrolyzed per minute/A_{600} of culture) using the following equation:

$$\text{Activity} = \frac{A_{420} - (1.75 \times A_{550})}{(A_{600})(t)} \cdot \frac{1}{0.0045} \cdot 5 \text{ ml}$$

    where t = time of reaction (min), 0.0045 = extinction coefficient (μM^{-1} cm^{-1}) for ONP under these conditions, ($1.75 \times A_{550}$) = correction factor for the light scattering of the culture at 420 nm, and 5 ml = total reaction volume.
15. Based on your knowledge of the regulation of the *lac* operon and catabolite repression, discuss whether or not each strain displayed a level of β-galactosidase activity that was expected under the various growth conditions.

Exercises

1. During the purification of β-galactosidase, a student found that 1 ml of a 1:10,000 dilution of crude extract gave A_{420} readings as follows:

Time (sec)	A_{420}
15	0.059
30	0.084
45	0.109
60	0.134
75	0.159

    The total reaction volume was 10 ml, containing 8.0 ml of buffer (pH 7.5), 1.0 ml of 7.5 mM ONPG solution, and 1.0 ml of the dilute β-galactosidase solution. A standard 10-mM solution of ONP, pH 7.5, has an A_{420} of 0.50 in the 1-cm pathlength colorimeter cuvettes used for these assays.

    a. One unit of β-galactosidase under these conditions should correspond to what increase in A_{420} per minute (1 unit = one micromole of ONPG hydrolyzed per minute)?
    b. If the protein concentration of the undiluted crude extract was 13.3 mg/ml, what is the specific activity of the extract in units per milligram?

2. Should the presence of IPTG in the growth medium have any effect on the expression of the *lac* operon in the *lacI*⁻ strain? Explain. Should the presence of cAMP in the growth medium have any effect on the expression of the *lac* operon in the *lacI*⁻ strain? Explain.
3. Should the presence of IPTG in the growth medium have any effect on the expression of the *lac* operon in the *cya*⁻ strain? Explain. Should the presence of cAMP in the growth medium have any effect on the expression of the *lac* operon in the *cya*⁻ strain? Explain.
4. Should the presence of IPTG in the growth medium have any effect on the expression of the *lac* operon in the *crp*⁻ strain? Explain. Should the presence of cAMP in the growth medium have any effect on the expression of the *lac* operon in the *crp*⁻ strain? Explain.
5. Can you think of a possible mutation in the operator sequence of the *lac* operon that would

produce a strain with the same level of expression of the *lac* operon as the *lacI*$^-$ strain in the presence of IPTG and/or cAMP?

6. Explain how the activity of adenylate cyclase affects expression of the *lac* operon. Describe situations (growth conditions) when the activity of adenylate cyclase would be high and when it would be low.
7. Explain how a mutation in the *lacY* gene could affect the regulation of the *lac* operon.
8. Why was it necessary to calculate a new conversion factor relating the ΔA_{420} to the number of micromoles of ONPG hydrolyzed for the kinetics experiments done on Day 3, despite the fact that the concentration of ONPG stock solution was the same as that on Day 1 and Day 2?

REFERENCES

Beckwith, J. R., and Zipser, D., eds. (1970). *The Lactose Operon.* Cold Spring Harbor, NY: Cold Spring Harbor Laboratories.

Lehninger, A. L., Nelson, D. L., and Cox, M. M. (1993). Regulation of Gene Expression. In *Principles of Biochemistry*, 2nd ed. New York: Worth.

Miller, J. H. (1972). *Experiments in Molecular Genetics.* Cold Spring Harbor, NY: Cold Spring Harbor Laboratories.

Perlman, R., Chen, B., DeCrombrugghe, B., Emmer, M., Gottesman, M., Varmus, H., and Pastan, I. (1970). The regulation of *lac* Operon transcription by cyclic adenosine-3′,5′-monophosphate. *Cold Spring Harbor Symp Quant Biol* **35**:419.

Stryer, L. (1995). Enzymes: Basic Concepts and Kinetics. In: *Biochemistry*, 4th ed. New York: Freeman.

Zubay, G., Schwartz, D., and Beckwith, J. (1970) The mechanism of activation of catabolite-sensitive genes. *Cold Spring Harbor Symp Quant Biol* **35**:433.

EXPERIMENT 8

Purification of Glutamate–Oxaloacetate Transaminase from Pig Heart

Theory

The detailed study of enzyme mechanisms requires the use of purified if not homogeneous enzymes. This experiment presents three procedures commonly used in protein purification: ammonium sulfate precipitation, heat denaturation, and ion-exchange chromatography. Although the purification procedure outlined in this experiment is useful in the isolation of glutamate–oxaloacetate transaminase (GOT), the same techniques can be modified to aid in the purification of many other proteins of interest.

One of the most important requirements for successful protein purification is the availability of an accurate, rapid, and quantitative assay that is specific for the protein of interest. If the protein has no known enzymatic activity, the amount of the protein present after each purification step must be analyzed by some general method such as sodium dodecyl sulfate–polyacrylamide gel electrophoresis (SDS-PAGE), Western blotting, or an enzyme-linked immunosorbent assay (ELISA). If the protein is known to bind a particular molecule, it may be possible to design a quantitative binding assay for the protein. If the protein of interest has enzymatic activity, an assay that is specific for the enzyme can be designed, and the *specific activity* of the enzyme can be monitored to assess the effectiveness of each purification step. Although you may find that the *total enzyme activity* decreases throughout the purification (percent yield less than 100), the specific activity of the enzyme should *increase* with each purification step. Recall that it is the specific activity of an enzyme that is the true measure of its purity (see Introduction to Section II).

GOT is a 90-kDa, homodimeric enzyme that catalyzes the reversible transfer of an amino group from aspartate to α-ketoglutarate (Fig. 8-1). Oxaloacetate and glutamate are the products of this reaction. GOT is one member of the family of transaminase enzymes. All members of this family catalyze the transfer of an amino group from an amino acid to an α-keto acid, which is nearly always α-ketoglutarate or pyruvate. In the process,

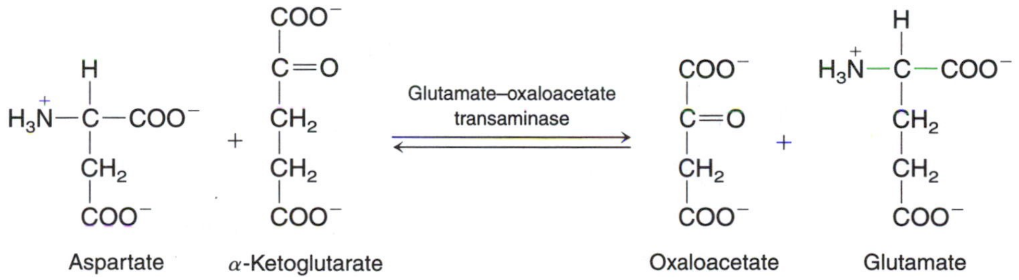

$$\underset{\text{Aspartate}}{H_3\overset{+}{N}{-}CH(CH_2COO^-){-}COO^-} + \underset{\alpha\text{-Ketoglutarate}}{^-OOC{-}C(=O){-}CH_2{-}CH_2{-}COO^-} \xrightleftharpoons[]{\text{Glutamate–oxaloacetate transaminase}} \underset{\text{Oxaloacetate}}{^-OOC{-}C(=O){-}CH_2{-}COO^-} + \underset{\text{Glutamate}}{H_3\overset{+}{N}{-}CH(CH_2CH_2COO^-){-}COO^-}$$

Figure 8-1 Reaction catalyzed by GOT.

glutamate (or alanine) and the corresponding α-keto acid of the amino acid substrate are produced. The specificity of each transaminase is determined by the amino acid it recognizes as a substrate. The transaminase enzymes provide an early and critical step in amino acid catabolism, funneling amino groups from different amino acids into the production of glutamate. The glutamate that is formed from these transaminase reactions has several fates: it can be converted to a neurotransmitter (γ-amino butyric acid, GABA), it can be used in the production of glutathione, it can be deamidated to produce NH_4^+ for the urea cycle, it can be converted to glutamine (a method of NH_4^+ assimilation in prokaryotes), or it can act as the precursor in the biosynthesis of other amino acids.

All of the transaminase enzymes rely on a common coenzyme, pyridoxal phosphate (PLP, Fig. 8-2). The PLP coenzyme is linked to the enzyme through the formation of a Schiff base between the ε-amino group of a lysine residue at the active site and the aldehyde carbon of PLP. The pyridoxal form of GOT is heat-stable, a property that can be exploited in its purification. As aspartate enters the active site, the PLP will form a new Schiff base with the α-amino group forming an aldimine intermediate. Addition of a water molecule will cleave the bond, producing oxaloacetate (the corresponding α-keto acid of aspartate) and pyridoxamine phosphate. This reaction will be reversed as α-ketoglutarate enters the active site and removes the amino group from pyridoxamine phosphate, forming glutamate and the pyridoxal form of the enzyme that is able to accept another aspartate molecule. The mechanism of the reaction catalyzed by GOT is outlined in Figure 8-3.

How do you measure GOT activity? The oxaloacetate produced by the enzyme can exist in either the keto or the enol form. At basic pHs favored by GOT (pH 8.3), the keto–enol equilibrium is displaced in favor of the enol form (Fig. 8-4). The enol form of oxaloacetate absorbs light at 256 nm, providing a direct means of measuring GOT activity (the $\Delta A_{256}/\Delta$time is directly proportional to the change in concentration of oxaloacetate over time). In this experiment, you will use a different method to measure GOT activity. You will quantify the rate of production of oxaloacetate (catalyzed by GOT) through the use of a coupled enzyme assay. In the presence of malic dehydrogenase (MDH), oxaloacetate and NADH will be converted to malate and NAD^+ (Fig. 8-5). Since NADH absorbs light at 340 nm, you can continuously measure the rate of oxidation of NADH to NAD^+ by following the *decrease* in absorbance at 340 nm over time. As long as the components of the coupled assay (MDH and NADH) are not rate-limiting, the $\Delta A_{340}/\Delta$time will be directly proportional to the change in concentration of oxaloacetate over time, providing an accurate measure of GOT activity. Since inexpensive spectrophotometers are not designed to give accurate absorbance readings in the short UV range of light, 340 nm is a much more convenient wavelength to use in the GOT assay.

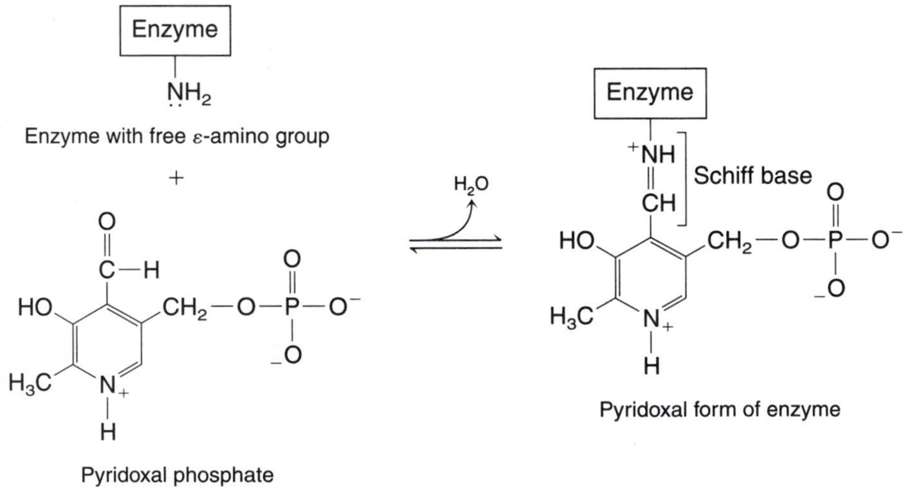

Enzyme with free ε-amino group + Pyridoxal phosphate $\xrightleftharpoons{H_2O}$ Pyridoxal form of enzyme (Schiff base)

Figure 8-2 PLP is a common coenzyme involved in transamination reactions.

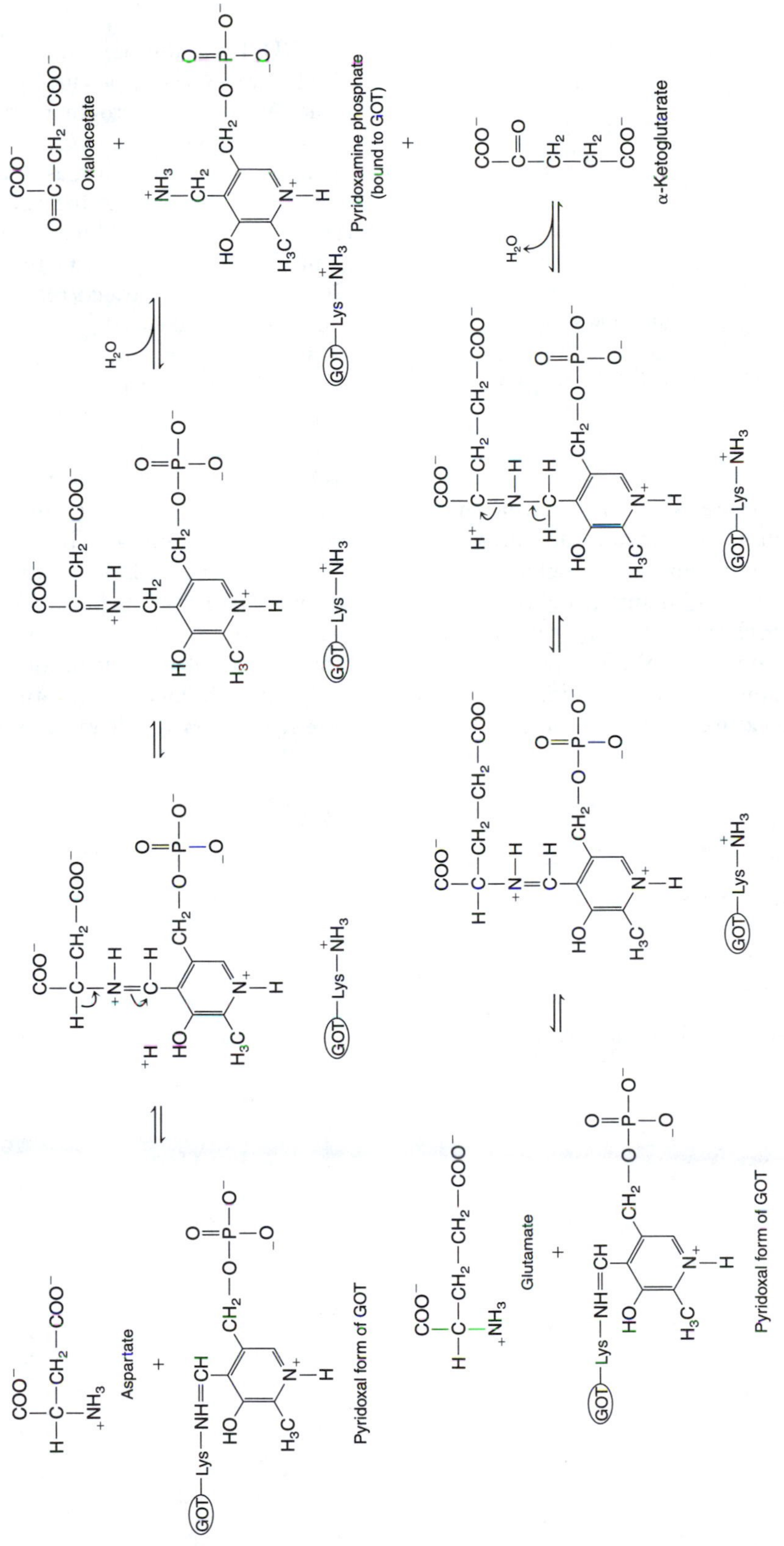

Figure 8-3 Mechanism of the reaction catalyzed by GOT.

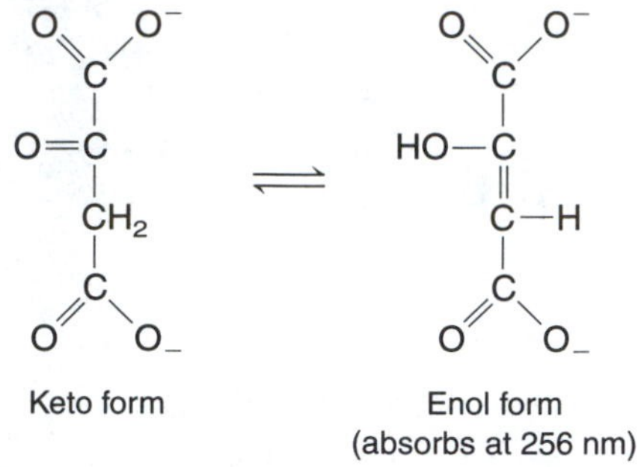

Keto form

Enol form (absorbs at 256 nm)

Figure 8-4 Keto–enol equilibrium of oxaloacetate.

Unfortunately, there is another enzyme in pig heart mitochondria, glutamate dehydrogenase (GDH), that also recognizes α-ketoglutarate and NADH as substrates. Glutamate dehydrogenase will convert α-ketoglutarate to glutamate, with the subsequent oxidation of an NADH molecule (Fig. 8-6). Since the change in concentration of NADH is what is being measured in the coupled GOT assay, GDH has the potential to interfere with the GOT activity assay, producing an artificially high value. Although α-ketoglutarate is a common substrate for both GOT and GDH, only GOT utilizes aspartate. Because of this, aspartate can be omitted from the GOT assay to measure GDH activity directly and accurately. Once GDH activity has been quantified, it can be subtracted from the apparent GOT activity to give a correct activity value for the transaminase enzyme.

In this experiment, you will perform a three-step purification of glutamate–oxaloacetate transaminase from pig heart. With each step in the purification, you will assay different fractions for GOT activity and total protein concentration. These values will then be used to calculate the total activity, specific activity, and percent yield of GOT following each purification step. The goal of the experiment is to achieve 80- to 100-fold purification of the enzyme with $\geq$5% yield. The overall effectiveness of the purification procedure will be analyzed by SDS-PAGE on the last day of the experiment. *Be sure that you determine three different*

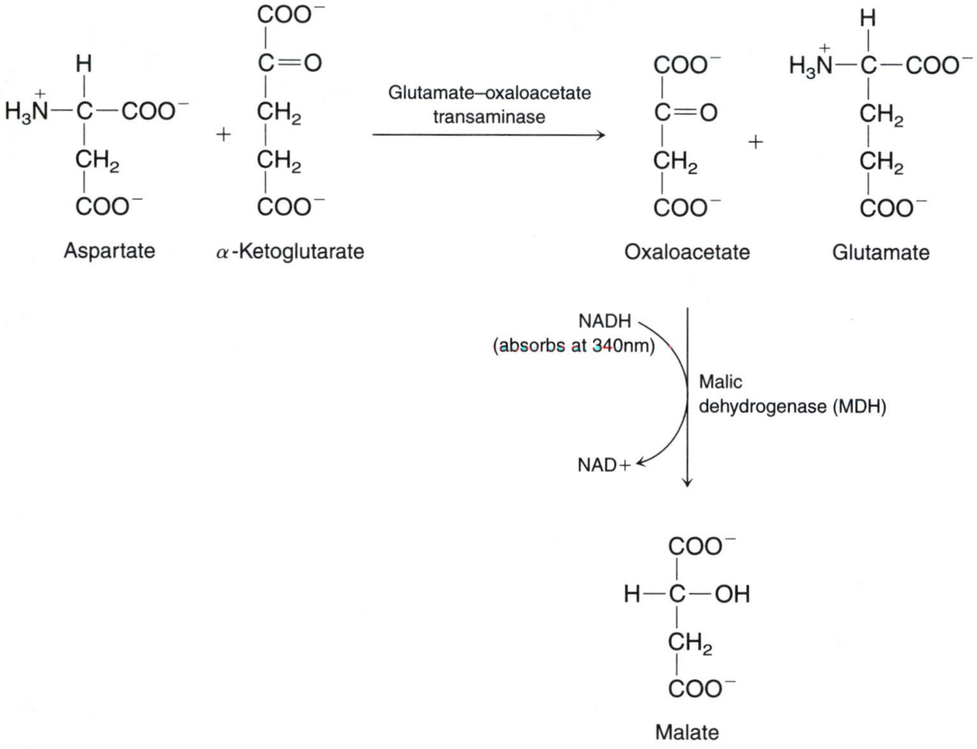

Aspartate + α-Ketoglutarate → (Glutamate–oxaloacetate transaminase) → Oxaloacetate + Glutamate

Oxaloacetate + NADH (absorbs at 340nm) → (Malic dehydrogenase (MDH)) → Malate + NAD+

Figure 8-5 Coupled enzymatic assay to measure GOT activity.

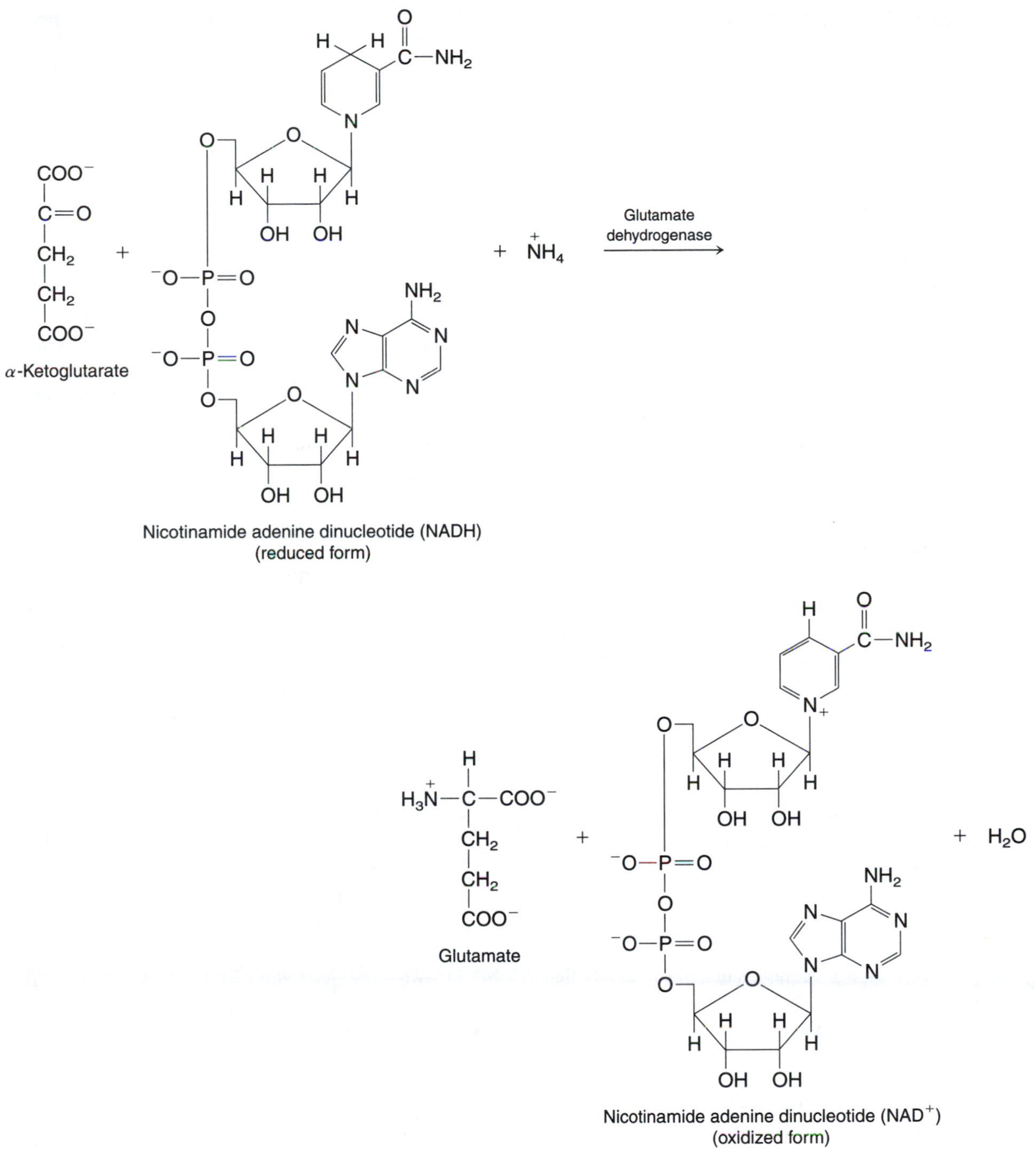

Figure 8-6 Reaction catalyzed by glutamate dehydrogenase (GDH).

values for each fraction following each purification step: The total volume of each fraction (in milliliters), the GOT activity in each fraction (in micromoles of oxaloacetate per minute per milliliter of enzyme), and the protein concentration in each fraction (in milligrams per milliliter).

Supplies and Reagents

P-20, P-200, P-1000 Pipetmen with disposable tips
0.5 M potassium malate buffer (pH 6.0) containing 5 mM EDTA
0.04 M α-ketoglutarate (adjusted to pH 6.0 with NaOH)
Saturated ammonium sulfate (reagent grade) solution
0.03 M sodium acetate buffer (pH 5.4)
0.03 M sodium acetate buffer (pH 5.0)
0.08 M sodium acetate buffer (pH 5.0)
Carboxymethylcellulose (equilibrated in 0.03 M sodium acetate, pH 5.0)
0.1 M potassium phosphate buffer (pH 7.5)
Standard bovine serum albumin solution (1 mg/ml)
Solutions for Folin-Ciocalteau protein assay:

- 2 N Folin–Ciocalteau reagent
- 2% Na_2CO_3 in 0.1 N NaOH
- 1% $CuSO_4 \cdot 5H_2O$
- 2% sodium tartrate

250-ml plastic centrifuge bottles
50-ml plastic centrifuge tubes
Bunsen burner
Pasteur pipettes
Spectrophotometer to read absorbance at 340 nm and 500 nm
Cuvettes or colorimeter tubes for spectrophotometer
Small and large glass test tubes
Cheesecloth
Glass wool
Chromatography column (20 ml)
Fresh pig hearts
Meat grinder
Blender
0.1 M potassium phosphate and 0.12 M aspartate buffer (pH 7.5)
3 mM NADH in 0.1 M potassium phosphate buffer (pH 7.5)—*prepared fresh daily*
Malic dehydrogenase (MDH), 200 units/ml in 0.1 M potassium phosphate buffer (pH 7.5)
0.1 M α-ketoglutarate solution (pH 7.5)
Dialysis tubing (8,000–12,000 Da molecular weight cut-off)
Reagents and supplies for SDS-PAGE (see Experiment 4):

- 10 × Buffer A (1.5 M Tris HCl, pH 8.8)
- 10 × Buffer B (1.0 M Tris HCl, pH 6.8)
- 30% acrylamide (292 g/liter acrylamide, 8 g/liter *N,N'*-methylenebisacrylamide)
- 10% (wt/vol) SDS in water
- TEMED (*N,N,N',N'*-tetramethylethylenediamine)
- 10% (wt/vol) ammonium persulfate in water—*prepared fresh*
- SDS-PAGE running buffer (25 mM Tris, 250 mM glycine, 0.1% (wt/vol) SDS, pH 8.3)

4× SDS sample buffer:

- 0.25 M Tris, pH 7.0
- 30% (vol/vol) glycerol
- 10% (vol/vol) 2-mercaptoethanol
- 8% (wt/vol) SDS
- 0.001% (wt/vol) bromophenol blue

Coomassie Blue staining solution (40% methanol, 10% glacial acetic acid, 0.25% (wt/vol) Coomassie Brilliant Blue R-250, 50% distilled water)
Protein molecular weight standards (see Experiment 4)
Destaining solution (40% methanol, 10% glacial acetic acid, 50% distilled water)
Power supply

GOT Activity Assay

The following procedure will be used for all of the GOT activity assays performed in this experiment.

1. Add the following, in order, to a small glass test tube:

Reagent	Volume (ml)
0.1 M potassium phosphate and 0.12 M aspartate buffer (pH 7.5)	2.5
3 mM NADH solution	0.1
MDH (200 units/ml)	0.1
GOT dilution (in 0.1 M potassium phosphate buffer, pH 7.5)	0.1

2. Blank your spectrophotometer to read zero absorbance at 340 nm against water.
3. Add 0.2 ml of 0.1 M α-ketoglutarate solution to the reaction tube, mix, and place the solution in the spectrophotometer.
4. Record the A_{340} of the solution at 30-sec intervals for 4 min. Determine whether the initial velocity of the reaction is linear over 4 min ($\Delta A_{340}/\Delta$time yields a straight line). If the kinetics are not linear over 4 min, dilute the enzyme twofold in 0.1 M potassium phosphate buffer, pH 7.5, and repeat the assay. Prepare a plot of A_{340} versus time and calculate the *absolute value* of the slope of the line.

5. Determine the activity of GOT (micromoles of oxaloacetate produced per minute per milliliter of enzyme) using the following equation:

$$\text{Activity} = \frac{\Delta A_{340}}{\Delta \text{time}} \cdot \frac{1}{6.22\ \text{mM}^{-1}\text{cm}^{-1}} \cdot (3\ \text{ml}) \cdot \frac{\text{fold dilution}}{0.1\ \text{ml of enzyme}}$$

where ($\Delta A_{340}/\Delta$time) is the absolute value of the slope of the line produced from a plot of A_{340} versus time, 6.22 $\text{mM}^{-1}\text{cm}^{-1}$ is the millimolar extinction coefficient for NADH at 340 nm, 3 ml is the total reaction volume, fold dilution is the extent to which the GOT sample was diluted *prior* to the assay, and 0.1 ml of enzyme is the volume of the dilute enzyme solution used in the assay.

GDH Activity Assay

The following procedure will be used for all GDH activity assays performed in the experiment. It is essential that GDH activity be assayed for and subtracted from the apparent GOT activity until you are sure that the GOT and GDH have been separated by one of the purification steps. Once this has occurred, you are no longer required to assay for GDH activity.

1. Add the following, in order, to a 13 × 100 mm glass test tube:

Reagent	Volume (ml)
0.1 M potassium phosphate (pH 7.5)—*no aspartate*	2.5
3 mM NADH solution	0.1
MDH (200 units/ml)	0.1
GOT dilution (in 0.1 M potassium phosphate buffer, pH 7.5)	0.1

2. Blank your spectrophotometer to read zero absorbance at 340 nm against water.
3. Add 0.2 ml of 0.1 M α-ketoglutarate solution to the reaction tube, mix, and place the solution in the spectrophotometer.
4. Record the A_{340} of the solution at 30-sec intervals for 4 min. Determine whether the initial velocity of the reaction is linear over 4 min ($\Delta A_{340}/\Delta$time yields a straight line). If the kinetics are not linear over 4 min, dilute the enzyme twofold in 0.1 M potassium phosphate buffer, pH 7.5, and repeat the assay. Prepare a plot of A_{340} versus time and calculate the *absolute value* of the slope of the line.
5. Determine the activity of GDH (micromoles of NADH oxidized per minute per milliliter of enzyme) using the following equation:

$$\text{Activity} = \frac{\Delta A_{340}}{\Delta \text{time}} \cdot \frac{1}{6.22\ \text{mM}^{-1}\text{cm}^{-1}} \cdot (3\ \text{ml}) \cdot \frac{\text{fold dilution}}{0.1\ \text{ml of enzyme}}$$

where ($\Delta A_{340}/\Delta$time) is the absolute value of the slope of the line produced from a plot of A_{340} versus time, 6.22 $\text{mM}^{-1}\text{cm}^{-1}$ is the millimolar extinction coefficient for NADH at 340 nm, 3 ml is the total reaction volume, fold dilution is the extent to which the GOT sample was diluted *prior* to the assay, and 0.1 ml of enzyme is the volume of the dilute enzyme solution used in the assay.

Folin–Ciocalteau Assay

The following procedure will be used for all protein assays performed in this experiment. Because of day-to-day variation in the assay, you must prepare a standard curve each day that you perform the Folin–Ciocalteau assay.

1. Set up the following reactions in 16 × 125 mm glass test tubes:

Tube	Volume of 1 mg/ml Bovine Serum Albumin (ml)	Volume of GOT Sample (ml)	Volume of Water (ml)
1	—	—	1.20
2	0.02	—	1.18
3	0.05	—	1.15
4	0.10	—	1.10
5	0.20	—	1.00
6	0.30	—	0.90
7	—	0.10	1.10
8	—	0.20	1.00

2. Prepare 100 ml of fresh alkaline copper reagent by mixing, in order:

   98 ml of 2% Na_2CO_3 in 0.1 M NaOH
   1 ml of 1% $CuSO_4 \cdot 5H_2O$
   1 ml of 2% sodium tartrate

   Add (and immediately mix in) 6 ml of this reagent to each tube and incubate at room temperature for 10 min.
3. Add (and immediately mix in) 0.3 ml of Folin–Ciocalteau reagent to each tube and incubate at room temperature for 30 min.
4. Blank your spectrophotometer to read zero absorbance at 500 nm against the solution in tube 1 (blank).
5. Read and record the A_{500} of the solutions in tubes 2 to 8.
6. Prepare a standard curve using the data from tubes 2 to 6 by plotting A_{500} versus milligrams of protein. Use the standard curve to determine the concentration of protein in the GOT sample (tubes 7 and 8). If the A_{500} of tubes 7 and 8 lies outside the range of the standard curve, dilute the GOT sample an additional twofold and repeat the assay.

Protocol

Day 1: Preparation of Pig Heart Homogenate and Heat Denaturation

We suggest that steps 1 to 3 be performed by the instructor prior to the beginning of the experiment.

1. In the cold room (4°C), trim the fresh pig hearts free of fat and auricles. Mince small portions of the heart tissue with a meat grinder. This minced heart tissue can be frozen at −20°C for up to 4 weeks.
2. Add 300 ml of ice cold 0.5 M potassium malate buffer (pH 6.0) containing 5 mM EDTA to each 200 ml portion of minced tissue.
3. Homogenize in a blender until no large tissue portions are present (about 1 min). Store the homogenate on ice until the beginning of the experiment.
4. Obtain 75 ml (record the exact volume in your notebook) of pig heart homogenate from the instructor and place in a 250-ml plastic centrifuge bottle. *Although you will only be assaying a small portion of the crude pig heart homogenate for GOT activity (see step 5 below), the total volume of this sample should be used in the calculations for the purification table at the end of the experiment.*
5. Remove 5 ml of the homogenate, place it in a clean 30-ml plastic centrifuge tube, and centrifuge at 8000 × *g* for 10 min. The supernatant is Fraction I (FR-I). Store it on ice. The pellet produced in this step may be discarded.
6. Prepare a water bath in a large beaker by heating over a Bunsen burner. Submerge the 250-ml centrifuge bottle (containing 70 ml of pig heart homogenate) into the water bath and bring the temperature of the homogenate to 60°C. *Place the clean thermometer directly into the homogenate (not the water bath) to monitor its temperature.*
7. Add 15 ml of 0.04 M α-ketoglutarate to the homogenate to convert the GOT to the heat-stable pyridoxal form.
8. Increase the temperature of the homogenate to 72°C and incubate it for exactly 20 min. *Do not exceed 75°C or incubate for longer than 20 min.* If the time of the incubation is too long or the temperature is too high, the GOT will denature.
9. Remove the homogenate from the water bath and chill on ice for 20 min.
10. Filter the heat-treated homogenate through cheesecloth into a clean 250-ml plastic centrifuge bottle. The solid material contains the denatured proteins and cell debris that precipitated out of solution at high temperature. The supernatant contains GOT, small molecules, and other heat-stable proteins.
11. Clarify the supernatant fraction further by centrifugation at 8000 × *g* for 10 min. The supernatant is Fraction II (FR-II). Record its exact volume and store it on ice. The pellet produced in this step (cell debris and denatured proteins) may be discarded.
12. Perform GOT and GDH assays on FR-I and FR-II using the following suggested dilutions in 0.1 M potassium phosphate buffer (pH 7.5):

    1:160 for FR-I
    1:80 for FR-II

    Calculate the GOT activity present in each fraction (micromoles of oxaloacetate produced per minute per milliliter of enzyme).

13. Place 0.5 ml of FR-I and FR-II in separate 1.5 ml microcentrifuge tubes labeled with your name. Return these to the instructor, who will store them at −20°C for SDS-PAGE analysis on the last day of the experiment.
14. Store the remainder of FR-I and FR-II on ice at 4°C for use on Day 2.

Day 2: Development of an Ammonium Sulfate Fractionation

1. Place 0.4 ml of FR-II into each of seven 1.5-ml microcentrifuge tubes. Place the tubes on ice.
2. *Slowly* (one drop at a time with gentle mixing in between) add the following volumes of saturated ammonium sulfate solution (SAS) to each tube:

Tube	Volume of SAS (ml)	% Ammonium Sulfate Saturation
1	—	0
2	0.40	50
3	0.49	55
4	0.60	60
5	0.74	65
6	0.93	70
7	1.20	75

Incubate all the tubes on ice for 20 min. *NOTE:* It is very important that the protein solution be brought up to the desired ammonium sulfate saturation *slowly*. If this step is not done correctly, it will be difficult to reproduce these results when the same procedure is carried out on the bulk preparation. Also, avoid frothing in the solution (a sign that some of the proteins are being denatured) by *gently* mixing the solution as the ammonium sulfate is added.

3. Centrifuge tubes 2 to 7 for 5 min at 4°C in a microcentrifuge. Remove and discard the supernatant from the precipitated protein pellet.
4. Resuspend the precipitated protein pellet from tubes 2 to 7 in 1.0 ml of 0.03 M sodium acetate buffer, pH 5.4. Add 0.6 ml of this same buffer to tube 1 (untreated) to bring the final volume of all of the samples to 1.0 ml.
5. Perform GOT assays (and GDH assays, if necessary) on all seven tubes using the following suggested dilutions in 0.1 M potassium phosphate buffer (pH 7.5):

Tube	Dilution
1	1:20
2	Undiluted
3	1:2
4	1:5
5	1:10
6	1:40
7	1:20

Calculate the GOT activity present in each fraction (micromoles of oxaloacetate produced per minute per milliliter of enzyme).

6. Perform Folin–Ciocalteau assays on FR-I, FR-II, and each of the solutions in tubes 1 to 7 (24 assays in all, including those for the standard curve) using the following suggested dilutions in water *(not phosphate buffer)*:

1:20 for FR-I
1:5 for FR-II
1:2 for tubes 1 to 7

Calculate the concentration of protein (in milligrams per milliliter) present in FR-I, FR-II, and each of the ammonium sulfate saturation trials.

7. Store FR-I and FR-II on ice at 4°C for use on Day 3.
8. Prepare a plot of percent GOT recovered versus percent ammonium sulfate saturation using the data from tubes 2 to 7 with the use of the following equation:

$$\% \text{ GOT recovered} = \frac{\text{total GOT activity in sample}}{\text{total GOT activity in Tube 1}} \times 100$$

where total GOT activity is the activity of GOT (micromoles of oxaloacetate produced per minute per milliliter of enzyme) times the total volume of the sample (1 ml).

9. On the same graph, prepare a plot of percent protein recovered versus percent ammonium sulfate saturation using the following equation.

$$\% \text{ Protein recovered} = \frac{\text{mass of protein in sample (mg)}}{\text{mass of protein in tube 1 (mg)}} \times 100$$

where mass of protein is the concentration (in milligrams per milliliter) of protein in the sample (determined from the standard curve) times the total volume of the sample (1 ml).

10. Determine the *highest* percent ammonium sulfate saturation that gave less than 20% GOT recovered. This will be a good percent ammonium sulfate saturation to use for the first cut on Day 3 (when the bulk of the GOT is still soluble). Next, determine the *lowest* percent ammonium sulfate saturation that gave more than 80% GOT recovered. This will be a good percent ammonium sulfate saturation to use for the second cut on Day 3 (when the bulk of the GOT will precipitate out of solution). *This information will be critical for the experiment performed on Day 3, so these calculations must be completed prior to the beginning of the next experiment.*

Day 3: Ammonium Sulfate Fractionation of FR-II (Bulk Preparation)

1. Based on your results from the ammonium sulfate precipitation trials performed on Day 2, determine what volume of SAS you will need to add to the remainder of FR-II to precipitate less than 20% of the total GOT in solution using the following formula:

$$\text{Volume of SAS} = \frac{[(\text{volume of sample at S1\%})(\text{S2\%} - \text{S1\%})]/100}{[(1 - (\text{S2\%} / 100)]}$$

   where S1% is the initial percent ammonium sulfate saturation (zero at this point for FR-II) and S2% is the desired final percent ammonium sulfate saturation.
2. *Slowly* (one drop at a time with gentle mixing in between) add the correct volume of saturated ammonium sulfate solution to the remainder of FR-II. Incubate on ice for 25 min.
3. Centrifuge the sample at 8000 × *g* for 20 min. Decant the supernatant into a *clean* 250-ml plastic centrifuge bottle.
4. Redissolve the precipitated protein pellet produced in step 3 in 3 ml of 0.03 M sodium acetate buffer (pH 5.0). This is ammonium sulfate fraction I (AS-I). Record the volume of this fraction and store AS-I on ice at 4°C for use on Day 4.
5. Determine what volume of SAS you will need to add to the remainder of the supernatant from the first cut to precipitate more than 80% of the total GOT in solution. Use the same formula shown in step 1, *but remember that S1% is now greater than zero and that the volume of FR-II has changed.*
6. Slowly (one drop at a time with gentle mixing in between) add the correct volume of SAS solution to the supernatant fraction from the first salt cut. Incubate on ice for 25 min.
7. Centrifuge the sample at 8000 × *g* for 20 min. Decant the supernatant into a clean, 250-ml plastic centrifuge bottle. This is AS-III. Store AS-III on ice at 4°C for use on Day 4 and record its volume.
8. Redissolve the precipitated protein pellet produced in step 7 in 3 ml of 0.03 M sodium acetate (pH 5.0). This is AS-II. Store this fraction on ice and record its *exact* volume.
9. Perform GOT assays (and GDH assays, if necessary) on AS-I, AS-II, and AS-III using the following suggested dilutions in 0.1 M potassium phosphate buffer (pH 7.5):

   1:100 for AS-I
   1:200 for AS-II
   1:10 for AS-III

   Calculate the GOT activity present in each fraction (micromoles of oxaloacetate produced per minute per milliliter of enzyme). Also, calculate the total GOT activity in each of these fractions (GOT activity times the total fraction volume). *Note that the sum of the total GOT activity in AS-I, AS-II, and AS-III should be roughly equal to the total GOT activity in FR-II.*
10. Obtain a 12-in. length of prepared dialysis tubing from the instructor (always wear gloves when handling the dialysis tubing and do not allow it to dry out). Tie two knots in one end of the tubing and fill the tubing with the entire AS-II fraction (~3 ml). Tie two more knots in the other end of the dialysis tubing and dialyze overnight against 6 liters of 0.03 M sodium acetate buffer (pH 5.0) to remove the residual ammonium sulfate from the sample. This is done to ensure that the ionic strength of the solution

(and the pH) will be low enough to allow GOT to bind the carboxymethylcellulose column. It is also necessary to remove the residual ammonium sulfate so that the NH_4^+ will not interfere with the Folin–Ciocalteau assay performed on AS-II.

Day 4: Folin–Ciocalteau Assay on Fractions and Preparation of Carboxymethylcellulose Column

1. Remove the dialysis bag containing AS-II and centrifuge at 8000 × *g* for 10 min. Remove and save the supernatant on ice. This is ammonium sulfate fraction II dialyzed (AS-IId). Remove 0.5 ml and place in a 1.5 ml microcentrifuge tube labeled with your name. Return this to the instructor, who will store it at −20°C for SDS-PAGE analysis at the end of the experiment.
2. Perform Folin–Ciocalteau assays on AS-I, AS-II, and AS-IId using the following suggested dilutions in water:

   1:20 for AS-I
   1:10 for AS-II
   1:10 for AS-IId

   Calculate the concentration of protein in (milligrams per milliliter) present in each fraction.
3. Obtain a 10-ml aliquot of carboxymethylcellulose (CMC) resin saturated with 0.03 M sodium acetate (pH 5.0). Close the outlet valve at the bottom of the column. Slowly pour the slurry into the chromatography column containing a small plug of glass wool at the bottom (consult the instructor). Allow the resin bed to settle with each small addition of the slurry. When the resin has settled to half of the column height (~5 ml), open the output valve and adjust the flow rate to ~1 ml/min. Equilibrate the column in 0.03 M sodium acetate (pH 5.0) by passing three column volumes (30 ml) of this buffer through the column. Leave about 10 ml of buffer above the resin bed and close the outlet valve at the bottom of the column.
4. Label the column with your name and store it at 4°C for use on Day 5.
5. Store the AS-IId fraction on ice at 4°C for use on Day 5.

Day 5: Carboxymethylcellulose Chromatography

1. Open the outlet valve on the CMC column and allow the level of the buffer to fall *even* with that of the resin bed (but not below it). During this time, adjust the flow rate to 1 ml/min.
2. Gently load *all but 0.1 ml* of AS-IId on the top of the bed and allow it to flow into the resin. Carefully layer (without disturbing the resin bed) the top of the resin bed with 20 ml of 0.03 M sodium acetate buffer (pH 5.0). Immediately begin collecting 2 ml fractions as they elute from the column. When this is complete, close the outlet valve at the bottom of the column.
3. Perform a quick GOT assay on every other fraction to ensure that the GOT adsorbed to the column. These rapid assays do not require any sample dilution. Simply add 0.1 ml of every other fraction to a GOT assay tube and determine whether you see any significant decrease in A_{340} after 2 min.
4. Elute GOT from the CMC column by passing 40 ml of 0.08 M sodium acetate buffer (pH 5.0) through the column. Immediately begin collecting 2-ml fractions as they elute.
5. Perform rapid GOT assays on every other fraction (see step 3) as well as absorbance readings at 280 nm for each fraction.
6. Prepare an elution profile by plotting relative GOT activity (ΔA_{340}/2 min) versus fraction number. Combine the three fractions that show the *highest* relative GOT activity.
7. Accurately assay the precolumn sample and pooled eluted samples for GOT activity using the following suggested dilutions in 0.1 M potassium phosphate buffer (pH 7.5):

   1:100 for pre-column fraction
   1:20 for pooled eluted fraction

   Calculate the GOT activity present in each fraction (micromoles of oxaloacetate produced per minute per milliliter of enzyme).
8. Perform a Folin–Ciocalteau assay on the pooled eluted fraction. No dilutions are needed at this point, since the total protein concentration is likely to be low. Calculate the concentration of protein (in milligrams per milliliter) present.

9. Label the pooled eluted fraction with your name and return it to the instructor, who will store it at −20°C for SDS-PAGE analysis on Day 6.
10. Prepare a purification table for GOT:

	Volume (ml)	[Protein] (mg/ml)	Total GOT Activity (μmol/min)	Specific Activity (μmol/min/mg)	% Yield	Fold Purification
Fraction I (crude homogenate)						
Fraction II (heat-treated)						
AS-II (predialysis)						
AS-IId (postdialysis)						
Pooled CMC fraction						

The total GOT activity of a fraction can be calculated by multiplying the GOT activity of the fraction (micromoles of oxaloacetate produced per minute per milliliter of enzyme) by the total fraction volume.

The specific activity of GOT of a fraction can be calculated by dividing the GOT activity of the fraction by the protein concentration (in milligrams per milliliter) for that fraction.

The percent yield for each fraction can be calculated by dividing the total GOT activity in a fraction by the total GOT activity in fraction I (crude homogenate) and multiplying the value by 100.

The fold purification for each fraction can be calculated by dividing the specific activity of GOT in a fraction by the specific activity of GOT in fraction I (crude homogenate).

11. What is the overall fold purification of GOT in your experiment?
12. What is the percent yield of GOT following this purification procedure?
13. Which step in the purification procedure afforded the greatest increase in the purity of GOT?
14. Which step in the purification procedure afforded the least increase in the purity of GOT?
15. Discuss the overall effectiveness of this GOT purification protocol in terms of fold purification and percent yield. Suggest possible modifications in the experiment that might improve the percent yield or fold purification. Describe specifically what the changes would be and how the new procedure would be performed (buffer composition, etc.).

Day 6: SDS-PAGE Analysis (See the Section II Introduction and Experiment 4)

1. Determine the concentration of protein in the FR-I, FR-II, AS-IId, and pooled CMC fractions from the Folin–Ciocalteau assays performed throughout the experiment. Remember that 0.5 ml of each of these fractions was saved for SDS-PAGE analysis.
2. For fractions that had a protein concentration greater than 1 mg/ml, determine how much each one would have to be diluted with water to bring the final concentration to 1 mg/ml. For example, if FR-I were at a concentration of 10 mg/ml, it could be diluted 10-fold with water to bring it to 1 mg/ml.
3. For fractions that had a protein concentration less than 1 mg/ml, determine what volume of that fraction would contain 10 μg of protein. For example, if AS-IId is at a concentration of 0.1 mg/ml, 100 μl of AS-IId contains 10 μg of protein. Add the appropriate volume of these fractions to a microcentrifuge tube and add

water to bring the final volume of each of these samples to 0.2 ml. Add 0.2 ml of 20% trichloroacetic acid to each of the samples, mix, and incubate on ice for 10 min. Centrifuge the samples at 4°C for 5 min and remove the supernatant (there should be a precipitated protein pellet at the bottom of the tube). Wash the protein pellet with 100 μl of acetone, centrifuge for 1 min at 4°C, and remove the supernatant. Allow the precipitated protein pellet to air dry and resuspend it in 10 μl of water.

4. Prepare the following samples for SDS-PAGE:

Tube	Volume of Sample at 1 mg/ml	Volume 4× SDS Sample Buffer
1	10 μl (FR-I)	5 μl
2	10 μl (FR-II)	5 μl
3	10 μl (AS-IId)	5 μl
4	10 μl (pooled CMC fraction)	5 μl
5	10 μl (protein molecular weight standard)	5 μl

The molecular weight protein standards (tube 5) contain 1 mg/ml each of phosphorylase (97,400 Da), bovine serum albumin (66,200 Da), ovalbumin (45,000 Da), carbonic anhydrase (31,000 Da), soybean trypsin inhibitor (21,500 Da), and lysozyme (14,400 Da) and will be provided by the instructor.

5. Incubate the samples for 5 min in a boiling water bath and centrifuge briefly to collect the sample at the bottom of the tube.
6. Load the samples on a 10-well, 12% SDS polyacrylamide gel with a 4% polyacrylamide stacking gel. Two groups can share a single SDS-PAGE gel. Apply 200 V (constant voltage), and allow the electrophoresis to continue until the bromophenol blue dye front reaches the bottom of the gel.
7. Remove the gel from the plates and place into a Coomassie Blue staining solution (50 ml per gel) for 1 hr. Pour off the staining solution and place in destaining solution for 1 hr, changing the destain solution every 15 min.
8. The proteins should appear as bright blue bands on the gel as the destain process progresses. At this point, the gel can be photographed or dried to serve as a permanent record.
9. Calculate the R_f values of the molecular weight protein standards using the following formula:

$$R_f = \frac{\text{distance traveled by protein band (cm)}}{\text{distance traveled by dye front (cm)}}$$

10. Prepare a standard curve by plotting the log of molecular weight versus R_f for each of the six protein molecular weight standards (see Fig. 4-8). Based on the R_f value of the protein band(s) in the lane containing the pooled CMC fraction, what is the molecular weight of the protein(s) present at the end of the purification? Can you identify GOT (GOT is a 90,000-Da homodimer composed of two identical subunits)? Are there any other proteins that copurified with GOT? What are their molecular weight(s)? Has the intensity of the protein band corresponding to GOT changed during the course of the purification? Explain what this means.

Exercises

1. What was being removed from FR-II during dialysis? Give two reasons why it was necessary to remove this component prior to continuing with the assay.
2. The pH of the buffer used to load FR-II on the CMC column was the same as that of the buffer used to elute GOT from the CMC column. How could two buffers of the same pH affect the ability of GOT to adsorb to the CMC resin?
3. Describe the difference between the activity and the specific activity of an enzyme. How is it possible that the total activity of an enzyme in a particular fraction could decrease while the specific activity of that same fraction increases?
4. Using the data shown below, answer questions a and b: The GOT assay was performed in a total volume of 3 ml. The value in parentheses indicates how much the fraction was diluted before using 0.1 ml in the assay. The protein assay was performed by diluting the samples as indicated in parentheses and determining an A_{280} value in a 1-cm pathlength cuvette. Assume

that the proteins in the solution have an extinction coefficient of 0.7 $(mg/ml)^{-1}$ cm^{-1}. The millimolar extinction coefficient for NADH is 6.22 mM^{-1} cm^{-1} at 340 nm.

Fraction	Volume (ml)	GOT assay (ΔA_{340}/min)	A_{280}
Crude cell lysate	100	0.57 (1:100)	0.464 (1:1000)
FR-II	80	0.40 (1:100)	0.186 (1:1000)
AS-IId	3	0.81 (1:1000)	0.440 (1:100)
CMC pooled	6	0.10 (1:1000)	0.200 (1:10)

a. Which single step in the purification protocol resulted in the greatest purification of GOT? Use specific numbers to justify your answer.
b. Calculate the overall fold purification and percent yield of GOT following this complete purification protocol.

5. Propose an experiment that would allow you to purify GOT with the use of a DEAE (diethylaminoethyl) cellulose column rather than with a CMC (carboxymethylcellulose) column. The pI of GOT is ~6.0. Include in your answer a description of the buffer system that you would use (buffering ions, ionic strength, pH) for loading GOT on the column and for eluting GOT from the column.

REFERENCES

Banks, B. E. C., Doonan, S., Lawrence, A. J., and Vernon, C. A. (1969). The Molecular Weight and Other Properties of Aspartate Aminotransferase from Pig Heart Muscle. *Eur J Biochem* **5**:528.

Christen, P., and Metzler, D. E. (1985). *Transaminases.* New York: Wiley-Interscience.

Jansonius, J. N., and Vincent, M. G. (1987). Structural Basis for Catalysis by Aspartate Aminotransferase. (Biological Macromolecules and Assemblies, Vol. 3.) New York: Wiley.

Lehninger, A. L., Nelson, D. L., and Cox, M. M. (1993). Amino Acid Oxidation and the Production of Urea. In: *Principles of Biochemistry,* 2nd ed. New York: Worth.

Nielsen, T. B., and Reynolds, J. A. (1978). Measurement of Molecular Weights by Gel Electrophoresis. *Methods Enzymol,* **48**:3.

Robyt, J. F., and White, B. J. (1987). Chromatographic Techniques. In: *Biochemical Techniques: Theory and Practice.* Prospect Heights, IL: Waveland Press.

Scopes, R. K. (1994). *Protein Purification: Principles and Practice.* New York: Springer-Verlag.

Sizer, I. W., and Jenkins, W. T. (1962). Glutamic Aspartic Transaminase from Pig Ventricles: Preparation and Assay of Enzymes. *Methods Enzymol* **5**:677.

Stryer, L. (1995). Amino Acid Degradation and the Urea Cycle. In: *Biochemistry,* 4th ed. New York: Freeman.

Voet, D., and Voet, J. G. (1990). Amino Acid Metabolism. In: *Biochemistry.* New York: Wiley.

EXPERIMENT 9

Kinetic and Regulatory Properties of Aspartate Transcarbamylase

Theory

The efficient and balanced functioning of the metabolism of an organism requires that various metabolic pathways be regulated. A biosynthetic pathway, for example, should be most active when the organism has a high demand for the product of that pathway and inactive when the organism has an outside source for the product. In biosynthetic pathways of bacteria, such regulation is accomplished by two kinds of mechanisms: control of the rate of synthesis of enzymes (generally referred to as repression) and control of the catalytic efficiency of enzymes by inhibition or activation. The latter mode of control will concern us in this experiment. It is commonly observed that the first enzyme in an unbranched pathway leading to the biosynthesis of a compound is subject to feedback inhibition by the end product of the pathway. Thus, the end product exerts feedback control over the role of its own synthesis.

In this experiment we will examine some of the properties of the aspartate transcarbamylase of *Escherichia coli*, which is typical of many enzymes subject to feedback inhibition and which has been studied extensively. Aspartate transcarbamylase (ATCase) catalyzes the first reaction unique to the biosynthesis of pyrimidine nucleotides. ATCase is subject to specific inhibition by quite low concentrations of one of its end products, cytidine 5′-triphosphate (CTP). This relationship and two other regulatory interactions important to the control of pyrimidine biosynthesis are summarized in Figure 9-1.

The study of the kinetic properties of the ATCase reaction and of its inhibition by CTP, some of which will be repeated in this experiment, reveals several unusual characteristics that cannot be explained by the simple Michaelis–Menten kinetic theory, presented in the Section II introduction. The inhibition of ATCase by CTP can be overcome by increased concentrations of the substrate aspartate, which suggests that these molecules compete for the active site of the enzyme. Yet the chemical structures of CTP and aspartate are very different and would not be expected to bind to the same site. Furthermore, inhibition by CTP is very specific; UTP does not inhibit. Another nucleotide, ATP, is a stimulator rather than an inhibitor of ATCase. Treatment of ATCase with mercurials results in an enzyme that is fully active (actually more active than the native enzyme) but which is insensitive to inhibition by CTP. These observations led to the proposal that aspartate and CTP bind at entirely separate sites—aspartate at the active or catalytic site and CTP at a separate *allosteric* site. The interactions between ligands at the active site and the allosteric site are negative; that is, binding of CTP lessens the ability of aspartate to bind. Hence, the effect of CTP can be overcome by high concentrations of aspartate. Clearly, if aspartate does not bind to ATCase, no product (carbamyl aspartate) can be formed, so CTP acts as an inhibitor. The allosteric nature of the CTP binding site on ATCase was proven by the demonstration that the active site and the CTP binding site actually reside on two different polypeptide chains (Fig. 9-2).

ATCase and many (but not all) enzymes subject to metabolic regulation have a second interesting property: their reaction velocity is not a hyperbolic function of substrate (aspartate) concentration, as

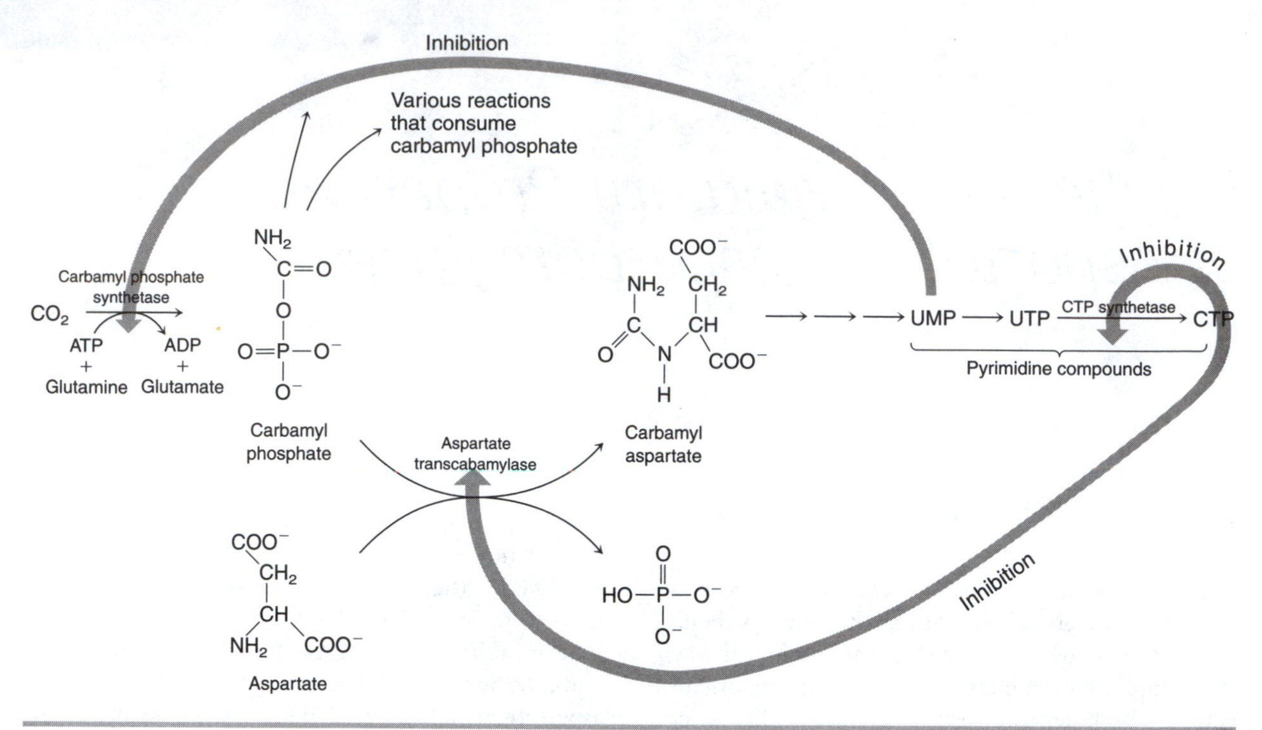

Figure 9-1 Sites of feedback inhibition in carbamyl phosphate metabolism of *E. coli*. Note that aspartate trascarbamylase is the first enzyme on the unique pathway to pyrimidine compounds.

would be predicted from simple steady-state kinetics (see Fig. II-8). Rather, the aspartate saturation curve is sigmoid (S-shaped). You can rationalize the results by suggesting that aspartate binds to more than one site on the enzyme and that once aspartate is bound to one of these sites, it binds more readily to a subsequent site. This is called *cooperative binding*. Such binding is similar to the interactions between aspartate and CTP binding, except that in this case interactions between aspartate sites are positive rather than negative. Simple enzyme kinetics generally assume single ligand sites or the absence of interactions between multiple sites. However, you must consider both positive and negative interactions between substrate and allosteric sites in order to understand many regulatory enzymes. Biochemists are now trying to learn the precise physical nature of such site–site interactions.

Some of the kinetic properties of ATCase described above will be examined in this experiment. The assay to be used is a fixed-time, colorimetric procedure. Carbamyl aspartate accumulated in the first step of the procedure is assayed in a second step (Fig. 9-3).

The source of the enzyme to be used in these studies is a mutant strain of *E. coli*, strain EK1104, in which the gene encoding ATCase *(pyrB)* has been deleted and has a defect in another enzyme of pyrimidine biosynthesis, orotidylic decarboxylase (encoded by the *pyrF* gene) (see Nowlan and Kantrowitz, 1985). The *pyrB* gene is supplied to the organism in many copies per cell on the pEK2 plasmid. However, because of the "leaky" defect in the orotidylate decarboxylase gene, the strain makes pyrimidine nucleotides very slowly. Thus, the pool of UTP, which is the repressing metabolite for the *pyrBI* that encodes ATCase, is very low, and the biosynthesis of ATCase is markedly derepressed. These observations underscore the fact that control of ATCase activity occurs in *E. coli* cells both by repression of enzyme synthesis and by inhibition of enzyme activity.

Supplies and Reagents

E. coli strain EK1104 bearing plasmid pEK2

Materials for growth medium and culturing bacteria (see Appendix)

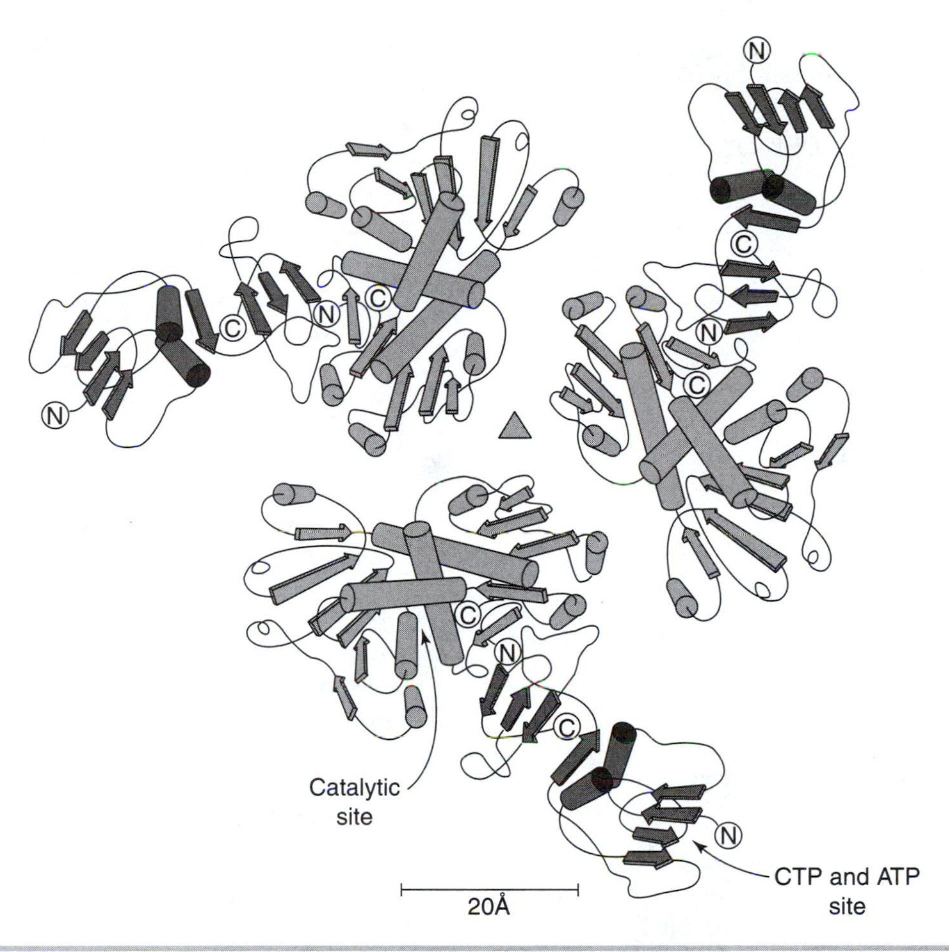

Figure 9-2 Three-dimensional structure of *E. coli* aspartate transcarbamylase. Half of the native c_6r_6 molecule (see Fig. II-4) is shown. The catalytic (c) submit, which binds the substrates aspartate and carbamyl phosphate is shown in light shading. The regulatory (r) subunit, which binds the allosteric effectors CTP and ATP, is shown in dark shading. From Kantrowitz, E.R., et al. (1980). *E. coli* Aspartate Transcarbamylase. Part II: Structure and Allosteric Interactions. *Trends Biochem Sci* **5**:150; and Stryer, L. (1995). *Biochemistry,* 4th ed. New York: Freeman, Figure 10-5, p. 240. Reprinted by permission.

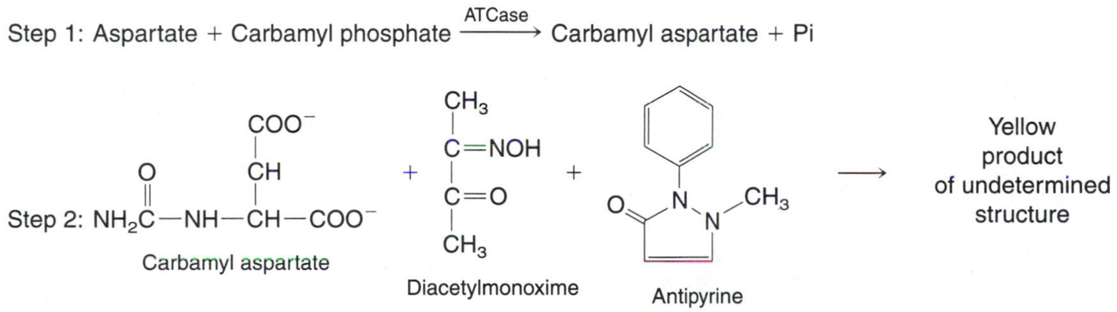

Step 1: Aspartate + Carbamyl phosphate $\xrightarrow{\text{ATCase}}$ Carbamyl aspartate + Pi

Step 2: $NH_2C(=O)-NH-CH(CH_2COO^-)-COO^-$ (Carbamyl aspartate) + $CH_3-C(=NOH)-C(=O)-CH_3$ (Diacetylmonoxime) + Antipyrine ⟶ Yellow product of undetermined structure

Figure 9-3 Colorimetric assay of aspartate transcarbamylase.

Sonicator
0.08 M sodium phosphate buffer, pH 7.0
0.125 M Na-L-aspartate
0.036 M carbamyl phosphate, freshly prepared (keep cold)
1.0 mM carbamyl aspartate standard solution
Antipyrine-H_2SO_4 reagent
Diacetylmonoxime (butanedione monoxime) reagent (refrigerate in foil covered bottle)
2% Perchloric acid
40 mM ATP (keep cold)
10 mM CTP (keep cold)
5 mM *p*-Mercuribenzoate
30 and 45°C water baths
Centrifuge
0.1 M Tris-hydrochloride buffer, pH 9.2

Protocol

Growth of *E. coli* EK1104/pEK2 (1-liter procedure)

(This portion of the experiment is usually performed in advance by an assistant.)

1. Inoculate 10 ml of sterile growth medium containing 36 μg/ml uracil and 100 μg/ml ampicillin (see Appendix) with strain EK1104 carrying the pEK2 plasmid, and grow for 16 to 18 hr (typically overnight) in a shaking incubator at 200 rpm and 37°C. The culture should be turbid with bacterial cell growth at this time (absorbance at 566 nm of about 1 or above).
2. Prepare 1 liter of sterile growth medium containing 12 μg/ml uracil and 100 μg/ml ampicillin as described in the Appendix. Cool to 37°C.
3. Using sterile technique, add 10 ml of the inoculum from step 1 to the 1 liter of culture medium. Cultivate the bacteria with shaking (300 rpm) at 37°C for 20 to 24 hr, whereupon the absorbance at 566 nm should be about 0.9.

Harvesting of Cells and Preparation of the Cell Extract

(This portion of the experiment is usually performed in advance by an assistant.)

1. Weigh two large centrifuge bottles and record their weight.
2. Pour the medium from the cell culture in step 3 of the growth procedure into the large bottles and centrifuge the cell-containing medium at 5000 × *g* for 10 min. Following centrifugation, decant the supernatant and discard.
3. The cell yield of the pellet, after allowing excess liquid to drain off, is determined by weighing the pellet-containing bottles and subtracting the weight of the empty bottles obtained in step 1. Record the cell yield, which should be about 5 to 6 g of wet cell paste per liter of culture.
4. Resuspend the cell pellets in ice-cold 0.1 M Tris chloride buffer, pH 9.2, using 4 ml of buffer per gram of cells. Transfer the resuspended cells to a small metal sonication container and pack in an ice-water bath. (The cells may be stored frozen at −20 or −70°C at this point for later use, if desired.)
5. To disrupt the cell walls, sonicate the cold breakage buffer-cell suspension for 30-sec intervals for a total of 20 to 30 min (30 sec sonication/ 30 sec off). *Do not operate the sonicator without immersing the tip.*
6. Pour the sonicated mixture into the medium-sized centrifuge tubes and centrifuge the mixture for 30 min at 10,000 × *g* and discard the pellet. The cell extract may be stored frozen at −20°C in small aliquots (0.5 ml) for subsequent use by the class.

Before it is used by the class, the extract should be tested for overproduction of ATCase by assay; see "General Procedure for the Assay of Aspartate Transcarbamylase Activity," step 9 "Range Finding" below. The extract can also be examined by SDS-PAGE for overproduction of the ATCase catalytic and regulatory subunits (34 kDa and 19 kDa, respectively). To do this, use the procedure described in Experiment 4 for 15% gels.

General Procedure for the Assay of Aspartate Transcarbamylase Activity

The standard assay procedure for ATCase is as follows. In the subsequent procedures, modifications and additions will be made as indicated. Read the entire experiment before planning your procedure. To obtain good results in the following experiment, it is *essential* to pipette enzyme and substrate solutions accurately and reproducibly and to measure absorbance values carefully. When you are comparing two sets of experimental conditions, you should assay the sets at the same time with the same enzyme dilution.

1. Add the following to each of a series of assay tubes: 0.5 ml 0.08 M sodium phosphate buffer, pH 7.0, varying amounts of 0.125 M aspartate (0.2 ml when range finding), 0.10 ml of an appropriate enzyme dilution (kept cold), and water to a final total volume of 0.90 ml.
2. Prepare a blank tube from which enzyme is omitted.
3. Incubate the tubes at 30°C for about 5 min to bring the contents to 30°C. At timed intervals (30 sec is convenient) add 100 μl of 0.036 M carbamyl phosphate to initiate the reaction and mix.
4. Allow the reaction to proceed at 30°C for 30 min. Then terminate each reaction at the same timed intervals as used in initiating the reactions by adding 1 ml 2% perchloric acid.
5. Set the tubes on ice until all are ready for the color test. (If necessary, remove any precipitate by centrifuging the cold tubes for 5 min in the clinical centrifuge, followed by decanting.) Develop the color in these tubes at the same time as the series of carbamyl aspartate standards described in the next step.
6. Prepare a series of carbamyl aspartate standards containing 0 (blank), 25, 50, 100, 150, 200, 250, and 375 μl of 1.0 mM carbamyl aspartate and water to a final volume of 2 ml.
7. Immediately before use, prepare the color reagent by mixing well two parts of antipyrine-H_2SO_4 reagent with one part of diacetyl-monoxime reagent.

> **WARNING**
>
> ***Warning: Sulfuric acid can burn clothing and skin. Wear protective gloves.***

Prepare the amount of color reagent you need—3 ml to be added to each tube containing 2 ml of carbamyl aspartate standard or enzyme assay mixture after addition of perchloric acid. Add 3 ml to each assay and standard tube and mix thoroughly. Cap the tubes with marbles or parafilm, cover them with aluminum foil, and store in a dark place at room temperature until the next lab period (15–48 hr is satisfactory).

8. After the incubation in the dark, place the tubes in a 60°C water bath exposed to room light, but not direct sunlight, for 70 to 75 min. Cool the tubes in cold water, and read the absorbance of each at 466 nm versus the blank containing all reagents except carbamyl aspartate. Read the tubes promptly, because the color will slowly fade.
9. *Range finding.* (If time is limited, this portion of the experiment should be performed in advance by an assistant.) In order to conduct the subsequent experiments under conditions that will yield valid assays, it is necessary to determine the amount of enzyme that will give a response proportional to the enzyme activity and within the range of the carbamyl aspartate color test. This corresponds to an amount of enzyme that forms about 0.05 to 0.10 μmol of carbamyl aspartate when 100 μl of a suitable dilution is assayed under standard assay conditions. Because the activity of your extract will depend on a number of variables, you must assay 25-, 50-, 75-, and 100-μl samples of a series of dilutions of the extract in 0.08 M sodium phosphate buffer, pH 7.0, to determine the appropriate enzyme level for use in the subsequent experiments. Suggested dilutions are: 1 to 5,000, 1 to 8,000, and 1 to 10,000.

Dependence of ATCase Activity on Aspartate Concentration in the Absence and Presence of Allosteric Effectors

1. Prepare a series of tubes that contain 0.5 ml 0.08 M sodium phosphate buffer in each and 10, 20, 40, 60, 100, 150, and 200 μl of 0.125 M aspartate. To each add 0.1 ml of an enzyme dilution shown to yield about 0.1 μmol of carbamyl aspartate under standard assay conditions above and shown to respond linearly to enzyme concentration (i.e., 50 μl gives half as much product as 100 μl). Add water sufficient to bring each to a volume of 0.9 ml. This is Series A. Use Table 9-1 to assist in preparation of the tubes.
2. Prepare two other series of tubes identical to the first series except that one series (Series B) also contains 50 μl 10 mM CTP, and the other series (Series C) contains 50 μl 40 mM ATP. Tables 9-2 and 9-3 describe these series.

Table 9-1 Series A: Activity of ATCase as a Function of Aspartate Concentration (All Volumes in Microliters)

Component	Tube Number							
	0	1	2	3	4	5	6	7
0.125 M Asparate	—	10	20	40	60	100	150	200
Water	300	290	280	260	240	200	150	100
0.08 M Na-Pi buffer, pH 7	500	500	500	500	500	500	500	500
Enzyme	100	100	100	100	100	100	100	100

Table 9-2 Series B: Activity of ATCase as a Function of Aspartate Concentration, with CTP Added (All Volumes in μl)

Component	Tube Number							
	0	1	2	3	4	5	6	7
0.125 M Asparate	100	10	20	40	60	100	150	200
Water	150	230	240	210	190	150	100	50
0.08 M Na-Pi buffer, pH 7	500	500	500	500	500	500	500	500
Enzyme	—	100	100	100	100	100	100	100
10 mM CTP	50	50	50	50	50	50	50	50

Table 9-3 Series C: Activity of ATCase as a Function of Asparate Concentration, with ATP Added (All Volumes in μl)

Component	Tube Number							
	0	1	2	3	4	5	6	7
0.125 M Aspartate	100	10	20	40	60	100	150	200
Water	150	240	230	210	190	150	100	50
0.08 M Na-Pi buffer, pH 7	500	500	500	500	500	500	500	500
Enzyme	—	100	100	100	100	100	100	100
40 mM ATP	50	50	50	50	50	50	50	50

3. Conduct the assays for Series A, B, and C, initiating each reaction with 100 μl 0.036 M carbamyl phosphate, terminating after 30 min of reaction with 1 ml of 2% perchloric acid, and developing the color by adding 3.0 ml of freshly prepared color reagent as described above. Include a series of carbamyl aspartate standards also as described above.
4. Using your carbamyl aspartate standard curve, convert your A_{466} readings to micromoles of carbamyl aspartate formed per hour per milliliter of diluted enzyme. Plot the rate of carbamyl aspartate formation as a function of aspartate concentration. Do the results obey Michaelis–Menten kinetics? What is the effect of adding CTP? Of adding ATP?
5. Prepare a Lineweaver–Burk plot of the data. Can you determine an accurate K_M value for aspartate? Prepare Hill plots of the data and calculate Hill coefficients (n_H) for each experimental series. (The Hill plot is a plot of log (v/V_{max}) versus log aspartate concentration,

where v is the reaction velocity at any aspartate concentration and V_{max} is the maximal velocity at saturating aspartate concentration (obtained by extrapolation on the Lineweaver–Burk plot). The slope of this plot is the Hill coefficient, n_H, and is a measure of the "sigmoidicity" of a velocity versus substrate concentration curve.)

Desensitization of the Allosteric Site of ATCase with Mercuribenzoate

1. Prepare two series of tubes similar to those described above. Series D should contain no nucleotides, and Series E should contain 50 μl of 10 mM CTP in each tube. Each tube should also contain 10 μl of 5 mM *p*-mercuribenzoate. You do not need to prepare a series containing ATP. The amount of enzyme added to each tube should be the same as that used in the previous series. Follow Tables 9-4 and 9-5.
2. Incubate the tubes at 30°C for 15 min. Then initiate the reactions with 100 μl of 0.036 M carbamyl phosphate. After 30 min of reaction, terminate each reaction (in the order they were initiated) with 1 ml 2% perchloric acid, and develop the color in the assay tubes (along with a set of carbamyl asparate standards) by adding 3.0 ml of freshly prepared color reagent as described above.
3. Analyze the data as in steps 4 and 5 above. What is the effect of 50 μM mercuribenzoate on the activity of ATCase and on its sensitivity to allosteric inhibition by CTP?

Exercises

1. Describe the current view of the molecular architecture of *E. coli* aspartate transcarbamylase. What techniques were used in arriving at this view?

Table 9-4 Series D: Activity of ATCase as a Function of Aspartate Concentration, after Treatment with Mercuribenzoate (All Volumes in μl)

Component	Tube Number							
	0	1	2	3	4	5	6	7
0.125 M Asparate	100	10	20	40	60	100	150	200
Water	290	280	270	250	230	190	140	90
0.08 M Na-Pi buffer, pH 7	500	500	500	500	500	500	500	500
Enzyme	—	100	100	100	100	100	100	100
5 mM Mercuribenzoate	10	10	10	10	10	10	10	10

Table 9-5 Series E: Activity of ATCase as a Function of Asparate Concentration, with CTP Added after Treatment with Mercuribenzoate (All Volumes in μl)

Component	Tube Number							
	0	1	2	3	4	5	6	7
0.125 M Aspartate	100	10	20	40	60	100	150	200
Water	240	230	220	200	180	140	90	40
0.08 M Na-Pi buffer, pH 7	500	500	500	500	500	500	500	500
Enzyme	—	100	100	100	100	100	100	100
10 mM CTP	50	50	50	50	50	50	50	50
5 mM Mercuribenzoate	10	10	10	10	10	10	10	10

2. In this experiment a sigmoid dependence of reaction velocity on substrate concentration was interpreted as evidence for cooperativity of binding of the substrate to the enzyme. What kinds of experiments would be necessary to confirm this proposal? Describe other possible situations, especially in a crude enzyme preparation, that would result in such a sigmoid curve; that is, what kinds of artifacts might lead you to conclude falsely that a sigmoid curve results from cooperativity in substrate binding?
3. How is aspartate transcarbamylase regulated in other bacteria? In fungi and yeasts? In mammalian systems?
4. List three other examples of enzymes that are made up of nonidentical subunits. List three enzymes that contain specific allosteric sites for regulatory molecules but that are made of one type of subunit only.
5. Is it possible to observe allosteric interactions in an enzyme made up of only one subunit? Is it possible to observe cooperativity in substrate binding in an enzyme made up of only one subunit? If so, under what circumstances; if not, why not?

REFERENCES

Monod, J., Changeux, J.-P., and Jacob, F. (1963). Allosteric Proteins and Cellular Control Systems. *J Mol Biol* **6**:306.

Nowlan, S. F., and Kantrowitz, E. R. (1985). Superproduction and Rapid Purification of *Escherichia coli* Aspartate Transcarbamylase and Its Catalytic Subunit under Extreme Derepression of the Pyrimidine Pathway. *J Biol Chem* **260**:14712.

Prescott, L. M., and Jones, M. E. (1969). Modified Methods for the Determination of Carbamyl Aspartate. *Anal Biochem* **32**:408.

Stadtman, E. R. (1966). Allosteric Regulation of Enzymic Activity. *Adv Enzymol* **28**:41.

EXPERIMENT 10

Affinity Purification of Glutathione-S-Transferase

Theory

As discussed in the introduction to Experiment 2, affinity chromatography is one of the most rapidly growing, new, and effective techniques in modern protein purification. A single protein or family of proteins that bind a particular ligand can often be purified in a single step, provided that the ligand of interest can be coupled to some type of solid support or matrix. As with other types of chromatography, affinity chromatography is carried out in a three-step process: a binding step that allows the protein of interest to interact with the immobilized ligand, a wash step to remove unbound proteins, and an elution step to recover the purified protein.

In this experiment, you will purify the C-terminal domain of the enzyme glutathione-*S*-transferase (GST) from an *E. coli* cell extract by exploiting the affinity that the enzyme has for its natural substrate, glutathione. Although the chromatography experiment used in the purification will not be performed in the column format (see Experiment 2), the "batch wash" format used here could easily be adapted for use in the column format.

Glutathione is a tripeptide that is involved in a number of different aspects of metabolism (Fig. 10-1). Found in nearly all cell types, the molecule can exist in either the oxidized or the reduced form. In the reduced form, it acts as a general reducing agent, protecting the cell from harmful peroxides, radicals produced from exposure to ionizing radiation, and maintaining the general sulfhydryl group status of other proteins in the cell. Some reductase enzymes utilize the sulfhydryl group on glutathione as an electron donor. Glutathione-*S*-transferase is an enzyme that catalyzes condensation reactions between glutathione and a number of other molecules. This reaction is often a necessary step in the detoxification of foreign organic compounds (xenobiotics) and toxic compounds produced by the cytochrome P-450 oxidase enzymes. Condensation of such compounds with glutathione converts them into harmless, water-soluble glutathione conjugates that can be filtered out of the blood in the kidneys.

The *E. coli* strain used in this experiment contains the pGEX-2T plasmid (Fig. 10-2). This plasmid contains the *bla* gene, which allows the cell to produce β-lactamase and grow in media containing ampicillin (a β-lactam antibiotic). It also contains the coding sequence for the C-terminal domain of GST from *Schistosoma japonicum*. This GST gene is under the control of the *tac* promoter, a modified form of the *lac* promoter that produces a high level of transcription of genes under its control when not bound by LacI (the *lac* repressor). Cells harboring this plasmid will not express GST until the inducer (isopropylthio-β,D-galactopyranoside, IPTG) is added, causing the *lac* repressor (LacI) to lose its affinity for the promoter sequence.

Following induction with IPTG and an incubation period to allow for the expression of GST, the cells are harvested and lysed by sonication in phosphate buffer containing 150 mM NaCl. Triton X-100 (a non-ionic or nondenaturing detergent) is added to help solubilize the membranes and release proteins that may be peripherally associated with them. The clarified cell extract is then incubated with the glutathione Sepharose 4B resin to allow GST to bind the immobilized ligand. The Triton

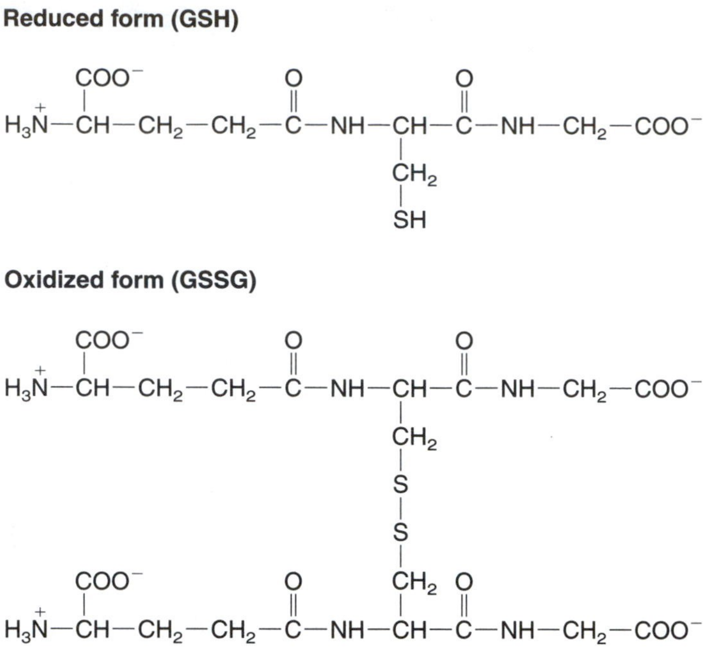

Reduced form (GSH)

$$H_3\overset{+}{N}-\underset{\displaystyle}{\overset{COO^-}{\overset{|}{CH}}}-CH_2-CH_2-\overset{O}{\overset{\|}{C}}-NH-\underset{\underset{\underset{\displaystyle SH}{|}}{\underset{\displaystyle CH_2}{|}}}{CH}-\overset{O}{\overset{\|}{C}}-NH-CH_2-COO^-$$

Oxidized form (GSSG)

$$H_3\overset{+}{N}-\overset{COO^-}{\overset{|}{CH}}-CH_2-CH_2-\overset{O}{\overset{\|}{C}}-NH-\underset{\underset{\displaystyle CH_2}{|}}{CH}-\overset{O}{\overset{\|}{C}}-NH-CH_2-COO^-$$
$$|$$
$$S$$
$$|$$
$$S$$
$$|$$
$$H_3\overset{+}{N}-\overset{COO^-}{\overset{|}{CH}}-CH_2-CH_2-\overset{O}{\overset{\|}{C}}-NH-\overset{CH_2}{\overset{|}{CH}}-\overset{O}{\overset{\|}{C}}-NH-CH_2-COO^-$$

Figure 10-1 The structure and forms of glutathione (γ-glutamyl cysteinyl glycine).

X-100 added earlier may help prevent nonspecific hydrophobic interactions between other proteins in the extract and the solid support matrix (or the GST bound to it), which could decrease the effectiveness of the purification. In the same fashion, the NaCl present in the binding buffer may prevent the formation of nonspecific electrostatic interactions between other proteins in the extract and the immobilized glutathione (remember that glutathione will carry a negative charge at pH 7.3 and may display some general cation exchange effects with other proteins in the extract). The glutathione resin is then washed several times in the binding buffer to remove loosely associated proteins.

To recover the purified GST from the resin, a buffer containing a high concentration of the ligand (free ligand) is introduced. This free ligand will compete with the immobilized ligand for the GST, eventually drawing the protein from the resin into the buffer. Remember that the enzyme–substrate interaction is a dynamic one. The enzyme does not bind the substrate irreversibly. Rather, the enzyme is in a rapid equilibrium between the free form and the ligand bound form, as determined by the dissociation constant (Equation 10-1).

(10-1)

$$E + S \underset{k_{-1}}{\overset{k_1}{\rightleftharpoons}} ES$$

The dissociation constant (K_s) = k_{-1}/k_1, while the equilibrium constant (K_{eq}) = [ES]/[E][S]. k_1 is equal to the rate of formation of the ES complex, while k_{-1} is equal to the rate of dissociation of the ES complex.

If a large concentration of *free* (non-immobilized) ligand is introduced, the enzyme will establish a new equilibrium with the immobilized ligand, releasing the protein from the resin into the buffer. The free glutathione in solution with GST following elution from the resin can be removed either by dialysis or by gel filtration chromatography. As stated in Experiment 2 (affinity chromatography), it may also be possible to elute GST from the solid support by altering the pH and/or the ionic strength of the elution buffer. Keep in mind, however, that the most *specific* method for eluting a protein of interest from an affinity-based chromatographic support is with the application of *excess free ligand.* If other proteins besides GST are bound to the glutathione resin after the numerous wash steps

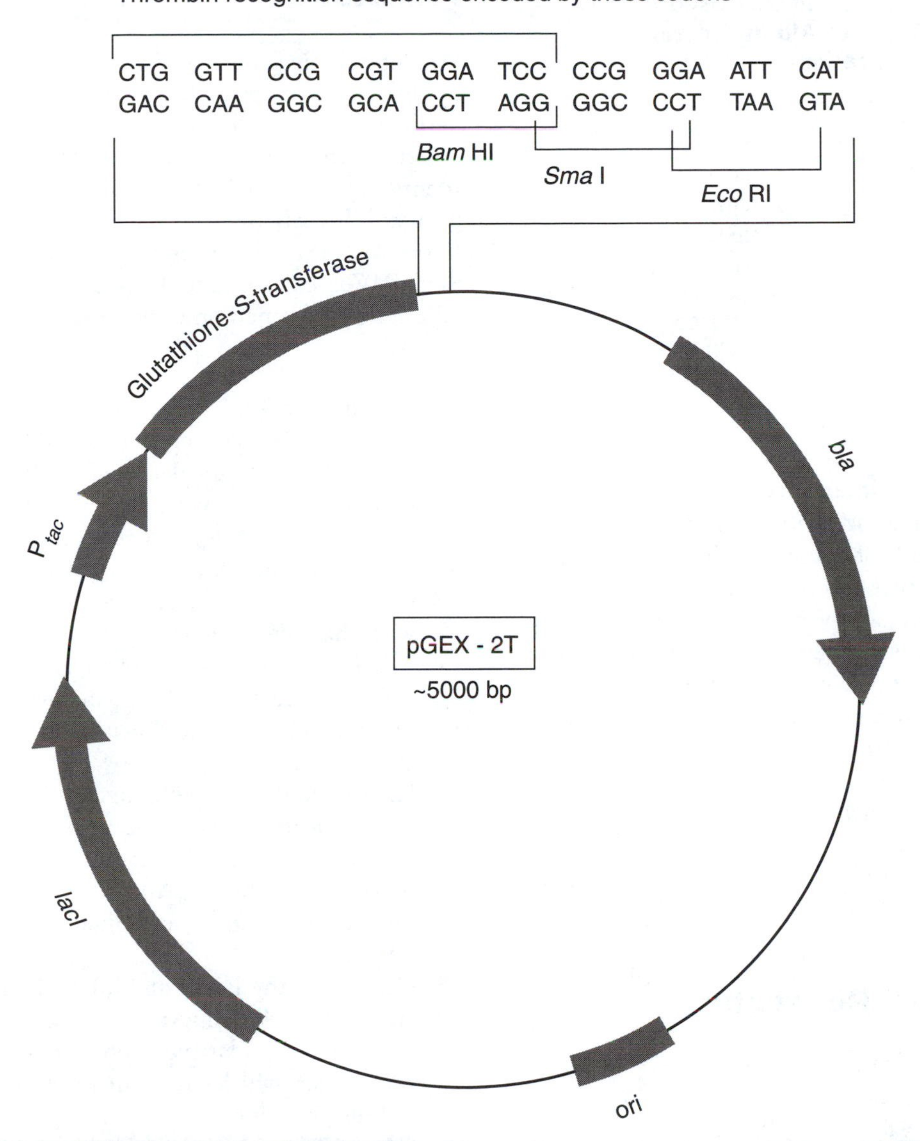

Figure 10-2 The pGEX-2T plasmid. Multiple-cloning site allows genes of interest to be inserted in frame with the GST coding sequence to express the gene as a GST fusion protein that is easily purified with affinity chromatography. The GST tag can be removed after purification by treatment with thrombin.

through some sort of *nonspecific* electrostatic or hydrophobic interactions, the addition of free glutathione will only cause the elution of the GST.

Notice that the pGEX-2T plasmid contains a multiple cloning site immediately 3′ to the GST coding sequence. If a gene of interest is cloned into this vector in the same reading frame as the GST gene, a fusion protein will be expressed as a hybrid including the GST domain and the protein encoded by the gene of interest. This protein "tagged" with the GST domain may then be purified using the same procedure described in this experiment. In

Table 10–1 Some Commonly Used Affinity "Tags" to Aid in Protein Purification

"Tag"	Affinity Ligand
Glutathione-*S*-transferase	Glutathione
Maltose binding protein	Amylose
Polyhistidine	Ni^{2+}
"Flag" epitope	"Flag" antibody (antibody recognizes the peptide sequence DYKDDDDK)

theory, a protein of interest could be "tagged" with any protein domain or peptide that has affinity for a particular ligand for use in affinity chromatography. Some examples of commercially available fusion protein expression vectors are shown in Table 10-1. In many cases, these vectors also contain a protease recognition sequence between the affinity tag coding sequence and the multiple-cloning site. This cleavage site (thrombin, factor Xa) will allow you to remove the affinity tag domain from the purified protein once it has been isolated. A graphical description outlining the technique of affinity chromatography used in this experiment is presented in Figure 10-3.

Supplies and Reagents

Cell extract from induced *E. coli* strain (TG1 or others; see Appendix) carrying pGEX-2T
Isopropylthio-β,D-galactoside (IPTG)
Triton X-100
YT broth (1% wt/vol bactotryptone, 0.5% (wt/vol) yeast extract, 0.5% (wt/vol) NaCl)
Glutathione sepharose 4B (Pharmacia Biotech, Inc.)
PBS buffer (150 mM NaCl, 16 mM Na_2HPO_4, 4 mM NaH_2PO_4, pH 7.3)
50 mM Tris, 10 mM reduced glutathione (pH 8.0)
4× SDS sample buffer (0.25 M Tris, pH 6.8, 8% SDS, 20% wt/wt glycerol, 10% β-mercaptoethanol, 0.1 % bromophenol blue)
Purified glutathione-*S*-transferase (prepared by instructor)
1.5-ml plastic microcentrifuge tubes
P-20, P-200, P-1000 Pipetmen with disposable tips

Protocol

We suggest that steps 1 through 6 be performed by the instructor prior to the beginning of the experiment. We perform this experiment in conjunction with Experiment 18 ("Western Blot to Identify an Antigen"). If this option is chosen, the protocol described above can be followed exactly. If the Western blot is not to be performed, the SDS-PAGE experiment described in Experiment 18 can be performed on a single set of samples (see step 16).

1. Prepare a 50-ml overnight culture of *E. coli* (strain TG1) containing pGEX-2T in YT broth with 100 μg/ml ampicillin. Incubate with shaking at 37°C overnight (~20 hr).
2. The next morning, inoculate 2 liters of fresh YT broth containing 100 μg/ml ampicillin with the 50 ml overnight culture. Incubate at 37°C with shaking for 1 hr.
3. Add IPTG to a final concentration of 0.5 mM and incubate at 37°C with shaking for 4 hr.
4. Harvest cells by centrifugation at 5000 × *g* for 10 min.
5. Resuspend the cell pellet in 20 ml of PBS buffer and sonicate to lyse the cells (four separate 30-sec bursts, with 30-sec incubations on ice in between each burst). Add Triton X-100 to 1% final concentration and incubate on ice for 30 min.
6. Centrifuge the lysate at 13,000 × *g* for 30 min and collect the supernatant (containing GST and other cytoplasmic proteins). This clarified cell extract will be used in the GST purification experiment.
7. Obtain 0.1 ml of a 50% slurry of Glutathione sepharose 4B resin in PBS buffer. Centrifuge the slurry for 30 sec in a 1.5 ml plastic microcentrifuge tube and discard the supernatant.
8. Obtain 0.5 ml of the *E. coli* cell extract from the instructor. *Save 15 μl of the crude cell extract for SDS-PAGE analysis.* Add the remainder of the cell extract to the resin. Flick the bottom of the tube gently to resuspend the resin in the extract.
9. Incubate for 20 min at room temperature. Make sure that the resin remains fully suspended in the cell extract by inverting the tube frequently during the incubation.

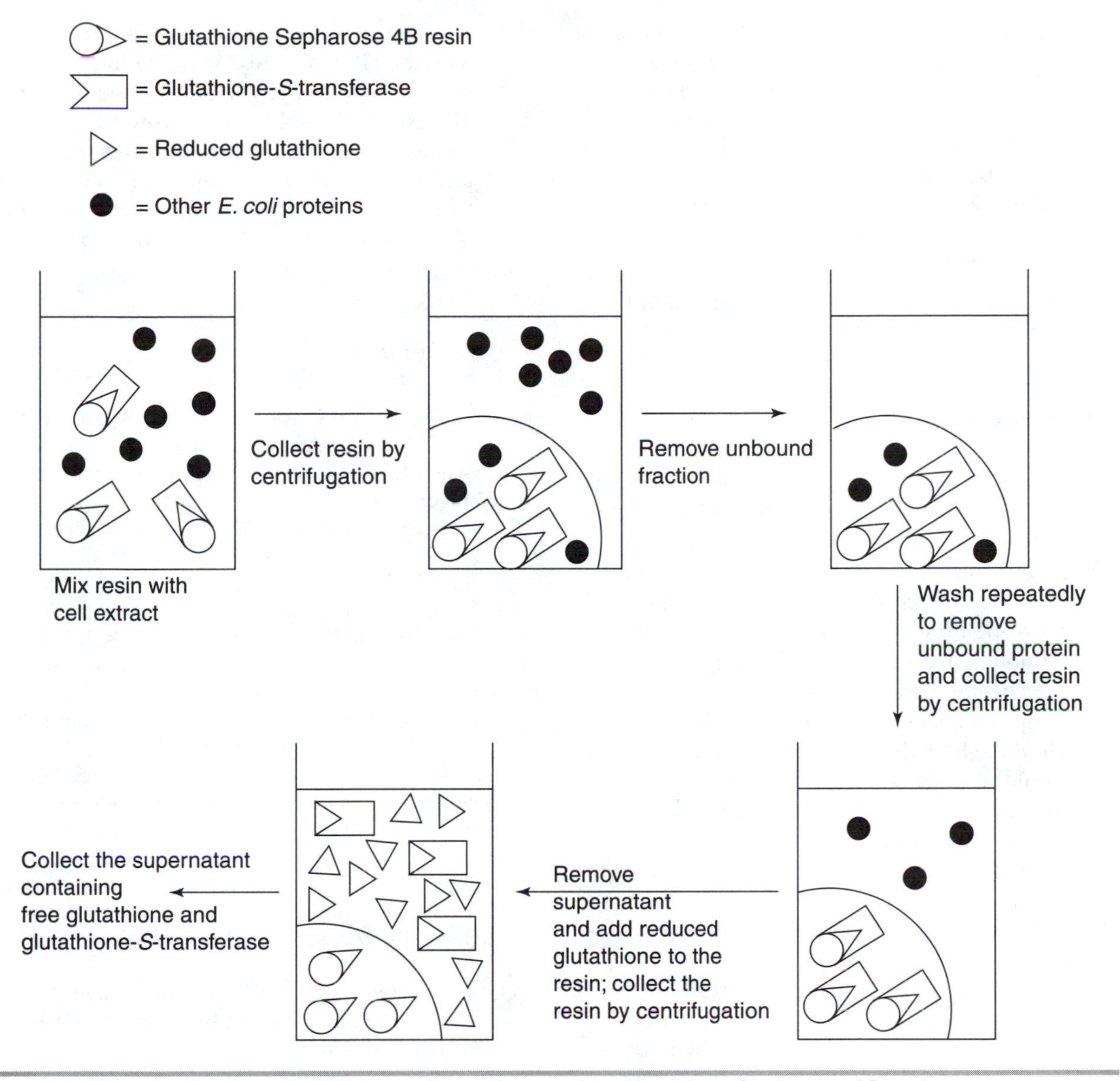

Figure 10-3 Purification of glutathione-*S*-transferase with glutathione Sepharose 4B resin.

10. Centrifuge the tube for 30 sec and collect the supernatant. Label the supernatant, "unbound fraction." Store this fraction on ice.
11. Add 0.5 ml of PBS buffer to the resin and invert the tube several times to completely resuspend the resin in the buffer. Centrifuge for 30 sec to collect the resin; discard the supernatant.
12. Repeat step 11 four more times.
13. Add 0.1 ml of 50 mM Tris/10 mM reduced glutathione (pH 8.0) to the resin and flick the bottom of the tube gently several times until the resin is fully resuspended.
14. Incubate the tube for 10 min at room temperature, inverting the tube frequently to keep the resin suspended in the buffer.
15. Centrifuge the tube for 30 sec to collect the resin. Save the supernatant and label it, "eluted fraction." Store this fraction on ice.
16. Prepare duplicate samples for SDS-PAGE and Western blot analysis (prepare in four tubes marked A to D):

Sample A: 4 μl of crude cell lysate + 11 μl of water + 5 μl of 4× SDS sample buffer
Sample B: 4 μl of "unbound fraction" + 11 μl of water + 5 μl of 4× SDS sample buffer
Sample C: 15 μl of "eluted fraction" + 5 μl of 4× SDS sample buffer
Sample D: 5 μl of purified GST (prepared by the instructor) + 10 μl of water + 5 μl of 4× SDS sample buffer

Proceed with SDS-PAGE (Experiment 18, Day 1).

Exercises

1. What characteristics of glutathione-*S*-transferase make it an attractive affinity "tag" for use in bacterial expression systems?
2. Is there another method that could be used to elute GST from the glutathione sepharose 4B resin besides the addition of reduced glutathione? Are there any advantages to eluting the enzyme with the addition of free substrate as opposed to these other methods?
3. Describe how the protocol used in this exercise would change if the experiment were performed in the column format rather than in the "batch wash" format.
4. How does the substrate affinity chromatography used in this experiment differ from the other types of affinity chromatography discussed in Experiment 2?
5. Why is it that fusion protein expression vectors contain thrombin and factor Xa recognition sequences rather than trypsin or chymotrypsin recognition sequences? Why would it sometimes be desirable to remove the affinity tag from the fusion protein?
6. You have isolated a protein of interest tagged with the GST domain. At this point, the fusion protein is bound to the glutathione Sepharose 4B resin and the unbound proteins have been washed away. Describe, in detail, two different techniques that you could use to isolate the protein without the affinity tag using only thrombin, glutathione Sepharose 4B, and reduced glutathione.

REFERENCES

Garrett, R. H., and Grisham, C. M. (1995). Recombinant DNA: Cloning and Creation of Chimeric Genes. In: *Biochemistry.* Orlando, FL: Saunders.

Lehninger, A. L., Nelson, D. L., and Cox, M. M. (1993). Biosynthesis of Amino Acids, Nucleotides, and Related Molecules. In: *Principles of Biochemistry,* 2nd ed. New York: Worth.

Mathews, C. K., and van Holde, K. E. (1990). Metabolism of Sulfur-Containing Amino Acids. In: *Biochemistry.* Redwood City, CA: Benjamin/Cummings.

Scopes, R. K. (1994). Purification of Special Types of Proteins. In: *Protein Purification: Principles and Practice,* 3rd ed. New York: Springer-Verlag.

Smith, D. B., and Johnson, K. S. (1988). Single-Step Purification of Polypeptides Expressed in *Escherichia coli* as Fusion Proteins with Glutathione-*S*-Transferase. *Gene* **67:**31.

Stryer, L. (1995). Pentose Phosphate Pathway and Gluconeogenesis: Biosynthesis of Amino Acids and Heme. In: *Biochemistry,* 4th ed. New York: Freeman.

Uhlen, M., and Moks, T. (1990). Gene Fusions for Purpose of Expression: An Introduction. *Methods Enzymol* **185:**129.

SECTION III

Biomolecules and Biological Systems

SECTION III

Biomolecules and Biological Systems

Introduction

Carbohydrates

Carbohydrates are the most abundant of all organic compounds in the biosphere. Many members of the carbohydrate class have the empirical formula $C_x(H_2O)_y$, and are literally "hydrates of carbon." The fundamental units of the carbohydrate class, the monosaccharides, are polyhydroxy aldehydes or ketones and certain of their derivatives. As with other classes of biologically important compounds, much of the function of the carbohydrates derives from the ability of the monosaccharides to combine, with loss of water, to form polymers: oligosaccharides and polysaccharides. The chemistry of carbohydrates is, at its core, the chemistry of carbonyl and hydroxyl functional groups, but these functional groups, when found in the same compound, sometimes exhibit atypical properties. The discussion that follows is designed to review the aspects of carbohydrate chemistry that are especially important for isolation, analysis, and structure determination of biologically important carbohydrates.

Structure and Stereochemistry of Monosaccharides

Drawing Monosaccharide Structures: The Fischer Projection. From the stereochemical point of view, monosaccharides are considered to be derived from the two trioses, D- and L-glyceraldehyde (see Fig. III-1). These two parent compounds differ only in the steric arrangement of the atoms about the central asymmetric carbon; they are mirror images (enantiomorphs, optical antipodes) of one another. The two compounds differ physically only in the orientation of their crystals and the direction in which their solutions rotate polarized light. These may seem like trivial differences, but when such optical isomers interact with other asymmetric molecules (especially enzymes) they may acquire very different chemical reactivities. This principle of optical selectivity is central to the biochemistry of carbohydrates (and other classes of naturally occurring compounds as well). The manner of drawing the two isomers as shown in the top part of Figure III-I is called the "Fischer projection." In this projection, the most oxidized (aldehyde) end of the molecule is drawn up and is considered to be behind the plane of the paper. The $-CH_2OH$ end is drawn down and is also behind the plane. Because the groups are

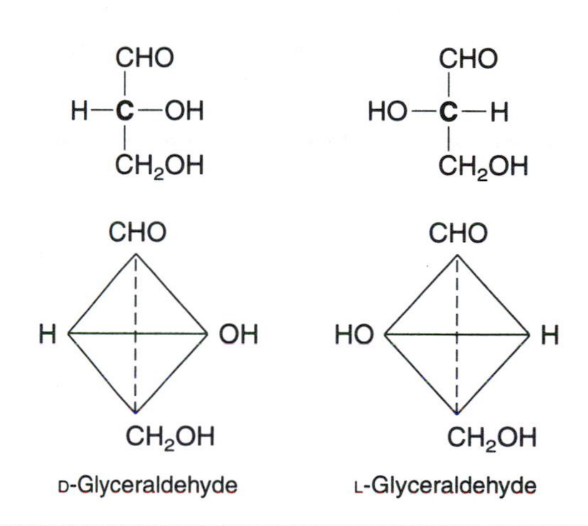

Figure III-1 The two forms of glyceraldehyde. The asymmetric carbon is shown in boldface.

arranged about the central carbon atom (which is in the plane of the paper) as at the corners of a tetrahedron, the H and OH group must project in front of the plane of the paper. The drawings at the bottom should make the arrangement of atoms clear. Although the arrangements of atoms in the D and L forms of glyceraldehyde were originally arbitrarily assigned by Emil Fischer, we now know from x-ray crystallographic analysis that the actual arrangement of atoms in these structures is the same as was originally assigned. You must always remember the three-dimensional meaning of the two-dimensional Fischer projection, especially when considering structures drawn in other arrangements.

If you have access to ball-and-stick models used for instruction in organic chemistry, use them to construct models corresponding to the drawings in Figure III-1 and determine the consequences of drawing flat projections of the structure in arrangements other than the Fischer convention. In this way, you will see why it is essential to use a generally accepted convention for drawing asymmetric structures in two dimensions.

When the carbohydrate molecule contains additional CHOH groups, the number of possible isomers increases (it will be equal to 2^n, where n is the number of asymmetric carbon atoms). By convention, all such sugars are considered to be members of two families, the D family and the L family. Members of the D family have the same arrangement at the next to last carbon atom (the penultimate carbon, the first asymmetric center from the bottom) as does D-glyceraldehyde. Likewise, the L sugars have the configuration of L-glyceraldehyde about the penultimate carbon. As before, the most oxidized end of the sugar (–CHO for aldo sugars and CH_2OH for keto sugars) is always drawn at the top.

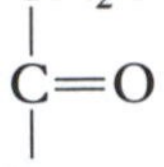

$$CH_2OH$$
$$|$$
$$C{=}O$$
$$|$$

The name of the sugar is determined by the arrangement of *all* the asymmetric centers; the family (D or L) is determined by the penultimate carbon only. (Note that D and L have no other meaning; they do *not* describe the direction of rotation of polarized light of solutions of the sugar.) D and L isomers of the same sugar (*enantiomorphs* or *enantiomers*) are mirror images. Isomers that are not mirror images, but differ in configuration about one or more carbon atom are called *diastereoisomers.* Two disastereoisomers differing in configuration about only one carbon atom are called *epimers.* Students should test their understanding of these concepts by stating how various pairs of the sugars in Figure III-2 are related.

A further point should be made about the three-dimensional meaning of the Fischer projection of sugars with more than three carbon atoms. The carbon atoms are arranged in a line (oxidized end up, as before) such that all H and OH groups project to the front and the C–C bonds curve back. You view the groups as they would project onto a plane curved toward you. The carbon chain is curved continuously in a single direction rather than staggered. This concept is illustrated in Figure III-3.

Monosaccharides Form Rings as a Consequence of Internal Hemiacetal or Hemiketal Formation. There is yet another level of isomerism of sugars, because aldehydes (and ketones) form freely reversible adducts with alcohols (hemiacetals and hemiketals) in aqueous solutions. In the presence of

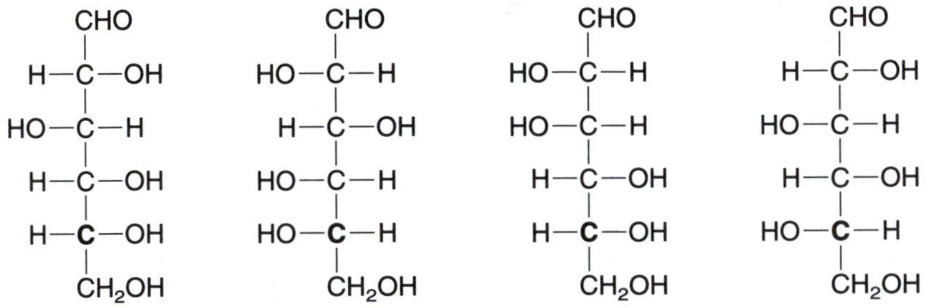

D-Glucose	L-Glucose	D-Mannose	L-Idose
CHO	CHO	CHO	CHO
H—C—OH	HO—C—H	HO—C—H	H—C—OH
HO—C—H	H—C—OH	HO—C—H	HO—C—H
H—C—OH	HO—C—H	H—C—OH	H—C—OH
H—**C**—OH	HO—**C**—H	H—**C**—OH	HO—**C**—H
CH_2OH	CH_2OH	CH_2OH	CH_2OH

Figure III-2 Which pairs of sugars are enantiomers? Diastereoisomers? Epimers? The stereochemistry of the penultimate carbon (boldface) determines whether the sugar belongs to the D family or the L family.

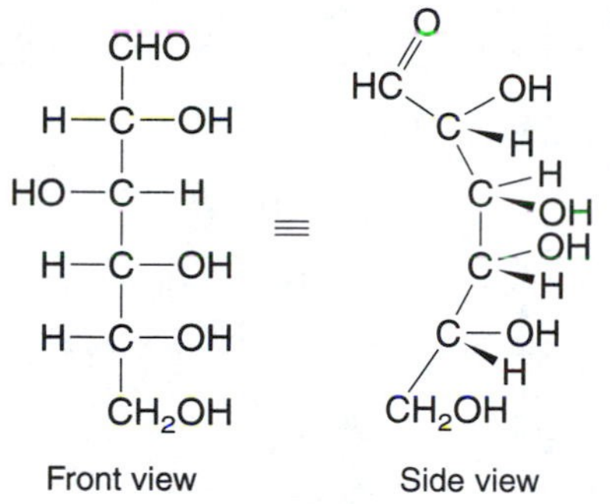

CHO
H—C—OH
HO—C—H
H—C—OH
H—C—OH
CH_2OH
Front view

≡

Side view

Figure III-3 Physical meaning of the Fischer projection of a hexose.

an acid catalyst, the hemiacetal can react with another alcohol molecule to form an acetal (or ketal) (see Fig. III-4*a*). Acetal formation is favored under dehydrating conditions (strong acid, anhydrous conditions); acetal hydrolysis is favored in dilute aqueous acid. Most sugars exist primarily as *internal hemiacetals*, in which intramolecular addition of a hydroxyl occurs to form a five-member (furanose) or six-member (pyranose) ring (see Fig. III-4*b*).

Note that formation of the hemiacetal ring form of a sugar creates a new optically active center. Sugars that differ in symmetry at this center are called *anomers*, and the aldehyde (or ketone) carbon involved is often called the "anomeric carbon." Anomers are named as derivatives of the parent aldehyde (or keto) sugar and are designated α- and β-. The best definition of these two designations relates them to the Fischer projection of the parent sugar. An α-form is the one in which the –OH on the anomeric carbon is on the same side of the carbon backbone in the Fischer projection as the –OH on the penultimate carbon. The examples in Figure III-5 illustrate this principle.

Drawing Monosaccharide Structures: The Haworth Projection. In an attempt to represent sugar structures in a manner more closely resembling the actual geometric arrangement of the atoms, W. N. Haworth proposed the system of perspective drawings generally used by biochemists. The Haworth formula is a natural consequence of turning the Fischer projection on its side and drawing the hemiacetal ring oxygen with more realistic bond dimensions as part of a ring. This idea should be made clear by the following step-by-step procedure for converting Fischer projections illustrated here with α-D-glucopyranose and β-D-fructofuranose.

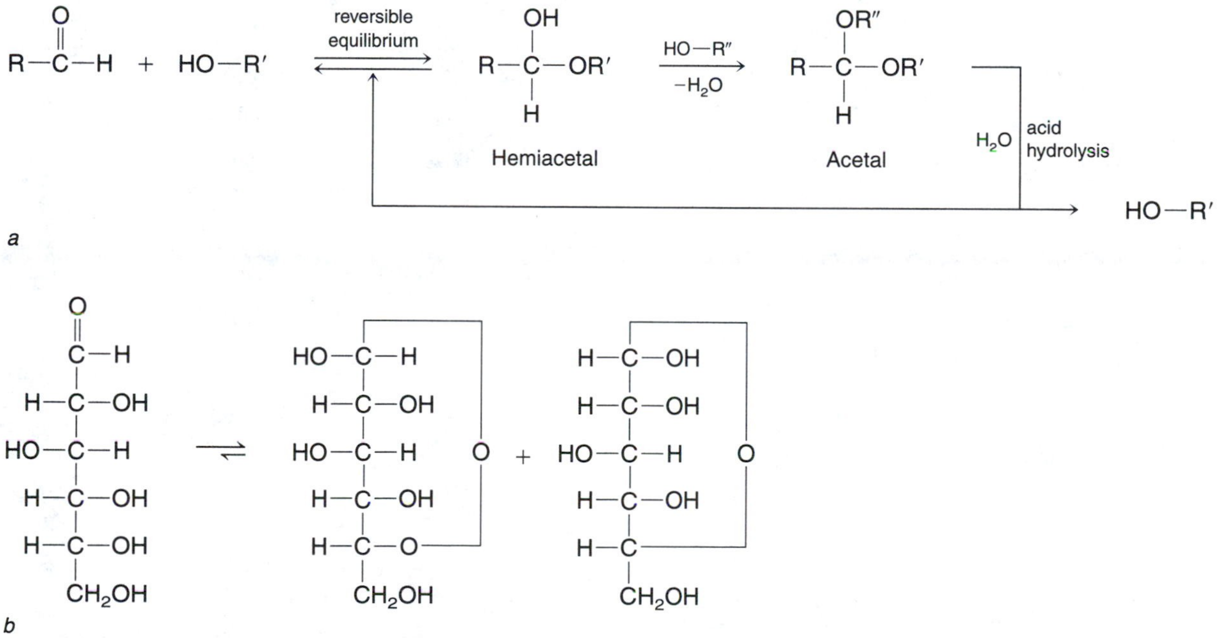

R—C(=O)—H + HO—R′ ⇌ (reversible equilibrium) R—C(OH)(H)—OR′ (Hemiacetal) —(HO—R″, $-H_2O$)→ R—C(OR″)(H)—OR′ (Acetal) —(H_2O, acid hydrolysis)→ HO—R″

a

b

Figure III-4 (*a*) Hemiacetal and acetal formation and acetyl hydrolysis. (*b*) Internal hemiacetal formation with the OH on carbon 5 of hexoses yields two six-membered rings.

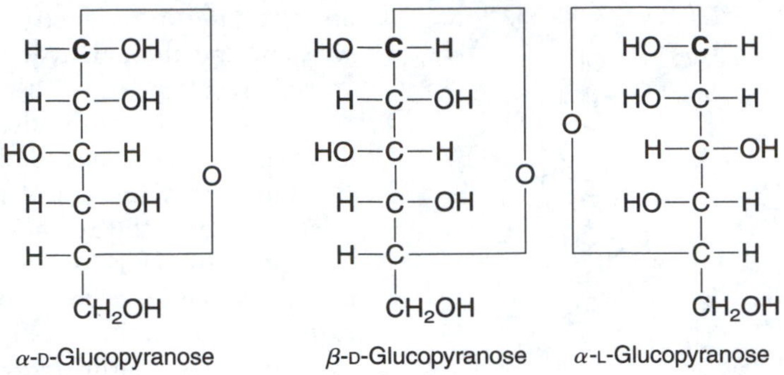

α-D-Glucopyranose β-D-Glucopyranose α-L-Glucopyranose

Figure III-5 α- and β-anomers of D- and L-glucose in Fischer projection. The anomeric carbons are shown in boldface.

1. Place the carbonyl carbon on the right and draw a potential five- or six-member ring clockwise (Fig. III-6). This rotation places hydroxyls that lie to the *right* on the Fischer projection *down* in Haworth representation; hydroxyls to the *left* will project *up*. (This is like placing the Fischer projection on its side.)
2. Draw the hemiacetal or hemiketal form of the sugar by making ring closure with the appropriate hydroxyl added to the carbonyl to form a furanose (five-member) or pyranose (six-member) ring with the ring O in the upper right-hand corner of the pyranose or at the top of a furanose (Fig. III-7). Ring closure will require a *rotation* of the carbon that bears the ring O so that remaining carbons of the "tail" are drawn opposite to the original direction (step 1) of the OH group involved in ring closure.
3. Establish the orientation of the asymmetric carbon created by hemiacetal or hemiketal formation. By definition, an α-anomeric OH is on the same side of the Fischer projection as the

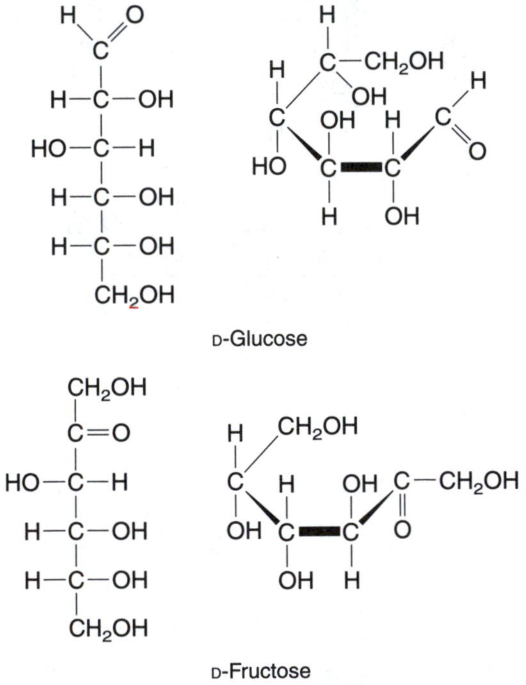

D-Glucose

D-Fructose

Figure III-6 Converting Fischer to Haworth projection—step 1.

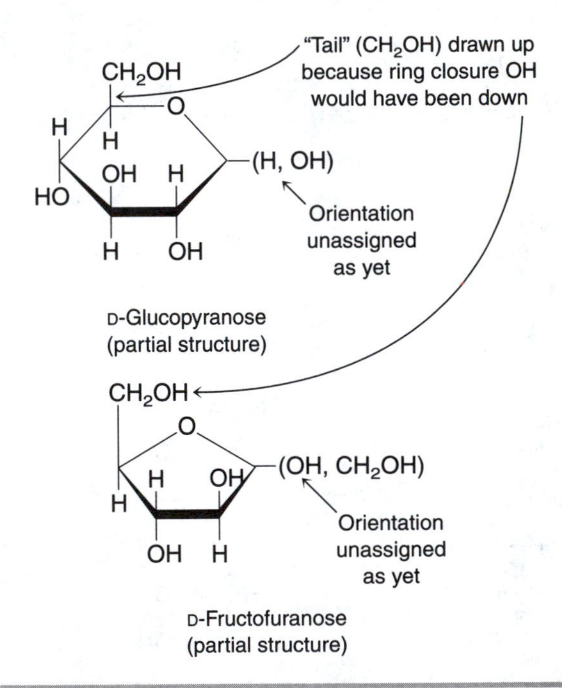

Figure III-7 Converting Fischer to Haworth projection—step 2.

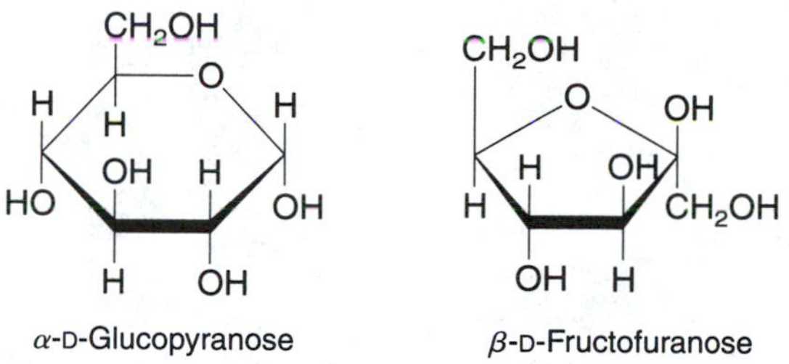

α-D-Glucopyranose

β-D-Fructofuranose

Figure III-8 Representation of α and β anomers of D sugars in Haworth projection.

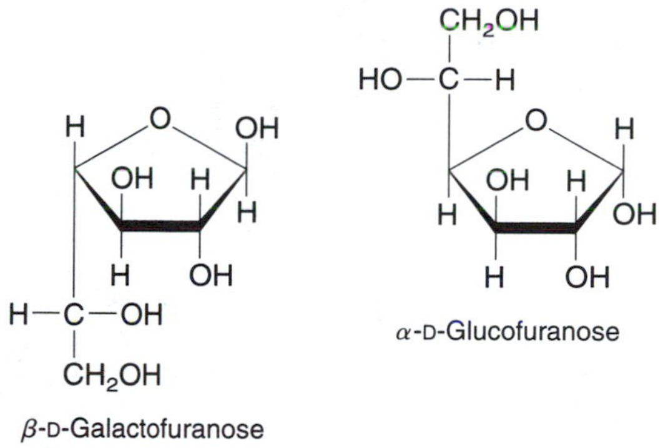

β-D-Galactofuranose

α-D-Glucofuranose

Figure III-9 Converting Fischer to Haworth projection—step 4.

hydroxyl on the penultimate carbon and will be *down* in D sugars (Fig. III-8). It follows that a β-anomeric hydroxyl will project *up*. It also follows that the opposite will be true for L sugars.

4. In cases in which ring formation involves an OH on a carbon other than the penultimate carbon (e.g., aldohexofuranoses), the remaining carbon atoms of the "tail" (e.g., CHOH—CH_2OH) are drawn in the Fischer projection convention (Fig. III-9).

Note that this last step becomes particularly confusing in cases in which the "tail" projects up, whereas the Fischer convention features the carbons of the "tail" down (e.g., α-D-glucofuranose). Correct application of step 4 to these cases requires that you rotate the plane of the paper to place the tail down and indicate the Fischer convention OH assignments (Fig. III-10). However, there still is unfortunate confusion over the correct way to draw such asymmetric centers. Consequently, it is a good practice to indicate in words the intended steroisomeric configuration on Haworth projections of sugars with such centers. It will be valuable to draw the Haworth formulae of α-D-galactofuranose and α-L-galactofuranose to test your ability to apply these rules.

Any sugar in solution should be considered a mixture of four ring forms and the straight-chain (aldehydo- or keto-) forms (see Fig. III-11). The mixture of forms existing in solution differs from sugar to sugar. Glucose, for example, has been shown to exist primarily as a mixture of α- and β-pyranose forms with only traces of the other forms present. Pure samples of single forms can often be obtained by crystallization under specific conditions.

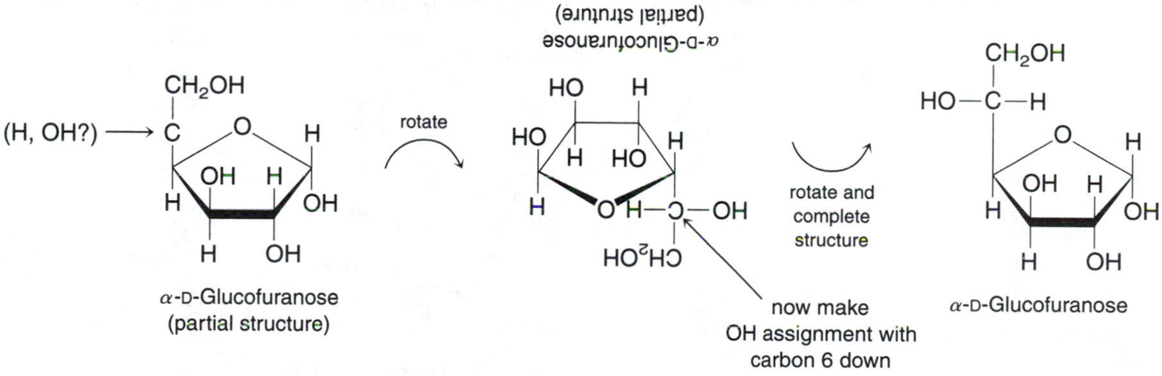

α-D-Glucofuranose (partial structure)

rotate

α-D-Glucofuranose (partial structure)

rotate and complete structure

now make OH assignment with carbon 6 down

α-D-Glucofuranose

Figure III-10 Haworth projections of some sugar hemiacetals.

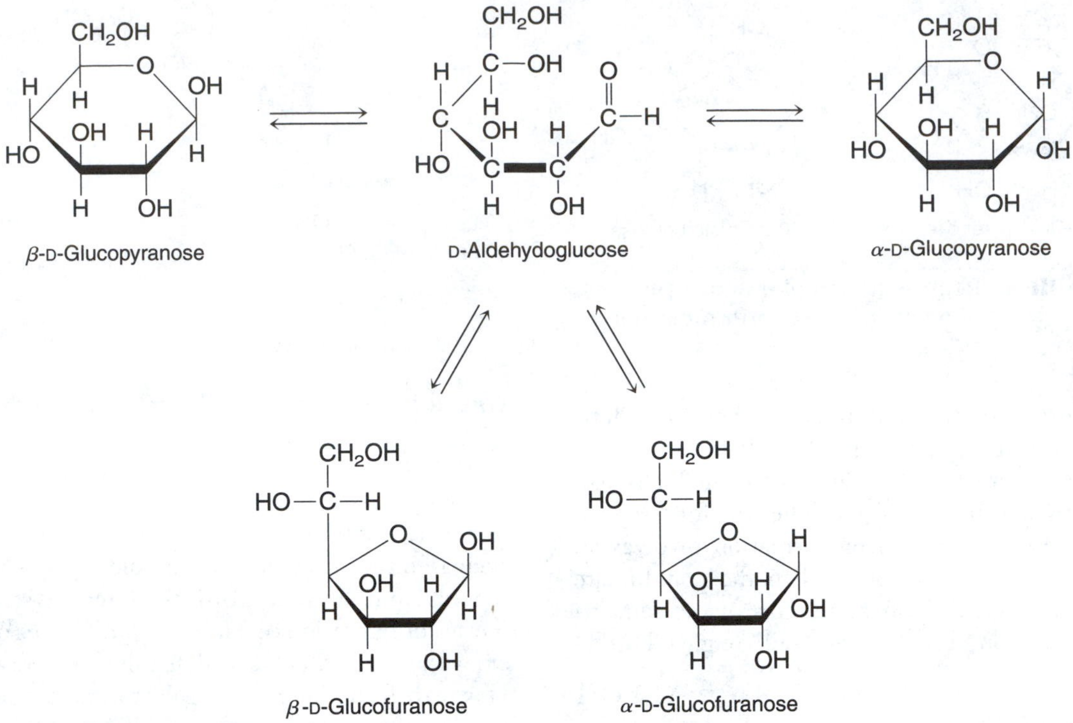

β-D-Glucopyranose

D-Aldehydoglucose

α-D-Glucopyranose

β-D-Glucofuranose

α-D-Glucofuranose

Figure III-11 Interconversion of hemiacetal forms of D-glucose.

Three-Dimensional Representation of Monosaccharide Rings. Even the Haworth projection is somewhat of an oversimplification. The most rational approach to carbohydrate chemistry requires us to consider the true conformation of pyranose rings. An examination of models using normal C–C bond angles makes it clear that six-membered carbon rings cannot assume a planar conformation. Two conformations with no bond strain are possible (Fig. III-12). The *chair* formation is favored over the *boat* conformation by about 1000 to 1 for pyranose rings, so conformational analysis of these carbohydrates is essentially limited to consideration of *two* different chair conformations (see Fig. III-13). To be certain that you have converted the Haworth projection into chair forms properly, imagine flattening the rings. When you do this, constituents that are above the ring in Haworth will be above the ring in a flattened chair. Consideration of space-filling models of such chair forms makes it clear that bulky groups in axial positions (perpendicular to the plane of the ring) will be crowded to distances smaller than their van der Waals radii, whereas such negative steric interactions are largely absent when such groups are in equatorial positions (projecting outward from the ring). Hence, the most stable conformation will favor equatorial arrangements of –OH and CH_2OH groups. In β-D-glucopyranose, shown in Figure III-13, it is clear that the C_1^4 conformation will be greatly favored over the C_4^1 conformation, which has five bulky axial groups. Even the difference in steric crowding between one axial OH group (α-D-glucopyranose in C_1^4 conformation) and no axial OH groups (β-D-glucopyranose in C_1^4 conformation) is sufficient to cause the β form of glucopyranose to predominate in solution over the α form by two to one.

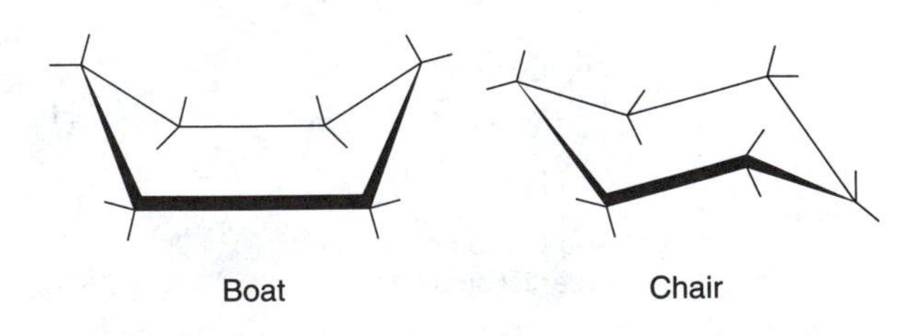

Figure III-12 Conformations of six-membered carbon rings.

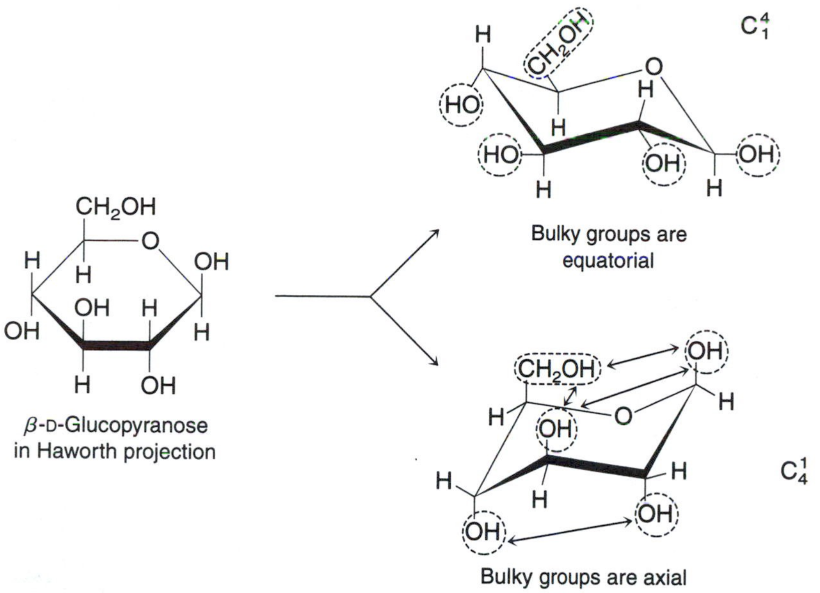

Figure III-13 Three-dimensional chair projections of β-D-glucopyranose.

Like pyranose rings, furanose rings are not actually planar as shown in Haworth projections, but exist in two conformations, called *envelope* (E) and *twist* (T): In the envelope conformation, four members of the ring lie in a plane with the fifth member bent above or below the plane. Only three members of the ring lie in a plane in the twist conformation; the other two members are twisted above and below the plane of the ring (see Fig. III-14). The differences in stability between conformations in furanose rings are not as great as in pyranose rings.

The use of conformational analysis is essential to a rational approach to carbohydrate chemistry, but for simply representing carbohydrate structures most biochemists continue to use the Haworth formulae.

Biologically Important Monosaccharides. A very large number of monosaccharides have been found in biological systems, but a relatively smaller number of monosaccharides is found very frequently in virtually all kinds of cells. Most biochemists find it convenient to memorize the structures of this smaller group. Consult a standard biochemistry textbook for the structures of the following monosaccharides. The hexoses, D-glucose and D-fructose (a 2-ketohexose), are especially important because of their central roles in metabolism, as are the pentoses, D-ribose and 2-deoxyribose, and the trioses, D-glyceraldehyde and dihydroxyacetone.

The rich variety of carbohydrate structures made possible by the presence of multiple chiral centers is further expanded in biology by the introduction of a number of modifications to the basic $C_x(H_2O)_y$ structure. These include formation of amino (usually 2-amino) sugars and *N*-acetylation of the amino group, oxidation of the terminal carbon atom to a COOH group (yielding uronic acids), and attachment of O-linked sulfate groups. Reduction of CHOH centers to CH_2 is also important, because it leads to deoxysugars; the formation of 2-deoxyribose, the constituent sugar of DNA, from ribose, the constituent sugar of RNA, is the most obvious example.

In addition to the monosaccharides listed above, the most commonly occurring monosaccharides in nature are mannose, galactose, glucosamine and *N*-acetylglucosamine, galactosamine and *N*-acetylgalactosamine, sialic acid (see Fig. III-41), glucuronic acid, and fucose (a 6-deoxyhexose). Most of these monosaccharides are found as components of oligosaccharides attached to proteins or lipids and polysaccharide polymers, which will be discussed next.

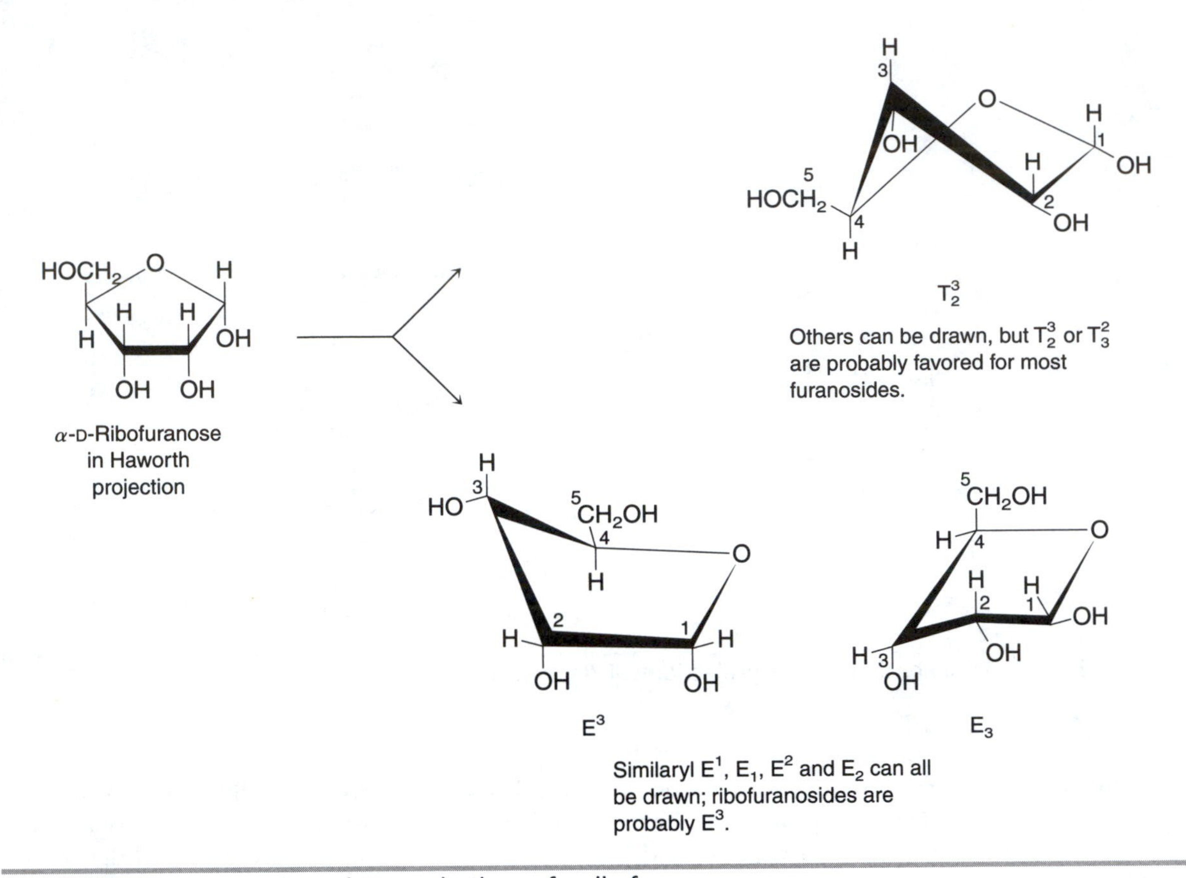

Figure III-14 Twist and envelope projections of D-ribofuranose.

Structure of Oligosaccharides and Polysaccharides

The Joining of Monosaccharide Units into Polymers. The fundamental linkage between monosaccharide units in oligosaccharides or polysaccharides (analogous to the peptide bond between amino acids in proteins) is the acetal or ketal linkage. You will recall that this linkage is formed by condensation of an alcoholic hydroxyl (usually the –OH of another monosaccharide) with the hemiacetal (or hemiketal) form of the carbonyl carbon of a sugar, accompanied by loss of water. Acetal or ketal linkages between the anomeric carbons of sugars and alcoholic hydroxyls (of sugars or other alcohols) are called *glycosidic linkages.* An example of such a linkage is shown in the disaccharide maltose (Fig. III-15). Such glycosidic linkages are quite stable at neutral and alkaline pH values, but are hydrolyzed by aqueous acid.

Oligosaccharides contain only a few monosaccharides. Most abundant natural oligosaccharides are *disaccharides,* which contain two monosaccharide units linked together. Biologically important disaccharides include the milk sugar, lactose; ordinary table sugar, sucrose; a derivative of starch common in plants, maltose; and the major circulating sugar in insects, trehalose. Oligosaccharides are extremely important in biology, where they are usually found

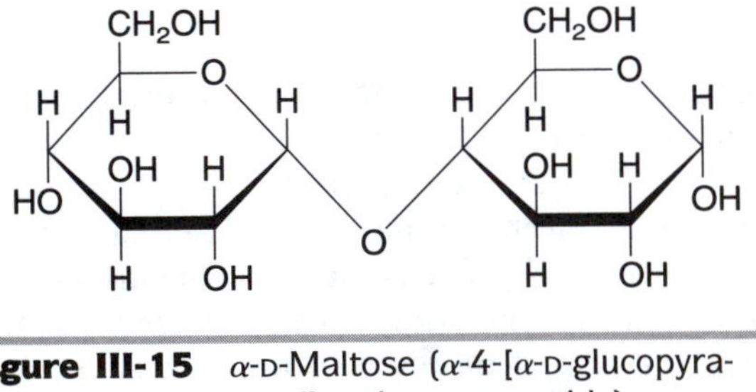

Figure III-15 α-D-Maltose (α-4-[α-D-glucopyranosyl]-D-glucopyranoside).

covalently attached to proteins or lipids. *Polysaccharides*, which are enormously abundant in nature, contain many monosaccharide units and may have molecular weights in the millions. Polysaccharides may contain only a single type of monosaccharide unit (homopolysaccharides, such as starch, glycogen, cellulose, chitin) or several different monosaccharide types (heteropolysaccharides).

The monosaccharide units of oligosaccharides and polysaccharides may be linked in a variety of isomeric arrangements. That is, glycosidic bonds may be formed by condensations between α- or β-OHs at the anomeric carbon of one sugar and alcoholic hydroxyl groups at carbon 2, 3, 4, or 6 of a second sugar (e.g., $1 \rightarrow 2$, $1 \rightarrow 3$, $1 \rightarrow 4$, and $1 \rightarrow 6$ for a hexopyranose). Furthermore, the polymer may be branched so that two sugars have glycosidic linkages to different alcoholic hydroxyls of a third unit (e.g., at a "branch point" in glycogen, one glucose residue is condensed with the C-4 hydroxyl of the branching residue and a second glucose is condensed with the C-6 hydroxyl of the same residue). Each glycosidic (acetal or ketal) linkage employs the *reducing carbon* (anomeric carbon) of one monosaccharide. Thus, all oligo- and polysaccharides have only a single reducing end (analogous to the C-terminal of a peptide). The only exceptions to this generalization are oligosaccharides, such as sucrose and trehalose, in which a sugar unit is linked by glycosidic linkage to the *anomeric* hydroxyl of another sugar. Such a sugar has no aldehyde or ketone in hemiacetal linkage and is consequently a *nonreducing* sugar. A linear polysaccharide also has only a *single* nonreducing end (analogous to the N-terminal of a peptide), but a branched polymer has one additional nonreducing end for each branch point. These elements of linkage isomerism, which are not found in proteins or nucleic acids, complicate the determination of the structure of polysaccharides, as will be discussed in following paragraphs.

Biological Functions of Oligosaccharides and Polysaccharides. Polysaccharides serve two important biological roles. Glycogen and starch are polymers of glucose units linked in $\alpha(1 \rightarrow 4)$ linkages that serve as carbohydrate reserves for animals, bacteria, and plants. Because these polymers are readily converted to intermediates for pathways that yield metabolic energy, they can also be thought of as "energy" storage materials. The mechanisms by which these reserves are mobilized by plants and animals are explored in Experiments 12 and 15, respectively. A second important role for polysaccharides is the formation of structural polymers for cell walls. Cellulose, a polymer of glucose units in $\beta(1 \rightarrow 4)$ linkages, is a major constituent of plant cell walls. Chitin, a polymer of $\beta(1 \rightarrow 4)$-linked *N*-acetylglucosamine units, is abundant in cell walls of yeasts and fungi, and in the exoskeletons of insects and crustaceans. A more complex class of carbohydrate-containing polymers are the *peptidoglycans* (muramylpeptides) and *teichoic acids* of bacterial cell walls, which contain peptides or phosphate residues, respectively, linked in polysaccharide structures.

As noted above, oligosaccharides are usually found linked to noncarbohydrate molecules, especially proteins and lipids. *Glycoproteins*, in which oligo- or polysaccharide chains are linked to polypeptide chains by N-glycosidic linkages to the amide nitrogen of asparagine, or by O-glycosidic linkage to the side chain OH of serine or threonine, are particularly abundant in blood and on cell surfaces. It is difficult to exaggerate the biological importance of the oligosaccharide components of these proteins. They affect the folding and stability of the proteins. They are important in cell–cell recognition, modulation of immunological and inflammatory responses, the invasion of host cells by pathogenic microorganisms, the biology and diagnosis of tumors, and a host of other phenomena. Modification of proteins with oligosaccharides governs their compartmentation into specific membranes or organelles within cells and their excretion from cells. Because these phenomena are highly sensitive to the precise structures of the oligosaccharides involved, it is crucial to develop methods for determining and modifying their complex structures. Characterization of the carbohydrate portion of a glycoprotein requires careful cleavage of the intact polysaccharide from the protein by digestion with proteolytic enzymes or glycosidases (or by mild alkaline degradation in the case of O-glycosides linked to serine) prior to structure determination. Two enzymes frequently used for release of oligosaccharides from glycoproteins are endo-β-*N*-acetylglucosaminidase H and peptide-N^4-(*N*-acetyl-β-D-glucosaminyl) asparagine amidase.

Glycolipids are a diverse class of compounds found throughout the living world. These compounds are classified according to the nature of the lipid portion of the molecule; for example, sphingolipids, abundant in nervous tissue and brain; glycero-glycolipids, found in plants and animals, and the very complex lipopolysaccharides of bacterial cell surfaces. Some glycolipids are conjugated to proteins, as for example, in glycosyl phosphatidylinositol structures that serve to anchor proteins to cell membranes. Glycolipids are discussed further in a subsequent section. Determination of the structure of the carbohydrate portion of these various complex polymers is obviously very important for understanding their biochemistry and biological functions. Such determination relies on the principles and techniques described in the following paragraphs.

Isolation of Carbohydrates

The isolation of polysaccharides presents problems like those encountered in the isolation of other biological polymers. Carbohydrates, with the exception of glycolipids, are insoluble in most organic solvents. Most are soluble in water, although a few, such as cellulose, are relatively insoluble. In contrast to proteins and nucleic acids, most polysaccharides are soluble at low pH, so proteins and nucleic acids can often be removed from them by precipitation with acids such as trichloroacetic acid. Proteins are denatured much more readily than polysaccharides and can be removed by heating or by treatment with denaturing solvents (such as chloroform:isoamyl alcohol [25:1]). Nucleic acids are sometimes removed by degradation with DNase or RNase, and proteins by digestion with proteases. Once other macromolecules have been removed or enzymatically degraded to low-molecular-weight fragments, most polysaccharides can be precipitated with water-miscible organic solvents (ethanol, acetone). Acidic polysaccharides or neutral polysaccharides rendered acidic when combined with borate can also be precipitated by the formation of water-insoluble complexes with cationic detergents. Other methods used in purification of polysaccharides resemble those used for purification of proteins, especially when the polysaccharide contains ionizable residues. Ion-exchange, gel-filtration, adsorption chromatography, and density-gradient centrifugation have all proven valuable.

Determination of the Structure of Polysaccharides

The problem of determining the structure of a polysaccharide is similar to that of determining the amino acid sequence of a protein, but it is complicated by the additional structural features unique to carbohydrates: branching and linkage isomerism. The elements of carbohydrate chemistry discussed in following paragraphs are those relevant to the problems of determination of carbohydrate composition and the structure of polysaccharides.

Determination of Carbohydrate Composition: Cleavage of Glycosidic Linkages. To analyze a polysaccharide, you must first determine the number and nature of monosaccharide units in the structure. For most oligo- and polysaccharides composed of neutral sugars, hydrolysis of the glycosidic linkages between sugars is complete in 3 to 6 hr in 0.5 to 1 N HCl or H_2SO_4 at 100°C (Fig. III-16). Furanoside linkages tend to be more labile to acid than pyranoside linkages. Glycosides of acidic sugar derivatives and amino sugars tend to

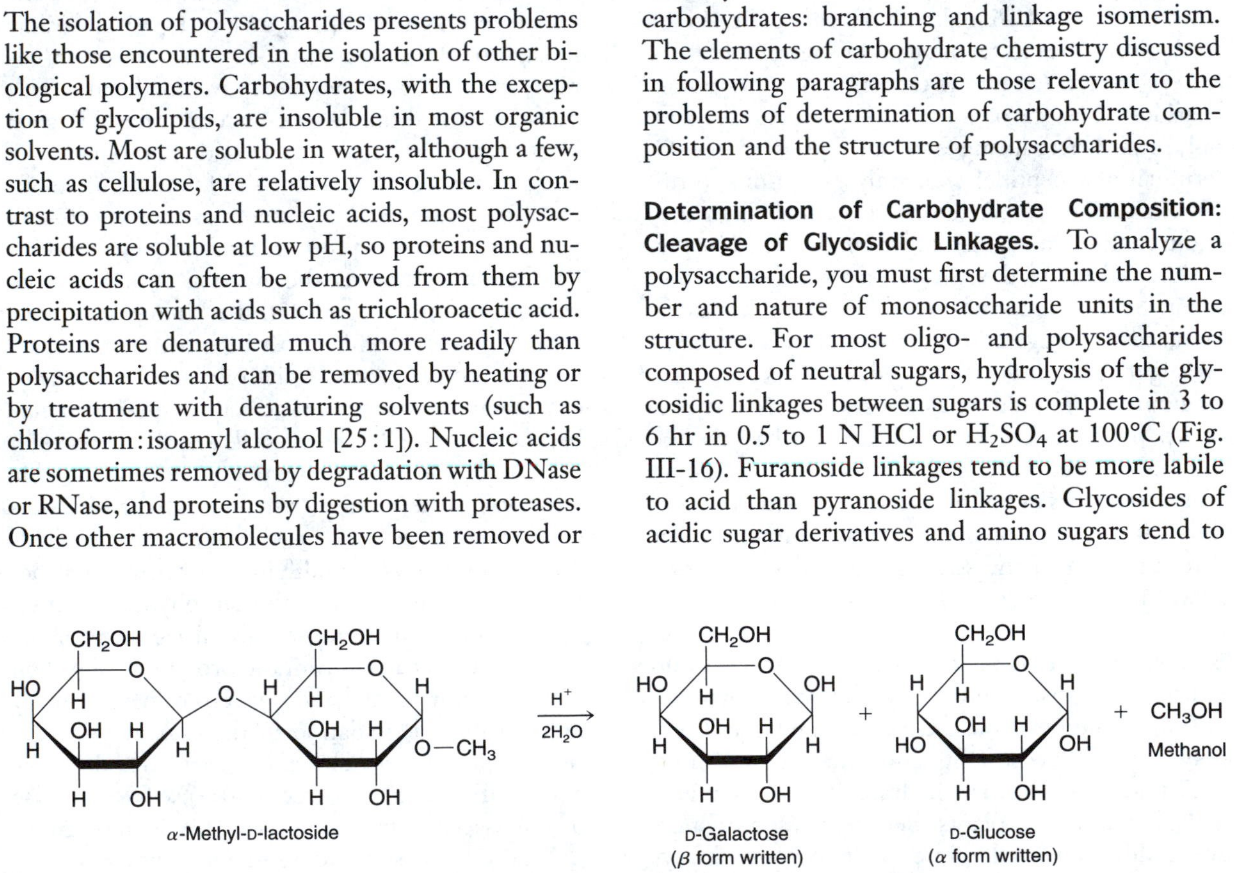

α-Methyl-D-lactoside $\xrightarrow[2H_2O]{H^+}$ D-Galactose (β form written) + D-Glucose (α form written) + CH_3OH Methanol

Figure III-16 Acid-catalyzed hydrolysis of acetal linkages in α-methyl-D-lactoside.

resist hydrolysis more strongly than corresponding neutral sugars. Only a few monosaccharides undergo degradative reactions under these mild acid conditions. Glycosidic linkages are generally stable under mild alkaline conditions.

Quantitative Determination of Monosaccharides by Reactions of Their Aldehyde or Ketone Functions: "Reducing Sugars." The quantitative determination of monosaccharides is often based on the reactivity of the hemiacetal (potential aldehyde or ketone) of the free sugar. (*NOTE:* This group is *not* present when the carbonyl is in acetal linkage.) The hemiacetal or "reducing group" is very reactive with dilute alkali (e.g., 0.035 N NaOH or saturated aqueous $Ca(OH)_2$), where the carbohydrate undergoes isomerization at room temperature. This interconversion, which probably involves an enolic intermediate (see Fig. III-17), is referred to as the Lobry de Bruyn–Alberda van Eckenstein rearrangement. Clearly, if the identity of individual sugars is important, you should avoid alkaline conditions during polysaccharide analysis. In the course of migration of double bonds in alkali, monosaccharides become very susceptible to oxidation by O_2 or by other electron acceptors (hence, the term *reducing sugar* for a free monosaccharide). Such oxidations are very complex and usually do not proceed in a strictly stoichiometric fashion. However, *if the conditions (temperature, alkali concentration, oxidant concentration, and especially time) are carefully controlled*, oxidation of monosaccharides by various oxidizing agents (e.g., Cu^{2+} or $Fe(CN)_6^{-3}$) can be used for quantitative analysis of sugar solutions. This is the basis of such methods as Nelson's and the Park and Johnson determinations, as well as many others not widely used today. The nonstoichiometric aspects of these oxidations show up as differences in the "reducing power" of sugars with very similar structures, which after a fixed time reduce different amounts of Cu^{2+} under identical conditions. Hence, different standard curves are obtained for different sugars. When you analyze an unknown sugar, you can expect substantial uncertainty in the analytical result. Any saccharide that has a reducing end reduces Cu^{2+}, $Fe(CN)_6^{3-}$, O_2, and like substances. Per unit weight, however, monosaccharides contain more reducing ends than polysaccharides. By measuring increases in reducing equivalents you can follow the cleavage of a polysaccharide to monosaccharides or oligosaccharides.

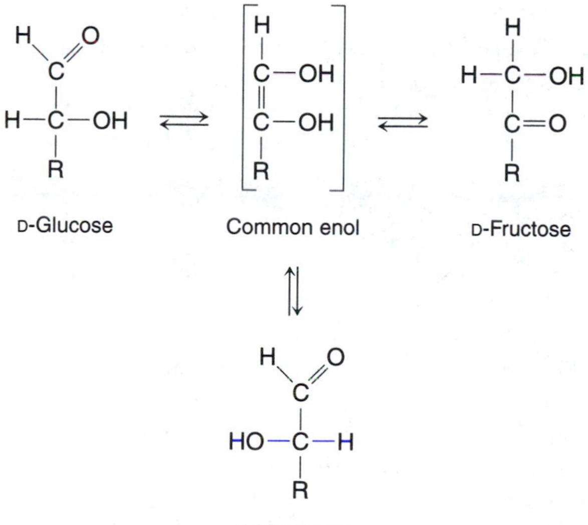

D-Glucose ⇌ Common enol ⇌ D-Fructose

Common enol ⇅ D-Mannose

Figure III-17 Alkaline interconversion of glucose, mannose, and fructose.

Quantitative Determination of Monosaccharides by Conversion to Furfurals. As described above, polysaccharides are cleaved in mild acid to monosaccharides, which are generally stable to acid. Strong mineral acids (4–6 N at 100°C, lower temperatures with more concentrated acids) not only cleave glycosidic bonds but cause dehydration of monosaccharides to yield furfurals and, subsequently, levulinic acid (Fig. III-18). Once again, the dehydration reaction is not stoichiometric, and different yields of furfurals and other products are obtained from very similar carbohydrates under identical conditions. Further, the yield of products is strongly dependent on the nature and strength of acid, temperature, and other variables. It is important to bear these complexities in mind when using one of many methods that utilize furfural formation for quantitative analysis of carbohydrates. The furfurals formed by acid dehydration from carbohydrates react in acid with any one of a variety of phenols (phenol, orcinol, carbazole, anthrone, etc.) or aromatic amines to give highly conjugated, colored products. The amount of color formation is a function of the amount and nature of the carbohyrate being analyzed. Because the rate of furfural

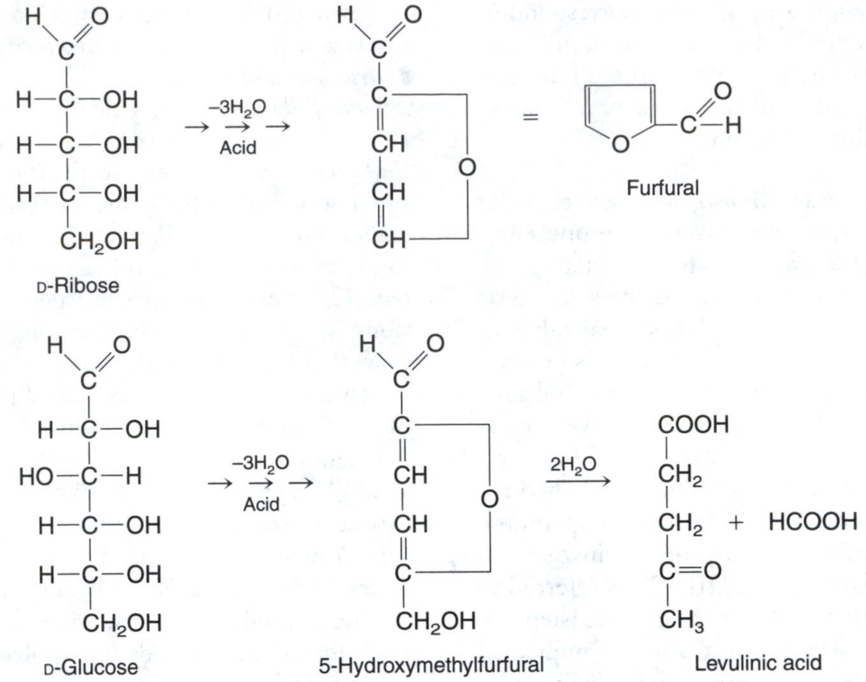

Figure III-18 Acid-catalyzed dehydration of monosaccharides.

formation is so dependent on the nature of the carbohydrate and the reaction conditions, careful standardization is required.

Chromatographic Analysis of Carbohydrate Mixtures. Much effort has been devoted in recent years to devising methods for analyzing carbohydrate mixtures quantitatively and qualitatively in a single process (as is done for amino acid analysis). Four methods that offer most promise of this goal are described here.

Gas–liquid chromatography of trimethyl-silyl (TMS) or acetyl derivatives of monosaccharides, of their methyl glycosides, or of their corresponding alditols (monosaccharides that have their carbonyl function reduced to the corresponding alcohol) is the most widely used. Some sample reactions illustrating the formation of these derivatives are shown in Figure III-19. An advantage of methanolysis is that it simultaneously cleaves the glycosidic bonds of a polysaccharide and forms the methyl glycoside derivatives. However, it suffers from the disadvantage that a single monosaccharide may give as many as four glycosidic derivatives (α- and β-pyranosides and α- and β-furanosides), a disadvantage not shared by the borohydride reduction–acetylation procedure. The volatile derivatives are identified and quantitated from the position and size of the elution peak on the gas chromatogram. They may also be analyzed by mass spectrometry when confirmation of identity is needed.

A second method of analyzing unknown carbohydrate mixtures is ion-exchange chromatography of the monosaccharides as their borate complexes. This method takes advantage of the fact that in alkaline solution, boric acid forms complexes with vicinal hydroxyl groups (Fig. III-20). The number and stability of the complexes that form on any given monosaccharide are a function of the structure and conformation of the carbohydrate, especially the spatial orientations of adjacent hydroxyls. Thus, the net negative charge of the borate complexes of various carbohydrates is different. This allows for their separation on anion-exchange resins and subsequent analysis by the colorimetric methods already described.

High-pH anion-exchange chromatography (HPAEC) of monosaccharides and oligosaccharides has been more recently developed as a useful analytical tool. At sufficiently alkaline pH (12 to 14)

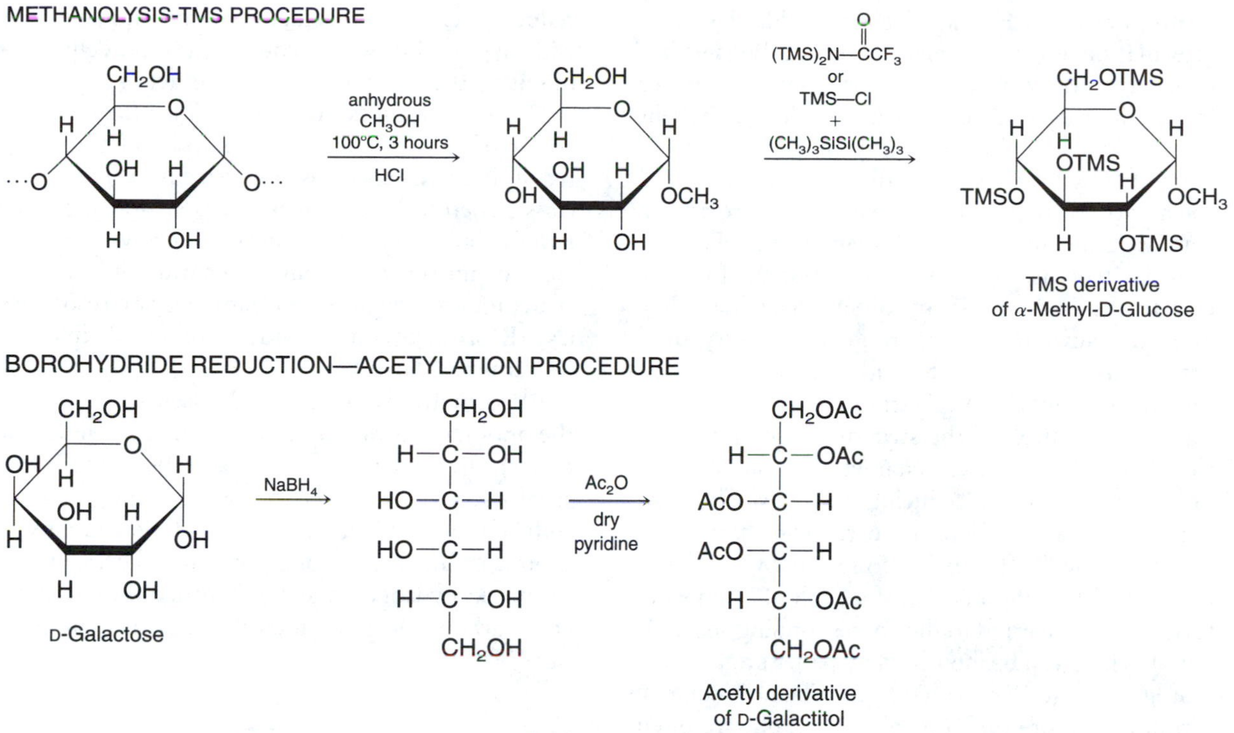

Figure III-19 Sugar derivatives commonly formed before gas-liquid chromatography.

the anomeric hydroxyl group of a monosaccharide loses its proton, and the carbohydrate acquires a negative charge. The acidity of the anomeric hydroxyl groups is a sensitive function of the structure of the monosaccharide from which it is formed, so many mono- and oligosaccharides can be resolved by gradient elution from an anion-exchange column (see Experiment 2). When this chromatographic separation is combined with pulsed amperometric detection, it provides a very sensitive method for analysis of the separated carbohydrates without derivatization or other chemical reactions.

A fourth method, analysis of sugar alcohols after reduction by $[^3H]$-$NaBH_4$ and separation by paper chromatography, is described in detail in Experiment 11.

Determination of Sequence and Linkage Isomerism of Polysaccharides. Structural characterization of polysaccharides is similar to protein characterization in that it is frequently necessary to degrade polysaccharides into more easily characterized subunits (oligosaccharides). This approach is, in fact, quite powerful, because very many polysaccharides are constructed of repeating or alternating oligosaccharide sequences rather than a very long unique sequence. Fragmentation can be carried out by partial acid hydrolysis or by use of specific glycosidases. Enzymes are especially useful because they are usually specific with respect to the monosaccharide units (e.g., glucosamine vs. glucuronic acid), position of the bond (e.g., $1 \rightarrow 4$ vs. $1 \rightarrow 6$), and anomeric configuration (α vs. β) of the

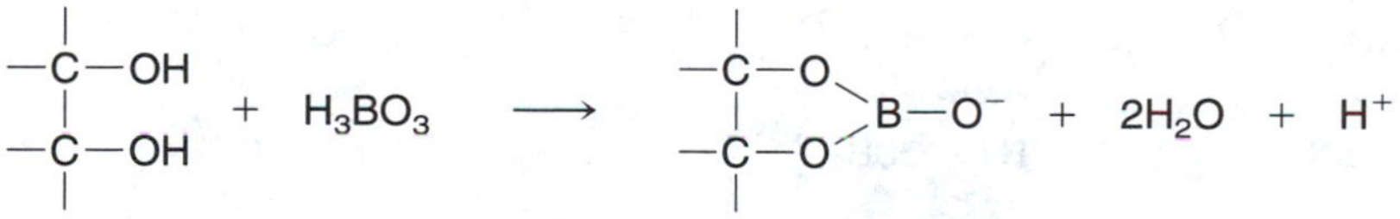

Figure III-20 Reaction of borate with vicinal hydroxyl groups.

residues cleaved. Hence, cleavage yields structural information about the nature of the linkage between the fragments, as well as releasing smaller fragments that can be analyzed further. In recent years, the availability of highly purified recombinant endoglycosidases with well-characterized specificity for their sites of cleavage has provided a very valuable set of tools for analyzing oligosaccharide structure by enzymatic digestion. In addition to the high specificity of such enzymes, they offer the advantage of cleavage under very mild conditions, eliminating the undesired side reactions that accompany chemical degradation.

Determination of the structure of oligosaccharides usually requires characterization of end groups, especially the reducing terminus. The most common means to determine reducing termini is reduction by [^{3}H]-$NaBH_4$ followed by total hydrolysis of the oligosaccharide. Only the reducing terminal is converted to the corresponding sugar alcohol, which can be identified by paper or gas chromatography (see Experiment 11). The nonreducing termini are more difficult to identify, but are often revealed by methylation. Exhaustive methylation of oligosaccharides or polysaccharides (conversion of all hydroxyl groups to the corresponding methyl ethers), followed by hydrolysis and identification of the constituent methyl sugars by chromatographic methods is a general means of determining structure. Only those OH groups on anomeric carbons and the OH groups that participate in glycosidic linkages will not be methylated (see Fig. III-21). (Note that in the example shown in Fig. III-21, the reducing terminus was identified by mild oxidation with hypoiodite, which converts the reducing terminus to the corresponding sugar acid.)

Recent advances in techniques for the determination of the structures of oligosaccharides by high-resolution mass spectroscopy make it likely that the classic chemical techniques discussed above may be largely replaced by mass spectroscopy. When sufficient quantities of an oligosaccharide of unknown structure are available, nuclear magnetic resonance (NMR) spectroscopy provides another valuable tool for structure analysis.

Two methods are generally used to determine the anomeric configuration of glycosidic linkages: cleavage by enzymes of known specificity and NMR spectroscopy. In the latter method, the chemical shift of the anomeric hydrogen and the coupling of proton spins (spin–spin coupling) between the anomeric hydrogen and the hydrogen on an adjacent carbon allow the identification of configuration.

Chemistry of Biological Phosphate Compounds

The chemistry of phosphate esters and anhydrides is so central to biochemistry that the subject might also be logically discussed in the sections on nucleic acids, lipids and membranes, or intermediary metabolism. However, because of the importance of sugar phosphates and their derivatives in Experiment 12, the subject is presented in this section. The background is important to more than Exper-

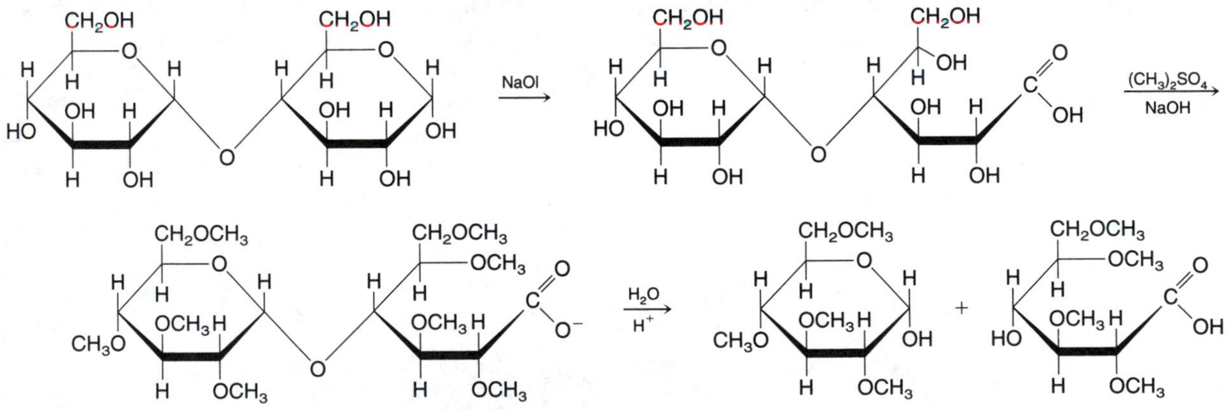

Figure III-21 Characterization of maltose by mild oxidation followed by exhaustive methylation and acetal hydrolysis.

iment 12, however, and you should keep these principles in mind when dealing with the other subjects mentioned in this paragraph.

Phosphate Esters. An ester is formed by elimination of H_2O and formation of a linkage between an acid and an alcohol (or phenol) (Fig. III-22). *Phosphomonoesters*, especially of monosaccharides, are very common (Fig. III-23). Because phosphoric acid is a tribasic acid, it can also form di- and triesters (Fig. III-24). Phosphotriesters are rarely found in nature, but *diesters* are extremely important, particularly as the fundamental linkage of the nucleic acid polymers, which are sequences of ribose (or deoxyribose) units linked by $3' \rightarrow 5'$ phosphodiester bonds (see Fig. III-25). Like phosphoric acid, which has three dissociable protons (Fig. III-26), phosphomono- and phosphodiesters are acidic and typically ionize as shown in Fig. III-27. Note the similarities between the pK_a values for phosphoric acid and the "primary" and "secondary" ionizations of phosphate esters. The ionization of phosphodiesters accounts for the very acidic properties of nucleic acids. Knowledge of the ionization of phosphate compounds and the consequent dependence of net charge on pH are essential to the use of ion-exchange chromatography and electrophoresis for the separation of these compounds (see Experiments 4 and 12).

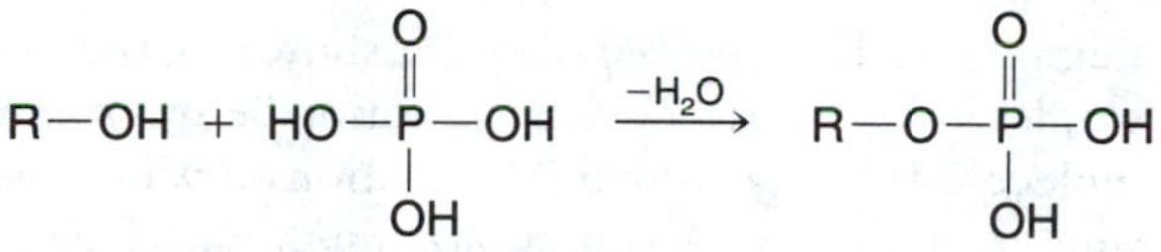

$$R{-}OH + HO{-}P(=O)(OH){-}OH \xrightarrow{-H_2O} R{-}O{-}P(=O)(OH){-}OH$$

Figure III-22 Formation of a phosphomonoester.

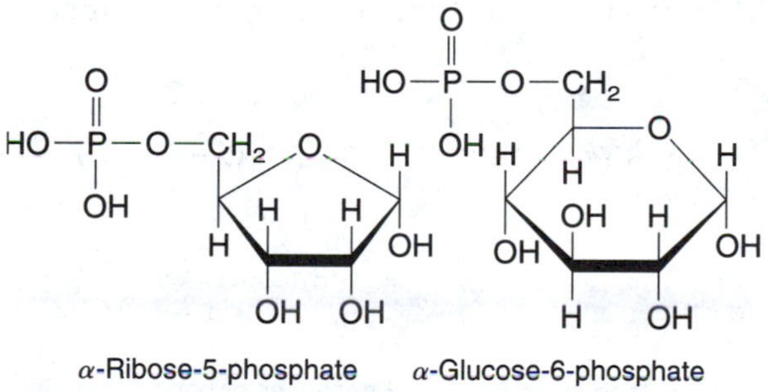

α-Ribose-5-phosphate α-Glucose-6-phosphate

Figure III-23 Examples of biologically important phosphomonoesters.

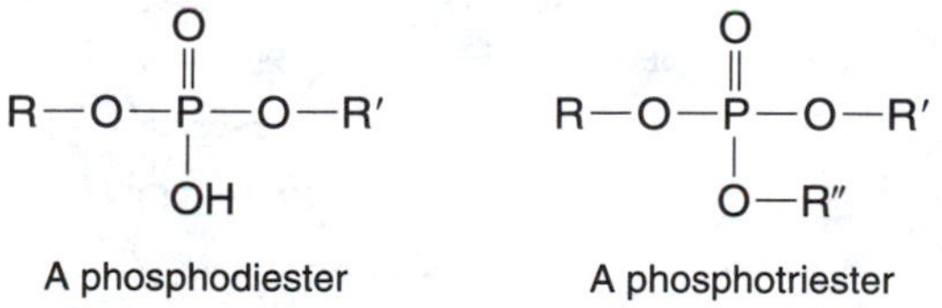

$$R{-}O{-}P(=O)(OH){-}O{-}R'$$

A phosphodiester

$$R{-}O{-}P(=O)(O{-}R''){-}O{-}R'$$

A phosphotriester

Figure III-24

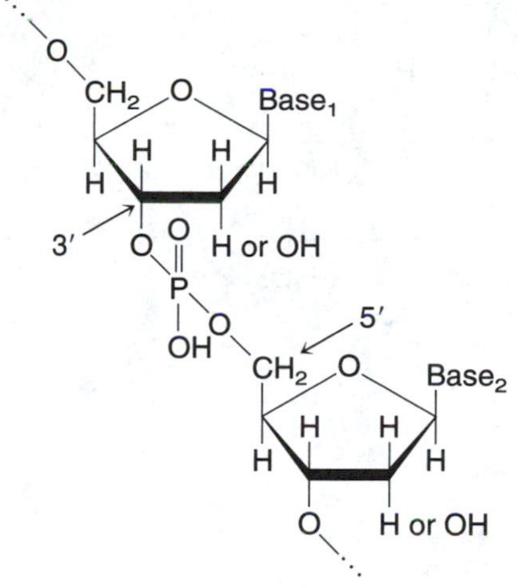

Figure III-25 Pentose phosphodiester polymer as found in nucleic acids.

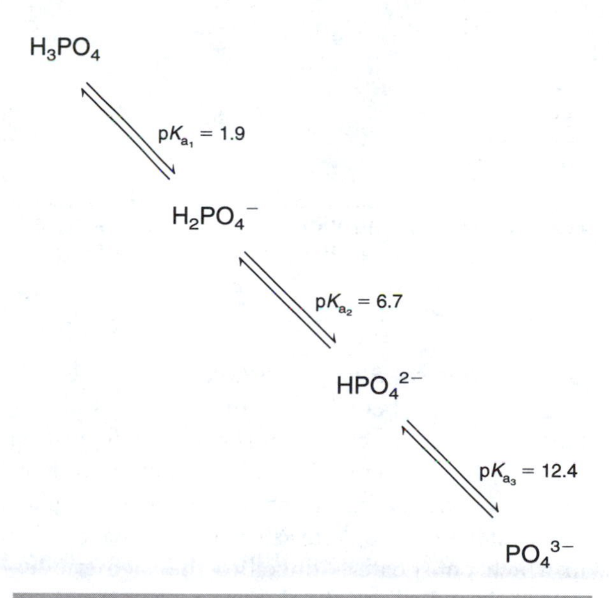

Figure III-26 Ionization of phosphoric acid as a function of pH.

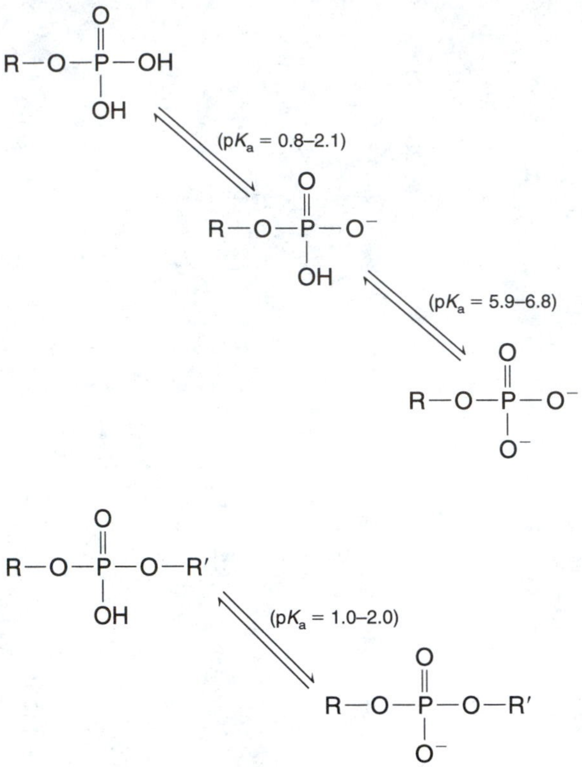

Figure III-27 Ionization of phosphomonoesters and phosphodiesters as a function of pH.

Other Biological Phosphate Compounds. Elimination of water between phosphoric acid and certain other types of compounds results in formation of a variety of phosphate compounds that have properties that are different from simple phosphomonoesters. A phosphate ester of a monosaccharide in which phosphate is linked to the anomeric hydroxyl is called a *phosphoacetal.* An example is α-D-glucopyranosyl-1-phosphate (glucose-1-phosphate) (Fig. III-28). A related group of compounds might be called *pyrophosphoacetals,* shown by the examples in Figure III-29. Condensation between two molecules of an acid with elimination of water results in the formation of an *acid anhydride.* Examples of such anhydride bonds are found in inorganic pyrophosphate and in the many pyrophosphoryl and tripolyphosphoryl esters of biological systems (Fig. III-30). Condensation between two different

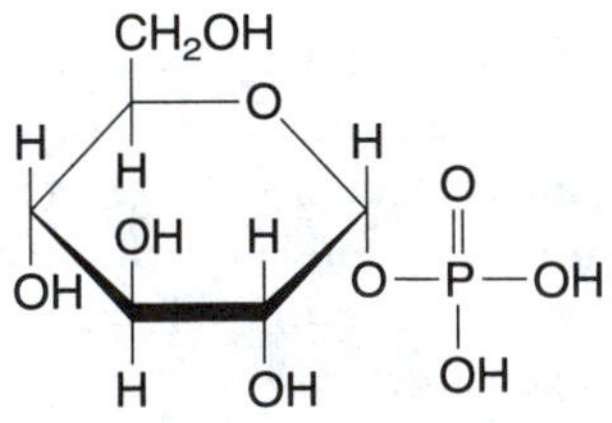

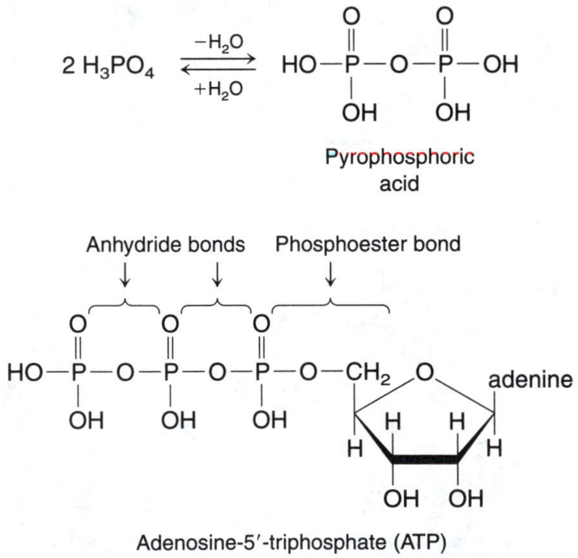

Figure III-28 An example of a phosphoacetal, glucose-1-phosphate.

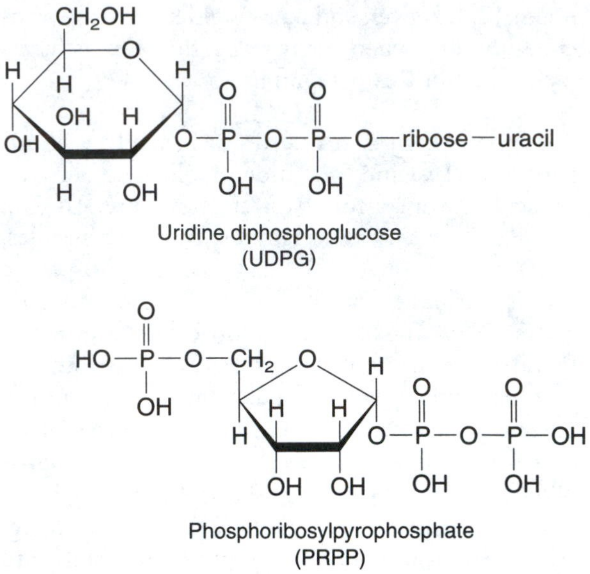

Figure III-29 Biologically important pyrophosphoacetals.

Figure III-30 Acid anhydride bonds in pyrophosphate and ATP.

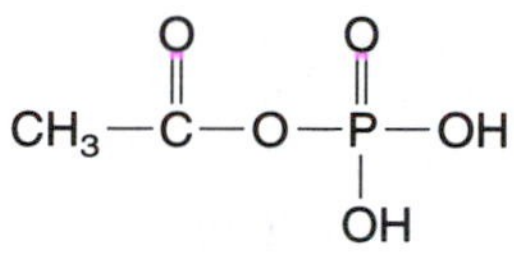

Acetyl phosphate

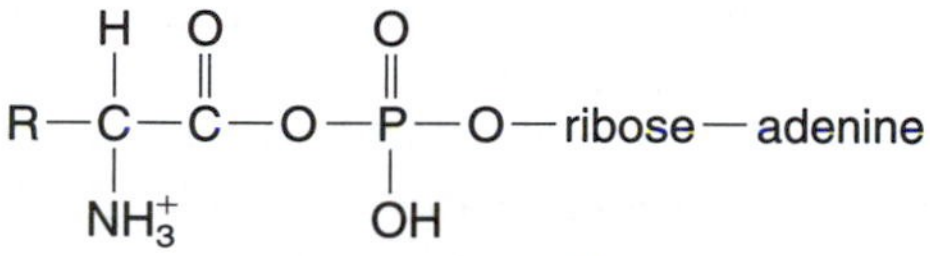

Aminoacyl-AMP

Figure III-31 Two biochemically important compounds that contain mixed anhydride bonds.

acids with elimination of water results in formation of *mixed anhydrides*, which are important as intermediates in biochemical reactions (Fig. III-31).

The chemical properties of an *enol phosphate* ester are quite different from simple phosphate esters. The only important example is the glycolytic intermediate, phosphoenolpyruvic acid (Fig. III-32).

A final group of biologically important phosphate compounds is the phosphoramidates, which are characterized by the structure shown in Figure III-33. Phosphocreatine, an important energy storage compound in muscle, and phosphohistidine, an intermediate in several enzyme reactions, are two examples (Fig. III-34).

Phosphorylation of Proteins. Phosphorylation of proteins is a central element in a large number of biological regulatory processes. Enzymes that catalyze this posttranslational modification are called *protein kinases*. They typically transfer the terminal (γ) phosphate group of ATP to a substrate protein or to a residue on their own polypeptide chain (autophosphorylating kinases) to yield phosphoserine, phosphothreonine, phosphotyrosine, phosphohistidine, or phosphoaspartate residues. Phosphorylation usually results in dramatic changes in the functional properties of the target proteins, which are often enzymes, receptors, or transcription factors. Removal of phosphoryl groups may be catalyzed by separate *phosphoprotein phosphatases* or may be catalyzed by the target proteins themselves. The combined action of protein kinases and phosphoprotein phosphatases renders protein phosphorylation highly reversible and susceptible to exquisitely sensitive regulation. Hundreds of examples are now known in which such processes play important roles in metabolic regulation and cell biology. An example is studied in Experiment 15.

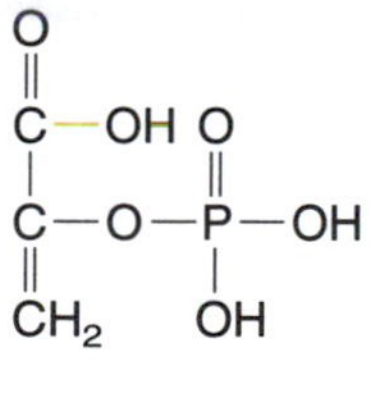

Figure III-32 Phosphoenolpyruvic acid.

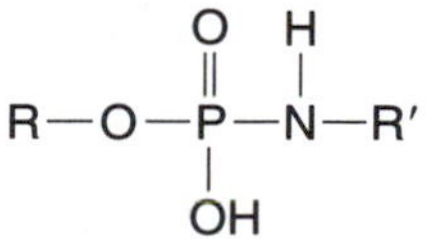

Figure III-33 A phosphoramidate.

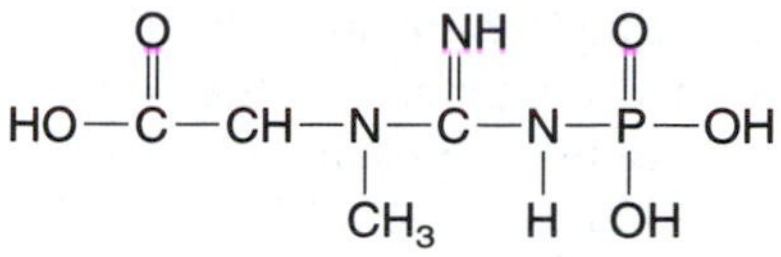

Phosphocreatine

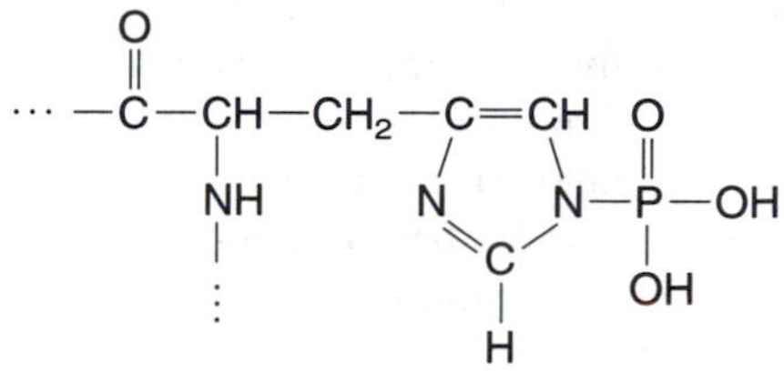

3-Phosphohistidine (in protein linkage)

Figure III-34 Examples of biologically important phosphoramidates.

Hydrolysis of Phosphate Compounds. From the point of view of the biochemist, the most important single reaction undergone by phosphate compounds found in biological systems is hydrolysis. An understanding of hydrolysis of phosphates is pertinent to the degradation and identification of compounds containing such bonds. It also provides an insight into the biological function of such molecules.

Although all of the compounds described here undergo hydrolysis with the release of inorganic phosphate during treatment with strong acid or alkali at elevated temperatures, they differ markedly in their rates of hydrolysis under milder conditions, especially in dilute acid (0.1–1 N, 20–100°C). These differences are often exploited for analysis or specific degradation (see Experiment 12). Table III-1 lists some examples from each class and compares their relative lability to acid hydrolysis. The compounds vary widely in their stability in mild acid. A useful generalization, based on the behavior of typical phosphate compounds in 1 N HCl at 100°C for 7 min is that most phosphomonoesters and diesters of alkyl alcohols are stable (no hydrolysis in 7 min at 100°C), whereas (with a few exceptions) members of the other classes are completely hydrolyzed under these conditions. This principle serves as the basis for the assays of Experiment 12.

Most of the compounds under discussion are quite labile in acid, but with two important exceptions they are resistant to mild alkaline hydrolysis (0.1–1 N base, 25°C). One exception is the mixed anhydrides of phosphate and carboxylic acids, such as acetyl phosphate. They are very susceptible to alkaline cleavage or cleavage by nucleophiles such as hydroxylamine. The other relatively alkali-labile phosphate compounds are phosphodiesters, in which the phosphate is attached to one of two vicinal hydroxyl groups (as in RNA or derivatives of phosphatidic acid). In these instances, the neighboring OH group participates in the hydrolysis of such compounds with intermediate formation of a cyclic diester. The resultant phosphomonoester is then generally very resistant to alkaline hydrolysis.

The susceptibility of different types of phosphoprotein linkages to chemical cleavage is often used to diagnose the nature of the attachment of phosphate to the protein in a case in which this is unknown. For example, O-phosphorylserine and O-phosphorylthreonine are very stable to acid, but are very rapidly hydrolyzed at alkaline pH. The reverse is true for most phosphoramidates (phosphorylhistidine, phosphoryllysine, phosphorylarginine), which are extremely acid-labile, but relatively stable to base. Acyl phosphates (phosphorylaspartate, phosphorylglutamate), as noted above, are quite labile to both acid and base. Finally, O-phosphoryltyrosine is extremely stable in alkaline. The phosphate of O-phosphoryltyrosine is not removed by hydrolysis in 5 N KOH at 155°C for 30 min, conditions that hydryolyze all the other types of phosphorylamino acid linkages.

The free energies of hydrolysis (phosphate group transfer potential) of the compounds in the classes we have been discussing are important to their roles as phosphate donors and acceptors, and as carriers of chemical energy in metabolism. Table III-2 lists some of these free energies under standard conditions (pH 7, 25°C, 1 M reactants and products). A

Table III-1 Rates of Hydrolysis of Phosphate Compounds in Dilute Acid (100°C, 1 N acid)

Hydrolysis Reaction	$t_{1/2}$ (time in minutes)	Percent of Reaction in 7 min
3-Phosphoglyceric acid $\xrightarrow{H_2O}$ glyceric acid + P_i	2140	0
Glucose-6-phosphate $\xrightarrow{H_2O}$ glucose + P_i	1300	0
Fructose-1-phosphate $\xrightarrow{H_2O}$ fructose + P_i	2.8	70
Glucose-1-phosphate $\xrightarrow{H_2O}$ glucose + P_i	1.05	100
Phosphoenolpyruate $\xrightarrow{H_2O}$ pyruvate + P_i	8.3	40
ATP $\xrightarrow{H_2O}$ ADP + P_i	8 (0.1 N acid)	100
Acetylphosphate $\xrightarrow{H_2O}$ acetate + P_i	11 (0.5 N acid, 40°)	100
Phosphocreatine $\xrightarrow{H_2O}$ creatine + P_i	4 (0.5 N acid, 25°)	100
UDPG $\xrightarrow{H_2O}$ UDP + glucose	20 (0.1 N acid)	100
PRPP $\xrightarrow{H_2O}$ ribose-5-P + PP_i	10 (pH 4, 65°)	100

Table III-2 Free Energy of Hydrolysis of Phosphate Compounds at pH 7, 25 °C. (Most of these free energy values are strongly dependent on pH and Mg^{2+} concentration.)

Reaction	$\Delta F°'$ of Hydrolysis (Phosphate Transfer Potential) kcal/mol
Phosphoenolpyruvate $\xrightarrow{H_2O}$ pyruvate + P_i	−14.8
Creatine phosphate $\xrightarrow{H_2O}$ creatine + P_i	−10.5
Acetylphosphate $\xrightarrow{H_2O}$ acetate + P_i	−10.1
PRPP $\xrightarrow{H_2O}$ ribose-5-P + PP_i	−8.2
ATP $\xrightarrow{H_2O}$ ADP + P_i	−7.3
UDPG $\xrightarrow{H_2O}$ UDP + glucose	−7.0
Glucose-1-P $\xrightarrow{H_2O}$ glucose + P_i	−5.0
Glucose-6-P $\xrightarrow{H_2O}$ glucose + P_i	−3.3
Fructose-1-P $\xrightarrow{H_2O}$ fructose + P_i	−3.1
3-Phosphoglycerate $\xrightarrow{H_2O}$ glycerate + P_i	−3.1

compound with a higher free energy of hydrolysis can spontaneously transfer its phosphate group to the conjugate acceptor of any phosphate compound with a lower free energy of hydrolysis under standard conditions and in the presence of an appropriate catalyst (enzyme). In this way, a table of group transfer potentials is analogous to a table of redox potentials; it can be used to predict the direction of many reactions and to analyze the bioenergetics of metabolic sequences. There is a general coincidence between the acid lability of these compounds and their free energies of hydrolysis, but you should not be deceived. There is no predictable relationship between the free energy change of a reaction and the rate at which that reaction will proceed.

Lipids

Because the universal liquid milieu of living systems is water, you might expect that the chemical components of cells would be polar substances that are readily soluble in water. This is generally the case, but there exists a rather heterogeneous class of biological molecules that are soluble in nonpolar solvents, such as benzene, ether, chloroform, or slightly more polar solvents, such as methanol or acetone. This class is called *lipids*. It is perhaps most useful to consider the lipids by their functions. A major function of the triglycerides (fats and oils) is storage and transport of metabolic energy. Waxes, hydrocarbons, and fatty alcohols, the most nonpolar lipids, function primarily in forming protective surfaces of plant and animal tissues, although waxes are used in place of triglycerides as energy storage forms in some aquatic animals. The most important lipids are those that serve as structural components of biological membranes. All of the members of this group (phospholipids, glycolipids, sphingolipids, and sterols—the latter two are usually absent from bacterial membranes) are *amphipathic*. That is, they possess a nonpolar tail, often containing long-chain fatty acyl groups, and a polar head group. Considerable attention will be paid in this section to these lipids and their role in membrane structure and function. Finally, there is a large group of lipids of relatively low molecular weight (steroids, terpenes, prostaglandins, and quinones) that have diverse and very important functions as hormones, vitamins, and photopigments, as well as other more specialized roles. As important as this group is, it will not be further discussed in this section. We will focus on lipids that are important in energy metabolism and membrane formation.

Structure and Classification of Lipids

Triacylglycerols and Related Simple Lipids. *Triacylglycerols* (triglycerides)—that is, glycerylesters of fatty acids (e.g., triolein and α-palmitoyl-α',β-diolein [Fig. III-35])—are by far the most abundant lipids in higher plants and animals. These lipids accumulate in fatty depots and seeds as a storage form of food energy. Natural fats generally contain a variety of fatty acids, and are best represented as "mixed" fats, as in the case of 1-palmitoyl-2,3-diolein in Fig. III-35. Diglycerides and monoglycerides exist in various tissues, usually in small amounts. Diglycerides play an important role in the biosynthesis of some of the more complex lipids and in signal transduction in eukaryotic cells.

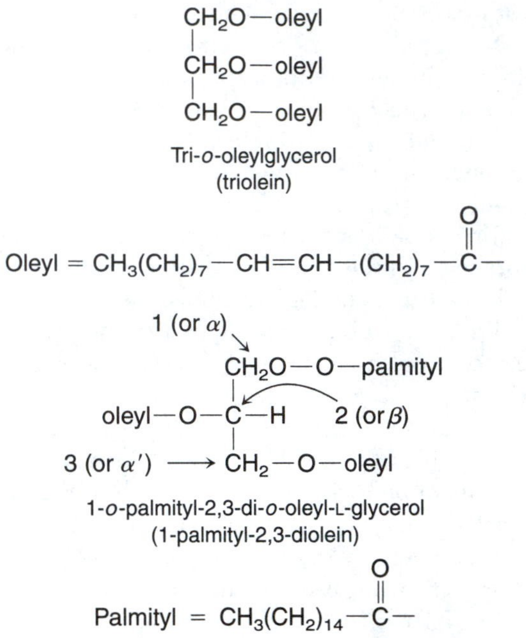
CH2O—oleyl
CH2O—oleyl
CH2O—oleyl
Tri-o-oleylglycerol
(triolein)
Oleyl = CH3(CH2)7—CH=CH—(CH2)7—C—
O
1 (or α)
CH2O—O—palmityl
oleyl—O—C—H
2 (or β)
3 (or α′)
CH2—O—oleyl
1-o-palmityl-2,3-di-o-oleyl-L-glycerol
(1-palmityl-2,3-diolein)
Palmityl = CH3(CH2)14—C—
O

Figure III-35 Structures of two triglycerides.

Sterols and Sterol Esters. The *sterols* are hydroxylated derivatives of the cyclopentanoperhydrophenanthrene (steroid) nucleus (Fig. III-36). All sterols are capable of forming esters with fatty acids, but in general, the free sterols are more abundant than their esters. The most abundant sterol is cholesterol, which is found in the membranes of most animal cells (Fig. III-36). When cholesterol occurs in the esterified forms, it usually is combined with common fatty acids, such as palmitic, oleic, or linoleic acids. Plant sterols are occasionally esterified with acids, such as acetic and cinnamic acids, or are incorporated into glycosides (saponins). Cholesterol is the precursor to many important steroid hormones that regulate metabolism and reproduction. Well-known examples include cortisol, aldosterone, progesterone, testosterone, and estradiol (see Fig. III-36), but there are many others.

Glycerophosphatides. These lipids are mainly acyl derivatives of α-glycerophosphoric acid and often are called "phospholipids." The simplest glycerophosphatides are the phosphatidic acids, which contain α-glycerophosphoric acid esterified with two fatty acids (Fig. III-37). Small quantities of phosphatidic acids have been isolated from a wide variety of plant and animal tissues. It is doubtful that these compounds exist in large amounts in tissues, because more complex glycerophosphatides are readily hydrolyzed by enzymes that are widely distributed in such tissues, yielding phosphatidic acids. Phosphatidic acid is a crucial intermediate in the biosynthesis of phospholipids.

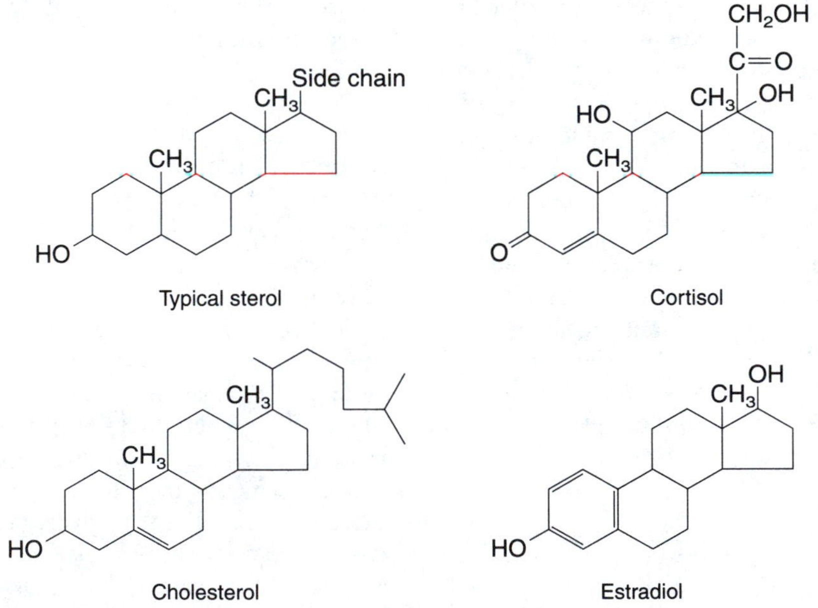
Side chain
CH3
CH3
HO
Typical sterol
CH2OH
C=O
CH3
OH
HO
CH3
O
Cortisol
CH3
CH3
HO
Cholesterol
OH
CH3
HO
Estradiol

Figure III-36 Structures of sterols.

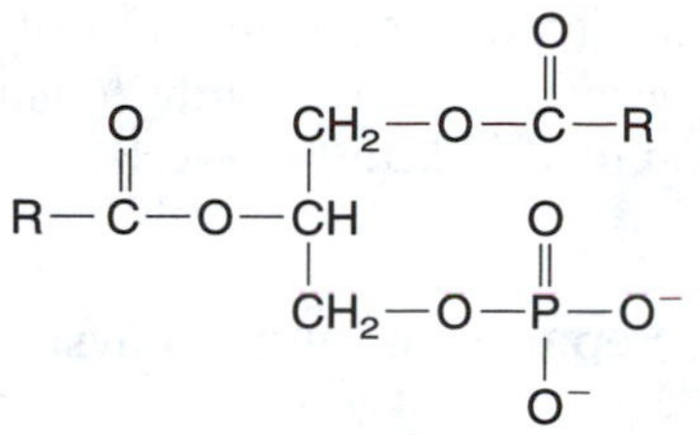

$$R{-}C(=O){-}O{-}CH(CH_2{-}O{-}C(=O){-}R)(CH_2{-}O{-}P(=O)(O^-){-}O^-)$$

Figure III-37 A phosphatidic acid.

The major glycerophosphatides are esters of phosphatidic acid that have the general formula shown in Figure III-38, in which R represents a long-chain alkyl group and R′ an amino alcohol or inositol. Additional phosphoryl groups are esterified to inositol in some phosphatidyl inositols. Phosphatidyl inositol diphosphate (PIP_2) and other phosphoinositols are important molecules in signal transduction.

Additional residues of glycerol are incorporated in the more complex structures of phosphatidyl glycerol, phosphatidyl 3′-*o*-aminoacylglycerol (lipamino acid), and cardiolipin (see Fig. III-38).

The plasmalogens represent an unusual type of glycerophosphatide that has been isolated from animal and plant tissues. Acid hydrolysis of these materials yields long-chain aldehydes and monoacylated α-glycerophosphate. The long-chain aldehydes originate from vinyl ether structures. The vinyl ether group appears usually (if not always) in

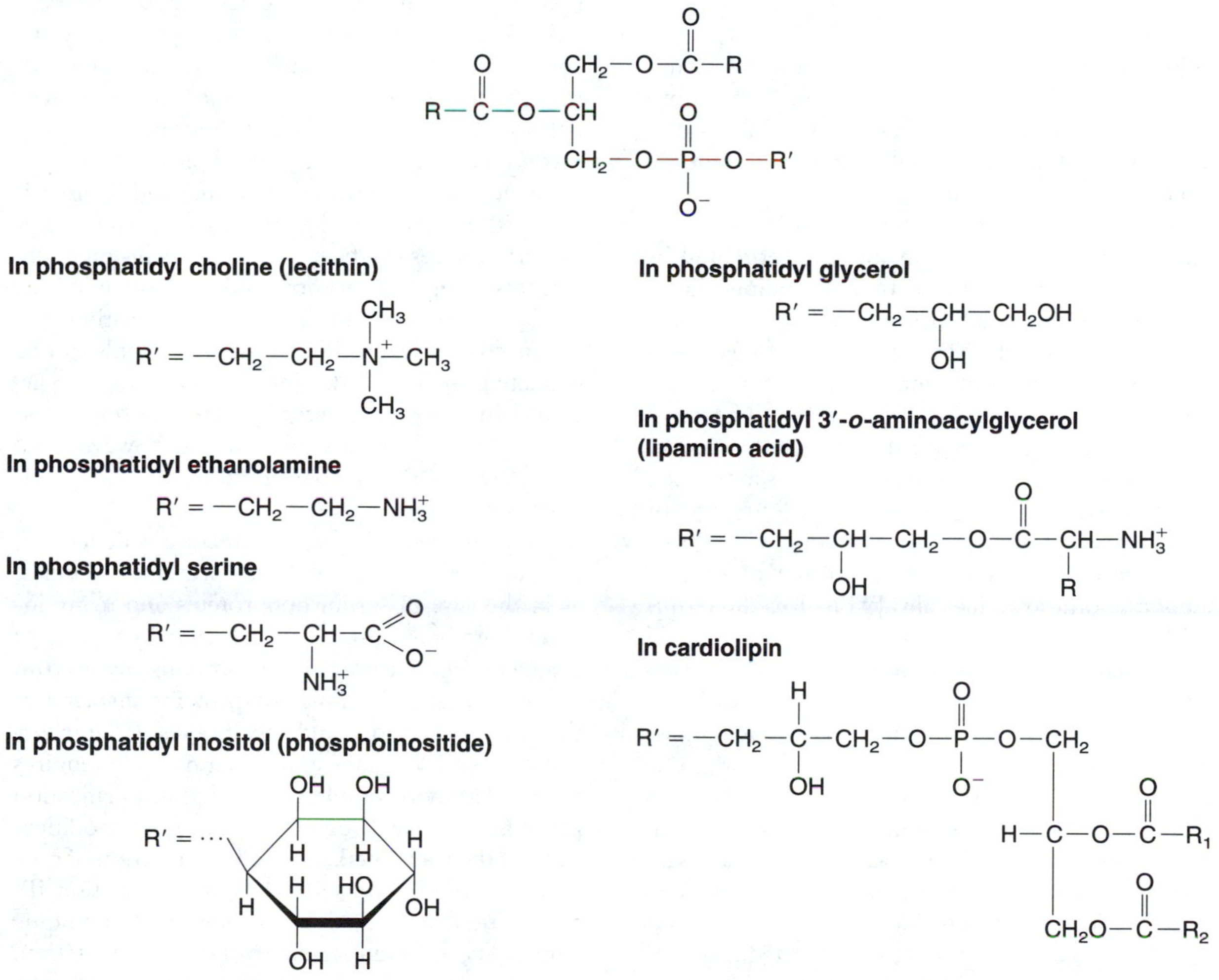

$$R{-}C(=O){-}O{-}CH(CH_2{-}O{-}C(=O){-}R)(CH_2{-}O{-}P(=O)(O^-){-}O{-}R')$$

In phosphatidyl choline (lecithin)

$$R' = {-}CH_2{-}CH_2{-}\overset{+}{N}(CH_3)_3$$

In phosphatidyl ethanolamine

$$R' = {-}CH_2{-}CH_2{-}NH_3^+$$

In phosphatidyl serine

$$R' = {-}CH_2{-}CH(NH_3^+){-}C(=O)O^-$$

In phosphatidyl inositol (phosphoinositide)

R′ = inositol ring (OH, OH, H, H, H, H, HO, H, OH, OH, H)

In phosphatidyl glycerol

$$R' = {-}CH_2{-}CH(OH){-}CH_2OH$$

In phosphatidyl 3′-*o*-aminoacylglycerol (lipamino acid)

$$R' = {-}CH_2{-}CH(OH){-}CH_2{-}O{-}C(=O){-}CH(R){-}NH_3^+$$

In cardiolipin

$$R' = {-}CH_2{-}C(H)(OH){-}CH_2{-}O{-}P(=O)(O^-){-}O{-}CH_2{-}C(H)(O{-}C(=O){-}R_1){-}CH_2O{-}C(=O){-}R_2$$

Figure III-38 Structures of various glycerophosphatides. The generic structure is shown at the top of the figure, and the classes differ in the structure of the R′ group as shown.

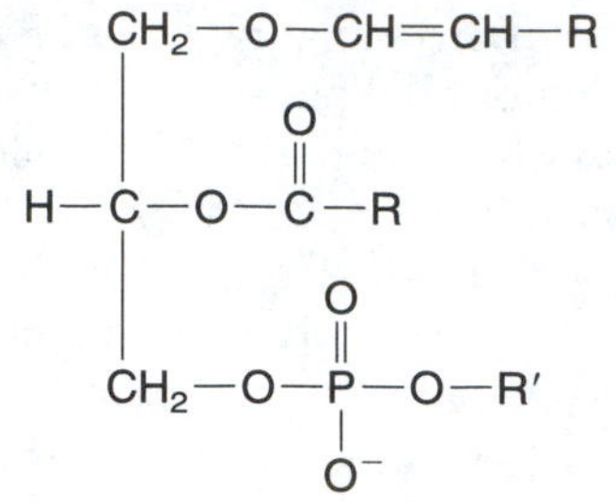

```
CH2—O—CH=CH—R
 |
 |        O
 |        ‖
H—C—O—C—R
 |
 |        O
 |        ‖
CH2—O—P—O—R′
          |
          O⁻
```

Figure III-39 Structure of plasmalogens.

the α position. The R′ phosphomonoester group may be either choline or ethanolamine, although ethanolamine usually predominates. In the structure of a plasmalogen shown in Figure III-39, R is an aliphatic chain and R′ is choline or ethanolamine.

Sphingolipids. The *sphingolipids* contain C_{18} hydroxyamino alcohols. Four such long-chain hydroxyamino alcohols have been isolated from natural lipids. Sphingosine and dihydrosphingosine are typical of animal tissues.

For simplicity we will concentrate on sphingosine, which is rarely seen in the free form, and its derivatives found in animal tissues. Sphingosine is formed biosynthetically from palmitoyl coenzyme A and serine (Fig. III-40). Attachment of a fatty acid to the amino group of sphingosine in amide linkage yields the *ceramides* (Fig. III-40). The most abundant sphingolipids are structures consisting of ceramides linked to a phosphate ester, as in the case of *sphingomyelin* or in glycosidic linkage to one or more carbohydrate residues to form *cerebrosides* or *gangliosides* (Fig. III-40). The linkage of the ceramide to other residues always involves the terminal hydroxyl group of sphingosine. The fatty acid components of the sphingolipids are usually saturated with an unusually high proportion of long-chain C_{24} acids such as lignoceric (n-tetracosanoic) and cerebronic (α-hydroxylignoceric) acid. The carbohydrate component of brain cerebrosides is galactose. Glucose is present in small amounts in spleen cerebrosides. Spleen also contains a disaccharide derivative made up of both glucose and galactose (ceramide lactoside).

The gangliosides, a more complicated group of ceramide oligosaccharides, contain (in addition to galactose and glucose) both N-acetylgalactosamine and sialic acid (N-acetylneuraminic acid) (Fig. III-41). Sialic acid is also frequently found in other protein- and lipid-linked oligosaccharides.

Isolation, Separation, and Analysis of Lipids

Solvent Extraction. The first step in the preparation of a lipid is extraction from the tissue. Since lipids are defined by their solubility in organic solvents (see above), whereas proteins, nucleic acids, carbohydrates and most of the low-molecular-weight metabolites of the cell are generally water-soluble, the isolation of lipids is, in principle, quite simple. Extraction of a biological sample with an organic solvent that is immiscible with water, followed by separation of the phases should yield the lipid components in the organic phase. The various classes of lipids can then be separated by the chromatographic techniques described below. Although such procedures are subject to the potential complications listed below, they are satisfactory if you wish to analyze the "free" lipids in a cell type or tissue—that is, lipids that are not tightly associated with proteins or oligosaccharides. The extraction procedures can be modified by inclusion of detergents or high concentrations of inorganic salts to disrupt noncovalent, but strong associations between lipids and proteins so that the lipids can be extracted and analyzed. Lipids that are *covalently* attached to proteins or carbohydrates are not necessarily extracted by such procedures, however, and specialized methods of separation must be developed for each case.

In many cases, biological function is dependent on specific associations between proteins and lipids, as in the case of serum lipoproteins and many integral membrane proteins, such as the enzymes of the electron transport chain, proteins involved in transmembrane signaling, or proteins involved in the transport of ions and metabolites. Characterization of such proteins in their native state requires their isolation in complexes with lipids. Purification procedures in such cases often require modifications of the usual methods used for isolation of proteins (see the Introductory Chapter to Section II). Such modifications might include use of nonionic detergents, differential centrifugal sedimentation, and chromatographic separations using lipophilic materials. Harsh extraction of the lipid components

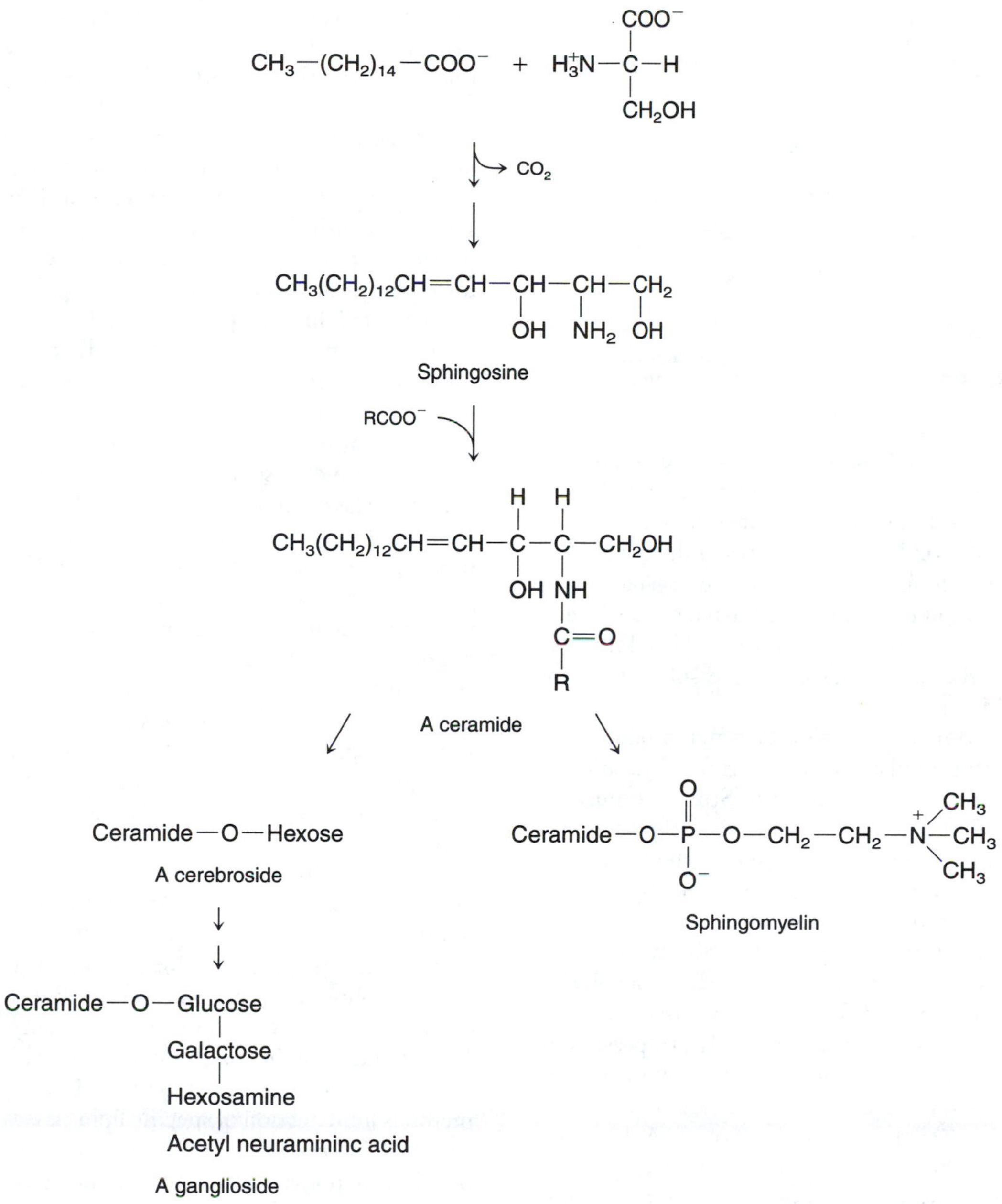

$$CH_3{-}(CH_2)_{14}{-}COO^- + H_3\overset{+}{N}{-}CH(COO^-){-}CH_2OH$$

$\downarrow$ CO_2

$$CH_3(CH_2)_{12}CH{=}CH{-}CH(OH){-}CH(NH_2){-}CH_2(OH)$$

Sphingosine

$RCOO^-$ $\downarrow$

$$CH_3(CH_2)_{12}CH{=}CH{-}CH(OH){-}CH(NH{-}C({=}O){-}R){-}CH_2OH$$

A ceramide

Ceramide—O—Hexose

A cerebroside

$\downarrow$

$\downarrow$

Ceramide—O—Glucose—Galactose—Hexosamine—Acetyl neuramininc acid

A ganglioside

$$\text{Ceramide}{-}O{-}P({=}O)(O^-){-}O{-}CH_2{-}CH_2{-}\overset{+}{N}(CH_3)_3$$

Sphingomyelin

Figure III-40 Origin and structure of the major sphingolipids.

from such complexes using organic solvents would allow their separation for analysis, but usually results in loss of biological activity and denaturation of the protein components.

Even in the case of extraction of "free" lipids by organic solvents, the experimenter must be aware of a number of potential problems.

1. Lipid mixtures have remarkable ability to carry nonlipid materials into organic solvents, partly by way of ionic interactions. Thus, hexane extracts of blood serum contain urea, sodium chloride, and amino acids, and hexane extracts of soybeans contain the sugars raffinose and stachyose.

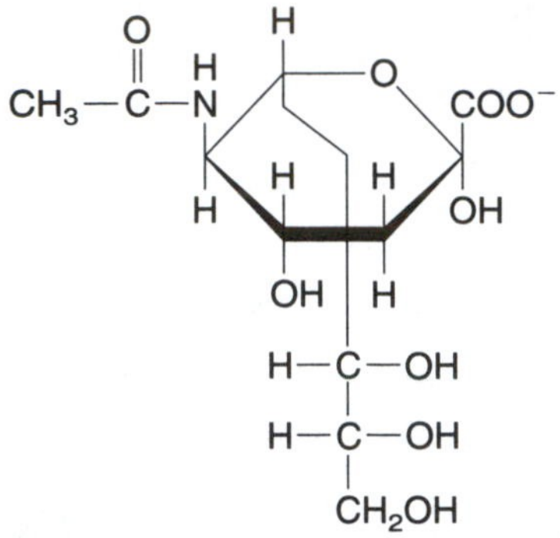

Figure III-41 Sialic acid (N-acetylneuraminic acid).

2. The glycerophosphatides usually contain substantial amounts of highly unsaturated fatty acids that undergo rapid autoxidation catalyzed by Cu^{2+} or Fe^{3+} salts. These salts tend to be carried along by lipids into the organic phase. Thus, lipid extractions may have to be made in a nitrogen atmosphere in the cold and may require the use of peroxide-free solvents or antioxidants.
3. Extraction of wet tissues often produces intractable emulsions; hence prior dehydration of the sample is often helpful. Still, you must realize that dehydration of certain tissues causes irreversible binding of some of the lipid components to proteins and consequently leads to decreases in the lipid yield.
4. Solvents vary widely in their ability to extract different types of lipids, and a series of solvents may be required for complete extraction.
5. Enzymes that hydrolyze lipids are present in various tissues. Certain of these are activated by organic solvents, and the rate of activation increases as the solvent is warmed. At least one such enzyme is active in an ether dispersion. Thus, consideration must be given to the possibility of enzymatic degradation, particularly when dealing with plant sources.

No single satisfactory procedure has been discovered that avoids all these complications. Thus, extraction conditions must be adapted to the tissue under study and to the quantity and type of lipid desired. As a point of departure, the total lipids of a wet tissue can usually be extracted with chloroform–methanol mixtures according to the procedure of Folch (or one of several variants of this procedure). Often, you can remove nonlipid materials present in the chloroform–methanol extract by extracting the organic phase with water.

Separation of Lipid Classes. Biochemists frequently obtain preparative scale separations of lipid mixtures by using step-by-step elution from silicic acid or Florosil columns with solvents of gradually increasing polarity. This is particularly applicable for mixtures of neutral lipids. Polar lipids also may be separated in this way, although ion-exchange separations on DEAE- or TEAE-cellulose are more powerful for large-scale separations of polar lipids. Lipophilic Sephadex (derivatives of the normal dextran-bearing alkyl groups) can be used to separate gross lipid classes and individual components within a class, such as various cholesterol esters.

Thin-layer chromatography (TLC), high-performance liquid chromatography (HPLC) and gas–liquid chromatography (GLC) are sensitive and powerful techniques for analytical scale separations of lipid mixtures into classes. These methods are also valuable for further resolution of the members of each class into single components or chemically very similar subgroups. TLC of lipids generally utilizes silica gel (silicic acid) as the adsorbant and the same nonpolar solvents used in preparative silicic acid chromatography. Like most TLC procedures, TLC resolution of lipids provides speed, sensitivity, and excellent resolving power. Many methods have been developed for detection of lipids on thin-layer chromatograms. Spraying the chromatogram with 50% H_2SO_4, followed by heating, reveals lipids (and all nonvolatile organic components) as charred spots. Most lipids contain unsaturated fatty acids, which can be visualized after reaction with I_2 vapor. Other methods for detection of specific lipid classes are also available. Incorporation of $AgNO_3$ into the silica gel of TLC or preparative column systems is a useful way of separating very similar lipids that contain fatty acids of varying degrees of unsaturation (i.e., differing numbers of double bonds). The Ag^+ ions form complexes with double bonds. The migration of lipids on the gel is therefore inversely related to the number of double bonds (i.e., Ag^+ ion complexes) per molecule of lipid.

GLC and HPLC are powerful techniques for analyzing lipids. GLC is particularly useful for analysis of fatty acid mixtures or fatty acid methyl esters derived from isolated lipid classes or specific lipids. GLC of intact nonpolar lipid classes, such as triglycerides or cholesterol esters, is also practical. In contrast, polar lipids must be converted to more volatile derivatives or be analyzed after degradation and derivatization. For this reason HPLC is more generally applicable to the analysis of polar lipids. The general principles of operation and detection used in GLC and HPLC have been described in Experiment 2. Identification and determination of unknown structures of components separated by chromatography can be greatly aided by use of mass spectrometry and NMR spectroscopy.

Degradation of Lipids and Analysis of Lipid Components. To determine the complete structure of an isolated lipid, usually you must degrade the lipid into its components, separate the components, and analyze them. This section reviews chemical and enzymic techniques for degradation of lipids, with an emphasis on common phospholipids.

Alkaline hydrolysis (e.g., the Schmidt–Thannhauser procedure, which employs 1 N KOH at 37°C for 24 hr) cleaves most ester bonds but leaves the amide groups in sphingolipids intact. Alcoholic alkali may be necessary with lipid mixtures that do not disperse readily into aqueous solutions. Potassium hydroxide is preferable because it is more soluble in alcohol than NaOH; similarly, potassium salts of fatty acids are more soluble than their equivalent sodium salts.

Acid hydrolyses are usually carried out by refluxing in 6 N aqueous hydrochloric acid (constant boiling) or 5 to 10% solutions of HCl in methanol (to promote solubility) for 4 to 30 hr, depending on the lipid in question. Most glycerophosphatides are hydrolyzed by acid to fatty acids, glycerophosphate, and the free base, just as with alkali. However, inositol phosphatides initially yield inositol phosphate and diglycerides on acid hydrolysis. Hydrochloric acid is easily removed by vacuum, which makes chromatographic examination of the hydrolysis products easier.

Because the sites of cleavage are often quite specific, enzymatic degradation of triglycerides and glycerophosphatides can provide useful structural information. Lipases generally cleave fatty acid residues from the 1 and 3 positions of glycerol, leaving the acyl groups on the secondary hydroxyl intact. The acyl groups can be removed subsequently by alkaline hydrolysis. The phospholipases (A_1, A_2, C, and D), which are abundant in bee and snake venoms, are valuable for structure determination because they selectively hydrolyze various phosphoglyceride components. As seen in Figure III-42, phospholipase A_2 removes the central fatty acid, phospholipase A_1 removes the terminal fatty acid, and phospholipase B (which is a mixture of phospholipase A_1 and phospholipase A_2 activities) removes both the central and terminal fatty acids. Phospholipases C and D cleave on either side of the phosphate, yielding a diacyl glycerol or a phosphatidic acid, respectively. The specificity of the various phospholipases for the site of cleavage is quite high. These enzymes have some specificity for the nature of the acyl group and (particularly the X group) of the phosphatide, placing some limitations on the use of these enzymes in structural studies of unknown phospholipids.

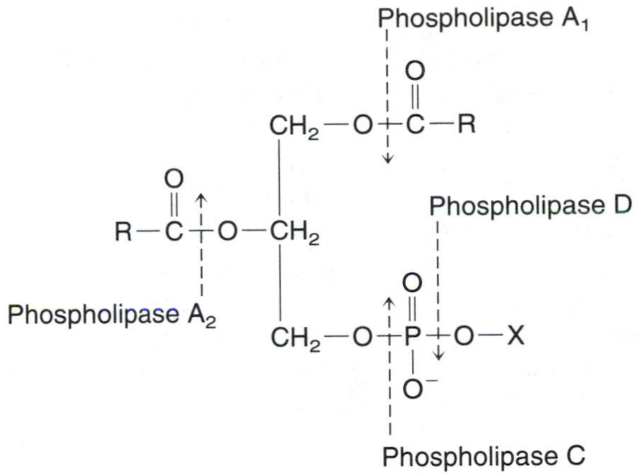

Phospholipase A_1
Phospholipase A_2
Phospholipase D
Phospholipase C

Figure III-42 Sites of hydrolytic cleavage by phospholipases.

Membranes

The most obvious function of a membrane is to serve as a physical barrier or boundary between the interior of a cell or subcellular organelle (nucleus, mitochondrion, chloroplast, etc.) and the surrounding medium. It has become evident, however, that this is far too static a view of the membrane. It is true that membranes act as envelopes and contain contents of cells or organelles, but all membranes possess machinery for selective movement of molecules across them, and frequently pump such molecules against a concentration gradient (active transport). Furthermore, the membrane is often the site of intense metabolic activity, particularly electron transport, oxidative and degradative reactions, biosynthesis of cell-surface components, and transformations of metabolic energy. Membranes are also important sites of cell-cell and organelle-cell interactions, such as hormone-cell interactions, contact inhibition, neurotransmitter action, recognition by the immune system and so on. In addition to being a physical barrier, the membrane is also a site of intense biochemical activity.

Most of these activities require both protein and lipid components of the membrane, as well as varying degrees of structural integrity of the membrane or cell. To obtain a chemical and physical understanding of membrane phenomena, it is necessary to isolate membranes and study their composition and structural arrangements. Ideally, you would reconstruct the phenomena of interest in isolated membranes or reconstituted systems made from membrane components. In some cases, this is difficult to carry out. An example of the study of biochemical functions in isolated membrane fractions is the characterization of electron transport performed in Experiment 14.

Isolation

One of the major problems encountered when isolating membranes is that they usually do not occur in sharply defined molecular sizes, so that differential centrifugation, while valuable, does not usually yield homogeneous preparations. Isopycnic density gradient centrifugation, which takes advantage of the generally low density of membranes or membrane-rich organelles, is therefore an invaluable adjunct to membrane isolation. When the membrane of a specific organelle is under study, it is usually desirable to isolate the organelle before separating the membrane. Contamination of plasma membranes with cellular organelles and organelle membranes is a serious problem in membrane isolation. The mammalian erythrocyte (red blood cell) is an especially useful source of homogeneous plasma membranes because these cells have no other organelles (see Experiment 13).

It is essential to have a set of criteria for assessing the purity of isolated membrane fractions. Two methods are most useful. Phase-contrast or electron microscopy gives a visual indication of the size and appearance of an isolated membrane preparation and reveals contamination by organelles and other material. If the membrane fraction in question is known to contain a specific chemical component or enzymic activity (such as Na^+,K^+-ATPase or 5′-nucleotidase for animal plasma membranes, or succinic dehydrogenase for mitochondrial membranes), assay for such is a valuable criterion of membrane homogeneity. Other criteria for identity of membranes include the presence of specific antigens, the phospholipid:protein ratio, and the specific protein composition revealed by sodium dodecyl sulfate–polyacrylamide gel electrophoresis (SDS-PAGE).

Composition of Membranes

The primary constituents of membranes are lipid and protein; carbohydrate residues are found in relatively small amounts attached to members of both major classes. Any organism has fairly constant membrane composition, but membranes vary widely from organism to organism and among various kinds of cells. For example, membranes vary from 18% protein, 79% lipid, and 3% other compounds in the myelin sheath, to 75% protein and 25% lipid in the cell membrane of gram-positive bacteria. A "typical" figure would be about 60% protein and 40% lipid.

The lipid composition of membranes can vary in the same manner. Major components of prokaryotic membrane lipids are phospholipids and glycolipids; eukaryotic membrane lipids typically contain these two classes and, in addition, sterols such as cholesterol, and sphingolipids. Table III-3 gives the lipid content of some membranes from various sources. Table III-3 does not list some of the unusual and less abundant lipids found in membranes, such as lipamino acids, glycolipids, and phosphatides containing unusual fatty acid residues.

The protein composition of a membrane is difficult to define precisely because proteins are bound to the membrane with highly varying degrees of avidity. Some proteins can be removed from a membrane by fairly mild treatments, such as washing with water or salt solutions. Such proteins are usually referred to as *peripheral membrane proteins*. In such cases, however, it may be difficult to say with certainty whether the protein in question is really a "membrane protein" or merely happens to associ-

Table III-3 Lipid Composition of Membranes from Various Sources

Lipids (given in percent of total lipid)	Animal				Bacteria	
	Myelin	Erythrocyte	Mitochondria	Microsome	*E. coli*	*Bacillus megaterium*
Cholesterol	25	25	5	6	0	0
Phosphatides						
Phosphatidylethanolamine	14	20	28	17	100	45
Phosphatidylserine	7	11	0	0	0	0
Phosphatidylcholine	11	23	48	64	0	0
Phosphatidylinositol	0	2	8	11	0	0
Phosphatidylglycerol	0	0	1	2	0	45
Cardiolipin	0	0	11	0	0	0
Sphingolipids						
Sphingomyelin	6	18	0	0	0	0
Ceramide	1	0	0	0	0	0
Cerebrosides	25	0	0	0	0	0
Others	11	1	0	0	0	10

Source: Modified from Wold, F. (1971). *Macromolecules: Structure and Function.* Englewood Cliffs, NJ: Prentice-Hall.

ate with the membrane during isolation. Complete release of protein from membrane preparations requires use of agents that disrupt the associations between lipids and proteins. Such procedures release *integral membrane proteins* that are deeply embedded in the membrane or have one or more transmembrane segments. Disruption of such protein associations with membranes requires agents rarely used in purifying soluble proteins, such as organic solvents, chaotropic agents, and detergents. Many membrane proteins are soluble in such agents and may require them (or phospholipids) for activity. Fractionation of membrane proteins follows conventional procedures (see Section II introduction), except that inclusion of detergents is frequently necessary. Analytical electrophoresis of membranes on polyacrylamide gels containing sodium dodecyl sulfate has proven to be a powerful tool. As we have discussed, many enzymes have been found to be associated with membranes. Glycoproteins are particularly abundant in animal-cell membranes, although they also occur in bacterial membranes.

Structure

The appreciation of the many important biological functions of membranes has led to intense research activity into the structural arrangement of lipids and proteins in membranes, the forces involved in forming such structures, and the relation between membrane structure and function. Our current view of membrane structure is based primarily on physical studies of natural membranes and of synthetic lipid monomolecular and bimolecular layers. *Amphipathic* molecules, particularly phospholipids, have a strong tendency to self-associate and to form layered structures when they are dispersed into aqueous medium. (Amphipathic molecules are molecules in which both highly polar, hydrophilic elements and nonpolar, hydrophobic elements are joined in the same molecule.) Monomolecular layers (monolayers) of amphipathic molecules tend to form at air–water interfaces. Bilayers form across an opening between two water-containing chambers. When a phospholipid is dispersed into water by ultrasonic oscillation, the lipids spontaneously form bilayered aggregates such as micelles and water-containing vesicles called "liposomes." These self-assembling systems provide models for studying the interactions of amphipathic lipids (phosphatidylcholine is a commonly used lipid) in aqueous environments. Such bilayers are thought to be models for the simplest or most primitive membrane structures (see Fig. III-43). The bilayer is

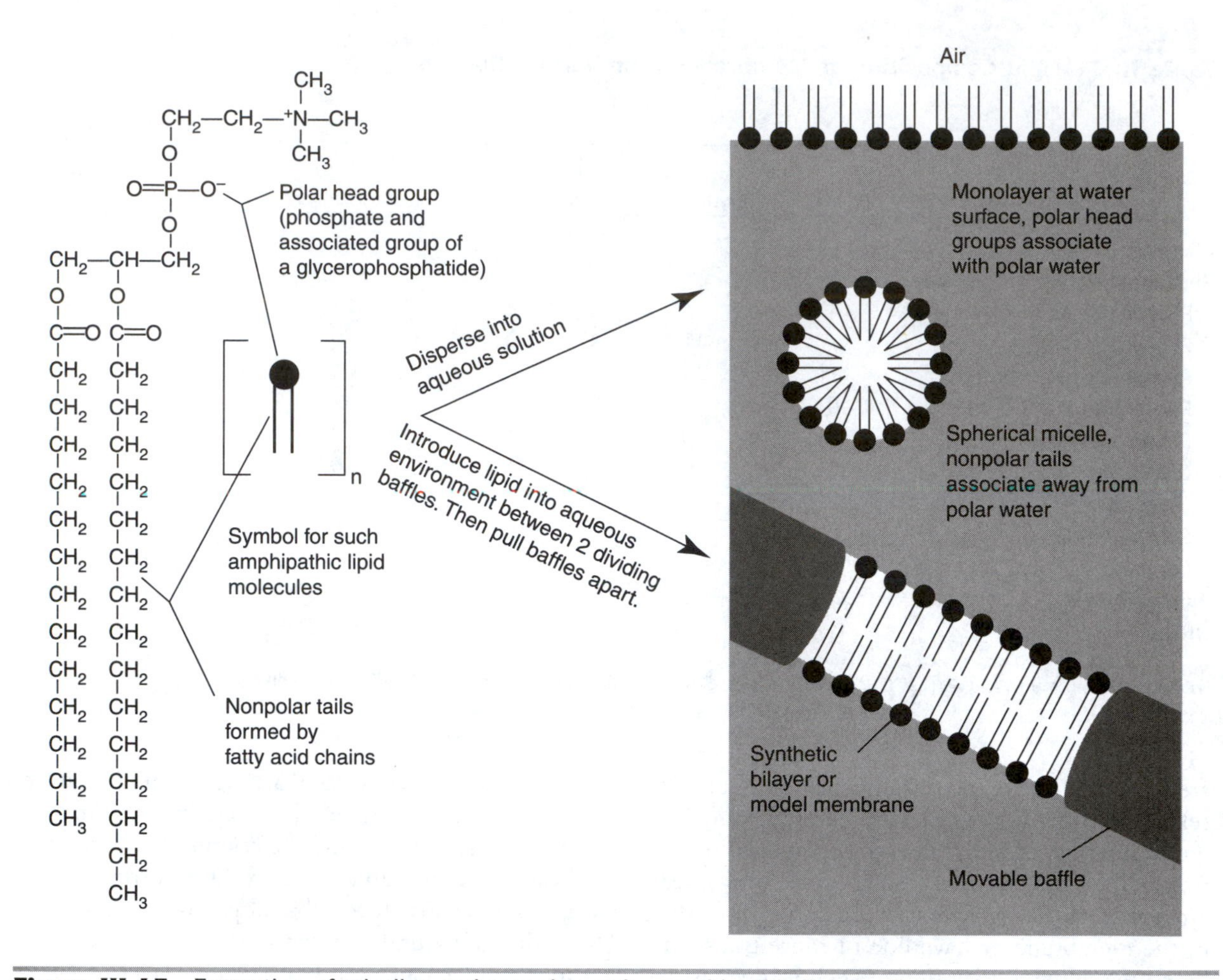

Figure III-43 Formation of micelles and monolayers by amphipathic lipid molecules in aqueous solution.

formed by hydrophobic interactions that lead to association of the nonpolar tails of the lipid molecule and interaction of the polar heads with water. Proteins are believed to associate with lipids in membranes by hydrophobic interactions as well as interactions with polar head groups, such as hydrogen bonds and electrostatic attractions. Synthetic lipid bilayer systems possess many of the physical properties of natural membranes.

The basic tenets of the fluid mosaic model of membrane structure shown in Figure III-44 are widely accepted. The phospholipid bilayer is the basic structural feature of the membrane, and proteins or multiprotein complexes are viewed as island embedded in a sea made up by the lipid bilayer. Membranes are *fluid* in the sense that lipids and proteins can diffuse freely in the plane of the membrane. Flip-flop of phospholipids, or other charged molecules, across the bilayer is unfavorable because of the energy required to move a charged or polar molecule through the hydrophobic interior of the membrane. There is also abundant evidence that membranes are asymmetric; that is, they have "inside" and "outside" surfaces with different proteins and lipids found on each side.

The relationship between the physical structure of membranes and their biological function is not yet well defined. Clearly, the interposition of a nonpolar layer between the spaces separated by a membrane prevents translocation of polar biological molecules, as a cell envelope must. Furthermore, the positioning of specific proteins at the surface of the membrane allows many specific "recognition"

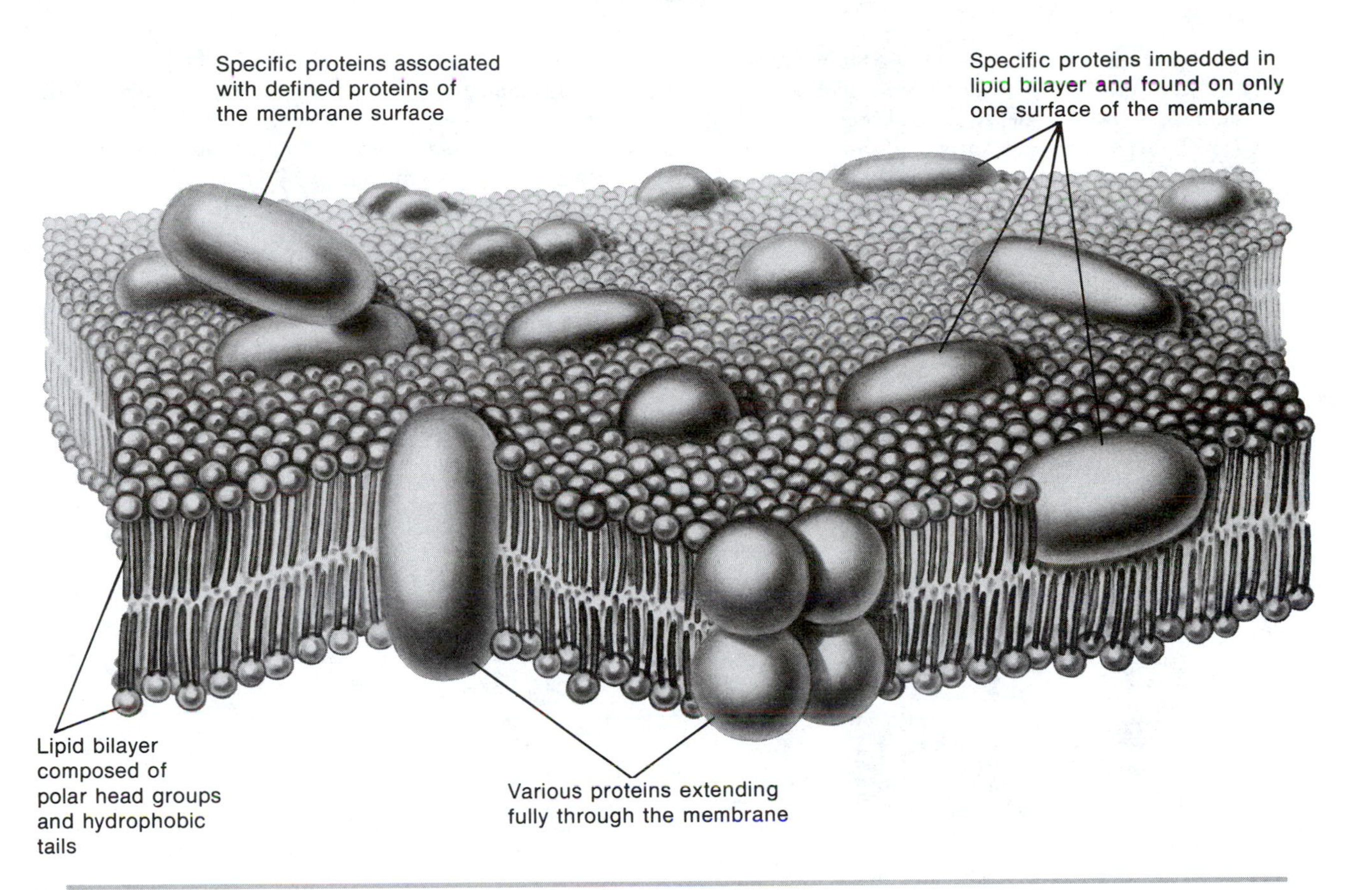

Figure III-44 The fluid mosaic model of membrane structure.

sites to be presented to other cells, hormones, and metabolites inside and outside the cell.

Studies have suggested that membrane proteins do not always undergo free diffusion in the membrane, and that they may be anchored by proteins of the cytoskelton, by the extracellular matrix or in other ways. Also, membranes are not uniform in their distribution of components, even on one side of the bilayer. They contain many specialized regions or domains, such as clathrin-coated pits, synapses in nerve cells, microvillae, and focal contacts. The lipid and protein composition of the domains are distinct and have been associated with specialized biological functions of the membrane.

The use of artificial lipid membranes and isolated membrane fragments and vesicles has been of great value in the study of membrane function, particularly transport and membrane-bound enzyme reactions. Current research in this area uses sophisticated techniques for determining the molecular architecture of these organelles at higher resolutions, including NMR, electron paramagnetic resonance, and fluorescence spectroscopy. Other methods for obtaining images of membranes and membrane components are x-ray crystallography of membrane proteins, Fourier analysis of electron microscopic images, and atomic force microscopy. Methods for isolating native membrane components in the functional state have also become much more effective. Important phenomena at the cell surface involve not only the cell membrane, but cell walls, capsular materials, and enzyme activities located outside the membranes of gram-negative bacteria (periplasmic proteins).

REFERENCES

Barker, R. (1971). *Organic Chemistry of Biological Compounds.* Englewood Cliffs, NJ: Prentice-Hall.

Casey, P. J., and Buss, J. E. (eds.) (1995). Lipid Modification of Proteins. *Methods Enzymol,* **250**.

Cummings, R. D., Merkle, R. A., and Stults, N. L. (1989). Separation and Analysis of Glycoprotein Oligosaccharides. *Methods Cell Biol* **32**:141.

Fukuda, M., and Kobata, A. (1994). *Glycobiology, A Practical Approach*. New York: Oxford University Press.

Gennis, R. B. (1989). *Biomembranes: Molecular Structure and Function*. New York: Springer Verlag.

Hardy, M. R., and Townsend, R. R. (1994). High-pH Anion-Exchange Chromatography of Glycoprotein-Derived Carbohydrates. *Methods Enzymol* **230**:208.

Hounsell, E. F. (ed.) (1998). *Glycoanalysis Protocols*. Totowa, NJ: Humana Press.

Lennarz, W. J., and Hart, G. W. (eds.) (1994). Guide to Techniques in Glycobiology. *Methods Enzymol,* **230**.

Kobata, A. (1992). Structures and Functions of the Sugar Chains of Glycoproteins. *Eur J Biochem* **209**:483.

Stryer, L. (1995). *Biochemistry*, 4th ed. chapters 11 and 18. W. H. New York: Freeman.

Vance, D. E., and Vance, J. E. (eds.) (1985). *Biochemistry of Lipids and Membranes*. Menlo Park, CA: Benjamin/Cummings.

Varki, A. (1993). Biological Roles of Oligosaccharides: All of the Theories Are Correct. *Glycobiology* **3**:97.

Wold, F., and Moldave, K. (1984). Posttranslational Modifications, Part B *Methods Enzymol,* **107**:3–23; 23–26.

Microanalysis of Carbohydrate Mixtures by Isotopic, Enzymatic, and Colorimetric Methods

Theory

Naturally occurring polysaccharides are made up of a variety of monosaccharides connected by glycosidic bonds. Often, these polysaccharides are linked to proteins and lipids. The function of such polysaccharides is often critically dependent on the composition and sequence of their monomeric units. Determination of the structure of polysaccharides is a multistep process. In the early stages, it is crucial to ascertain both the relative amount and the chemical identity of the monomers. After a polysaccharide has been broken down to its monomeric units by hydrolytic or enzymatic digestion, a variety of analytical techniques can be employed to identify and quantify each monomer. Three of these techniques will be demonstrated in this laboratory exercise.

In this experiment, you will be given a sample containing D-glucose, D-ribose, and D-glucosamine. Your objective will be to determine the concentration of each of these carbohydrates in the sample using three independent techniques:

1. Reduction with tritiated $NaBH_4$ (NaB^3H_4), followed by analysis of the radioactive sugar alcohols by paper chromatography and liquid scintillation counting.
2. Quantification of D-glucose with the use of a coupled enzyme assay.
3. Quantification of D-glucosamine with the use of the Elson–Morgan colorimetric reaction.

Reaction of Reducing Sugars with Sodium Borohydride

A characteristic property of reducing sugars is that they can be reduced to their corresponding sugar alcohols with sodium borohydride (Fig. 11-1). This is one of the few analytical reactions of carbohydrates that is known to proceed stoichiometrically to give a quantitative yield of product. Notice that *one* mole of NaB^3H_4 can reduce *four* moles of sugar to its corresponding sugar alcohol. If the specific activity of the NaB^3H_4 used in the reaction and the amount of carbohydrate in the sample are known, you can predict the amount of radioactivity that will be present in the product. In the same sense, the amount of product in a sample can be determined if the specific activity of the NaB^3H_4 used in the reaction and the amount of radioactivity present in the product are known. For example, suppose you perform a carbohydrate reduction reaction with 40 μCi/μmol NaB^3H_4, and you determine that the sugar alcohol contains 10 μCi of ^3H. Based on the stoichiometry of the reaction, you know that you have then reduced 1 μmol of the sugar to its sugar alcohol. If you know the volume of the carbohydrate solution used in the reaction, you can then determine the concentration of carbohydrate present in the unknown sample.

When utilized in conjunction with paper chromatography, this technique can be effective in identifying and quantifying a number of different carbohydrates in an impure solution. Commonly, polysaccharides are hydrolyzed in 1 N acid to produce their constituent monomers. The solution is then adjusted to pH 8.5 with the addition of 1 volume of 2 N sodium carbonate and reduced with NaB^3H_4. Following the completion of the reduction reaction, the excess (unreacted) NaB^3H_4 is removed from the solution through the addition of a strong acid:

$$NaB^3H_4 + 3H_2O + H^+ \longrightarrow Na^3H_2BO_3 + 4\,^3H_2\uparrow$$

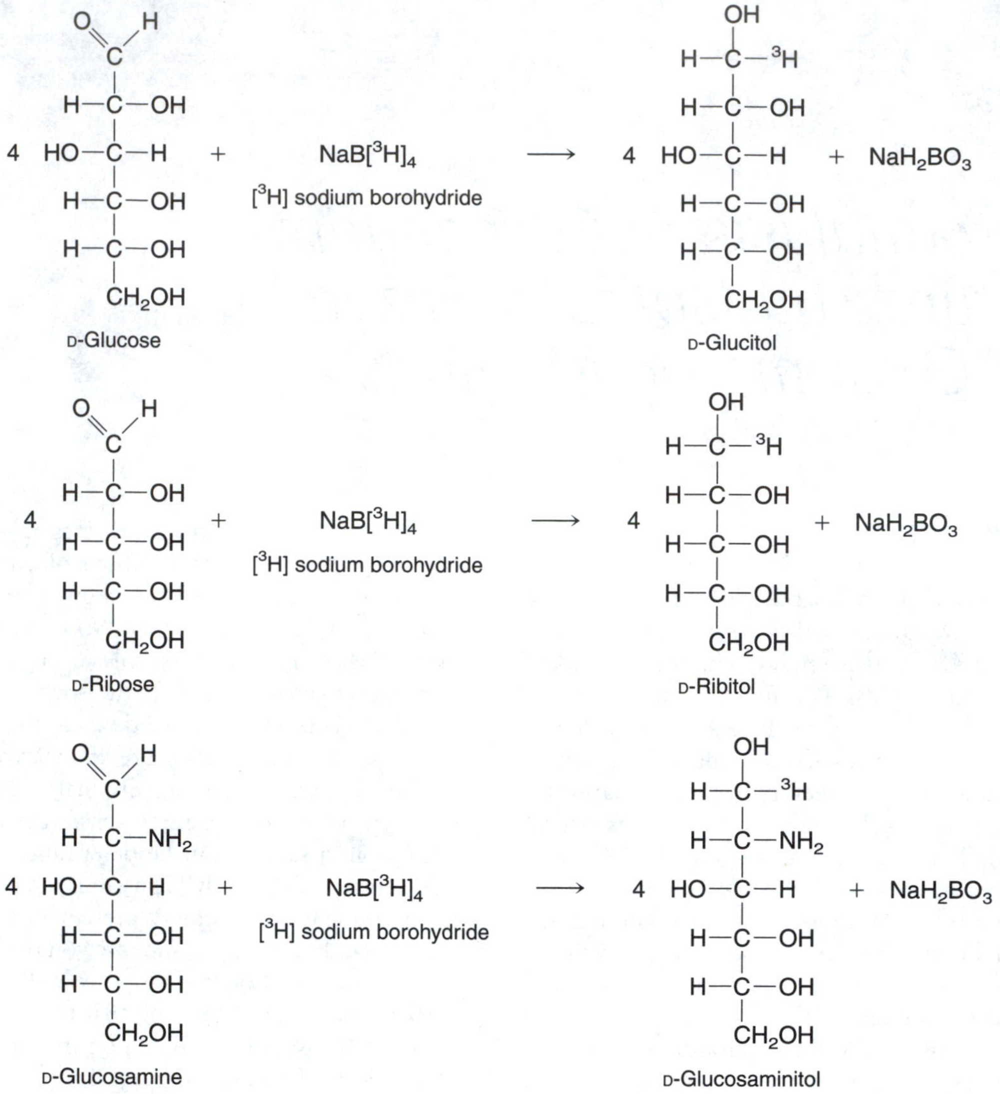

4 D-Glucose (CHO–$CHOH$–$HOCH$–$CHOH$–$CHOH$–CH_2OH) + $NaB[^3H]_4$ ([^{3}H] sodium borohydride) $\longrightarrow$ 4 D-Glucitol (HO–CH^3H–$CHOH$–$HOCH$–$CHOH$–$CHOH$–CH_2OH) + NaH_2BO_3

4 D-Ribose (CHO–$CHOH$–$CHOH$–$CHOH$–CH_2OH) + $NaB[^3H]_4$ ([^{3}H] sodium borohydride) $\longrightarrow$ 4 D-Ribitol (HO–CH^3H–$CHOH$–$CHOH$–$CHOH$–CH_2OH) + NaH_2BO_3

4 D-Glucosamine (CHO–$CHNH_2$–$HOCH$–$CHOH$–$CHOH$–CH_2OH) + $NaB[^3H]_4$ ([^{3}H] sodium borohydride) $\longrightarrow$ 4 D-Glucosaminitol (HO–CH^3H–$CHNH_2$–$HOCH$–$CHOH$–$CHOH$–CH_2OH) + NaH_2BO_3

Figure 11-1 Reaction of reducing carbohydrates with sodium borohydride.

It is very important to perform this reaction uncapped in a fume hood to prevent the room from filling with tritium gas. The mixture of tritiated sugar alcohols is then separated by paper chromatography on thin strips. These strips are divided into half-inch segments and analyzed by liquid scintillation counting. The total counts per minute of ^3H in each sugar alcohol peak can then be used to calculate the amount of each carbohydrate present in the sample.

In addition to the tritium isotope present in the sodium borohydride, the carbohydrate sample that you are given will also contain a small amount of high specific activity ^{14}C-glucose, which will serve as an *internal control* or an *internal standard* for the reaction. Inclusion of ^{14}C-glucose in the reaction will allow you to verify three important parameters:

1. Whether the reduction reaction with NaB^3H_4 went to completion.
2. How much of the total reaction mixture was transferred to the chromatogram.
3. Which of the ^3H-containing peaks on the chromatogram corresponds to glucitol, the product of the reduction of glucose.

If the reduction reaction was complete, there will be a single ^{14}C peak superimposed on the ^3H-glucitol peak. If the reduction was incomplete, there will be an additional ^{14}C peak that corresponds to unreduced ^{14}C-glucose. The ^{14}C-glucose internal control will also indicate what percent of the total reaction was analyzed on the chromatogram. If the reaction contained 1 μCi of ^{14}C-glucose, and

0.25 μCi total are present on the chromatogram, then you know that only 25% of the sample was analyzed. Finally, the ^{14}C-glucose internal control will allow you to identify the glucitol peak. Since D-glucose is the only carbohydrate in the unknown sample that is labeled with ^{14}C, the glucitol peak will be the only sugar alcohol peak that contains this isotope. It will be important to identify the glucitol peak, since the migration of all of the other sugar alcohols will be measured relative to glucitol (an $R_{glucitol}$ value will be calculated for each sugar alcohol). By using *high specific activity* ^{14}C-glucose as the internal control component, one can follow the fate of D-glucose in the reaction without significantly affecting the total mass of glucose present in the sample (it will add identifiable counts to the reaction while contributing negligible mass).

Elson–Morgan Determination of Hexosamines

The most specific and frequently used assay to quantify 2-deoxy-2-amino sugars employs the condensation of the carbohydrate with 2,4-pentanedione in basic solution. The product (a pyrrole derivative) is then reacted with *p*-dimethylaminobenzaldehyde to form a chromogen that has a maximum absorbance at 530 nm (Fig. 11-2). Although both 6-deoxy-6-amino and 3-deoxy-3-amino sugars can also be analyzed using the Elson–Morgan reaction, the chromogens

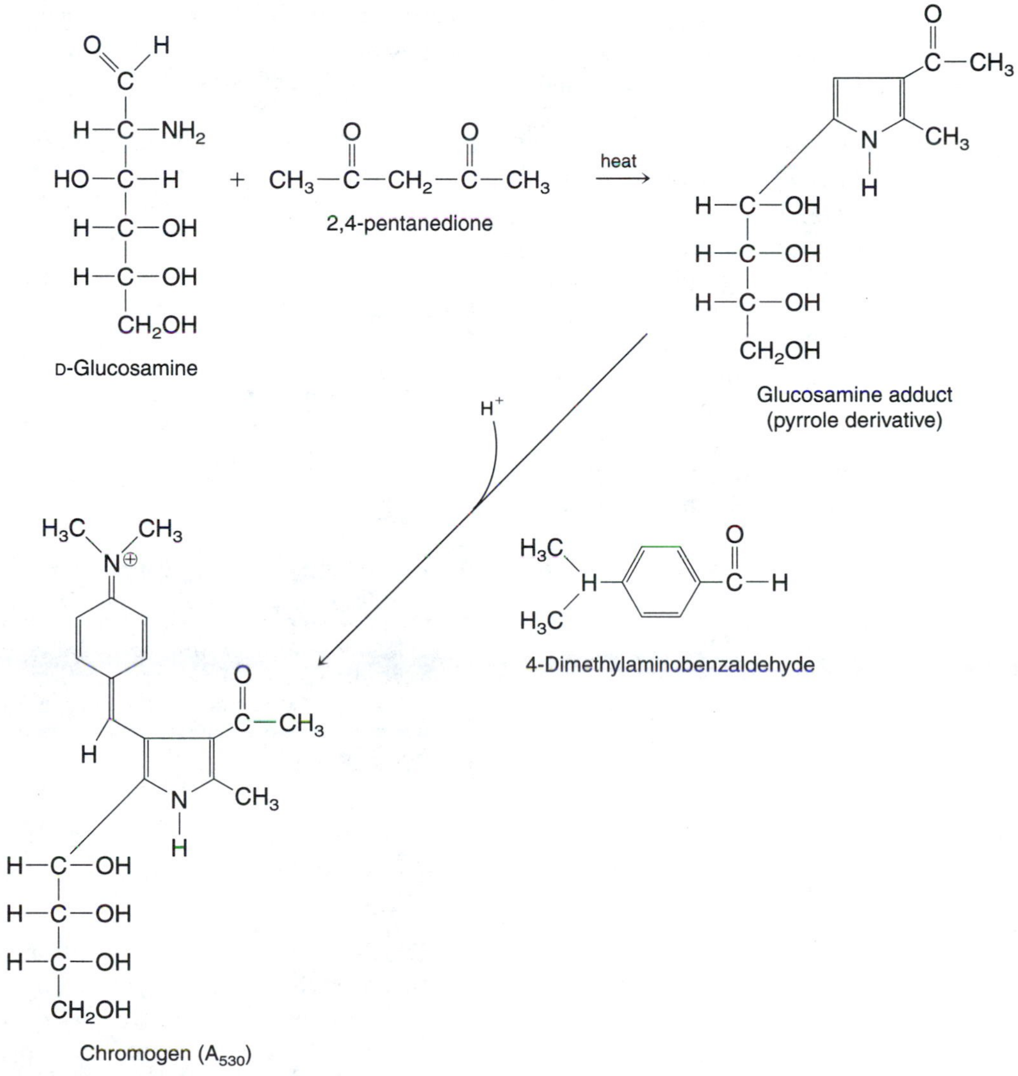

D-Glucosamine + 2,4-pentanedione —heat→ Glucosamine adduct (pyrrole derivative)

4-Dimethylaminobenzaldehyde, H^+ → Chromogen (A_{530})

Figure 11-2 The Elson–Morgan reaction.

that they produce absorb light maximally at different wavelengths.

Determination of Glucose with a Coupled Enzyme Assay

In the presence of *excess* ATP and NAD^+, and hexokinase plus glucose-6-phosphate dehydrogenase, the concentration of glucose in an unknown sample can be determined in an *end-point assay* that follows the reduction of NAD^+ to NADH + H^+ (Fig. 11-3). In the presence of ATP, hexokinase will convert glucose to glucose-6-phosphate. This reaction *is not specific for glucose* in that hexokinase will recognize many different six-carbon sugars as a substrate. It is the second reaction that makes this overall assay specific for glucose. Glucose-6-phosphate dehydrogenase will convert the glucose-6-phosphate produced in the first reaction to 6-phosphogluconate. During this process, one molecule of NAD^+ will be reduced to NADH. This enzyme will only recognize glucose-6-phosphate as a substrate, ignoring all other six-carbon sugar phosphates that may have been produced in the first reaction. Since the production of NADH can be followed spectrophotometrically at 340 nm, you can determine the number of moles of NADH produced. This number will be equal to the number of moles of glucose present in the reaction. This information can then be used to determine the *concentration* of glucose in the unknown solution, provided you know the *volume* of the unknown sample that was present in the assay.

Supplies and Reagents

P-20, P-200, P-1000 Pipetmen with disposable tips
Heating block
Microcentrifuge tubes (0.5 ml and 1.5 ml)
^{14}C-glucose, 200 mCi/mmol (2,000 dpm/μl)
0.5 M NaB^3H_4, 15–25 mCi/mmol
Solutions of unknown concentration containing D-glucose, D-ribose, and/or D-glucosamine
0.75 N H_2SO_4
Fume hood
Whatman 3 MM chromatography paper
Descending chromatography tank
Ethyl acetate
Acetic acid
Formic acid
Scintillation fluid (5 g of 2,5-diphenyloxazole/L-toluene)
Scintillation vials
Scintillation counter
Glucose reagent (For large laboratory groups, this reagent can be purchased in a mixed dry form from Sigma Chemical Co.)

- NAD^+ (1.5 mM)
- ATP (1 mM)
- Yeast hexokinase (1 unit/ml)
- Glucose-6-phosphate dehydrogenase (1 unit/ml)
- $MgCl_2$ (2.1 mM)
- Buffer at pH 7.5

Spectrophotometer
4-ml glass vials with Teflon-lined screw caps
Glucosamine standard solution (0.5 μmol/ml)
Acetylacetone reagent (prepared fresh daily); 2 ml of 2,4-pentanedione mixed with 98 ml of 1 M Na_2CO_3
95% ethanol
p-Dimethylaminobenzaldehyde reagent (dissolve 677.5 mg of *p*-dimethylaminobenzaldehyde in 25 ml of a 1:1 mixture of ethanol:concentrated HCl)

Protocol

Day 1: Reduction of Carbohydrates with Sodium Borohydride and Enzymatic Determination of Glucose

WARNING
Safety Precaution: This laboratory exercise utilizes radioisotopes (^3H and ^{14}C). Use caution when handling these compounds. Avoid ingesting and skin contact with these radioisotopes. Wear gloves and safety glasses at all times. Consult the instructor for information on how to properly dispose of the radioactive material produced in this experiment.

Reduction of Carbohydrates with Sodium Borohydride

1. Obtain a numbered 1.5-ml microcentrifuge tube containing 1 ml of your unknown carbohydrate mixture. *Record the number of your unknown solution in your notebook.* This solution will be used on Day 1 and Day 2.

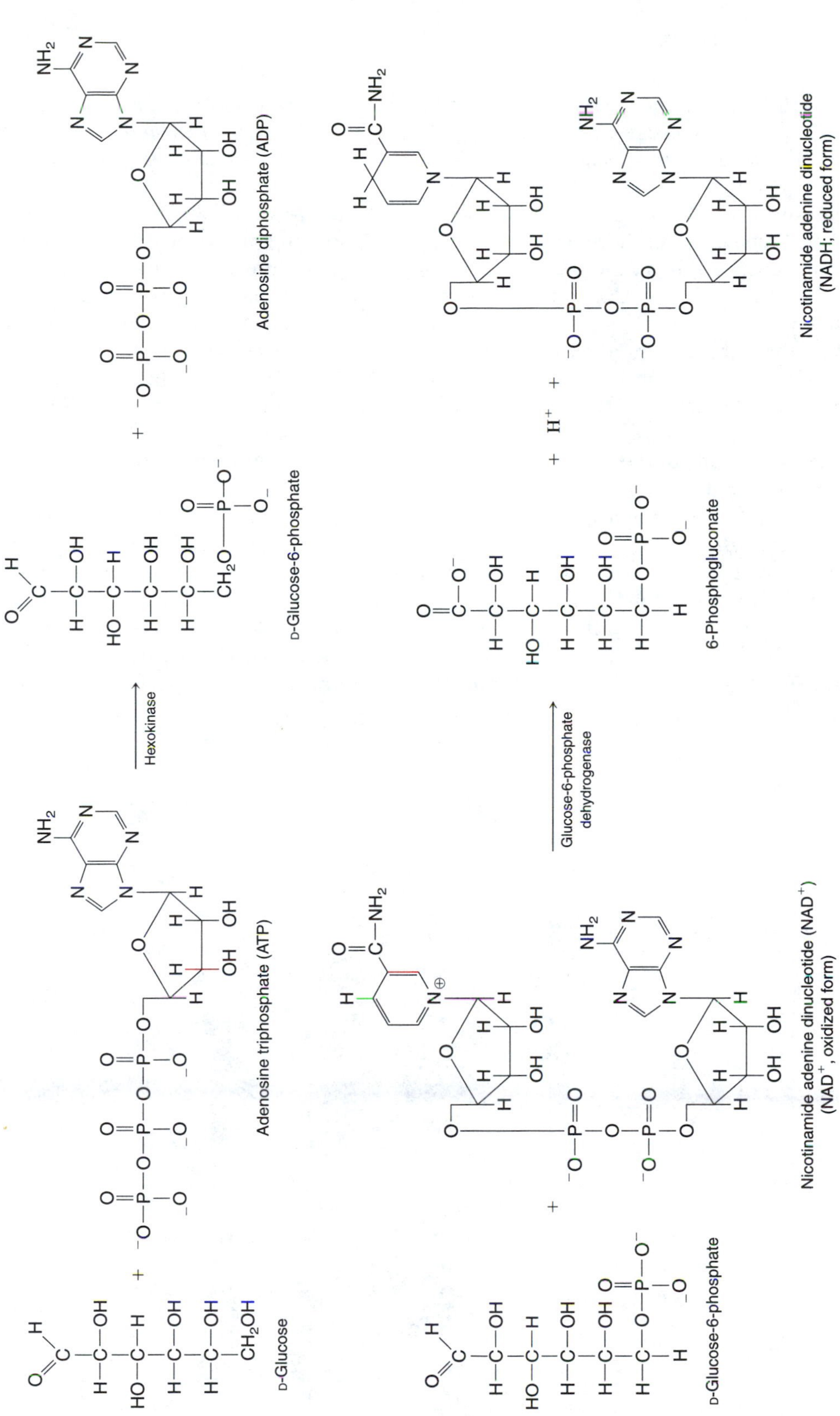

Figure 11-3 Coupled enzyme assay to determine glucose concentration.

2. Obtain a 0.5-ml microcentrifuge tube containing 5 μl of ^{14}C-glucose. *The reduction reaction will be done in this tube, so label it with your name.*
3. Add 5 μl of your unknown carbohydrate solution to this tube and mix thoroughly.
4. Bring this reaction tube to the instructor, who will add 5 μl of NaB^3H_4. Mix all components thoroughly.
5. Incubate the reaction in a 50°C oven for 30 min. During this incubation, you may proceed with the "Enzymatic Determination of Glucose," below.
6. *In the fume hood,* add 10 μl of 0.75 N H_2SO_4 to the reaction. Mix thoroughly and allow the reaction to stand for 30 min uncapped *in the fume hood* to remove all $^3H_2$ gas.
7. Obtain a strip of Whatman 3 MM chromatography paper (1 in. by 22 in). Draw a line with a soft lead pencil about 7 cm from one end of the strip. This line will mark the origin where the sample will be spotted. *Keep in mind that the carbohydrate sample is radioactive. Wear gloves and safety goggles at all times when handling the sample. Be sure that all radioactive waste is disposed of properly.*
8. Apply a 2-μl aliquot of the sample on the origin and allow the sample to dry. Repeat this procedure until a total of 8 μl of the sample has been applied. Successive 2-μl aliquots should be spotted on top of one another to give a single, 8-μl sample at the origin.
9. Using a soft lead pencil, label the strip at the end opposite the origin with your name.
10. Place the strip at the top of the descending chromatography tank so that the origin is just above the level of the mobile phase (Fig. 11-4). The mobile phase for this experiment is ethyl acetate:acetic acid:formic acid:water (18:3:1:4). Allow the chromatogram to develop for ~18 to 20 hr.
11. When the development is complete, remove the strip from the chromatography tank and allow it to air dry in the fume hood.

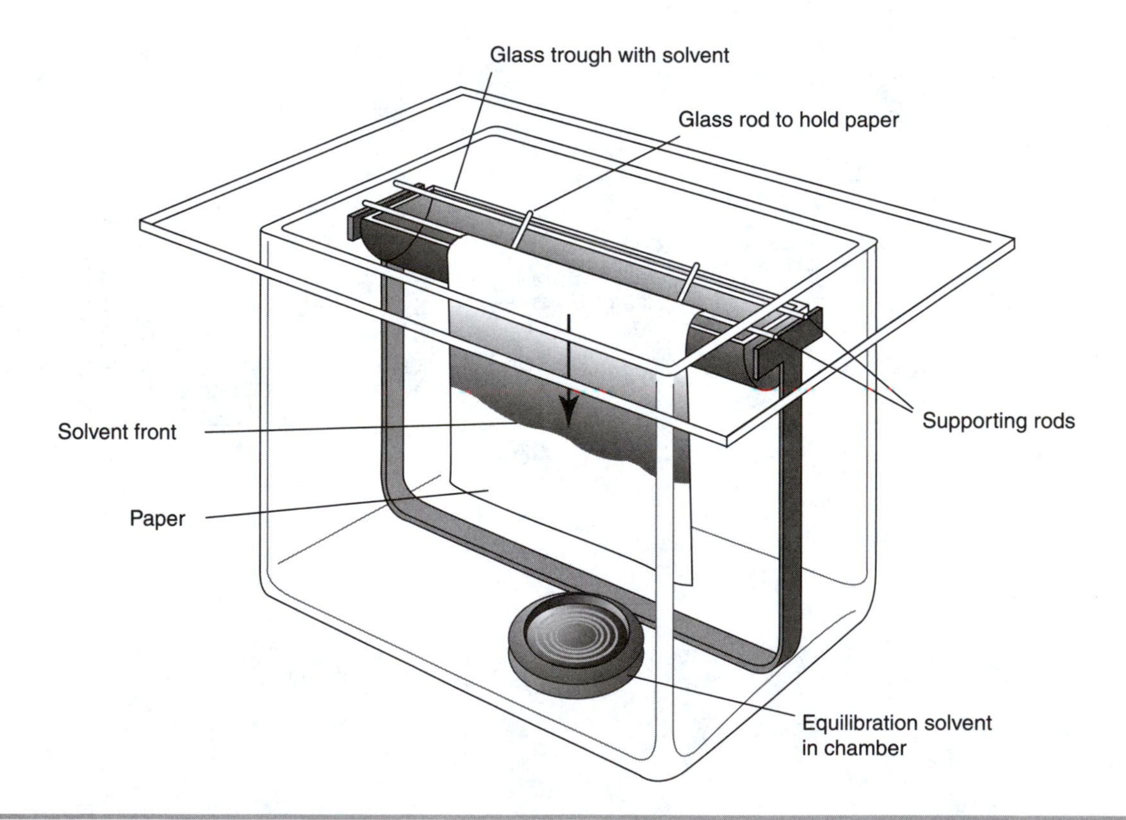

Figure 11-4 Descending paper chromatography.

Enzymatic Determination of Glucose

1. Dilute your unknown carbohydrate sample 1:20 with distilled water to give a total of 40 μl of dilute carbohydrate solution. This diluted sample will be used in the following assay.
2. Set up the following reactions in two 1.5-ml microcentrifuge tubes:

Tube	Volume of Glucose Reagent (ml)	Volume of Diluted Carbohydrate Solution (μl)	Volume of Water (μl)
1	1.0	—	20
2	1.0	20	—

3. Mix thoroughly and allow the tubes to incubate at room temperature for 15 min.
4. Read the absorbance of both samples at 340 nm in the spectrophotometer. Set the instrument to read zero absorbance using water as a blank.
5. Subtract the A_{340} reading of tube 1 (control) from the A_{340} reading from tube 2 (sample). This corrected value is the true A_{340} reading for your carbohydrate sample.
6. Using Beer's law, calculate the concentration of NADH in this 1.02-ml sample. The millimolar extinction coefficient at 340 nm for NADH is 6.22 mM^{-1} cm^{-1}.
7. Using this information, calculate the number of moles of NADH that are present in the 1.02-ml sample.
8. Using the information obtained above, calculate the concentration of glucose present in the unknown carbohydrate solution in units of micromoles per milliliter and milligrams per milliliter. *Remember that the original unknown carbohydrate solution was diluted for use in this assay.*

Day 2: Elson–Morgan Determination of Glucosamine and Analysis of Paper Chromatogram

Elson–Morgan Determination of Glucosamine

1. Dilute your original unknown carbohydrate solution 1:100 with distilled water to give a total of 1 ml of dilute carbohydrate solution. This diluted sample will be used in the following assay.
2. Set up eight 4-ml glass vials containing the following:

Vial	Volume of 0.5 μmol/ml Glucosamine Standard (μl)	Volume of Dilute Carbohydrate Solution (μl)	Volume of Water (μl)
1	—	—	500
2	50	—	450
3	100	—	400
4	150	—	350
5	250	—	250
6	—	50	450
7	—	150	350
8	—	250	250

3. Add 0.5 ml of acetylacetone reagent to all eight vials. Mix thoroughly and loosely seal the vial with a Teflon-lined screw cap to allow the fumes to escape.
4. Place the vials in a 90°C heating block and incubate for 45 min. While you are waiting, you may complete "Analysis of Paper Chromatogram," below.
5. Allow the vials to cool to room temperature (~5 min).
6. Add 2 ml of 95% ethanol to each vial. Slowly add 0.5 ml of *p*-dimethylaminobenzaldehyde reagent to each vial. *The reaction may bubble vigorously as this is done, so use caution so as not to lose your samples.*
7. When the solutions have stopped bubbling, cap the vials and incubate for 1 hr at room temperature.
8. Blank your spectrophotometer to read zero absorbance at 540 nm against the solution in vial 1 (blank).
9. Measure the absorbances of vials 2 to 8 at 540 nm and record these values in your notebook.
10. Prepare a standard curve by plotting the number of micromoles of glucosamine versus A_{540} using the data obtained from vials 2 to 5 (these contain known amounts of the glucosamine standard solution). Does this plot yield a straight line through the data points?
11. Determine which of vials 6 to 8 had absorbance readings that lie in the *linear portion* of the standard curve. Based on these absorbance values,

determine the number of micromoles of glucosamine present in these vials.

12. Taking into account the dilution of the original unknown carbohydrate sample and the *volume* of this diluted sample present in vials 6 to 8, determine the *concentration* of glucosamine in the original unknown carbohydrate solution. Express the concentration in units of micromoles per milliliter *and* milligrams per milliliter.

Analysis of Paper Chromatogram

1. Using a soft lead pencil, mark off half-inch segments on the chromatography strip, starting from the origin and extending to the end of the strip. *As before, remember that the chromatogram is radioactive. Wear gloves and safety glasses at all times.*
2. Cut along each line with a scissors and place the half-inch segments into scintillation vials labeled 1 to 30 (with 1 corresponding to the segment nearest the origin). Be sure that the segments are placed in the vials in the correct order so that you will be able to identify each peak by its $R_{glucitol}$ value.
3. Fill each scintillation vial with enough scintillation fluid to completely cover each paper segment.
4. Place the vials (in order) into a scintillation counter rack and count each vial in the ^3H channel (channel 1) and the ^{14}C over ^3H channel (channel 2). Consult the instructor for the proper scintillation counter channel settings.
5. *Data analysis*

   a. Identify the segment that has the lowest cpm value for channel 1 (^3H, background counts per minute for channel 1).

   b. Subtract this value from all other channel 1 cpm values for the segments.

   c. Identify the segment that has the lowest cpm value for channel 2 (^{14}C, background cpm for channel 2).

   d. Subtract this value from all other channel 2 cpm values for the segments.

   e. Prepare a plot of ^{14}C counts per minute versus segment number and a plot of ^3H counts per minute versus segment number.

   f. Identify the glucitol peak, which is (are) the segment(s) with ^{14}C and ^3H cpm.

   g. Calculate the $R_{glucitol}$ values for the other ^3H sugar alcohol peaks on the chromatogram:

$$R_{glucitol} = \frac{\text{distance from } ^3\text{H peak to the origin (cm)}}{\text{distance from glucitol peak to the origin (cm)}}$$

   Using this mobile phase for development, $R_{glucitol}$ values for the various compounds are typically:

Ribitol	~1.4
Glucose	~0.85
Glucosaminitol	~0.3

   Which sugar alcohols can you identify on your chromatogram? Was the reduction reaction with sodium borohydride complete? Explain.

   h. Determine the *true* number of ^3H cpm and ^{14}C cpm in the glucitol peak:

$$^3\text{H cpm} = (\text{channel 1 cpm}) - (\text{channel 1/channel 2 ratio})(\text{channel 2 cpm})$$

$$^{14}\text{C cpm} = (\text{channel 2 cpm}) + (\text{channel 1/channel 2 ratio})(\text{channel 2 cpm})$$

   The channel 1/channel 2 ratio is a value that will be provided by the instructor. It is a correction factor that takes into account the fact that some ^{14}C counts per minute were counted in channel 1. The instructor determined this ratio by scintillation counting of a sample containing a known number of ^{14}C disintegrations per minute in channel 1 and channel 2 under conditions identical to your experiment. *Refer to Section I, Experiment 3 for further discussion of the channel ratio method of quantifying radioactivity.*

   i. Calculate the number of micromoles of glucose, ribose, and glucosamine in your carbohydrate sample:

$$\text{Micromoles of sugar} = \frac{(^3\text{H cpm in peak})(\mu\text{mol sugar}/^3\text{H cpm})}{\text{fraction recovered}}$$

   where

$$\text{Fraction recovered} = \frac{\text{total } ^{14}\text{C cpm in glucitol peak}}{\text{total } ^{14}\text{C cpm in the reaction}}$$

The number of micromoles of sugar/^{3}H cpm or the specific activity of sodium borohydride are values that will be provided by the instructor. The total number of ^{14}C counts per minute present in the reaction will also be provided by the instructor.

j. From the number of micromoles of each carbohydrate on the chromatogram and the volume of the original unknown carbohydrate solution used in the reduction reaction, can you determine the *concentrations* of all three carbohydrates in your unknown? Express your answer in units of micromoles per milliliter and milligrams per milliliter.

6. How do the concentrations of D-glucose and D-glucosamine determined in this sodium borohydride reduction experiment compare with the values that you obtained for these two carbohydrates in the hexokinase and Elson–Morgan assays, respectively? Discuss the quantitative differences as well as possible explanations for the differences arising from the various assays.

Exercises

1. Acetyl derivatives of amino sugars (e.g., *N*-acetyl glucosamine) will undergo an Elson–Morgan reaction in basic solution. Write an equation to describe this reaction and draw the structure of the likely intermediate.
2. The following data were obtained in a sodium borohydride reduction reaction as performed in this experiment. The reaction contained a single carbohydrate and the ^{14}C-labeled internal control:

Data

Sample	Measured Channel 1 cpm	Measured Channel 2 cpm
^{3}H (10,000 dpm)	5,800	0
^{14}C (10,000 dpm)	4,700	3,600
Unknown (^{14}C and ^{3}H)	49,000	7,500

Questions

a. Calculate the counting efficiencies for ^{3}H and ^{14}C both in channel 1 and channel 2.
b. Determine how many ^{14}C disintegrations per minute are in the unknown sample.
c. Determine how many ^{3}H disintegrations per minute are in the unknown sample.
d. If the specific activity of NaB^3H_4 used in this reaction was 15 mCi/mmol, and 6 μl of the original carbohydrate solution was used in the reaction, calculate the concentration of the carbohydrate in the unknown solution in micromoles per milliliter. Assume that the fraction recovered was 40%.

REFERENCES

Conrad, H. E., Varboncouer, E., and James, M. E. (1973). Qualitative and Quantitative Analysis of Reducing Carbohydrates. *Anal Biochem* **51**:486.

Elson, L. A., and Morgan, W. T. J. (1933). A Colorimetric Method for Determination of Glucosamine and Chondrosamine. *Biochem J* **27**:1824.

Iyer, R. S., Kobierski, M. E., and Salomon, R. G. (1994). Generation of Pyrroles in the Reaction of Levuglandin E2 with Proteins. *J Org Chem* **59**:6038.

Lehninger, A. L., Nelson, D. N., and Cox, M. M. (1993). Structure and Catalysis of Carbohydrates. In: *Principles of Biochemistry*, 2nd ed. New York: Worth.

Mathews, C. K., and van Holde, K. E. (1990). Carbohydrates. In: *Biochemistry*. Redwood City, CA: Benjamin/Cumming.

Morrison, R. T., and Boyd, R. N. (1987). *Organic Chemistry*, 5th ed. Boston: Allyn & Bacon.

Robyt, J. F., and White, B. J. (1990). Chromatographic Techniques. In: *Biochemical Techniques: Theory and Practice*. Prospect Heights, IL: Waveland Press.

Spiro, R. G. (1966). Analysis of Sugars Found in Glycoproteins. *Methods in Enzymol* **8**.

Stryer, L. (1995). Carbohydrates. In: *Biochemistry*, 4th ed. New York: Freeman.

Wheat, R. W. (1966). Analysis of Hexosamines. *Methods Enzymol* **8**:60.

Glucose-1-Phosphate: Enzymatic Formation from Starch and Chemical Characterization

Theory

Starch, a complex carbohydrate found in most plants, is a mixture of two polysaccharides: *amylose*, a straight-chain polymer of glucose units joined by α-1,4 linkages, and *amylopectin*, a branched-chain glucose polymer that differs from glycogen primarily in its larger number of α-1,4-linked glucose units between the α-1,6 branch points. Both glycogen (which is similar to amylopectin and is found in animals and bacteria) and starch are substrates for the enzyme *phosphorylase*. Phosphorylase catalyzes the *phosphorolysis* of glycogen or starch to glucose-1-phosphate (Fig. 12-1).

The primary function of starch is to serve as a storage form for carbohydrate. Mobilization of starch for metabolism or synthesis of other polysaccharides requires cleavage of monosaccharide units from the polymer and formation of phosphorylated sugar derivatives, which are the substrates for glycolysis or for formation of nucleoside diphosphosugars that are used for biosynthetic reactions. Phosphorolysis of starch fulfills both requirements in a single step. If starch were cleaved by *hydrolytic* reactions (e.g., by amylases), subsequent formation of the sugar phosphate from the released glucose would require the expenditure of metabolic energy in the form of ATP. Thus, the phosphorylase reaction increases the efficiency of starch or glycogen utilization by conserving the energy of the acetal bonds between glucose units through formation of the phosphoacetal bond of glucose-1-phosphate.

The regulation of glycogen phosphorylase in mammalian muscle and liver has been studied intensively. As might be expected for a pathway governing use of a major energy reserve, this enzyme (and the corresponding synthetic enzyme, glycogen synthetase) is under complex metabolic and hormonal control. The regulation of glycogen phosphorylase by covalent modification (reversible phosphorylation) and by allosteric effectors will be studied further in Experiment 15.

In this experiment, soluble starch is incubated with inorganic phosphate (Pi) and a phosphorylase preparation from potatoes. After 24 to 48 hr, the reaction is stopped by heating the mixture to destroy the enzyme. After removal of the denatured enzyme, the unreacted inorganic phosphate is removed from the filtrate by precipitation as magnesium ammonium phosphate. The glucose-1-phosphate is then isolated by ion-exchange chromatography and is subsequently crystallized. The ion-exchange purification consists of two steps. First, cations in the solution (Mg^{2+}, NH_4^+, K^+) are adsorbed to the cation exchange resin, Dowex 50, with displacement of the H^+ initially adsorbed to the resin. The H^+ ions released into the solution lower its pH and convert glucose-1-phosphate to the monanionic form (G-1-P^{-1}). Then, the glucose-1-phosphate is adsorbed onto an anion exchange resin, Amberlite IR-45, which allows uncharged materials (starch, acetic acid) to pass through. The glucose-1-phosphate is then eluted as the dipotassium salt from the IR-45 column with the addition of a concentrated solution of OH^- ions.

Glucose-1-phosphate is determined by measurement of the inorganic phosphate released from the molecule following acid hydrolysis. Phosphoacetals of all sugars can be hydrolyzed by treatment with 1 N acid for 7 min at 100°C, but most

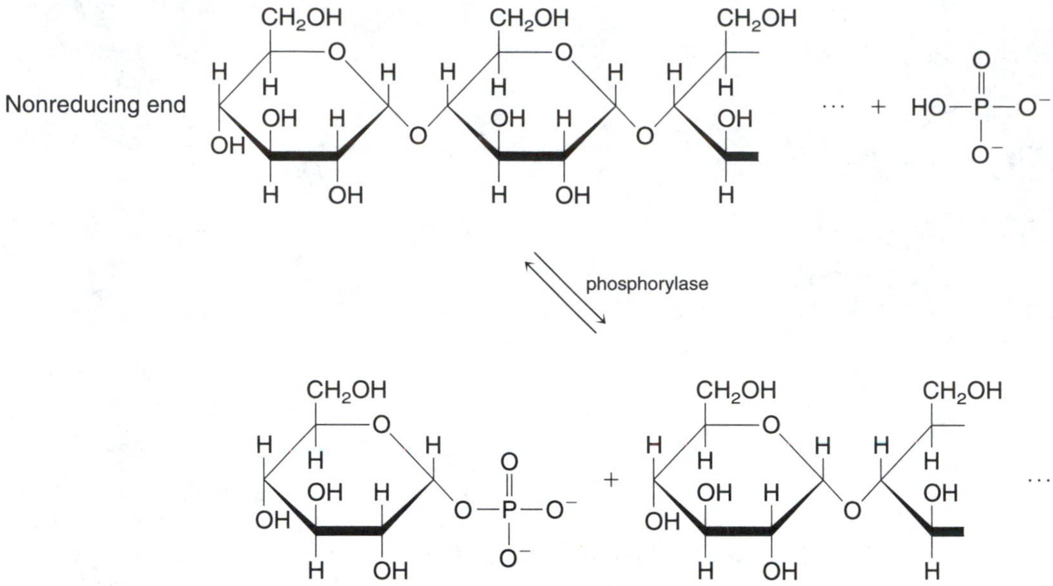

Figure 12-1 Phosphorylase-catalyzed phosphorolysis of glycogen or starch to glucose-1-phosphate.

other sugar phosphate esters (such as glucose-6-phosphate) ordinarily require longer heating times or more concentrated acid for complete hydrolysis. The inorganic phosphate released by acid hydrolysis is measured by the colorimetric method of Fiske and Subbarow, which is specific for inorganic phosphate; that is, inorganic phosphate can be analyzed with no interference from organic phosphate compounds such as phosphate esters or phosphoacetals that may also be present in the solution. Hence, the increase in inorganic phosphate after 7-min hydrolysis of a sample (compared to an unhydrolyzed sample) is a measure of the phosphoacetal content of the sample. Stability of various other biologically important phosphate compounds to acid hydrolysis has been discussed in the Section III introduction. The color developed in the Fiske–Subbarow reaction is dependent on the formation of a phosphomolybdic acid complex, which forms an intense blue color when reduced by a mixture of bisulfite and *p*-methylaminophenol. (Note that a variant of this assay is used in Experiment 13.)

It is necessary to establish both the identity and purity of the newly isolated glucose-1-phosphate. In this experiment, the first step consists of showing that the product presumed to be glucose-1-phosphate is, in fact, composed of *equal quantities* of glucose and phosphate. If the ratio of glucose to total phosphate is greater than 1, the results may indicate the presence of phosphate-containing contaminants. Second, it is necessary to distinguish between the known isomers of glucose phosphate. This step is particularly pertinent in this experiment because we know that glucose-1-phosphate may be converted to glucose-6-phosphate by the enzyme phosphoglucomutase, another enzyme present in the crude potato extracts used for synthesis of glucose-1-phosphate from starch.

The characterization of glucose-1-phosphate is based on the relative stability of various sugar phosphates in acid solution. Because sugar phosphoacetals are nonreducing sugars, they release equal amounts of reducing sugar and phosphate on 7-min hydrolysis in 1 N acid at 100°C. In contrast, glucose-6-phosphate and other phosphate esters require more concentrated acid or more prolonged heating for complete hydrolysis. Therefore, it is possible to characterize glucose-1-phosphate by qualitative (chromatographic) and quantitative analysis of materials present before and after 7-min hydrolysis. It is also possible to estimate the purity of the product by careful evaluation of the results.

Chromatographic characterization of sugar phosphates can be achieved by two different methods. First, you can usually detect the sugar portion of sugar phosphates by using sugar-specific reagents, such as the aniline–acid–oxalate spray reagent. Detection of sugars with this reagent involves the formation of colored glycosylamines when reducing sugars react with aniline. (Note that glucose-1-phosphate is a nonreducing sugar, but it is readily hydrolyzed to form glucose under the acidic conditions used to develop the color.) Organic phosphates, including sugar phosphates, can be detected with phosphate-specific methods. The modified Hanes–Ischerwood spray reagent used in this experiment hydrolyzes any organic phosphate so that inorganic phosphate is released. The inorganic phosphate then combines with the molybdic acid and is reduced to yield a blue spot (phosphomolybdous acid). The alternate iron–sulfosalicylic acid complex assay involves binding of Fe^{3+} by organic phosphates. The bound Fe^{3+} cannot then form a red-brown complex with sulfosalicylic acid, so phosphate-containing areas appear as white spots on a colored background.

Supplies and Reagents

Phenylmercuric nitrate
0.8 M Potassium phosphate buffer, pH 6.7
Soluble starch
Filter aid
14% NH_4OH
2 N NaOH
1 N NaOH
2 N HCl
1 N HCl
Potatoes (fresh)
Blender
Cheesecloth
$Mg(OAc)_2 \cdot 4\ H_2O$
Dowex 50 (H^+ form)
IR-45 (OH^- form)
Absolute methanol
Cylindrical glass tubes, or equivalent chromatography columns
Glass wool
Reducing reagent (3% $NaHSO_3$, 1% *p*-methylaminophenol)
Acid molybdate reagent (see Appendix)
Phosphate standard (1 mM)
5% KOH
0.01 M KI, 0.01 M I_2
Charcoal (Norit)
Vacuum desiccator
P_2O_5
Centrifuges
300-ml Polypropylene centrifuge bottles
Nelson's reagents A and B
Arsenomolybdate reagent (see Appendix)
1.0 mM glucose standard (1.0 mM glucose in 1.0 M NaCl)
Whatman No. 1 paper
1% Glucose-1-phosphate
1% Glucose-6-phosphate
1% Glucose
1% Inorganic phosphate
10% $Mg(NO_3)_2$ in ethanol
Methanol
88% Formic acid
Aniline–acid–oxalate spray reagent (see Appendix)
Modified Hanes–Ischerwood spray reagent [(4% $(NH_4)_6MoO_{24} \cdot 4\ H_2O$: 1 N HCl : 70% $HClO_4$: H_2O (5 : 2 : 1 : 12)]
Acid–$FeCl_3$ in acetone (Dissolve 150 mg $FeCl_3 \cdot 6\ H_2O$ in 3 ml 0.3 N HCl and mix with 97 ml acetone—prepare fresh before use.)
1.25% Sulfosalicylic acid in acetone
UV lamp
100°C oven

Protocol

The isolation and characterization of glucose-1-phosphate can be completed within four laboratory periods. The enzyme incubation is begun in the first period. The removal of cations with Dowex 50 should be completed in the second period, and the product should be stored at 0 to 5°C. The ion-exchange chromatography can then be completed and the crystallization begun in the third period. Characterization of the isolated product will be conducted in the fourth period, after collecting and drying the crystals of dipotassium glucose-1-phosphate dihydrate in between the third and fourth periods. Store all solutions at 0 to 5°C between laboratory periods to avoid bacterial or chemical degradation. Use distilled or deionized water throughout the experiment.

Day 1: Incubation of Starch and Pi with Potato Phosphorylase, Preparation of the IR-45 Column

1. Using about 50 ml of H_2O, make a smooth slurry of 10 g of soluble starch. Add this *slowly*

to 100 ml of vigorously boiling H_2O, and stir continuously with a glass rod until the solution is nearly clear. The solution may be cloudy, but should not be milky. Further heating may be required to dissolve the starch, but avoid prolonged heating. Add 100 ml of cold H_2O, stirring continuously, to help cool the solution to room temperature. *Do not add the enzyme until the solution has cooled to room temperature* because high temperatures inactivate the enzyme. Step 2 can be conducted while the solution is cooling.

2. Prepare a 500-ml filter flask to receive the potato extract by pouring 250 ml of H_2O into it and marking the position of the top of the fluid in the filtration reservoir with a felt-tipped pen. Pour out the H_2O. Suspend about 100 mg of phenylmercuric nitrate in a few milliliters of H_2O and pour it into the filter flask. Phenylmercuric nitrate is added to inhibit other enzymes and to prevent bacterial growth during the incubation of potato phosphorylase with starch.
3. When the starch solution prepared in step 1 has cooled, add the solution (250 ml) to 250 ml of 0.8 M phosphate buffer in a 1-liter Erlenmeyer flask.
4. Cut a fresh, medium-sized potato (precooled for 24 hr at 1 to 5°C; you need not peel the potato, but it should be washed to remove dirt and dust) into half-inch cubes. Blend 150 g of these cubes, added over a 30-sec period, with 150 ml of H_2O for 2 min in a blender. Then *quickly* pour the resultant slurry onto a Buchner funnel lined with two to four layers of cheesecloth, and filter with vacuum, collecting the filtrate in the filter flask you prepared in step 2. Wash the crude pulp with 50 ml of H_2O to ensure thorough enzyme extraction; that is, add the water to the pulp, stir the mixture, and pass the mixture through the filter as before. *Failure to complete these operations within 2 min of blending may result in loss of enzyme activity.* Adjust the extract to a volume of 250 ml with H_2O.
5. *Immediately* add the 250 ml of enzyme solution (potato filtrate) to 500 ml of the solution of starch plus phosphate buffer prepared in step 3. Record the total volume, and store the solution in a stoppered Erlenmeyer flask at room temperature in your desk. During the incubation period (24–48 hr), the reaction mixture will turn red and then dark blue or purple because of the action of tyrosinases present in the crude extract that catalyze melanin formation from proteins. These colored materials will be removed during later procedures.
6. Prepare the IR-45 column for use on Day 3 as follows: Prepare a column about 4 cm in diameter and 20 to 30 cm long using a rubber stopper, screw clamp, and glass wool plug (Fig. 12-2). Before assembling the column, rinse all parts thoroughly with distilled H_2O. Insert the rubber stopper tightly at the bottom, and tape the edges to prevent leaks. Fill the column about one-third full with H_2O, and insert the glass wool plug, pushing it gently down to the rubber stopper at the bottom. Mix 250 ml of

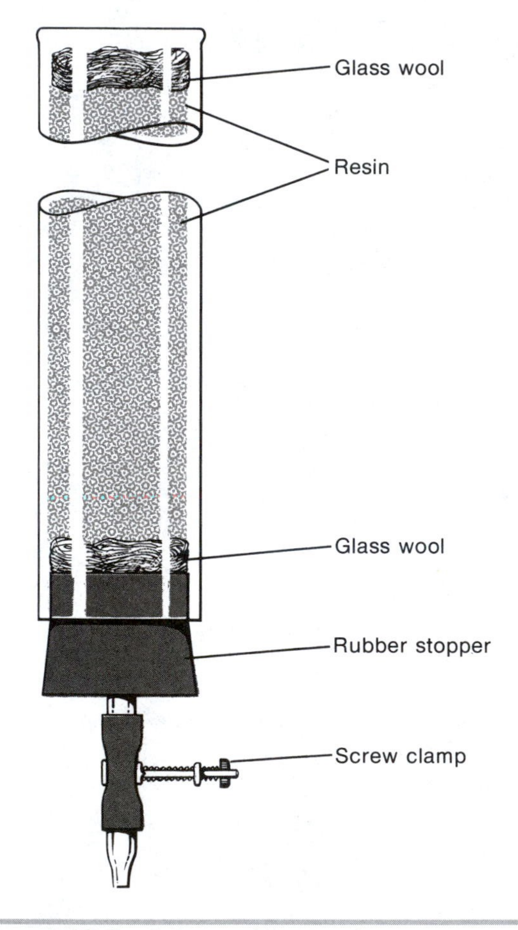

Figure 12-2 Anion-exchange column.

H_2O with 250 ml of moist IR-45 (OH^-) (prepared by the instructor), and pour the resultant slurry into the column so that there are no air pockets in the settled resin. For best results, pour the slurry of resin completely into the column before allowing the resin to settle. Then, wash the resin with H_2O until the effluent is about pH 9 or lower (use a pH meter). Drain or pipette off the excess fluid until the fluid surface just covers the top of the resin bed. Cover the surface of the resin with a layer of glass wool and cover the top of the column with Parafilm. The column can stand at room temperature until use on Day 3, but *be sure that it does not leak.*

Day 2: Precipitation of Pi, Cation Exchange with Dowex 50, and Initial Glucose-1-Phosphate Assays

1. After 24 to 48 hr of incubation, stop the enzymatic reaction by rapidly heating the potato extract/starch solution to 95°C and then slowly cooling it to room temperature over a 30-min period or longer. *If possible, come to the laboratory 4 to 6 hr before the usual laboratory period to carry out the heating, so that the solution will slowly cool before the next step.* Rapid cooling of this solution (e.g., in an ice bath) may cause the residual starch to form a gel and should be avoided.
2. Remove the coagulated protein by centrifuging the fluid in large (300 ml) bottles at room temperature for 15 min at 4000 × *g*. Decant the supernatant fluid into a large beaker.
3. Remove the excess inorganic phosphate from the solution by dissolving 0.2 mol of magnesium acetate (44 g of $Mg(OAc)_2 \cdot 4\,H_2O$) in the solution and adjusting the pH to 8.5 with 14% NH_4OH (use pH paper first, then a pH meter; about 25 to 30 ml of 14% NH_4OH will be required, but avoid adding excess NH_4OH). Cool the solution in a salted ice bath for 10 min, and remove the precipitated magnesium ammonium phosphate by suction filtration. Use Whatman No. 1 filter paper covered with a thin layer of Celite filter aid. If filtration is very slow, use more than one Buchner funnel or change the filter as frequently as needed. Filter the solution a second time using two layers of Whatman No. 1 filter paper.
4. Collect the filtrate and record its volume. Remove *duplicate* 0.05, 0.1, 0.2, and 0.5 ml samples of the filtered solution for the inorganic phosphate and 7-min phosphate assays described in steps 5 through 9. Place these duplicate samples in small, labeled glass test tubes.

   NOTE: Most laboratory detergents contain large amounts of phosphate, which may contaminate your glassware. All tubes used in phosphate analyses should be thoroughly cleaned with a sodium dodecyl sulfate detergent that does not contain phosphate and rinsed with deionized water.
5. *Assay for inorganic phosphate and glucose-1-phosphate:*

   In the Fiske–Subbarow colorimetric method for the determination of phosphate, *the color yield is directly proportional to the amount of inorganic phosphate only when the sample analyzed contains between 0.1 and 1.0 μmol of phosphate.* The efficiency of the enzymatic formation of glucose-1-phosphate on Day 1 will vary somewhat depending on the sources of the enzyme and the starch, as well as the length of the incubation period. Accordingly, the sample sizes suggested may not lie within the range in which the assay is valid. In this event, use the results you obtained to decide on several different sample sizes to analyze until you find at least one that falls within the accurate range of the assay. For example, if the suggested sample sizes gave too little color for accurate measurement, use larger sample sizes in the next analysis. In this experiment, as in all isolation procedures, you must obtain an accurate measurement of the amount of the desired compound in each fraction. Therefore, you will need to determine and use appropriate aliquot ranges for analysis before proceeding to the next steps in the experiment. Further, you must keep accurate records of aliquot sizes and protocols in order to evaluate your data correctly.
6. Carry out the 7-min hydrolysis on *one* of the set of duplicate samples of the magnesium ammonium phosphate supernatant fluid from step 4 by adding an equal volume of 2 N HCl to each and heating at 100°C for 7 min. That is, set up *tubes 1 through 4*, which contain 0.05, 0.1, 0.2, and 0.5 ml of the supernatant fluid, respectively. Add 0.05, 0.1, 0.2, and 0.5 ml 2 N

HCl, respectively, to tubes 1 through 4. Place the tubes in a boiling water bath for exactly 7 min. Remove the tubes, cool them in cold water, and add 0.05, 0.1, 0.2, and 0.5 ml of 2 N NaOH, respectively, to tubes 1 through 4. It is important that the solutions be nearly neutral; use pH paper to check them. Bring the volume in each tube to 3 ml by adding 2.85, 2.7, 2.4, and 1.5 ml of H_2O, respectively, to tubes 1 through 4.

7. Prepare the other (unhydrolyzed) set of duplicate samples of the magnesium ammonium phosphate supernatant fluid from step 4 by bringing their volumes to 3.0 ml with H_2O. That is, set up *tubes 5 through 8*, which contain 0.05, 0.1, 0.2, and 0.5 ml of the supernatant fluid, respectively. Add 2.95, 2.9, 2.8, and 2.5 ml of H_2O, respectively to tubes 5 through 8.
8. Prepare a phosphate standard curve and blank as follows. Set up *tubes 9 through 15*, which contain 0, 0.1, 0.2, 0.4, 0.6, 0.8, and 1.0 ml of the 1 mM (1 μmol/ml) inorganic phosphate standard, respectively. Add 3.0, 2.9, 2.8., 2.6, 2.4, 2.2, and 2.0 ml of H_2O, respectively, to tubes 9 through 15.
9. Add, in order, 1 ml of acid molybdate reagent, 1 ml of reducing reagent (3% $NaHSO_3$, 1% *p*-methylaminophenol), and 5 ml of H_2O to all of tubes 1 through 15. Mix all of the solutions by inverting the tubes, and allow the color to develop for 20 min before reading the absorbance at 660 nm *using the solution in tube 9 as the blank.* Calculate the quantity of inorganic phosphate and glucose-1-phosphate in each sample and in the entire reaction mixture as follows. Prepare a standard curve that plots absorbance at 660 nm versus number of micromoles of Pi based on the data obtained from tubes 10 through 15. From the absorbance at 660 nm of your hydrolyzed and unhydrolyzed samples, determine the number of micromoles of Pi and the concentration of glucose-1-phosphate in your original sample. *Remember that the number of micromoles of glucose-1-phosphate in the sample is equal to the total Pi found in the 7-min hydrolyzed sample minus the number of micromoles of Pi found in the unhydrolyzed sample of the same volume.* You have made these determinations in four different volumes of the magnesium ammonium phosphate supernatant fluid from step 4. How do the different determinations agree? Which determinations are most reliable? Why?

   In subsequent phosphate analyses on Days 3 and 4, you should prepare a fresh standard curve according to step 8 along with the samples being analyzed for that day's experimental work.
10. If the phosphate assay reveals that an excess of inorganic phosphate is still present in the incubation filtrate (i.e., if an intense blue color forms in the unhydrolyzed 0.1 ml sample), add 1 g of $Mg(OAc)_2 \cdot 4\,H_2O$, adjust to pH 8.5 with 14% ammonia, cool the solution, and filter it again. Then repeat the phosphate assays. (Step 10 is not usually necessary.)
11. Calculate the number of micromoles of Pi and 7-min phosphate in the entire volume of filtered solution. If the inorganic phosphate (in the unhydrolyzed sample) in the filtrate is less than 15% of the phosphate found after 7-min hydrolysis, you may proceed with the rest of the experiment. (*NOTE*: When students are working in pairs, some instructors prefer to have one student proceed directly with steps 12 through 14 while the other student conducts the assays of steps 5 through 9. This assumes that the magnesium ammonium phosphate precipitation effectively removes most of the inorganic phosphate from your preparation, as occurs in most (but not all) cases. Consult your instructor.)
12. Decolorize the solution of glucose-1-phosphate (separated from excess inorganic phosphate) by stirring 2 g of charcoal into the solution and then removing the charcoal by vacuum filtration using filter aid. This procedure should yield a clear or yellowish solution containing glucose-1-phosphate, unreacted starch, and many salts that were present either in the original potato extract or in the reagents added during the course of the experiment.
13. Remove the cations by treatment with Dowex 50 in the following manner: Add 350 ml of moist Dowex 50 in the H^+ form (prepared by the instructor) to the decolorized solution and stir gently for 5 min before separating by vacuum filtration. *Do not use filter aid when removing the Dowex;* use only Whatman filter paper.
14. Determine the pH of the filtrate produced in step 13. If the pH of the filtrate is not acidic

(pH 1 to 3), add 100 ml of moist Dowex 50 in the H^+ form, then stir and filter the solution as before. Repeat this procedure until the pH of the solution is between 1 and 3. Record the volume of the resultant solution, and remove duplicate 0.05-, 0.1-, 0.2-, and 0.5-ml samples. Place these samples in four small, phosphate-free, labeled glass test tubes.

15. Assay the samples from the Dowex supernatant solution for inorganic phosphate and 7-min phosphate exactly as described under steps 5 to 9. If necessary, the remaining solution from step 14 may now be stored at 0 to 5°C for several days. There may be small losses of glucose-1-phosphate during storage in acidic solution, but these are minimized at 0 to 5°C. Alternatively, if time is sufficient, proceed directly to the protocol for Day 3.
16. At the end of the period return the Dowex to the container provided by your instructor. *(NOTE: Do not allow the Dowex to become mixed with the Amberlite resin used in Day 3.)*

Day 3: Isolation of Glucose-1-Phosphate by Anion Exchange

The next step in the purification procedure is the column chromatography of the Dowex 50 filtrate on Amberlite IR-45 in the OH^- form. This involves removing the anions in the acidic solution from all other contaminating materials. Thus, when the acidic filtrate (pH 1 to 3) contacts the IR-45 in OH^- form, the glucose-1-phosphate ($G\text{-}1\text{-}P^{-1}$) is electrostatically adsorbed on the resin, whereas un-ionized materials in the solution (such as acetic acid from $Mg(OAc)_2$ and unreacted starch) pass through the column. When the resin is eluted with strong alkali (5% KOH), the adhering anions, including glucose-1-phosphate, are displaced by the excess OH^- ions and are obtained in the eluate from the column.

1. To cause adsorption of the glucose-1-phosphate by the IR-45, gently pour the acidic solution treated with Dowex 50 (steps 13 and 14 of the protocol for Day 2) onto the column, avoiding the introduction of air bubbles into the resin bed. Open the screw clamp and adjust the flow rate to about 15 ml/min. Pass the entire solution through the IR-45 resin without permitting air to enter the column.
2. Collect the entire effluent in one fraction, determine its volume, and assay duplicate 0.5-ml samples for inorganic phosphate and 7-min phosphate as described for tubes 4 and 8 under steps 5 to 9 of Day 2.
3. If appreciable 7-min phosphate appears (>30% of total glucose-1-phosphate present in the Dowex 50 filtrate), and if time permits, add an additional 50 g of IR-45 resin to the column, and pass the effluent through the column again.
4. After the original solution has passed into the column and the effluent has been satisfactorily freed of the 7-min phosphate (i.e., glucose-1-phosphate should have bound to the column), wash the column with deionized H_2O until the effluent no longer gives a positive test for starch (i.e., a blue color forms on mixing a drop of effluent with a drop of 0.01 M I_2 in 0.01 M KI). About 500 ml of water will be required to remove all of the starch from the column.
5. To elute the glucose-1-phosphate from the resin, pass 5% KOH *(Remember: KOH is caustic and can cause painful skin burns)* through the column, adjusting the flow rate to 20 ml/min. Collect separate, successive 100 ml fractions in clean 250 ml Erlenmyer flasks. Continue collecting fractions for about 500 ml after the pH of the effluent becomes markedly alkaline (pH 11 to 13) as observed with pH paper. (The usual total volume of all fractions equals 700–800 ml.)
6. Test 0.1-ml samples from each 100-ml fraction for 7-min phosphate as described for tube 2 under steps 5 to 9 of Day 2. Some fractions may require smaller samples to fall within the linear range of the standard curve, but 0.1-ml samples will serve to find the "peak" for glucose-1-phosphate. *It is not necessary to repeat analyses for samples that give too much color to fall on the standard curve.*
7. Combine the fractions containing 80 to 90% of the recovered glucose-1-phosphate, and adjust the solution to pH 8 (or higher) by adding a few drops of 5% KOH, if necessary. *Do not add KOH if the pH is already above 8.* Determine the volume of the combined KOH effluent (glucose-1-phosphate "peak") fractions. Save *duplicate* 0.05-, 0.1-, 0.2-, and 0.5-ml samples for inorganic phosphate and 7-min phosphate analysis in clean, phosphate-free, labeled small glass test tubes.

8. Add 3 volumes of absolute methanol to the combined fractions, and leave them at 0 to 5°C for at least 12 hr for crystallization of the dipotassium glucose-1-phosphate dihydrate. In this temperature range, this compound will remain stable indefinitely.
9. Collect the crystals by centrifugation, after pouring off the bulk of the clear supernatant fluid. A clinical centrifuge or refrigerated laboratory centrifuge at low speed may be used. Wash the crystals with 5 ml of absolute ethanol and centrifuge again. Then dry the crystals in a tared container in a vacuum desiccator over P_2O_5 at 5°C. Weigh the crystals to determine the final yield of glucose-1-phosphate and save them for analysis on Day 4 of this experiment. *It will save much time on Day 4 if you return to the laboratory and carry out step 9 on the day preceding the experiments of Day 4.*

Report of Results for Days 1 through 3

1. Prepare a flow sheet of the steps in the isolation procedure, indicating the purpose of each step.
2. Prepare a table (see Table 12-1) showing the percentage of the glucose-1-phosphate recovered at each step in the procedure. Account for any poor recoveries.
3. Prepare a graph of the elution pattern of the glucose-1-phosphate from IR-45, plotting micromoles of glucose-1-phosphate/100 ml as the ordinate and the milliters of effluent as the abscissa.
4. Calculate the maximal yield of glucose-1-phosphate that would be expected if the phosphorylase reaction had reached equilibrium. The quantity of starch is altered only slightly during the incubation; therefore, the starch concentrations in the numerator and denominator are roughly equivalent and cancel out. Use the following equilibrium constant for your calculation:

$$K_{eq} = \frac{\text{glucose-1-phosphate}}{\text{inorganic phosphate}} = 0.088$$

Remember that the initial concentration of inorganic phosphate (0.8 M × 250 ml/750 ml = 0.267 M) has been decreased at equilibrium by the amount of glucose-1-phosphate formed. Compare your yield (glucose-1-phosphate found in the $MgNH_4PO_4$ supernatant fluid) with the theoretical yield expected. Suggest reasons for any discrepancy between your results and the theoretical value calculated from the phosphorylase equilibrium constant.

Day 4: Chemical Characterization of the Isolated Glucose-1-Phosphate

You should begin this day's experiments by preparing tubes 1 through 6 and Tubes 1′ through 5′ as described in steps 1 through 6 below. Then you should prepare the chromatograms and start their development (steps 7 and 8). Next you should perform the phosphate and reducing sugar analyses as described in steps 12 and 13. Finally, spray the developed chromatograms to detect the separated products (steps 9 through 11).

Table 12-1 Percentage of Glucose-1-Phosphate Recovered during Purification

Step	Vol of Solution	μmol of 7-min Phosphate/ml	Total μmol of 7-min Phosphate	Percent Recovered
$MgNH_4PO_4$ supernatant				100
Dowex 50 supernatant				
Original IR-45 (OH^-) washes (before eluting with H_2O or KOH)				
Combined KOH effluent fractions				
Glucose-1-phosphate crystals*	—	—		

*Assume that the crystals are dipotassium glucose-1-phosphate dihydrate (MW = 372).

1. You will evaluate your sample of isolated glucose-1-phosphate for purity by determining the following quantities (tubes with prime values are duplicates). Instructions for preparing these tubes follow the list below.
   a. Reducing sugar equivalents before 7-min hydrolysis (tubes 1 and 1′)
   b. Reducing sugar equivalents after 7-min hydrolysis (tubes 2 and 2′)
   c. Inorganic phosphate present before 7-min hydrolysis (tubes 3 and 3′)
   d. Inorganic phosphate present after 7-min hydrolysis (tubes 4 and 4′)
   e. Phosphate present after total hydrolysis (tubes 5 and 5′)
   f. Chromatographic characterization after 7-min hydrolysis (tube 6)
2. Weigh out a 20- to 40-mg sample of your isolated glucose-1-phosphate with an analytical balance, determining the weight of the sample to the nearest milligram, and dissolve the sample in 10 ml of H_2O. Place 0.1-ml samples of this solution in each of 10 small glass tubes, which should be numbered tubes 1, 1′, 2, 2′, 3, 3′, 4, 4′, 5, and 5′. Save the remaining solution for chromatographic characterization (step 7, below).
3. To another tube (tube 6), add 0.1 ml of a separately prepared 10 mg/ml solution of your isolated glucose-1-phosphate and add 0.1 ml 2 N HCl.
4. Add 1.0 ml of 1 N HCl to tubes 2, 2′, 4, and 4′. Heat them for 7 min in a boiling water bath; then cool them at once in a beaker of cold water. Finally, neutralize the contents by adding 1 ml of 1 N NaOH to each of the tubes. At the same time as tubes 2, 2′, 4, and 4′ are heated, heat tube 6 at 100°C, but do not add additional HCl or NaOH to this tube. Cool tube 6 with the other tubes and save the contents for chromatographic analysis (step 7 below).
5. Add 1 drop of 10% $Mg(NO_3)_2$ in ethanol to tubes 5 and 5′. Cautiously evaporate the fluid to dryness over a flame until brown fumes disappear and only a white ash remains in the tubes. Cool these two tubes, then add 1.0 ml 1 N HCl to each and heat them in a boiling water bath for 15 min to hydrolyze any pyrophosphates that may have formed during the heating with $Mg(NO_3)_2$. Cool the tubes, neutralize with 1 ml of 1 N NaOH, and analyze for inorganic phosphate as with the other tubes (step 13).
6. Add 2 ml of H_2O to tubes 1, 1′, 3, and 3′ and save them for the assay of unhydrolyzed product (steps 12 and 13).
7. Prepare two identical chromatograms using Whatman No. 1 paper squares, 20 cm on a side, which carry the following spots, 5 to 7 mm in diameter, spotted at 1-in. intervals on a line drawn lightly with a pencil 1 in. from the bottom of the paper:

   20 μl of the hydrolyzed material (from tube 6)
   10 μl of 1% glucose-1-phosphate standard
   10 μl of 1% glucose-6-phosphate standard
   20 μl of the isolated, suspected glucose-1-phosphate (from step 2)
   10 μl of 1% glucose standard
   20 μl of 1% inorganic phosphate standard

   Keep the spots small by spotting about 3 μl at a time and allowing drying of the spots between applications.
8. Develop each chromatogram in ascending fashion (see Fig. 6-8, Experiment 6), using freshly mixed solvent, which consists of 80 ml absolute methanol, 15 ml 88% formic acid, and 5 ml H_2O. (If time permits, you can improve the separation by suspending the chromatogram in the chromatography jar with a fine wire for 2 hr before allowing the paper to dip into the solvent.) *Since the chromatography requires about 3 hr for development, you should carry out steps 12 and 13 during this time and complete Steps 9 through 11 afterward.*
9. When the chromatograms have developed, dry them, and spray *one* of them with aniline–acid–oxalate spray to detect sugars. Spray the chromatogram with the reagent until it is fully dampened, but avoid excessive application of the spray, as this may cause diffusion of the spots on the paper. Heat the paper at 100 to 105°C for 5 to 15 min to develop the colored spots. Note that this spray reagent requires reducing sugars to develop color. Explain how it can be used to detect glucose-1-phosphate, which is not a reducing sugar.
10. Subject the other dried chromatogram to one of the following phosphate-detection proce-

dures. Either the modified Hanes–Ischerwood procedure or the iron-sulfosalicylic acid complex method may be used to detect organic phosphate compounds on chromatograms.

a. *Modified Hanes–Ischerwood Spray Procedure.* Spray the chromatogram lightly with freshly prepared modified Hanes–Isherwood reagent (25 ml 4% $(NH_4)_6Mo_7O_{24} \cdot 4\,H_2O$, 10 ml 1 N HCl, 5 ml 70% perchloric acid, 60 ml H_2O). Dry the paper in a 100°C oven for a few minutes, but do not allow the paper to darken. The paper will become fragile, so handle it very carefully. Irradiate the paper with a UV lamp held at a distance of 10 cm for 1 to 10 min. Phosphate-containing compounds will appear as blue spots on a light background.

WARNING
UV will damage eyes: wear glasses and avoid direct irradiation.

b. *Iron–Sulfosalicylic Acid Detection of Phosphates.* Dip the chromatogram paper containing more than 0.05 μmol phosphate per spot in a bath of acidic $FeCl_3$ in acetone and immediately hang it up in a fume hood to dry. When dry, dip the paper in 1.25% sulfosalicylic acid in acetone and again immediately hang it up in a fume hood to dry. Organic phosphates and inorganic phosphate appear on the dry chromatogram as white spots in a red-brown field.

11. Determine the R_f values for the various standard and experimental spots on both chromatograms and record them in a table. What conclusions can you draw from your observations?
12. *Nelson's Test for Equivalents of Reducing Sugar.* (*NOTE:* A simple and specific alternative method to assay the glucose present in tubes 1, 1′, 2, and 2′ is the enzymatic assay for glucose described in Experiment 11. This assay is specific for glucose (i.e., will not detect glucose-6-phosphate or other reducing sugars) and is optimal in the range from 0.01 to 0.10 μmol of glucose. If you choose this option, follow the steps in the section on "Enzymatic Determination of Glucose" in Experiment 11, except that 100- and 200-μl samples from tubes 1 and 2 should be assayed, and equivalent volumes of H_2O should be added to the analysis of the glucose standard. In calculating the amount of glucose determined, remember that the final volume in the assay will be 1.1 or 1.2 ml.) Analyze tubes 1 and 1′ (untreated) and tubes 2 and 2′ (hydrolyzed 7 min) for their reducing sugar content as follows: Prepare a blank sample containing 2 ml of H_2O and standards containing 0.1, 0.2, 0.4, 0.6, and 0.8 μmol of glucose in 2-ml final volume (i.e., 0.1, 0.2, 0.4, 0.6, and 0.8 ml of a 1.0 mM glucose standard plus 1.9, 1.8, 1.6, 1.4, and 1.2 ml of H_2O, respectively). Mix 0.5 ml of Nelson's reagent B with 12.5 ml of Nelson's reagent A. Add 1 ml of the combined reagent to each tube (i.e., the blank, the five standards, and tubes 1, 1′, 2, and 2′). Place the tubes simultaneously in a vigorously boiling water bath (500-ml beaker or larger), and heat for exactly 20 min. Remove the tubes simultaneously and place them in a beaker of cold water to cool. When the tubes are cool (25°C), add 1 ml of arsenomolybdate reagent to each and shake well occasionally during a 5 min period to dissolve the precipitated Cu_2O and to reduce the arsenomolybdate. Dilute the contents of each tube to a final volume of 10 ml with H_2O. Read the absorbance of all of the solutions at 540 nm. Calculate the number of micromoles of glucose-reducing equivalents in tubes 1, 1′, 2, and 2′ from your glucose standard curve.
13. *Determination of Inorganic Phosphate.* Using the modified Fiske–Subbarow method as described above, assay the following for inorganic phosphate: tubes 3 and 3′ (no hydrolysis), tubes 4 and 4′ (7-min hydrolysis), tubes 5 and 5′ (total hydrolysis), five standards (0.1, 0.3, 0.5, 0.7, and 0.9 ml of 1.0 mM standard phosphate plus 1.9, 1.7, 1.5, 1.3, and 1.1 ml H_2O, respectively), and a 2.0-ml H_2O blank. Add 1 ml of acid molybdate reagent and 1 ml of reducing reagent to all of the tubes, and bring the final volume of each tube to 10 ml with H_2O. Allow them to stand for 20 min, then read the absorbance of each solution at 660 nm against the blank. Determine the number of micromoles of inorganic phosphate in tubes 3, 3′, 4, 4′, 5, and 5′ from your standard curve.

Report of Results for Day 4

1. Prepare a table similar to Table 12-2. From the weight of isolated glucose-1-phosphate dissolved in H_2O for analysis, calculate the theoretical values for micromoles per milliliter reducing equivalents and inorganic phosphate in each sample, assuming a molecular weight of 372 and 100% purity. Then, determine the observed values from your analyses in steps 12 and 13. In all cases, report the values as micromoles per milliliter of the original glucose-1-phosphate solution analyzed.
2. From the analysis of tubes 3 and 3′ in step 13, determine how many micromoles per milliliter of inorganic phosphate contaminate your sample. How many micromoles per milliliter of glucose-6-phosphate contaminate your sample? (You have *two independent determinations* of this value. What are they? *Hint:* One determination uses the determinations with tubes 1 and 1′ in step 12, and the other requires you to compare the determinations in tubes 4 and 4′ to those of tubes 5 and 5′ in step 13.) What do you assume in using these values as assays for glucose-6-phosphate? Do the two determinations agree? If not, suggest possible reasons why not. (Note that if you used the enzymatic determination of glucose, you have only a single determination of glucose-6-phosphate. Explain why.) Do your chromatographic results agree with the conclusions drawn from the chemical analyses?

Table 12-2 Analysis of Glucose-1-Phosphate Isolated from Potato Phosphorylase Reaction

	μmol of Reducing Equivalents/0.1 ml		μmol of Inorganic Phosphate/0.1 ml	
Sample	Theory	Observed	Theory	Observed
Initial sample (μg/0.1 ml)				
7-min hydrolyzed sample				
Total hydrolysis sample				

3. Determine the extent of hydration of your isolated compound by performing the following operations:
   a. Calculate the percentage of 7-min *phosphorus* (not phosphate) in your sample (e.g., micrograms of phosphorus per 100-μg sample), assuming that the sample is pure.
   b. Calculate the expected percentage of 7-min phosphorus from dipotassium glucose-1-phosphate (no water of hydration).
   c. Calculate the expected percentage of 7-min phosphorus from dipotassium glucose-1-phosphate dihydrate.
   d. Compare your value for the percentage of phosphorus (a) with the two theoretical values (b and c), and decide which formula best fits your data. Comment on the reliability of your conclusion.
4. Weigh and label the remaining glucose-1-phosphate, and turn it in to the instructor.

Exercises

1. Consider the following set of data for inorganic phosphate determinations on unhydrolyzed and 7-min hydrolyzed aliquots taken from a 700-ml volume of the $MgNH_4PO_4$ supernatant obtained in this experiment.

	Absorbance at 660 nm	
Volume of Sample	Unhydrolyzed	Hydrolyzed for 7 min
0.1 ml	0.004	0.225
0.2 ml	0.010	0.435
0.5 ml	0.022	0.860
0.4 μmol Pi standard	0.150	
0.8 μmol Pi standard	0.300	

   a. Which absorbance values can be used for further calculations?
   b. Assume 100% recovery of the glucose-1-phosphate and inorganic phosphate at the beginning of the anion exchange step. How many milliequivalents of anion exchange resin must be used to adsorb a solution of these anions at pH 3.0? ($pK_{a1} = 1$, and $pK_{a2} = 6$ for glucose-1-phosphate.)

c. Assume 80% recovery of the glucose-1-phosphate as crystals. How many grams of dipotassium glucose-1-phosphate dihydrate would be isolated?

2. Propose a series of steps using ion-exchange resins for the isolation of glucose-6-phosphate from a solution containing methylamine, sodium acetate, glucose, and glucose-6-phosphate.
3. What steps would you perform to convert the rather insoluble barium salt of glucose-6-phosphate to the dipotassium salt?
4. Describe chemical tests for determining whether a given pure sample of unknown is (a) glucose-6-phosphate, (b) glucose-1-phosphate, (c) β-methylglucoside, (d) glucose, (e) fructose-1,6-diphosphate, or (f) sorbitol.
5. Which of the following compounds yield inorganic phosphate on 7-min hydrolysis: acetyl phosphate, 3-phosphoglyceraldehyde, ribose-5-phosphate, adenosine-5′-phosphate, ribose-1-phosphate, and pyrophosphate?
6. Point out the chemical similarities between Nelson's test and the Fiske–Subbarow test for phosphate, as well as between these tests and the Folin–Ciocalteau (Lowry) protein determination.

REFERENCES

Bandurski, R. S., and Axelrod, B. (1951). The Chromatographic Identification of Some Biologically Important Phosphate Esters. *J Biol Chem* **193**:405.

Cori, C. F., Colowick, S. P., and Cori, G. T. (1937). The Isolation and Synthesis of Glucose-1-Phosphoric Acid. *J Biol Chem.* **121**:465.

Fiske, C. H., and Subbarow, Y. (1925). The Colorimetric Determination of Phosphorus. *J Biol Chem* **66**:375.

Hanes, C. S., and Isherwood, F. A. (1949). Separation of the Phosphoric Esters. *Nature* **164**:1107.

McCready, R. M., and Hassid, W. Z. (1957). Preparation of α-D-Glucose-1-Phosphate by Means of Potato Phosphorylase. *Methods Enzymol* **3**:137.

Nelson, N. (1944). A Photometric Adaptation of the Somogyi Method for the Determination of Glucose. *J Biol Chem* **153**:375.

Runeckles, V. C. and Krotkov, G. (1957). The Separation of Phosphate Esters and Other Metabolites. *Arch Biochem Biophys* **70**:442.

Isolation and Characterization of Erythrocyte Membranes

Theory

In modern biochemical research it is often necessary to isolate and characterize biological membranes. Erythrocytes (red blood cells) are a popular cell type for these studies since they are abundant, they contain no organelles, they conduct relatively little metabolism, and they offer a homogeneous source of easily obtainable membranes. This experiment features a three-stage isolation of plasma membranes from porcine (pig) erythrocytes. First, the red blood cells will be separated from the plasma by low-speed centrifugation and subsequent washing. Second, the isolated red blood cells will be lysed by osmotic shock following transfer from an isotonic to a hypotonic buffer. In the final stage, the membranes will be isolated following a series of washes to remove hemoglobin and other cytoplasmic contaminants. After the erythrocyte membranes have been purified, you will perform a series of assays to characterize and quantify the different protein and lipid components present in these membranes.

Remember that biological membranes contain a great number of different types of proteins, which typically account for about 60% of their total mass. There are about 20 prominent proteins found in erythrocyte membranes. The chloride–bicarbonate anion exchanger accounts for about 30% of the total protein content of the red blood cell membrane. As carbonic anhydrase converts CO_2 to the more water-soluble form of HCO_3^-, this membrane protein will transfer the ion across the erythrocyte membrane. At the same time, chloride ions are transported in the opposite direction to prevent the formation of an electrochemical gradient. Another protein that is abundant in the erythrocyte membrane is the Na^+/K^+ pump. This transport protein uses the energy of ATP to *import* K^+ ions and *export* Na^+ ions. The net effect of this transport system is to establish an electrochemical gradient that can act as an energy source to drive the transport of other molecules against a concentration gradient. Glycophorin is another protein abundant in erythrocyte membranes. This highly charged transmembrane glycoprotein is the target (docking site) of the influenza virus. It is believed that this protein plays a role in preventing the aggregation of red blood cells in narrow capillaries. Finally, red blood cells contain a significant concentration of a transport protein called glucose permease. This transport protein spans the membrane 12 times and allows passive transport of glucose into the cell at a rate 50,000 times faster than would occur through the lipid bilayer in the absence of this transport protein.

All of the proteins just described are examples of integral membrane proteins. As such, they are firmly embedded in the phospholipid bilayer and actually span the membrane; one side of the protein faces the cytoplasmic side of the membrane, the other side faces the outer surface of the cell. The erythrocyte membrane also contains a number of peripheral-membrane proteins. Unlike the membrane-spanning integral-membrane proteins, these peripheral-membrane proteins are tightly associated only with the cytoplasmic side of the phospholipid bilayer. Together, these peripheral-membrane proteins form a meshlike matrix or skeleton on the inner surface of the membrane that gives the erythrocyte its distinctive and unusual biconcave (disklike) shape.

The majority of this matrix is composed of a protein called spectrin, a heterodimeric protein containing a 220-kDa α subunit and a similar but slightly larger β subunit. The highly repetitive amino acid sequences of both the α and β subunits give them a filamentous three-dimensional structure. As the α and β subunits of spectrin associate with one another, they form the flexible monomeric units that are used to create the membrane skeleton.

Two additional peripheral membrane proteins anchor the spectrin filaments to the cytoplasmic side of the erythrocyte membrane. One of these polypeptides, the 210-kDa ankyrin protein, binds both a single spectrin molecule and the chloride–bicarbonate anion-exchange protein discussed previously. The second of these polypeptides, actin, is capable of binding several molecules of spectrin. Since actin is able to associate with more than a single spectrin monomer, it acts as a branch point for the spectrin protein as the membrane skeleton or matrix is assembled (see Fig. 13-1). In this experiment, you will determine the concentration of total protein in the erythrocyte membrane through the use of the Folin–Ciocalteau assay.

Lipids account for about 40% of the total mass of the typical mammalian cell membrane. One class of lipids, the sterols, acts as the precursor to steroid hormones, affects membrane fluidity, and constitutes a major component of bile salts. In this experiment, you will quantify the amount of cholesterol (a sterol) using a coupled enzymatic assay containing cholesterol esterase, cholesterol oxidase, peroxidase, and a chromophore that absorbs light at 500 nm in the oxidized form (see Fig. 13-2).

Phospholipids, a second class of lipids, are the major components of the lipid bilayer of biological membranes. In this experiment, you will separate and identify phosphatidylcholine, phosphatidylethanolamine, phosphatidylserine, and phosphatidylinositol using thin-layer chromatography. Cholesterol and sphingomyelin (a sphingolipid) will also be identified in this part of the experiment. In addition, you will quantify the amount of total phospholipid present in the erythrocyte membrane using a colorimetric assay that produces a reduced phosphomolybdate complex that absorbs light at 660 nm (see Fig. 13-3). Note the relationship between the phosphate assay and the Folin–Ciocalteau protein assay (Experiment 1). What is the reducing agent in each case? What is the limiting reagent in each assay? A flow diagram for this three-period experiment is presented in Figure 13-4 to help guide you through the experiment.

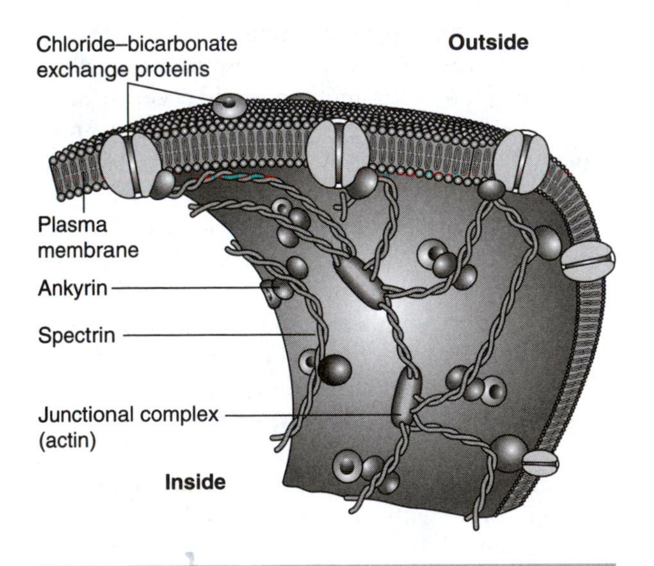

Figure 13-1 The membrane skeleton of erythrocytes.

Supplies and Reagents

Pig blood (if you do not have access to blood from this source, out of date whole human blood from a blood bank can also be used)
Isotonic phosphate buffer (310 mM sodium phosphate)
Hypotonic phosphate buffer (20 mM sodium phosphate)
250-ml plastic centrifuge bottles
30- to 50-ml plastic centrifuge tubes with screw caps
P-20, P-200, and P-1000 Pipetmen with disposable tips
15-ml plastic conical tubes with screw caps
Microcentrifuge tubes
Heparin sulfate
Pasteur pipettes
Reagents for Folin–Ciocalteau protein assay:

- 2% sodium tartrate solution
- 1% $CuSO_4 \cdot 5\ H_2O$ solution
- 2% Na_2CO_3 in 0.1 N NaOH
- Folin–Ciocalteau Reagent (2 N)

Lysozyme solution (1 mg/ml) in water
4-ml glass vials with Teflon-lined screw caps
Chloroform

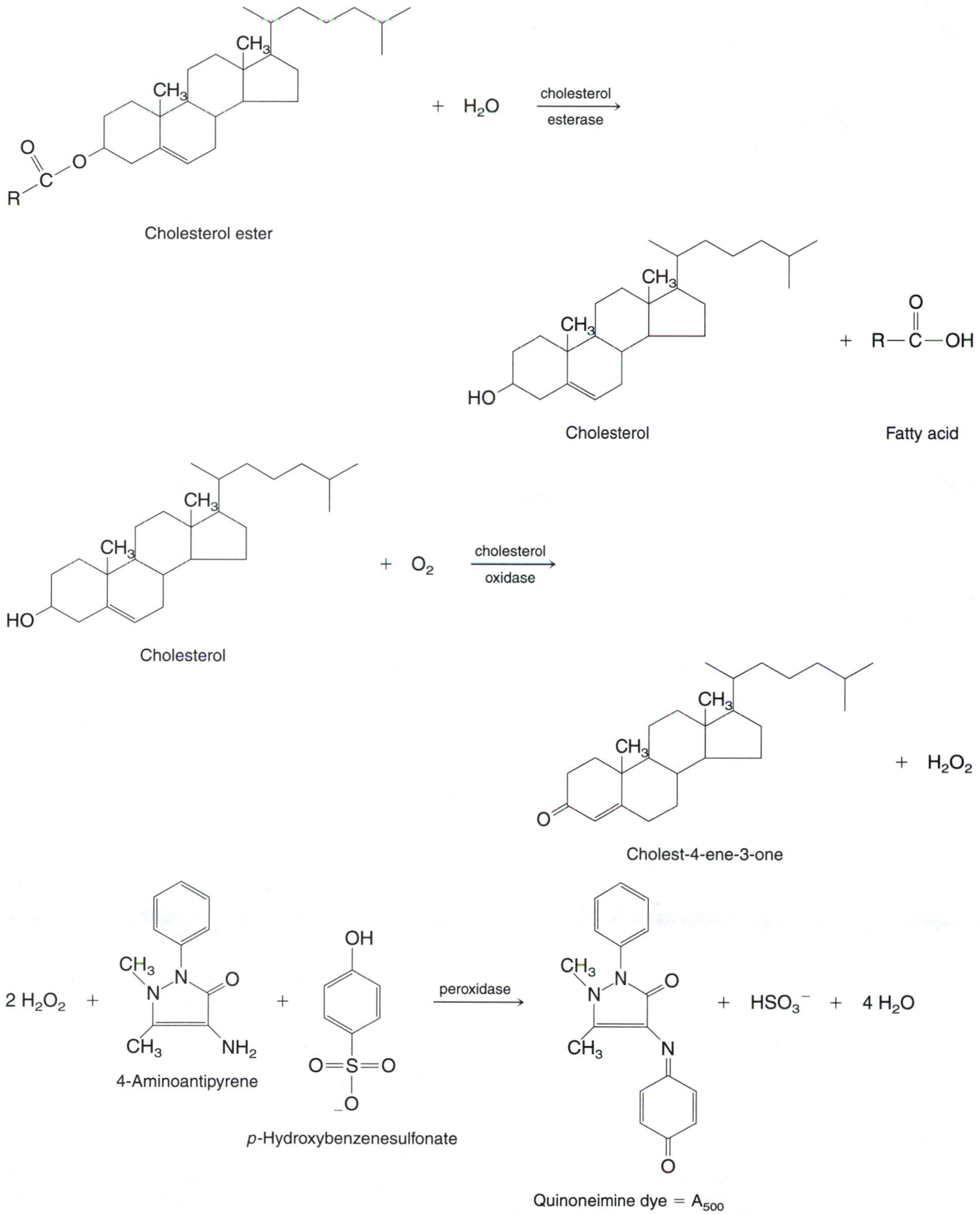

Cholesterol ester + H_2O —(cholesterol esterase)→ Cholesterol + $R-C(=O)-OH$ (Fatty acid)

Cholesterol + O_2 —(cholesterol oxidase)→ Cholest-4-ene-3-one + H_2O_2

$2\ H_2O_2$ + 4-Aminoantipyrene + *p*-Hydroxybenzenesulfonate —(peroxidase)→ Quinoneimine dye = A_{500} + HSO_3^- + $4\ H_2O$

Figure 13-2 Quantitative assay for total cholesterol.

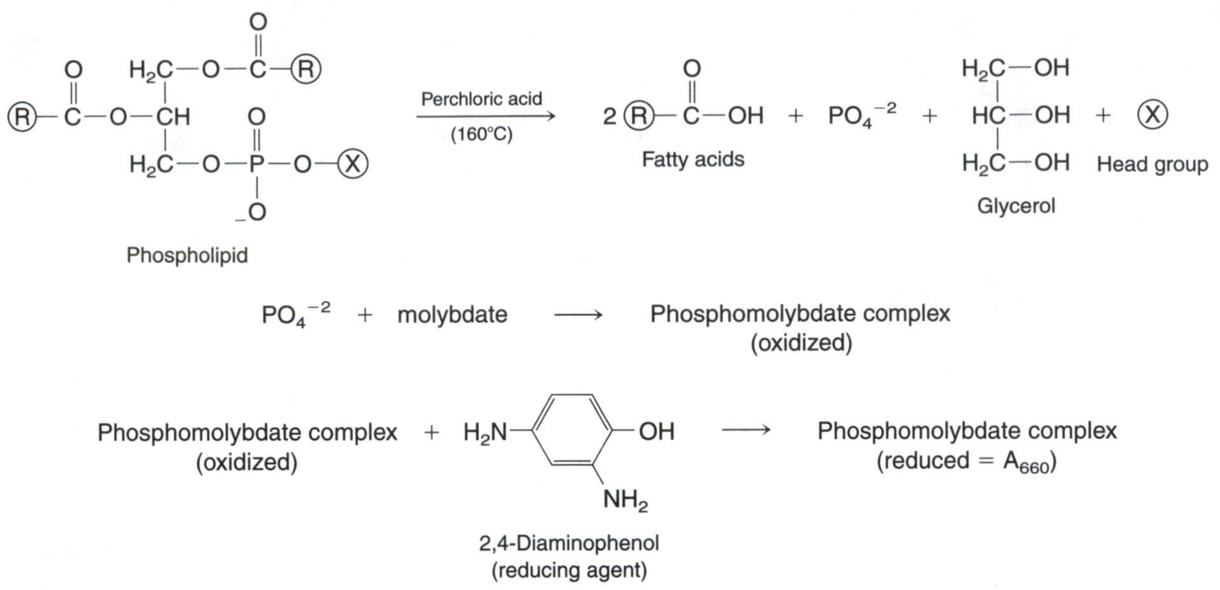

Phospholipid $\xrightarrow[\text{(160°C)}]{\text{Perchloric acid}}$ 2 Ⓡ—C(=O)—OH (Fatty acids) + PO_4^{-2} + Glycerol + Ⓧ (Head group)

PO_4^{-2} + molybdate ⟶ Phosphomolybdate complex (oxidized)

Phosphomolybdate complex (oxidized) + 2,4-Diaminophenol (reducing agent) ⟶ Phosphomolybdate complex (reduced = A_{660})

Figure 13-3 Quantitative phosphate assay to quantify phopholipids.

Methanol
Glacial acetic acid
Silica gel thin-layer chromatography plates (Fisher catalog #05–713–317)
16 × 125-mm glass test tubes
Phospholipid standard solutions in $CHCl_3$ (10–50 mg/ml of sphingomyelin, phosphatidylcholine, phosphatidylethanolamine, phosphatidylserine, phosphatidylinositol, and cholesterol)
I_2 chamber
Cholesterol standard solutions (1 mg/ml, 2 mg/ml, 4 mg/ml) in isopropanol
Cholesterol reagent (buffered to pH 6.5) (This reagent is supplied in a mixed dry form by Sigma Chemical Company. It is convenient for use with large laboratory groups.)

Cholesterol esterase (0.1 unit/ml)
Cholesterol oxidase (0.3 unit/ml)
Peroxidase (1 unit/ml)
Chromophore (4-aminoantipyrine, 0.3 mM)
p-Hydroxybenzenesulfonate (30 mM)

Phosphate standard solution (1 μmol/ml)
Pressurized N_2 tank
Perchloric acid (70%)
5% ammonium molybdate solution in water
2,4-diaminophenol reagent (1 g of sodium bisulfite and 40 mg of 2,4-diaminophenol per 4 ml of water)
Heating block
Preparative centrifuge
Clinical tabletop centrifuge (low speed)
Spectrophotometer to read absorbance at 500 and 660 nm with cuvettes
100°C oven

Protocol

Day 1: Isolation of Erythrocyte Membranes and Lipid Extraction.

We suggest that steps 1 to 6 be performed by the instructor prior to the beginning of the exercise.

Isolation of Erythrocyte Membranes

1. Obtain 3 liters of fresh pig blood (see "Supplies and Reagents" section for an alternative supply of blood, if necessary) and *immediately* mix with heparin sulfate (30 mg/liter final concentration) to prevent clotting. Place the blood in an ice bucket and proceed. *All additional steps through step 6 should be performed at 4°C.* We suggest that steps 1 to 6 be performed by the instructor prior to the beginning of the experiment. If you plan to perform Experiment 16 (Experiments in "Clinical Biochemistry and Metabolism") in conjunction with this experiment, a

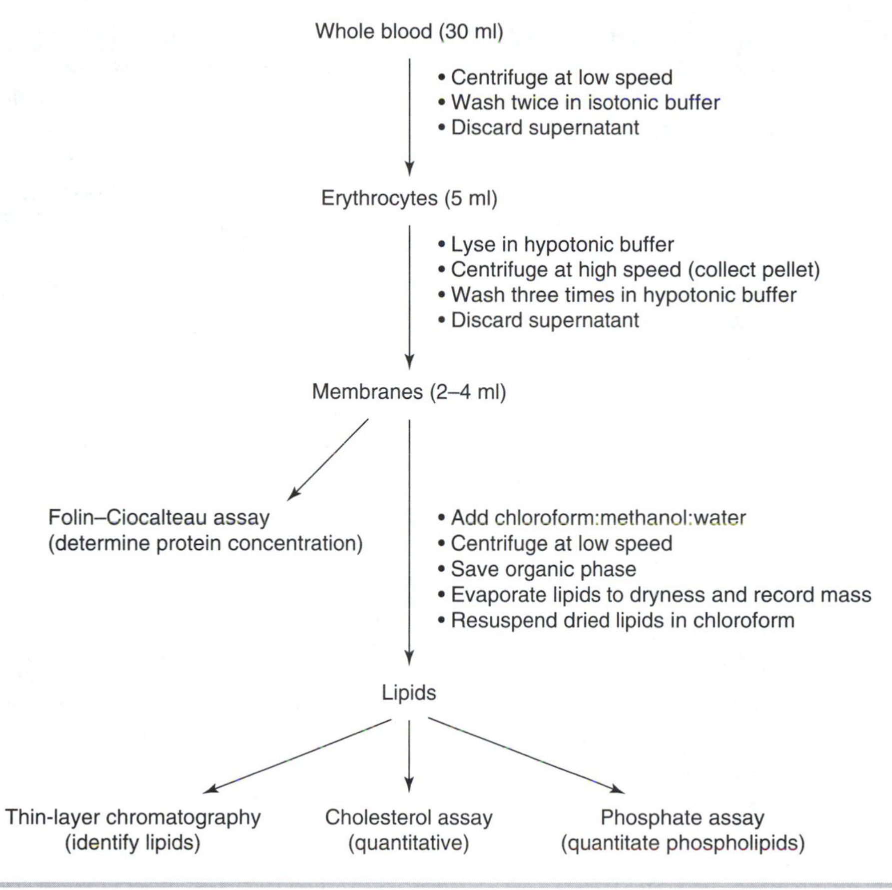

Figure 13-4 Flow diagram for erythrocyte membrane isolation and characterization.

portion of the pig blood can be centrifuged at 400 × g for 10 min to obtain the plasma (no phosphate buffer added).

2. Add 32 ml of isotonic phosphate buffer to a number of 250-ml plastic centrifuge bottles. Add 180 ml of whole pig blood to each bottle, cap, and invert several times to mix.
3. Centrifuge at 1000 × g for 10 min. Using a Pasteur pipette, carefully remove the plasma and "buffy coat" from the pelleted red blood cells.
4. Gently resuspend the erythrocyte pellet from each bottle in 120 ml of isotonic phosphate buffer. Centrifuge at 1000 × g for 10 min and again remove the supernatant.
5. Repeat step 4 one more time.
6. Resuspend the red blood cell pellet from each centrifuge bottle in isotonic buffer to a final volume of 30 ml. If these erythrocytes are prepared in advance, they can be stored on ice at 4°C for several days. DO NOT FREEZE THE CELLS AT THIS POINT. *Each 30 ml suspension of erythrocytes (obtained from 180 ml of whole blood) will be enough material for six groups of students to perform the experiment. We always collect extra blood (3 liters total) in case problems arise in the preparation of the erythrocytes.*
7. Add 5 ml of the erythrocyte suspension to 25 ml of hypotonic phosphate buffer in a 50-ml plastic centrifuge tube. Cap the tube and invert several times to mix (do not use a Vortex mixer).
8. Centrifuge at 20,000 × g for 40 min. The membrane pellet at the bottom of the tube will be dark red in color and *loosely* packed. *Carefully* remove the supernatant (hemoglobin, cytoplasmic contaminants) with a Pasteur pipette.
9. *Gently* resuspend the membrane pellet in the centrifuge tube in 25 ml of hypotonic phosphate buffer. Mix well by inversion and centrifuge at 20,000 × g for 20 min. Carefully

remove and discard the supernatant above the membrane pellet with a Pasteur pipette.

10. Repeat step 9 two more times. Note the change in color of both the membrane pellet and the supernatant with each wash.
11. After the last wash step, remove as much of the supernatant as possible and gently agitate the tube to resuspend the loosely packed, milky-looking membrane pellet in the residual buffer at the bottom of the tube. At this point, you should have 2 to 4 ml of concentrated membrane solution. Transfer this solution to a 15-ml plastic conical tube with a screw cap. Label the tube with your name and record the *exact volume* of the membrane solution in your notebook. This membrane solution can be stored at 4°C.

Lipid Extraction

1. Transfer 0.8 ml of the membrane suspension into two separate 15-ml plastic conical tubes with screw caps. *Avoid polystyrene and any other types of plastics that are not compatible with this solvent.* Add 3 ml of chloroform:methanol (1:2) to each tube. Cap and swirl each tube vigorously with a Vortex mixer for 30 sec.
2. Add 1 ml of chloroform to each tube, cap the tube, and swirl vigorously for 30 sec.
3. Add 1 ml of water to each tube, cap the tube, and swirl vigorously for 30 sec.
4. Separate the methanol:water phase (top) from the chloroform phase (bottom) in each tube by centrifugation at low speed ($\sim 500 \times g$, 5 min) in a tabletop centrifuge. You will see a small disk of white precipitate (denatured protein) at the interface between these two phases. If this disk extends down too far into the chloroform layer (more than halfway into the chloroform phase), add 0.1 ml of 0.1 M HCl, mix, and repeat the last low-speed centrifugation step.
5. Remove as much of the upper methanol:water phase as possible with a Pasteur pipette and discard.
6. With a clean Pasteur pipette, carefully puncture through the denatured protein disk and remove the chloroform layer at the bottom of each tube. Transfer the chloroform layer from both tubes to a single, *preweighed* 4-ml glass vial with a Teflon-lined screw cap. *Take care not to transfer any of the denatured protein layer (white precipitate) to these glass vials.*
7. Evaporate the combined chloroform phase containing the erythrocyte membrane lipids to dryness by passing a stream of N_2 over the top of the vial in a fume hood (see Experiment 6, "Sequence Determination of a Dipeptide" Fig. 6-7).
8. When the lipids have been dried completely, weigh the vial again and determine the *mass* of lipid that you have extracted from the erythrocyte membrane. *Record the mass of lipids in your notebook.*
9. Based on the mass of lipid present in 1.6 ml of the membrane solution, determine the number of milligrams of lipid present in the *entire* erythrocyte membrane solution (determined in step 11 of "Isolation of Erythrocyte Membranes").
10. Resuspend your dried lipids in 1 ml of chloroform. Cap the vial *tightly* (chloroform is very volatile), label it with your name, and store it at room temperature.

Day 2: Thin-Layer Chromatography of Lipids, Quantitative Cholesterol Assay, Folin–Ciocalteau Assay, and Preparation for Phosphate Assay

Thin-Layer Chromatography of Lipids

1. Obtain a silica-gel thin-layer chromatography plate that has been activated in a 100°C oven for 5 to 10 min. On either side of the plate, make a small scratch in the silica matrix about 2 cm from the bottom of the plate. This will mark the position of the origin, along which the samples will be spotted. Do not mark on the plate with pencil.
2. Remove 0.3 ml of your extracted lipids in chloroform prepared on Day 1 and place in a 1.5-ml microcentrifuge tube. Dry the sample under a stream of N_2 and resuspend the lipids in 40 μl of chloroform.
3. Using your P-20 Pipetman or a thin capillary tube, apply a total of 10 μl of this lipid solution on the origin. Keep the area of application small by repeated spotting and drying of 2-μl aliquots. Try to keep each sample spot as con-

centrated as possible. This will give you a much less diffuse lipid sample following development, which will make the identification of the different phospholipids much easier.

4. Using the same technique, spot 10 μl of each of the lipid standard solutions in chloroform (phosphatidylcholine, phosphatidylserine, phosphatidylethanolamine, phosphatidylinositol, cholesterol, and sphingomyelin) on different positions along the origin. Spot an additional 20 μl of your extracted lipid sample using the same technique (in case the concentration of lipid in your prepared sample is low).
5. When all the spots have dried, place the plate (origin side toward the bottom) in an ascending chromatography tank containing 1 cm of mobile-phase solvent on the bottom of the chamber. The mobile phase for this experiment is chloroform:methanol:acetic acid:water (25:15:4:1). Be sure that the mobile phase is in contact with the plate, but that the origin along which your samples are spotted lies *above* the level of the mobile phase (See Fig. 6-8). *While your thin-layer chromatography plate is developing, you may complete the quantitative cholesterol assay and the Folin–Ciocalteau protein assay.*
6. Allow the mobile phase to travel through (up) the stationary phase until the solvent front is about 1 in. from the top of the plate (~1.5–2.0 hr). Remove the plate from the tank, mark the position of the solvent front with a pencil, and allow the plate to air dry in the fume hood for about 5 min.
7. Place the plate in a sealed chamber saturated with I_2 vapor and incubate for 5 min. *The I_2 will react with unsaturated carbon bonds present in the lipids and produce yellow spots on the plate.*
8. Remove the plate from the I_2 chamber and circle the position of the spots with a pencil *(the yellow spots will disappear in about 1 hr).*
9. Measure the R_f values of the lipid standards:

$$R_f = \frac{\text{Distance traveled by sample spot (cm)}}{\text{Distance traveled by solvent front (cm)}}$$

Based on the R_f values of the lipids present in your sample, what lipids appear to be present in the erythrocyte membrane? Based on the size and intensity of the spots present in your sample compared to those of the lipid standards, which lipid(s) are most abundant in the pig erythrocyte membrane?

Quantitative Cholesterol Assay

1. Set up the reactions described in the table below in 1.5-ml microcentrifuge tubes: *The extracted lipids that are currently dissolved in chloroform cannot be used directly in this assay, since the chloroform will denature the enzymes that will eventually produce a colored product (see Fig. 13-1). To prepare the lipids in your sample for this assay, add the indicated volume of lipids in $CHCl_3$ to tubes 4 and 5, dry them under a stream of N_2, and resuspend each sample in 10 μl of* ***isopropanol.*** *The standard cholesterol solutions (tubes 1–3) can be used directly in this assay by adding the indicated volumes to the appropriate tubes.* (See chart below.)
2. Incubate the tubes at room temperature for 15 min.
3. Read the absorbance of the solution in each tube at 500 nm within 30 min of the completion of the 15-min incubation. Use the solution in the blank tube to zero your spectrophotometer at 500 nm. Record the absorbance values in your notebook.
4. Prepare a standard curve by plotting A_{500} versus milligrams of cholesterol for tubes 1 to 3.
5. Based on the absorbance readings of tubes 4 and 5 at 500 nm, determine the mass of cholesterol (in milligrams) present in these tubes.

Component	Blank	Tube 1	Tube 2	Tube 3	Tube 4	Tube 5
Water	100 μl	90 μl	90 μl	90 μl	90 μl	90 μl
1 mg/ml cholesterol std.	—	10 μl	—	—	—	—
2 mg/ml cholesterol std.	—	—	10 μl	—	—	—
4 mg/ml cholesterol std.	—	—	—	10 μl	—	—
Lipids (in $CHCl_3$)	—	—	—	—	20 μl	100 μl
Cholesterol reagent	1.00 ml	1.00 ml	1.00 ml	1.00 ml	1.00 ml	1.00 ml

6. Based on the volume of lipid sample (in $CHCl_3$) present in tubes 4 and 5, determine the concentration of cholesterol present in your lipid sample in units of milligrams per milliliter.
7. Using the molecular weight of cholesterol (386.7), calculate the concentration of cholesterol in your entire membrane sample in units of micromoles per milliliter.

Folin–Ciocalteau Assay to Determine Membrane Protein Concentration

1. Dilute your *membrane solution* prepared on Day 1 (*not the extracted lipid solution*) by adding 0.2 ml of the membrane solution to 0.8 ml of hypotonic phosphate buffer. This diluted erythrocyte membrane solution will be used in the Folin–Ciocalteau assay described below.
2. Set up the following reactions in 16 × 125-mm glass test tubes:

Tube	Volume of Hypotonic Buffer (ml)	Volume of 1 mg/ml Lysozyme Solution (μl)	Volume of Diluted Membrane Solution (μl)
1	1.20	—	—
2	1.18	20	—
3	1.15	50	—
4	1.10	100	—
5	1.00	200	—
6	0.90	300	—
7	1.15	—	50
8	1.10	—	100
9	0.90	—	300

3. Separately prepare 100 ml of fresh alkaline copper reagent by mixing, in order, 1 ml of 1% $CuSO_4 \cdot 5H_2O$ and 1 ml of 2% sodium tartrate into 98 ml of 2% Na_2CO_3 in 1 N NaOH.
4. Add 6 ml of this alkaline copper reagent to each of the nine tubes and mix immediately after each addition to avoid precipitation. Incubate the tubes at room temperature for 10 min.
5. Add (and immediately mix in) 0.3 ml of Folin–Ciocalteau reagent to each of the nine tubes. Incubate for 30 min at room temperature.
6. Read the absorbance of the solution in each tube at 500 nm. *Record the absorbance values in your notebook.* Use the solution in tube 1 (blank) to zero your spectrophotometer at 500 nm.
7. Prepare a standard curve by plotting A_{500} versus milligrams of protein for tubes 2 to 6.
8. Determine which of tubes 7 to 9 gave an A_{500} that lies in the *linear* portion of the standard curve. What mass of protein (in milligrams) is indicated by its position on the standard curve?
9. Based on the *volume* of the membrane solution present in tubes 7 to 9 (remember the dilution at the beginning of the assay), calculate the *concentration* of protein present in the membrane solution in units of milligrams per milliliter.
10. Based on the total *volume* of the entire membrane solution isolated on Day 1, how many milligrams of protein are present in the total membrane fraction?
11. Using the *total mass* of lipid in the membrane solution (calculated at the end of Day 1), what is the *mass ratio* of protein to lipid in the erythrocyte membrane?

Preparation for Phosphate Assay

The glass vials used in this part of the experiment should be cleaned with a 0.5% solution of sodium dodecyl sulfate (SDS), since many commercially available detergents contain a significant concentration of phosphate.

1. Transfer 20 μl, 50 μl, and 100 μl of your extracted lipid sample in $CHCl_3$ to three 4-ml glass vials with Teflon-lined screw caps. Transfer 0.30 ml of a 1 μmol/ml phosphate standard solution to a fourth 4-ml glass vial.
2. Dry each of the samples under a stream of N_2 in the fume hood.
3. Add 0.5 ml of 70% perchloric acid to each vial and resuspend each of the dried lipid samples with gentle agitation.
4. Cap the vials tightly with a Teflon-lined screw cap and incubate in a 160°C oven or heat block overnight (~12 hr). After cooling to room temperature, these solutions will be used on Day 3. *The 1 μmol/ml phosphate solution in perchloric acid does not have to be heated overnight. This standard solution can simply be stored at room temperature for use on Day 3.*

Day 3: Completion of Phosphate Assay

1. Prepare a series of phosphate standard solutions in 4-ml glass vials using the 0.3-ml sample of phosphate (1 μmol/ml) prepared on Day 2.

Tube	Volume of Phosphate Solution in Perchloric Acid (μl)	Volume of Perchloric Acid (ml)
1	—	0.50
2	20	0.48
3	50	0.45
4	100	0.40
5	150	0.35

2. Add and mix the following reagents, in order, to these five standard vials as well as to the three vials containing your lipid sample in perchloric acid that were hydrolyzed on Day 2:

   2.4 ml of water
   0.1 ml of 5% ammonium molybdate in water
   0.1 ml of 2,4-diaminophenol (amidol) reagent

3. Seal all eight vials with a Teflon-lined screw cap and incubate at 110°C for 30 min in the heat block.
4. Allow the vials to cool to room temperature and read the absorbance of all of the solutions at 660 nm. Use the solution in vial 1 (blank) to zero your spectrophotometer.
5. Prepare a plot of A_{660} versus micromoles of phosphate. Based on the A_{660} of your three lipid samples, how many micromoles of phosphate are present in each vial?
6. Based on the fact that there is a 1:1 ratio of phosphate to phospholipid, and considering the total volume of the lipid fraction recovered on Day 1, how many micromoles of phospholipid are present in your extracted lipid sample?
7. Based on the average molecular weight of the four phospholipids (~780 g/mol), how many milligrams of phospholipid are present in your total extracted lipid sample?
8. Based on the number of milligrams and the number of micromoles of cholesterol present in your total extracted lipid sample (calculated on Day 2), what is the mass ratio and mole ratio of cholesterol:phospholipid in the erythrocyte membrane?

Exercises

1. Prior to the beginning of this experiment, the erythrocytes were washed twice in isotonic phosphate buffer. What was the purpose of these washes?
2. The erythrocytes were then washed several times in hypotonic phosphate buffer and subjected to centrifugation. What purpose did the hypotonic phosphate buffer serve in these washes? What components were present in the pellet and supernatant fractions following each wash step?
3. State what components were present in the water:methanol phase and the chloroform phase following the water:methanol:chloroform extraction of the erythrocyte membranes.
4. The lipid sample that was used in the cholesterol assay underwent a special preparation procedure before it could be used in the assay. What was this procedure and why was it necessary?
5. List the four phospholipids that are commonly found in biological membranes. Explain how the single assay performed in this experiment could be used to quantify this entire group of lipids in the erythrocyte membrane.
6. Explain why the protein:lipid ratio may vary between different cell types. Explain why there may be variations in the relative amounts of each type of lipid from one cell type to another.
7. Why are erythrocytes commonly used as a source of membranes for biochemical analysis?
8. Analyze your thin-layer chromatography plate. What lipids displayed the lowest and highest R_f values? Explain the reason for this result in terms of the chemical nature of the stationary and mobile phases used in this experiment and the nature of the different molecules being separated in this experiment (which lipids would you expect to show the greatest and the least mobility on the plate and why).

REFERENCES

Allain, C. C., Poon, L. S., Chan, C. S. G., Richmond, W., and Fu, P. C. (1974). Enzymatic Determination of Total Serum Cholesterol. *Clin Chem* **20**:470.

Bennett, V. (1985). The Membrane Skeleton of Human Erythrocytes and Its Implications for More Complex Cells. *Annu Rev Biochem* **54**:273.

Dittmer, J. C., and Wells, M. A. (1969). Quantitative and Qualitative Analysis of Lipids and Lipid Components. *Methods Enzymol* **14**:482.

Garrett, R. H., and Grisham, C. M. (1995). Lipids and Membranes. In: *Biochemistry*. Orlando, FL: Saunders.

Gurr, M. I., and Harwood, J. L. (1990). *Lipid Biochemistry. An Introduction*, 4th ed. London: Chapman & Hall.

Kates, M. (1986). *Techniques of Lipidology: Isolation, Analysis, and Identification*, 2nd ed. (Laboratory Techniques in Biochemistry and Molecular Biology, Vol. 3.) New York: Elsevier Science Publishing Co.

Lehninger, A. L., Nelson, D. L., and Cox, M. M. (1993). Lipids. In: *Principles of Biochemistry*, 2nd ed. New York: Worth.

Op den Kamp, J. A. F. (1979). Lipid Asymmetry in Membranes. *Annu Rev Biochem* **48**:47.

Skipski, V. P., and Barclay, M. (1969). Thin-Layer Chromatography of Lipids. *Methods Enzymol* **14**:530.

Stryer, L. (1995). Membrane Structure and Dynamics. In: *Biochemistry*, 4th ed. New York: Freeman.

EXPERIMENT 14

Electron Transport

Theory

Most of the metabolites that serve as energy sources for aerobic organisms are broken down (catabolized) by various pathways to yield substrates for the central, energy-yielding citric acid cycle (also known as the tricarboxylic acid cycle or the Krebs cycle). The citric cycle is a sequence of enzymatic steps whose net reaction consists of the oxidation of acetate to carbon dioxide.

(14-1)

$$CH_3COOH + 2\ O_2 \longrightarrow 2\ CO_2 + 2\ H_2O$$

Oxygen is not reduced directly during oxidation of acetate. Rather, pairs of electrons and pairs of protons are removed from the citric acid cycle substrates and transferred to intermediate redox carriers, as shown by the half reactions of Equation 14-2.

(14-2)

$$AH_2 \longrightarrow A + 2H^+ + 2\ e^-$$

$$2\ e^- + 2\ H^+ + B \longrightarrow BH_2$$

Here AH_2 represents the *reduced* substrate, A is the *oxidized* (dehydrogenated) substrate, B is the *oxidized* carrier and BH_2 is the *reduced* carrier. That is, the oxidation of the substrate A is coupled with the reduction of electron carrier B (Equation 14-3).

(14-3)

$$AH_2 + B \longrightarrow A + BH_2$$

Several electron carriers are found in nature; the most important examples include pyridine nucleotides, flavins, quinones, and cytochromes. Electron transport consists of a series of reactions in which electrons and protons are passed through a cascade of these electron carriers.

Natural electron transport involves consecutive reduction and oxidation of a series of electron carriers, which must include electron transfer, but does not always include transfer of the protons. In these cases, the protons become a part of the aqueous medium (e.g., in the reduction of ferric iron associated with a cytochrome, as shown in Equation 14-4).

(14-4)

$$\underset{\text{(in protein)}}{CH_2 + 2\ Fe^{3+}} \longrightarrow \underset{\text{(in protein)}}{C + 2\ Fe^{2+}} + 2\ H^+$$

Eventually, the electrons are passed to molecular oxygen, and protons are reclaimed from the aqueous medium through the formation of water (half reaction shown in Equation 14-5).

(14-5)

$$1/2\ O_2 + 2\ H^+ + 2\ e^- \longrightarrow H_2O$$

This final reaction accounts for most of the known oxygen consumption by aerobic organisms. The cascade of redox reactions that couples the oxidation of organic substrates to reduction of molecular oxygen in biological systems is called *electron transport*, and is often presented schematically as shown in Figure 14-1. When substrates are oxidized by such a system, the rate and extent of substrate oxidation is directly dependent on, and can be measured by, the decrease in concentration of molecular oxygen, as will be done in this experiment.

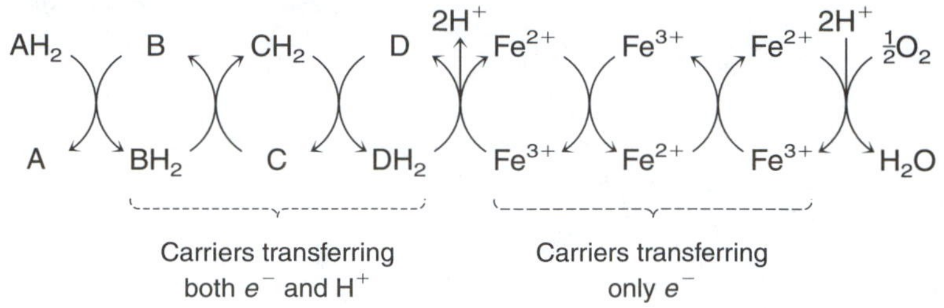

AH_2 B CH_2 D $2H^+$ Fe^{2+} Fe^{3+} Fe^{2+} $2H^+$ $\frac{1}{2}O_2$

A BH_2 C DH_2 Fe^{3+} Fe^{2+} Fe^{3+} H_2O

Carriers transferring both e^- and H^+

Carriers transferring only e^-

Figure 14-1 Schematic representation of electron transport.

The general features of electron transport are similar throughout nature. The processes take place in highly structured environments within the cell membranes of bacteria and within specialized subcellular particles—the mitochondria—of higher plants and animals. The electron carrier molecules are also similar throughout nature: the diphosphopyridine nucleotide NAD^+, proteins that contain flavin (FMN and/or FAD) and iron–sulfur (Fe-S) clusters, quinones that are soluble in the lipid component of membranes, and several heme-containing proteins called cytochromes. Each of these electron-carrying molecules has a characteristic optical absorption spectrum that differs between the oxidized and reduced states of the component. You have seen the absorption spectra of reduced and oxidized NAD in Experiment 8 and the spectrum of oxidized riboflavin, the chromophore of FAD, in Experiment 1. Several cytochromes are involved in the electron transport scheme. These cytochromes all possess an iron-containing heme, which is Fe-protoporphyrin IX in the case of the various members of the cytochrome *b* family (Fig. 14-2). The heme is covalently linked to the protein via addition of cysteinyl residues to the vinyl groups of Fe-protoporphyrin IX in the cytochromes of the *c* class (Fig. 14-2). Further modifications of the heme lead to heme *a*, which is found in cytochromes of the *a* class (Fig. 14-2). In all of the cytochromes, electron transfer reactions involve the reversible oxidation and reduction of the heme Fe between the Fe^{2+} and Fe^{3+} states. In their reduced forms cytochromes demonstrate a characteristic three-peak heme absorption spectrum with α, β, and γ (or Soret) bands similar to those shown in Figure 14-3.

Biochemists use the changes in the ultraviolet-visible absorption spectra of the individual components of the electron transport chain to follow the order of the steps in electron transport and the rates of individual steps in the electron transport chain. This has been a particularly useful approach in analyzing the cytochrome components of the chain. Small differences in the heme structures of different cytochromes, together with major differences in their protein components, create characteristic spectral differences among them. These spectral differences, as observed in the visible spectra of reduced cytochromes, originally served as the primary tool of cytochrome classification (see Table 14-1). Functional assays also serve to classify cytochromes. For example, the protein conformation of most cytochromes buries the cytochrome's heme within the protein and makes it inaccessible to molecular oxygen. Those cytochromes therefore cannot react with O_2 or its analogs (CO, CN^-, and azide). In contrast, the cytochromes of the terminal oxidase contain a heme iron that is accessible to O_2 and its analogs. Thus, cytochrome a_3 of Table 14-1, for example, forms a spectrally distinct complex with CO, CN^-, and azide, whereas the other cytochromes in Table 14-1 do not.

The components of the electron transport chain are organized into a series of very complex, highly ordered integral-membrane–multiprotein complexes. It is essential that the electron-transport apparatus be integrated into an enclosed membrane system that enables the large negative free energy of the oxidation steps to be captured in the form of an *electrochemical gradient.* That is, the electron-transport components are arranged so that the protons which are released during oxidation of the hydrogen-carrying electron carriers (see, for example, Equation 14-4) are released on the *outside* of the membrane, creating a *gradient* of protons. The reverse flow of these protons to the interior of the membrane-enclosed cell or organelle discharges the

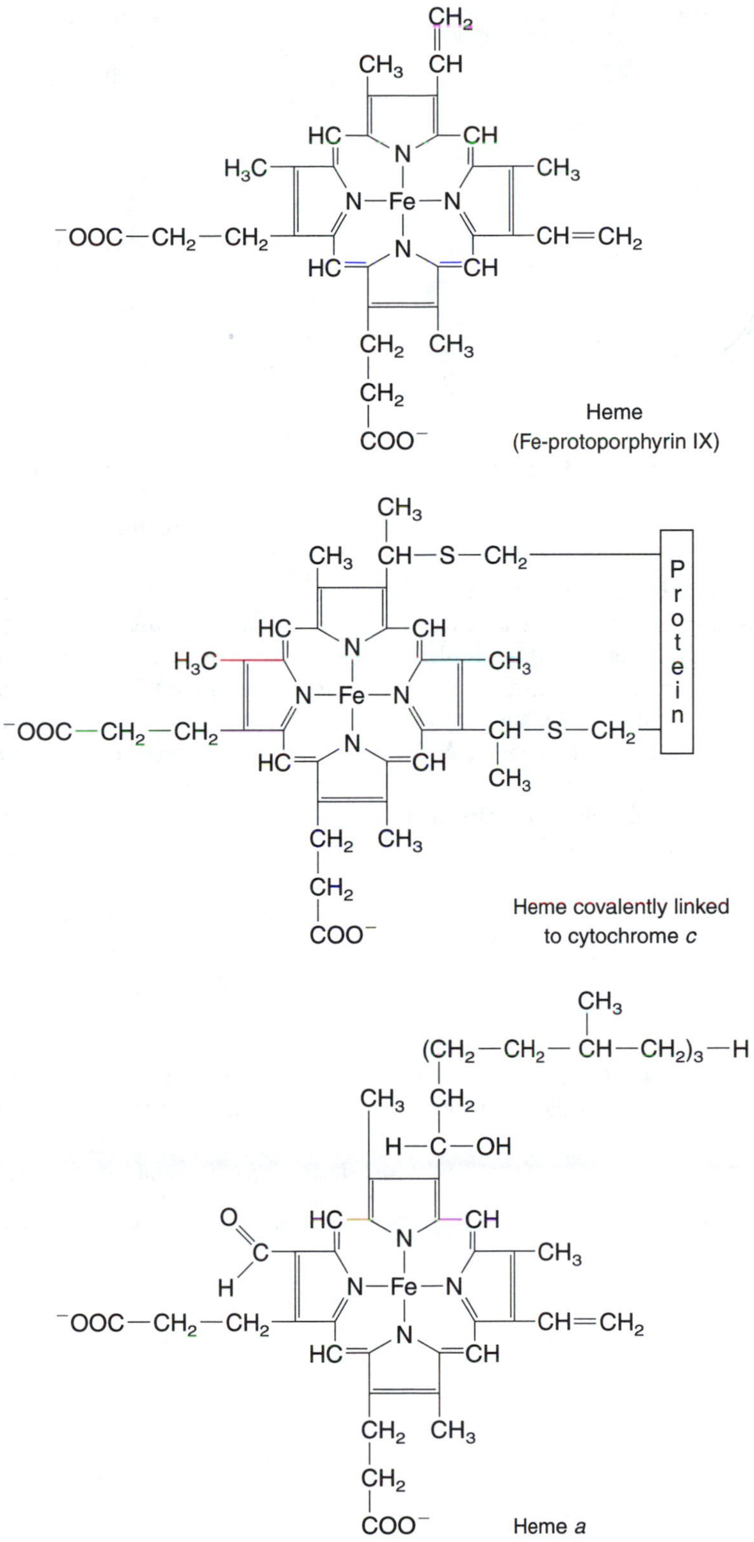
Heme
(Fe-protoporphyrin IX)
Heme covalently linked
to cytochrome *c*
Protein
Heme *a*

Figure 14-2 Structures of the heme prosthetic groups of the cytochromes. In all cytochromes the heme Fe undergoes reversible oxidation and reduction between Fe^{2+} and Fe^{3+}.

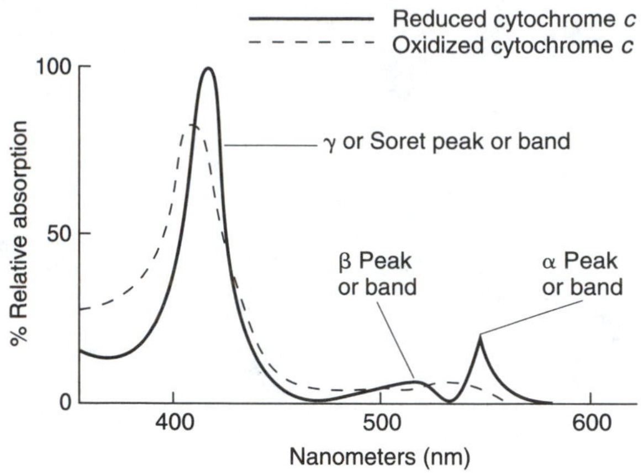

Figure 14-3 Absorption spectra of cytochrome *c*.

gradient and drives the synthesis of ATP from ADP and Pi by an integral membrane enzyme known as the F_0F_1-ATPase or ATP synthase. Although electron transport can occur in membrane fragments without an enclosed membrane space, such a system cannot use oxidation reactions to generate ATP. Such a system is said to be *uncoupled.*

In higher plants and animals, electron transport occurs within the mitochondria. Biochemists have devised methods to isolate intact mitochondria containing all of the functional citric-acid-cycle enzymes and electron-transport components. Such intact mitochondria provide useful information about the overall process. Yet, such intact mitochondria are too complex to allow the intricate details of electron transport to be studied. For this reason, the integral-membrane–multienzyme complexes of electron transport have been carefully dissected into separate functional components that catalyze a portion of the overall electron-transport chain, and these components have been characterized separately. For example, the electron transport chain of mammalian mitochondria has been resolved into three fundamental multienzyme complexes: NADH-ubiquinone reductase, cytochrome *c* reductase, and cytochrome *c* oxidase (see Fig. 14-4 and Table 14-2). Two components of the electron-transport chain, ubiquinone (Fig. 14-5) and cytochrome *c*, are not tightly associated with these complexes. Oxidation of the citric-acid-cycle intermediate succinate involves yet a fourth integral membrane enzyme complex, succinate–ubiquinone reductase, which bypasses NADH–ubiquinone reductase and contributes reducing equivalents (electrons) from succinate oxidation to ubiquinone, which is dissolved in the membrane (Fig. 14-4 and Table 14-2). An exciting area of current research is the determination of the detailed structure of the multienzyme complexes of electron transport; the structure of the 13-subunit cytochrome *c* oxidase from bovine mitochondria has been determined at better than 3 Å resolution, for example.

Table 14-1 Absorption Bands (λ_{max}) of Cytochromes (position of bands in nanometers, cytochromes in reduced state)

Cytochrome	Name of Band	Absorption Maxima of Pigment in Reduced Form	Maxima of CO Complex Formed on Addition of CO to Reduced Cytochrome
a	α	605	—
	γ	452	—
a_3	α	600	590
	γ	445	432
b	α	564	—
	β	530	—
	γ	431	—
c	α	550	—
	β	520	—
	γ	415	—

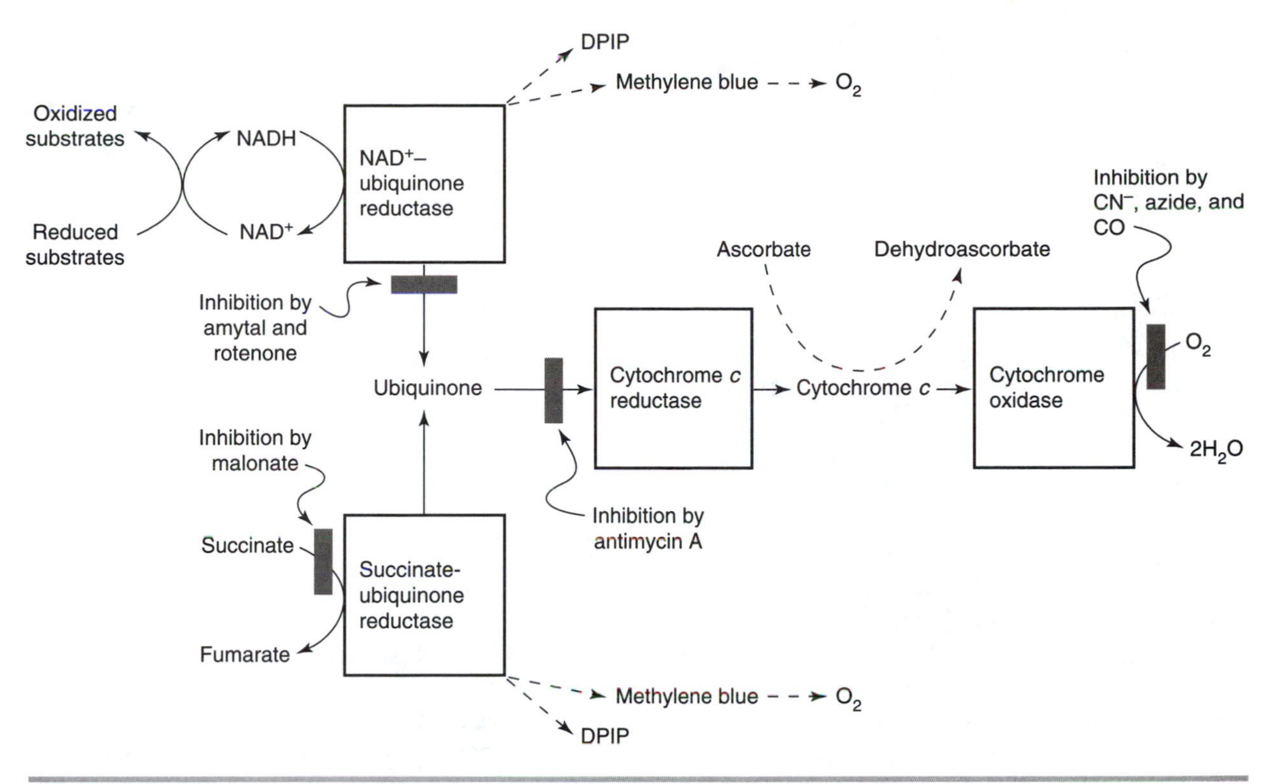

Figure 14-4 Schematic representation of electron transport in mitochondria.

The sum of a large number of studies with intact mitochondria, submitochondrial membrane fractions, and purified electron-transport–multienzyme complexes has led to the somewhat simplified scheme for the functional organization of the electron-transport system in mammalian mitochondria shown in Figure 14-4. You should use this scheme to aid in your interpretation of your results from this experiment. Note that different substrates are oxidized by different components of the system

Table 14-2 Multienzyme Complexes of the Electron-Transport Chain in Mammalian Mitochondria

Complex	Mass (kDa)	Number of Subunits	Prosthetic Groups
NADH-ubiquinone reductase	880	34	FMN, Fe-S centers
Succinate-ubiquinone reductase	140	4	FAD, Fe-S centers
Cytochrome *c* reductase	250	10	Heme *b*-562
			Heme *b*-566
			Heme c_1
			Fe-S centers
Cytochrome oxidase	160	13	Heme *a*
			Heme a_3
			Cu_A and Cu_B

SOURCE: Stryer, L. (1995). *Biochemistry,* 4th ed. p. 534 (Table 21-2), New York: Freeman. Reprinted by permission.

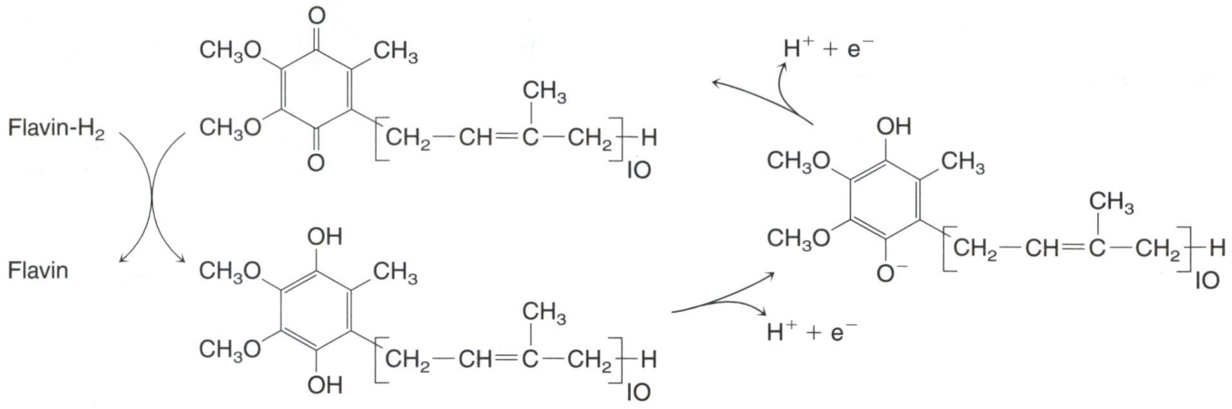

Figure 14-5 Oxidation–reduction reactions of ubiquinone.

and thus feed their reducing equivalents into the chain at different points. Most citric-acid-cycle components are oxidized by soluble dehydrogenases of the mitochondria matrix with the concomitant reduction of NAD^+ to NADH. NADH is then oxidized by the NADH–ubiquinone reductase. Succinate, on the other hand, is oxidized by a membrane-bound succinate dehydrogenase complex (a flavoprotein containing iron–sulfur centers), which reduces ubiquinone without involvement of pyridine nucleotides. Ascorbate (vitamin C) is not a normal substrate of the electron-transport chain; it readily reduces cytochrome *c*, but not the electron carriers upstream (such as flavoproteins, ubiquinone, or cytochrome *c* reductase), so it can be used to study the properties of the portion of the electron-transport chain from cytochrome *c* to O_2. As noted below, artificial electron-accepting dyes can accept electrons from the flavin-containing components of the electron transport chain (see Fig. 14-4) and can thus be used to study the portions of the chain upstream from the flavins. Finally, various inhibitors have been shown to block the electron-transport chain at specific points; use of such inhibitors played an important role in developing the scheme shown in Figure 14-4. For example, malonate is a structural analog of succinate and competitively inhibits succinate dehydrogenase. CN^-, azide ions, and CO, as noted above, bind specifically to the heme a_3 of the terminal cytochrome *c* oxidase. The antibiotic antimycin A specifically inhibits cytochrome *c* reductase. The use of such specific inhibitors is an important part of this experiment.

For less sophisticated studies, preparations of electron-transport components in mitochondrial membranes are useful. Keilin–Hartree heart particles, a preparation of mitochondrial fragments and damaged mitochondria used in this experiment, provide a useful submitochondrial system for the assay of many general properties of electron transport. These particles have lost most of the citric-acid-cycle enzymes, which are soluble in the mitochondrial matrix and are washed away from the membrane fragments during purification. Keilin–Hartree particles are also not capable of oxidative phosphorylation, because the integrity of an enclosed membrane system has been lost. However, the membrane fragments still contain functional succinate–ubiquinone reductase, ubiquinone, cytochrome *c* reductase, cytochrome *c*, and cytochrome *c* oxidase complexes. Thus, these particles will catalyze oxidation of succinate and normal electron transport to oxygen. In this experiment, some properties of the electron-transport chain from succinate to O_2 will be characterized using Keilin–Hartree particles from pig heart mitochondria.

Two methods will be used to assay electron transport catalyzed by these Keilin–Hartree particles. In most of the experiments, an oxygen electrode will be used to measure the rate of oxygen consumption. Oxygen electrodes are described in the next paragraph. A spectrophotometric assay of electron transport will also be used. This assay replaces O_2 with an organic dye that acts as the terminal electron acceptor. Some reducible dyes, such as methylene blue, are "autoxidizable." That is, they are reduced as the result of accepting electrons from a reduced electron transport component and they, in turn, pass on electrons by reducing O_2 to H_2O. Such autoxidizable dyes therefore provide a shunt, or second means of electron transfer to oxygen, but they are not reduced in a stoichiometric manner. Because of this, they are of little use in aerobic spectrophotometric assays (Fig. 14-6). In contrast, a dye that is not autoxidizable, such as 2,6-dichlorophenol-indophenol (DPIP), will accept electrons from a biological electron transport system but will not pass these electrons on to O_2. Thus, the reduction of DPIP, which is accompanied by a loss of visible color, can be used to assay biological oxidation reactions (Fig. 14-6).

Commercially available electrodes that measure O_2 in aqueous systems are called oxygen electrodes, dissolved oxygen meters, or oxygen polarographs (although few of these systems involve true oxygen polarography; i.e., a dropping Hg electrode under controlled diffusion conditions). All of the systems are based on an oxygen detector called an oxygen electrode similar to that illustrated in Figure 14-7. Oxygen electrode units contain two chambers—a sample chamber and an electrode chamber—separated by a synthetic membrane. The electrode chamber is filled with an electrolyte system that minimizes artifacts at the electrode surfaces. The membrane is placed over the electrode chamber,

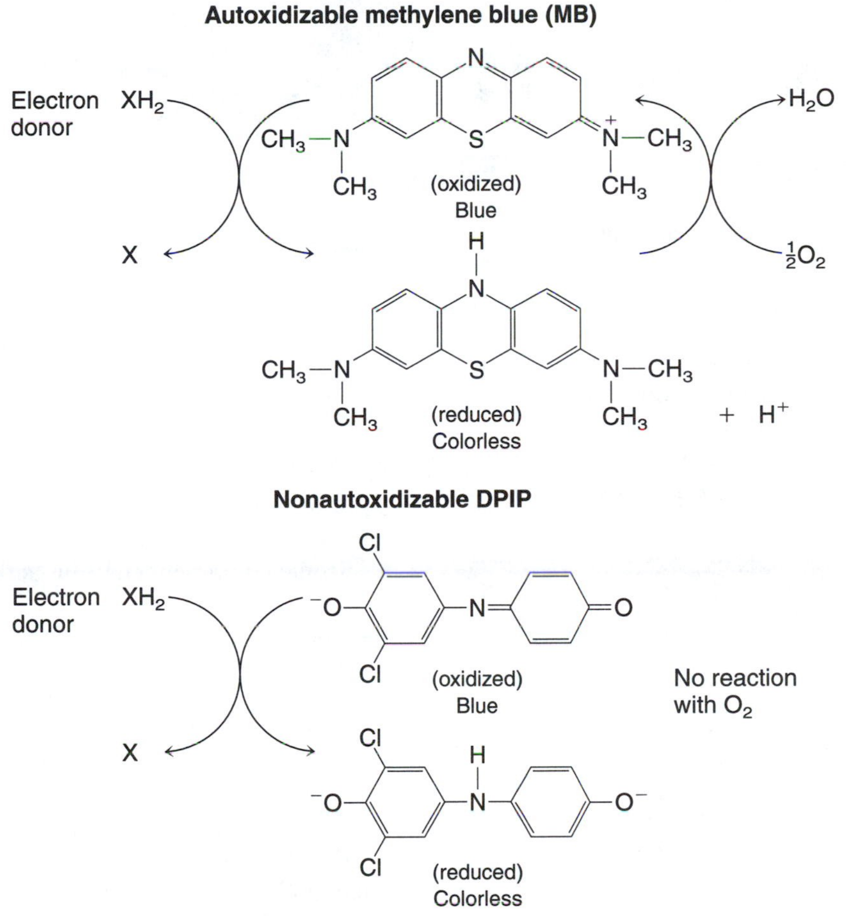

Autoxidizable methylene blue (MB)

Electron donor XH_2 → X

(oxidized) Blue

(reduced) Colorless + H^+

H_2O ← $\frac{1}{2}O_2$

Nonautoxidizable DPIP

Electron donor XH_2 → X

(oxidized) Blue

(reduced) Colorless

No reaction with O_2

Figure 14-6 Comparison of autoxidizable and nonautoxidizable dyes.

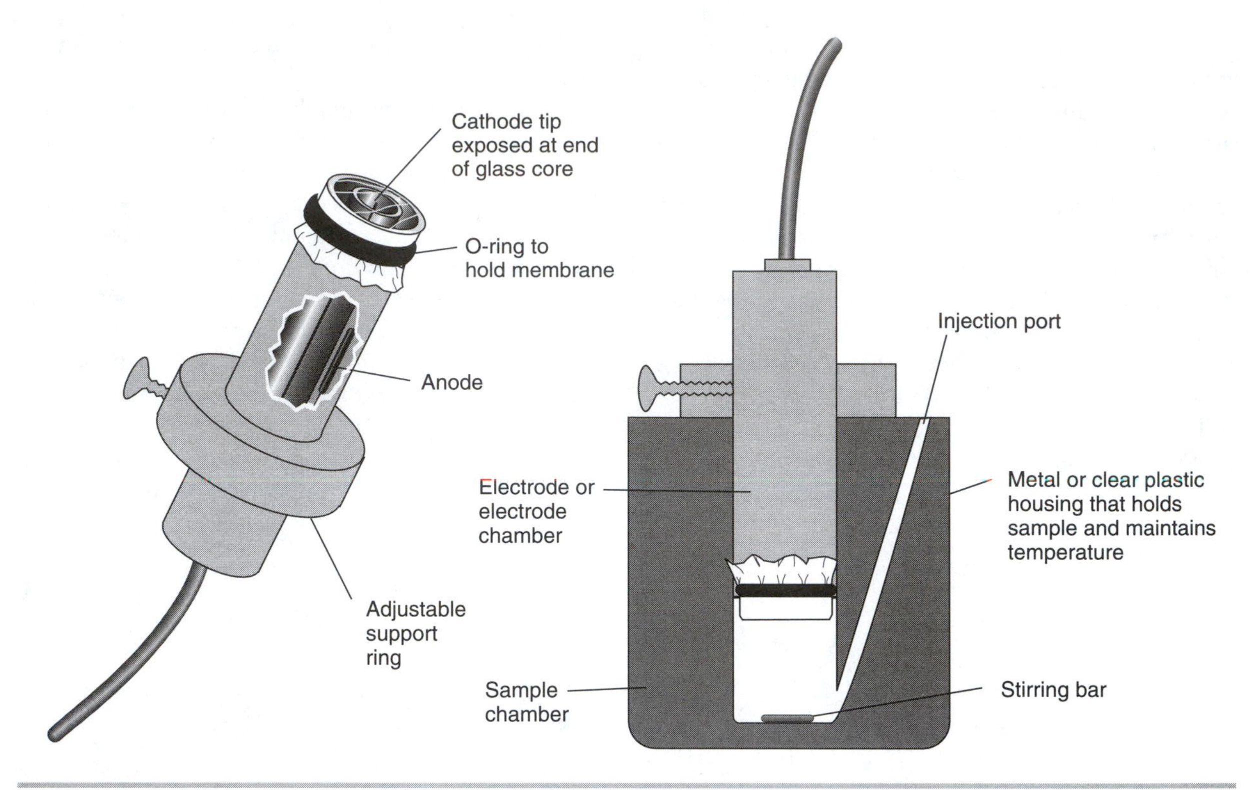

Figure 14-7 Features of a typical oxygen electrode system.

and the electrode chamber is placed in contact with the sample chamber. Next, the sample chamber is filled with an aqueous solution containing dissolved O_2 such that the fluid is in complete contact with the membrane. The laws governing diffusion dictate that dissolved oxygen from the aqueous system in the sample chamber will diffuse through the membrane into the electrode chamber at a rate proportional to the concentration of dissolved O_2 in the sample. Thus, the system will establish an initial rate of diffusion of O_2 into the electrode chamber that is defined by the geometry of the sample chamber and the initial concentration of dissolved O_2. The electrode system of the electrode chamber measures the O_2 as it enters the chamber in one of two ways. Amperometric systems use a constant voltage between electrodes. This voltage causes the reduction of O_2 at the cathode according to Equations 14-6 through 14-8, which results in a flow of current in the electrolyte solution.

(14-6)

$$O_2 + 2\,H_2O + 2\,e^- \longrightarrow H_2O_2 + 2\,OH^-$$

(14-7)

$$H_2O_2 + 2\,e^- \longrightarrow 2\,OH^-$$

(14-8)

$$\text{Sum: } O_2 + 2\,H_2O + 4\,e^- \longrightarrow 4\,OH^-$$

Potentiometric systems, which are less commonly used, measure changes in electric potential caused by interaction of O_2 with the cathode. Appropriate electronic equipment transduces the amperometric output into a voltage, which is amplified and recorded on a conventional strip chart recorder.

As long as the solution is adequately mixed so that the solution at the membrane surface has a dissolved O_2 content that is representative of the bulk solution, the voltage output from an oxygen electrode is proportional to the rate of diffusion of O_2 across the membrane, which is in turn proportional to the O_2 concentration in the bulk solution. Changes in amounts of dissolved O_2 in the sample chamber, due to O_2 reduction by electron transport, alter the rate of O_2 diffusion into the electrode chamber and result in a proportional alteration of the recorder readings. Thus, you can easily measure relative rates of O_2 consumption (or production, as in studies of photosynthesis) with an oxygen electrode. If you have previously calibrated the oxygen electrode with an O_2 standard, as will be done in this experiment, you can interpret these rates and overall extents of reaction in units of microliters or micromoles of O_2 (22.4 μl of a gas = 1 μmol under standard conditions).

Supplies and Reagents

Oxygen electrode with recorder, electrolyte solution, membranes, magnetic stirrers and sample injection syringes (see Appendix)
Fresh pig heart
Meat grinder
1 mM EDTA (pH 7.4)
Blender
0.02 M potassium phosphate containing 1 mM EDTA (pH 7.4)
High-speed centrifuge
0.25 M potassium phosphate containing 1 mM EDTA (pH 7.4)
1 M acetic acid
Oxygen electrode system (e.g., Yellow Springs Instruments Model 53)
1 mM NAD^+
0.3 M potassium phosphate buffer (pH 7.4)
0.05 M succinate
0.5 M succinate
0.25 M malate
0.25 M β-hydroxybutyrate
0.6 mM cytochrome *c*
Sodium ascorbate (198 mg sodium ascorbate in 10 ml H_2O, or 175 mg ascorbic acid plus 1 ml 1 N NaOH and diluted to 10 ml with H_2O—*freshly prepared immediately before use*)
0.5 M malonate
0.1 M KCN—*Poison!*
0.01 M methylene blue
Antimycin A (180 μg/ml in 95% ethanol)
95% Ethanol
6 M KOH
2,6-Dichlorophenol indophenol (DPIP) solution (13.5 mg dissolved in 50 ml H_2O)
Spectrophotometer

Protocol

Preparation of Keilin–Hartree Particles from Heart Muscle

We suggest that steps 1 through 5 of the experiment be performed by the instructor or teaching assistant a day or two before the beginning of the experiment.

1. Working at 0 to 4°C (cold room), free a pig heart *from a freshly slaughtered animal* of visible fat and fibrous tissue. Pass the muscle twice through a precooled (2°C) meat grinder fitted with a fine-grind plate. (The minced muscle may be stored at −20°C for months without risking loss of activity.)
2. Wash the minced heart tissue by stirring in 10 volumes of cool (less or equal to 20°C) 1 mM EDTA (pH 7.4) for about 4 hr, replacing the buffer six to eight times during this interval. Squeeze excess fluid from the minced muscle with four layers of cheesecloth after each washing.
3. Homogenize the washed tissue in a chilled blender for 7 min at 0 to 4°C with 0.02 M potassium phosphate and 1 mM EDTA (pH 7.4), using 500 ml per 200 g of original, trimmed heart muscle. Centrifuge the homogenate at 800 × *g* for 15 to 20 min. Decant the cloudy supernatant solution into a fresh centrifuge bottle, and save it on ice. Homogenize the precipitate again in a chilled blender for 2 min with 300 ml of cold 0.02 M potassium phosphate and 1 mM EDTA (pH 7.4), and centrifuge as before. Add this supernatant liquid to that from the first centrifugation. (Usually 500 to 600 ml of suspension is obtained from 200 g of heart.)
4. Reduce the pH of the cold (<5°C) combined supernatant suspension to 5.6 by cautiously adding (with stirring) cold 1.0 M acetic acid. Centrifuge at 4000 to 5000 × *g* for 15 min. The clear, reddish supernatant from the centrifugation contains cytochrome *c* and hemoglobin and can be discarded. Wash the precipitate once by suspending it in an equal volume of ice water and then centrifuge for 5 min at 1500 × *g*. Save the precipitate and discard the supernatant fluid.

5. Suspend the precipitate in an equal volume of 0.25 M potassium phosphate and 1 mM EDTA (pH 7.4). The final volume should be about 70 ml from 200 g of heart. Disperse any lumps of precipitate by shaking gently in a flask or by gentle homogenization in a hand-operated glass homogenizer. Store this preparation at 1 to 4°C until used. (It is stable for about 1 week, although some losses in activity may be observed by the end of a week.)

Calibration of the Oxygen Electrode

1. Slowly remove your O_2 electrode from the sample chamber and check to see that the membrane is intact and that the electrolyte solution within the electrode is free of gas bubbles. If necessary, remove the membrane holder, refill the electrode with electrolyte so a bulging pool of electrolyte is left, and then slip on and reseal a membrane over the electrode so that no air bubbles are trapped inside. Finally, slowly place the O_2-electrode into the sample holder so that the electrode contacts the desired quantity of fluid to be assayed (~3 ml). Leave the sample chamber empty.
2. Turn on the O_2-electrode system and allow the circuitry to warm up or stabilize for 5 min. Check the operation of the recorder during this interval. Then set the recorder on "pen" and set the zero point adjustment control of the recorder to achieve a zero setting on the recorder during zero output from the O_2-electrode system. This usually requires that you disconnect the O_2-electrode system from the recorder or switch a zero output control on the O_2-electrode system.
3. Set the O_2-electrode system to "amplifier zero," and adjust the amplifier zero control until the recorder registers zero. This establishes the reading that occurs in the absence of O_2. You can confirm this setting by later using an O_2-free water solution, such as water that has been boiled vigorously and cooled or sparged with an inert gas.
4. Establish the other end of the recorder scale by carrying out the following steps *in order.* First, open the sample chamber of the O_2 electrode and place a small stirring bar into the sample chamber of the O_2-electrode system. Second, using the sample injection syringe containing a flexible end of tubing, carefully fill the sample chamber with 2.9 ml (or more, no remaining air bubbles) of air-saturated water. (Saturate the water with air by squirting a water sample repeatedly into a beaker.) Third, set the sample chamber and O_2 electrode over a magnetic stirrer, turn on the stirrer, and adjust to a slow (~100 rpm) stirring rate. Last, set the O_2-electrode system to "air" or "air-saturated H_2O" and adjust the 100% adjust control of the O_2-electrode system so that the recorder registers 85 units, at which full deflection equals 100 units. This may require a minute or two to reach a stable reading. Since the sample chamber contains 3 ml, and air-saturated H_2O contains 28.3 μl/ml of O_2, this setting calibrates the recorder so that *each unit or division on the strip chart equals 1 μl of O_2.*
5. These steps will give you a fully adjusted and calibrated system. To proceed directly with the O_2 measurements in this experiment, return the O_2-electrode system to the amplifier zero state, turn off the magnetic stirrer, replace the air-saturated water sample with a magnetically stirred (~100 rpm) experimental sample from one of the series tables as described below, reassemble the system (no air bubbles!) and make appropriate readings with the O_2-electrode system at the "air" or "air saturated H_2O" state. At the conclusion of your O_2 measurements, turn off the O_2-electrode system, the recorder, and the magnetic stirrer. Empty the sample chamber and rinse all components gently with distilled water. Repeat this brief calibration procedure before performing an experiment in a following lab period.

Characterization of Electron Transport using Keilin–Hartree Particles and the Oxygen Electrode

The following tables describe four series of assays. Series I and II characterize the substrate specificity of Keilin–Hartree particles, whereas Series III and IV employ the autoxidizable dye, methylene blue, and various inhibitors to characterize the electron-transport system in these particles. It may be practical to divide the class into groups that will only perform one or two of the series of experiments and for the groups to share their results to obtain a full data set. Consult your instructor about the distribution of the assays among the class members.

1. Prepare separate tubes containing all of the "sample chamber" components of the desired assays of Series I through IV *except for the addition of KCN solution where it is called for in Series III and IV.* Add the components to the tubes in the order in which they are listed in the tables. Prevent plugging of micropipette tips used to transfer the heart particle suspension by cutting about 2 mm off the tips.
2. Using an injection syringe tipped with a flexible tubing end, load the "sample chamber" contents of an individual assay tube into the sample chamber of a 2.9-ml calibrated oxygen electrode. When the assay tube contents include KCN, add it as the last component just before mixing them and adding them to the sample chamber with the injection syringe.

WARNING

KCN releases dangerous HCN at the pH of the system and must be added just before starting the system. Do not pipette by mouth! Use a propipetter.

Set the oxygen electrode control to "air" or "air-saturated H_2O" and turn on the recorder.

3. Two areas to which you should pay particular attention are sample volume and potential interfering air bubbles. Most of the assays involve initiating the reaction by addition of only 50 or 100 μl of substrate. Thus, the contents of the sample chamber in these assays essentially fills the O_2-electrode sample chamber before the substrate is added. This allows you to establish a base value of O_2 concentration in the main vessel contents before adding potential substrate. Therefore, if the assay is initiated by addition of 100 μl or less of substrate, turn on the magnetic stirrer and the recorder and record base-line value readings for 2 min.
4. During this interval, load an injection syringe with the substrate to be used in initiating the reaction, withdrawing the syringe barrel just enough to take up the desired volume of substrate. *Avoid excess air in the syringe because this will introduce bubbles into the O_2-electrode sample chamber.*
5. After recording base-line O_2 levels for 2 min, lower the flexible tip of the injection syringe into the sample chamber, inject the substrate (no air bubbles, let any excess fluid overflow out of the system), and withdraw the syringe. Continue recording data over 2 to 4 min; that is, as long as a linear reaction occurs. Then, turn off the system, disassemble the O_2 electrode, and rinse out the sample chamber and injection syringe with distilled H_2O prior to further assays or storage.
6. If the total volume of the main vessel components of an assay does not equal or exceed 2.9 ml—that is, if the substrate volume exceeds 100 μl—you cannot establish a good base-line reading before adding substrate because air bubbles in the sample chamber will distort the base-line values. In such instances, transfer the sample chamber components into the O_2-electrode sample chamber, but do not turn on the recorder or the magnetic stirrer—do not attempt to establish a base-line value. Instead, load the injection syringe with just the desired volume of substrate (no air bubbles), and inject the substrate into the bottom of the O_2-electrode sample chamber. Immediately turn on the magnetic stirrer and recorder. With the O_2-electrode set on "air" or "air saturated H_2O," begin to monitor the reaction with the recorder. Continue readings for 2 to 4 min, or for as long as a linear reaction occurs. Then, turn off the system, disassemble the O_2 electrode, and rinse out the sample chamber and injection syringe with distilled H_2O prior to further assays or storage.
7. Using the linear assay range from each assay studied, calculate a rate of O_2 consumption in units of microliters of O_2 consumed per minute. Explain your results for each tube in all four series in terms of our current knowledge of electron transport (see Fig. 14-4). Estimate a K_M for succinate from the data of Series I, tubes 1 through 3. (If you do not remember how to do this, see Experiment 7.) Which tubes represent controls for other tubes? Are the sites of entry of the various electron donors and the modes of action of the inhibitors depicted in Figure 14-4 consistent with

your findings? Discuss possible reasons for any unexpected results.

Spectrophotometric Assay of Electron Transport

1. Prepare a series of tubes containing the components outlined in Series V below, *but without the specified amounts of heart particles and succinate.* Add the components in the order listed in the table. *Add the KCN* carefully *to tube 5 just before adding the heart particle preparation to tube 5—not before.*
2. Add the heart-particle suspension to tube 1 (the blank) and to tube 2. Mix the contents of each of these two tubes, and read the optical density of tube 2 against the blank (tube 1) at 600 nm at 30-sec intervals for 2 min to check for oxidation of any endogenous substrate.
3. Add the succinate to tube 2. Mix the contents, and read the optical density at 600 nm against the blank at 30-sec intervals for 5 min.

Series I Assays of Citric-Acid-Cycle Substrates

Addition (all volumes in μl)	Assay Tube Number					
	1	2	3	4	5	6
Sample chamber						
H_2O	950	990	950	650	500	500
0.3 M K-phosphate, pH 7.4	1500	1500	1500	1500	1500	1500
1 mM NAD^+	—	—	—	300	300	300
Heart-particle suspension	500	500	500	500	500	500
Substrate (added last)						
0.05 M succinate	50	—	—	—	—	—
0.5 M succinate	—	10	50	50	—	—
0.25 M malate	—	—	—	—	200	—
0.25 M β-hydroxybutyrate	—	—	—	—	—	200

Series II Assays with Succinate and Ascorbate as Substrates

Addition (all volumes in μl)	Assay Tube Number				
	1	2	3	4	5
Sample chamber					
H_2O	950	650	400	700	400
0.3 M K-phosphate, pH 7.4	1500	1500	1500	1500	1500
0.6 mM cytochrome *c*	—	300	300	—	300
Heart-particle suspension	500	500	500*	500	500
Substrate (added last)					
0.5 M succinate	50	50	—	—	—
0.1 M ascorbate†	—	—	300	300	300

*For tube 3 use heart particles that have been heated at 100°C for 10 min.

†198 mg sodium ascorbate/10 ml or 175 mg ascorbic acid plus 1 ml 1 N NaOH made up to 10 ml with H_2O. Prepare *immediately* before use.

Series III **Sites of Action of Electron Transport Inhibitors**

Addition (all volumes in μl)	Assay Tube Number 1	2	3	4	5	6	7
Sample chamber							
H_2O	950	650	350	650	850	500	550
0.3 M K-phosphate, pH 7.4	1500	1500	1500	1500	1500	1500	1500
0.5 M malonate	—	—	—	—	100	100	100
10 mM methylene blue	—	—	300	300	—	—	300
Heart-particle suspension	500	500	500	500	500	500	500
0.1 M KCN	—	300	300	—	—	—	—
Substrate (added last)							
0.5 M succinate	50	50	50	50	50	400	50

Series IV **Sites of Inhibitor Action and Ascorbate Oxidation**

Addition (all volumes in μl)	Assay Tube Number 1	2	3	4	5	6
Sample chamber						
H_2O	920	920	890	370	370	70
0.3 M K-phosphate, pH 7.4	1500	1500	1500	1500	1500	1500
0.6 mM cytochrome *c*	—	—	—	300	300	300
10 mM methylene blue	—	—	30	—	—	—
Antimycin A in 95% ethanol (180 μg/ml)	—	30	30	—	30	—
95% Ethanol	30	—	—	30	—	30
Heart-particle suspension	500	500	500	500	500	500
0.1 M KCN	—	—	—	—	—	300
Substrate (added last)						
0.5 M succinate	50	50	50	—	—	—
0.1 M ascorbate*	—	—	—	300	300	300

*198 mg sodium ascorbate/10 ml or 175 mg ascorbic acid plus 1 ml 1 N NaOH made up to 10 ml with H_2O. Prepare *immediately* before use.

4. Repeat steps 2 and 3 with each of the remaining 4 tubes (tubes 3, 4, 5, and 6), one at a time, reading the absorbance at 600 nm of each solution against the tube 1 blank.
5. Prepare a plot of the decrease in absorbance at 600 nm versus time in minutes for the results obtained from tubes 2 through 6. Explain the results for each reaction in terms of our current knowledge of electron transport, the site in the electron transport chain at which reducing equivalents are "shunted" to the DPIP dye, and the proposed sites of action of the inhibitors used (see Fig. 14-4).

Series V Spectrophotometric Assays of Electron Transport

Addition (all volumes in ml)	Assay Tube Number					
	1 (blank)	2	3	4	5	6
H_2O	4.4	3.9	3.8	3.4	3.0	3.9
0.3 M K-phosphate, pH 7.4	4.5	4.5	4.5	4.5	4.5	4.5
DPIP solution	—	0.4	0.4	0.4	0.4	0.4
0.5 M malonate	—	—	0.1	0.1	—	—
Antimycin A (180 μg/ml in 95% ethanol)	—	—	—	—	—	0.01
0.1 M KCN	—	—	—	—	0.9	—
Heart-particle suspension (next to last)	0.1	0.1	0.1	0.1	0.1	0.1
0.5 M succinate (added last)	—	0.1	0.I	0.5	0.1	0.1

Exercises

1. Define: (a) dehydrogenase, (b) oxidase, (c) cytochrome.
2. Ascorbic acid reduces cytochrome *c* but has little impact on cytochrome *b*. Explain this observation in terms of the reduction potentials (redox potentials) of these substances.
3. The following parts describe a series of separate assays that use the heart particle system described in this experiment. Assuming each component (substrate, dye, or inhibitor) is present in sufficient concentration to assert its function, predict whether O_2 will be consumed or not consumed in each of these heart-particle assays.

   a. Succinate, DPIP, and cyanide
   b. Succinate, antimycin A, cyanide, and methylene blue
   c. Ascorbate, antimycin A, DPIP, and malonate
   d. Succinate, methylene blue, cytochrome *c*, cyanide, and ascorbate

4. Four electron carriers (A, B, C, and D) are required in the electron transport system of a recently discovered aerobic bacterium. You find that in the presence of substrate and O_2, four inhibitors (I, II, III, and IV) block respiration at four different sites. From differential spectrophotometry of the electron carriers, you find that these inhibitors yield the patterns of oxidation state of the carriers shown below. What is the order of electron transfer to the electron carriers in the respiratory chain of this bacterium? Explain your answer by writing the reaction sequence from reduced substrate to O_2 and indicate the site of action of each inhibitor.

Oxidation State of Electron Carrier in the Presence of Substrate and Inhibitor (+ means fully oxidized, – means fully reduced)

Inhibitor	A	B	C	D
I	+	–	–	+
II	+	+	–	+
III	–	–	–	+
IV	–	–	–	–

REFERENCES

Iwata, S., et al. (1998). Complete Structure of the 11-Subunit Bovine Mitochondrial Cytochrome bc_1 Complex. *Science* **281**:64.

Keilin, D., and Hartree, E. F. (1947). Activity of the Cytochrome System in Heart Muscle Preparations. *Biochem J* **41**:500.

Lehninger, A. L. (1964). *The Mitochondrion: Molecular Basis of Structure and Function.* Menlo Park, CA: Benjamin.

Stryer, L. (1995). *Biochemistry,* 4th ed. chapters 20 and 21. New York: Freeman.

Trumpower, B. L., and Gennis, R. B. (1994). Energy Transduction by Cytochrome Complexes in Mitochondrial and Bacterial Respiration: The Enzymology of Coupling Electron Transfer Reactions to Transmembrane Proton Translocation. *Annu Rev Biochem* **63**:675.

Tsukihara, T. et al. (1995). Structures of Metal Sites of Oxidized Bovine Heart Cytochrome c Oxidase at 2.8 Å. *Science* **269**:1069.

Xia, D. et al. (1997). Crystal Structure of the Cytochrome bc_1 Complex from Bovine Heart Mitochondria. *Science* **277**:60.

Yoshikawa, S. et al. (1998). Redox-Coupled Crystal Structural Changes in Bovine Heart Cytochrome c Oxidase. *Science* **280**:1723.

Study of the Phosphoryl Group Transfer Cascade Governing Glucose Metabolism: Allosteric and Covalent Regulation of Enzyme Activity

Theory

A living cell contains hundreds of enzymes involved in many very different yet equally important metabolic processes. As the environment of the cell and the demands on it change, so too must the activity of many of these biological pathways to allow the cell to cope with each new challenge. This general concept is easy to understand if you consider how cellular activity following a large meal is different than that during strenuous physical activity. In the resting state, when energy supplies are plentiful, the cell will begin to switch from a catabolic (energy-yielding metabolism) mode to a more anabolic (energy-storing or biosynthetic metabolism) mode. This situation will be reversed with increases in physical activity; energy stores replenished during the resting state will be metabolized to produce large quantities of the cell's energy currency, ATP.

How does the cell regulate the enzymes involved in catabolic and anabolic metabolism? The cell has several options: it can control the *expression* of enzymes involved in these processes, or it can control the *activity* of the enzymes involved in these processes. In Experiment 7 ("Study of the Properties of β-Galactosidase"), you investigated a common strategy utilized by prokaryotes to regulate metabolic enzyme activity at the level of transcription. Expression of the gene for β-galactosidase (*lacZ*) is *induced* in the presence of the enzyme's substrate, lactose. The cell will *repress* expression of the enzymes involved in lactose metabolism until it finds itself in the presence of this carbohydrate. In some cases, however, the cell is required to respond rapidly to its changing metabolic needs. It may not have time to produce the enzymes required to perform a specific reaction. Therefore, cells are also capable of regulating the *activity* of enzymes that are expressed or maintained at a constant level. The net result of either form of regulation is the same: enzymes can be regulated according to the ever-changing metabolic needs of the cell.

Two mechanisms that are commonly employed in altering enzyme activity are covalent modification and allosteric regulation. *Covalent modification* is an enzymatically catalyzed reaction that involves the *reversible* formation of a *covalent* bond between a small molecule and a specific amino acid side chain(s) on an enzyme that affects its activity. *Allosteric regulation* of an enzyme's activity involves *noncovalent* binding of a small molecule at a site other than the active site that alters the enzyme's activity. Unlike the limited examples of covalent modification that have been discovered (see Table 15-1), a wide variety of small molecules have been found to regulate the activity of particular enzymes allosterically.

Glycogen phosphorylase was one of the first enzymes shown to be allosterically regulated. It is also one of the first enzymes shown to be covalently modified by what has been shown to be a very common mechanism in enzyme regulation, phosphorylation. Glycogen phosphorylase is the enzyme responsible for cleaving the $\alpha(1 \rightarrow 4)$ linkage between the glucose subunits that comprise glycogen, a common energy store for some mammalian cells (Fig. 15-1). The glucose-1-phosphate units that result from this *phosphorolysis* reaction can eventually enter the glycolytic pathway and produce ATP. Glycogen phosphorylase can exist in the *a* form or

Table 15-1 Established Forms of Covalent Modification in the Regulation of Enzyme Activity

Modification	Group Donor	Attached Group	Byproduct	Modified Residues	Examples
Phosphorylation	ATP	PO_4^{-2}	ADP	Ser Thr His Tyr	Phosphorylase *b*, glycogen synthetase, EGF receptor, insulin receptor, protein kinase C, protein kinase G, protein kinase A
ADP-ribosylation	NAD^+	ADP-ribose	Nicotinamide	Cys Arg Gln	α subunit of G protein (cholera toxin), i subunit of G protein (pertussis toxin)
Adenylylation	ATP	AMP	PPi	Tyr	glutamine synthetase
Uridylylation	UTP	UMP	PPi	Tyr	P_{II} protein (another form of glutamine synthetase regulation)
Methylation	*S*-adenosyl-methionine	CH_3	*S*-adenosyl-homocysteine	Glu	Methyl-accepting chemotaxis proteins

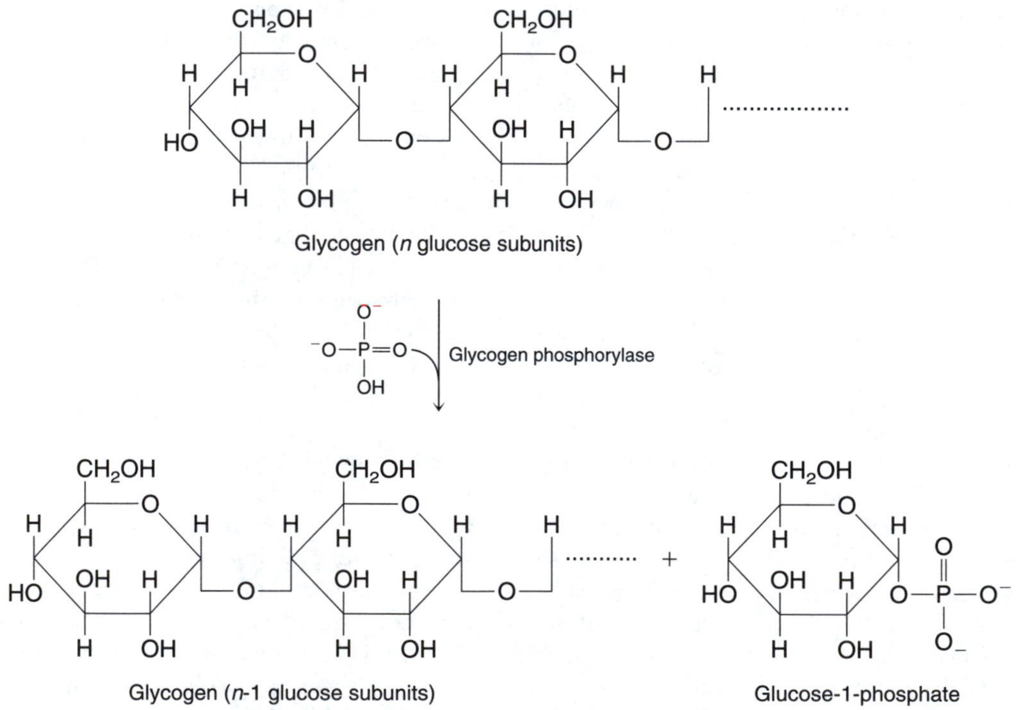

Glycogen (*n* glucose subunits)

Glycogen phosphorylase

Glycogen (*n*-1 glucose subunits) + Glucose-1-phosphate

Figure 15-1 Reaction catalyzed by glycogen phosphorylase.

the *b* form, each made up of a dimer of identical 94.5-kDa subunits. Glycogen phosphorylase *b* is the less active form of the enzyme. Glycogen phosphorylase *a* is the active form of the enzyme that has covalently bound phosphoryl groups on the Ser-14 side chain of each subunit. The phosphoryl group transfer from ATP to phosphorylase *b* to produce phosphorylase *a* is catalyzed by an enzyme called *glycogen phosphorylase kinase*.

Like glycogen phosphorylase, phosphorylase kinase is also subject to covalent modification (regulation) by phosphorylation. In response to hormones that signal an increase in physical activity, adenylate cyclase is activated to produce an increase in the intracellular cAMP concentration. In turn, a cAMP-dependent protein kinase is activated to cause an increase in the level of phosphorylated glycogen phosphorylase kinase. This activated (phosphorylated) kinase will convert phosphorylase *b* to phosphorylase *a*, causing an increase in the rate of glycogen breakdown to glucose-1-phosphate for the production of ATP. Similar types of phosphorylation cascades have been shown to be critical in many signal transduction pathways.

In addition to being activated by phosphorylation, glycogen phosphorylase *b* is also subject to allosteric activation by AMP. As physical activity increases, so too does the intracellular AMP:ATP ratio. This increase in AMP concentration will allow immediate activation of glycogen phosphorylase *b* prior to its hormonally mediated conversion to phosphorylase *a* by cAMP-dependent protein kinase and glycogen phosphorylase kinase. This dual regulation of glycogen phosphorylase ensures that active muscle cells will have an adequate supply of glucose for the production of ATP. A schematic diagram of the signal transduction process just described is presented in Figure 15-2.

In this laboratory exercise, you will study the effects of phosphorylation and allosteric regulation on the activity of glycogen phosphorylase. In the first experiment, you will phosphorylate glycogen phosphorylase *b* in vitro using γ-[^{32}P]ATP and glycogen phosphorylase kinase. In the second experiment, you will study the effect of phosphorylation on glycogen phosphorylase activity, as well as the effect of AMP on glycogen phosphorylase *b* activity with the use of a coupled enzymatic (kinetic) assay (Fig. 15-3).

Although phosphorylation is the type of covalent modification that will be investigated in this experiment, it is not the only form of covalent modification that has been adopted by cells as a means of regulating enzyme activity. Table 15-1 lists a number of other types of covalent modifications that have been confirmed in a number of different biological systems and pathways.

Supplies and Reagents

P-20, P-200, and P-1000 Pipetmen with sterile tips
1.5-ml Microcentrifuge tubes
Stock solution of 1 M Tris, pH 8.0
Stock solution of 1 M glycerophosphate, pH 8.0
Stock solution of 60 mM Mg acetate, pH 7.3
Concentrated enzyme storage buffer:

- 50 mM Tris, pH 8.0
- 50 mM glycerophosphate, pH 8.0
- 40% vol/vol glycerol
- 2 mM EDTA, pH 8.0
- 0.1% β-mercaptoethanol

Reaction buffer:

- 50 mM Tris, pH 8.0
- 50 mM glycerophosphate, pH 8.0

2 mM ATP solution (2 mM ATP prepared in 60 mM Mg acetate, pH 7.3)
Glycogen phosphorylase *b* stock solution (Sigma catalog #P-6635):

Dissolve lyophilized enzyme to ~1000 units/ml in concentrated enzyme storage buffer. Divide into 20-μl aliquots and store at −20°C.

Glycogen phosphorylase kinase stock solution (Sigma catalog #P-2014):

Dissolve lyophilized enzyme to ~2,500 units/ml in concentrated enzyme storage buffer. Divide into 20-μl aliquots and store at −20°C.

4X SDS/EDTA buffer:

- 0.25 M Tris, pH 7.0
- 30% vol/vol glycerol
- 10% vol/vol β-mercaptoethanol
- 8% wt/vol sodium dodecyl sulfate
- 250 mM EDTA
- 0.01% Bromophenol blue (tracking dye)

γ-[^{32}P]ATP (4000 Ci/mmol)—ICN Biochemicals, catalog #35001X. *Dilute this solution 1:80 for use in the in vitro phosphorylation assay*
Prestained high-molecular-weight protein markers (Gibco BRL cat no. 26041-020)

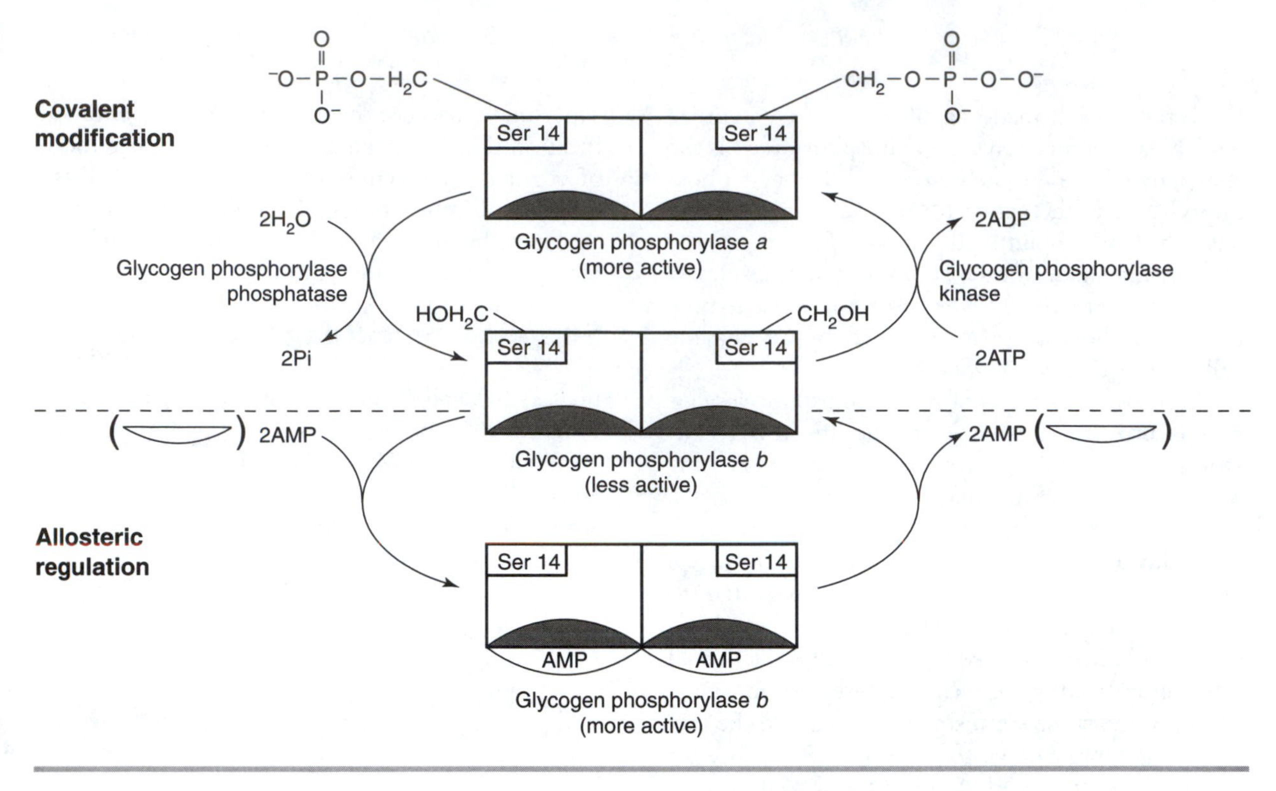

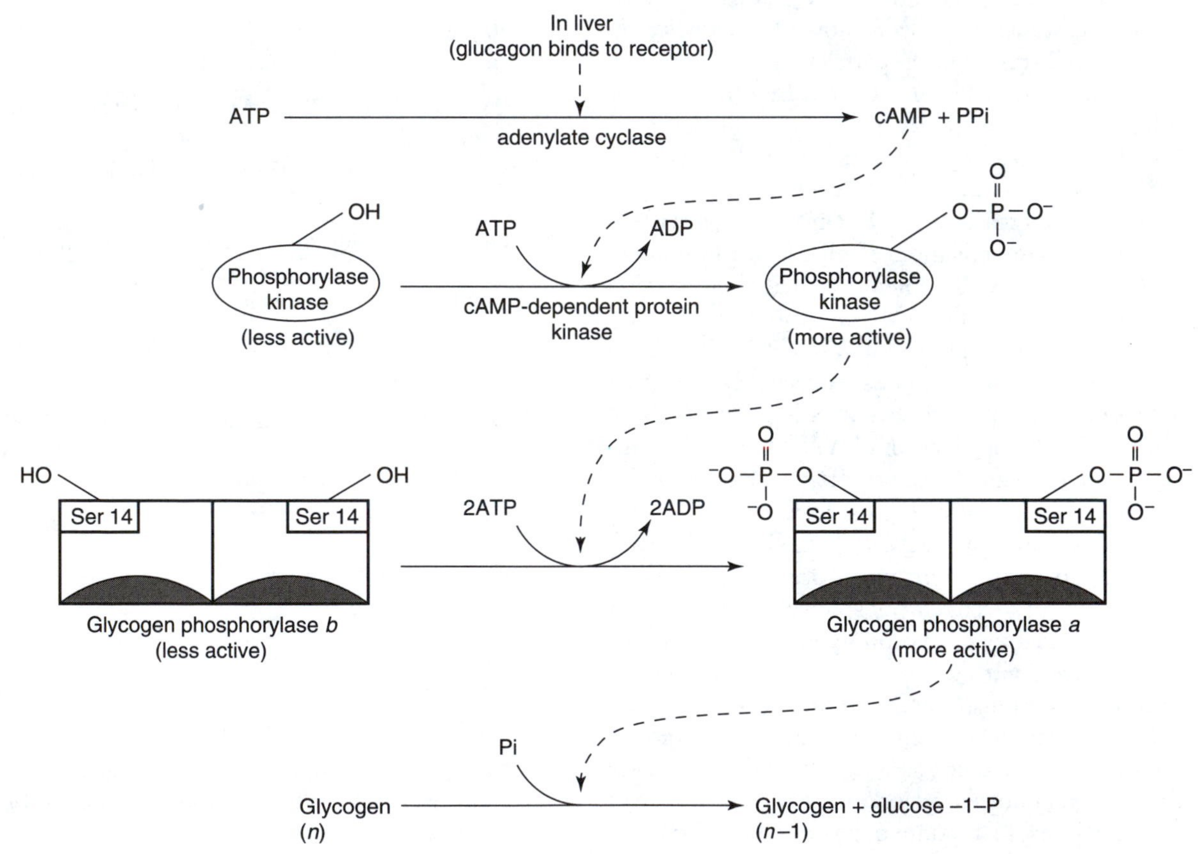

Figure 15-2 Allosteric and covalent regulation of glycogen phosphorylase.

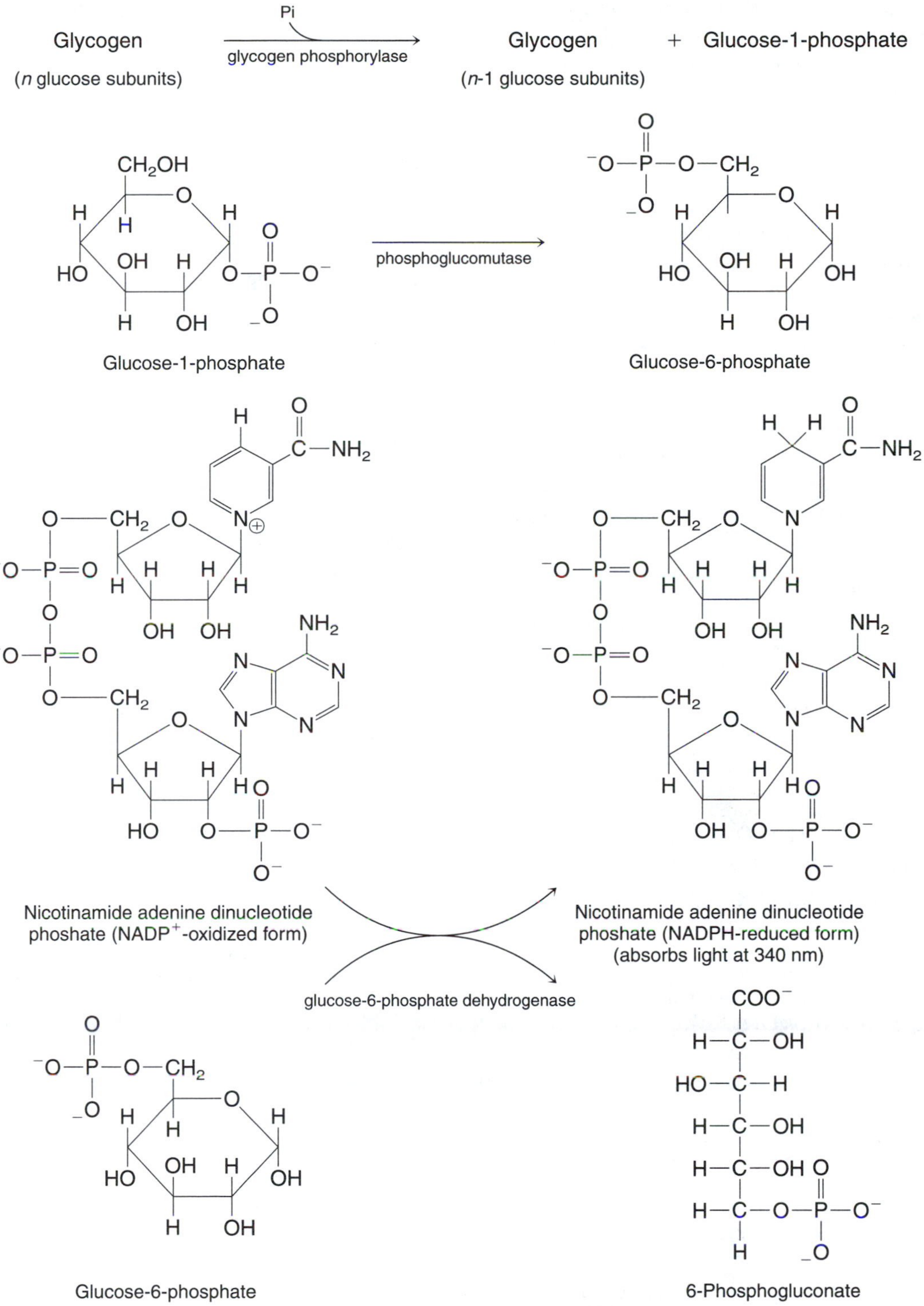

Glycogen (n glucose subunits) $\xrightarrow[\text{glycogen phosphorylase}]{\text{Pi}}$ Glycogen (n-1 glucose subunits) + Glucose-1-phosphate

Glucose-1-phosphate $\xrightarrow{\text{phosphoglucomutase}}$ Glucose-6-phosphate

Nicotinamide adenine dinucleotide phoshate (NADP$^+$-oxidized form)

Nicotinamide adenine dinucleotide phoshate (NADPH-reduced form) (absorbs light at 340 nm)

glucose-6-phosphate dehydrogenase

Glucose-6-phosphate

6-Phosphogluconate

Figure 15-3 Coupled enzyme assay to determine glycogen phosphorylase activity. As long as the activity of phosphoglucomutase and glucose-6-phosphate dehydrogenase are not rate-limiting, the $\Delta A_{340}/\Delta t$ is directly proportional to the Δ[glucose-1-phosphate]/Δtime.

Reagents and supplies for SDS-PAGE (see Experiment 4)
Glass or plastic plates for casting SDS–polyacrylamide gel with a 10-well comb
Plastic wrap
Whatman 3 MM filter paper
Film cassette
Kodak X-omat film
Dark room
Film developer or Kodak developing solutions
1.5-ml Microcentrifuge tubes
Glass test tubes (13 × 100 mm)
10-ml Pipette with bulb
Spectrophotometer to read absorbance at 340 nm
30 mM cysteine, pH 7.0 (prepare using L-cysteine hydrochloride, monohydrate, and pH to 7.0 using sodium bicarbonate)
80 mM cysteine, pH 6.8 (prepare using L-cysteine hydrochloride, monohydrate, and pH to 6.8 using sodium bicarbonate)
250 mM Tris buffer, pH 7.7
250 mM glycerophosphate (prepare in 250 mM Tris, pH 7.7)
40 mM glycerophosphate (prepare in 80 mM cysteine, pH 6.8 and adjust pH to 6.8 with 1 M HCl or 1 M NaOH if necessary)
Glycogen phosphorylase *b* solution (100 units/ml)

Dilute enzyme stock solution (see Day 1 reagents) 1:10 in 30 mM cysteine, pH 7.0

30 mM ATP solution (prepare in deionized water and pH to 7.7 with 1 M NaOH)
Glycogen phosphorylase kinase solution

Dilute enzyme stock solution (see Day 1 reagents) 1:125 in 30 mM cysteine, pH 7.0

100 mM Mg acetate solution (prepare in deionized water with Mg acetate tetrahydrate and adjust pH to 7.7 with 1 M NaOH)
500 mM potassium phosphate buffer, pH 6.8
4% (wt/vol) glycogen solution (Sigma catalog #G-8876)
300 mM $MgCl_2$
100 mM EDTA, pH 8.0
6.5 mM β-nicotinamide adenine dinucleotide phosphate solution (β-NADP)

Prepare in deionized water *immediately before use*

0.1% (wt/vol) α-D-glucose-1,6-diphosphate solution (prepare in deionized water)
Phosphoglucomutase solution (10 units/ml)—Sigma catalog #P-3397

Prepare in deionized water *immediately before use*

Glucose-6-phosphate dehydrogenase solution (10 units/ml)—Sigma catalog #G-6378

Prepare in deionized water *immediately before use*

Glycogen phosphorylase a enzyme solution (2 units/ml)—Sigma catalog #P-1261

Dilute a 1000-units/ml stock in concentrated enzyme storage buffer 1:500 in 40 mM glycerophosphate solution

Glycogen phosphorylase *b* control solution (2 units/ml)

Dilute enzyme stock 1:500 in 40 mM glycerophosphate solution

Glycogen phosphorylase *b* control solution (2 units/ml) with AMP

Dilute enzyme stock 1:500 in 40 mM glycerophosphate solution containing 60 mM AMP

Remember to keep all enzyme solutions on ice.

Protocol

WARNING

Caution: This experiment requires the use of significant amounts of ^{32}P-ATP. Wear gloves and safety goggles at all times during the experiment. Plexiglas shields should be used to screen students from excessive exposure to the radioisotope. Consult the instructor for proper handling and disposal of radioactive waste.

Day 1: In Vitro Phosphorylation of Glycogen Phosphorylase *b*

1. Prepare an ~250-units/ml phosphorylase kinase solution by diluting the stock enzyme 1:10 with reaction buffer.
2. Set up the following reactions in three 1.5-ml microcentrifuge tubes:

Tube	Volume of ~250 U/ml Phosphorylase Kinase	Volume of ~1,000 U/ml Phosphorylase *b*	Volume of Reaction Buffer
1	—	10 μl	34 μl
2	10 μl	—	34 μl
3	10 μl	10 μl	24 μl

3. To tubes 1 and 2, add 6 μl of an ATP solution that contains 20 μl of 2 mM ATP in 60 mM Mg acetate, pH 7.3 and 4 μl of γ-[^{32}P]ATP (provided by the instructor). Pipette up and down rapidly several times to mix. After exactly 10 min, remove a 10-μl aliquot from each tube and add it to 10 μl of 4X SDS/EDTA buffer to stop the reaction.
4. To tube 3, add 6 μl of the same ATP solution described above and pipette up and down rapidly several times to mix. At exactly 5 min and 10 min, remove a 10-μl aliquot and add it to 10 μl of 4X SDS/EDTA buffer to stop the reaction. Separate 5-min and 10-min samples should now have been taken from tube 3.
5. Load each 20-μl sample in the following order on a 12% SDS–PAGE gel (prepared by the instructor). Two groups will share one SDS–PAGE gel. Obtain the prestained protein molecular weight standards from the instructor.

   Tube 1—10 min
   Tube 2—10 min
   Tube 3—5 min
   Tube 3—10 min
   Standards
   Standards
   Tube 1—10 min
   Tube 2—10 min
   Tube 3—5 min
   Tube 3—10 min

6. Connect the negative electrode (cathode) to the upper buffer chamber and the positive electrode (anode) to the lower buffer chamber. The SDS bound to the proteins will give them a negative charge, causing them to migrate toward the positive electrode. Apply 200 V (constant voltage) and continue the electrophoresis until the bromophenol blue dye front has migrated off the bottom of the gel. *The unreacted ^{32}P-ATP present in the samples will migrate off the gel with the dye front. At this point, the solution in lower buffer chamber is radioactive. Wear gloves and safety goggles at all times.*
7. When the electrophoresis is complete, turn off the power supply and disconnect the electrodes. Remove from the unit the glass plates, which contain the gel. Carefully separate the plates (the gel will adhere to one of the two plates) and carefully transfer the gel to a small piece of Whatman 3 MM filter paper, avoiding any wrinkles in the gel.
8. Enclose the gel in a piece of plastic wrap and place it in a film cassette. Using a felt-tipped pen, mark the film in a manner that will allow you to determine position of gel on the film after it is developed. Expose the gel to a piece of Kodak X-omat film for ~18 hr at −70°C. By exposing the film at low temperature, radioactive scattering will be minimized, giving you well-defined radioactive protein bands on the autoradiogram.
9. Develop the film the next morning. Align the autoradiogram with the gel and mark the positions of the wells and the prestained protein molecular weight standards on the film.
10. Prepare a plot of log molecular weight versus distance migrated on the gel using the molecular weight protein standards. Are there any radiolabeled protein bands present on the autoradiogram? What is the molecular weight of this protein? What is the identity of this protein? Support your answer in terms of what you know about the molecular weight of the proteins present in the assay. Are the radiolabeled protein bands of the same intensity on the autoradiogram? Describe and discuss your results in terms of the kinetics of the kinase present in this assay.
11. Would you expect to see any radiolabeled bands present in the lanes containing samples from reaction tubes 1 and 2? Why or why not?

Day 2: In Vitro Assay for Phosphorylase Kinase and Glycogen Phosphorylase Activity

1. Set up the reactions shown in the table on the next page in 1.5-ml microcentrifuge tubes. The stock glycogen phosphorylase and the stock glycogen phosphorylase kinase solutions should be diluted to the proper concentrations with 30 mM cysteine buffer, pH 7.0.

   To start the phosphorylase kinase assay, add 25 μl of 30 mM ATP solution to each tube and pipette rapidly up and down several times to mix. All three reactions can be run simultaneously, staggering the start of each reaction by about 30 sec. At exactly 5 min and 10 min, remove 100 μl from each reaction and stop it by adding it to 1.9 ml of 40 mM glycerophosphate,

Tube	Volume of 30 mM Cysteine, pH 7.0	Volume of 250 mM Glycerophosphate	Volume of 100 mM Mg Acetate	Volume of 100 units/ml Phosphorylase *b*	Volume of 20 units/ml Phosphorylase Kinase
1	122.5 μl	62.5 μl	25 μl	—	15 μl
2	37.5 μl	62.5 μl	25 μl	100 μl	—
3	22.5 μl	62.5 μl	25 μl	100 μl	15 μl

pH 6.8 in a 13-by-100-mm test tube. *Each of these "stop" tubes (six of them) should be set up in advance of starting the reaction and each tube should be labeled with the reaction tube number and the time (e.g., tube 1—5 min).* These stopped, 2.0-ml reactions will be used in the glycogen phosphorylase *a* assay described below. *Place all of these tubes on ice until you are ready to proceed with the next assay.*

3. Prepare a reaction cocktail containing the following stock components (this quantity will be sufficient for at least four groups):

Stock Solution	Volume (ml)
Deionized water	99.5
500 mM potassium phosphate buffer, pH 6.8	15.00
4% (wt/vol) glycogen	7.50
300 mM $MgCl_2$	0.67
100 mM EDTA	0.15
6.5 mM β-NADP	10.0
0.1% (wt/vol) glucose-1,6-diphosphate	0.50

4. Dispense 2.7-ml of this reaction cocktail in 10 small glass test tubes (13 × 100 mm), labeled A through J. To each of the 10 test tubes, add 100 μl of the 10-units/ml glucose-6-phosphate dehydrogenase solution and 100 μl of the 10-units/ml phosphoglucomutase solution. The total volume in each test tube after this step will be 2.9 ml. *Refer to Table 15-2 to aid you in setting up the rest of the assay tubes.*
5. To tube A, add 100 μl of water and shake the test tube gently for about 10 sec to mix. At 1-min intervals, measure and record the A_{340} of the sample. Continue taking absorbance readings until the reaction achieves a linear change in A_{340} versus time (at least 5 min). This is a negative control reaction that contains no glycogen phosphorylase or glycogen phosphorylase kinase.
6. Perform the same procedure using tubes B to G by adding 100 μl of your different phosphorylase kinase assay reactions to these tubes. *Be sure that you know which phosphorylase kinase reaction corresponds to which test tube (e.g., tube B is phosphorylase kinase reaction 1—5 min, tube C is phosphorylase kinase reaction 1—10 min, etc.).* Measure and record the A_{340} of each reaction at 30-sec intervals for at least 5 min, or until each reaction achieves a linear change in A_{340} versus time.
7. Perform the same procedure using tube H by adding 100 μl of the 2-units/ml phosphorylase *a* solution (this is a positive control for the glycogen phosphorylase *a* assay). Measure and record the A_{340} of the reaction at 30-sec intervals for at least 5 min, or until the reaction achieves a linear change in A_{340} versus time.
8. Perform the same procedure using tube I by adding 100 μl of the 2-units/ml phosphorylase *b* solution (this is a negative control for the glycogen phosphorylase *a* assay). Measure and record the A_{340} of the reaction at 30-sec intervals for at least 5 min, or until the reaction achieves a linear change in A_{340} versus time.
9. Perform the same procedure using tube J by adding 100 μl of the 2-units/ml phosphorylase *b* solution in 60 mM AMP. This assay is designed to show *allosteric activation* of glycogen phosphorylase *b* by AMP. Measure and record the A_{340} of the reaction at 30-sec intervals for at least 5 min, or until the reaction achieves a linear change in A_{340} versus time.
10. Construct a plot of A_{340} versus time for each of the 10 glycogen phosphorylase reactions (A–J). Determine the slope of the line for each plot (ΔA_{340}/min). It may be helpful to construct four plots for tubes B–J: one graph will contain the data for tubes B and C, another will contain the data for tubes D and E, another will

Table 15-2 Assay for Glycogen Phosphorylase Activity

Components Added (ml)	Tubes									
	A	B	C	D	E	F	G	H	I	J
Reaction cocktail	2.7	2.7	2.7	2.7	2.7	2.7	2.7	2.7	2.7	2.7
Glucose-6-phosphate Dehydrogenase (10 units/ml)	0.1	0.1	0.1	0.1	0.1	0.1	0.1	0.1	0.1	0.1
Phosphoglucomutase (10 units/ml)	0.1	0.1	0.1	0.1	0.1	0.1	0.1	0.1	0.1	0.1
H_2O	0.1	—	—	—	—	—	—	—	—	—
From Step 2:	—	**Tube 1, 5**	**Tube 1, 10**	**Tube 2, 5**	**Tube 2, 10**	**Tube 3, 5**	**Tube 3, 10**	—	—	—
Volume from step 2 (ml)	—	0.1	0.1	0.1	0.1	0.1	0.1	—	—	—
Phosphorylase *a* (2 units/ml)	—	—	—	—	—	—	—	0.1	—	—
Phosphorylase *b* (2 units/ml)	—	—	—	—	—	—	—	—	0.1	—
Phosphorylase *b* (2 units/ml in 60 mM AMP)	—	—	—	—	—	—	—	—	—	0.1

contain the data for tubes F and G, and a fourth will contain the data for tubes H to J.

11. The goal of the analysis is to determine the concentration (units per milliliter) of phosphorylase kinase used in these reactions as well as the concentration of glycogen phosphorylase *a* (units per milliliter) present in each of the reactions after 5 min and 10 min. One unit of phosphorylase kinase is defined as the amount of the enzyme that will convert 1 μmol of phosphorylase *b* to phosphorylase *a* per min. One unit of phosphorylase *a* is defined as the amount of the enzyme that will produce 1 μmol of glucose-1-phosphate from glycogen per minute.
12. Determine the glycogen phosphorylase activity present in tubes D to J using the following equation:

$$\text{units/ml enzyme} = \frac{(\Delta A_{340}/\text{min test} - \Delta A_{340}/\text{min blank})(20)(3\ \text{ml})}{(6.22\ \text{mM}^{-1}\ \text{cm}^{-1})(\text{l})(0.1\ \text{ml enzyme})}$$

where test = tubes other than A to C whose activity is being tested; blank = the blank for the reactions in tubes H to J is tube A (no glycogen phosphorylase). The blank for the reactions in tubes D and F is tube B (5-min time point, no glycogen phosphorylase *b*). The blank for the reactions in tubes E and G is tube C (10-min time point, no glycogen phosphorylase *b*); 20 = dilution factor (how much the glycogen phosphorylase enzyme solution being tested was diluted before being assayed); 3 ml = total volume of glycogen phosphorylase assay; 6.22 $\text{mM}^{-1}\ \text{cm}^{-1}$ = millimolar extinction coefficient for β-NADPH at 340 nm; l = pathlength through the spectrophotometer (1.3 cm for benchtop spectrophotometers); and 0.1 ml enzyme = volume of the phosphorylase kinase assay or enzyme solution used in the glycogen phosphorylase assay.

NOTE: No dilution factor (20) is required for the calculation of glycogen phosphorylase activity in tubes H to J, since the enzyme stock solution was added directly to the assay. Otherwise, the same formula can be used for tubes H to J.

13. Determine the concentration of glycogen phosphorylase kinase present in reaction tube 3 (from step 2) using the following equation:

$$\text{Units/ml enzyme} = \frac{(\text{units glycogen phosphorylase test/ml}) - (\text{units glycogen phosphorylase blank/ml})(20)}{(\text{time})(0.015\ \text{ml})}$$

where test = the 5-min and 10-min time points for phosphorylase kinase reaction 3 (tubes F and G, respectively); blank = the 5-min and 10-min time points for phosphorylase kinase reaction 2 (no glycogen phosphorylase kinase; tubes D and E, respectively); 20 = dilution factor (how much the glycogen phosphorylase enzyme solution being tested was diluted before being assayed); time = number of minutes (5 or 10) of the phosphorylase kinase assay; and 0.015 ml = Volume (ml) of phosphorylase kinase used in reactions 1 and 3.

14. After you have completed these calculations, you will have obtained two values for the concentration of glycogen phosphorylase kinase present in this assay, one based on the 5-min time point and one based on the 10-min time point. From these data, determine the average value of glycogen phosphorylase kinase present in this assay (units/ml of enzyme).
15. Compare the glycogen phosphorylase activity between the reactions in tubes I and J. What effect does AMP have on the activity of glycogen phosphorylase *b*? Can you estimate how much the activity of the enzyme is increased or decreased in the presence of this allosteric effector?
16. Compare the glycogen phosphorylase activity between the reactions in tubes H and I. What effect does phosphorylation of Ser 14 have on the activity of glycogen phosphorylase *b*? Can you estimate how much the activity of the enzyme is increased or decreased with this type of covalent modification?

Exercises

1. What do you think is the advantage of having an enzyme such as glycogen phosphorylase *b* controlled both by an allosteric effector and by covalent modification?
2. Would you expect the glycogen phosphorylase assay data from phosphorylase kinase reaction 3 to be any different if the concentration of the kinase used in this reaction was increased by a factor of 5? Explain.
3. Would you expect the glycogen phosphorylase assay data from phosphorylase kinase reaction 3 to be any different if the concentration of ATP used in this reaction was increased by a factor of 10? Explain.
4. What must be true about the concentrations of glucose-6-phosphate dehydrogenase and phosphoglucomutase used in the glycogen phosphorylase assay for the change in A_{340} to be directly proportional to the change in concentration of glucose-1-phosphate? Explain.
5. You have obtained a liver cell line that is mutant in glycogen metabolism. The mutation is such that glycogen phosphorylase activity is always high (glycogen phosphorylase activity is no longer regulated). This phenotype could be the result of a mutation in a number of different genes involved in glycogen metabolism. For each of the enzymes listed below, state whether the phenotype of this cell line is consistent with the activity of the enzyme being constitutively high or constitutively low. Support your answer using your knowledge of the control of glycogen phosphorylase activity.

   a. Adenylate cyclase
   b. Glycogen phosphorylase kinase
   c. Glycogen phosphorylase phosphatase
   d. cAMP-dependent protein kinase

REFERENCES

Antoniw, J. F., Nimmo, H. G., Yeaman, S. J., and Cohen, P. (1977). Comparison of the Substrate Specificities of Protein Phosphatases Involved in the Regulation of Glycogen Metabolism in Rabbit Skeletal Muscle. *Biochem J* **162**:423.

Cohen, P. (1983). Phosphorylase Kinase from Rabbit Skeletal Muscle. *Methods Enzymol.* **99**:243.

Garrett R. H. and Grisham, C. M. (1995). Gluconeogenesis, Glycogen Metabolism, and the Pentose Phosphate Pathway. In: *Biochemistry.* Orlando, FL: Saunders.

Hemmings, B. A., and Cohen, P. (1983). Glycogen Synthase Kinase-3 from Rabbit Skeletal Muscle. *Methods Enzymol.* **99**:337.

Lehninger, A. L., Nelson, D. L., and Cox, M. M. (1993). Glycolysis and Catabolism of Hexoses. In: *Principles of Biochemistry*, 2nd ed. New York: Worth.

Sigma Quality Control Test Procedure (EC 2.7.1.38). St. Louis: Sigma Chemical Co.

Stryer, L. (1995). Glycogen Metabolism. In: *Biochemistry*, 4th ed. New York: Freeman.

Voet, D. and Voet, J. G. (1990). Glycogen Metabolism. In: *Biochemistry*. New York: Wiley.

Experiments in Clinical Biochemistry and Metabolism

Theory

Medical technology has made tremendous advances in the past several decades, due in part to the advances made in clinical chemistry. Before the development of clinical chemistry, doctors were forced to make biochemistry diagnoses based on macroscopic observations and symptoms. Today, medical professionals have an extensive arsenal of biochemical tests available to aid in the early detection of a particular disease or condition. These tests were made possible through improved knowledge of how individual cells and tissues function at the molecular level.

How do you determine whether a particular organ, such as the liver, heart, lung, or kidney is functioning properly? Since tissues consist of specialized cells that perform specific biological processes, information about the integrity of a particular tissue can be obtained by monitoring the activity of enzymes present in that cell type, or the level of substrates or products used or produced by these enzymes. Since it is not always feasible or desirable to invade an organ and remove cells for analysis, an alternative means of quantifying the level of biological molecules in an individual had to be developed.

What is the next best thing to studying the activity present in a particular cell type directly? A great deal of information can be obtained from analysis of the fluid surrounding a particular tissue or cell type. Extracellular body fluid accounts for nearly 20% of the total weight of an individual. Of this, about 75% is found around and between tissues and cells (interstitial fluid). The remainder is in the form of vascular bound fluid, such as blood and lymph. Cerebrospinal fluid (in the spinal cord and around the brain) and the fluid between joints account for only a trace of the total body fluid of an individual.

Undoubtedly, blood plasma or serum is the most common source of material for experiments in clinical biochemistry. This makes logical sense if one considers the role of the blood in the body. Nearly every cell comes into contact with capillary blood for the purpose of acquiring oxygen, obtaining nutrients, receiving signals via hormones, and expelling the products of metabolism. If a particular cell is undergoing some unusual type of metabolism due to infection, exposure to a toxin, or other condition, evidence for this will be present in the identity and/or quantity of the various molecules that the cell secretes into the bloodstream. In the same sense, dying cells associated with tissue damage or necrosis will lyse and secrete enzymes into the bloodstream. Identifying and quantifying the level of particular enzymes in the blood has become a useful tool in the diagnosis of specific conditions and diseases. In this laboratory exercise, you will become familiar with several assays that are currently used in the field of clinical biochemistry.

Glucose

Although high blood glucose levels are commonly associated with diabetes, it has also been found to be associated with individuals possessing a hyperactive thyroid or adrenal gland. Conversely, a low blood glucose level (hypoglycemia) is usually associated with cases of insulin overdose, an underactive endocrine system, and some forms of liver disease. In general, blood glucose levels are useful in assessing the overall integrity of carbohydrate metabolism in an individual.

The assay used in the quantitation of blood glucose is one that you have become familiar with from an earlier experiment. It is a coupled assay that relates the reduction of NAD^+ (ΔA_{340}) to the concentration of glucose in the sample (see Experiment 11, Figure 11-3).

Lactate

Lactate is formed from the degradation of carbohydrates via glycolysis. During times of oxygen limitation, cells will turn toward anaerobic metabolism. In this process, lactate dehydrogenase will convert pyruvate to lactate, with the concomitant oxidation of NADH to NAD^+. This, in turn, will lead to an increase in the concentration of blood lactate. Since lactate is metabolized by the liver to produce glucose via the gluconeogenesis cycle, high blood lactate levels are usually found in individuals with poor liver function. The liver acts as the target organ for many drugs and toxins; therefore, the assay is also useful in diagnosing a person's exposure to these agents.

The assay used to quantify lactate in the blood allows you to relate the production of a colored dye (chromophore) to the level of lactate in the sample. This chromophore will absorb light at 540 nm, allowing its concentration to be measured spectrophotometrically (Fig. 16-1).

Urea Nitrogen

One of the major products of amino acid metabolism is ammonia (NH_3), a molecule known to be highly toxic to higher organisms. In the liver, ammonia and carbon dioxide are used to produce a water-soluble form of nitrogen, urea, via the urea cycle. The liver passes this urea to the blood, which carries it to the kidneys to be filtered out and excreted in the urine. Since one function of the kidney is to collect and excrete urea, increases in the concentration of this compound in the blood are an indicator of poor kidney function. Since urea is formed in the liver, low blood urea nitrogen is often the consequence of impaired liver function due to disease or as the result of infection (hepatitis).

The assay used to quantify blood urea nitrogen relies on coupling the production of ammonia and carbon dioxide from urea to the transamination of α-ketoglutarate to glutamate and the subsequent oxidation of NADH (A_{340}) (Fig. 16-2).

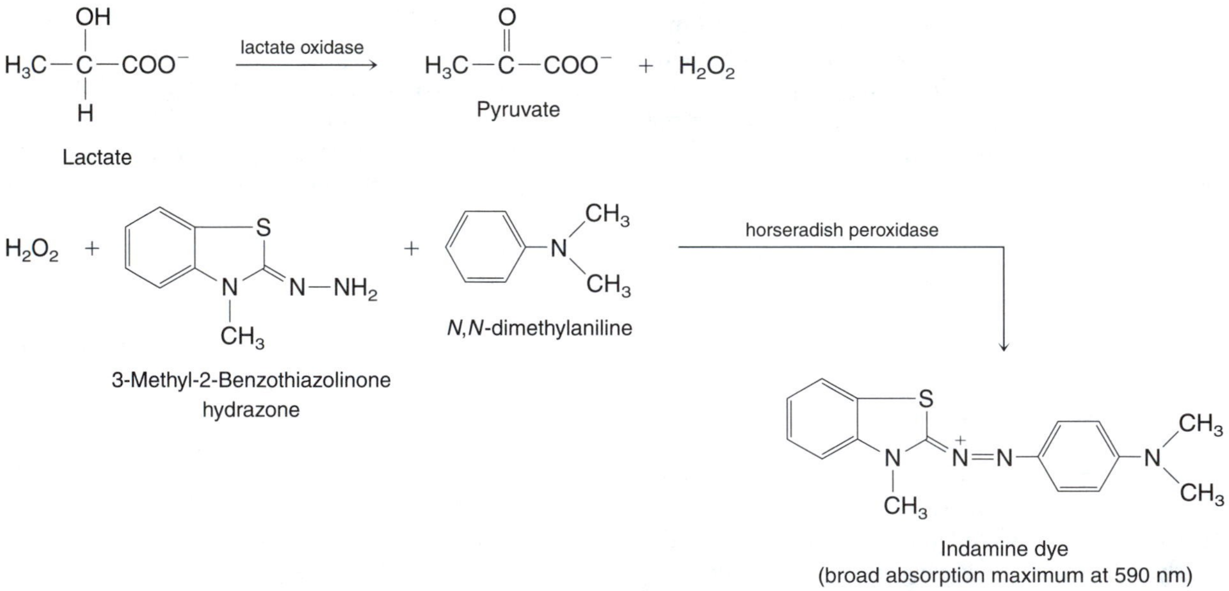

Lactate $\xrightarrow{\text{lactate oxidase}}$ Pyruvate + H_2O_2

H_2O_2 + 3-Methyl-2-Benzothiazolinone hydrazone + *N,N*-dimethylaniline $\xrightarrow{\text{horseradish peroxidase}}$ Indamine dye (broad absorption maximum at 590 nm)

Figure 16-1 Quantitative assay for serum lactate.

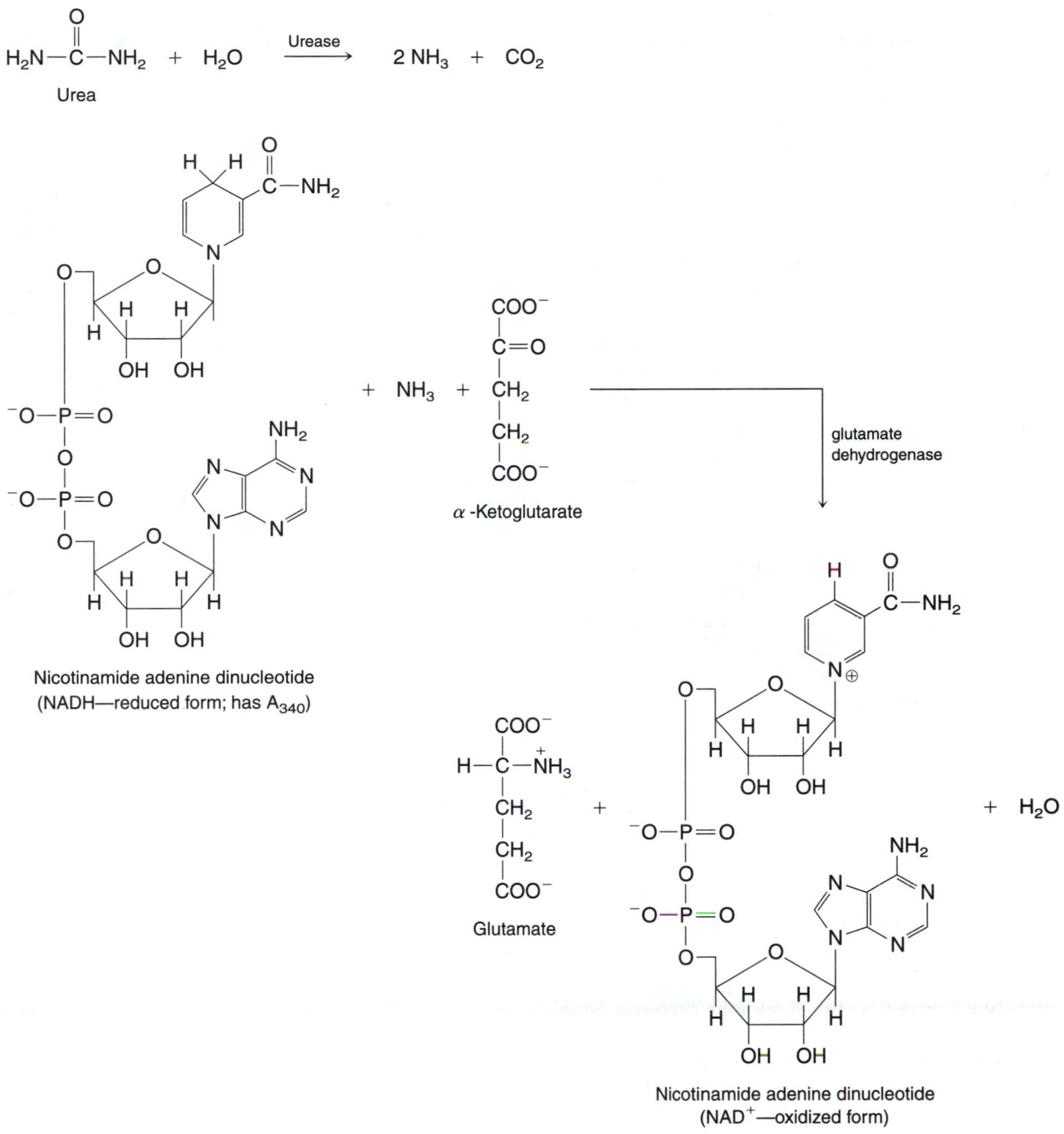

$$H_2N{-}\overset{\overset{\Large O}{\|}}{C}{-}NH_2 \;+\; H_2O \;\xrightarrow{\text{Urease}}\; 2\,NH_3 \;+\; CO_2$$

Urea

Nicotinamide adenine dinucleotide
(NADH—reduced form; has A_{340})

$$+\; NH_3 \;+\; \begin{array}{c} COO^- \\ | \\ C{=}O \\ | \\ CH_2 \\ | \\ CH_2 \\ | \\ COO^- \end{array}$$

α-Ketoglutarate

glutamate dehydrogenase

$$\begin{array}{c} COO^- \\ | \\ H{-}C{-}\overset{+}{N}H_3 \\ | \\ CH_2 \\ | \\ CH_2 \\ | \\ COO^- \end{array} \;+\; \text{NAD}^+ \;+\; H_2O$$

Glutamate

Nicotinamide adenine dinucleotide
(NAD$^+$—oxidized form)

Figure 16-2 Quantitative assay for serum urea nitrogen.

Creatine Kinase (CK)

Phosphocreatine is a high-energy compound that can be used by cells to drive energy-requiring reactions. This compound is formed from creatine and ATP by the action of creatine kinase (CK). This enzyme is also often called creatine phosphokinase (CPK). The reaction performed by this enzyme is readily reversible: ADP and phosphocreatine can be converted to creatine and ATP. Surprisingly, increased levels of this enzyme in the blood have been found to be associated with a number of serious yet seemingly unrelated conditions including heart attacks, alcoholism, muscular dystrophy, stroke, epilepsy, and other neurological and endocrine disorders. In general, elevated serum CK levels are associated with conditions leading to tissue damage and cell death (necrosis).

Creatine kinase functions as a dimer. The dimer can consist of combinations of two different subunits, M and B. Different cell types produce or express the MM, MB, or BB forms of CK. Although the amino acid composition of the M and B subunit differ, all three dimeric forms can catalyze the reaction described above. Different forms of an enzyme that catalyze the same reaction (such as the MM, MB, and BB forms of CK) are referred to as *isoenzymes.*

It has been determined that different tissues produce or express different CK isoenzymes. Skeletal muscle cells express the CK-MM isoenzyme almost exclusively, while brain cells and cells of the gastrointestinal tract express the CK-BB isoenzyme almost exclusively. Cardiac muscle cells, however, express an approximate 1:3 ratio of the CK-MB and CK-MM isoenzymes, respectively. If you had a means of determining which isoenzyme is present at elevated levels in the serum, you might be able to determine what type of tissue the enzyme originated from.

One common method of determining the identity of the three isoenzymes makes use of the difference in charge between the two subunits that comprise the dimeric protein. While the M subunit carries a slightly positive charge at neutral to slightly alkaline pH, the B subunit carries a large negative charge. When subjected to an electric field (electrophoresis, see Experiment 4), the CK-BB isoenzyme will migrate rapidly toward the anode (positive electrode), while the CK-MM isoenzyme will display a slow migration toward the cathode (negative electrode). The CK-MB isoenzyme, carrying less of a net negative charge than the CK-BB isoenzyme, will migrate toward the anode at a slower rate.

Another method of determining the identity of the various CK isoenzymes that is gaining widespread popularity involves immunochemical techniques such as immunoprecipitation and Western blotting (see Section IV). It is now possible to produce and screen for monoclonal antibodies that recognize the M subunit or B subunit exclusively.

The assay used to quantify blood concentrations of this enzyme is nearly identical to that used in the assay described for determining blood glucose concentrations (Fig. 16-3). Whereas the ATP in the latter assay is supplied in excess, the ATP in the creatine kinase assay is provided by the action of the enzyme (creatine kinase) in the presence of ADP and creatine phosphate. In both assays, you are measuring the change in absorbance of the sample at 340 nm. Since glucose is the limiting reagent in the glucose assay, the A_{340} will be directly proportional to the glucose concentration. In the case of the assay for creatine kinase, glucose is in excess, and the rate of change in absorbance at 340 nm will be limited by the ATP that is being produced by the activity of creatine kinase. This demonstrates an important feature common to all coupled assay systems: *for the assay to be valid, the molecule or enzyme under study must be the limiting reagent in the assay.*

Supplies and Reagents

Plasma or serum from pig
P-20, P-200, and P-1000 Pipetmen with sterile tips
5-ml and 10-ml pipettes
13 × 100-mm glass test tubes
Heparin sulfate solution (1000 units/ml)
0.5 M EDTA, pH 8.0
B/L spectrophotometer tubes or cuvettes
Spectrophotometer to read absorbance at 340 nm and 540 nm
Glucose assay reagent* consisting of:

- 50 mM Tris, pH 7.5
- 1.5 mM NAD^+
- 1.0 mM ATP
- 1.0 units/ml hexokinase
- 1.0 units/ml glucose-6-phosphate dehydrogenase
- 5 mM $MgCl_2$

D-Glucose standard solutions in water: 3.00 mM, 11.10 mM, and 16.65 mM
Lactate assay reagent* consisting of:

- 50 mM Tris, pH 7.2
- 0.5 units/ml lactate oxidase

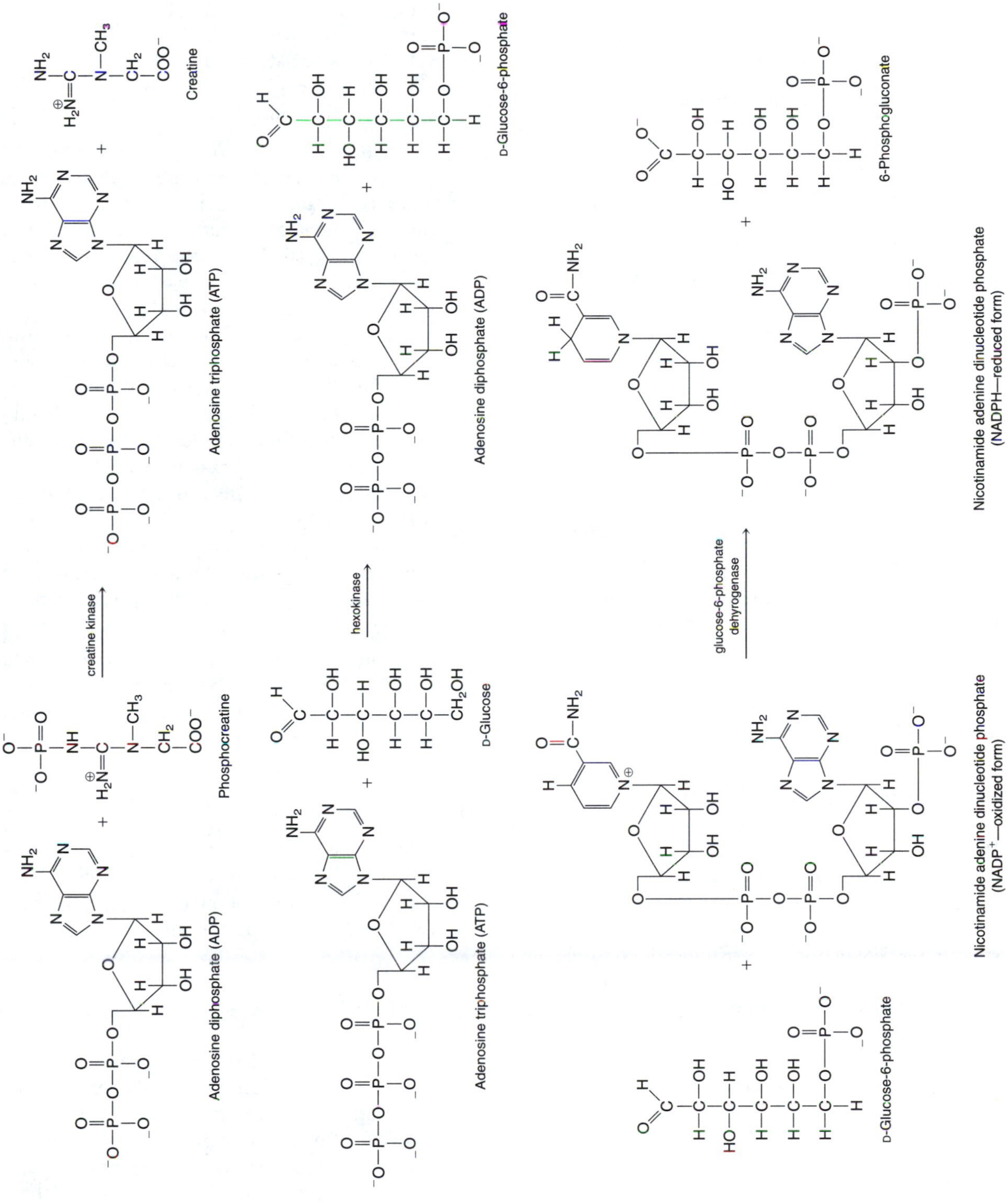

Figure 16-3 Assay for serum creatine kinase (CK) activity.

2.5 units/ml horseradish peroxidase
5 mM 3-methyl-2-benzothiozolinone hydrazone
5 mM *N,N*-dimethylaniline

Urea nitrogen assay reagent* consisting of:

50 mM Tris, pH 8.0
0.25 mM NADH
5.0 mM α-ketoglutarate
15 units/ml urease
25 units/ml glutamate dehydrogenase

Urea standard solutions in water: 3.57 mM, 8.92 mM, 10.71 mM, and 35.70 mM

Creatine kinase assay reagent consisting of:

50 mM Tris, pH 7.5
10 mM phosphocreatine
1.5 mM ADP
1 mM NADP
5.0 mM D-glucose
1.5 units/ml hexokinase
0.5 units/ml glucose-6-phosphate dehydrogenase
20 mM dithiothreitol

*(Reagent is sold in a mixed dry form by Sigma Chemical Co. This is convenient for use with a large laboratory class.)

Protocol

Preparation of Blood Plasma

Pig Blood. Collect a 100–200-ml blood sample in a sterile container with cap containing 750 units of heparin sulfate (anticoagulant) and 25–50-μl of 0.5M EDTA, pH 8.0 (preservative). As soon as possible after collection, centrifuge the sample at $400 \times g$ for 10 min to separate the plasma or serum from the blood cells (erythrocytes, etc.). Remove the clear plasma from the blood cells and store in a fresh, sterile container at 4°C. *For best results, the assays described below should be performed within 36 hr after the serum is isolated.*

Blood Glucose Assay

1. Add 2.0 ml of glucose reagent to five 13-by-100-mm test tubes labeled G1 through G5.
2. Add 20 μl of the following to each of these tubes and shake gently to mix:

G1 Water
G2 Plasma sample
G3 Glucose standard (3.00 mM)
G4 Glucose standard (11.10 mM)
G5 Glucose standard (16.65 mM)

3. Incubate the tubes at room temperature for 10 min. It is not critical that the incubation time be exact. Rather, it is important that the reaction be allowed to run to completion (no change in A_{340}). When the reaction is complete, add 1 ml of water to each of the five tubes before taking a final A_{340} reading.
4. Blank your spectrophotometer at 340 nm against a water reference. Record the absorbance of all of your samples, G1 through G5.
5. Subtract the absorbance value of G1 (blank) from the rest of the samples. Construct a standard curve of A_{340} versus micromoles of glucose for your three glucose standard solutions (G3, G4, and G5).
6. Based on the A_{340} of your plasma sample and the *volume* of plasma used in the assay, determine the *concentration* of glucose in your plasma sample.

   NOTE: The normal or expected concentration of blood glucose in an adult human is ~4.0 mM to 6.5 mM. How does the concentration of glucose in the pig serum sample compare to that expected in human serum?

Blood Lactate Assay

1. Add 2.0 ml of lactate reagent to three 13-by-100-mm test tubes labeled L1 through L3.
2. Add 20 μl of the following to each of these tubes and shake gently to mix:

L1 Water
L2 Plasma sample
L3 Lactate standard (4.44 mM)

3. Incubate the tubes at room temperature for 10 min. It is not critical that the incubation time be exact. Rather, it is important that the reaction be allowed to run to completion (no change in A_{540}). When the reaction is complete, add 1 ml of water to each of the three tubes before taking a final A_{540} reading.
4. Blank your spectrophotometer at 540 nm against the sample blank (L1). Record the absorbance of your samples, L2 and L3.

5. Use the following equation to determine the concentration of lactate in your plasma sample:

$$[\text{Lactate}]\ (\text{mM}) = \frac{A_{540}\ \text{plasma sample (L2)}}{A_{540}\ \text{lactate standard (L3)}} \times 4.44\ \text{mM}$$

NOTE: The normal or expected concentration of blood lactate in an adult human is ~0.3 mM to 1.3 mM. The assay is reported to be accurate for plasma samples containing as much as ~13.0 mM lactate. How does the concentration of lactate in the pig serum sample compare to that expected in human serum?

Blood Urea Nitrogen Assay

1. Add 2.0 ml of urea nitrogen reagent to five 13-by-100-mm test tubes labeled U1 through U5.
2. Add 20 μl of the following to each of these tubes and shake gently to mix:

U1 Water
U2 Plasma sample
U3 Urea standard (3.57 mM)
U4 Urea standard (8.92 mM)
U5 Urea standard (10.71 mM)
U6 Urea standard (35.70 mM)

3. Incubate the tubes at room temperature for 5 min. It is not critical that the incubation time be exact. Rather, it is important that the reaction be allowed to run to completion (no change in A_{340}). When the reaction is complete, add 1 ml of water to each of the five tubes before taking a final A_{340} reading.
4. Blank your spectrophotometer at 340 nm against a water reference. Record the absorbance of all of your samples, U1 through U6.
5. Subtract the absorbance value of U1 (blank) from the rest of the samples. Construct a standard curve by plotting the *absolute value* of A_{340} versus micromoles of urea nitrogen for your four urea standard solutions (U3, U4, U5, and U6).
6. Based on the A_{340} of your plasma sample and the *volume* of the plasma sample used in the assay, determine the *concentration* of urea nitrogen in your plasma sample.

NOTE: The normal or expected concentration of blood urea nitrogen in an adult human is ~2.0 mM to 6.0 mM. How does the concentration of urea nitrogen in the pig serum sample compare to that expected in human serum?

Blood Creatine Kinase Assay

This assay is different from that of the glucose, lactate, and urea nitrogen assays in that you are attempting to measure the activity of an enzyme rather than the concentration of a molecule. *Remember that in an enzymatic activity assay, it is critical that the time between absorbance readings be recorded exactly.*

1. Add 2.5 ml of creatine kinase reagent to two 13-by-100-mm test tubes labeled CK1 and CK2.
2. Dilute your plasma sample 1:5 in distilled water. This dilute plasma sample will be used in the assay described below.
3. Add 100 μl of distilled water to CK1 and 100 μl of the dilute plasma sample to CK2. Shake gently to mix and allow the reaction to incubate at room temperature for 5 min.
4. Blank your spectrophotometer at 340 nm against a water reference. Record the absorbance of CK1 and CK2 at *exactly* 5 min and 10 min.
5. Determine the ΔA_{340} of your plasma sample (CK2) and the blank (CK1) by subtracting the A_{340} for the 5-min reading from the A_{340} for the 10-min reading for each of these tubes.
6. To determine the concentration of creatine kinase in your plasma sample (units/liter), use the following equation:

$$[\text{Creatine kinase}]\ (\text{units/liter}) = \frac{([\Delta A_{340}\ \text{CK2} - \Delta A_{340}\ \text{CK1}]\ \text{per 5 min})(2.6)(1000)}{(5\ \text{min})(6.22\ \text{mM}^{-1}\ \text{cm}^{-1})(\text{l})(0.02)}$$

where 2.6 = total volume of reaction (ml); 5 min = total reaction time; 1000 = conversion factor to convert micromoles per milliliter to micromoles per liter; 6.22 $\text{mM}^{-1}\ \text{cm}^{-1}$ = millimolar extinction coefficient for NADPH at 340 nm; l = pathlength of spectrophotometer (1.3 cm for benchtop colorimeters); and 0.02 = volume of plasma sample used in the assay (in milliliters), taking into account the dilution.

NOTE: The normal or expected concentration of blood creatine kinase in an adult human is

~10 to 50 units/liter. One unit of creatine kinase is defined as the amount of the enzyme that will produce 1 μmol of ATP from phosphocreatine and ADP per minute under these reaction conditions. How does the activity of creatine kinase in the pig serum sample compare to that expected in human serum?

Exercises

1. You are presented with two individuals: one of whom completed a marathon earlier that day and one who has been watching television all afternoon. What test would you use to determine which individual had completed a marathon? Describe in your answer for what molecule you would assay, the relevant features of the test, and how the results for each individual would be different.
2. The assays for serum glucose and creatine kinase were similar in many respects. List the chemical reactions that were common to both assays. Also, describe how these two similar assay systems could be used to quantify these two very different biomolecules.
3. Two of the molecules that were assayed for in pig serum are often used to assess overall liver function. Name these two molecules. Using your knowledge of the gluconeogenesis pathway and the urea cycle, explain how increased or decreased levels of these two molecules could be used to assess the overall function of the liver.
4. Why do you think that serum creatine kinase levels might be elevated in patients with seemingly unrelated conditions such as alcoholism and epilepsy?
5. Why was it necessary to know the exact incubation times for the creatine kinase assays performed in this experiment, while it was less important in the blood urea nitrogen, glucose, and lactate assays?

REFERENCES

Barhan, D., and Trinder, P. (1972). An Improved Color Reagent for the Determination of Blood Glucose by the Oxidase System. *Analyst* **97**:142.

Bondar, R. J. L., and Mead, D. C. (1974). Evaluation of Glucose-6-Phosphate Dehydrogenase from *Leuconostoc mesenteroides* in the Hexokinase Method for Determining Glucose in Serum. *Clin Chem* **20**::586.

Gotchman, N., and Schmitz, J. M. (1972). Application of a New Peroxide Indicator Reaction to the Specific, Automated Determination of Glucose with Glucose Oxidase. *Clin Chem* **18**:943.

Hess, J. W., Murdock, K. J., and Natho, G. J. W. (1968). Creatine Phosphokinase—A Spectrophotometric Method with Improved Sensitivity. *Am J Clin Pathol* **50**:89.

Kaplan, A., Szabo, L. L., and Opheim, K. E. (1988). *Clinical Chemistry: Interpretation and Techniques*, 3rd ed. Philadelphia: Lea & Febiger.

Lehninger, A. L., Nelson, D. L., and Cox, M. M. (1993). Bioenergetics and Metabolism. In: *Principles of Biochemistry*, 2nd ed. New York: Worth.

Martinek, R. G. (1969). Review of Methods for Determining Urea Nitrogen in Biological Fluids. *J Am Med Technol* **31**:678.

Rosalki, S. B. (1967). An Improved Procedure for Serum Creatine Phosphokinase Determination. *J Lab Clin Med* **69**:696.

Sigma Chemical Co., Procedure No. 16-UV (1997).

Sigma Chemical Co., Procedure No. 735 (1997).

Sigma Chemical Co., Procedure No. 66-UV (1997).

Sigma Chemical Co., Procedure No. 45-UV (1997).

Stryer, L. (1995). Metabolic Energy. In: *Biochemistry*, 4th ed. New York: Freeman.

Trinder, P. (1969). Determination of Glucose in Blood Using Glucose Oxidase with an Alternative Oxygen Acceptor. *Ann Clin Biochem* **6**:24.

SECTION IV

Immunochemistry

SECTION IV

Immunochemistry

Introduction

Antibodies, or *immunoglobulins* (Ig), are proteins produced by cells of the immune system that allow the body to resist viral and bacterial infection, as well as cells that have undergone malignant transformation. Antibodies act by binding with *high specificity* to foreign molecules, called *antigens*. The cells that produce and secrete antibodies are called B cells. An individual B cell can differentiate to produce a *single* type of antibody that will recognize a *single* epitope (region) on a *single* antigen molecule. Because of their high degree of specificity and often high affinities for antigens ($K_D \sim 10^{-6}$ to 10^{-10} M), antibodies have proven to be powerful tools for biochemical studies that require detection and/or quantitation of a specific antigen.

The structural characteristics common to all immunoglobulins is shown in Figure IV-1. An antibody is comprised of two identical heavy chain and two identical light chain subunits that are linked by several disulfide bonds. Both the heavy and the light chains possess a *constant region* at their C-termini and a *variable region* at their N-termini. Depending on the type of antibody (see below), one or more N-linked carbohydrates may be found at different positions along the heavy chains. *It is the N-termini of the heavy and light chains that are variable in amino acid sequence. Together, these variable (V) regions form a binding site that interacts with the antigen and thereby determines the specificity of the antibody* (see Fig. IV-1). Because each "arm" of the immunoglobulin is made up of identical heavy and light chain subunits, a single antibody molecule has the ability to bind the identical epitope or antigenic determinant on two individual molecules of the antigen. In other words, a single antibody can bind two molecules of the antigen.

Once the epitope of a single molecule of antigen is bound by an antibody molecule, other antibodies that bind the same epitope are unable to bind (Fig. IV-2). Antibodies that bind to different epitopes on the same antigen may still be able to bind. These properties underlie the basis of competition and capture assays, as developed by many diagnostic testing laboratories. Such assays are frequently used to determine the amount of a specific antigen that is present in clinical samples (e.g., shed tumor-associated antigens that might be diagnostic for a particular type of cancer).

Types of Antibodies

The process of stem-cell differentiation into a mature, antibody-secreting B cell is complex. The first part of this process, which takes place in the bone marrow, occurs before the B-cell precursor is exposed to any potential antigens, and is therefore termed *antigen independent* differentiation. As the B-cell precursor migrates to the lymph and continues to mature, it eventually reaches a stage in its differentiation at which further development will be guided by a particular antigen that it happens to encounter (*antigen-dependent* differentiation). The terminal step in differentiation will give rise to a plasma B cell that secretes a single immunoglobulin. The secreted immunoglobulin is classified as one of the five isotypes described below.

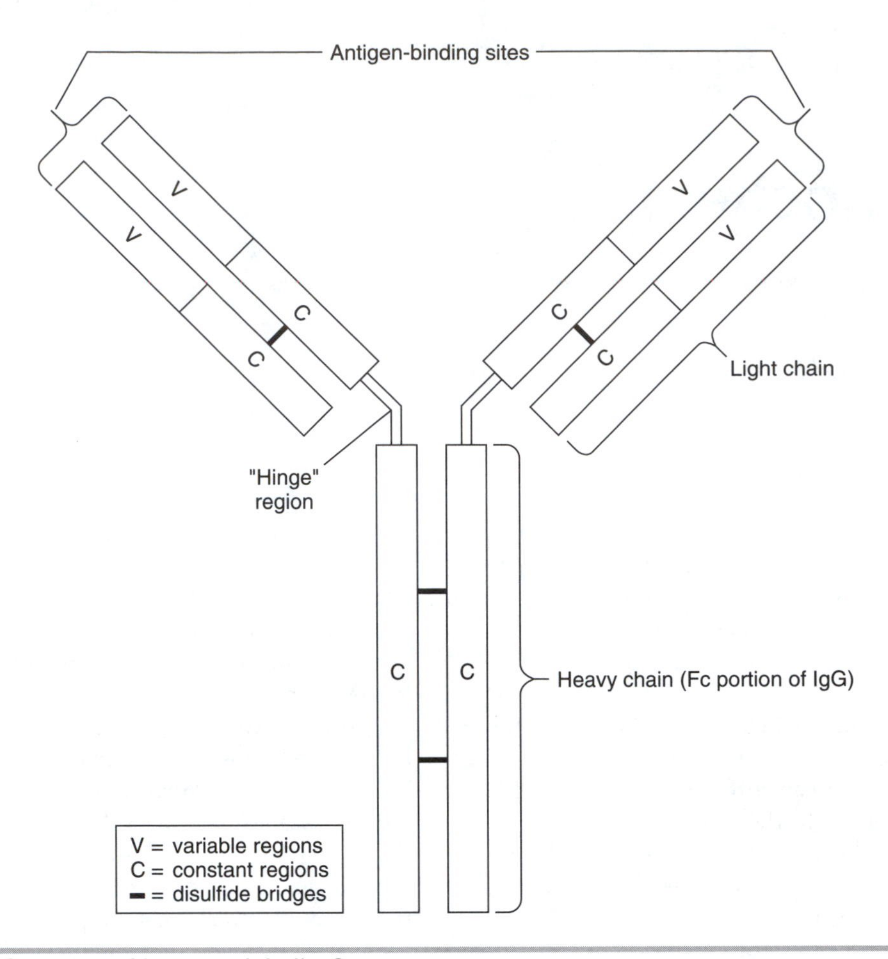

Figure IV-1 Structure of immunoglobulin G.

Immunoglobulin M (IgM)

Immunoglobulin M is the first antibody isotype to be expressed on the surface of an immature B-cell precursor as the cell leaves the bone marrow and enters the lymph to undergo antigen-dependent differentiation. Unlike IgG (see Fig. IV-1), IgM has no hinge region between the heavy- and light-chain junction, and it is heavily glycosylated on the most C-terminal regions of the heavy chain. Immature B cells that recognize self-antigens on tissue surfaces with their membrane-bound IgM will quickly undergo programmed cell death (apoptosis), preventing the onset of various autoimmune diseases. B cells that do continue to differentiate into IgM-secreting cells (those that do not recognize surface self-antigens) will produce a *pentameric* form of the antibody (5 immunoglobulins with 10 identical antigen binding sites, Fig. IV-3). This pentameric form of IgM resides almost exclusively in the vascular bound fluid, and is the major component required for activation of the complement system. IgM antibodies typically have low intrinsic binding affinities (i.e., at a single antigen-binding site) but high avidities (i.e., multiple sites on the IgM bind very well to multivalent ligands).

Immunoglobulin D (IgD)

Although the role of IgD in the immune system has not yet been fully elucidated, it appears to be involved in determining if a particular B-cell precursor will continue to develop and differentiate. While an immature B cell expresses only IgM on the cell surface, a mature B cell expresses both IgM and IgD on the cell surface. A B cell that recognizes any *soluble* self-antigen with its surface IgM or IgD will quickly halt production of IgM and increase expression of surface IgD. This high concentration of surface IgD will place the B cell in a

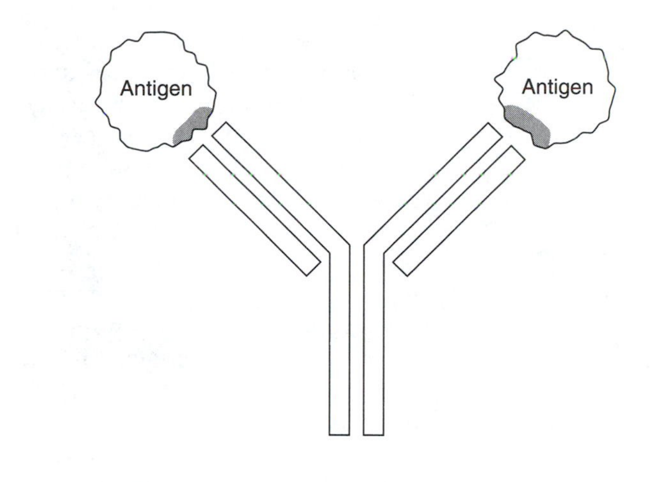

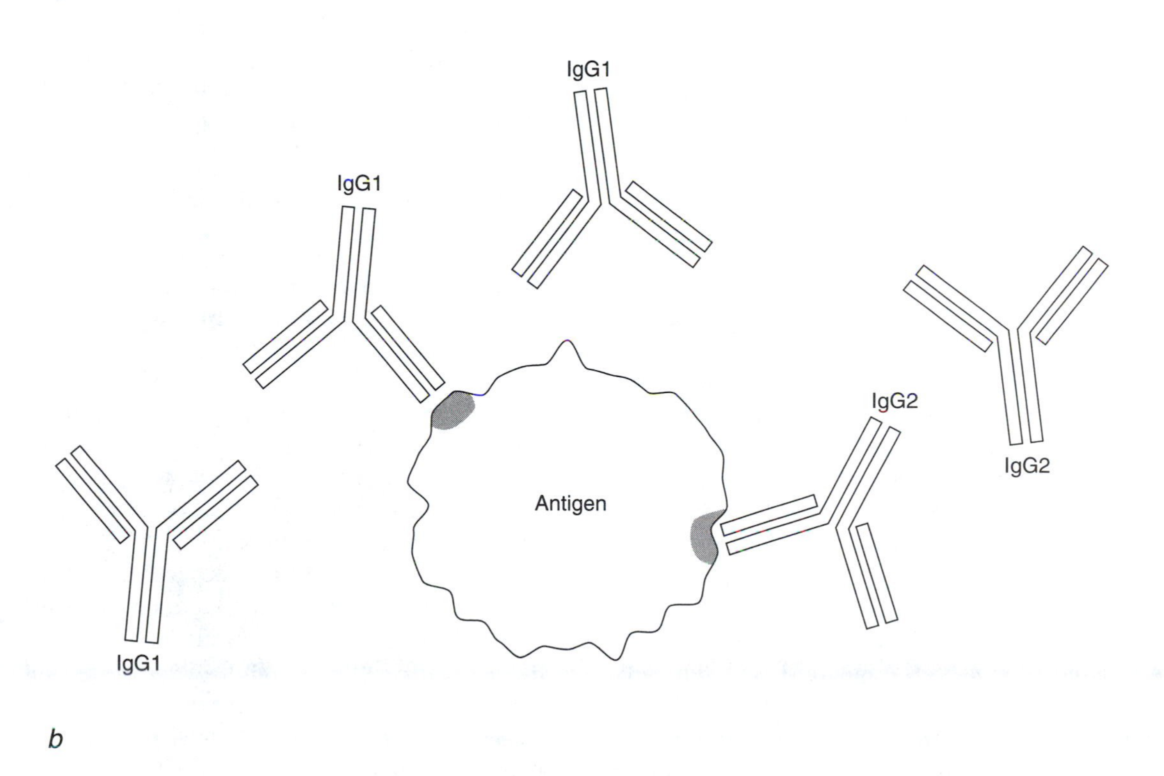

Figure IV-2 Interaction of an antibody with an antigen. (*a*) One molecule of antibody can bind the same epitope or region on two individual molecules of the antigen. (*b*) Once a single epitope of an antigen is bound by a single molecule of the antibody (IgG1), another molecule of the same antibody will not be able to bind the same molecule of antigen. A different antibody (IgG2) may still be able to bind the same molecule of antigen at a different site or epitope.

state of *anergy*, in which it will no longer be allowed to develop into a fully mature, antibody-secreting B cell. As is the case with surface IgM, selection against B-cell precursors that recognize soluble self-antigens with their surface IgD prevents the onset of various autoimmune diseases. Structurally, IgD is similar to IgG (see Fig. IV-1), but it contains a number of N-linked carbohydrate groups near the

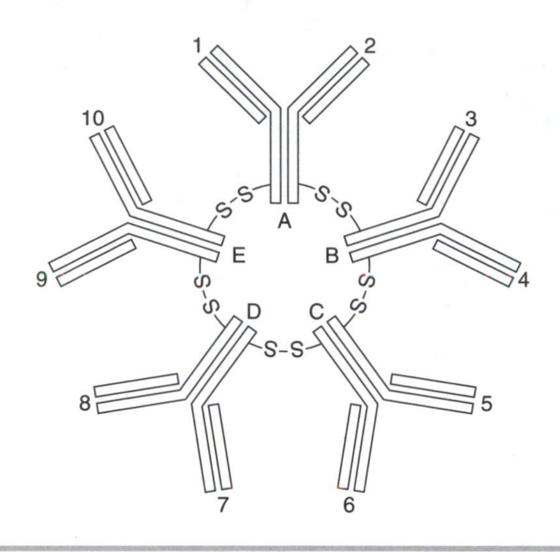

Figure IV-3 The pentameric form of IgM: five disulfide-linked immunoglobulins (A to E) with 10 identical antigen-binding sites.

junction between the variable and constant regions of the heavy chain.

Immunoglobulin A (IgA)

Immunoglobulin A is the isotype responsible for conferring protective immunity to the mucousal membranes that line the intestinal, respiratory, and urogenital tracts. IgA acts by *neutralizing* foreign bacteria and viruses that enter the body via these pathways. Possessing a high affinity (and avidity) for surface antigens on these types of pathogens, IgA binds them and prevents their attachment to, and subsequent infection of, the epithelial cells that line these tissues. Unlike IgD and IgM, IgA functions mainly as a *dimer* in these areas of the body, although very low levels of the *monomeric* form of IgA may be found in vascular bound fluids. The structure of the IgA isoform most closely resembles that of the IgD isoform, containing a heavily glycosylated stretch of amino acids at the junction between the variable and constant regions of the heavy-chain peptides.

Immunoglobulin E (IgE)

The immunoglobulin E isotype is associated with *mast cells* of the skin, mucosal membranes, and vessels that line connective tissue. IgE appears to play a more *indirect* role in protecting the body from infection, particularly by parasitic organisms. While IgE has affinity for surface antigens found on foreign pathogens that may enter these tissues, it also has high affinity for receptors present on the mast cells. As IgE binds foreign antigens, it signals the mast cells to secrete powerful compounds that induce the contraction of these tissues. The vomiting, coughing, and sneezing that are induced by these contractions function to rid the body of the invading pathogen or agent. The IgE-mediated response can also give rise to allergic reactions toward dust, pollen, mold, spores, and other environmental agents that the immune system becomes sensitive to through the development of IgE antibodies specific for these antigens (allergens). Like IgM, IgE is larger than the other isotypes and lacks the hinge region between the heavy- and light-chain junction. Like IgM, IgD, and IgA, IgE is heavily glycosylated throughout the heavy-chain subunit.

Immunoglobulin G (IgG)

Immunoglobulin G is the most abundant isotype found in the blood. Unlike the pentameric form of IgM, which provides protection due largely to its ability to activate the complement system, IgG offers protection through *opsonization* of the invading pathogen or agent. After IgG binds the antigen, it can induce macrophages, neutrophils, and other phagocytes to engulf and destroy it. This process occurs after the macrophage or other cell binds the antigen–antibody complex through IgG receptors (Fc receptors) on their cell surface. IgG is the major immunoglobulin isotype induced with the repeated introduction of a foreign antigen into the bloodstream (hyperimmunization), and generally exhibits the highest intrinsic binding affinities for antigens. Thus, IgG is the most common isotype used in biochemical studies (see below).

Polyclonal and Monoclonal Antibodies

An *immunogen* is defined as an agent that can elicit an immune response. An *antigen* is defined as any molecule that a particular antibody has affinity for, or recognizes. While it is possible to raise antibodies against virtually any protein, hormone, or chemical, not all potential antigens are themselves immunogenic. There are two methods that can be employed to elicit an immune response to an inherently nonimmunogenic compound in an attempt to raise antibodies that will recognize it.

The first method involves the use of an *adjuvant*, an immunostimulant that will increase the immunogenicity of other compounds that are mixed with it. If a protein is mixed with heat-killed mycobacteria in an oil/water emulsion (Freund's adjuvant) and injected into a host animal, the host's immune system will mount a response to the protein in the mixture and produce antibodies that will have affinity for it. By delivering the protein antigen in an oil emulsion, the protein is slowly released into the bloodstream so that it can be engulfed by macrophages and recognized by mature B cells expressing surface IgM and IgD. The components present in the mycobacteria are also crucial in that they stimulate macrophages and proliferation of other immune system cell types that will eventually be required for B cell activation. Another adjuvant containing aluminum hydroxide and heat-killed *Bordetella pertussis* (the pathogenic bacteria that causes whooping cough) is becoming more frequently used because of its ability to enhance the immune response to immunogenic antigens.

The second method involves chemically conjugating a small molecule (*hapten*) to an immunogenic protein carrier. This procedure is often used to raise antibodies against small organic molecules such as nitrophenols, antibiotics, and hormones, which are normally not able to elicit an immune response on their own. Although this method will lead to the production of antibodies that recognize both the protein carrier and the hapten–carrier conjugate, a subpopulation of antibodies will also be able to recognize only the hapten.

As the antigen of interest is injected into the bloodstream of the host animal, it will encounter B cells expressing both surface IgD and IgM. Three events will take place that will ultimately lead to the activation and proliferation of B cells that recognize the injected antigen. First, the surface IgM on the B cell will become cross-linked as it binds the antigen, triggering a phosphorylation cascade that will release intracellular Ca^{2+}, activate Ca^{2+}-dependent kinases, and eventually lead to the activation (phosphorylation) of various transcription factors that promote cell division (proliferation). Second, some of the IgM-bound antigen will be engulfed by the B cell, degraded, and presented again on the surface of the B cell as small peptides bound by the *major histocompatibilty complex* class II (MHC II). Eventually, this MHC II complex is bound by *T helper cells* that secrete interleukins-4, 5, and 6. The net effect of T-helper-cell binding and interleukin secretion is that the B cell will undergo rapid differentiation and proliferation. Third, *follicular dendritic cells* in the lymph nodes will bind the B cell and stimulate its growth and development via a separate phosphorylation cascade. Remember that *each B cell produces a single antibody that recognizes a single epitope on the antigen with a defined affinity.* A population of antibodies derived from many different B cells will therefore recognize different epitopes on the antigen with different affinities and with potentially different attractive forces (electrostatic, hydrogen bonding, hydrophobic, etc.).

Polyclonal Antibodies

Polyclonal antibodies are a population of different immunoglobulins derived from different B cells that recognize an antigen. As such, they frequently bind to different epitopes on the antigen with different affinities. The process of raising polyclonal antibodies against an antigen of interest is relatively simple (Fig. IV-4). The antigen–adjuvant mixture is first injected into the bloodstream of the host animal. About 7 days after this *primary* injection, the concentration of serum IgG that is able to recognize the antigen will begin to rise. About 13 to 17 days following the primary immunization, the serum level of IgG that recognizes the antigen of interest will be at a maximum. The host organism will then begin to "clear" the antigen from its system, causing the concentration of antigen-specific serum IgG to decrease back to pre-immunization levels.

If the same host animal is again immunized with the antigen (a "booster" injection), the immune response mounted against it will be both greater in magnitude and more rapid than the response seen following the primary immunization. Approximately 4 days after the *secondary* immunization, the serum levels of antigen-specific IgG will be at a maximum (as compared to 13 to 17 days following the primary immunization). In addition to being more rapid, the response to the secondary

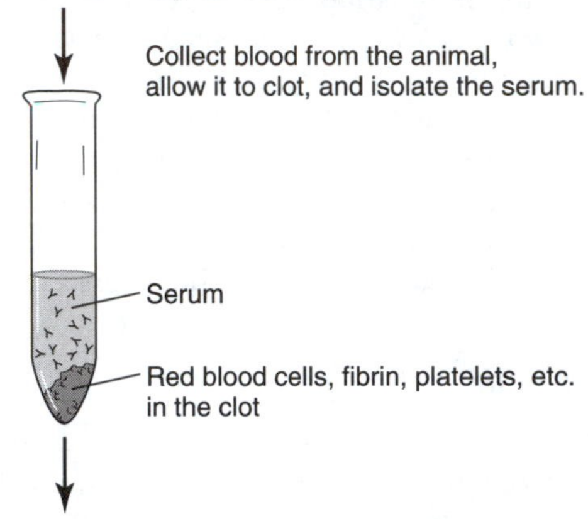

Figure IV-4 Raising polyclonal antibodies against an antigen of interest.

immunization will give rise to a higher concentration of antigen-specific IgG in the serum. This population of IgG molecules may also have a higher intrinsic binding affinity for antigen. The immune response mounted against the antigen after repeated injection is more intense and more rapid due to the fact that the primary immunization gave rise to a number of *memory* B cells that will allow the host organism to recognize the antigen more quickly and efficiently upon future exposures to it. This process of *hyperimmunization* (multiple rounds of antigen injection), blood collection, and serum isolation may be repeated many times on an individual animal to obtain large quantities of antibodies that will recognize the antigen of interest.

There are several things you must keep in mind when attempting to raise a polyclonal antibody against an antigen. First, you must consider what host animal should be used in raising the antibody. The immune system naturally selects against B cells that produce surface immunoglobulins that recognize "self" antigens in their early stages of development. Therefore, it may be difficult to raise antibodies against a mouse protein by injecting it into another mouse. This same mouse antigen, however, may prove to be quite immunogenic when introduced into another species, such as a rat, horse, goat, guinea pig, or rabbit. In general, it is best to use a host animal for antibody production from which the antigen of interest was not derived.

Second, you must remember that the serum from an immunized animal contains thousands of different IgG molecules, of which only a small fraction (perhaps less than 1%) actually recognize the antigen of interest. The IgG present in the serum prior to the introduction of the antigen is still present. The immunization process merely attempts to increase the relative concentration of antigen-specific antibodies in the serum with respect to the numerous other nonspecific antibodies that are also present.

Finally, the heterogeneity of the population of antibodies in the serum that do recognize the antigen make polyclonal antibodies very useful in some biochemical applications and less useful in others. This stems from the fact that the different antibodies can interact with different epitopes on the antigen with a wide variety of prevailing attractive forces (electrostatic, hydrophobic, hydrogen bonding, etc.). This topic will be discussed more thoroughly later in this introduction.

Monoclonal Antibodies

Monoclonal antibodies are a population of *identical* immunoglobulins all derived from a *single* B cell. Because of this, they all recognize a single epitope on the antigen with the same affinity. The process of producing monoclonal antibodies is more complex than that used to produce polyclonal antibodies (Fig. IV-5). Following hyperimmunization and several days of incubation, the animal is killed to obtain its spleen, which contains a large population of B cells. These B cells are then fused with myeloma cells (transformed or immortalized cells of the immune system) with the addition of polyethylene glycol to the cell mixture. To select for re-

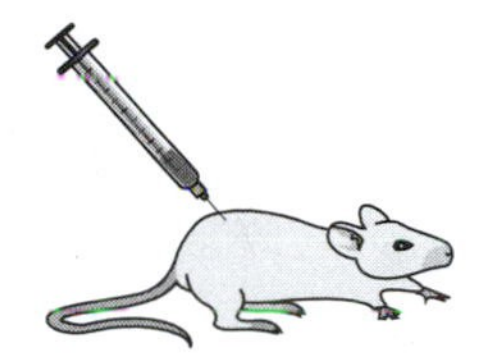

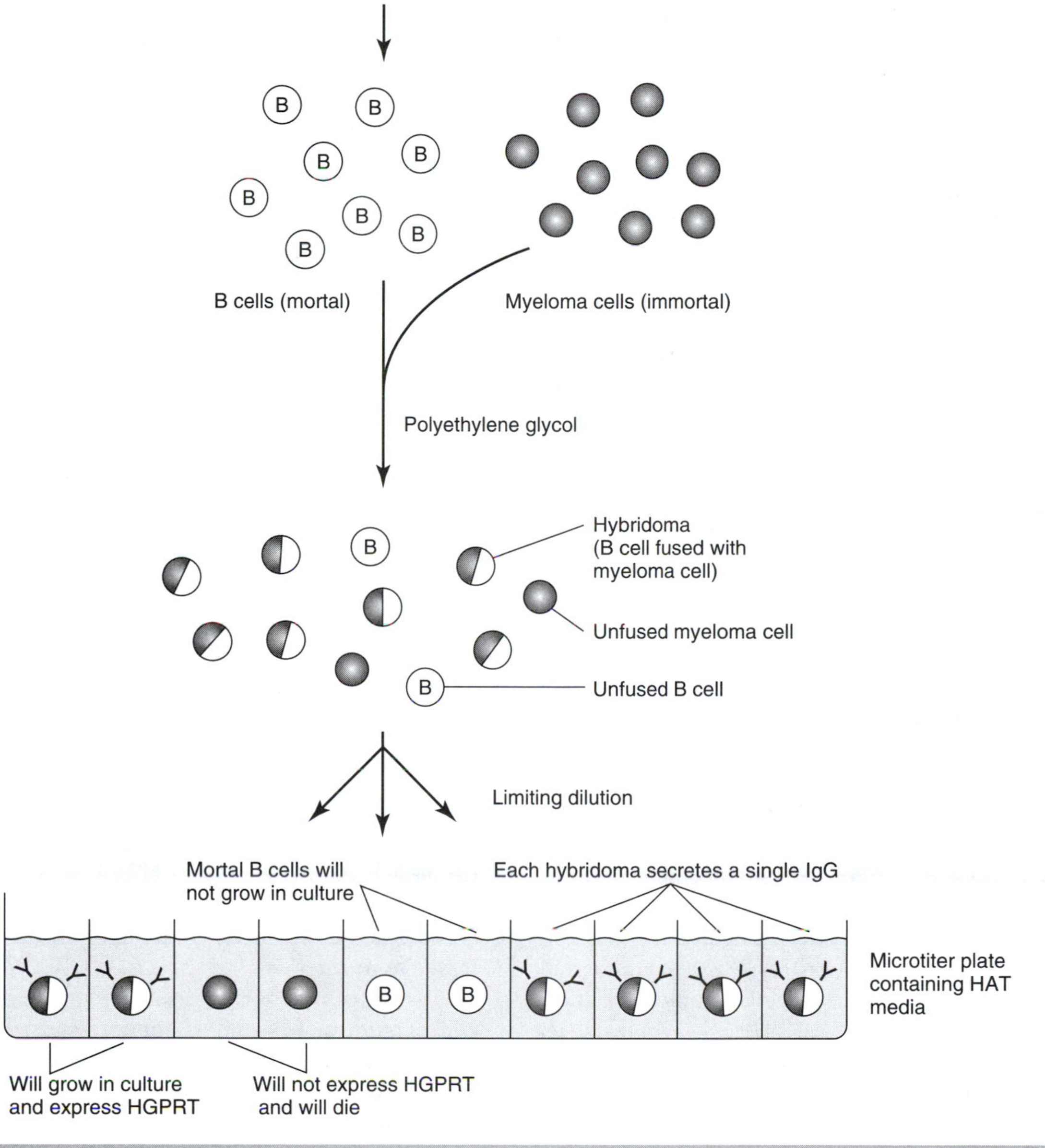

Figure IV-5 Production of a hybridoma cell line that produces a monoclonal antibody against an antigen of interest.

sultant hybridoma cells in culture (*mortal* B cells fused with *immortal* myeloma cells), the mixture is added to a medium that contains hypoxanthine-aminopterin-thymidine (HAT). Aminopterin is an antifolate that blocks de novo purine nucleotide biosynthesis, making cells dependent on hypoxanthine and thymidine for growth. B cells normally express an enzyme called *hypoxanthine guanosine phosphoribosyl transferase* (HGPRT), which allows cells to take up hypoxanthine. The myeloma fusion partner does not express this enzyme. A cell that does not express HGPRT, therefore, will not be able to grow in the presence of HAT. Thus, any B cell that has fused with a myeloma cell (a hybridoma) will be immortal, will produce HGPRT derived from the B cell, and will be able to grow in the HAT-containing medium. Myeloma cells that did not fuse with a B cell will not express HGPRT and will die in HAT medium. Although unfused B cells do express HGPRT, their lifespan in culture is sufficiently short that they too will be unable to grow effectively in the HAT-containing medium. In effect, only hybrid cells that are immortal and which express HGPRT will grow effectively.

Once a collection of hybridoma cells has been isolated, a long screening and cloning process lies ahead. After identification of cultures that contain the desired antibody (the hybridoma cell will secrete antibody into the growth medium), *limiting dilutions* of the hybridoma population are made in HAT media in a 96-well microtiter plate. This limiting dilution is performed in an effort to obtain one hybridoma cell per well. These hybridomas are allowed to grow, and the medium is again analyzed for the presence of antibody that binds to the antigen of interest. After one or several positive clones are identified, an additional round of limiting dilution and screening is performed. *Each hybridoma clone represents a single, immortalized B cell that will produce a single (monoclonal) antibody.* A number of different selective criteria may be employed during the screening and cloning process. For instance, you may wish to clone a hybridoma that produces an antibody with a defined affinity ($K_D = 10^{-8}$, 10^{-6}, etc.). You may wish to raise an antibody that recognizes a particular epitope on the antigen of interest. You may wish to raise an antibody that binds the antigen with a certain type of prevailing attractive force (e.g., electrostatic as opposed to hydrophobic). If the screening process is designed and performed properly, it is possible to isolate a hybridoma that produces an antibody that fits any selective criterion.

Purification of Antibodies

A hybridoma cell line will secrete a monoclonal antibody into the culture medium in relatively pure form. Because of this, monoclonal antibodies require less purification than polyclonal antibodies. A serum sample containing polyclonal antibodies contains both antigen-specific and nonspecific immunoglobulins, as well as multiple other serum proteins. For example, a large percentage of the serum (80–90%) is comprised of albumin. Unlike the serum immunoglobulins, serum albumin is soluble in a solution at 50% ammonium sulfate saturation. Thus, if a serum sample is slowly brought to 50% saturation with ammonium sulfate, the immunoglobulins will selectively precipitate out of solution, leaving the albumin in the supernatant. This very simple procedure, completed in less than an hour, can increase the purity of a polyclonal antibody solution by an order of magnitude (see Experiment 17).

If you wish to separate the antibodies in a polyclonal antibody solution that recognize the antigen of interest from those that do not, the specificity of the antigen–antibody interaction must be exploited. Because all IgGs possess the same general structure, differing only in amino acid composition within the variable region, you cannot effectively isolate specific IgGs within the population by differences in size, shape, and surface charge on which general techniques in protein purification are based (see introduction to Experiment 2). The one and perhaps only property that will distinguish one IgG from another is the specificity of the antigen with which it is able to interact.

If you are able to chemically conjugate the antigen of interest to an inert support or matrix, the antigen will "capture" antibodies in the polyclonal solution that recognize it as it passes through the matrix (Fig. IV-6). After extensive washing, a population of immunoglobulins that recognize epitopes on the antigen will remain bound to the antigen on the matrix. To isolate the antigen-specific antibodies, you have to experiment with different mobile-phase compositions (alter pH, ionic strength, de-

Figure IV-6 Immunoaffinity chromatography to purify antigen-specific antibodies.

tergent concentration, etc.) that will weaken the antigen–antibody interaction enough to allow the antibodies to elute from the column. Ideally, the mobile-phase composition change will disrupt the antigen–antibody interaction without denaturing the antibody. Typical elution procedures involve low pH buffers (0.1 M glycine, pH ~ 2.0) or 5 M $MgCl_2$. Fortunately, IgG molecules are quite stable, and can often withstand the relatively harsh conditions that are often required to elute them from affinity columns. Although this type of immunoaffinity chromatography may be difficult to optimize, it has been used successfully in the purification of antigen-specific antibodies from polyclonal antibody solutions. Many of the secondary antibody reagents that are commercially available (e.g., used in Experiments 17 and 18) have been purified in this manner.

Use of Antibodies in Biochemical Studies

Western Blotting

Western blotting is one of the most widely used immunochemical techniques in modern biochemistry. In general, it is used to detect the presence of a particular protein antigen in an impure mixture. Western blots can provide information that is (1) qualitative: Is the protein present? (2) quantitative: How much of the protein is present? and (3) structural: What is the size of the protein? The general protocol for Western blotting is as follows (see Fig. 18-1):

1. A mixture of proteins is separated by SDS–PAGE.
2. The proteins are electrophoretically transferred from the gel to a membrane and incubated with a high concentration of nonspecific protein to "block" or "mask" the areas of the membrane not already bound with protein.
3. A *primary* antibody specific for the antigen of interest is applied to the membrane to allow the antigen–antibody interaction to take place.
4. The membrane is washed extensively to remove unbound or nonspecifically bound primary antibody.
5. An enzyme-conjugated (alkaline phosphatase or horseradish peroxidase) *secondary* antibody, which has a strong affinity for the primary antibody, is incubated with the membrane.

6. The unbound and nonspecifically bound secondary antibody is removed with extensive washing.
7. A substrate for the enzyme-conjugated secondary antibody is applied to the membrane. The product of this reaction, a colored *precipitate* or light, is detected on the membrane to reveal the location and relative amount of the antigen of interest.

The technique of Western blotting is demonstrated and described more thoroughly in Experiment 18.

Enzyme-Linked Immunosorbent Assay (ELISA)

Like Western blotting, ELISA is a very popular tool used in biochemistry. This technique relies on the ability of a particular antibody to recognize or bind a specific antigen that is adsorbed to the wells of a microtiter plate. Like the Western blot, an enzyme-conjugated secondary antibody that has affinity for the primary antibody is utilized to detect the presence and relative quantity of primary antibody from one well to the next (see Fig. 17-2). When the enzyme is supplied with the appropriate substrate, it will convert it to a *soluble* colored product, the relative concentration of which can be determined easily from well to well with the use of a specialized, 96-well plate spectrophotometer. By monitoring the change in absorbance of a collection of wells containing different dilutions of the antibody, the *titer* of an antibody solution (the inverse dilution of the antibody required to give a half-maximum signal) can be determined.

The main advantage of the ELISA over Western blotting is its ability to analyze many samples at one time. In addition, modifications to the general ELISA protocol can be used to quantify the exact concentration of antigen in an unknown solution, the concentration of a specific antibody in an unknown solution, the dissociation constant for a particular antigen–antibody interaction, or the relative affinity of two different antibodies for the same antigen. An example of an indirect ELISA, and its use in the determination of the concentration of an antigen in an unknown solution, is the subject of Experiment 17.

Radioimmunoassay (RIA)

In many respects, the *radioimmunoassay* is similar to the ELISA. The two techniques differ only in their method of detection of antibody when bound to the antigen. While the traditional ELISA detects the binding of the antibody by monitoring the appearance of a soluble colored product after the enzyme-conjugated antibody acts on the appropriate substrate, the radioimmunoassay detects the antigen–antibody complex with a radioactive atom covalently attached to either of the two components. For instance, if the antibody is labeled with ^{125}I and added to a microtiter plate well coated with the antigen, the amount of radioactivity present in the well after washing will be directly proportional to the amount of antibody that bound the antigen. This assay can be altered by coating the well with the antibody of interest and incubating it with a ^{125}I-labeled antigen. The amount of antigen "captured" by the antibody, then, is directly proportional to the amount of antigen in the unknown solution (Fig. IV-7).

In addition to using radiolabeled antibodies in the ELISA format, you can also use them in the

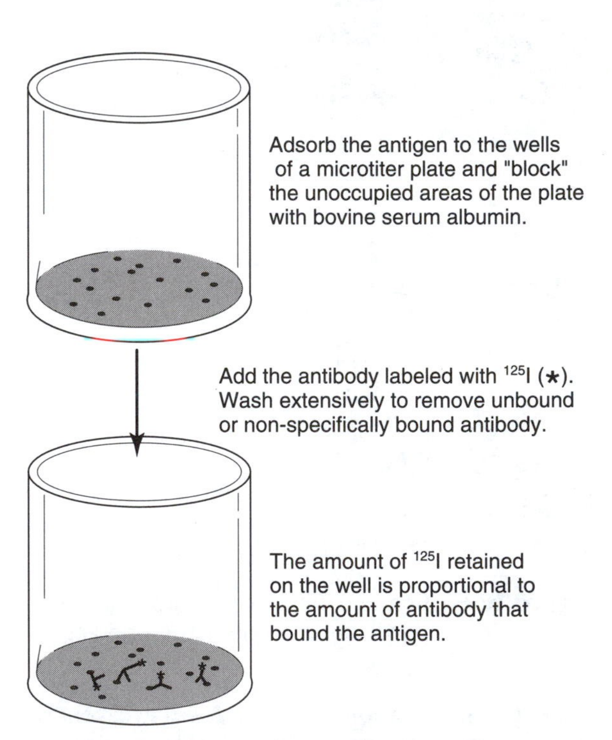

Figure IV-7 The radioimmunoassay (RIA).

Western blot format. In general, radiolabeled antibodies alleviate the need for enzyme-conjugated secondary antibodies in both the ELISA and the Western blot. The major disadvantage to using ^{125}I-labeled antibodies is the hazard associated with this isotope (^{125}I is volatile in its unconjugated form and emits very high-energy γ rays).

Immunoprecipitation

As discussed earlier, a single IgG antibody has two identical antigen-binding sites or "arms." In the case of polyclonal antibodies, the antigen-binding sites and the epitopes on the antigen that they recognize differ among members of the population. If the antigen and antibody are mixed together in correct proportions, the different polyclonal antibodies can bind to different regions on a number of different antigen molecules to create a large complex (Fig. IV-8*a*). These large protein aggregates are often insoluble, precipitating out of solution. If a fixed amount of either the antigen or antibody is incubated with increasing concentrations of the other component, you will be able to determine an optimal antigen:antibody ratio for use in future immunoprecipitation experiments.

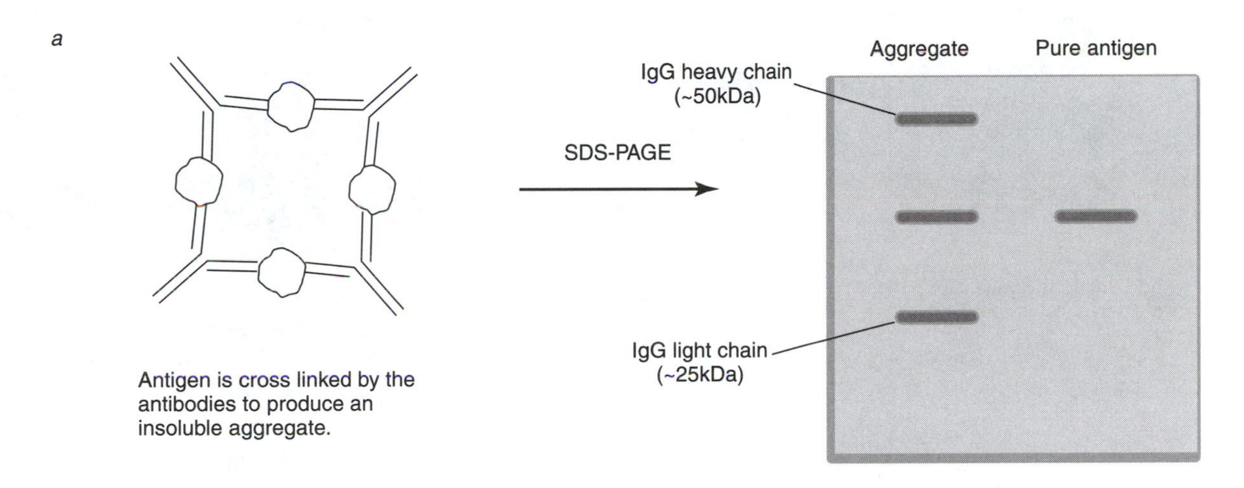

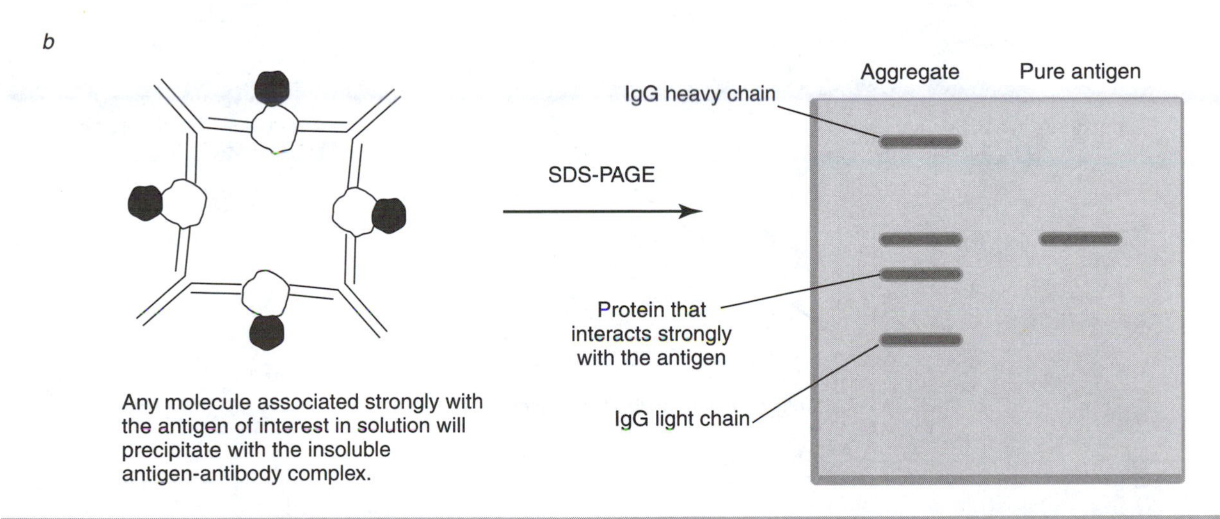

Figure IV-8 Use of immunoprecipitation to purify antigens and antigen associated proteins.

Although immunoprecipitation is not used as widely in modern biochemistry, it has been used in the past to study protein–protein interactions. If an antigen of interest has a strong affinity for another molecule present in the solution, the associated molecule will precipitate in the complex along with the antigen that is recognized directly by the antibody (Fig. IV-8*b*). This technique is often used to identify new members of physiologically important multiprotein complexes when one component of the complex has been purified, but others are unknown.

If the antibody is unable to cross-link or aggregate the antigen of interest to the extent needed to induce precipitation of the complex (as is often the case with monoclonal antibodies), an alternative method may be employed. *Protein A* from *Staphylococcus aureus* is known to have a high affinity for immunoglobulins. If a soluble antigen–antibody complex is incubated with an insoluble matrix covalently conjugated with Protein A, the complex will be captured by the Protein A covalently attached to the matrix (Fig. IV-9). Because Protein A

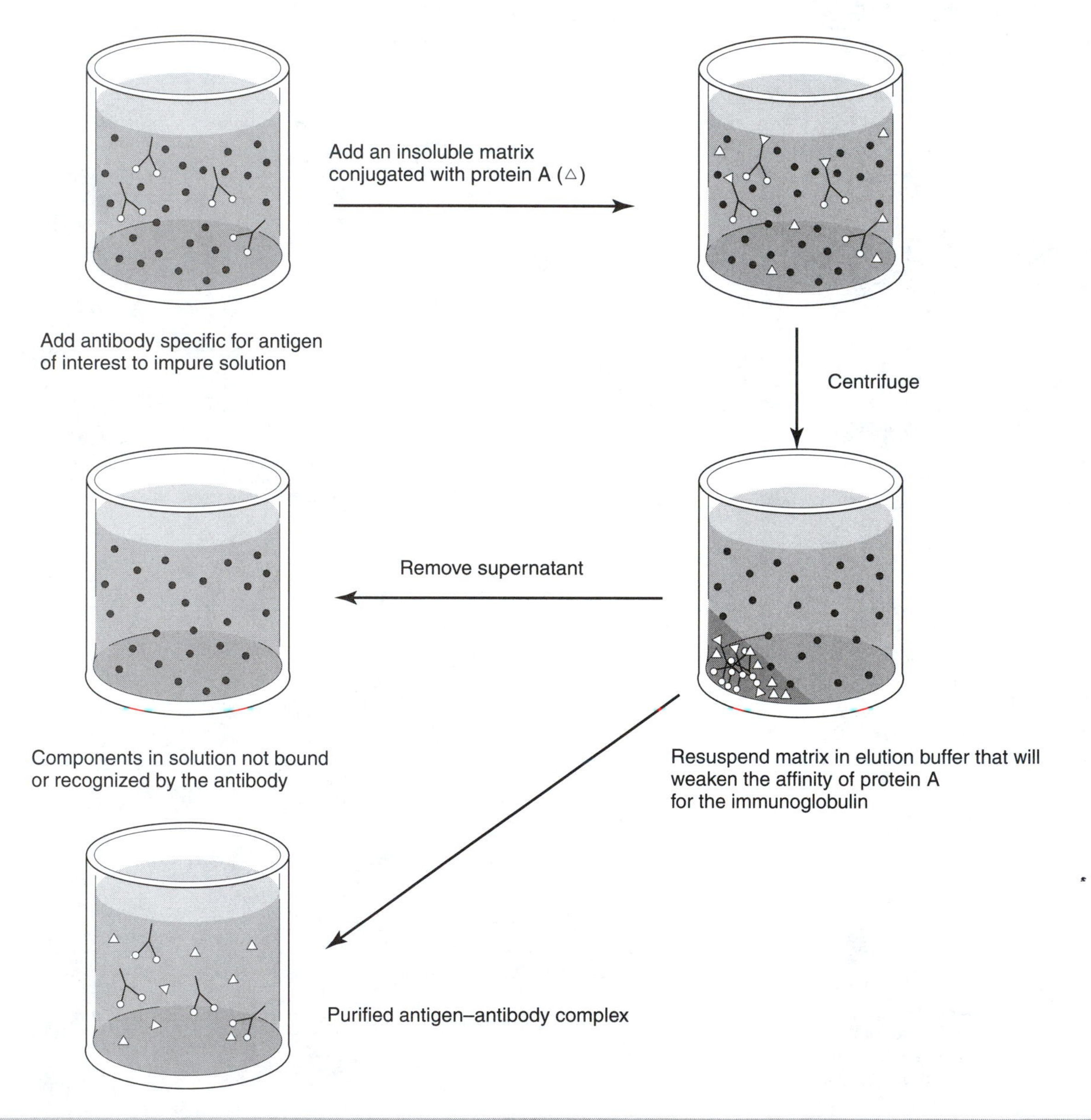

Figure IV-9 Using Protein A to purify antigen–antibody complexes.

has affinity for most immunoglobulins (not just those that recognize the antigen of interest), it is of little use in purifying antigen-specific antibodies from a polyclonal antibody solution. Immobilized Protein A is used, however, to purify immunoglobulins from serum samples.

Immunofluorescence Microscopy and Immunohistochemistry

These two techniques are used to identify the *location* of a particular antigen within a cell or tissue. In addition to covalent coupling of antibodies to enzymes for Western blots and ELISAs, antibodies can also be conjugated to *fluorescent dyes*. After a tissue section or cell has been isolated, ruptured, and fixed to a coverslip, it can be labeled with a fluorescently "tagged" antibody that is specific for a particular antigen believed to reside in it. As the sample is subjected to incident light, the antibodies bound to the antigen of interest will emit light in the visible spectrum and be detectable with a fluorescence microscope. Because there are a number of different dyes available that excite at different wavelengths and emit light of different colors within the visible range of wavelengths, a number of different fluorescent antibodies can be applied to a single cell or tissue section simultaneously. The results of this type of experiment can provide insight into the cellular location of numerous proteins and small molecules, providing the investigator with an understanding of the in vivo environment of the molecule under study.

The technique of *immunohistochemistry* is very similar to fluorescence microscopy. This technique differs only in the method of detection or localization of the antibody and can be performed with a conventional light microscope. As with the ELISA and Western blot, the antibody used in this experiment is covalently conjugated to an enzyme, such as horseradish peroxidase. This enzyme is then incubated with a substrate that is converted to an *insoluble* colored product that will precipitate or deposit at the site of enzyme activity. The distribution and location of the colored product is readily detected with an ordinary light microscope.

Another related technique utilizes antibodies that are conjugated to large, electron-dense atoms, such as gold. When used in conjunction with electron microscopy, these probes can be used to determine the subcellular localization of an antigen of interest with high resolution.

Flow Cytometry

In addition to the technique of fluorescence microscopy, fluorescently labeled antibodies are currently used in *flow cytometry*. This technique allows the identification and isolation of individual cells in a complex mixture *according to the antigens they express on their surface*. For example, among a diverse population of cells, some may express an antigen, X, on their surface. After incubating the cells with a fluorescently labeled anti-X antibody, the cells are passed in a narrow tubing through a laser beam. As the laser hits the surface of a single cell, it will excite the fluorescently labeled antibody bound to antigen X on the surface of the cells that express it. Each time a fluorescently "tagged" cell passes through the laser, a sensitive photomultiplier tube (PMT) will detect and count that event (Fig. IV-10). More than 100,000 cells can be passed through the laser beam in less than a minute. Cells that do not express antigen X on their surface will not fluoresce (although they will scatter the incident light to some extent) and will not be detected by the PMT. This approach can determine the exact *number* or *percentage* of cells in a diverse population that express antigen X on their surface. In addition, the PMT can distinguish between differences in fluorescent *intensity* from one cell to another, providing insight into variations in the level of expression of antigen X from cell to cell. For example, if cell 1 is five times greater in fluorescent intensity than cell 2 as it passes through the laser, then the concentration of antigen X on the surface of cell 1 is five times greater than that on cell 2.

A flow cytometer may be equipped with more than one PMT, each capable of detecting fluorescence of a particular wavelength. Thus, a cell population can be incubated with more than one antibody, provided that the fluorescent dyes conjugated to them emit light of different wavelengths. This "double label" experiment will allow you to distinguish cells expressing both antigens from those expressing either of the two antigens individually.

A *fluorescence-activated cell sorter* (FACS) is a flow cytometer capable of isolating individual cells within a diverse population that are expressing a

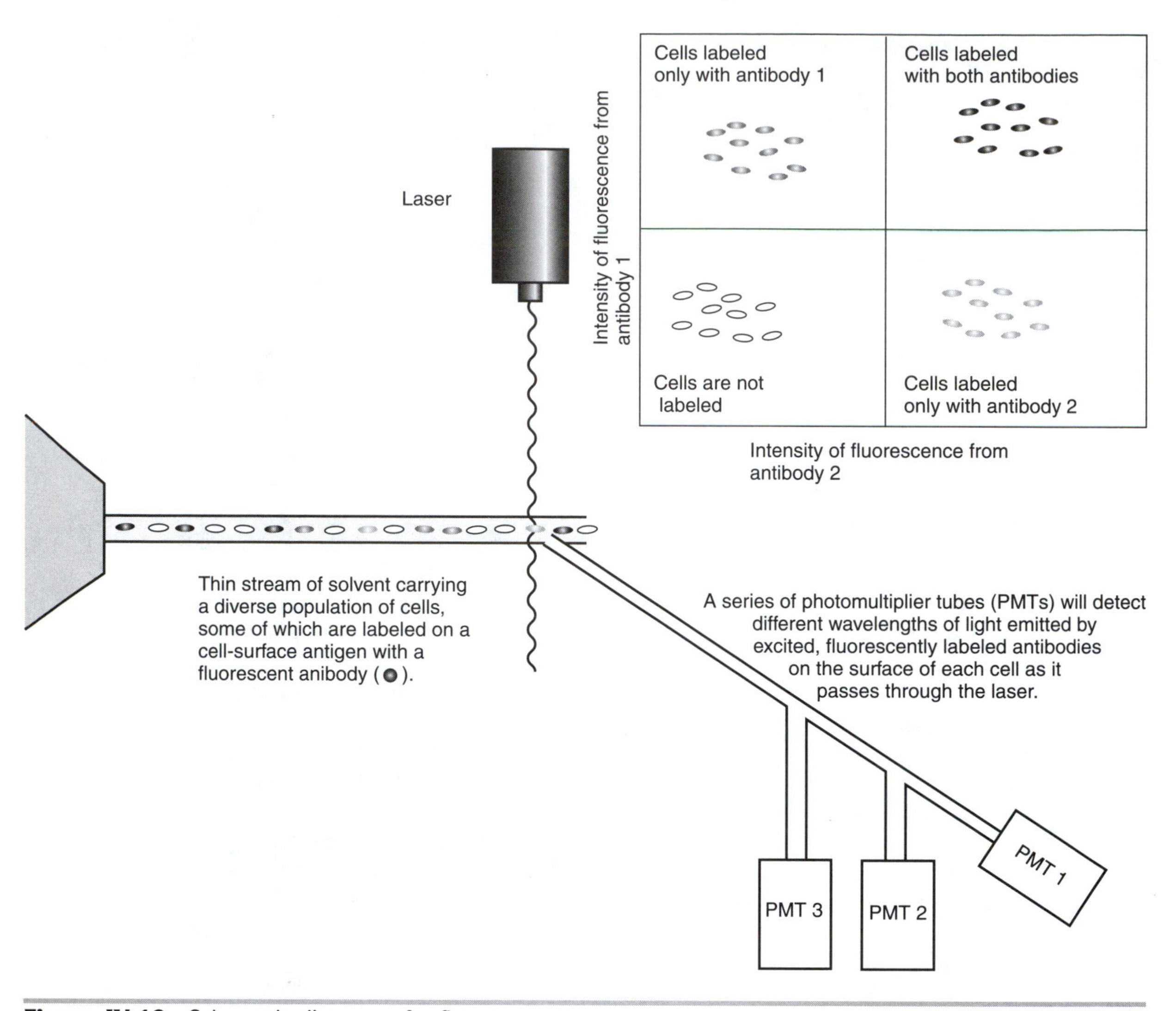

Figure IV-10 Schematic diagram of a flow cytometer.

surface antigen of interest. The detection mechanism of the FACS is identical to that of the conventional flow cytometer. The difference lies in the computer-generated feedback system of the FACS that is absent in the flow cytometer. As the PMT of the FACS identifies a fluorescent cell, the instrument divides the population into a fine mist that carries a single cell per droplet. Each *fluorescent* droplet is *charged* by the PMT and passed through an electrode that carries a charge *opposite* that of the droplet. Any cell that expresses antigen X (fluorescently labeled) can thereby be separated from the remaining cell population as the mist passes across the charged plate (Fig. IV-11). Cell populations isolated by this technique can be propagated in culture and utilized in future studies.

Advantages and Disadvantages of Polyclonal and Monoclonal Antibodies

Both polyclonal and monoclonal antibodies have proven to be very effective in a number of different biological applications, but there are distinct advantages and disadvantages to employing each type of antibody for a particular study. First, polyclonal antibodies are generally more stable than mono-

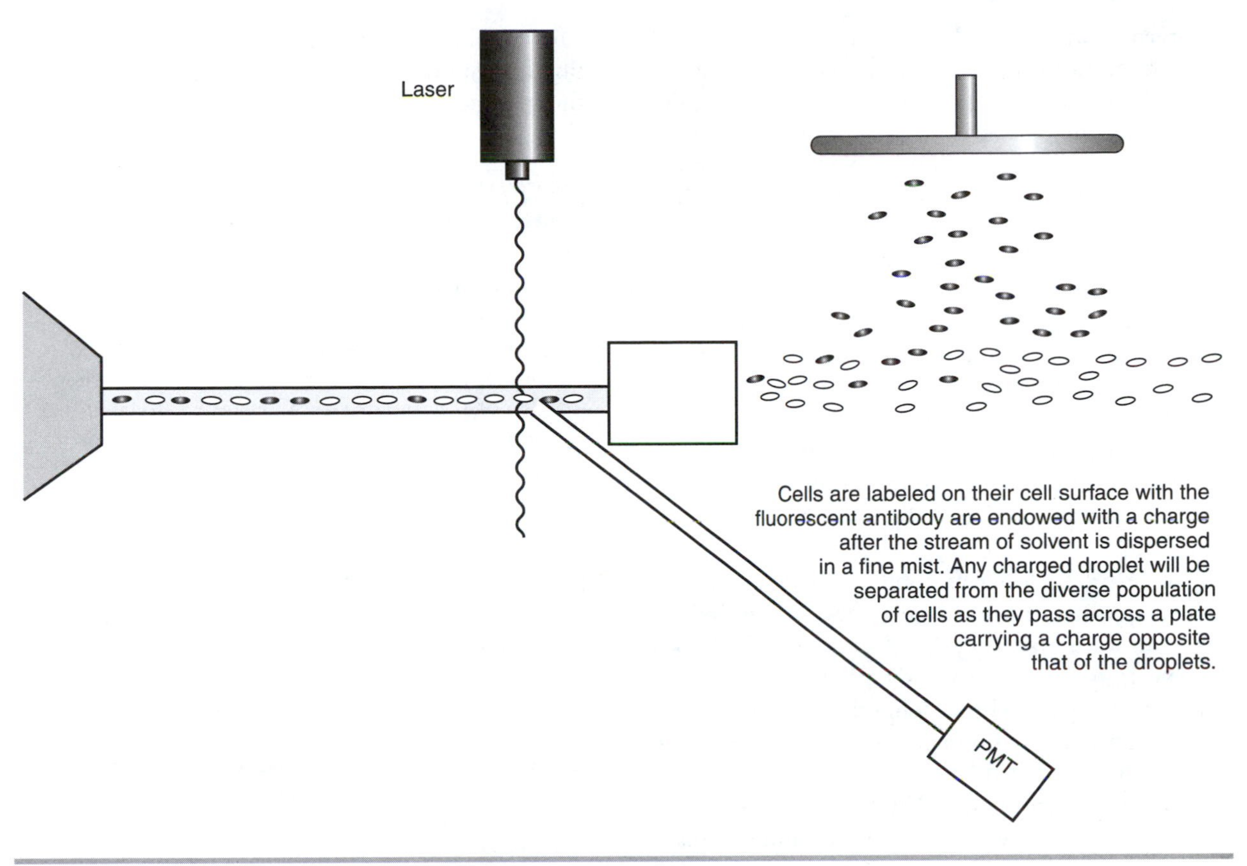

Figure IV-11 The fluorescence-activated cell sorter (FACS).

clonal antibodies. While monoclonal antibodies may begin to denature after several weeks of storage at −80°C, polyclonal antibody serum can often be stored at 4°C for years before any significant loss of titer is observed. While polyclonal antibodies are more stable, they differ from monoclonal antibodies in that their supply is limited. Once the host animal dies, the source of a particular preparation of polyclonal antibody is gone. While the immunization process can be repeated on another host animal, a different polyclonal antibody will be produced that requires its own quality control and testing. In contrast, monoclonal antibodies are produced from a single immortalized B cell. The cell line can be stored indefinitely (as a frozen culture stock) and grown whenever more of the antibody is needed.

Polyclonal antibody production requires the use of relatively large quantities of *pure* antigen for immunization. Remember that the goal of polyclonal antibody production is to increase the relative concentration of antigen-specific serum IgG. If an impure antigen is used for immunization, the relative concentration of serum antibodies that recognize the impurities in the antigen mixture will be increased along with the relative concentration of antigen-specific antibodies. In contrast, monoclonal antibodies can often be raised with the injection of lower concentrations of impure antigen. Since the thousands of hybridomas produced will be screened for antigen-specific antibody production, the B cells that differentiated to produce antibodies that recognize antigens other than the one of interest will eventually be eliminated during the cloning process. In effect, you must find the few hybridomas that are producing the antibodies of interest among the thousands that are screened.

Since polyclonal antibodies recognize a variety of epitopes on the antigen, they generally work well in the technique of immunoprecipitation (see below). Many polyclonal antibody preparations have also proven to be effective in the techniques of

Western blotting and ELISA. The main concern with the use of polyclonal antibodies for these latter techniques is the potential for cross-reactivity of some of the antibodies with proteins other than the antigen of interest. The problem of cross-reactivity is easier to control in Western blotting than in ELISAs, since a Western blot (see Experiment 18) will identify the molecular weight of any cross-reacting protein and allow you to evaluate which of the signals may be due to nonspecific interactions. Because an ELISA analyzes the extent of interaction between all of the antibodies in a solution with all of the potential antigens present in the microtiter plate (see Experiment 17), it is more difficult to determine which of the potentially numerous interactions are specific for the antigen of interest.

Since monoclonal antibodies recognize a particular epitope on the antigen, they tend to be less cross-reactive than polyclonal antibodies in Western blotting and ELISAs. Also, because the interaction takes place with a defined affinity (dissociation constant), monoclonal antibodies are frequently preferred in the technique of immunoaffinity chromatography (see Introduction to Experiment 2). The wide variety of antigen–antibody interactions that are present in a polyclonal antibody preparation makes it difficult to define a mobile-phase buffer that will promote the elution of the antigen of interest in good yield from an immunoaffinity column prepared with polyclonal antibodies.

Regardless of the type of antibody you decide to produce, it is likely that you will be able to develop effective applications for their use in the study of a particular biochemical process or pathway. A survey of scientific literature indicates that at least 25% of all published research articles in the discipline of biochemistry utilize antibodies and immunochemistry in one form or another. In addition to providing scientists with a powerful tool in the study of biochemistry, natural and recombinant antibody technology promises to be of great use in the treatment and prevention of numerous diseases.

REFERENCES

Alzari, P. M., Lanscombe, M., and Poljak, R. J. (1988). Three-Dimensional Structure of Antibodies. *Annu Rev Immunol* **6**:555.

Burkhardt, A. L., Brunswick, M., Bolen, J. B., and Mond, J. J. (1991). Anti-immunoglobulin Stimulation of B Lymphocytes Activates src-Related Protein-Tyrosine Kinases. *Proc Natl Acad Sci USA* **88**:7410.

Chapus, R. M., and Koshland, M. E. (1974). Mechanisms of IgM Polymerization. *Proc Natl Acad Sci USA* **71**:657.

Harlow, E., and Lane, D. (1988). *Antibodies, A Laboratory Manual.* Cold Spring Harbor, NY: Cold Spring Harbor Laboratory Press.

Janeway, C. A., Rosen, F. S., Merler, E., and Alper, C. A. (1969). *The Gamma Globulins*, 2nd ed. Boston: Little, Brown.

Janeway, C. A., Jr., and Travers, P. (1996). *Immunobiology: The Immune System in Health and Disease*, 2nd ed. New York: Garland.

Lehninger, A. L., Nelson, D. L., and Cox, M. M. (1993). An Introduction to Proteins. In: *Principles of Biochemistry*, 2nd ed. New York: Worth.

Natvig, J. B., and Kunkel, H. G. (1973). Human Immunoglobulin: Classes, Subclasses, Genetic Variants, and Idiotypes. *Adv Immunol* **16**:1.

Nemazee, D., and Burki, K. (1989). Clonal Deletion of B Lymphocytes in a Transgenic Mouse Bearing Anti-MHC Class I Antibody Genes. *Nature* **337**:562.

Paul, W. E. (1993). *Fundamental Immunology*, 3rd ed. New York: Raven Press.

Stryer, L. (1995). Exploring Proteins. In: *Biochemistry*, 4th ed. New York: Freeman.

Tedder, T. F., Zhou, L. J., and Engel, P. (1994). The CD19/CD21 Signal Transduction Complex of B Lymphocytes. *Immunol Today* **15**:437.

Partial Purification of a Polyclonal Antibody, Determination of Titer, and Quantitation of an Antigen Using the ELISA

Theory

As discussed in the introduction to Section IV, the *e*nzyme-*l*inked *i*mmuno*s*orbent *a*ssay (ELISA) has become a widely used and powerful technique in the field of medical diagnostics. If an individual has been exposed to a pathogenic organism (virus, bacteria), antibodies that recognize particular antigens specific to these organisms will be present in the blood plasma. Because of this, a blood test can often reveal the source of an infection and dictate a proper course of treatment.

In addition to being used as a method of identifying and quantifying protein antigens in a sample, the ELISA is also gaining widespread popularity as a tool for quantifying low-molecular-weight compounds (haptens) such as drugs, chemical toxins, and steroid hormones. Although many of these haptens are not themselves immunogenic (able to elicit an immune response), antibodies specific for them can often be raised by coupling them to a foreign protein prior to being injected into the host animal. The following is an example of how ELISA technology is currently being used: the human chorionic gonadotropic hormone is produced by the placenta several days after the onset of pregnancy. A modified ELISA using an antibody against this hormone is used to determine whether an individual is pregnant based on the quantity of this hormone in the urine. The assay is a sensitive, reliable, an immediate method to verify pregnancy in its very early stages.

In this experiment, you will purify *polyclonal* antibodies against β-galactosidase from the serum of rabbits immunized with this antigen. Using the ELISA, you will determine the titer of this antibody, as well as that of a *monoclonal* antibody against β-galactosidase, called J1, that is derived from a mouse hybridoma cell line. Based on the titer of the J1 antibody, you will perform a *competitive* ELISA to determine the concentration of β-galactosidase in an unknown solution.

In many respects, the ELISA performed in this experiment is similar to the Western blot described in Experiment 18. The antigen of interest (β-galactosidase) is first adsorbed to the bottom of a 96-well polyvinyl chloride (PVC) or polystyrene plate. Next, the sites on the bottom of each well not already bound with protein are "blocked" with a buffered solution containing a high concentration of bovine serum albumin (BSA). This blocking step is done to ensure that the primary and secondary antibodies added later in the assay will bind only to the antigen of interest, rather than to the bottom of the well. As with the Western blot, if the blocking step is not performed, the background would be too high to obtain any useful information (see Experiment 18). Although you will use BSA as the blocking protein in this assay, any protein that does not have affinity for the antibodies could be used.

At this point, the anti-β-galactosidase antibody (primary antibody) diluted in blocking buffer is added to the well. During the incubation, it will bind to the β-galactosidase adsorbed to the bottom of the well. Several washes are then performed to remove primary antibody that may have bound nonspecifically to proteins on the bottom of the well other than β-galactosidase. As in the Western blot experiment, the wash solution contains a small amount of the non-ionic detergent, Tween 20, and NaCl to disrupt any weak ionic or hydrophobic interactions that may be involved in these nonspecific interactions. A second antibody specific for the anti-β-galactosidase antibody is then added (secondary antibody) to allow it to bind the primary antibody. As with the Western blot experiment, the secondary antibody is covalently conjugated to the enzyme horseradish peroxidase (HRP). This enzyme will eventually take part in a color-producing reaction that will allow you to determine the titer of the primary antibody. Remember that you will be using two different anti-β-galactosidase primary antibodies in this experiment. For assays performed using *rabbit* polyclonal antibodies, HRP-goat-anti*rabbit* IgG must be used as the secondary antibody. For assays performed using the *mouse* monoclonal antibody as the primary antibody, HRP-goat-anti*mouse* IgG must be used as the secondary antibody.

After a series of wash steps to remove any secondary antibody bound nonspecifically to proteins on the bottom of the well other than the primary antibody, a mixture of 3,3′,5,5′-tetramethylbenzidene (TMB) and hydrogen peroxide is added. The HRP enzyme conjugated to the secondary antibody will use these substrates to produce a soluble, blue-colored product. The reaction is stopped with the addition of 4 N H_2SO_4, which will also convert the blue-colored product to a form that absorbs light at 450 nm (yellow in color, Fig. 17-1). The intensity of the yellow color in each well (A_{450}) will be proportional to the amount of primary antibody that has bound to the antigen at the bottom of the well. The ELISA procedure just described is depicted in Figure 17-2.

Once the titer of the mouse-anti-β-galactosidase monoclonal antibody has been determined, a competitive ELISA can be performed to determine the concentration of the antigen in an *unknown* solution. The only variation in the competitive ELISA from that of the standard ELISA is that the primary antibody incubation will be performed in the presence of *soluble* β-galactosidase (antigen that is not adsorbed to the bottom of the well). The theory underlying the competitive ELISA is shown in Figure 17-3. If the primary antibody-binding step is performed in the presence of soluble antigen, the soluble antigen will *compete* with the plate-bound antigen for antibody binding. The more soluble antigen that is present during the primary antibody incubation, the less antibody will bind to the antigen on the plate. The lower the concentration of antibody

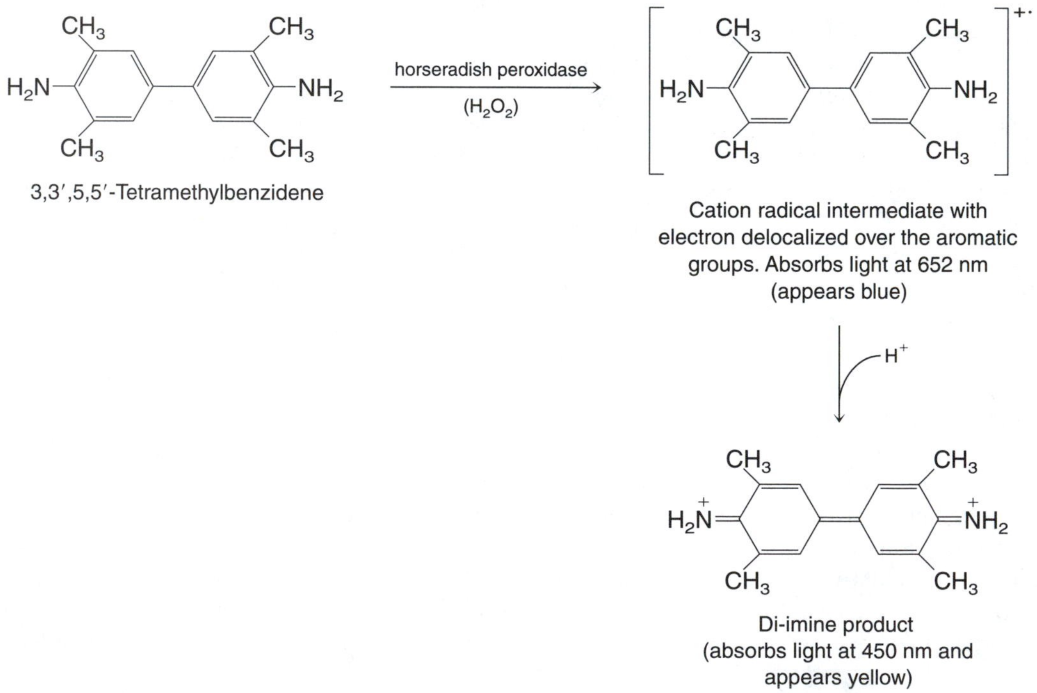

3,3′,5,5′-Tetramethylbenzidene

horseradish peroxidase (H_2O_2)

Cation radical intermediate with electron delocalized over the aromatic groups. Absorbs light at 652 nm (appears blue)

H^+

Di-imine product (absorbs light at 450 nm and appears yellow)

Figure 17-1 Oxidation of 3,3′,5,5′-tetramethylbenzidene by horseradish peroxidase.

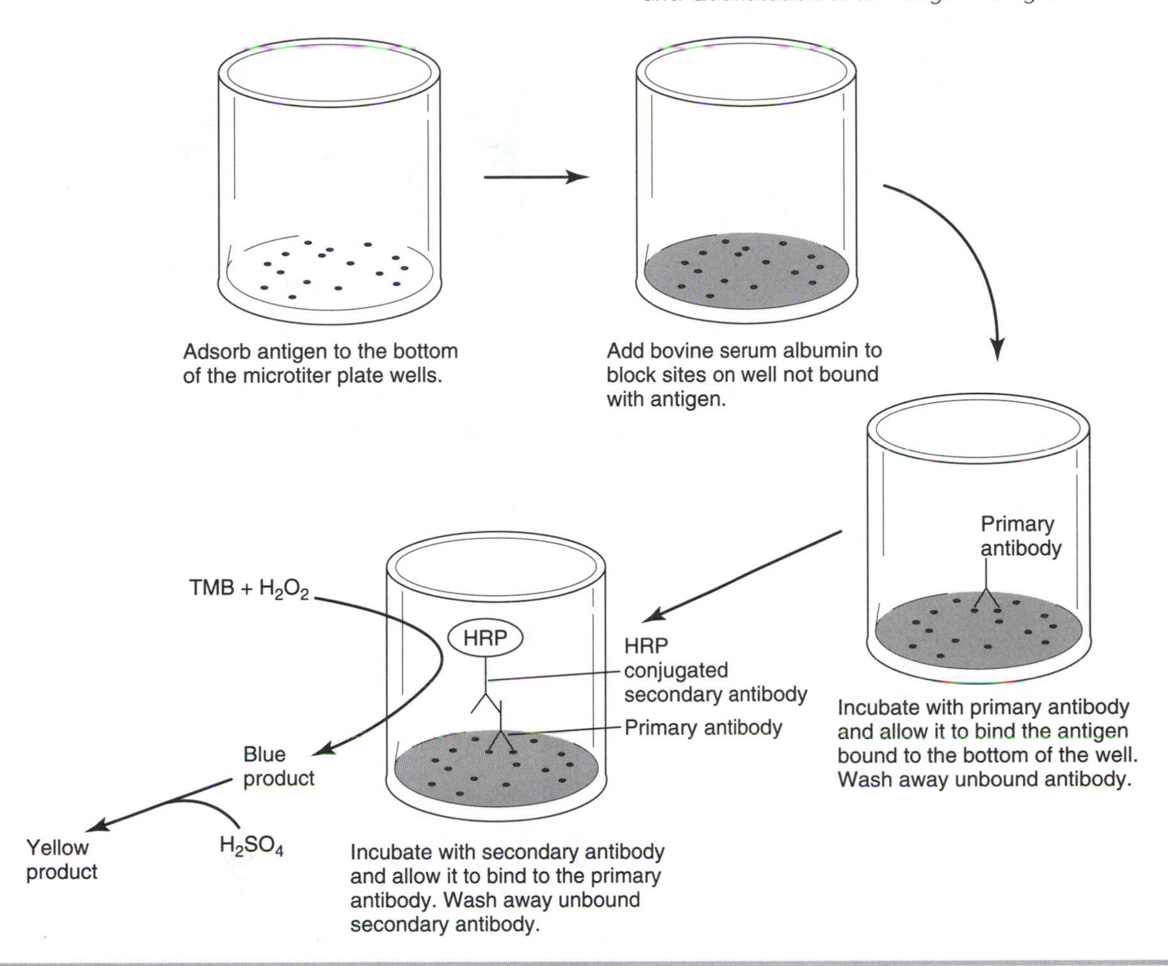

Figure 17-2 Principle of two-antibody (indirect) ELISA.

bound to the plate, the lower the A_{450} value that each well will display at the end of the assay. As the concentration of soluble antigen decreases, more antibody will be free to bind the antigen on the plate, increasing the A_{450} value for each well.

It is important that the competitive ELISA experiment be performed at a primary antibody concentration (dilution) that is *most* sensitive to changes in the concentration of soluble β-galactosidase. This dilution can be determined from a plot of A_{450} versus the log of the inverse dilution of the primary antibody as described in the experimental protocol (inverse dilution is the same as the degree to which the sample was diluted; for example, if the sample was diluted 1:1000, the inverse dilution is 1000). If the primary antibody concentration is too high, there will be enough antibody to bind both the soluble and plate-bound antigen, and changes in the concentration of soluble β-galactosidase may go undetected. If the antibody concentration is too low, addition of *any* soluble β-galactosidase will prevent binding of the antibody to antigen on the plate, and the A_{450} values will never increase above the background value. As with Western blotting, the ideal dilutions of the primary and secondary antibodies must often be determined empirically (optimized) for each experimental situation.

Horseradish peroxidase is one of several enzymes that are commonly used for the detection of antigens in the ELISA. Like HRP-conjugated antibodies, secondary antibodies conjugated with these other enzymes are commercially available. A list of ELISA detection enzymes, the substrates that they use, and the wavelengths of detection for the soluble colored products that they produce is presented in Table 17-1.

Still another variation of the ELISA exploits the strong interaction between biotin and streptavidin,

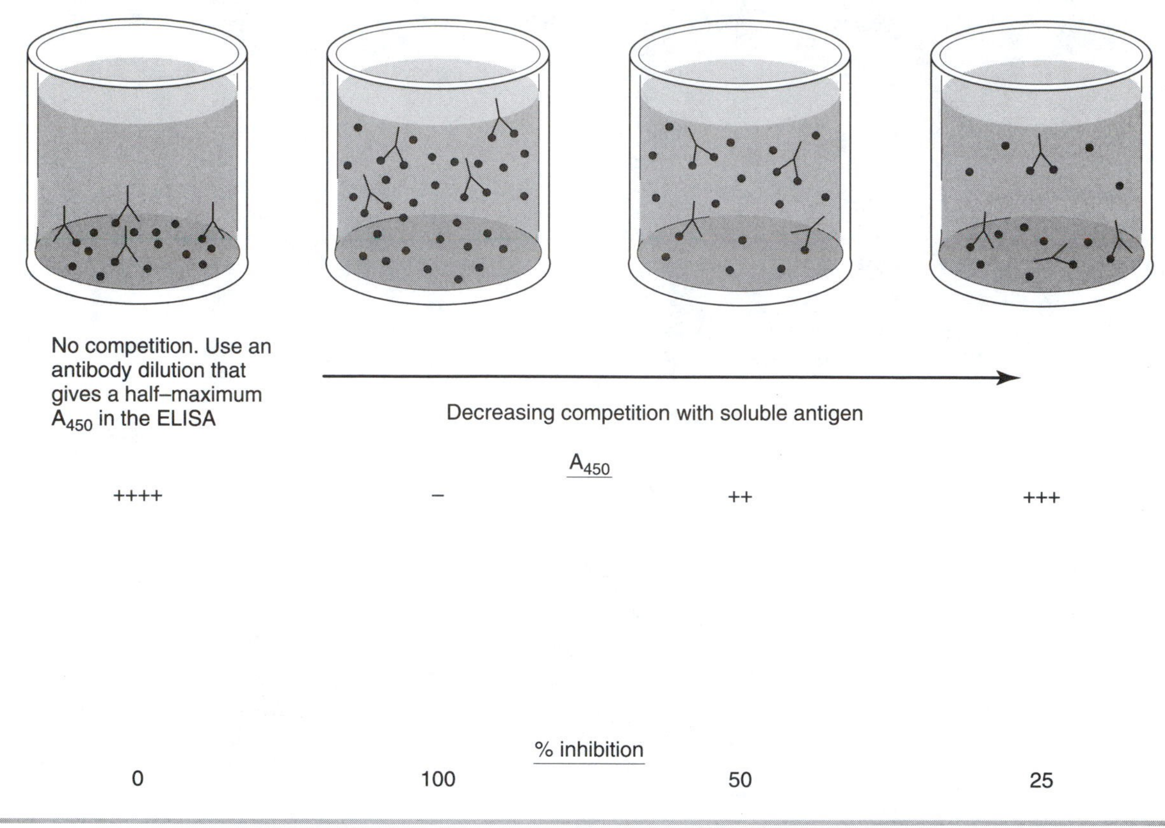

Figure 17-3 The competitive ELISA. If *soluble* antigen is included during the primary antibody incubation, the soluble antigen will compete for antibody binding with the plate-bound antigen.

one of the strongest noncovalent interactions found in nature ($K_D = 10^{-15}$ M). Many commercially available reagents allow investigators to cross-link (conjugate) biotin to antibodies. If a biotinylated antibody is used as the primary antibody in an ELISA, it is detected with the use of streptavidin that has been conjugated to one of the enzymes listed in Table 17-1. For more information on biotin, strep-

Table 17-1 Soluble Colorimetric Enzyme/Substrate Systems for Detection of Antigens in the ELISA

Enzyme	Substrate	Wavelength for Detection (nm)
Alkaline phosphatase	*p*-Nitrophenyl phosphate (PNPP)	405
Horseradish peroxidase	2,2′-Azino[3-ethylbenzo-thiazoline-6-sulfonic acid]-diammonium salt (ABTS)	410 or 650
	o-Phenylenediamine (OPD)	490
	3,3′,5,5′-Tetramethyl-benzidene (TMB)	450
β-Galactosidase	*o*-Nitrophenyl-*β*,D-galactopyranoside (ONPG)	405

tavidin, and their use in immunochemical assays, see Figure 18-4 and the introduction to Experiment 18.

Supplies and Reagents

P-20, P-200, P-1000 Pipetmen with disposable tips
Pasteur pipettes
Parafilm
Saturated ammonium sulfate solution
1.5-ml microcentrifuge tubes
10× Phosphate-buffered saline (10X PBS, pH 7.3): *This solution should be diluted 1:10 with distilled water to produce the 1X PBS (working solution) described in the protocol.*

2.56 g/liter $NaH_2PO_4 \cdot H_2O$
22.48 g/liter $Na_2HPO_4 \cdot 7H_2O$
87.6 g/liter NaCl

Rabbit serum (normal or pre-immune rabbit serum and anti-β-galactosidase serum)*
Spectrophotometer to read absorbances at 280 nm
Cuvettes for spectrophotometer
Polyoxyethylenesorbitan monolaurate (Tween 20, Sigma Catalog #P-5927)
β-Galactosidase solutions (40 μg/ml and 10 μg/ml) in 1X PBS (Sigma product #G-6008)
Blocking solution (1X PBS containing 1.0% wt/vol bovine serum albumin (BSA) and 0.05% vol/vol polyoxyethylenesorbitan monolaurate (Tween 20))
Wash solution (1X PBS containing 0.05% vol/vol Tween 20)
Mouse-anti-β-galactosidase monoclonal antibody (J1) (Sigma Product #G-8021)*
Secondary antibodies:

Goat-antirabbit IgG (HRP conjugated) (Kirkegaard & Perry product #074–1506)
Goat-antimouse IgG (HRP conjugated) (Kirkegaard & Perry product #074–1802)

Horseradish peroxidase (HRP) substrate (3,3′5,5′-tetramethylbenzidene and H_2O_2) (Kirkegaard & Perry product #50–7600)

*The dilutions of the anti-β-galactosidase antibodies described in this experiment are approximate. The titers of particular antibody samples (even if they are from commercial sources) may vary significantly. Because of this, we strongly recommend that the instructor test different dilutions of the primary antibodies that are to be used in the experiment beforehand (using this same general protocol) to optimize the results. We find that there is less variability with the secondary antibody dilutions than with the primary antibody dilutions from semester to semester.

Immulon-2 96 well flat-bottomed microtiter plate (Fisher product #1424561)
96-Well microplate spectrophotometer
β-galactosidase solution (unknown concentration) in 1X PBS
4 N H_2SO_4

Protocol

Day 1: Partial Purification of Antibodies (IgG) from Normal Rabbit Serum and Anti-β-Galactosidase Rabbit Serum

1. Obtain a 0.2-ml sample of normal or pre-immune rabbit serum (NS) and anti-β-galactosidase rabbit serum (immune serum, IS) from the instructor.
2. Transfer 0.1 ml of each sample to two separate 1.5-ml microcentrifuge tubes labeled NS and IS.
3. Slowly add 0.1 ml of saturated ammonium sulfate solution to each tube, cap, and invert the tubes several times to mix *(do not use a Vortex mixer).*
4. Incubate the tubes for 15 min on ice. Centrifuge the samples at 4°C for 10 min in the microcentrifuge.
5. Remove the supernatant from the precipitated IgG pellet in each tube and discard. The serum consists of 80 to 90% albumin, which is soluble at 50% ammonium sulfate saturation. The IgG is insoluble at 50% ammonium sulfate saturation and will precipitate.
6. Resuspend the protein (IgG) pellet in 0.1 ml of 1X PBS, pH 7.3, and repeat steps 3 to 5.
7. Resuspend the protein pellet in 1.0 ml of PBS, pH 7.3. Make the following dilutions of the NS and IS samples in 1X PBS as follows:

Sample	Volume of PBS (ml)	Volume of Sample (ml)
NS (untreated)	0.990	0.010
IS (untreated)	0.990	0.010
NS [$(NH_4)_2SO_4$ purified]	0.667	0.333
IS [$(NH_4)_2SO_4$ purified]	0.667	0.333

8. Read the absorbance of each 1.0-ml solution at 280 nm, using 1X PBS to zero your spectrophotometer. *Estimate* the total protein concentration in each sample using Beer's law

($\varepsilon_{280} = 0.80\ (\text{mg/ml})^{-1}\ \text{cm}^{-1}$). This milligram per milliliter extinction coefficient is based on the *average* value of aromatic amino acids found in most proteins. Remember to take into account the *dilution* of each sample in your calculation when determining the concentration of protein in each sample. How does the total protein concentration compare between the untreated and ammonium sulfate–treated NS and IS samples? Explain your results. Store all four of these rabbit serum samples at 4°C for use on Day 2.

9. Obtain a 96-well Immulon-2 flat-bottomed ELISA plate. *Cover rows F, G, and H tightly with a piece of parafilm or with tape.* To each well in rows A through E, add 50 μl of the 10 μg/ml β-galactosidase solution in 1X PBS. Tap the plate gently to ensure that the 50-μl sample *completely covers* the bottom of each well. Store the plate at 4°C covered loosely with a piece of plastic wrap for use on Day 2. During the overnight incubation, the β-galactosidase in the solution will absorb noncovalently to the bottom of the wells.

Day 2: Determination of Titer

1. Remove the ELISA plate from the cold room and vigorously shake out the β-galactosidase solution added to rows A to E. Fill each well in row A through E completely (to the top of each well) with wash buffer and flick the solution out as before. Invert the plate and tap vigorously several times on a piece of paper towel to remove all of the wash buffer.
2. Fill the well in rows A through E completely with *blocking* buffer. Incubate for 30 min at room temperature. During this incubation, the BSA in the blocking buffer will bind to all of the sites on the bottom of the well not already bound with β-galactosidase. During the 30-min incubation, prepare the antibody dilutions described in step 4.
3. Remove the blocking buffer as before and tap extensively on a piece of paper towel to remove all of the excess blocking solution.
4. Dilute the various rabbit serum antibody solutions prepared on Day 1 *in blocking buffer* as follows:

Sample	Dilution
NS (untreated)	1:20
IS (untreated)	1:1000
NS [$(NH_4)_2SO_4$ purified]	1:5
IS [$(NH_4)_2SO_4$ purified]	1:1000

Prepare at least 0.5 ml of each antibody sample at the appropriate dilution. Also, obtain an aliquot of mouse-anti-β-galactosidase monoclonal antibody (J1) diluted 1:1000 *in blocking buffer* (prepared by the instructor).

5. Add 75 μl of blocking buffer to all of the wells in rows A through E *except for* 2A, 2B, 2C, 2D, and 2E. Once this is done, you are ready to make serial, twofold dilutions of the different primary antibody solutions across each row (see Fig. 17-4).
6. Add 75 μl of normal rabbit serum (NS-untreated) to wells 2A and 3A. Remove 75 μl from well 3A and place into well 4A. Pipette up and down several times to mix.
7. Remove 75 μl from well 4A and place into well 5A. Pipette up and down several times to mix. Continue the twofold serial dilutions of the NS sample all the way down row A to well 12A. Remove 75 μl from well 12A and discard.
8. *Using a fresh micropipette tip*, follow the same procedure (steps 6 and 7) to do twofold serial dilutions of the *untreated IS sample* across *row B*. Remember that well 1B will contain no primary antibody.
9. *Using a fresh micropipette tip*, follow the same procedure (steps 6 and 7) to do twofold serial dilutions of the *ammonium sulfate–purified NS sample* across *row C*. Remember that well 1C will contain no primary antibody.
10. *Using a fresh micropipette tip*, follow the same procedure (steps 6 and 7) to do twofold serial dilutions of the *ammonium sulfate–purified IS sample* across *row D*. Remember that well 1D will contain no primary antibody.
11. *Using a fresh micropipette tip*, follow the same procedure (steps 6 and 7) to do twofold serial dilutions of the *J1 monoclonal antibody* across *row E*. Remember that well 1E will contain no primary antibody. *NOTE: At this point, all of the wells in rows A through E should contain 75 μl of either blocking buffer or blocking buffer plus a particular antibody dilution.*

Dilution of antibody across each row

Samples (Day 2)		1	2	3	4	5	6	7	8	9	10	11	12
Normal rabbit serum stock (untreated)	A	Blank No 1° or 2° Ab	No dilution of stock	1:2	1:4	1:8	1:16	1:32	1:64	1:128	1:256	1:512	1:1024
Immune rabbit serum stock (untreated)	B	No 1° Ab	No dilution of stock	1:2	1:4	1:8	1:16	1:32	1:64	1:128	1:256	1:512	1:1024
Normal rabbit serum stock (ammonium-sulfate –purified)	C	No 1° Ab	No dilution of stock	1:2	1:4	1:8	1:16	1:32	1:64	1:128	1:256	1:512	1:1024
Immune rabbit serum stock (ammonium-sulfate –purified)	D	No 1° Ab	No dilution of stock	1:2	1:4	1:8	1:16	1:32	1:64	1:128	1:256	1:512	1:1024
JI monoclonal (mouse stock)	E	No 1° Ab	No dilution of stock	1:2	1:4	1:8	1:16	1:32	1:64	1:128	1:256	1:512	1:1024

Dilution of soluble β-galactosidase solution across each row

Day 3		1	2	3	4	5	6	7	8	9	10	11	12
40 μg/ml β-galactosidase standard	F	Blank No 1° or 2° Ab	No dilution of stock	1:2	1:4	1:8	1:16	1:32	1:64	1:128	1:256	1:512	1:1024
β-galactosidase at unknown concentrations	G	No 1° Ab	No dilution of stock	1:2	1:4	1:8	1:16	1:32	1:64	1:128	1:256	1:512	1:1024
	H	No 1° Ab	No soluble β-galacto-sidase	No soluble β-galacto-sidase	No soluble β-galacto-sidase								

Use goat-antirabbit IgG secondary antibody for rows A to D. Use goat-antimouse IgG secondary antibody for rows E to H.

Figure 17-4 96-Well microtiter plate ELISA. Rows A to E will be used to determine the titer of antibody samples (dilute antibody across each row). Rows F to H will be used to determine the concentration of soluble β-galactosidase in an unknown solution (dilute antigen across each row).

12. Incubate the primary antibody solutions at room temperature for 30 min. During the incubation, the anti-β-galactosidase antibodies in the various samples will bind to the antigen on the bottom of the wells.
13. Invert and shake the plate to remove the primary antibody solutions from rows A through E. Fill each well completely with wash buffer. Discard the wash buffer as before and repeat the wash step three more times. Remove the last traces of wash buffer from the wells after the final wash by inverting the plate and tapping the plate extensively on a piece of paper towel.
14. Dilute the HRP conjugated secondary antibodies as follows *in blocking buffer:*

Secondary Antibody	Dilution
HRP-goat-anti-rabbit IgG	1:9000 (2.5 ml total)
HRP-goat-anti-mouse IgG	1:10,000 (0.8 ml total)

15. Add 50 μl of the HRP-goat-anti*rabbit* IgG antibody to all of the wells in rows A to D *except well 1A.* Add 50 μl of blocking buffer to well 1A to prevent it from drying out.
16. Add 50 μl of the HRP-goat-anti*mouse* IgG antibody to all of the wells in row E.
17. Incubate the secondary antibody solutions at room temperature for 30 min. During this incubation, the secondary antibody will bind the primary antibody bound to the antigen at the bottom of the wells.
18. Remove the secondary antibody solutions from rows A to E and wash each well four times with wash buffer as described in step 13. Be sure to remove all of the wash buffer from these wells following the last wash step.
19. Prepare a fresh 1:1 mixture of the 3,3′,5,5′-tetramethylbenzidene (TMB) solution and the H_2O_2 solution (5 ml total, 2.5 ml of each solution mixed together). Add 100 μl to all of the wells in rows A to E. *Note the time that the TMB solution is added to the first well.*
20. Carefully monitor the wells visually for the development of blue color. When you begin to see the first signs of blue color in any of the wells in columns 11 or 12 (~2 min), the reaction in that row is ready to be stopped. Consult the instructor if you are not sure when to stop the reaction. *NOTE:* It is important that the wells *within* a particular row be incubated with the substrate for the same period of time. It is not necessary to have the incubation time be the same *between* each row. For instance, the wells in row A may require a 5-min incubation time before color development, while the wells in row E may require only a 2-min incubation for color development. Before you begin to stop the reactions in a particular row, the intensity of blue color in wells 2 and 3 of that row should appear to be about equal. If the reactions in a particular row are stopped too quickly, the determination of the titer will be difficult (see step 24 and Fig. 17-5).
21. Stop the reaction in each well by adding 100 μl of 4 N H_2SO_4. The solution will immediately change from a blue to yellow color. *It is important that the reactions are stopped in the same order in which the TMB substrate was added to start the reaction. For instance, if row A was started 1A to 12A, then row A should be stopped 1A to 12A.*
22. Using well 1A as a blank to zero the instrument at 450 nm (no primary or secondary antibody), read the A_{450} of the wells in rows A to E in the

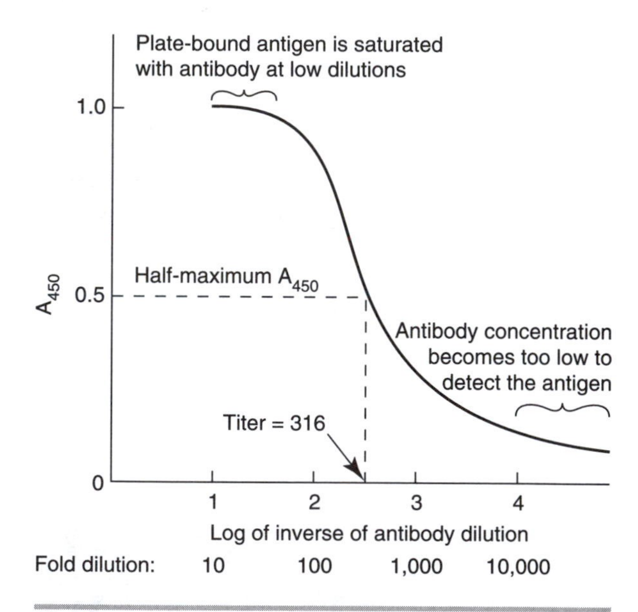

Figure 17-5 Sigmoidal titration curves will allow you to determine the titer of an antibody.

microplate spectrophotometer. Consult the instructor for information on how to operate the microplate spectrophotometer.

23. Subtract the A_{450} value of well 1C from all of the other A_{450} values in rows A and C. Subtract the A_{450} value of well 1B from all of the other A_{450} values in row B. Subtract the A_{450} value of well 1D from all of the other A_{450} values in row D. Subtract the A_{450} value of well 1E from all the other A_{450} values in row E. These corrected absorbance values will be used to produce the graphs described below. *This is done to account for the fact that the secondary antibodies may have bound nonspecifically to other proteins on the bottom of the wells besides the different IgG samples (background). The value in well 1 for each row represents the background A_{450} value produced from the secondary antibody bound nonspecifically to the bottom of the well (no primary antibody present).*
24. Prepare a plot of A_{450} versus the log of the inverse antibody dilution (a 1:10 dilution = 1, a 1:100 dilution = 2, etc.). Plot the data from all five primary antibody samples on a single graph for comparison. If the assay was successful, each antibody solution should produce a sigmoidal curve in this type of plot (see Fig. 17-5). *The various primary antibody stock solutions were diluted prior to performing twofold serial dilutions across each row (see step 4). Keep this in mind when you construct your graph.*
25. Identify the position on each curve where the A_{450} value is exactly *half of the maximum A_{450} value for each primary antibody sample.* From these points on the graph, determine the *titer* of each antibody sample: *For purposes of this experiment, the titer will be defined as the inverse of the dilution of the antibody that yields a half-maximum A_{450} value in the ELISA.* For example, an antibody that displayed a half maximum A_{450} value at a 1:1000 dilution has a titer of 1000.
26. Remove the parafilm from rows F to H. Add 50 μl of the 10 μg/ml β-galactosidase solution in 1× PBS to all the wells in rows F and G, as well as wells 1H through 4H. Cover the plate loosely with plastic wrap and store it at 4°C for use on Day 3.
27. Compare the titer of the untreated NS and IS samples. Which is larger? Explain your observation.
28. Compare the titers of the untreated NS and IS samples with those of their corresponding ammonium-sulfate–treated samples. Has the titer of NS or IS changed at all following treatment with ammounium sulfate? Explain.
29. Using the total protein concentrations (in milligrams per milliliter) for each primary antibody sample determined on Day 1, calculate the "specific activity" of the four rabbit serum samples using the following equation:

$$\text{“Specific activity” (titer} \cdot \text{ml/mg)} = \frac{\text{titer of antibody sample}}{\text{total protein concentration (mg/ml)}}$$

(Note that this differs from the traditional definition of specific activity as used in the discussion of enzymes.)

Is this value the same for the untreated NS and IS samples and their corresponding ammonium-sulfate–treated samples? Explain your results.

30. Determine the titer of the J1 monoclonal antibody. The half maximum A_{450} value should be somewhere between 0.5 and 0.6 if the reactions in the wells were stopped at the appropriate time. This information will be used on Day 3 in the competitive ELISA assay to determine the concentration of β-galactosidase in an unknown solution.

Day 3: Determination of the Concentration of β-Galactosidase in an Unknown Solution

1. Follow steps 1 to 3 from the protocol described for Day 2 (blocking steps) for all of the wells in rows F to H (see Fig. 17-4).
2. Prepare 1 ml of a J1 monoclonal antibody solution in blocking buffer that is *twice as concentrated* as the dilution that yielded a half maximum A_{450} value in the ELISA performed on Day 2 ($A_{450} \approx 0.5$ to 0.6). For example, if a 1:10,000 dilution yielded a half-maximum A_{450} value on Day 2, prepare a solution that is diluted 1:5000 in blocking buffer. *Because you will be adding an equal volume of soluble β-galactosidase to the wells during the primary antibody incubation, you will need a primary antibody dilution that is half that used on Day 2. This*

will bring the working dilution of the J1 monoclonal antibody in the well to its titer value, where it will be most sensitive to changes in the concentration of soluble β-galactosidase.

3. Fill all the wells in rows F and G with 37.5 μl of blocking buffer *except* wells 1F, 2F, 1G, and 2G. Also, add 37.5 μl of blocking buffer to wells 2H, 3H, and 4H. Add 75 μl of blocking buffer to wells 1F, 1G, and 1H.
4. Add 37.5 μl of the *40 μg/ml β-galactosidase solution* in PBS to wells 2F and 3F. Remove 37.5 μl from well 3F and add it to well 4F. Pipette up and down several times to mix.
5. Remove 37.5 μl from well 4F and add it to well 5F. Pipette up and down several times to mix. Continue the twofold serial dilution of the 40 μg/ml β-galactosidase solution across row F until you reach 12F. Remove 37.5 μl from well 12F and discard.
6. Add 37.5 μl of the *β-galactosidase solution at an unknown concentration* to wells 2G and 3G. Perform twofold serial dilutions of this unknown β-galactosidase solution across row G from 4G to 12G. Remove 37.5 μl from well 12G and discard.
7. Add 37.5 μl of the J1 monoclonal antibody solution prepared in step 2 to all the wells in rows F and G except for 1F and 1G. Also, add 37.5 μl of the J1 monoclonal antibody solution prepared in step 2 to wells 2H, 3H, and 4H. *Wells 2H, 3H, and 4H contain no soluble β-galactosidase. The A_{450} values of these wells will be important in the calculations performed at the end of the assay.*
8. Incubate the primary antibody for 30 min at room temperature. Remove the primary antibody solution from the wells and wash the wells four times with wash buffer as described in the protocol for Day 2.
9. Add 50 μl of the HRP-conjugated goat-anti-*mouse* IgG antibody to all of the wells in rows F and G except for 1F (using the same dilution of the secondary antibody as used on Day 2). Also, add 50 μl of the secondary antibody solution to wells 1H to 4H.
10. Incubate the secondary antibody for 30 min at room temperature. Wash the wells four times with wash buffer as described in the protocol for Day 2.
11. Add 100 μl of a 1:1 mixture of 3,3′,5,5′-tetramethylbenzidene (TMB) and H_2O_2 to all of the wells in rows F and G, as well as wells 1H to 4H. *Note the time at which you add the substrate to the first well. It is critical for the success of this experiment that the total reaction time be the same for all wells in the competitive ELISA.* Allow the blue color to develop until you begin to see any blue color in wells 2F or 3F (~2–4 min). Stop the reactions by adding 100 μl of 4 N H_2SO_4. *Again, it is very important that the reaction time for all of the wells is the same in this experiment.*
12. Using well 1F as a blank to zero the instrument at 450 nm (no primary or secondary antibody), read the A_{450} of all of the wells in rows F to H in the microplate spectrophotometer.
13. Determine the *average* A_{450} value of wells 1G and 1H and subtract this value from all of the other A_{450} values in rows F to H.
14. Prepare a plot of percent inhibition versus the log of the inverse of the dilution of the 40 μg/ml and unknown β-galactosidase solutions. Use the following formula to determine the percent inhibition for *each well* in rows F and G:

$$\%\ \text{inhibition} = \frac{(A_{450}\ \text{of J1}) - (A_{450}\ \text{of J1} + \beta\text{-gal})}{A_{450}\ \text{of J1}} \times 100$$

where A_{450} of J1 is the *average* absorbance value of wells 2H, 3H, and 4H and A_{450} of J1 + β-gal are the values of *each individual well* in rows F and G. Plot the data for rows F and G on the same graph for comparison. *NOTE:* A percent inhibition value can be calculated for each well in rows F and G.

15. Identify the dilution of each β-galactosidase solution (40 μg/ml solution and the unknown solution) that gave 50% inhibition. If the unknown solution showed 50% inhibition at a larger dilution than the 40 μg/ml β-galactosidase solution, then the unknown solution is more concentrated than 40 μg/ml. If the unknown solution showed 50% inhibition at a lower dilution than the 40 μg/ml β-galactosidase solution, then the unknown solution is less concentrated than 40 μg/ml. Based on the dilution required for the unknown solution to show 50% inhibition compared to the 40 μg/ml solution (0.25 times, 0.5 times, 2 times, 4 times, etc.), determine the concentration of β-galactosidase in the unknown solution. For instance, if the 40 μg/ml

β-galactosidase solution showed 50% inhibition at a 1:8 dilution (well 5F), and the unknown β-galactosidase solution showed 50% inhibition at a 1:16 dilution (well 6G), then the unknown solution is *twice* as concentrated as the 40 μg/ml solution (80 μg/ml β-galactosidase).

Exercises

1. Explain why it was necessary to incubate the TMB and H_2O_2 with the HRP-conjugated secondary antibody for the same time in each well in the competitive ELISA performed on Day 3. Why was the incubation time between rows less of a consideration in the ELISA experiment performed on Day 2 (to determine the titer of each antibody sample)?
2. Discuss the advantages and disadvantages of using the ELISA to determine the concentration of an antigen in an unknown solution as opposed to performing Western blots (see Experiment 18).
3. You have obtained several monoclonal antibodies that all recognize Protein X. You would like to determine whether any of these antibodies recognize Protein X only in the native form as opposed to the denatured form. Describe an ELISA experiment that would allow you to do this, as well as the predicted results of the experiment.
4. Why is it a good idea to test the serum of an animal for the presence of antibodies against a protein of interest *before* the animal is injected with the antigen?
5. The production of polyclonal antibodies usually requires a relatively large amount of a pure protein, whereas the production of monoclonal antibodies may require smaller amounts of an impure protein. Explain this in terms of the method of production of these two types of antibodies.

REFERENCES

Harlow, E., and Lane, D. (1988). *Antibodies, a Laboratory Manual.* Cold Spring Harbor, NY: Cold Spring Harbor Laboratory.

Kaplan, A., Szabo, L. L., and Opheim, K. E. (1988). Immunochemical Techniques. In: *Clinical Chemistry: Interpretation and Techniques.* Philadelphia: Lea & Febiger.

Lehninger, A. L., Nelson, D. L., and Cox, M. M. (1993). An Introduction to Proteins. In: *Principles of Biochemistry.* New York: Worth.

Mathews, C. K., and van Holde, K. E. (1990). Tools of Biochemistry. In: *Biochemistry.* Redwood City, CA: Benjamin/Cummings.

Western Blot to Identify an Antigen

Theory

A protein has been purified using the latest techniques in chromatography. How can you prove that the protein band visualized on a Coomassie Blue–stained sodium dodecyl sulfate-(SDS)-polyacrylamide gel is the protein of interest? If the protein is an enzyme, it can be subjected to an activity assay and total protein determination to assess its purity (specific activity). If the protein has no enzymatic activity, immunological studies may be employed. Western blotting is a technique commonly used to verify the presence of a protein of interest in a crude mixture. In addition to its use as a qualitative tool (to determine if a particular protein is present), Western blotting can also be used to quantify the amount of protein present in a pure or impure solution. Western blotting can be used to detect any protein of interest, provided that an antibody raised against (that recognizes) the protein of interest is available.

In this experiment, you will identify the C-terminal domain of glutathione-*S*-transferase (GST) from *Schistosoma japonicum* (purified from an *Escherichia coli* cell extract in Experiment 10) through the use of a Western blot. After the proteins present in the crude cell extract, unbound fraction, and the fraction eluted from the glutathione Sepharose resin are separated by SDS–polyacrylamide gel electrophoresis (SDS-PAGE), the proteins will be electrophoretically transferred to a polyvinylidene fluoride (PVDF) membrane (Fig. 18-1). Although membranes composed of nitrocellulose or nylon can also be used for Western blotting, PVDF membranes are particularly effective for use with the chemiluminescent detection system employed in this experiment.

After the proteins have been transferred to the PVDF membrane, a blocking step is performed with powdered milk to saturate all of the sites on the membrane not already bound with protein. This blocking step is critical to the success of the technique: if the unoccupied sites are not blocked with protein, the primary and secondary antibodies will bind non-specifically to sites throughout the membrane, making it very difficult to localize the protein of interest to a single location (protein band) on the membrane. In addition to using powdered milk as a source of protein for the blocking step, bovine serum albumin and casein are commonly used as blocking agents. *If an avidin/biotin system is being utilized for detection of the antigen (see below), nonfat powdered milk or casein should not be used as the blocking agent.* The biotin in the milk may adhere to the membrane and cause high background.

After the proteins have been transferred to the PVDF membrane and the unoccupied sites have been blocked with protein, the antibody directed against the protein of interest (primary antibody) is added. During the incubation, the primary antibody will bind to the protein of interest. The amount of primary antibody to be added (the dilution) must be determined empirically for each experimental situation. If the antibody concentration is too high, it is likely to bind to other proteins on the membrane in addition to the protein of interest. If the primary antibody concentration is too low, it may not be able to bind the protein of interest effectively. A compromise must be reached between the sensitivity of detection of the protein of interest (signal) and the

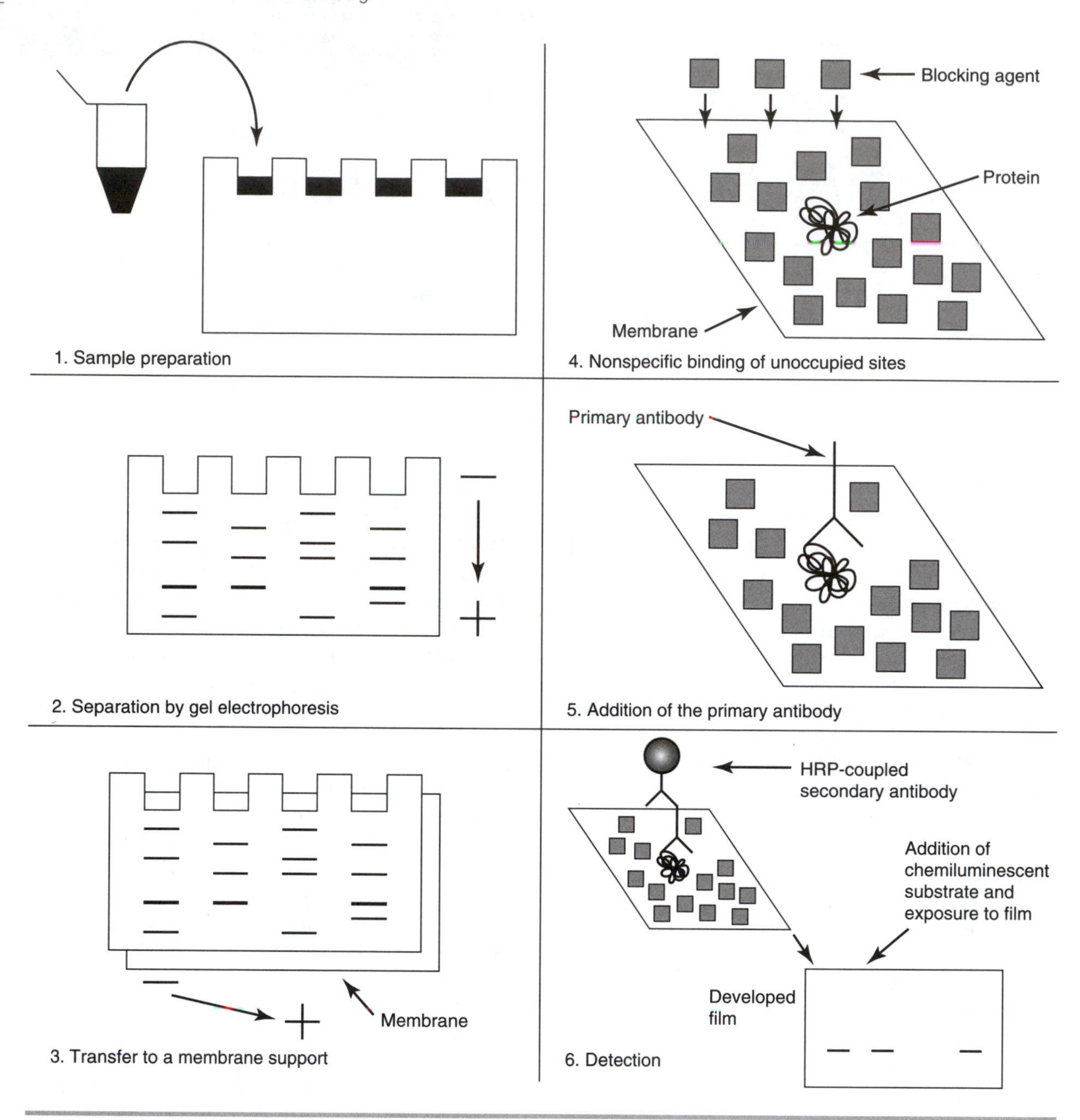

Figure 18-1 The technique of Western blotting.

extent of nonspecific binding of the antibody to other proteins on the membrane ("noise").

After the membrane has been incubated with the primary antibody (diluted in blocking buffer), the membrane is washed in blocking buffer containing polyoxyethylenesorbitan monolaurate (Tween 20). This non-ionic (nondenaturing) detergent will disrupt hydrophobic interactions between the primary antibody and other proteins on the membrane (besides the protein of interest) that formed during the incubation. The membrane is then incubated with a secondary antibody that will recognize and bind to the primary antibody. The anti-GST antibody (primary) used in this experiment was raised in a mouse. The secondary antibody used in this experiment is an antimouse IgG antibody that was raised in a goat (goat-antimouse IgG). This secondary antibody is conjugated to an enzyme called horseradish peroxidase (HRP), the critical component necessary for the light-producing (luminescent) chemical reaction. As with the primary antibody, the dilution of the secondary antibody must be determined empirically for each experimental situation to optimize the results.

The PVDF membrane is again washed with blocking buffer containing Tween 20 to remove any secondary antibody that has bound nonspecifically to proteins on the membrane other than the mouse IgG. At this point, the PVDF membrane is ready to be treated with the SuperSignal reagents. One of these reagents contains hydrogen peroxide, which will convert the HRP to its oxidized form. The second reagent contains a component called *luminol.* As HRP converts luminol from the reduced form to the oxidized form (Fig. 18-2), 428-nm light is given off. If the PVDF membrane is exposed to photographic film during the course of this reaction, a dark spot will appear on the film where the primary and secondary antibodies are localized. If care is taken to optimize the system under study, a single protein of interest can be identified among thousands of others.

Before the introduction of chemiluminescence to the technique of Western blotting, the secondary antibodies were localized on the membrane with the use of compounds and chemical reactions that formed *colored precipitates* on the membrane. For instance, if the membrane used in this experiment were treated with a solution of 3,3′-diaminobenzidene tetrahydrochloride dihydrate (Fig. 18-3) and H_2O_2, a brown precipitate would form at the site of action of the HRP conjugated to the secondary antibody. If the secondary antibody used in this experiment were conjugated to *alkaline phosphatase* rather than HRP, the membrane could be treated with a solution containing 5-bromo-4-chloro-3-indolyl phosphate and nitroblue tetrazolium chloride (Fig. 18-3) to form a blue precipitate at the site of alkaline phosphatase action on the membrane. A list of enzymes used for detection of antigens in Western blotting, their substrates, and the color of the insoluble products that they produce is shown in Table 18-1. Although these latter methods are still in wide use today, they are less sensitive than the chemiluminescent technique demonstrated in this experiment. For lab-

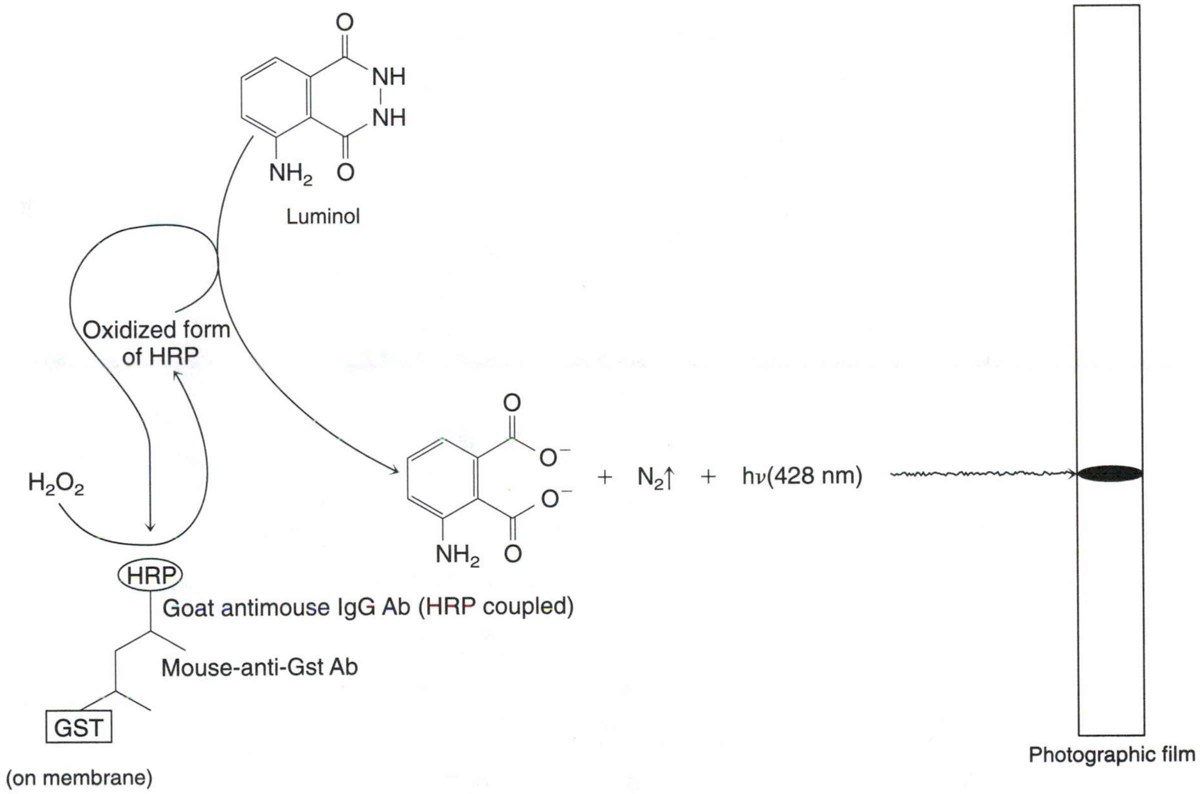

Figure 18-2 Reaction of luminol and H_2O_2 to produce light in the presence of horseradish peroxidase.

For use with Antibodies Coupled to Horseradish Peroxidase

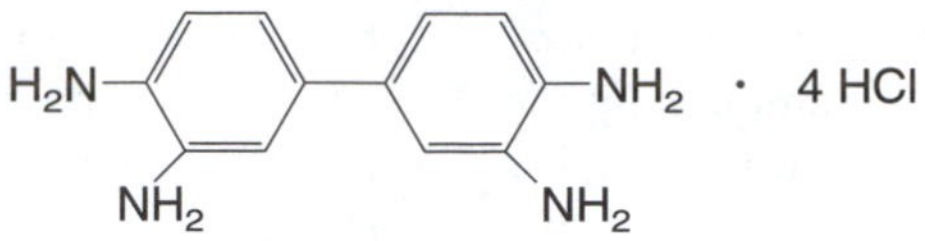

3,3′-Diaminobenzidene tetrahydrochloride

For use with Antibodies Coupled to Alkaline Phosphatase

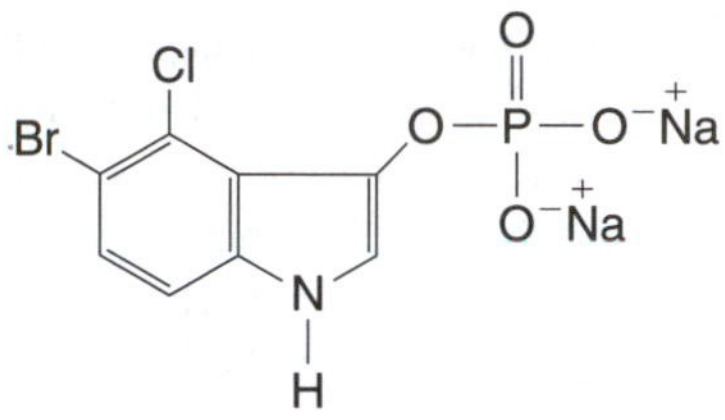

5-Bromo-4-chloro-3-indolyl phosphate

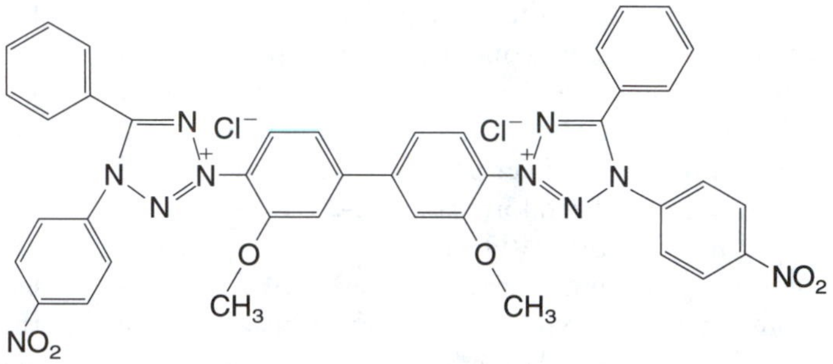

Nitroblue tetrazolium chloride

Figure 18-3 Other commonly used substrates that form colored precipitates on the membrane when acted on by enzymes conjugated to secondary antibodies.

oratories that do not have access to darkrooms and/or developing solutions, we offer an alternative protocol that uses a precipitating HRP substrate for the detection of GST at the end of the experiment.

Why has the "two-antibody" (primary and secondary) technique become so popular for use in immunochemistry experiments such as the Western blot? Enzyme-conjugated antibodies that recognize

Table 18-1 Precipitating Colorimetric Enzyme/Substrate Systems for Detection of Antigens in Western Blots

Enzyme	Substrate	Color of Precipitate
Alkaline phosphatase	5-Bromo-4-chloro-3′-indolyl-phosphate with nitroblue tetrazolium chloride (BCIP/NBT)	Black/purple
	Naphthol AS-MX phosphate with Fast Red TR	Bright red
Horseradish peroxidase	4-Chloro-1-naphthol (CN)	Blue/purple
	3-Amino-9-ethyl carbazole (AEC)	Red/brown
	3,3′-Diaminobenzidene tetrahydrochloride (DAB) *(carcinogenic)*	Brown
Glucose oxidase	Phenazine methosulfate with nitroblue tetrazolium chloride	Black/purple
β-Galactosidase	5-Bromo-4-chloro-3-indolyl-β,D-galactopyranoside (X-Gal)	Blue

species-specific IgG are commercially available and relatively inexpensive. For instance, an HRP-conjugated goat-antirabbit IgG antibody will recognize *any* antibody that has been raised against *any* antigen, so long as the latter antibody has been raised in a rabbit. It is far more cost effective for biotechnology companies to develop and conjugate *secondary* antibodies that recognize antibodies obtained from a *limited* number of hosts than it is for them to develop and conjugate *primary* antibodies that recognize the seemingly *unlimited* number of antigens that are currently under study. Still, you should realize that it is possible to conjugate a particular primary antibody of interest to an enzyme and alleviate the need for a secondary antibody in Western blotting and in the ELISA (see Experiment 17).

Yet another variation of the Western blot exploits the strong interaction between biotin and streptavidin. Biotin is a 244-Da vitamin found in all cells (Fig. 18-4). Streptavidin is a 75-kDa, tetrameric protein isolated from *Streptomyces avidinii.* Each streptavidin monomer is capable of binding a single molecule of biotin. The dissociation constant of the streptavidin/biotin complex is on the order of 10^{-15} M, one of the strongest noncovalent interactions found in nature. Many reagents are commercially available that allow investigators to cross-link (conjugate) antibodies with biotin. Antibodies can be conjugated to biotin through a number of different reactive groups on the protein, including primary amine groups (lysines), sulfhydryl groups (produced after reduction of the antibody to break disulfide bonds), carboxyl groups (glutamate and aspartate), carbohydrate groups (after oxidation of *cis*-diols to reactive aldehyde groups), etc. As is the case with the enzyme-conjugated secondary antibodies described above, horseradish peroxidase and alkaline-phosphatase–conjugated streptavidin are also commercially available. When the enzyme-conjugated streptavidin is incubated with a membrane containing a biotinylated antibody bound to the antigen of interest, this antigen can be localized to a particular portion of the membrane with the use of luminol or any other of the substrates described in Table 18-1 (Fig. 18-5).

Supplies and Reagents

Reagents and apparatus for SDS–PAGE (See Experiment 4)

P-20 Pipetman with disposable tips

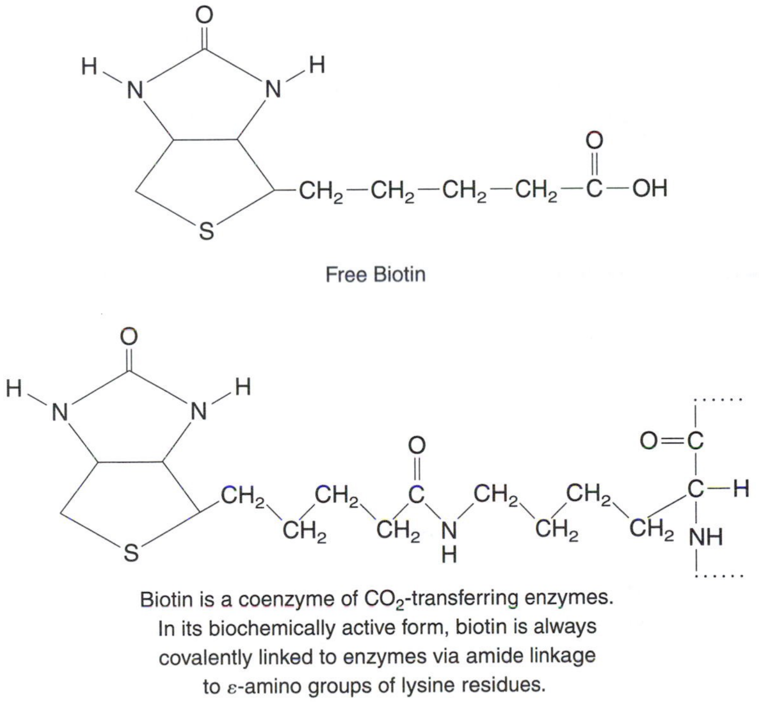

Free Biotin

Biotin is a coenzyme of CO_2-transferring enzymes. In its biochemically active form, biotin is always covalently linked to enzymes via amide linkage to ε-amino groups of lysine residues.

Figure 18-4 The chemical structure of biotin.

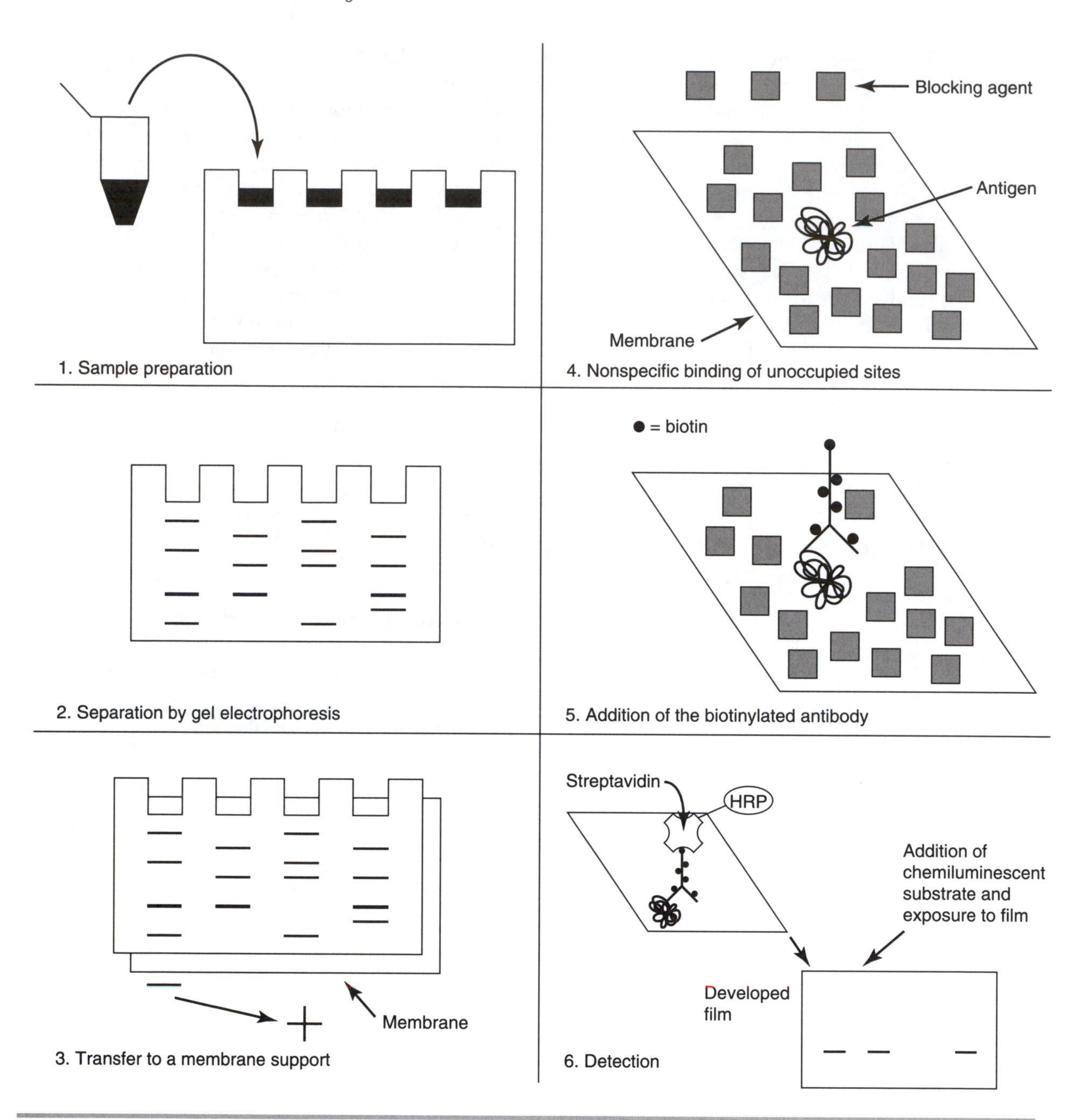

Figure 18-5 Use of streptavidin–enzyme conjugates and biotinylated antibodies in the detection of antigens on Western blots.

Prestained high molecular-weight protein standard (Gibco BRL, Catalog #26041-020)

Coomassie Blue staining solution (40% methanol, 10% glacial acetic acid, 0.25% wt/vol Coomassie Brilliant Blue R-250)

Destaining solution (40% methanol, 10% glacial acetic acid)

Methanol

Transfer buffer (48 mM Tris, 39 mM glycine, 20% methanol, 0.004% SDS, pH 9.2)

Immobilon-P PVDF membrane, 0.45-μm pore size (Millipore, Catalog #IPVH 000 10)

Whatman 3 MM filter paper

Tris/NaCl/powdered milk solution (20 mM Tris–HCl, pH 8, 14.62 g/liter NaCl, 86.22 g/liter nonfat powdered milk)

Boiling water bath
Razor blade
Aluminum foil
Small pans or containers for membrane incubations and washes
Power supply
Tween 20 (polyoxyethylenesorbitan monolaurate, Sigma Catalog #P-5927)
Mouse–anti-glutathione-*S*-transferase monoclonal antibody (Sigma Product #G-1160)
SuperSignal West Pico chemiluminescent substrate for Western blotting (Pierce Catalog #34080)
HRP-conjugated goat-antimouse IgG antibody (Pierce Catalog #31430)
Plastic wrap
Kodak X-omat film
Film cassette
Dark room
Transfer apparatus (such as BIORAD Transblot SD®)

Reagents required for the colorimetric detection of HRP (optional):

10 mM Tris-buffered saline (TBS buffer), pH 7.4 (Dissolve 1.8 g of Trizma base and 13 g of NaCl in 1.5 liters of distilled water. Adjust the pH to 7.4 with 1 N HCl.)
4-Chloro-1-naphthol (Pierce Catalog #34010)
Absolute ethanol
3% (vol/vol) Hydrogen peroxide in distilled water—*prepare immediately before use*

Protocol

Day 1: SDS-PAGE and Preparation of Blot for Western Blot Analysis

1. Heat the eight samples prepared for SDS–PAGE (performed in step 16 of Experiment 10) in a boiling water bath for 5 min, along with a 40 μl sample of molecular weight protein standards in 4× SDS sample buffer (prepared by the instructor).
2. Centrifuge the samples for 10 sec to collect the material at the bottom of the tubes.
3. Load a 10-well, 12% SDS polyacrylamide gel with 20 μl per well of each sample as follows:

Lane	Sample
1	Crude cell extract
2	Unbound fraction
3	Eluted fraction
4	Purified GST
5	Molecular weight protein standards
6	Crude cell extract
7	Unbound fraction
8	Eluted fraction
9	Purified GST
10	Molecular weight protein standards

4. Apply current (~120–180 V, constant voltage, ~30 mA) and continue electrophoresis until the bromophenol blue dye front has migrated to the bottom of the gel.
5. Turn off the power supply and remove the gel from the electrophoresis apparatus. Separate the plates (the gel will adhere to one of the two plates), and remove the stacking gel. Be careful not to tear the resolving gel during this process.
6. Cut the gel in half between lanes 5 and 6 with a razor blade. The gel will tear if the razor blade is dragged across it, so it is better to make several cuts holding the blade perpendicular to the plate and pressing straight down to cut.
7. Cut a notch in the upper corner of each half of the gel, next to lane 1 and lane 6. This will help you determine the orientation of the gel following Coomassie staining and transfer to the PVDF membrane.
8. Place the portion of the gel containing lanes 1 to 5 in a small pan. Add enough Coomassie Blue staining solution to completely cover the gel and incubate at room temperature for 1 hr.
9. Place the portion of the gel containing lanes 6 to 10 in a small pan containing transfer buffer. Incubate the gel for 10 to 15 min to allow the gel to equilibrate in the transfer buffer.
10. Cut six pieces of Whatman 3 MM filter paper to the *exact size* of the gel portion containing lanes 6 to 10.
11. Cut a piece of PVDF membrane *(wear gloves when handling the membrane)* to the exact size of the gel portion containing lanes 6 to 10. Wet the membrane completely by dipping it in a small pan containing methanol. It is critical that the PVDF membrane be fully wetted in methanol prior to being equilibrated in transfer buffer (see below). *It is important that the filter paper and membrane be cut to the exact size of*

the membrane so that the current applied to the transfer unit (see below) be forced through the gel rather than around it. If the top and bottom layers of filter paper on either side of the gel come in contact with one another, the current will take the path of least resistance and flow around rather than through the gel. The result of this is that the proteins will not transfer to the PVDF membrane, and the subsequent Western blot will not be successful.

12. Place the methanol-treated PVDF membrane and the four pieces of Whatman 3 MM filter paper in a small pan containing transfer buffer and incubate for 5 to 10 min.
13. Assemble the blot as described below. Make sure that all components are stacked directly on top of one another, *with no overhanging edges and no air bubbles in between the layers:*

    ANODE (+)
    Whatman 3 MM filter paper
    Whatman 3 MM filter paper
    Whatman 3 MM filter paper
    PVDF membrane
    Gel
    Whatman 3 MM filter paper
    Whatman 3 MM filter paper
    Whatman 3 MM filter paper
    CATHODE (−)

14. Connect the electrodes and apply 15 V (constant voltage) to the unit for 45 min.
15. Disconnect the power supply from the transfer unit, disassemble the layers, and cut a notch on the PVDF membrane to mark the position of lane 6. Stain the gel following the transfer with Coomassie Blue as described in step 8. This is done to ensure that the proteins on the gel were transferred to the PVDF membrane.
16. Place the PVDF membrane in a small pan containing 100 ml of Tris/NaCl/powdered milk solution (blocking buffer). Incubate at 4°C overnight with gentle shaking. Recall that this step is performed to block any portion of the membrane that does not already contain protein.
17. After each half of the gel has been incubated with the Coomassie Blue staining solution for 1 hr, they may be destained with destain solution by incubating for 1 hr at room temperature. The destain solution should be changed four times during the course of the 1-hr incubation.
18. Calculate the R_f values of the molecular weight protein standards using the following equation:

$$R_f = \frac{\text{distance traveled by protein band (cm)}}{\text{distance traveled by dye front (cm)}}$$

    Prepare a standard curve by plotting log molecular weight vs. R_f.
19. What is the molecular weight of the protein(s) in the eluted fraction? Do you believe that the purification of GST was successful? Explain.

Day 2: Western Blot

1. Drain the 100 ml of blocking buffer solution from the pan containing the PVDF membrane. Rinse the membrane with a small amount of distilled water to remove the last traces of blocking buffer.
2. Add 20 ml of fresh blocking buffer to the PVDF membrane, along with 2.5 μl of the mouse–anti-GST monoclonal antibody (a 1:8000 dilution of the antibody). Swirl the pan continuously and gently for 45 min at room temperature.
3. Pour out the antibody solution and rinse the PVDF membrane with a small amount of distilled water.
4. Add 100 ml of fresh blocking buffer containing 0.05% Tween 20 (wash buffer) and incubate at room temperature for 10 min with continuous gentle swirling. Pour out the wash solution and repeat step 4 two more times.
5. Add 20 ml of fresh blocking buffer to the PVDF membrane. Add 5 μl of a 1:10 dilution (in water) of the goat-antimouse IgG (HRP-conjugated). The final dilution of this antibody is 1:40,000. Cover the pan with aluminum foil and incubate at room temperature for 45 min with continuous gentle swirling. *NOTE: If you will be developing the Western blot with a colorimetric substrate (see below), the final dilution of the HRP-goat-antimouse IgG should be between 1:5000 and 1:20,000. This results from the fact that the 4-chloro-1-naphthol substrate is much less sensitive than the chemiluminescent substrate (i.e., you will need more HRP at the site of the antigen to obtain a dark precipitate). Consult the instructor for information concerning the final dilution of the secondary antibody if the colorimetric detection option is chosen.*

6. Pour out the antibody solution and rinse the PVDF membrane with a small amount of distilled water.
7. Add 100 ml of wash buffer and incubate at room temperature for 10 min with continuous gentle swirling. Pour out the wash solution and repeat step 7 two more times. *If you will be using colorimetric detection with a precipitating HRP substrate, proceed to the section entitled "Colorimetric Detection of HRP."*
8. Mix 5 ml of each of the two SuperSignal reagents (peroxide solution and luminol/enhancer solution) and add them to the top of the PVDF membrane (the side containing the proteins). Incubate at room temperature for 5 min.
9. Remove the PVDF membrane from the pan, allow the excess SuperSignal substrate to drip from the membrane, and wrap the membrane in plastic wrap.
10. Expose the PVDF membrane *(protein side up)* to a piece of film for 1 to 5 min. Mark the film with a notch corresponding to the notch made next to lane 6 on the SDS gel to aid in orientation. Remove the film from the cassette and develop. *The time of the exposure will depend on how quickly the blot is exposed to film after the chemiluminescence reaction has been performed. The longer the time between these two events, the longer the exposure time will be needed to visualize the bands.*
11. Using the notch made on the film, align it with the notch on the PVDF membrane. Based on the position of the prestained molecular weight protein standards present on the PVDF membrane, determine the molecular weight of the protein band(s) present on the film. Do you believe that it is GST? Explain. Are there any other protein bands besides GST that are present on the film (in the crude cell lysate and unbound fraction lanes)? If so, explain what this means. Propose how you could alter the conditions of the experiment to prevent these other protein bands from cross-reacting with the antibodies used in the Western blot.

Colorimetric Detection of HRP

1. Dissolve 150 mg of 4-chloro-1-napthol (CN) in 50 ml of ethanol. The stock solution can be prepared in advance and stored at −20°C. This will be enough reagent for 50 groups to perform the experiment.
2. Add 1 ml of the 3 mg/ml CN solution prepared in step 1 to 10 ml of Tris-buffered saline, pH 7.4.
3. Immediately before use, add 100 μl of a 3% (vol/vol) hydrogen peroxide solution to the 10 ml of CN/H_2O_2 reagent prepared in step 2, and pour the entire solution over the side of the membrane containing your protein.
4. Incubate the membrane with the CN/H_2O_2 reagent for approximately 30 min at room temperature. Monitor the membrane visually during this incubation for the appearance of blue or purple protein band(s). If the membrane does not contain a large amount of GST and/or HRP-conjugated goat-antimouse secondary antibody, an incubation time longer than 30 min may be required.
5. When you are satisfied with the color development on your Western blot, rinse the membrane gently several times with distilled water to stop the reaction. *NOTE: The blue color will begin to fade several days after development. To obtain a permanent record of the experiment, we advise that you take a picture or prepare a photocopy of the membrane after it has been allowed to dry.*

Exercises

1. Where does the term "Western" blot come from?
2. You are performing a Western blot on a purified sample of protein X. You know that protein X is composed of three subunits, one of 50 kDa, one of 30 kDa, and one of 15 kDa. You have two monoclonal antibody samples raised against native protein that you will use as the primary antibodies in two separate experiments. Describe what you could expect to see in two different Western blots performed with these two antibodies. Explain your answer.
3. Describe how monoclonal and polyclonal antibodies are made.
4. Why do you think that it is more cost effective for companies to sell enzyme-conjugated antibodies that recognize antibodies from a number of animals rather than to sell enzyme-

conjugated antibodies that recognize the antigens themselves (why is the two-antibody system commonly used rather than a primary antibody that is already enzyme conjugated)?

5. Describe how a chemiluminescent Western blot could be used to quantify the amount of an antigen in an unknown sample.
6. How would a Western blot be affected if the blocking step were omitted? How would a Western blot be affected if the blocking step were carried out, but the wash steps following incubation with the primary and secondary antibodies were omitted?

REFERENCES

Harlow, E., and Lane, D. (1988). *Antibodies, A Laboratory Manual.* Cold Spring Harbor, NY: Cold Spring Harbor Laboratory.

Hines, K. (1995). *Immunoperoxidase Assay Systems: Visualization of Immunoblots with SuperSignal® CL-HRP Substrate System.* Rockford, IL: Pierce Chemical.

Lehninger, A. L., Nelson, D. L., and Cox, M. M. (1993). An Introduction to Proteins. In: *Principles of Biochemistry,* 2nd ed. New York: Worth.

Stryer, L. (1995). Exploring Proteins. In: *Biochemistry,* 4th ed. New York: Freeman.

Thorpe, G. H. G., and Kricka, L. J. (1986). Enhanced Chemiluminescence Reactions Catalyzed by Horseradish Peroxidase. *Methods Enzymol* **133**:331.

Voet, D., and Voet, J. G. (1990). Techniques of Protein Purification. In: *Biochemistry.* New York: Wiley.

Weir, D. M. (1986). *Handbook of Experimental Immunology.* London: Oxford University Press.

SECTION V

Nucleic Acids

Experiment 19	**Isolation of Bacterial DNA**
Experiment 20	**Transformation of a Genetic Character with Bacterial DNA**
Experiment 21	**Constructing and Characterizing a Recombinant DNA Plasmid**
Experiment 22	**In Vitro Transcription from a Plasmid Carrying a T7 RNA Polymerase–Specific Promoter**
Experiment 23	**In Vitro Translation: mRNA, tRNA, and Ribosomes**
Experiment 24	**Amplification of a DNA Fragment Using Polymerase Chain Reaction**

SECTION V

Nucleic Acids

Introduction

Structural Features of Nucleic Acids

Nucleic acids are polymers containing nitrogenous bases attached to sugar–phosphate backbones. The common nitrogenous bases of nucleic acids are the bicyclic *purines*, adenine and guanine, and the monocyclic *pyrimidines*, cytosine, uracil, and thymine (Fig. V-1).

These purines and pyrimidines join to the sugar–phosphate backbones of nucleic acids through repeating β-linked N-glycosidic bonds involving the N^9 position of purines and the N^1 position of pyrimidines. There are two classes of nucleic acids: ribonucleic acid (RNA) and deoxyribonucleic acid (DNA). DNA and RNA differ in one of their nitrogenous base components (uracil in RNA, thymine in DNA) and in their sugar (ribose) moiety, as indicated in Fig. V-2.

Repeating $3',5'$-phosphodiester bonds join the D-ribose units of RNA and the 2-deoxy-D-ribose units of DNA. A *nucleoside* is a nitrogenous base plus a sugar, while a *nucleotide* is a nitrogenous base, plus a sugar, plus a phosphate. Shorter polymers of 2 to 50 nucleotides are referred to as *oligonucleotides*, while larger polymers of nucleotides are referred to as *polynucleotides*.

Because nucleic acids contain a large number of nucleotides, biochemists have devised an abbreviation system to indicate the nucleotide sequence of nucleic acids (see Fig. V-2). This system assumes the $3',5'$-phosphodiester linkages and lists, in a $5'$ to $3'$ order, the nucleotide sequence of the nucleic acid with p (phosphate) between each nucleotide. Using this latter system, the RNA nucleotide sequence shown in Figure V-2 is . . . ApGpCpUp . . . . You may also encounter an even simpler abbreviation system where the phosphate (p) between each nucleotide is omitted (e.g., . . . AGCU . . .).

Many of the physical and biological properties of nucleic acids arise from secondary structures created by a regular stacking of hydrogen-bonded pairs of nitrogenous bases. Only certain pairs of bases are capable of hydrogen bonding: Adenine hydrogen bonds with thymine and guanine hydrogen bonds with cytosine in DNA, while adenine hydrogen

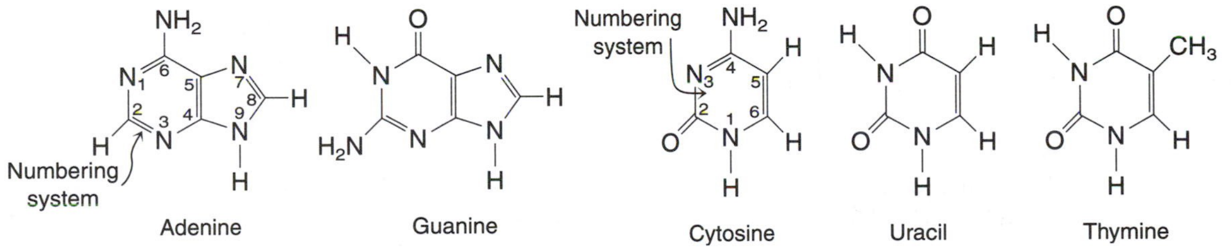

Adenine　Guanine　Cytosine　Uracil　Thymine

Figure V-1 The most common nitrogenous bases of nucleic acids.

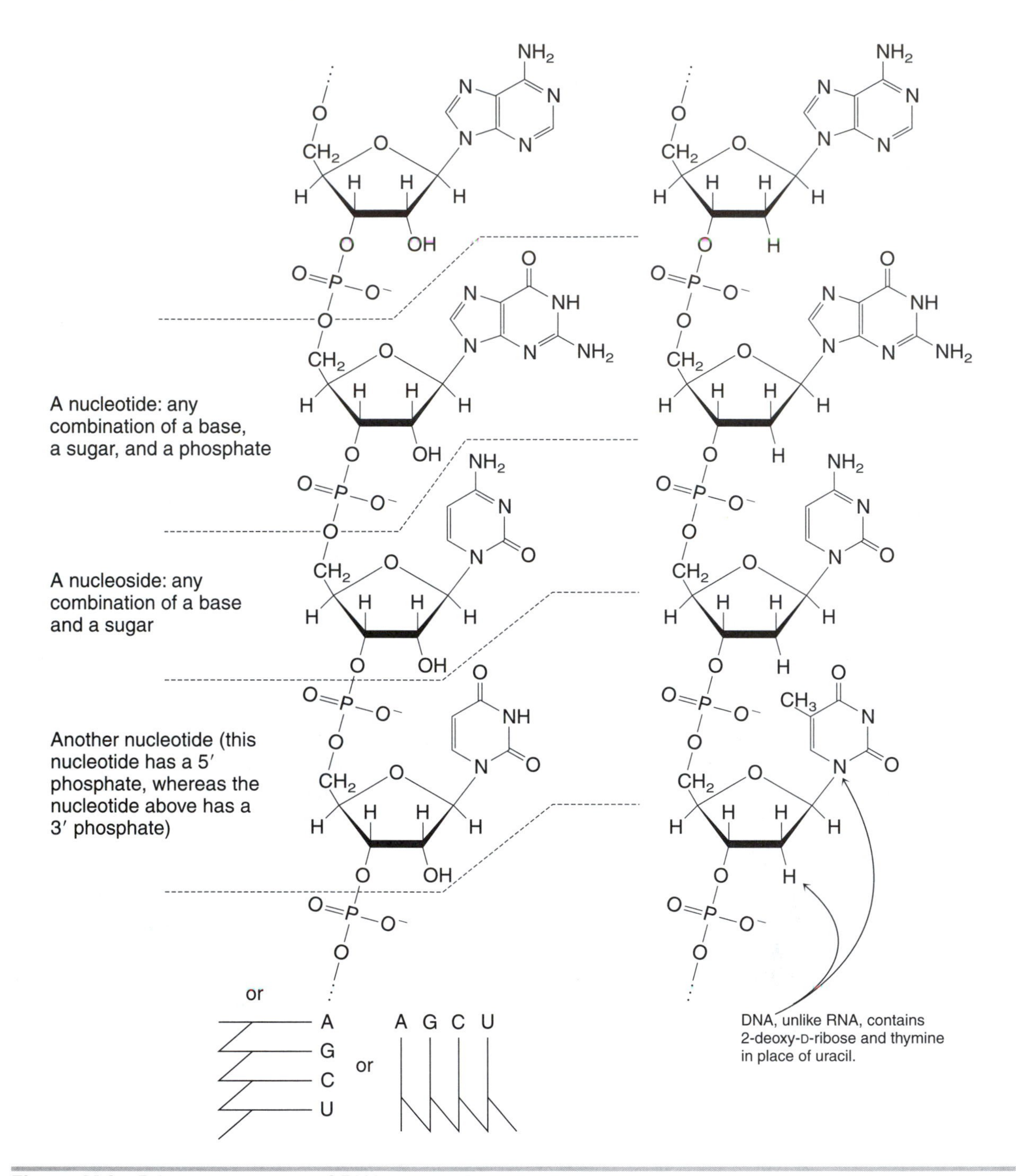

Figure V-2 Primary structures of RNA and DNA.

bonds with uracil and guanine hydrogen bonds with cytosine in RNA and in DNA:RNA duplexes. It should be noted, however, that noncanonical base pairing (e.g., G═U) is also commonly found in stretches of RNA. The overlapping π electron orbitals of the stacked planar base pairs add to the stability of the polynucleotide (Figs. V-3, and V-4).

Most DNAs contain extensive secondary structure, existing as large double-stranded helixes containing two separate, hydrogen-bonded

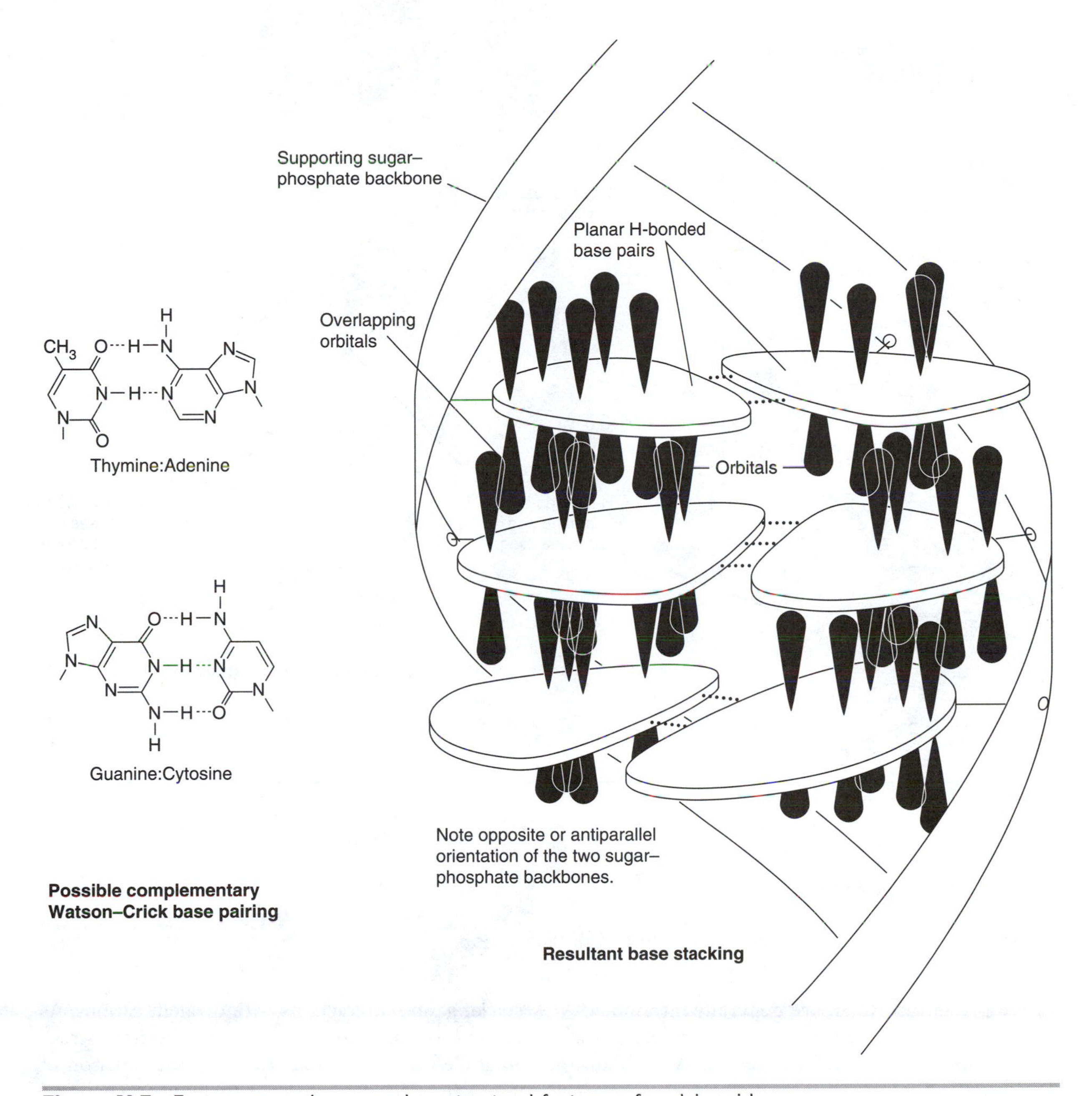

Figure V-3 Factors governing secondary structural features of nucleic acids.

complementary sequences. These separate primary strands exist together in opposite or antiparallel orientation (one strand running 5′ to 3′ and the other running 3′ to 5′), while being stabilized by intramolecular base stacking and by intermolecular hydrogen-bonded base pairing. This yields the long, fibrous "double helix" characteristic of most DNAs (Fig. V-4). This lengthy structure contributes to the pronounced physical characteristics of DNA: extensive viscosity in solution, the potential to rupture or "shear" on agitation, ease of fiber formation for x-ray analysis, and so forth. Double-stranded DNA will denature or "melt" (uncoil to form single strands) when subjected to high temperatures, strong bases, and denaturants such as urea and formamide. Secondary structure also plays a role in the function of various RNAs. Hydrogen-bonding and base pairing,

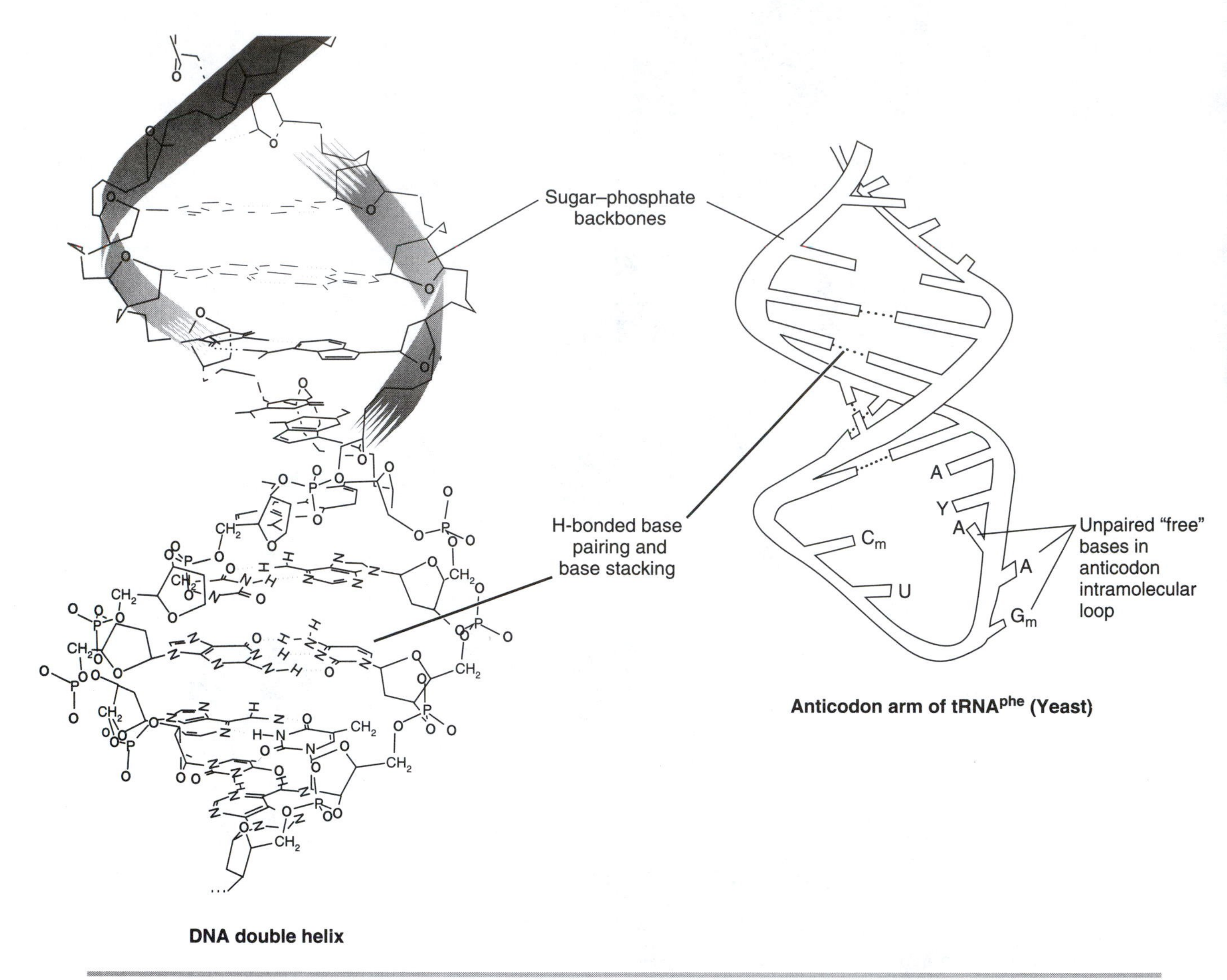

Figure V-4 Examples of secondary structure in nucleic acids.

where present in RNA, are generally intramolecular and generate helical hairpin loops within a primary nucleotide sequence (see Fig. V-4). Thus, precipitated RNAs are more amorphous than fibrous. Accordingly, most RNAs do not markedly increase solution viscosity and can be difficult to crystallize for subsequent x-ray analysis.

Certain RNAs also possess substantial amounts of tertiary structure. Tertiary structure in RNA refers to the folding of secondary structural elements, such as double helical regions or hairpin stem-loops, into discrete three-dimensional structures. Forces involved in stabilizing such interactions are diverse, involving hydrogen-bonding, base stacking, and interactions with divalent cations. As is the case for secondary structure, tertiary structural elements are often critical to the function of a particular RNA molecule, which may include activating or "charging" amino acids for translation, proper ribosome function, or messenger RNA splicing (see below).

Cellular Localization and Roles of Nucleic Acids

DNA contains the basic genetic information of all living cells. As such, cellular DNA is located at the site of primary genetic activity within cells. In *prokaryotic* cells (i.e., cells lacking a nucleus) genetic activity occurs throughout the cytoplasm. Thus, the various molecules of circular DNA (chromosome and plasmids) residing in prokaryotic cells are not localized to a specific compartment of the cell. In contrast, the DNA of *eukaryotic* cells (i.e., cells with defined nuclei) lies within discrete, membrane-enclosed organelles. The majority of the DNA of a eukaryotic cell exists in the nucleus. In the "resting" cell not undergoing DNA replication and cell division, the nuclear DNA is complexed with a number of histone proteins. This compact form of DNA is referred to as *chromatin* (Fig. V-5). During mitosis, the nuclear DNA will "unwind" from its compact structure into individual chromosomes that can be replicated by the action of DNA polymerase. The remaining DNA of eukaryotic cells exists in the partially self-duplicating chloroplasts and mitochondria. These nonnuclear DNAs have densities different from the nuclear DNA, and are dispersed throughout these membrane-bound organelles.

There are three types of RNAs found within cells that participate in some aspect of the transfer of genetic information from DNA to protein: ribosomal RNAs, messenger RNAs, and transfer RNAs. Ribosomal RNAs (rRNAs) participate directly in this genetic information transfer process in that they constitute approximately 60% of the mass of the ribosome. Ribosomes differ slightly in their physical properties and component rRNAs depending on their prokaryotic or eukaryotic origin (Fig. V-6). Because cells contain a large number of these ribosomes, rRNAs account for about 70% of the total cellular RNA. Because of their high concentration and uniform size, rRNAs are also easy to detect and isolate.

Messenger RNAs (mRNAs) and transfer RNAs (tRNAs) participate directly in protein synthesis. The nucleotide sequence of one of the two DNA strands in the double-helix is *transcribed* by RNA polymerase to produce a mRNA that is complementary to it (see Experiment 22). A portion or all of this mRNA nucleotide sequence will then serve as the template for the ribosome, which will *translate* the mRNA molecule into a polypeptide or protein (see Experiment 23). Messenger RNAs, then, carry information contained in the DNA into the production of protein.

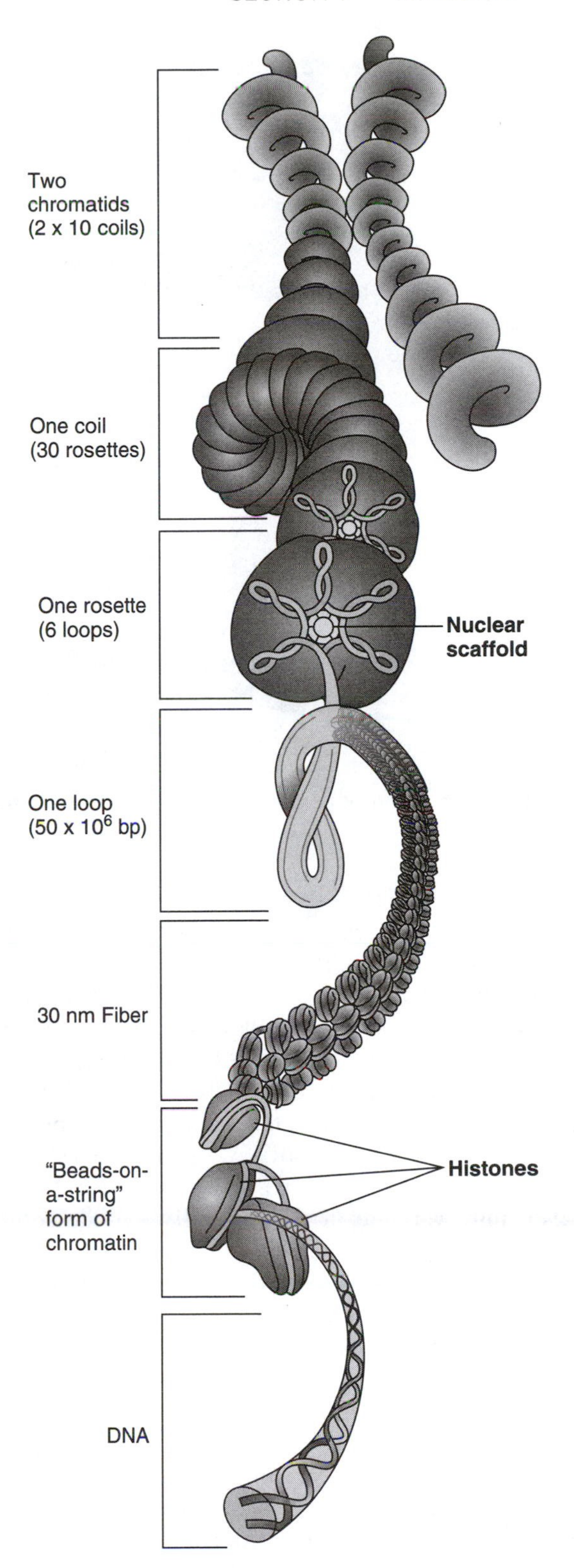

Figure V-5 The highly ordered structure of DNA in chromatin.

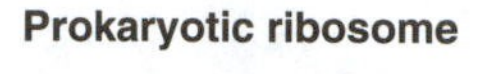

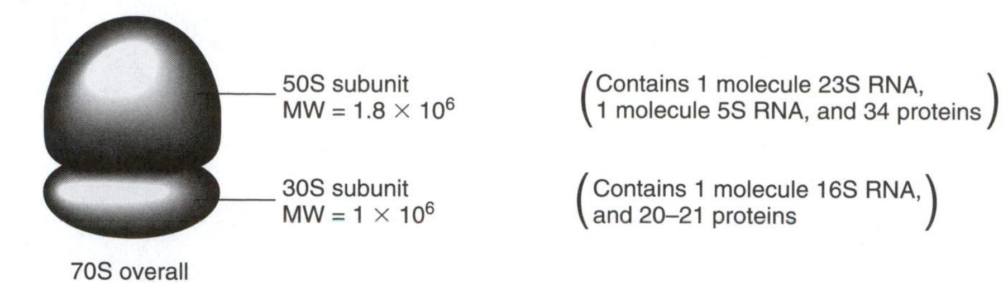

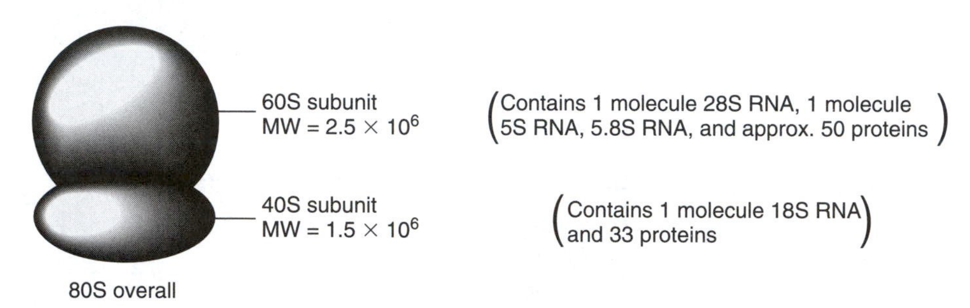

Figure V-6 Component parts of natural ribosomes.

Messenger RNAs constitute approximately 10% of the total cellular RNA. This mRNA can be either *monocistronic* (governing the synthesis of a single polypeptide) or *polycistronic* (governing the synthesis of two or more polypeptide chains). Since the ribosomes bind the mRNA and initiate translation at defined positions (the *ribosome-binding sites*), translation may begin from more than one site on a single molecule of mRNA. As suggested above, mRNAs can be quite variable in size. mRNAs are also quite variable in their stability. While some mRNAs may have a half-life of several hours, others may have a half-life on the order of minutes. In general, prokaryotic mRNA is nearly one hundred times less stable than the average eukaryotic mRNA in terms of half-life. This difference in mRNA stability may be a function of the faster growth rate of prokaryotic cells compared to eukaryotic cells (the typical prokaryotic cell can divide every 20 min, while the typical eukaryotic cell divides once every 30 hr). The ribonuclease activity of both types of cells ensures that the composition and concentration of the total mRNA "pool" within the cell will be constantly changing to meet the changing needs of the cell. Eukaryotic cells have developed specific mechanisms to control the stability of individual mRNAs as a means of regulating gene expression.

Transfer RNAs (tRNAs) participate in translation by transferring free amino acids into the growing peptide chain. Unlike mRNAs, tRNAs are uniform in size (25 to 30 kDa) and constitute approximately 20% of the total cellular RNA. Normal cells will have a variety of different tRNAs, one or more that will recognize each of the 20 amino acids and deliver them to the growing polypeptide chain at the appropriate time. Although there are only 20 amino acids, the degeneracy of the genetic code dictates that a single cell must contain at least 32 different tRNAs with the capability of recognizing different codons (three-base sequence) on the mRNA.

One final class of eukaryotic RNA that deserves mention are the *small nuclear RNAs* (snRNAs). As their name suggests, this class of RNA is found in the nucleus and participates in the splicing of introns from certain types of eukaryotic mRNAs. Five snRNAs (U1, U2, U4, U5, and U6) take part in these splicing reactions when combined with a number of nuclear proteins. Together, the snRNAs and the proteins form a functional splicing unit referred to as *small nuclear ribonuclear proteins* (snRNPs or "snurps"). Although the snRNAs are very important in eukaryotic RNA processing, their small size (between 100 and 200 nucleotides) contributes a very small percentage to the total cellular RNA. The role of snRNAs and the mechanism of intron splicing are currently a topic of intensive research. In addition to snRNAs, a number of other small nuclear and cytoplasmic RNAs have been discovered and shown to serve a variety of biochemical functions.

Chemical Properties of Nucleic Acids

Action of Alkali

The *N*-glycoside linkages of DNA and RNA are stable in mildly alkaline conditions. However, the phosphodiester linkages of DNA and RNA demonstrate markedly different chemical properties in mild alkali. For example, 0.3 M KOH (37°C) rapidly (~1 hr) catalyzes the cleavage of all phosphodiester bonds in RNAs and yields intermediate cyclic-2′,3′-phosphonucleosides that hydrolyze to nucleoside-2′-phosphates and nucleoside-3′-phosphates on further (12–18 hr) incubation (Fig. V-7). The phosphodiester bonds of DNA are stable under these conditions because they lack the adjacent 2′ hydroxyl group on D-ribose required to form the cyclic-2′,3′-phosphonucleoside intermediates.

Action of Acid

Very mild acid, or dilute acid at or below room temperature, has little effect on nucleic acids. Indeed, dilute acids (e.g., trichloroacetic, $HClO_4$, HCl) at 0 to 25°C are routinely used to precipitate (isolate) nucleic acids and other polar macromolecules. Somewhat stronger acidic conditions (prolonged exposure to dilute acid or dilute acid at increased temperature or acid strength) cause *depurination*, i.e., acid-catalyzed hydrolysis of purine *N*-glycosides. Thus, 15 min of treatment with 1 N HCl at 100°C removes most of the purines from DNA and RNA. Much harsher acidic conditions (several hours at 100°C or more concentrated acid) are required to remove pyrimidine *N*-glycosides. Some limited

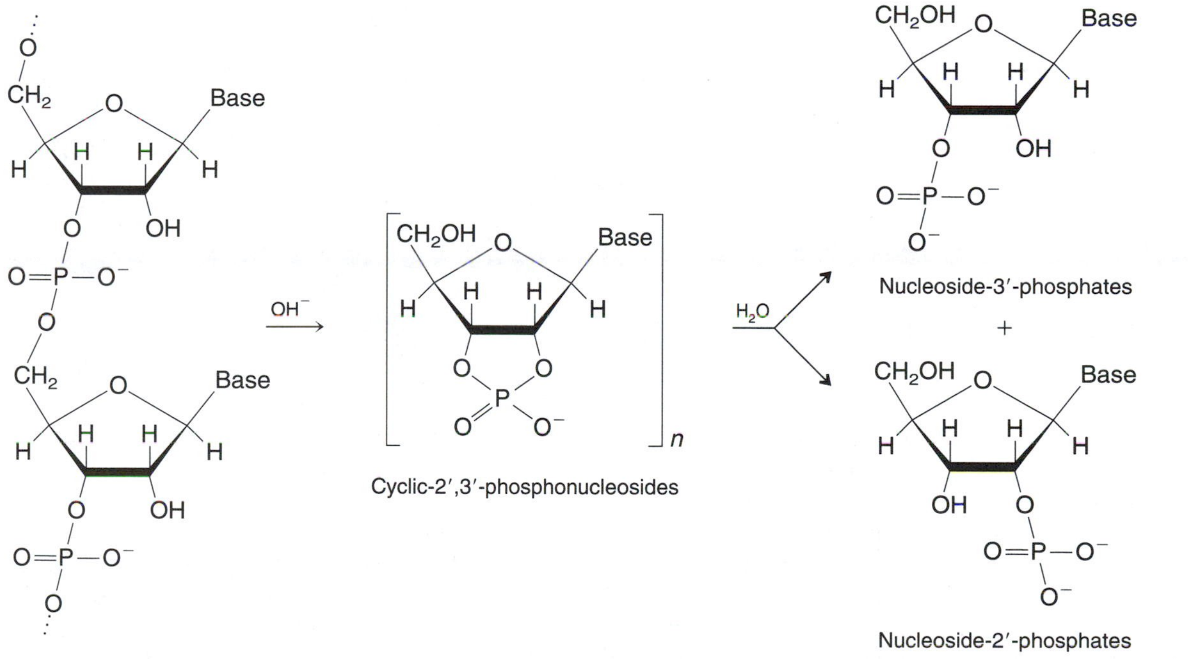

Figure V-7 Alkaline hydrolysis of RNA.

phosphodiester bond cleavage occurs during acid-catalyzed depurination. Even more phosphodiester cleavage occurs during acid-catalyzed depyrimidination.

Physical Properties of Nucleic Acids

Several of the physical properties of nucleic acids have already been mentioned in the discussion of nucleic acid structure. The following properties play important roles in the isolation, detection, and characterization of nucleic acids.

UV Absorption of Nucleic Acids

All nucleic acids, nucleosides, and nucleotides strongly absorb ultraviolet (UV) light. This UV absorption arises from the nitrogenous bases, each of which possesses its own unique absorption spectrum (Fig. V-8). The sum of these individual nitrogenous base spectra generates the overall UV spectrum of DNA and RNA. Several features of these spectra deserve further comment. First, the nucleoside spectra are a function of pH. This property reflects the titration of specific functional groups within the nitrogenous bases. Such pH-dependent spectral shifts provide an additional tool for characterizing unknown solutions containing these nucleosides. Second, the ratios of the absorption at two different wavelengths (e.g., A_{280}/A_{260}) at any pH provide criteria for identification of individual nitrogenous bases or nucleosides, and for the estimation of the relative proportions of individual nucleosides in an oligonucleotide. Third, a nucleic acid spectrum is the sum of the individual nitrogenous base spectra present in the nucleic acid. Such summation yields a spectrum with a λ_{max} of 260 nm. It follows, then, that the absorption of a solution containing a nucleic acid is proportional to the nucleic acid concentration. In general, a 1-mg/ml solution of single-stranded DNA (lacking any degree of secondary structure) has an A_{260} of 25 at neutral pH in a 1-cm light path ($\varepsilon_{260} = 25$ (mg/ml)$^{-1}$ cm^{-1}). Extensive secondary structure, as found in native tRNA, rRNA, and DNA, reduces the UV absorption of the stacked, hydrogen bonded bases so that a 1-mg/ml solution of these types of nucleic acids has an A_{260} of 20 at neutral pH in a 1-cm light path. It follows, then, that the disruption of secondary structure within DNA, rRNA, and tRNA increases the A_{260} of a solution containing these molecules.

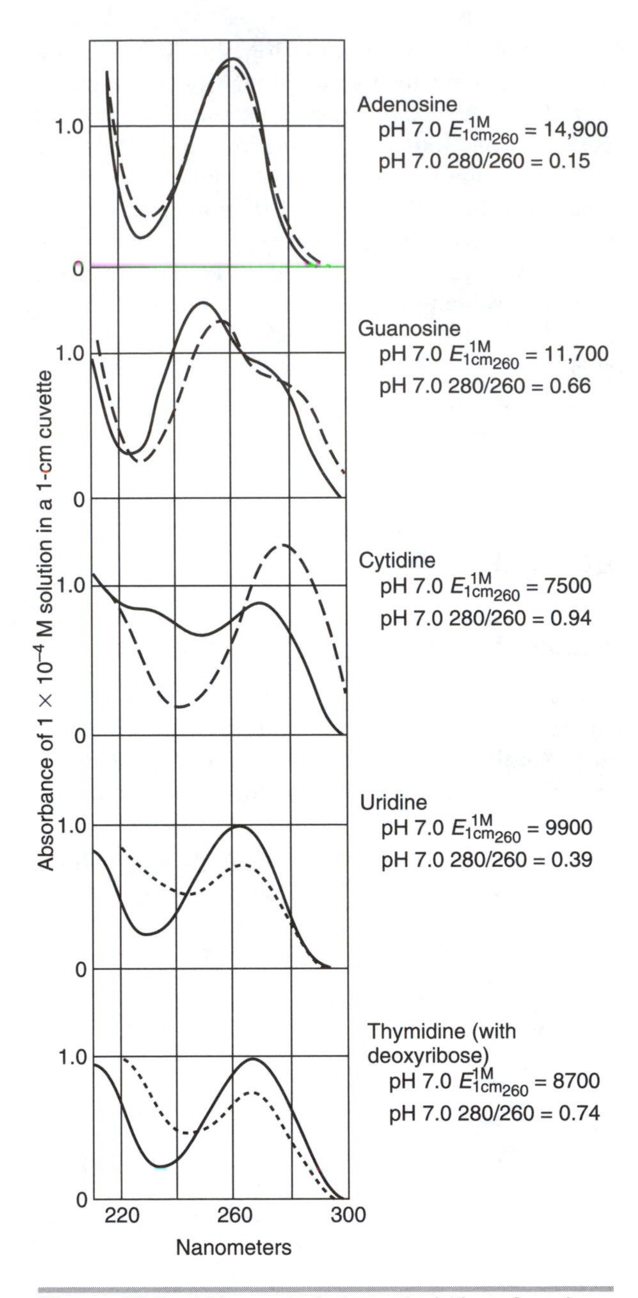

Figure V-8 UV spectral characteristics of nucleosides. (——) = pH 7.0, (— — —) = pH 1.0, and (- - -) = pH 13.0.

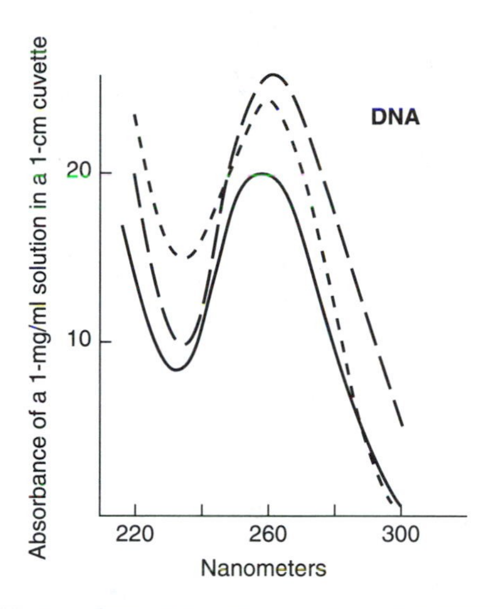

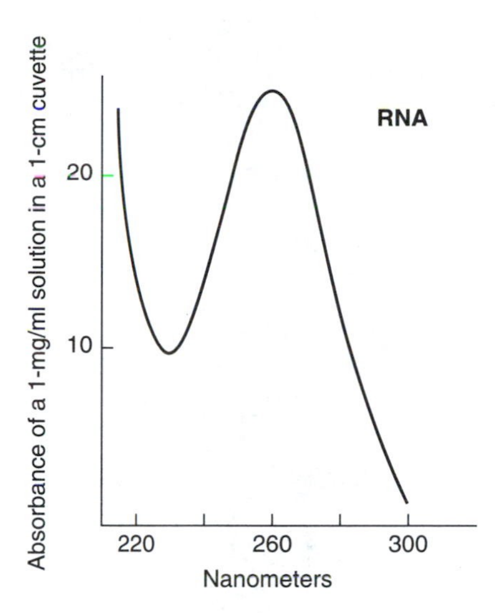

Figure V-9 UV characteristics, right, of single-stranded RNA and, left, of DNA. (———) = pH 7.0, (— — —) = pH 1.0, (- - -) = pH 13.0. The denaturation of double-stranded DNA at pH 1 and 13 causes an increase in UV absorbance, an illustration of the hyperchronic effect.

This increase in the absorption spectrum following denaturation (destruction of secondary structure) is termed the "hyperchromic effect" (Fig. V-9). Conversely, the decrease in the absorption spectrum on renaturation of these types of nucleic acids (restoration of secondary structure) is termed the "hypochromic effect." These effects are observed in Experiment 19.

Acid–Base Titration of Nitrogenous Bases, Nucleotides, Nucleosides, and Nucleic Acids

The nitrogenous bases and phosphates of nucleic acids contain groups that titrate in aqueous media (Table V-1). Thus, nucleic acids have titration curves and act as buffers in solution and in biological extracts. Extremes of acid or alkali will introduce charges on nitrogenous bases and yield structures that no longer participate in base stacking and hydrogen bonding. Extremes of acid or alkali destroy (denature) secondary structure in nucleic acids. Using Table V-1 and the Henderson–Hasselbach equation, you can calculate the net charge on any nucleoside or nucleotide at any pH. These charges, and their effect on the overall polarity of the molecule, serve as the basis for resolution of nucleotides and nucleosides in chromatographic and electrophoretic systems (see Experiments 2 and 4).

Hybridization

Secondary structure in nucleic acids can be disrupted by extremes of pH or by increasing the temperature of the nucleic acids residing in a solution of low ionic strength. Disruption of double-stranded nucleic acids, which contain stacked bases and hydrogen-bonded base pairs (e.g., DNA), yields separate, single-stranded, complementary strands of polynucleotides. At favorable temperatures, pHs, and ionic strength conditions, these single-stranded molecules will reassociate with themselves or *hybridize* with other polynucleotides containing a complementary sequence. Biochemists make great use of this phenomenon. Specifically, if hybridization is allowed to take place in the presence of an *isotopically labeled*, single-stranded DNA molecule with a known sequence, then a fragment of DNA containing the complementary sequence can be identified in a complex DNA sample. This process of carrying out DNA:DNA hybridization on a solid support (Fig. V-10) was developed by Edward Southern. Because of this, the technique is called *Southern blotting*. The technique has proven

Table V-1 Titration Properties of Nitrogenous Bases and Phosphates of Nucleic Acids and Their Subunits

Name of Ionizing Group	Primary Ionization	Secondary Ionization
Phosphodiesters of internucleotide linkages of RNA and DNA	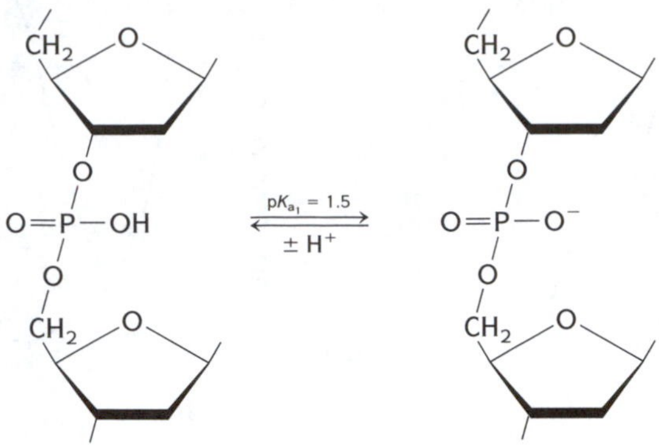 $pK_{a_1} = 1.5$, $\pm H^+$	
Phosphomonoesters of nucleotides or terminal phosphates	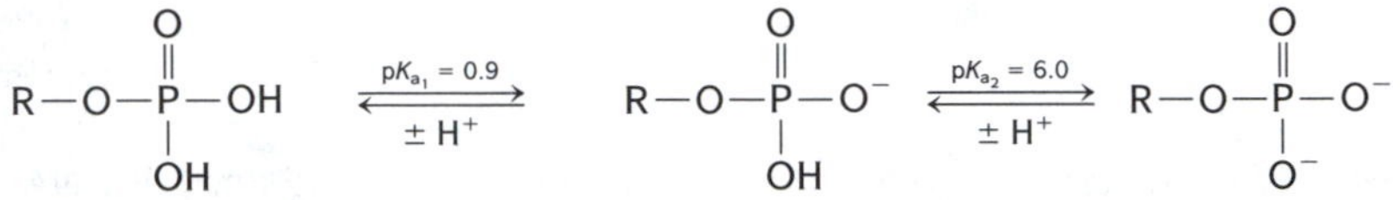 $R{-}O{-}P(=O)(OH){-}OH$ $\underset{\pm H^+}{\overset{pK_{a_1} = 0.9}{\rightleftharpoons}}$ $R{-}O{-}P(=O)(OH){-}O^-$	$\underset{\pm H^+}{\overset{pK_{a_2} = 6.0}{\rightleftharpoons}}$ $R{-}O{-}P(=O)(O^-){-}O^-$
Adenine in adenosine, adenylic acids, AMP, etc., and RNA and DNA	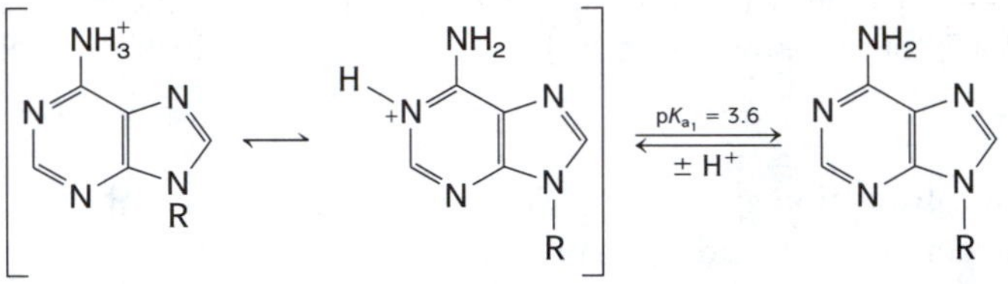 $pK_{a_1} = 3.6$, $\pm H^+$	
Guanine in guanosine, guanylic acids, GMP, etc., and RNA and DNA	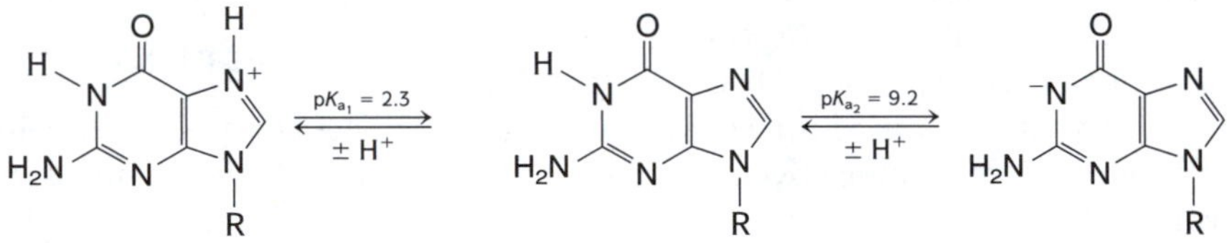 $pK_{a_1} = 2.3$, $\pm H^+$	$pK_{a_2} = 9.2$, $\pm H^+$
Cytosine in cytidine, cytidylic acids, GMP, etc., and RNA and DNA	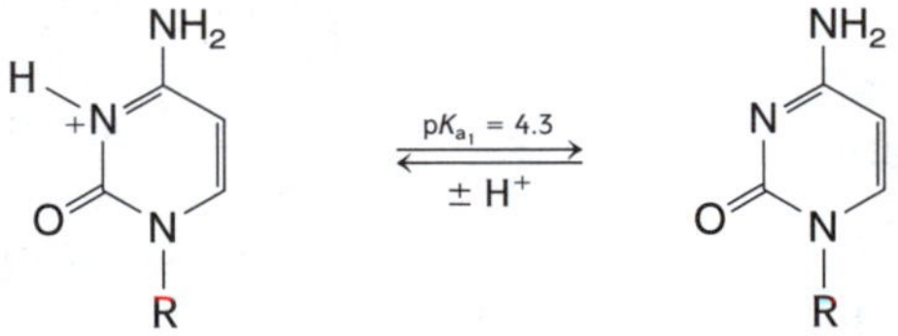 $pK_{a_1} = 4.3$, $\pm H^+$	
Uracil in uridine, uridylic acids, UMP, etc., and RNA	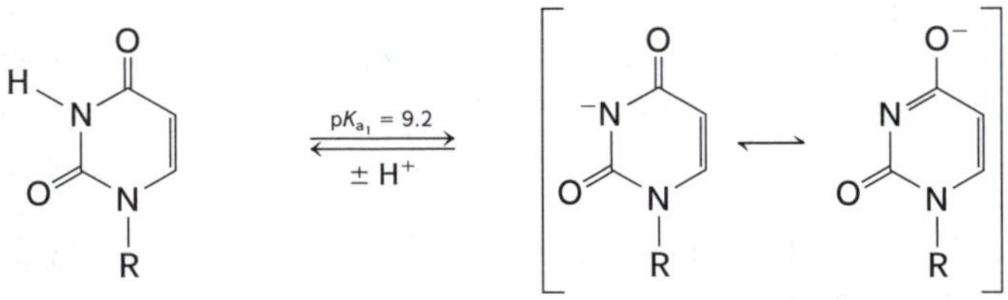 $pK_{a_1} = 9.2$, $\pm H^+$	
Thymine in thymidine, thymidylic acids, TMP, etc., and DNA	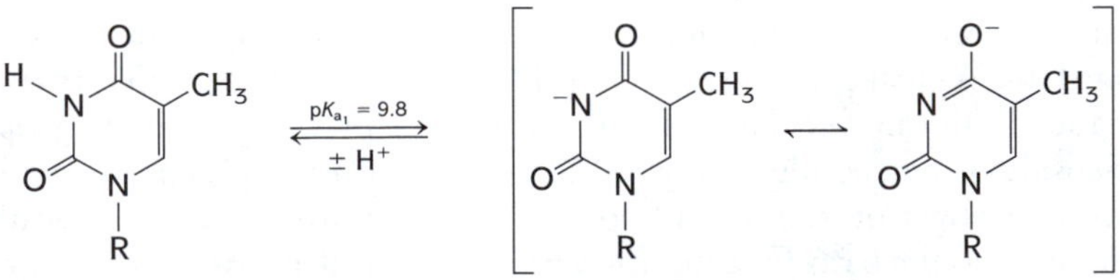 $pK_{a_1} = 9.8$, $\pm H^+$	

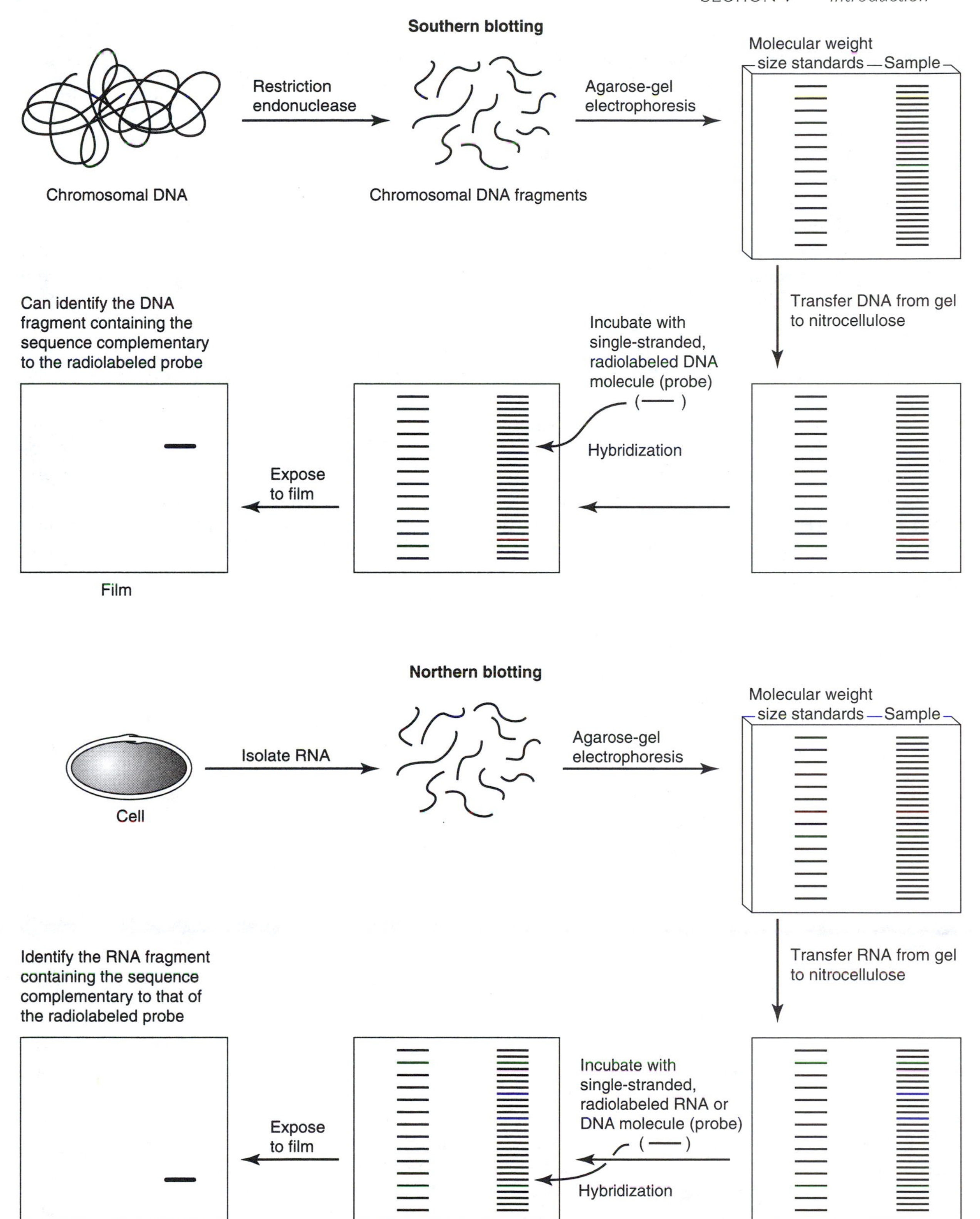

Figure V-10 Southern blotting and Northern blotting.

to be a very powerful tool in molecular biology. You can identify a DNA fragment of interest in any complex mixture of nucleic acids (such as from a chromosomal digest), and attempt to isolate and clone it. Southern blotting can also be used to identify the position of various restriction endonuclease cleavage sites within the nucleotide sequence of interest, a technique that has been put to good use in forensic biochemistry and genetic screening for mutations associated with various diseases and conditions (see Experiment 24, Fig. 24-5).

A similar technique, in which RNA is bound to a solid support and hydridized to an isotopically labeled single-stranded RNA or DNA molecule, is called *Northern blotting* (Fig. V-10). As with the technique of Southern blotting, Northern blotting can be used to determine if an RNA molecule with a particular sequence is present in a complex mixture of nucleic acids. You can imagine how this technique may be useful in studies involving the regulation of gene expression: the mRNA transcribed from a particular gene can be identified and quantified at any time during the growth cycle, and changes in the level of transcription of particular genes can be monitored as specific alterations are made to the growth media. With this technique, it has been possible to study the effects of specific biological componds (nutrients, drugs) on gene expression, as well as the roles of various protein and nucleic acid structural elements that control the process of transcription. Hybridization is also a critical element in the powerful technique known as the polymerase chain reaction (PCR), described in detail in Experiment 24.

Plasmids and Other Cloning Vectors

Plasmids

Plasmids are naturally occurring, extrachromosomal, circular DNA elements that endow the host organism with potentially advantageous capabilities. Genes carried on these plasmids may confer resistance to specific antibiotics, allow the cell to metabolize harmful organic compounds, and/or allow the cell to ward off infection by bacteriophage. Some plasmids also carry genes involved in bacterial conjugation (mating) and virulence. The discovery of plasmids has revolutionized the study of biochemistry. Although the recombinant plasmids used in modern molecular biology have been engineered for a number of different applications, there are several features that are common to most plasmids (Fig. V-11).

Recall that DNA replication initiates at a specific DNA sequence termed the *origin of replication.* It is the specificity of this sequence on a plasmid that will determine the host organism in which a particular plasmid will be able to be propagated (replicated). An *Escherichia coli* plasmid contains an *E. coli* origin of replication, a *Bacillus subtilis* plasmid contains a *B. subtilis* origin of replication, and so forth. Some recombinant plasmids contain the origin of replication for more than one organism. These "shuttle" plasmids can therefore be propagated in more than one organism. Specific elements within the origin of replication will determine the frequency of replication of the plasmid residing in the host cell. "Low-copy" plasmids have origins of replication that are under the same stringent control as the chromosome, which is ultimately regulated by the cell cycle. Hence, one or two molecules of a "low-copy" plasmid will be present in an individual host cell. "High-copy" plasmids contain a less stringent or "relaxed" origin of replication. Since a high-copy plasmid is not under the same replication control as the chromosome during cell division, hundreds of molecules of a high copy plasmid can be present in a single cell.

All recombinant plasmids contain a genetic element that provides a means of selecting for those cells that harbor it. Most commonly, the plasmid will contain a gene coding for an enzyme that will allow the cell to metabolize or inactivate a specific antibiotic. For instance, the β-lactamase gene (*bla*) on many *E. coli* plasmids will allow the cell to cleave the lactam ring structure common to the family of β-lactam antibiotics, thereby inactivating them. A list of antibiotics, their mode of action, and the genes involved in their inactivation are listed in Table V-2. Antibiotic resistance genes are not the only genetic selection markers that can be present in plasmids. Genes encoding enzymes involved in essential biosynthetic pathways can also provide a means of selection, provided that the host strain into which the plasmid is introduced contains a null mutation (inactivation) of

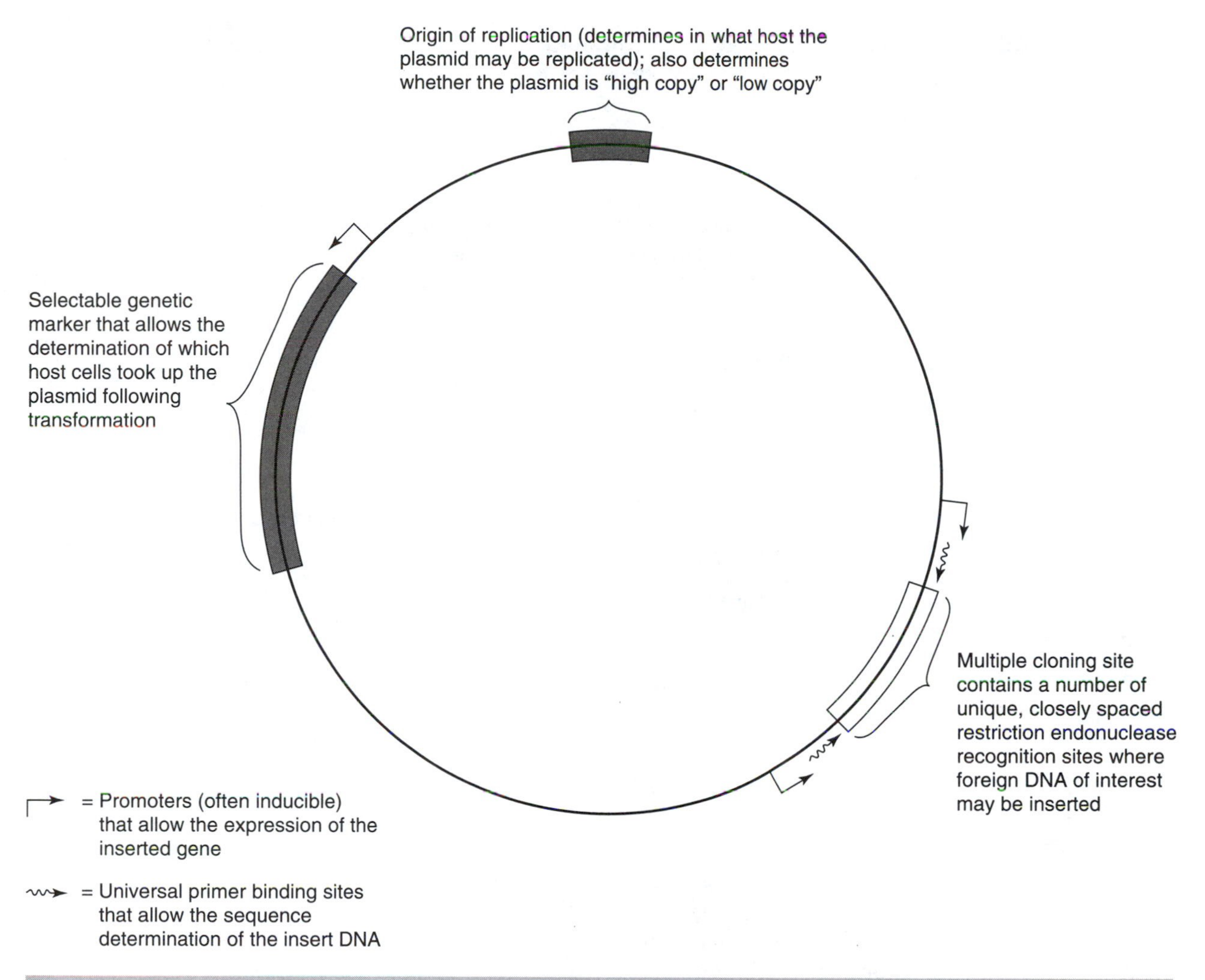

Figure V-11 Features found on many popular plasmids used for different molecular biology applications.

that same gene on the chromosome. This idea of selection for transformed cells is discussed later in this section.

Recombinant plasmids have been engineered to "accept" fragments of foreign DNA. Most plasmids contain a multiple cloning site (MCS) into which foreign DNA can be inserted. The MCS is a short segment of DNA on the plasmid that contains a number of unique restriction enzyme recognition sequences found nowhere else on the plasmid. The plasmid and the DNA insert can be digested with one or more of these restriction endonucleases and ligated together to produce a new recombinant plasmid that carries the DNA insert of interest (Fig. V-12). Once this recombinant plasmid has been isolated, it can be introduced into and faithfully propagated (cloned) by the host organism.

This process of inserting genes into plasmids is the cornerstone of molecular biology. Once a gene of interest has been separated from the chromosome and inserted into a plasmid, the gene can be subjected to a number of different manipulations that provide insight into its biological function. In biochemical studies, it is often necessary to express the gene carried on the plasmid, isolate the protein that it encodes, and study its function. To aid in this process, many recombinant plasmids have been engineered to include powerful promoter sequences 5′ to the MCS. Genes cloned into the MCS downstream (3′) to the promoter can be expressed at high

Table V-2 Some Commonly Used Antibiotics, Their Modes of Action, and Their Inactivation by Resistant Host Cells

Antibiotic	Mode of Action	Inactivation
Chloramphenicol	Binds to 23S rRNA in the 50S ribosomal subunit and inhibits translation	Chloramphenicol acetyl transferase acetylates the antibiotic
Ampicillin	Inhibits cross-linking of peptidoglycan chains in bacterial membrane synthesis	β-lactamase cleaves the ring in the antibiotic
Erythromycin	Binds to 23S rRNA in the 50S ribosomal subunit and inhibits translation	An enzyme methylates a specific adenine residue in the rRNA so that it is unable to bind the antibiotic
Tetracycline	Binds to the 30S ribosomal subunit and prevents the tRNA from interacting with the ribosome	A membrane protein actively expels or transports the antibiotic out of the cell
Kanamycin and Neomycin	Bind to the 30S ribosomal subunit and inhibits translation	Aminoglycoside phosphotransferase phosphorylates the antibiotic
Streptomycin	Binds to the 30S ribosomal subunit and inhibits translation	A deletion of the S12 ribosomal protein prevents binding of the antibiotic

levels in the host cell, and the proteins that they encode can be purified. Often, you would like to control the expression of genes cloned into plasmids. This becomes critical if the high level expression of a particular gene product is found to be lethal to the host cell. Many of the promoter elements that flank the MCS, therefore, are able to be *induced* through the addition of a small molecule or some other change in the condition of the growth medium. In this way, the host strain can be grown in culture, and the gene of interest can be expressed at any particular time during the growth cycle. A list of inducible promoters found in recombinant plasmids and their means of induction are listed in Table V-3.

Expression of a gene of interest is by no means the only manipulation that can be carried out once the gene of interest has been inserted into a plasmid. Current techniques in molecular biology allow you to make virtually an unlimited number of mutations in the gene-carrying recombinant plasmid: 1 specific base pair can be changed, 2 specific base pairs can be changed, 50 base pairs can be deleted, 20 base pairs can be added, and so forth. The information obtained from the analysis of these mutant genes can provide a great deal of insight into the various domains, structure, and function of the proteins that they encode.

Bacteriophage λ

Although plasmids are the most popular cloning vectors used in modern molecular biology, they are generally limited to use with DNA inserts less than 10 kb in size. If the DNA fragment of interest is between 10 kb and 20 kb, you may wish to use a *viral* cloning vector. One commonly used viral DNA vector that can be propagated in *E. coli* is the double-stranded, linear chromosome of bacteriophage λ. This bacteriophage chromosome, about 48 kb in size, encodes a number of proteins required for replication of the viral chromosome, head and tail assembly of the phage particle, and lysis of the host cell. The genes that encode these proteins are known to lie on either end of the linear chromosome, and constitute about 65% of the total phage genome. The remaining 35% of the bacteriophage genome contains nonessential genes that can be

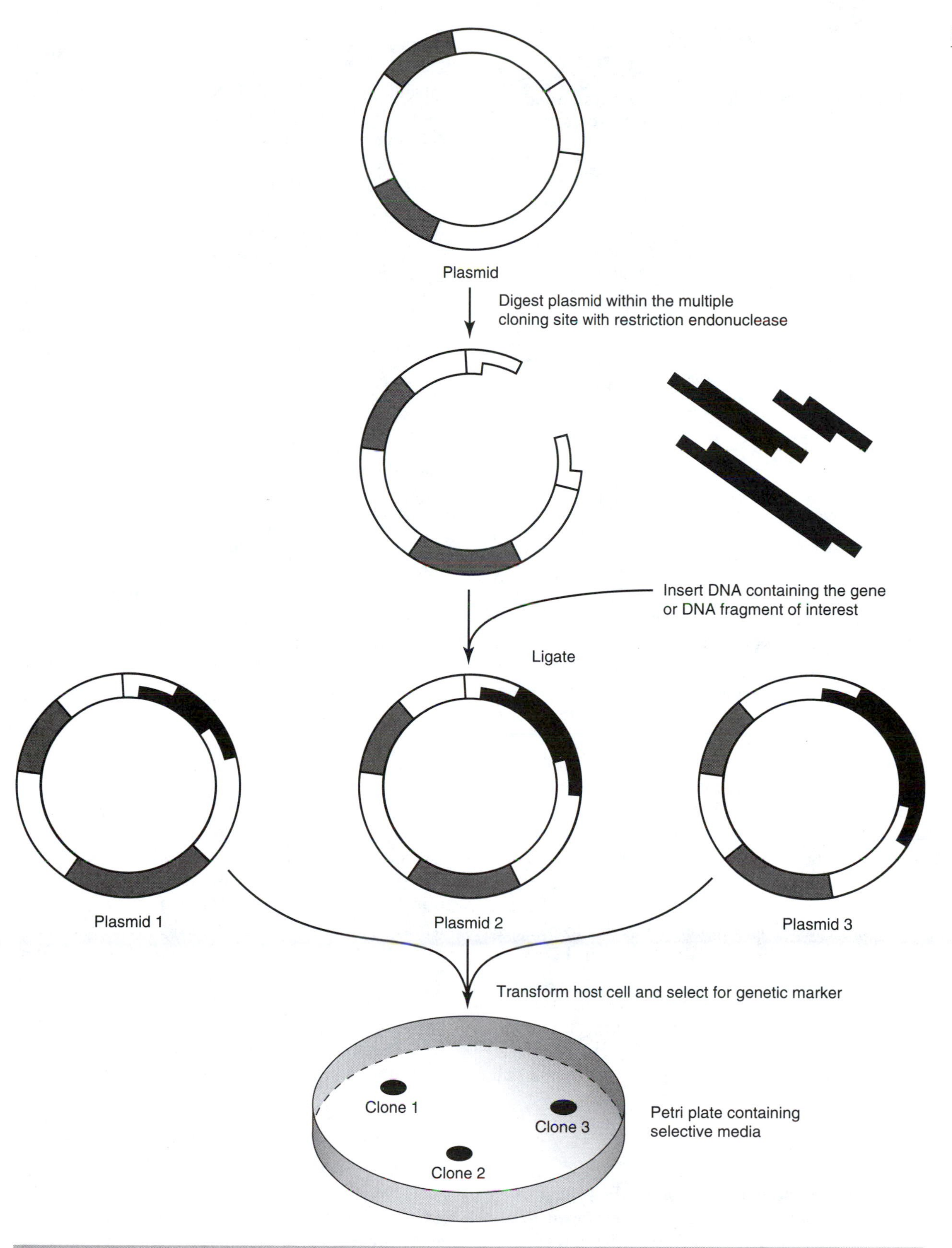

Figure V-12 General cloning strategy utilizing plasmids.

Table V-3 Some Inducible Promoters Found in Commercially Available Vectors

Promoter*	Inducer
lac	Isopropylthiogalactoside (IPTG)
tac	Isopropylthiogalactoside (IPTG)
λ	Increased temperature (42°C)
BAD	Arabinose
MMTV (mouse mammary tumor virus)	Dexamethasone
GAL1	Galactose
AOX1	Methanol
MT (metallothionein)	$CuSO_4$
Δ*HSP*	Muristerone A

*Lower-case letters denote prokaryotic origin, upper-case letters denote eukaryotic origin.

excised and replaced with any foreign DNA fragment that is 10 to 20 kb in size (Fig. V-13). Once the DNA fragment of interest has been inserted into the bacteriophage λ chromosome, the chimeric chromosome can be packaged in vitro into phage particles that will deliver the DNA into the *E. coli* host. Since the chimeric genome still contains the genes required for replication, phage assembly, and cell lysis, the DNA fragment that is inserted into the chromosome will be replicated (cloned), packaged, and used to infect neighboring *E. coli* cells. The *plaque* produced when cells infected with the recombinant phage are plated on a lawn of uninfected *E. coli* cells contains progeny (clones) of a single phage. Since bacteriophage λ is able to package only DNA between 40 and 50 kb into the phage head, DNA inserts smaller than 10 kb and larger than 20 kb in size cannot be cloned using this vector. Other *E. coli* bacteriophages, such as the single-stranded M13 phage, have smaller chromosomes that can accept DNA fragments smaller than 10 kb in size.

Cosmids

The *cosmid* is another variation of the plasmid and viral DNA vectors that has been developed to allow the cloning of DNA fragments that range from 40 to 50 kb. As the linear bacteriophage λ DNA is introduced into *E. coli*, DNA ligase will convert it to a closed, circular form by joining together the cohesive ends (12 bases in length) at either end of the molecule (Fig. V-14). The region at which these two complementary strands are joined is the *cos* site. Following replication and packaging of this now circular chromosome into the phage head, the chromosome is again cleaved at the *cos* site to produce the linear form of the chromosome that will be used to infect the next host cell. By placing the *cos* site from bacteriophage λ into a plasmid, the phage will be able to package and process the vector in vitro, *provided that the DNA fragment inserted in the multiple cloning site is 40 to 50 kb in size.* After the cosmid is cleaved (linearized) and introduced into the *E. coli* host, it will be closed or circularized to reform the *cos* site. Unlike the viral DNA vector, this circular DNA molecule contains none of the phage genes required for replication, packaging, or cell lysis. Therefore, the cosmid will continue to be replicated from the bacterial origin of replication, just as with a plasmid. As stated earlier, the in vitro packaging system requires that the total size of the DNA molecule containing the *cos* site falls within a specified range. Therefore, DNA fragments smaller than 40 kb and larger than 50 kb cannot be cloned with cosmid vectors.

Yeast Artificial Chromosomes (YACs)

This type of DNA cloning vector, designed to accept DNA fragments up to several hundred kilobases in size, contains the three elements that will allow propagation in yeast: an origin of replication—the *autonomously replicating sequence*—a centromere, and a telomere at either end of the linear DNA molecule. The centromere is the region on the chromosome that will attach to the spindle during mitosis. The telomeres are the small (100-base-pair) repetitive sequences found at the end of linear eukaryotic chromosomes that are crucial in the DNA replication process.

Restriction Endonucleases and Other Modifying Enzymes

The discipline of molecular biology requires the ability to manipulate the gene or DNA sequence under study. In the early 1970s, an enzyme was isolated

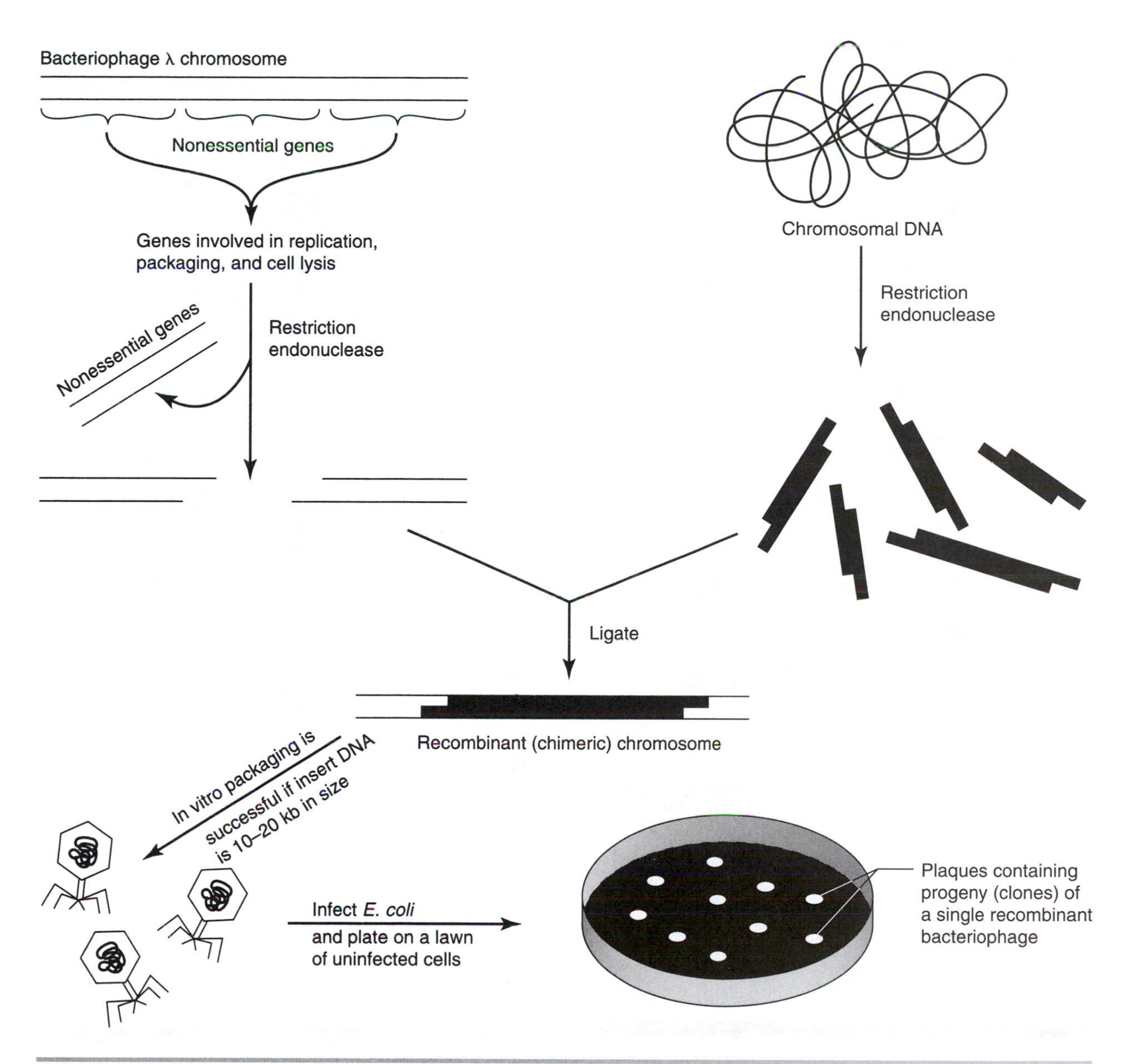

Figure V-13 Cloning genes using bacteriophage λ.

that catalyzes the hydrolysis of the phosphodiester backbone in both strands of a double-stranded DNA molecule at a highly specific sequence. The enzyme was termed a *restriction endonuclease*. Currently, over 700 restriction endonucleases have been isolated from a variety of different prokaryotic organisms. Why would an organism produce enzymes that cleave DNA molecules? It is thought that restriction endonucleases act as a defense mechanism to block or *restrict* infection by bacteriophage. If the host cell that the phage is invading is able to destroy the phage chromosome before the genes carried on it are expressed, the phage will be unable to replicate and lyse the host cell. In a sense, you might consider restriction endonucleases to be a crude immune system for organisms that produce them.

How does the cell prevent these same restriction endonucleases from destroying its own chromosomal and plasmid DNA? It has been determined that restriction endonucleases are one

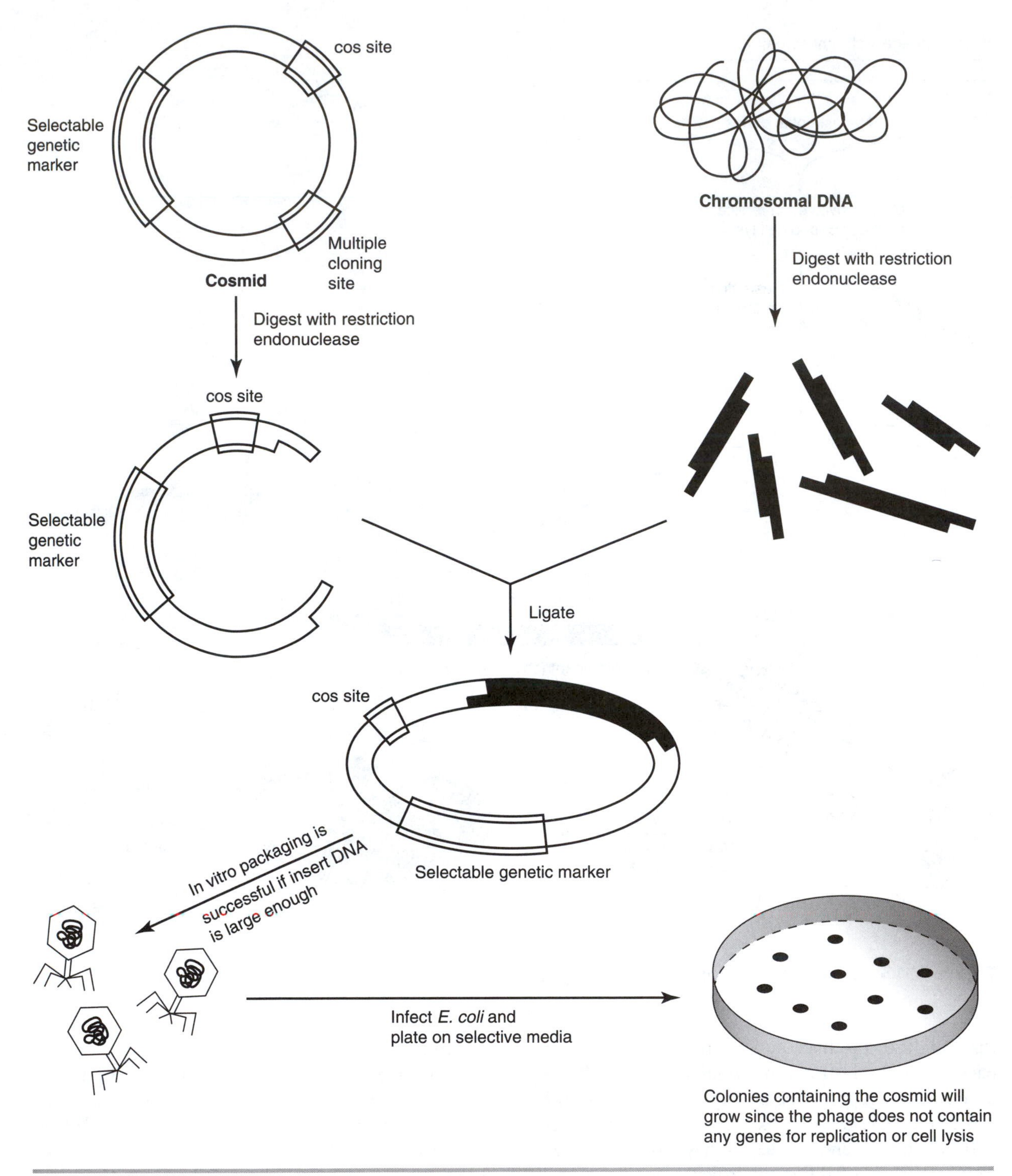

Figure V-14 Cloning genes with the use of cosmids.

component of a two-component system collectively referred to as the *restriction-modification system*. The restriction endonuclease component will recognize a specific base-pair sequence in the DNA and catalyze the hydrolysis of the phosphodiester backbone in both strands of the DNA duplex. The second component of this system, the *DNA methylase*, catalyzes the transfer of a methyl group from S-adenosylmethionine to a specific adenine or cytosine base *within* the same DNA sequence recognized by the endonuclease (Fig. V-15). Once the recognition sequence has been methylated, the DNA is resistant to the action of the restriction endonuclease. As the bacterial chromosome is replicated, the DNA methylase methylates the appropriate sequences and protects the DNA from hydrolysis. Since the invading phage DNA is not protected by methylation, it will be destroyed by bacterial restriction endonucleases and exonucleases (see below) before it can cause damage.

There are three types of restriction-modification systems. Type I and type III systems consist of a single, multisubunit enzyme that conducts both the endonuclease and methylase activities. In an ATP-dependent mechanism, type I restriction endonucleases scan the DNA, bind to a specific recognition sequence, and cleave both strands of the DNA duplex at a site 1 to 10 kb from the recognition sequence. Type III restriction endonucleases do not require ATP and usually cleave both strands of the DNA duplex within 100 base pairs of the recognition sequence. For reasons that are not entirely clear, *S*-adenosylmethionine is required for the activity of both the type I and type III endonucleases, despite the fact that it is not consumed during the hydrolysis of the phosphodiester backbone. While the type III DNA methylases modify (methylate) a single strand of the DNA duplex within the recognition sequence, type I DNA methylases methylate both strands of the duplex within the recognition sequence.

Type II restriction-modification systems differ from their type I and type III counterparts in that the endonuclease and DNA methylase activities are conducted by two separate enzymes (not a single multisubunit complex). The restriction endonuclease cleaves both strands of the DNA duplex within a defined recognition sequence, while the companion DNA methylase methylates a specific base within the same recognition sequence. In contrast to the type III endonucleases, the hydrolysis of the phosphodiester backbone does not require the presence of ATP or *S*-adenosylmethionine. All of the restriction endonucleases discussed above are similar in that they utilize a divalent cation as a cofactor, most commonly Mg^{2+} or Mn^{2+}.

Because type II restriction endonucleases cleave the DNA at a specified position within (rather than adjacent to) a defined recognition sequence, they are the enzymes of choice in most practical applications in molecular biology. Some commercially available type II restriction endonucleases, their recognition sequences, and their cleavage sites are listed in Table V-4. Note that some of these enzymes make *staggered* cuts across the often palindromic recognition sequence, producing 5′ single-stranded DNA overhangs, while others cleave straight across the DNA duplex, producing *blunt-ended* DNA fragments.

In addition to the very useful type II restriction endonucleases discussed above, a number of other commercially available enzymes can be utilized in various applications in molecular biology (Table V-5). Exonucleases have been discovered that hydrolyze DNA and/or RNA from either the 3′ or 5′ terminus. Some of these enzymes act on single-stranded nucleic acids, others on both strands of the DNA duplex. The majority of these modifying enzymes are used to remove nucleic acids from cell extracts during the early phases of protein purification. Other applications include the preparation of single-stranded DNA templates for sequence analysis, preparation of blunt-ended DNA fragments for cloning, DNA footprinting (see Experiment 22), removal of phosphoryl groups from the 5′ terminus of cleaved vector DNA to improve cloning efficiency, and radiolabeling of the 5′ hydroxyl group of chemically synthesized oligonucleotides for DNA sequencing and Southern blotting.

Ligation of DNA Fragments

The ability of a cell to join (ligate) DNA fragments is essential for its survival. This reaction is performed by an enzyme called *DNA ligase*. Recall that DNA ligase plays a critical role in DNA replication, linking together the Okazaki fragments synthesized by DNA polymerase on the lagging strand of the replication fork. DNA ligase is also a key

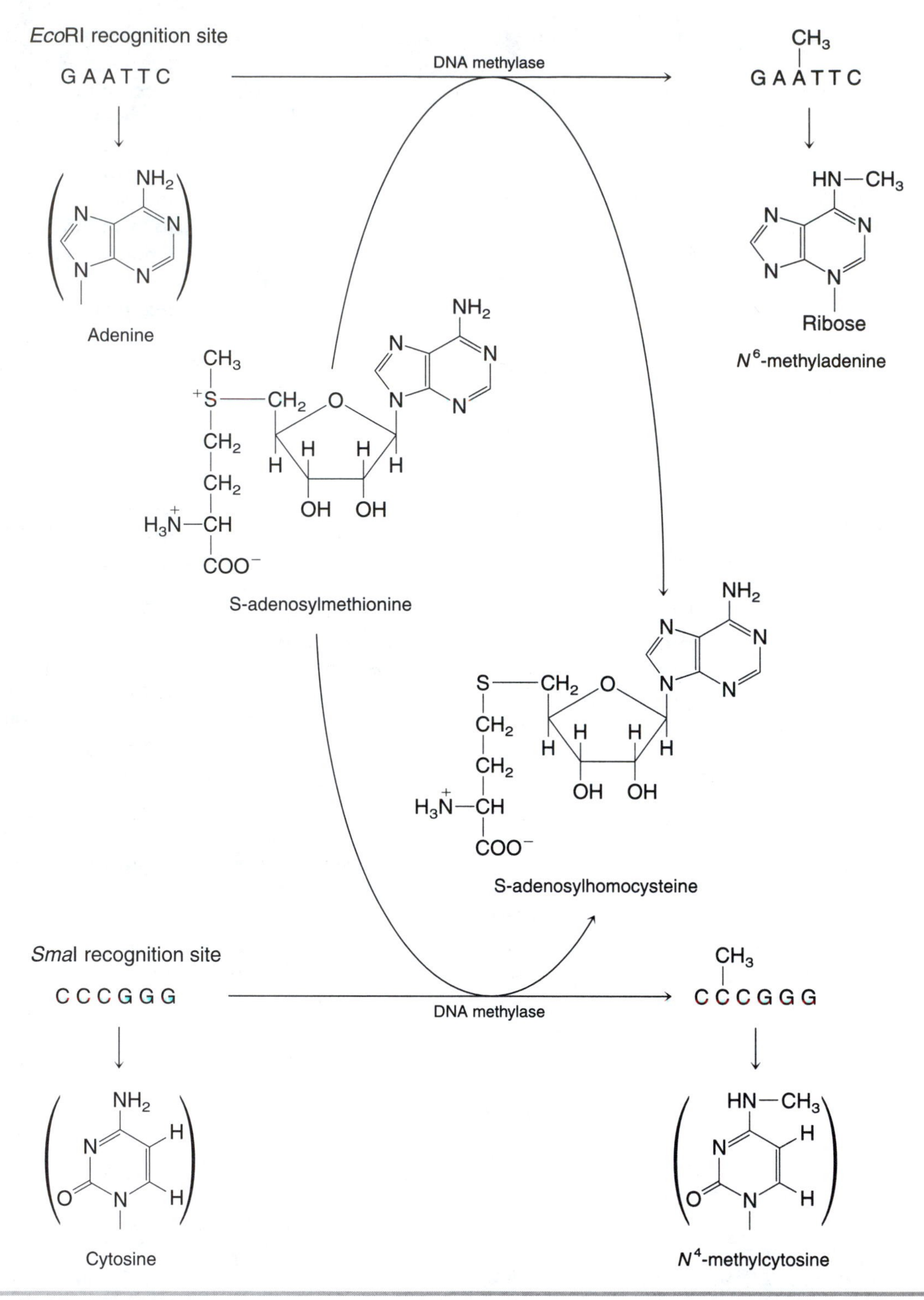

Figure V-15 The DNA methylase component of the restriction modification system methylates a specific cytosine or adenine residue, making it incapable of being acted on by the companion restriction endonuclease.

Table V-4 Some Commercially Available Type II Restriction Endonucleases

Restriction Endonuclease	Recognition Sequence
*Bam*HI	5′-G ↓ GATTC-3′ 3′-CCTAA ↑ G-5′
*Bgl*II	5′-A ↓ GATCT-3′ 3′-TCTAG ↑ A-5′
*Cla*I	5′-AT ↓ CGAT-3′ 3′-TAGC ↑ TA-5′
*Eco*RI	5′-G ↓ AATTC-3′ 3′-CTTAA ↑ G-5′
*Eco*RV	5′-GAT ↓ ATC-3′ 3′-CTA ↑ TAG-5′
*Hind*III	5′-A ↓ AGCTT-3′ 3′-TTCGA ↑ A-5′
*Kpn*I	5′-GGTAC ↓ C-3′ 3′-C ↑ CATGG-5′
*Nde*I	5′-CA ↓ TATG-3′ 3′-GTAT ↑ AC-5′
*Not*I	5′-GC ↓ GGCCGC-3′ 3′-CGCCGG ↑ CG-5′
*Sac*I	5′-GAGCT ↓ C-3′ 3′-C ↑ TCGAG-5′
*Sal*I	5′-G ↓ TCGAC-3′ 3′-CAGCT ↑ G-5′
*Sau*3AI	5′- ↓ GATC-3′ 3′-CTAG ↑ -5′
*Sma*I	5′-CCC ↓ GGG-3′ 3′-GGG ↑ CCC-5′
*Ssp*I	5′-AAT ↓ ATT-3′ 3′-TTA ↑ TAA-5′
*Xho*I	5′-C ↓ TCGAG-3′ 3′-GAGCT ↑ C-5′

component in four different DNA repair mechanisms: base-excision repair, base mismatch repair, nucleotide excision repair, and recombinational (SOS) repair. Homologous recombination, or exchange of DNA fragments within the genome that have similar sequences, also requires DNA ligase. This latter type of recombination, which occurs frequently during meiosis, is responsible for the genetic and phenotypic diversity seen within populations of higher eukaryotes.

Although DNA ligase is essential for numerous in vivo processes, it has also proven to be a very powerful tool in the growing field of recombinant DNA technology. As discussed earlier in this section, a gene of interest can be excised from the chromosome with restriction endonucleases, isolated, and inserted into a vector. The ligation of the DNA fragment of interest into the vector is catalyzed in vitro by DNA ligase (Fig. V-16). The reaction catalyzed by DNA ligase is energy-requiring. The enzyme first attaches an adenyl group (from ATP or NAD^+) to the 5′-phosphoryl group present on one of the two DNA strands. The phosphoamide adduct produced in this reaction is then subject to nucleophilic attack at the 5′ phosphoryl group by the 3′ hydroxyl group present on the other

Table V-5 Some Commercially Available Modifying Enzymes

Enzyme	Activity
Deoxyribonuclease I (DNaseI)	Cleaves both single- and double-stranded DNA
Exonuclease III	Removes nucleotides from the 3′ ends of double-stranded DNA
Lambda exonuclease	Degrades the 5′ phosphorylated strand of double-stranded DNA
Nuclease P1	Completely hydrolyzes DNA and RNA to form nucleotide monophosphates
Nuclease S1	Hydrolyzes single-stranded nucleic acids to form nucleotide monophosphates or 5′-phosphoryl oligonucleotides
Mung bean nuclease	Hydrolyzes single-stranded nucleic acids to form nucleotide monophosphates or 5′-phosphoryl oligonucleotides
Snake venom phosphodiesterase I	Removes nucleotides from the 3′-hydroxyl group terminus of DNA and RNA
Ribonuclease H	An endonuclease that degrades the RNA strand in a DNA–RNA duplex
Ribonuclease I	Hydrolyzes the phosphodiester bonds in RNA to form oligonucleotides
Polynucleotide kinase	Catalyzes the phosphorylation of the 5′-hydroxyl group terminus of DNA or RNA. Also catalyzes an exchange reaction between the γ-phosphate group of ATP and the 5′-phosphoryl group on DNA and RNA
Alkaline phosphatase	Removes 5′-phosphoryl groups from nucleic acid fragments

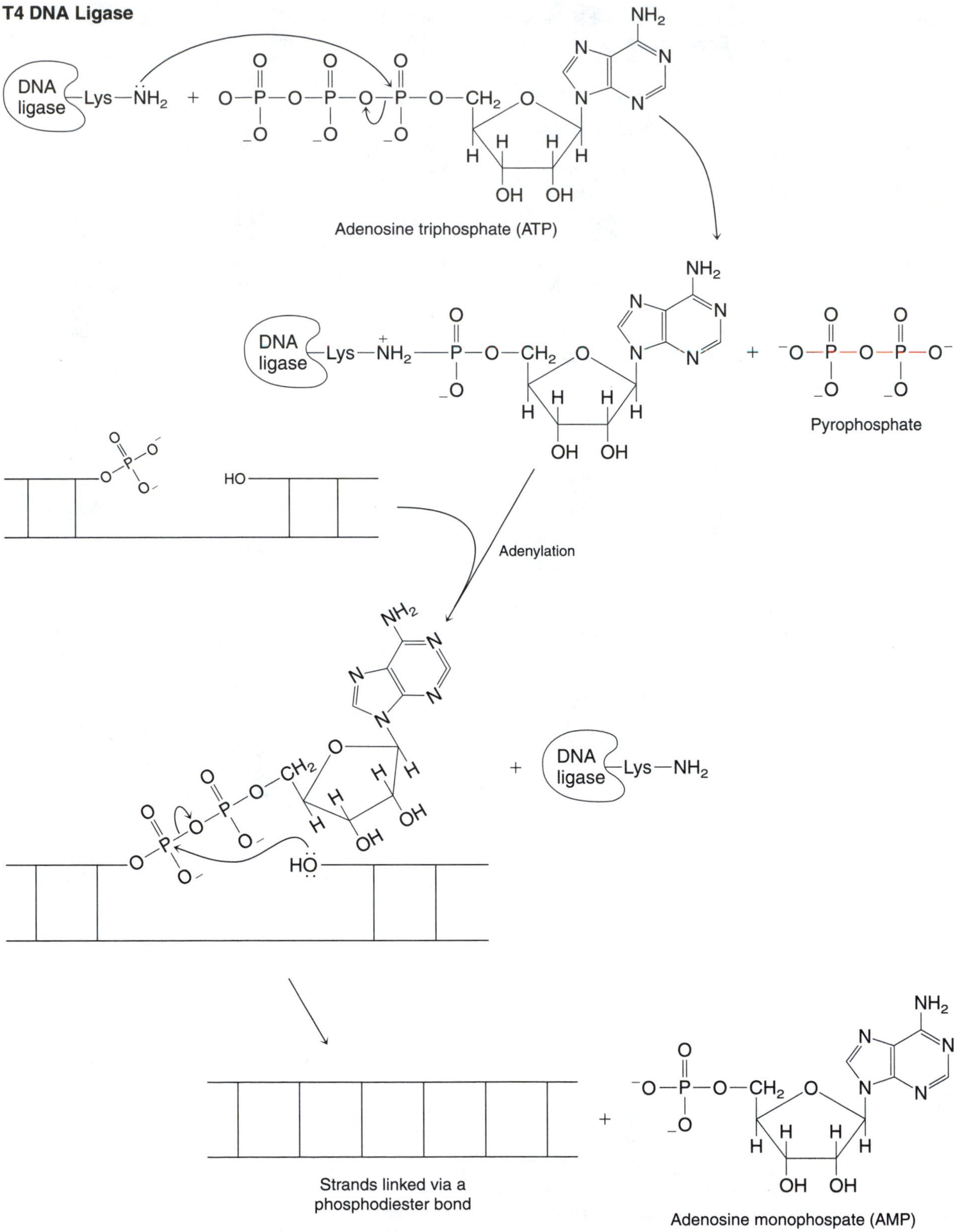

T4 DNA Ligase

DNA ligase — Lys — NH_2 + ATP

Adenosine triphosphate (ATP)

DNA ligase — Lys — $\overset{+}{N}H_2$ — AMP + Pyrophosphate

Pyrophosphate

Adenylation

+ DNA ligase — Lys — NH_2

Strands linked via a phosphodiester bond

\+ Adenosine monophospate (AMP)

Figure V-16 The reaction catalyzed by DNA ligase (T4).

***E. coli* DNA Ligase**

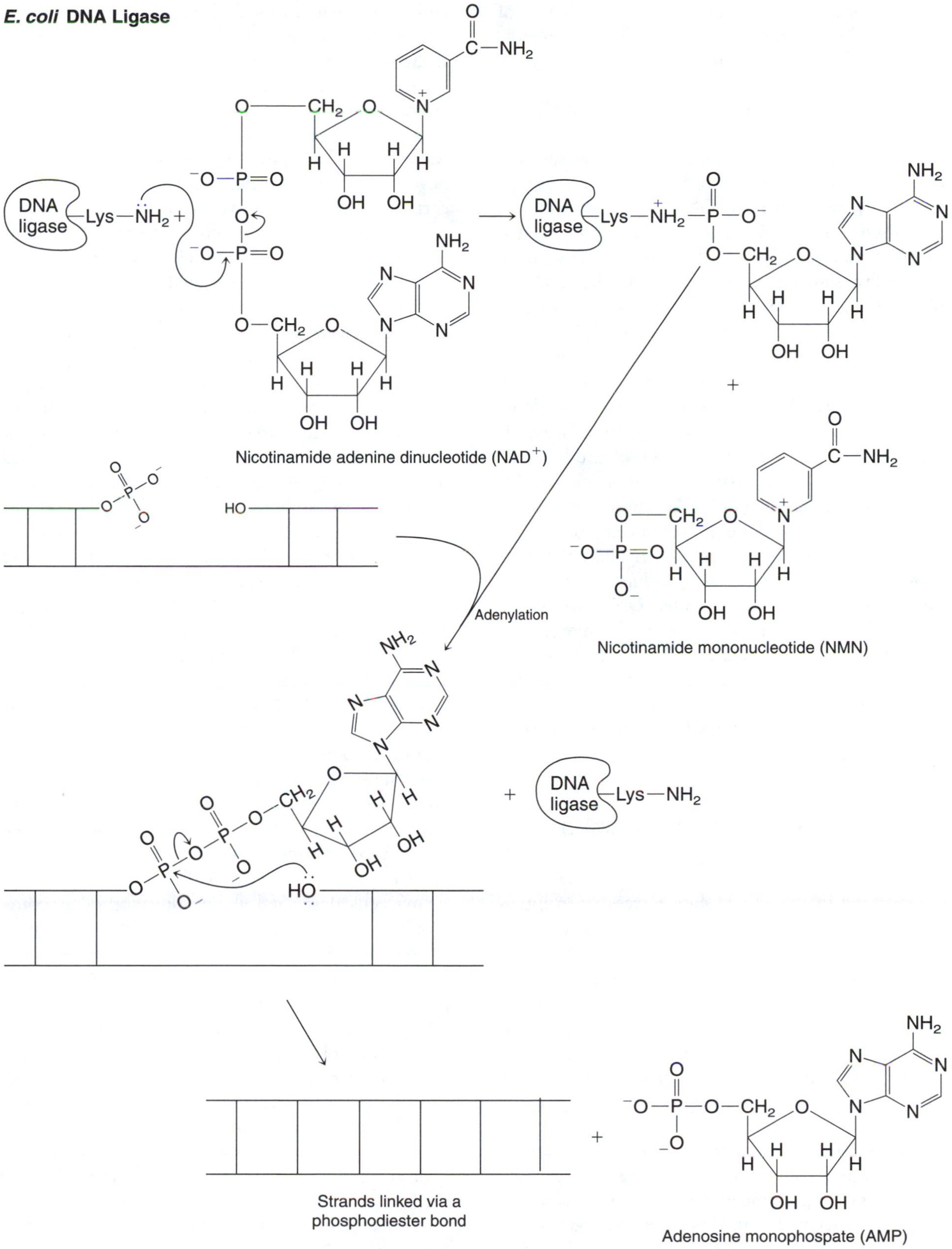

Nicotinamide adenine dinucleotide (NAD^+)

Adenylation

Nicotinamide mononucleotide (NMN)

Strands linked via a phosphodiester bond

Adenosine monophospate (AMP)

Figure V-16 (continued) The reaction catalyzed by DNA ligase (*E. coli*).

DNA strand. At the expense of one molecule of ATP, two DNA strands are joined via a phosphodiester bond. Although eukaryotic and T4 phage DNA ligase use ATP as the adenyl group donor in this reaction, *E. coli* DNA ligase utilizes NAD^+ (Fig. V-16).

Since the type II restriction endonucleases produce DNA fragments with 5′ phosphoryl groups and 3′ hydroxyl groups, DNA fragments isolated after digestion can be easily joined together in vitro in the presence of T4 DNA ligase and ATP. Since many of these same restriction endonucleases cleave the DNA at staggered positions within a specific recognition sequence, single-stranded DNA overhangs are present on the ends of these fragments. As long as there is complementarity between these single-stranded DNA overhangs on two DNA fragments, DNA ligase will be able to connect them. If the vector DNA and the insert DNA are both digested with the same restriction endonuclease, this will be the case (Fig. V-17). You may even be able to join two DNA fragments produced from digestion with two different restriction endonucleases, provided that both enzymes produce complementary (compatible) single-stranded DNA overhangs. Although the reaction usually requires higher concentrations of DNA ligase and longer incubation times, the enzyme can even join together blunt-ended (no single stranded overhangs) DNA fragments produced from digestion with some type II restriction endonucleases.

Transformation of Host Cells

Once the desired recombinant vector has been constructed in vitro, it is then necessary to introduce the vector into the appropriate host cell. If a viral cloning vector such as bacteriophage λ is used, this process becomes trivial. After the recombinant bacteriophage chromosome containing the gene of interest is packaged in vitro into an infectious phage particle, the genetic material packaged in the phage head will be delivered into the host cell via the normal, phage-directed, transfection mechanism.

Some bacterial species, such as *Bacillus subtilis*, have a genetically specified system that will allow them to take up extracellular DNA. Expression of the genes that make up this *competence system* (the ability to take up extracellular DNA) can be induced as the bacteria are starved for particular nutrients. If the bacteria are exposed to a recombinant vector after induction of the competence system, the host cell will take up the DNA and become *transformed* (eukaryotic cells that have taken up a recombinant vector are referred to as being *transfected*). It is believed that the competence system has evolved in certain bacterial species as a means of obtaining potentially advantageous genes from other organisms during times of environmental stress. Transformation by this process is illustrated in Experiment 20.

Other commonly used prokaryotic and eukaryotic host cells do not posess a genetically controlled competence system. Despite this fact, these types of cells can become competent after a series of chemical and/or physical treatments. For instance, if *E. coli* cells are first treated with a solution of $CaCl_2$ or $RbCl_2$ (see Table V-6) and heat shocked in the presence of a recombinant vector, a certain population or percentage of the cells ($\sim 10^6$ per μg of DNA) will become *transformed.* The hypotonic cation solution used in these procedures is believed to swell the outer membrane, allow expulsion of periplasmic proteins into the surrounding solution, and allow the negatively charged DNA to bind to the membrane. On heat shock, transient pores are created in the membrane, which allows the DNA to enter the cell. Although the host cells are near death following this heat shock treatment, many of them will soon recover and faithfully propagate the introduced vector. Transformation of yeast cells involves a similar protocol utilizing lithium acetate, polyethylene glycol (PEG), and heat shock (see Table V-6). Although the detailed mechanism of yeast-cell transformation is different from that of *E. coli* cells described above, the method is commonly used to produce stable yeast transformants.

Nearly any type of cell (prokaryotic or eukaryotic) can be transformed by the technique of *electroporation.* Protoplasts are first prepared by enzymatic or chemical disruption of the host-cell membrane polysaccharides. Next, the recombinant vector is introduced to the protoplast suspension residing in a *very low* ionic strength buffer (or distilled water). This DNA–protoplast suspension is then subjected to one or several 250-V pulses delivered from a cathode and anode placed directly into the solution. This applied voltage gradient will cause a certain population of the cells ($\sim 10^{10}$ per

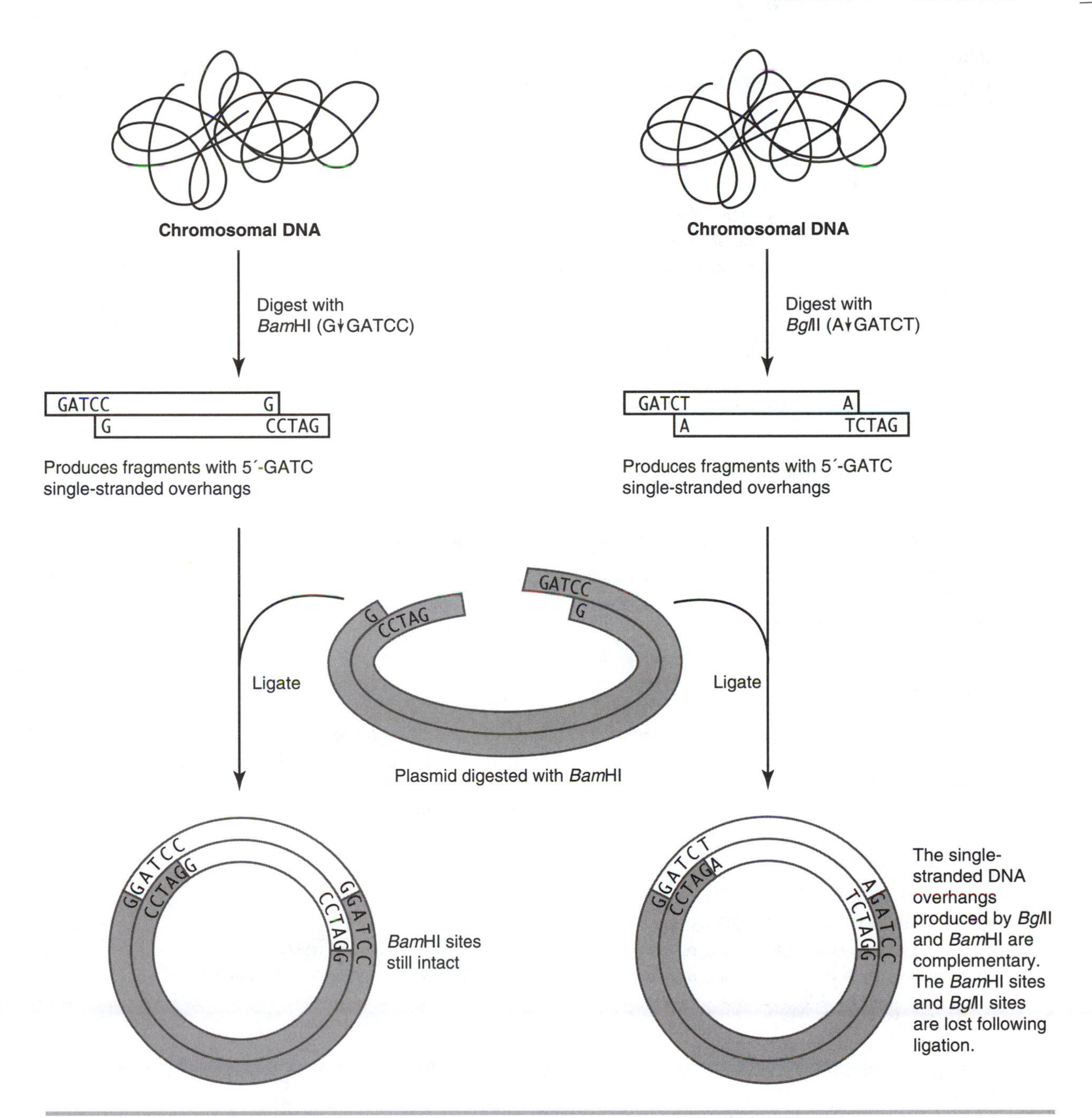

Figure V-17 Cloning genes using a single or compatible enzymes (*Bam*HI/*Bgl*II).

microgram of DNA) to take up the vector. *It is extremely important when using this technique to remove all of the salt from the solution.* If this is not done, the solution will arc (a spark will appear), heat, and kill all of the host cells. Even with a very-low-ionic-strength solution, a mortality rate of up to 80% is not uncommon with electroporation. Its major advantage is the nearly 10,000-fold increase in *transformation efficiency* (number of transformed cells produced per microgram of vector DNA) compared to the traditional *E. coli* transformation method described earlier.

Evidence indicates that transformation efficiency can be improved with yeast transformation

Table V-6 Preparation of *E. coli* and Yeast Competent Cells

E. coli ($CaCl_2$ method)

1. Grow 100 ml of *E. coli* cells to an OD_{600} of 0.35.
2. Harvest cells by centrifugation.
3. Resuspend cell pellet in 20 ml of 0.1 M $CaCl_2$. Incubate on ice for 10 min.
4. Harvest cells by centrifugation.
5. Resuspend cell pellet in 4 ml of ice cold 0.1 M $CaCl_2$.
6. Use 200–300 μl of these cells immediately in the transformation reaction.
7. Incubate the cells on ice for 30 min and heat shock at 42°C for 1–2 min.
8. Add 800 μl of rich media and incubate at 37°C for 1 hr.
9. Plate on selective media.

E. coli ($RbCl_2$ method)

1. Grow 50 ml of *E. coli* cells to an OD_{600} of 0.35.
2. Harvest cells by centrifugation.
3. Resuspend the cell pellet in 16 ml of transformation buffer 1:
   12 g of $RbCl_2$
   1.2 g of $MnCl_2 \cdot 2H_2O$
   1 ml of 1 M potassium acetate, pH 7.5
   1.5 g of $CaCl_2 \cdot 2H_2O$
   150 g of glycerol
   Bring total volume of solution to 100 ml with distilled water and *filter sterilize*
4. Incubate the cells on ice for 15 min.
5. Harvest cells by centrifugation.
6. Resuspend cell pellet in 4 ml of transformation buffer 2:
   1.2 g of $RbCl_2$
   20 ml of 0.5 M MOPS buffer, pH 6.8
   11 g of $CaCl_2 \cdot 2H_2O$
   150 g of glycerol
   Bring total volume of solution to 100 ml with distilled water and *filter sterilize*
7. Use 200–300 μl of these cells immediately in the transformation reaction. Cells prepared by this method can also be flash frozen in a dry ice ethanol bath and stored at −70°C for later use.
8. Incubate the cells on ice for 30 min and heat shock at 42°C for 1–2 min.
9. Add 800 μl of rich media and incubate at 37°C for 1 hr.
10. Plate on selective media.

Yeast

1. Grow 300 ml of yeast cells to an OD_{600} of 0.5.
2. Harvest the cells by centrifugation and resuspend the cell pellet in 10 ml of distilled water.
3. Harvest the cells by centrifugation and resuspend the cell pellet in 1.5 ml of Tris/EDTA/lithium acetate:
   10 mM Tris, pH 7.5
   1 mM EDTA
   100 mM lithium acetate
4. Add 1–5 μg of plasmid DNA plus 100 μg of single-stranded carrier (salmon sperm) DNA to a sterile 1.5-ml microcentrifuge tube.
5. Add 200 μl of the cells prepared in step 3 to the same tube.
6. Add 600 μl of the following solution to the same tube:
   40% wt/vol polyethylene glycol (PEG-8000)
   10 mM Tris, pH 7.5
   1 mM EDTA
   100 mM lithium acetate
7. Incubate the cells at 30°C for 30 min.
8. Add 90 μl of dimethyl sulfoxide (DMSO), and heat shock the cells at 42°C for 15 min.
9. Harvest the cells by centrifugation and resuspend the cell pellet in 1 ml of Tris/EDTA buffer (10 mM Tris, pH 7.5, 1 mM EDTA).
10. Plate cells on selective media.

and electroporation if a defined concentration of single-stranded *carrier DNA* is included in the host-cell suspension during transformation. This carrier DNA is usually from a *nonhomologous* source (such as salmon sperm) to ensure that the genetic integrity of the host genome is maintained. Carrier DNA containing homologous gene sequences may have the potential to recombine with sequences on the host cell genome and alter the genotype of the host cell. For reasons that are not entirely understood, a larger percentage of host cells will take up the desired vector in the presence of this carrier DNA.

Finally, a new transformation technique has been developed that literally "fires" or "shoots" the vector DNA into the desired host cell. This "gene gun" technique works particularly well on plant cells, which contain a rather tough and impermeable cell wall. The technique involves coating microscopic heavy metal particles (such as tungsten or gold) with vector DNA and projecting them at high speed toward the host cell. These particles actually pierce the cell wall and physically carry the DNA into the cell. Another advantage of this technique is that different cell types can be transformed in situ, or directly in the tissue where that cell type is normally found. By doing this, you can determine the effect of the gene in the developing organ or tissue.

Selection and Screening for Transformed Cells

Once a recombinant DNA vector has been introduced into the appropriate host cell, it is then time to *select* for those cells that received the vector. In general, you do this by growing the transformed cell population in conditions under which only those that received the plasmid will be able to survive. For instance, suppose that an *E. coli* strain is transformed with a plasmid carrying the *bla* (β-lactamase) gene and plated on rich media containing ampicillin. Since *E. coli* does not carry the *bla* gene on the chromosome, only those cells that took up the plasmid will express β-lactamase, metabolize (inactivate) the antibiotic, and form colonies on the plate. This type of drug selection is feasible for any cell type provided that the selectable marker (gene) is not normally expressed by the host cell.

Antibiotic resistance genes are not the only genetic markers that can be carried on a vector that will allow a powerful selection for transformed cells. Selection can also be made with the use of various *nutritional* genetic markers. Suppose that a population of yeast cells with a null mutation (deletion) of the *leu2* gene on the chromosome is transformed with a vector carrying the *leu2* gene. The *leu2* gene encodes β-isopropylmalate dehydrogenase. This enzyme oxidizes β-isopropylmalate to α-ketoisocaproate, the last intermediate in the leucine biosynthetic pathway. *Because of the null mutation in the leu2 gene, this strain will not be able to grow in media that does not contain leucine.* If the yeast cells are transformed with a vector carrying the *leu2* gene and plated on media that does not contain leucine, only the cells that took up the vector will be able to synthesize leucine and form colonies on the plate. Obviously, this same vector could not be selected for in a yeast strain that did not contain a deletion of the *leu2* gene on the chromosome. Similar selection techniques have been devised for different types of host cells that contain chromosomal deletions of genes involved in the biosynthesis of histidine, methionine, tryptophan, lysine, isoleucine, phenylalanine, threonine, and valine. *The only requirement for selection is that the vector must endow the host cell with a phenotype that will allow you to distinguish it from those cells that did not take up the vector.*

Although it is essential that one be able to *select* for cells that took up the vector following transformation, it is also desireable to *screen* for cells that harbor the *desired recombinant* vector. Suppose, for instance, that you have isolated a gene of interest by cleaving the parent DNA with *Bam*HI (see Fig. V-18). You digest the pBR322 plasmid with the same restriction endonuclease, mix it with the gene-carrying DNA fragment, and perform an in vitro ligation reaction. The ligation mixture is used to transform *E. coli* cells, which are then plated on rich media containing ampicillin. As can be seen in Figure V-18, two plasmids (or more) may result from this procedure: the pBR322 plasmid may re-ligate without the gene of interest, or the gene of interest may ligate into pBR322 at the *Bam*HI site. Since both plasmids will give rise to ampicillin resistant colonies, how would you go about determining which of these colonies harbor the pBR322 plasmid containing the gene of interest?

The pBR322 plasmid carries genes that confer resistance to both ampicillin and tetracycline. The *Bam*HI site lies *within* the tetracycline resistance gene. If the gene of interest has been inserted into

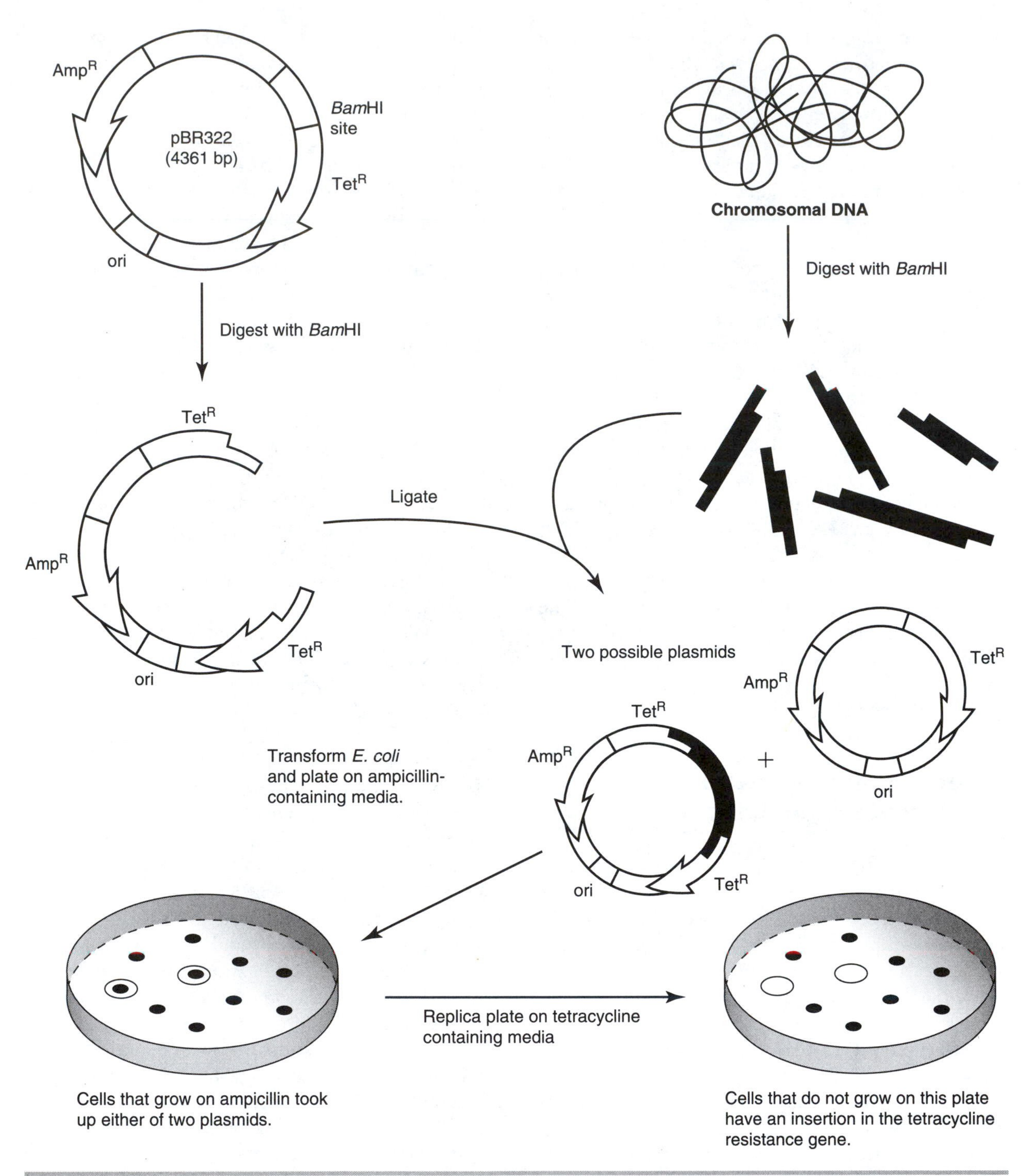

Figure V-18 Insertional inactivation of the tetracycline resistance gene following cloning in pBR322. *amp*^{R} = ampicillin-resistant; *tet*^{R} = tetracycline-resistant; *ori* = origin of plasmid replication

the plasmid at the *Bam*HI site, the tetracycline resistance gene will be disrupted *(insertional inactivation)*, and cells harboring the plasmid will be unable to grow in the presence of tetracycline. Therefore, if ampicillin-resistant colonies are replica plated on a tetracycline containing plate and do not grow, then they contain the pBR322 plasmid with an insertion (most likely of the gene of interest) at the *Bam*HI site. Replica plating is the process of lifting individual colonies from one plate and transferring them to the exact same position on a new plate. For a plate that contains a large number of colonies, this can be performed by placing a sterilized felt patch across the surface of the plate and pressing firmly to allow the felt to come into contact with the bacteria. This felt patch can then be placed over the surface of a fresh plate and pressed firmly to obtain an exact replica of the original plate.

A similar type of *insertional inactivation* screening process for recombinant vectors is described in the introduction to Experiment 21. Many modern *E. coli* cloning vectors contain multiple cloning sites within the *lacZ* α-peptide coding sequence. The LacZ α-peptide consists of the first 150 amino acids of β-galactosidase. It is known that the LacZ α-peptide can be removed (cleaved) from the enzyme and added back to the C-terminal domain of β-galactosidase to produce the fully active enzyme. If this plasmid is introduced into an *E. coli* strain in which the chromosomal copy of the *lacZ* gene has the region of the α-peptide deleted, the host strain will be *complemented* (restored) for β-galactosidase activity. If this same plasmid has a gene inserted in the multiple cloning site within the *lacZ* α-peptide sequence, the *lacZ* α-peptide reading frame will be disrupted, and the host cell will not display any β-galactosidase activity. Any cell that is not complemented for β-galactosidase activity (contains a plasmid with an insertion at the multiple cloning site) will give rise to white colonies on an IPTG/Xgal plate, while cells that received the unmodified plasmid will give rise to blue colonies. This screening process involving *α complementation* is described in greater detail in Experiment 21.

REFERENCES

Balbas, P., *et al.* (1986). Plasmid Vector pBR322 and Its Special-Purpose Derivatives—A Review. *Gene* **50**:3.

Chu, G., Hayakawa, H., and Berg, P. (1987). Electroporation for the Efficient Transfection of Mammalian Cells with DNA. *Nucleic Acids Res* **15**:1311.

Collins, J., and Hohn, B. (1978). Cosmids: A Type of Plasmid Gene-Cloning Vector That Is Packageable *in Vitro* in Bacteriophage λ Heads. *Proc Natl Acad Sci* **75**:4242.

Dagert, M., and Ehrlich, S. D. (1979). Prolonged Incubation in Calcium Chloride Improves the Competence of *Escherichia coli* Cells. *Gene* **6**:23.

Davidson, J. N. (1972). *The Biochemistry of Nucleic Acids*, 7th ed. New York: Academic Press.

Davies, J., and Smith, D. I. (1978). Plasmid Determined Resistance to Antimicrobial Agents. *Annu Rev Microbiol* **32**:469.

Doerfler, W. (1983). DNA Methylation and Gene Activity. *Annu Rev Biochem* **52**:93.

Dugaiczyk, A., Boyer, H. W., and Goodman, H. M. (1975). Ligation of *Eco*RI Endonuclease-Generated Fragments into Linear and Circular Structures. *J Mol Biol* **96**:171.

Friedberg, E. C. (1985). *DNA Repair.* New York: Freeman.

Gerard, G. F. (1983). Reverse Transcriptase. In: Jacob, S. T., ed. *Enzymes of Nucleic Acid Synthesis and Modification: Volume I. DNA Enzymes.* Boca Raton, FL: CRC Press.

Hohn, B. (1979). *In Vitro* Packaging of λ and Cosmid DNA. *Methods Enzymol* **68**:299.

Hung, M. C., and Wensink, P. C. (1984). Different Restriction Enzyme-Generated Sticky DNA Ends Can Be Joined *in Vitro*. *Nucleic Acids Res* **12**:1863.

Ish-Horowicz, D., and Burke, J. F. (1981). Rapid and Efficient Cosmid Cloning. *Nucleic Acids Res* **9**:2989.

Kornberg, T., and Kornberg, A. (1974). Bacterial DNA Polymerases: In: Boyer, P. D., ed. *The Enzymes*, 3rd ed. Vol. 10, pp. 119–144. New York: Academic Press.

Lehninger, A. L., Nelson, D. L., and Cox, M. M. (1993). Nucleotides and Nucleic Acids. In: *Principles of Biochemistry*, 2nd ed. New York: Worth.

Mathews, C. K., and van Holde, K. E. (1990). Nucleic Acids. In: *Biochemistry*. Redwood City, CA: Benjamin/Cummings.

McClelland, M., and Nelson, M. (1987). The Effect of Site-Specific DNA Methylation on Restriction Endonucleases and DNA Modification Methyltransferases—A Review. *Gene* **74**:291.

Messing, J. (1983). New M13 Vectors for Cloning. *Methods Enzymol* **101**:20.

Minton, N. P., Chambers, S. P., Prior, S. E., Cole, S. T., and Garnier, T. (1988). Copy Number and Mobilization Properties pf pUC Plasmids. *Bethesda Res Lab Focus* **10**(3):56.

Oishi, M., and Cosloy, S. D. (1972). The Genetic and Biochemical Basis of the Transformability of *Escherichia coli* K12. *Biochem Biophys Res Comm* **49**:1568.

Patterson, T. A., and Dean, M. (1987). Preparation of High Titer Lambda Phage Lysates. *Nucleic Acids Res* **15**:6298.

Roberts, R. J. (1988). Restriction Enzymes and Their Isoschizomers. *Nucleic Acids Res (suppl)* **16**:r271.

Saenger, W. (1984). *Principles of Nucleic Acid Structure.* New York: Springer-Verlag.

Sambrook, J., Fritsch, E. F., and Maniatis, T. (1989). *Molecular Cloning: A Laboratory Manual,* 2nd ed. Plainview, NY: Cold Spring Harbor Laboratory Press.

Stoker, N. G., Fairweather, N. F., and Spratt, B. G. (1982). Versatile Low-Copy-Number Plasmid Vectors for Cloning in *Escherichia coli. Gene* **18**:335.

Sykes, R. B., and Mathews, M. (1976). The Beta-Lactamases of Gram-Negative Bacteria and Their Role in Resistance to Beta-Lactam Antibiotics. *J Antimicrob Chemother* **2**:115.

Ullmann, A., Jacob, F., and Monod, J. (1967). Characterization by *in Vitro* Complementation of a Peptide Corresponding to an Operator-Proximal Segment of the β-Galactosidase Structural Gene of *Escherichia coli. J Mol Biol* **24**:339.

Watson, J. D., Hopkins, N. H., Roberts, J. W., Steitz, J. A., and Weiner, A. M. (1987). *Molecular Biology of the Gene,* 4th ed. Menlo Park, CA: Benjamin/Cummings.

Isolation of Bacterial DNA

Theory

To isolate a functional macromolecular component from bacterial cells, you must accomplish three things. First, you must efficiently disrupt the bacterial cell wall and cell-membrane system to facilitate extraction of desired components. Second, you must work under conditions that either inhibit or destroy the many degradative enzymes (nucleases, proteases) released during cell disruption. Finally, you must employ a fractionation procedure that separates the desired macromolecule from other cellular components in satisfactory yield and purity.

The isolation of bacterial DNA described in this experiment, patterned after the work of Marmur (1961), accomplishes these objectives. Bacterial cells are disrupted by initial treatment with the enzyme, egg-white lysozyme, which hydrolyzes the peptidoglycan that makes up the structural skeleton of the bacterial cell wall. The resultant cell walls are unable to withstand osmotic shock. Thus, the bacteria lyse in the hypotonic environment. The detergent, sodium dodecyl sulfate, (SDS, sodium dodecyl sulfate) then completes lysis by disrupting residual bacterial membranes. SDS also reduces harmful enzymatic activities (nucleases) by its ability to denature proteins. The chelating agents, citrate and EDTA (ethylenediamine tetraacetic acid), also inhibit nucleases by removing divalent cations required for nuclease activity.

This experiment employs a variety of fractionation methods to purify the bacterial DNA. Perchlorate ion is used to dissociate proteins from DNA. Chloroform–isoamyl alcohol is used to denature and precipitate proteins by lowering the dielectric constant of the aqueous medium. The precipitated material is removed by centrifugation. DNA is then isolated from the solution by ethanol precipitation, which allows spooling of the long, fibrous DNA strands onto a glass rod. Although the residual RNA and proteins in the solution also coagulate, they do not form fibers and are left in the solution as a granular precipitate. The resultant crude DNA is dissolved and precipitated again with isopropanol under conditions that favor specific DNA precipitation. These steps form a general procedure for the isolation of chromosomal DNA from virtually any bacterium. If you intend subsequently to assay the biological activity of the isolated DNA in a transformation assay (as in Experiment 20), it is necessary to start this isolation with the appropriate species and strain of bacteria possessing the genotype of interest.

This experiment will also provide you with experience in characterizing DNA samples by determining their ultraviolet absorption spectra and observing their thermal denaturation by means of the *hyperchromic effect*. These topics were discussed in the introduction to Section V.

Supplies and Reagents

For growth of wild-type (Trp^+) *Bacillus subtilis* (if this experiment is to be done together with Experiment 20. Other kinds of frozen bacterial cells can be used if Experiment 20 will not be performed.)

- Stock culture of wild-type *B. subtilis* (Bacillus Genetic Stock Center strain 1A2)
- Inoculating needle

Luria broth (see Appendix)
37°C Incubator-shaker
Fermenter
Centrifuges, continuous and laboratory
Sterile saline solution (8.5 g NaCl/liter)
0.15 M NaCl, 0.1 M Na_2EDTA
Lysozyme solution, 10 mg/ml
25% (wt/vol) SDS solution
5.0 M $NaClO_4$
Chloroform–isoamyl alcohol (24:1, vol/vol)
95% Ethanol
Dilute (1/10X) saline-citrate (0.015 M NaCl, 0.0015 M Na_3-citrate)
Standard (1X) saline-citrate (0.15 M NaCl, 0.015 M Na_3-citrate)
Concentrated (10X) saline-citrate (1.5 M NaCl, 0.15 M Na_3-citrate)
3.0 M Sodium acetate, 1 mM EDTA, pH 7.0
Isopropanol
Chloroform
Spectrophotometer and quartz cuvettes

Protocol

Growth of Wild-Type *Bacillus subtilis* for DNA Isolation

This portion of the experiment may be performed by an instructor in advance.

1. Inoculate (sterile technique) 100 ml of sterile Luria broth (200- to 400-ml flask) with a sample of wild-type *Bacillus subtilis* obtained from a stock culture slant. Incubate this solution on a shaker at 37°C for 15 to 24 hr (until the culture is densely turbid and is entering the stationary phase of growth).
2. Transfer the starter culture (sterile technique) into a fermenter containing 10 to 12 liters of sterile Luria broth and continue the bacterial growth at 37°C while stirring (~400 rpm) and aerating at 1 liter of air per liter of fluid per minute. Monitor the growth periodically by determining the turbidity of the culture (absorbance at 660 nm) during the 3- to 5-hr growth until this culture just enters the stationary phase of growth.
3. Stop the growth by cutting off both air and agitation. Use a continuous centrifuge (e.g., Sharples) to harvest the cells as quickly as possible.
4. Suspend the cell paste in an equal volume of cold (1 to 4°C) sterile saline solution. Complete the cell-washing procedure by harvesting the *B. subtilis* cells (10 min, 20,000 × *g*, 1 to 3°C) with a high-speed centrifuge.
5. Decant the supernatant fluid from the cell pellet after centrifugation and store the washed cells in Parafilm or waxed-paper-wrapped lots at −20°C until use (2-g lots are convenient). The expected yield is 35 to 45 g of cells per 10 liters.

Isolation of Bacterial DNA

1. Suspend 2 g of bacterial cell paste in 25 ml of 0.15 M NaCl, 0.1 M Na_2EDTA solution. The bacterial paste is most easily suspended by first adding a small amount of the NaCl–EDTA solution, making a slurry, and then slowly adding the rest of the NaCl–EDTA solution. After suspension, add 1 ml of a 10-mg/ml lysozyme solution and incubate the suspension at 37°C for 30 min, gently agitating occasionally.
2. After the 30-min incubation, complete the lysis of the bacteria by adding 2 ml of a 25% SDS solution and heating this preparation for 10 min in a 60°C water bath. Cool the solution to room temperature in a bath of tap water.
3. Add 7.5 ml of 5.0 M $NaClO_4$ to the 25 ml of lysed bacteria cells and mix gently. Then add 32.5 ml of $CHCl_3$–isoamyl alcohol (24:1) (i.e., a volume equal to the volume of lysed cell preparation containing 1.0 M perchlorate). Slowly shake the solution (30–60 oscillations per minute, by hand or on a low-speed shaker) in a tightly stoppered flask for 30 min at room temperature.
4. Transfer the suspension to a 250-ml polypropylene centrifuge bottle, and separate the resulting emulsion by centrifuging for 5 min at 10,000 × *g* at room temperature.
5. During the centrifugation step, remove any residual enzymes from a glass rod by heating one end of the rod in a flame and allowing the sterile end to air cool in a thoroughly washed 250-ml beaker.
6. After the centrifugation step, carefully pipette the clear aqueous phase (top layer) away from the coagulated protein emulsion at the interface between the aqueous and organic phases of the centrifuged solution. Do not attempt to

remove all of the aqueous phase; rather, avoid contaminating the aqueous phase with material from the other layers. Place the aqueous phase, which contains the extracted nucleic acids, in the 250-ml beaker. *Dispose of the $CHCl_3$-containing waste in the special container provided by your instructor.*

7. Gently stir the nucleic acid solution with the sterilized rod while *slowly and gently* adding 2 volumes (about 60 ml) of 95% ethanol. *Pour the ethanol gently down the side of the beaker so that it is layered over the viscous aqueous phase.* Continue to stir this preparation *gently* with the sterilized glass rod so that the ethanol is very gradually mixed throughout the entire aqueous phase. Avoid stirring vigorously, which would shear the DNA strands and make the DNA more difficult to isolate. The DNA will form a white fibrous film at the water–ethanol interface, and you should be able to spool all of the gelatinous, threadlike, DNA-rich precipitate onto the glass rod. Continue gentle stirring until the aqueous and ethanol phases are completely mixed and no more DNA precipitate can be collected on the glass rod. *Drain off excess fluid from the spooled crude DNA by pressing the rod against the walls of the beaker until no further fluid can be squeezed from the spooled preparation.*
8. Dissolve the crude DNA by stirring the glass rod with its spool of material in 9 ml of dilute (1/10X) saline-citrate in a test tube. Difficulty in dissolving the sample indicates failure to remove sufficient alcohol from the sample during step 7. If such difficulty is encountered, continue working the sample within the solution until you obtain as uniform a suspension as possible.
9. To the DNA solution add 1 ml of 3.0 M sodium acetate, 1 mM EDTA, pH 7.0 solution and transfer the preparation to a 100-ml beaker. Gently swirl the sample while slowly dripping in 6.0 ml of isopropanol. If fibrous DNA is readily apparent, collect the DNA threads by stirring and spooling with a sterilized glass rod as described in step 7. If a gel-like preparation develops, add an extra 1.0 ml of isopropanol and stir to spool the DNA onto the sterilized glass rod. Remove excess fluid from the spooled DNA by pressing the sample against the walls of the beaker.
10. Wash the sample (by submersion and brief swirling of the DNA on the rod) in test tubes containing, in order, 10 ml of 70% ethanol (made by adding 2.6 ml H_2O to 7.4 ml of 95% ethanol), and then 10 ml of 95% ethanol.
11. If you plan to isolate the DNA only for use in the genetic transformation outlined in Experiment 20, or if a day or longer will elapse before you will characterize the DNA as described below, store the DNA in a stoppered tube (2°C) as a spool submerged on the rod in 95% ethanol. Dissolve the DNA as described in step 1 of the following section when you are ready to use it.

Characterization of DNA

1. Remove the rod and the spooled DNA from the 95% ethanol, blot away all obvious residual fluid with a clean piece of filter paper, and then dissolve the DNA by stirring the glass rod, with its spooled DNA, in a test tube containing 10 ml of dilute (1/10X) saline-citrate *(sterile solution if the DNA is to be used in Experiment 20)*. This solution can be stored at 2°C.

Spectral Characterization of DNA

2. Dilute a 0.5 ml sample of the DNA solution with 4.5 ml of standard (1X) saline-citrate and determine the absorbance of this diluted sample at 260 nm against a dilute saline-citrate blank containing no DNA. If the absorbance at 260 nm is greater than 1.0, dilute the sample with a precisely known volume of standard (1X) saline-citrate until you obtain an absorbance of between 0.5 and 1.0.
3. Determine the absorbance of this solution at 5-nm intervals between 240 and 300 nm, so as to obtain an absorption spectrum for the isolated DNA. (Remember that *quartz cuvettes* must be used in this wavelength range.) *Blank the spectrophotometer to read zero absorbance with 1X saline-citrate at each wavelength.* (If equipment for the automated collection of this absorbance spectrum is available, follow the instructions for that instrument.) Include this absorption spectrum in your report.
4. Calculate the A_{260}:A_{280} ratio for your sample, and compare it with the value expected for pure DNA (2.0).

5. Assuming that a 1-mg/ml solution of native DNA has an absorbance at 260 nm of 20 and that all of your absorbance at 260 nm is due to native DNA, calculate the concentration (in milligrams per milliliter) of the DNA in your *undiluted* DNA solution and the total yield (in micrograms or milligrams) of DNA you obtained.
6. Assuming 100% recovery of the cellular DNA and assuming the wet bacterial cell paste was 80% water, calculate the percent of the dry weight in the bacterial cells that is DNA. How do your results agree with the DNA content of a typical bacterium, which is about 3% of the cell's dry weight? If your results do not agree with this value, discuss possible reasons for the discrepancy.

Melting of DNA—The "Hyperchromic Effect"

1. If your laboratory is equipped with a spectrophotometer with a thermoprogrammer designed for determination of melting curves of nucleic acid samples (Gilford, Beckman Instruments), determine a melting curve of a sample of your DNA dissolved in the required volume (usually 1 ml or less) of standard (1X) saline-citrate so as to give an initial absorbance of about 1 at 260 nm. Use the results from the preceding section to decide how much you need to dilute the stock DNA solution. Follow the instructions provided by the instrument's manufacturer in programming the rate of temperature increase, protecting the sample against evaporation, etc.
2. If your laboratory does not have access to a spectrophotometer with a thermoprogrammer, use the "quick cool" method described in steps 3 through 6 below to obtain a melting curve for your DNA. This method does not observe a true equilibrium between native and denatured ("melted") DNA and is appreciably less accurate than the method described in step 1, but it will suffice to illustrate the principles underlying the thermal separation of DNA strands.
3. Using your stock DNA solution from step 11, prepare 10 ml of a diluted DNA solution in the dilute saline-citrate (1/10X) that has a final absorbance at 260 nm of about 0.5. Use the results from the preceding section to decide how much you need to dilute the stock DNA solution. Determine the absorbance at 260 nm of this dilute solution at room temperature (about 20°C).
4. Transfer 1.0 ml of the same dilute DNA solution into each of six microcentrifuge tubes and cap them tightly. Incubate one tube at each of the following temperatures for 20 min: 50°C, 65°C, 75°C, and 85°C. In addition, incubate two of the tubes at 100°C (i.e., a boiling water bath).
5. After 20 min of incubating the tubes at the indicated temperatures, *quickly* cool all of them by placing them in an ice-water bath and gently agitating them. *Allow one of the tubes heated at 100°C to cool slowly to room temperature by placing it in a rack at your bench (30 to 40 min will be required).*
6. As quickly as possible after they have cooled, determine the absorbance at 260 nm of the solution in each tube. For a blank, use a sample of dilute saline-citrate (1/10X) that contains no DNA.
7. During heating, the double-stranded structure of DNA "melts"; that is, the strands separate. The nucleotide bases in the separated strands are no longer stacked in a regular fashion, which results in an increase in their absorbance of UV light. This is called the *hyperchromic effect* or *hyperchromicity*. When melted DNA is quickly cooled in solutions with very low salt concentrations, the strands are very slow to reassociate into their original base-paired, double-stranded form. Thus, you can estimate the degree of melting at each temperature even though you are measuring the absorbance at 260 nm at room temperature. The DNA sample that was heated to 100°C and slowly cooled to room temperature, on the other hand, is expected to form normal double-stranded DNA again (reanneal) and to exhibit little hyperchromic effect. Do your data support this expectation?
8. Prepare a plot of the absorbance at 260 nm at each temperature for the unheated sample and each of the rapidly cooled samples. Estimate the melting temperature (T_m) for your DNA. (T_m is the temperature at which the DNA sample is half melted. Assume that the DNA is

completely melted at 100°C. Calculate the hyperchromic effect for your DNA as the percent increase in absorbance at 260 nm on complete melting. Calculate the percent guanine plus cytosine (% GC) base pairs in your DNA using the following formula

$$\% \text{ GC} = 2.44\ (T_m - 81.5°\text{C} - 16.6\ \log_{10}[\text{Na}^+])$$

For dilute (1/10X) saline-citrate, this becomes % GC = 2.44 (T_m − 53.9°C). *B. subtilis* DNA is known to contain about 43% GC. How does this agree with your results? Discuss the results of this experiment.

Exercises

1. The DNA of cells containing nuclei (eukaryotic cells) is tightly associated with proteins called histones, most of which are rather basic proteins (i.e., they have high isoelectric points). Considering the basic character of most histones and the amino acids that contribute to this basic character, what enzymes would probably be most effective in disrupting histones from eukaryotic cellular DNA? Would you expect the procedure used in this experiment for the isolation of bacterial DNA to be useful for the isolation of DNA from eukaryotic cells?
2. This isolation of bacterial DNA utilizes reagents in the neutral pH range. What effects, if any, would you expect extremes in acid or alkali to have on the fibrous character of the DNA isolated in this procedure? Why?
3. Examine the procedure described in Experiment 21 for the isolation of plasmid DNA from bacteria. It is very different from the procedure used to isolate chromosomal DNA in this experiment. Could it be used to isolate chromosomal DNA from bacteria? Why or why not? Explain why such different procedures are used for isolating these two types of DNA.
4. What other physical and chemical treatments besides heating would be expected to allow you to demonstrate the hyperchromicity of single-stranded DNA? Explain how these treatments bring about strand separation.
5. Most naturally occurring RNA molecules are singled-, not double-stranded duplexes, yet RNA also shows a hyperchromic effect on melting. Explain.
6. The equation used for calculating % GC bases pairs in this experiment predicts that increasing % GC and increasing Na^+ concentration increases the T_m of DNA. Explain the physical basis for these effects.

REFERENCES

Mandel, M., and Marmur, J. (1968). Use of ultraviolet absorbance-temperature profiles for determining the guanine plus cytosine content of DNA. *Methods Enzymol* **12B**:195.

Marmur, J. (1961). A procedure for the isolation of desoxyribonucleic acid from microorganisms. *J Mol Biol* **3**:208.

Puglisi, J. D., and Tinoco, I. (1989). Absorbance melting curves of RNA. *Methods Enzymol* **180**:304.

EXPERIMENT 20

Transformation of a Genetic Character with Bacterial DNA

Theory

Transformation is the process by which exogenous DNA is taken up by bacterial cells and permanently alters the cells' inherited characteristics. Distinction is made between "natural" transformation and "artificial" transformation. Numerous bacterial species are naturally transformable; that is, they have the genetic capacity to develop *competence* for transformation naturally under certain growth conditions as one of their normal physiological responses. Examples of naturally competent bacteria include members of the genera *Bacillus*, *Streptococcus*, *Haemophilus*, and *Neisseria*. Many other bacterial species and eukaryotic cells do not develop competence naturally, but they can be made artificially competent for transformation by laboratory procedures that render them capable of taking up exogenous DNA, such as a combination of osmotic shock and treatment with $CaCl_2$, electroporation, or removal of cell walls to form protoplasts. The development of such techniques has played an extremely important role in the development of modern molecular biology. (See the introduction to Section V and Experiment 21 for a further discussion of artificial competence.)

Artificial transformation of *Escherichia coli* cells will be performed as part of Experiment 21. In this experiment, you will study natural transformation using strains of the bacterium *Bacillus subtilis*. Natural transformation has been subjected to extensive study by genetic and biochemical analysis. *B. subtilis* cells are not able to take up DNA and undergo transformation under all growth conditions, but generally develop competence during periods of nutrient limitation (along with other responses such as motility, the ability to degrade and utilize alternative metabolites, and endospore formation). The development of competence involves expression of a large number of *com* genes and is subject to a complex regulatory system. Many details have yet to be learned, but a number of steps in transformation have been identified. First, double-stranded DNA is bound by specific proteins found on the cell surface of competent, but not noncompetent, cells. There appears to be little sequence specificity in the binding of DNA, but only DNA molecules with a high degree of sequence homology to the DNA in the *B. subtilis* chromosome will eventually recombine and lead to genetic changes in the recipient cells. Very large DNA molecules undergo one or more cleavages to yield smaller double-stranded DNAs. The DNA is then transported into the cell in a process that requires metabolic energy. Then, the DNA is converted to the single-stranded form in a process that probably involves the random degradation of one of the two strands of the imported DNA. The single-stranded DNA forms a heteroduplex with homologous sequences of the chromosomal DNA. Subsequent integration, recombination, and repair of mismatches in the DNA proceed by steps that are not fully understood, but which involve a number of the genes required for other forms of genetic recombination.

Transformation of *B. subtilis* cells is easily studied in the laboratory if the genes under observation lead to a readily selectable phenotype. For example, consider competent recipient cells that are *auxotrophic* for a required metabolite, such as the amino acid tryptophan; that is, they have a genetic defect in a gene (*trp* gene) that specifies one of the enzymes required for the biosynthesis of tryptophan (such cells are called Trp^-). Trp^- cells cannot grow on a medium lacking tryptophan. If, however, Trp^- cells are grown under conditions leading to the development of competence, and they are treated with DNA isolated from a *B. subtilis* strain that had a normal copy of the *trp* gene that was defective in the recipient cells, some of the recipient cells will take up the normal gene, recombine it into their chromosome, and inherit a normal *trp* gene. Such cells will now be able to grow on a medium lacking tryptophan. Cells that did not receive the *trp* gene will remain unable to grow on a medium that lacks tryptophan. The Trp^+ cells will grow as individual colonies on Petri dishes containing agar medium without tryptophan, but the Trp^- cells that were not transformed will not grow.

In this experiment you will use DNA isolated from a wild-type (Trp^+) strain of *B. subtilis* as described in Experiment 19 to transform a competent Trp^- mutant strain of *B. subtilis* (strain 168) from tryptophan auxotrophy (Trp^-) to tryptophan prototrophy (Trp^+). The results will also enable you to determine the transformation efficiency obtained with this protocol.

Supplies and Reagents

For growth of competent recipient *Bacillus subtilis* strain 168 (Trp^-):

- Stock culture of *B. subtilis* strain 168 (Bacillus Genetic Stock Center strain 1A1)
- Inoculating needle
- 0.8% Nutrient broth, 0.5% glucose, 50 μg/ml tryptophan
- 37°C incubator-shaker
- Competence growth medium I (SPI)*
- Competence growth medium II (SPII)*
- Minimal medium agar Petri plates*
- Minimal medium agar Petri plates supplemented with 50 μg/ml tryptophan*
- Centrifuge
- Sterile centrifuge bottles or tubes with caps
- Sterile glass rod
- Sterile synthetic medium*
- Sterile 10-ml pipettes
- Sterile saline (8.5 g NaCl/liter)
- Sterile microcentrifuge tubes
- Sterile 100% glycerol

For transformation:

- Sterile test tubes with caps (preferably 13-×-100-mm tubes)
- Sterile 0.1-ml and 1.0-ml pipette tips
- Sterile minimal medium*
- DNA from wild-type *B. subtilis* (from Experiment 19)
- Standard (1X) saline-citrate (0.15 M NaCl, 0.015 M Na_3-citrate) (sterile)
- Pancreatic DNase (1 mg/ml in sterile 1X saline-citrate)
- Recipient *B. subtilis* strain 168 cells in sterile saline
- Sterile 0.1 M EGTA, pH 7.9
- 40 to 45°C water bath
- 37°C incubator-shaker
- 95% Ethanol
- Bent glass rods
- Minimal medium agar Petri plates*
- Minimal medium agar Petri plates supplemented with 50 μg/ml tryptophan*
- Tubes containing 9.9 ml of sterile saline
- Tubes containing 0.9 ml of sterile saline
- Masking tape

*Additional details in the Appendix.

Protocol

Growth of Competent Recipient *Bacillus subtilis* Strain 168 (Trp^-)

This portion of the experiment may be performed by an instructor in advance.

NOTE: Sterile technique is essential in the following transfers, incubations, and centrifugations to avoid contamination with other bacteria.

1. Inoculate 5 ml of 0.8% nutrient broth, 0.5% glucose, and 50 μg/ml tryptophan with a sample of *Bacillus subtilis* strain 168 (Trp^-) obtained from a stock culture slant.
2. Incubate this culture on a shaker at 37°C for 12 hr overnight.
3. After this incubation, use a sterile inoculating needle to streak out a few droplets of the growth medium on one Petri plate containing minimal medium agar and another plate containing minimal medium agar supplemented with 50 μg/ml tryptophan. The results of these platings are necessary to confirm the Trp^- phenotype of the cells. Incubate the plates at 37°C for 24 hr and examine the plates. Many colonies

should appear on the plates containing tryptophan, and *no* colonies should be found on the plates lacking tryptophan.

4. After removing samples for plating to confirm the Trp^- phenotype, use the remaining culture to inoculate 100 ml of SPI medium in a 500-ml flask. Incubate the culture at 37°C overnight with shaking. The solution should be turbid in the morning.
5. Transfer 5 to 10 ml of this culture to 100 ml of fresh SPI medium in a 500-ml sidearm flask so that the final cell suspension has an absorbance at 500 nm of about 0.1 when viewed against an SPI medium blank. (*NOTE:* If you do not use a Klett colorimeter and a sidearm flask, remove 1-ml samples for absorbance readings in a spectrophotometer. Use sterile technique for all samplings. Do not return samples from the cuvettes to the cell suspension.) Incubate this culture with shaking at 37°C.
6. Follow the growth of the cells in the culture by taking readings from the sidearm with a Klett colorimeter or using the sampling technique described in step 5. Plot the absorbance as a function of time of growth on semilog paper.
7. When the turbidity of the culture (cell growth) stops increasing exponentially (usually 4 to 5 hr after inoculation), transfer 50 ml of the culture to a 2-liter flask containing 450 ml of fresh SPII medium, which is the same as SPI medium except that it contains 0.5 mM $CaCl_2$ and 2.5 mM $MgCl_2$. Incubate this culture with shaking at 37°C for 90 min.
8. Cells from the culture in step 7 should have developed maximal competence and can be used directly for transformation. However, it is usually much more convenient to store the cells as frozen cell cultures and to thaw and use them when needed. Frozen competent cells are prepared as follows:

   Centrifuge the cell culture in a high-speed centrifuge (10 min, 20,000 × *g*, 1 to 4°C) using sterilized centrifuge bottles. Working aseptically and quickly, decant all but about 1/10 of the original volume (about 50 ml) of the SPII growth medium. Resuspend (sterile glass rod) the cell pellet in the remaining growth medium. Add 5 ml of sterile glycerol and mix thoroughly. The resultant cell suspension should have an absorbance at 660 nm of roughly 0.5 (i.e., $\sim 10^8$ cells per milliliter). Store these cells in 0.6-ml aliquots in sterile 1.5-ml microcentrifuge tubes at −70°C until use in the transformation experiments below. This procedure yields sufficient cells for about 80 pairs of students.

Transformation

The following procedures require sterile technique; that is, use of sterilized glassware, pipettes, Petri plates, and other equipment. Do not use equipment that has lost a covering cap and consequently has been exposed to the air for long periods. If you are unfamiliar with sterile techniques for transferring bacterial cultures, consult your instructor for details of the techniques required in this experiment.

1. Start by carrying out the special treatments of DNA required for tubes 4 and 6 in Table 20-1.
2. If frozen competent cells are to be used, thaw the entire 0.6-ml sample by incubation for

Table 20-1 Protocol for Transformation of Competent *B. subtilis* Cells with DNA

	Tube Number					
Component (volumes in ml)	1	2	3	4	5	6
Minimal growth medium	0.8	0.8	0.8	0.8	0.8	0.8
B. subtilis DNA from wild type*	—	0.1	0.2	0.1	0.1	0.1†
Standard (1X) saline-citrate	0.2	0.1	—	—	0.2	0.1
DNase solution	—	—	—	0.1‡	—	—
Competent *B. subtilis* cells (Trp^-)	0.1	0.1	0.1	0.1	—	0.1

*DNA dissolved in standard (1X) saline-citrate as described in Experiment 19.

†Denature this DNA sample before use in transformation by heating in a boiling water bath for 5 min and cooling rapidly in an ice bath.

‡Incubate the growth medium + DNA + DNase at 37°C for 15 min before adding the competent cells.

5 min at 40 to 45°C. Add 6 μl of sterile 0.1 M EGTA, pH 7.9, and mix gently. Use these cells for step 3.

3. Using sterile technique, pipette the following ingredients into six empty sterile test tubes in the order shown in Table 20-1. Note that tubes 4 and 6 require special treatment of the DNA before initiating the incubations. Label the tubes clearly with their numbers and your initials.
4. Cover the tubes with caps and incubate them for 2 hr at 37°C, preferably in a slowly shaking water bath, but alternatively in a 37°C incubator with occasional gentle hand agitation. During the 2-hr incubation, set up and familiarize yourself with the details of the following dilution and plating operations.

You will determine the number of cells in each bacterial suspension by use of the agar plating method. In this method, you spread a small sample of known volume (or a known dilution of such a sample) over the surface of a sterile Petri plate containing an agar-stabilized growth medium (Fig. 20-1). The sample dif-

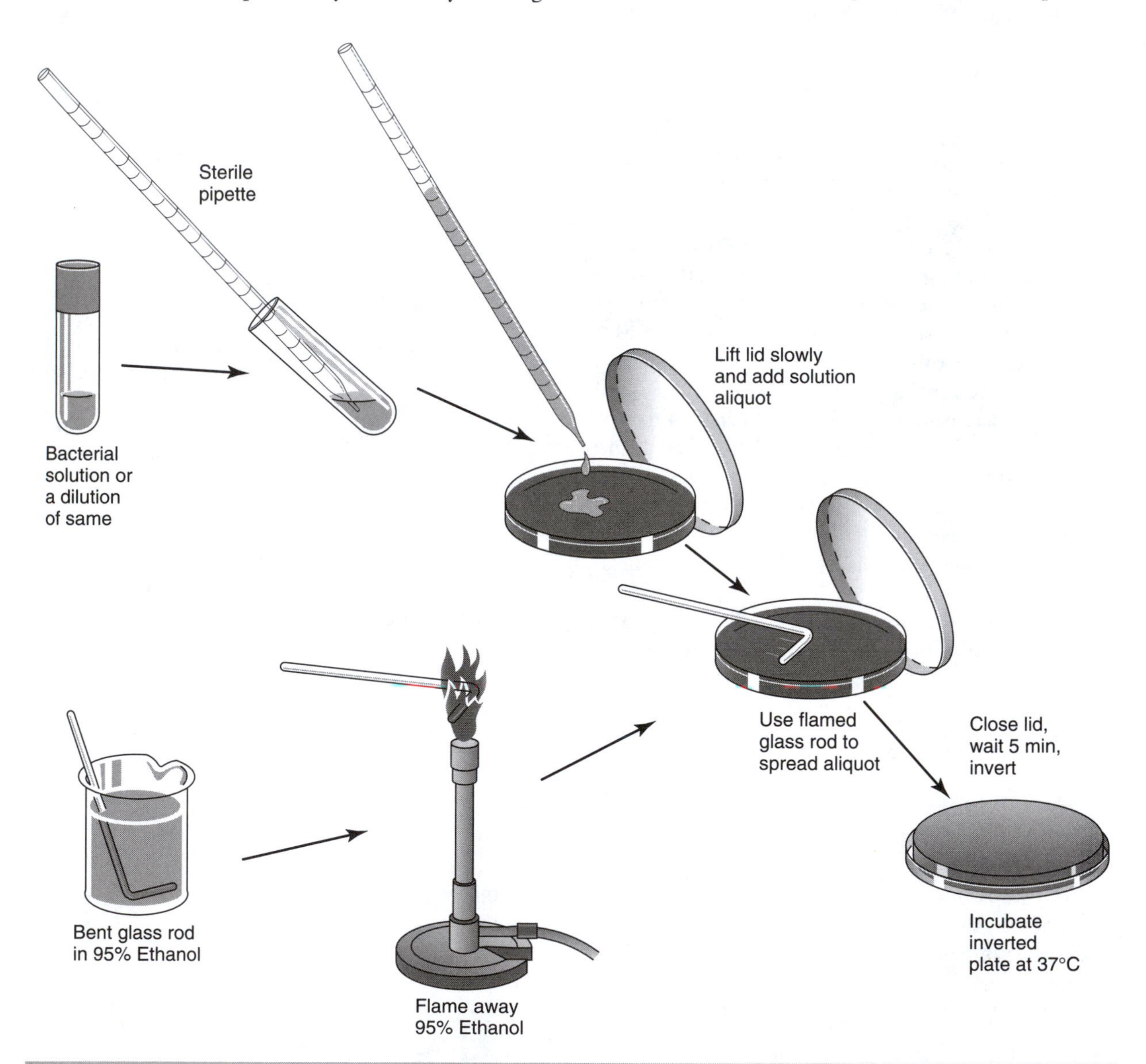

Figure 20-1 Steps of the procedure for plating bacteria.

fuses into the growth medium, embedding single bacterial cells in the agar. On incubation, the bacterial cells take up nutrients and divide. Each original bacterial cell capable of growth on the growth medium eventually yields a visible colony of cells growing in the same spot on the agar where the original single bacterial cell was embedded. Thus, the number of colonies obtained from the sample placed on the plate equals the number of viable cells originally present in the sample. This assumes that all of the cells can grow on the growth medium provided in the Petri plate. If the growth medium is altered such that only cells of a certain phenotype can grow on the plate, the number of colonies will reflect only the cells of that subpopulation.

Analysis of the transformation performed in this experiment requires agar plating to determine the number of transformed cells (Trp^+ cells) in each of the six tubes of the experiment. The Petri plates used in this determination obviously do not contain tryptophan in the growth medium. In a separate experiment, you will determine the total number of cells (Trp^+ *plus* Trp^- cells) in the transformation mixture by plating a diluted sample of known volume and known dilution on a Petri plate in which the agar growth medium *does* contain tryptophan. The number of transformants (Trp^+) cells divided by the total number of cells (Trp^+ plus Trp^-) is the *efficiency* of that transformation experiment.

The number of transformed cells in each tube will be small compared with the total number of cells (perhaps 0.1 to 1%). You can therefore determine the number of transformed cells in each tube by plating a 0.1-ml sample (or a limited dilution of a 0.1-ml sample) on minimal medium agar (i.e., growth medium containing sugars and salts but lacking tryptophan). In contrast, analysis of the total number of cells in each tube requires extensive dilution of the contents of each tube before plating on minimal medium agar plates supplemented with tryptophan.

5. After the 2-hr incubation at 37°C, use sterile technique to perform the following steps for dilution and plating, labeling all tubes and plates with the dilutions and medium used. (*NOTE:* Once you obtain a specified bacterial dilution, conserve pipettes by using the same pipette both for plating and for subsequent dilutions.) Use the technique illustrated in Figure 20-1 to plate the samples and spread them evenly over the agar surface.
6. Determine the *number of transformed (Trp^+) cells* in each of the six assay tubes by plating a 0.1-ml sample from each tube on labeled plates containing minimal medium agar (without tryptophan).
7. The Trp^+ colonies obtained from tubes 2 and 3 may be too numerous to count. For this reason you must confirm the number of transformed (Trp^+) cells in tubes 2 and 3 by transferring a 0.1-ml sample from each of these tubes into separate marked tubes containing 9.9 ml of sterile saline. Mix these solutions thoroughly and plate a 0.1-ml sample of the 100-fold diluted samples on separate marked plates containing minimal medium agar. Save the diluted culture samples for later use.
8. Determine the *total number of cells (Trp^+ plus Trp^-)* in tubes 1 and 2 by preparing a 100-fold dilution of tube 1 by adding 0.1 ml of culture from tube 1 to 9.9 ml of sterile saline. The same dilution was already performed for tube 2 in step 5. Take a 0.1-ml sample from the dilutions of tubes 1 and 2 and transfer each to separate marked tubes containing 9.9 ml of sterile saline. Plate a 0.1-ml aliquot of each of these latter 10,000-fold dilutions on minimal medium plus tryptophan agar plates. Finally, transfer separate 0.1-ml samples of both 10,000-fold diluted samples to marked tubes containing 0.9 ml of sterile saline, mix the contents, and plate separate 0.1 ml samples of these 100,000-fold dilutions on minimal medium plus tryptophan agar plates. Two dilutions of the cultures in tubes 1 and 2 are prepared in case the number of colonies on the plate containing the 10,000-fold dilution are too numerous to count. In this case, the 100,000-fold dilution of the same culture should produce a number of colonies that you can count readily.
9. Immediately after placing each 0.1-ml sample of culture on a Petri plate, spread the sample uniformly over the surface of the plate with a *sterile* bent glass rod (see Fig. 20-1). Be sure that you sterilize the glass rod between spreadings

of each sample by dipping it in ethanol and flaming it.

10. Allow all the plated samples to diffuse into the agar for 5 to 10 min. Then, invert each Petri plate (agar side up), write your name or initials on each, combine the plates in stacks held by masking tape, and incubate all plates for 48 to 72 hr at 37°C. Count the colonies on each plate.
11. Use the dilution factors and the number of colonies detected on each plate to determine the number of Trp^+ cells per milliliter of undiluted growth medium in each of the six transformation tubes. How do the results obtained for tubes 2 and 3 with the 100-fold diluted samples agree with those obtained when 0.1 ml of undiluted cells was plated? Explain your findings.
12. Use the dilution factors and colony counts to determine the total number of viable cells (Trp^+ plus Trp^-) present per milliliter of undiluted culture in tubes 1 and 2. Assuming that all of the Trp^+ cells are transformants and assuming that the total number of viable cells in tubes 2 through 6 is the same, calculate the efficiency of transformation in tubes 2 to 6. Interpret all your results in terms of the contents of each tube and your knowledge of the process of transformation. Which tubes served as controls for the experiment? Explain.

Exercises

1. What biological advantage do bacteria derive from having the capacity to develop natural competence? Suggest why competence develops optimally in nutrient-limited cultures.
2. A student performing this experiment detects 2×10^7 total viable cells (Trp^+ plus Trp^-) per milliliter in tube 1 of this experiment, yet the student cannot detect a significant number of viable cells in tube 2 of this experiment. Suggest an explanation for this result.
3. Assume that indoleglycerolphosphate synthetase, an enzyme essential for tryptophan biosynthesis in *B. subtilis*, has a molecular weight of 50,000, that amino acids in proteins have an average molecular weight of 100, and that nucleotides in nucleic acids have an average molecular weight of 333. Calculate the *minimal* molecular weight of a transforming fragment of double-stranded DNA that could carry the complete genetic information necessary for synthesis of indoleglycerolphosphate synthetase.

REFERENCES

Anagnostopoulos, C., and Spizizen, J. (1961). Requirements for transformation in *Bacillus subtilis*. *J Bacteriol* **81**:741.

Dubnau, D. (1993). Genetic exchange and homologous recombination. In: Sonenshein, A. L., Hoch, J. A., and Losick, R., eds. *Bacillus subtilis and Other Gram Positive Bacteria: Biochemistry, Physiology and Molecular Genetics.* Washington, DC: American Society for Microbiology, pp. 555.

Mahler, I. (1968). Procedures for *Bacillus subtilis* transformation. *Methods Enzymol* **12B**:846.

Sadaie, Y., and Kada, T. (1983). Formation of competent *Bacillus subtilis* cells. *J Bacteriol* **153**:813.

Wilson, G. A., and Bott, K. F. (1968). Nutritional factors influencing the development of competence in the *Bacillus subtilis* transformation system. *J Bacteriol* **95**:1439.

Constructing and Characterizing a Recombinant DNA Plasmid

Theory

The discovery of plasmids and restriction endonucleases is largely responsible for the enormous technological advances made in the field of molecular biology over the past 20 years. A desired DNA fragment or gene sequence can be identified, excised from the chromosome, isolated, and cloned into a number of different plasmids or vectors. This gene can then be sequenced, mutated, and/or expressed to gain insight into the function of the protein that it encodes. The plasmid carrying the gene of interest can be introduced into a host cell and faithfully propagated during the cell's normal cycle of DNA replication and cell division. In this experiment, you will subclone the gene for aminoglycoside-3′-phosphotransferase from plasmid pUC4K into pUC19. This gene will confer resistance to the antibiotic, kanamycin, a phenotype that can be exploited in the selection process for cells that obtained the desired plasmid. The resulting pUC19/4K recombinant plasmid will be selected for, and the construction of the plasmid will be verified using restriction endonuclease digestion and agarose-gel electrophoresis.

Plasmid pUC4K is a commercially available *Escherichia coli* vector that contains the kanamycin resistance (kan^R) gene from transposon (Tn) 903 (Fig. 21-1). The gene is flanked on the 5′ and 3′ sides by four different restriction endonuclease recognition sites. Thus, the kan^R gene can be excised from pUC4K and inserted into virtually any other existing plasmid to create a new vector that confers kan^R to its host. The kan^R gene can also be inserted into the open reading frame of a gene of interest, allowing you to study the effect of deletion of the gene (null mutation) on the organism. This type of gene disruption experiment has proven to be very useful in efforts to determine the function of particular gene products in various biological processes.

Plasmid pUC19 (and its derivative, pUC18) is one of the most popular and widely used *E. coli* cloning vectors in molecular biology (see Fig. 21-1). Like pUC4K, pUC19 contains the gene that confers ampicillin resistance (amp^R) to its host (*bla*, β-lactamase). This vector also contains a multiple cloning site (MCS) within the sequence of the LacZα peptide. The LacZα peptide, which encodes the N-terminal 150 amino acids of β-galactosidase, will function in *trans* to complement (restore) β-galactosidase activity in a host strain deleted for the LacZα peptide. The position of the MCS on pUC19 will play an important role in the selection process to identify cells that harbor the desired pUC19/4K plasmid. If cloned into pUC19 in the same reading frame as the LacZα peptide sequence, a gene of interest can be expressed as a fusion protein with the N-terminus of the LacZα peptide. Like the LacZα peptide, the expression of the fusion protein will now be under the control of the *lac* promoter, which is inducible with the addition of isopropylthio-β-galactoside (IPIG) to the growth medium. Recall that the expression of genes under the control of the *lac* promoter is normally low in the presence of the *lac* repressor protein (LacI), the gene for which is also contained in pUC19 (see Fig. 21-1) (See introduction to Experiment 7).

On Day 1 of the experiment, you will digest pUC4K with *Eco*RI and *Xho*I. You will also digest

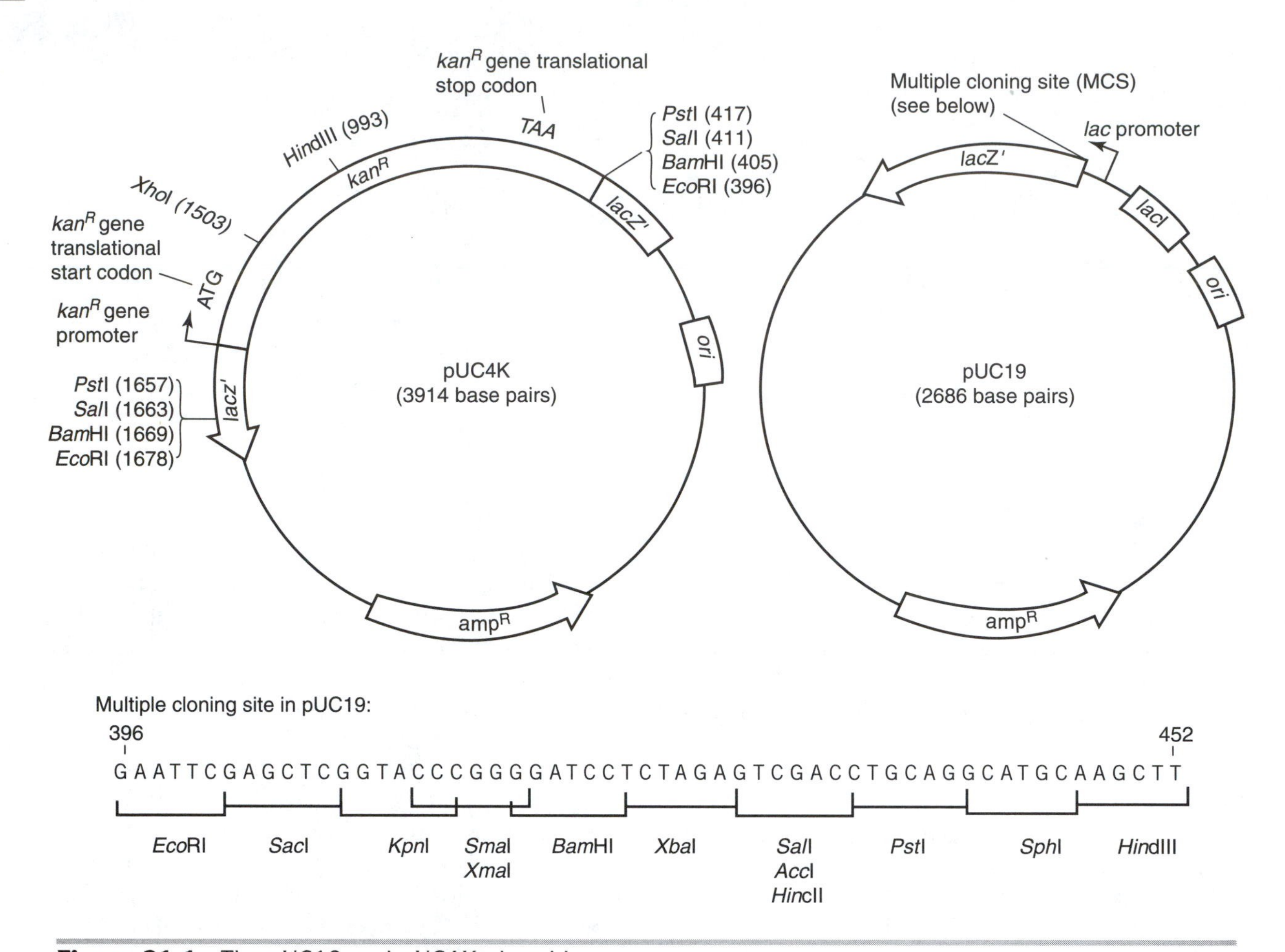

Figure 21-1 The pUC19 and pUC4K plasmids.

pUC19 with *Eco*RI, *Bam*HI, and *Sal*I. Following digestion, the DNA fragments resulting from these reactions (Fig. 21-2) will be mixed together and precipitated with ethanol. Next, T4 DNA ligase and ATP will be added, and the ligation reaction will be allowed to proceed overnight. As shown in Fig. 21-2, the 5′ single-stranded DNA overhang produced in the *Xho*I digest on pUC4K will be compatible with the 5′ single-stranded DNA overhang produced in the *Sal*I digest on pUC19. Although DNA ligase will be able to join these two fragments together, the resulting sequence will no longer be recognized by *Xho*I or *Sal*I. The 5′ single-stranded DNA overhangs produced in the *Eco*RI digests on pUC4K and pUC19 will also be compatible, and will be joined by DNA ligase to regenerate the *Eco*RI site. This type of *directional*, or *forced*, *cloning* ensures that the kan^R gene will ligate into pUC19 in a specific orientation (see Fig. 21-2). If both the kan^R gene and pUC19 were digested with *Eco*RI only, this would not be the case (see Fig. 21-3).

Also note that the *Xho*I digest on pUC4K cleaves off the promoter and start codon (ATG) for the kan^R gene, rendering it incapable of being transcribed or translated by the host cell (see Fig. 21-1). However, the *Xho*I/*Sal*I ligation of the kan^R gene into pUC19 will put it into the same reading frame as the LacZα peptide sequence. What will be produced is a *lacZ/kan^R* gene fusion protein that will confer kan^R to the host cell *and* that is inducible in the presence of IPTG. This differs from the case in pUC4K, where the kan^R gene is *constitutively* expressed under the control of a different promoter.

Since the ligation reaction is performed in the presence of a number of different DNA fragments

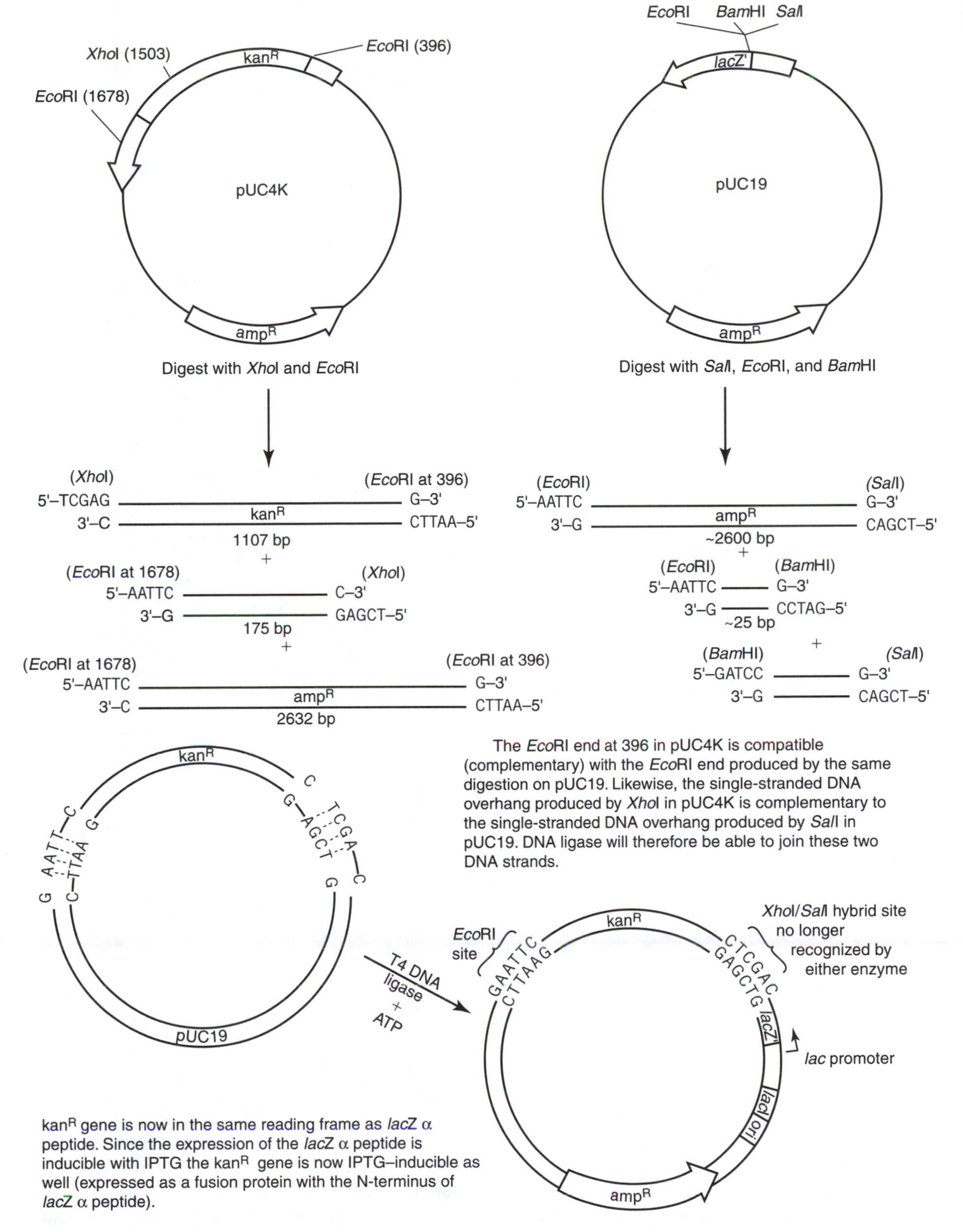

Figure 21-2 DNA fragments produced from pUC4K and pUC19 following restriction enzyme digestion.

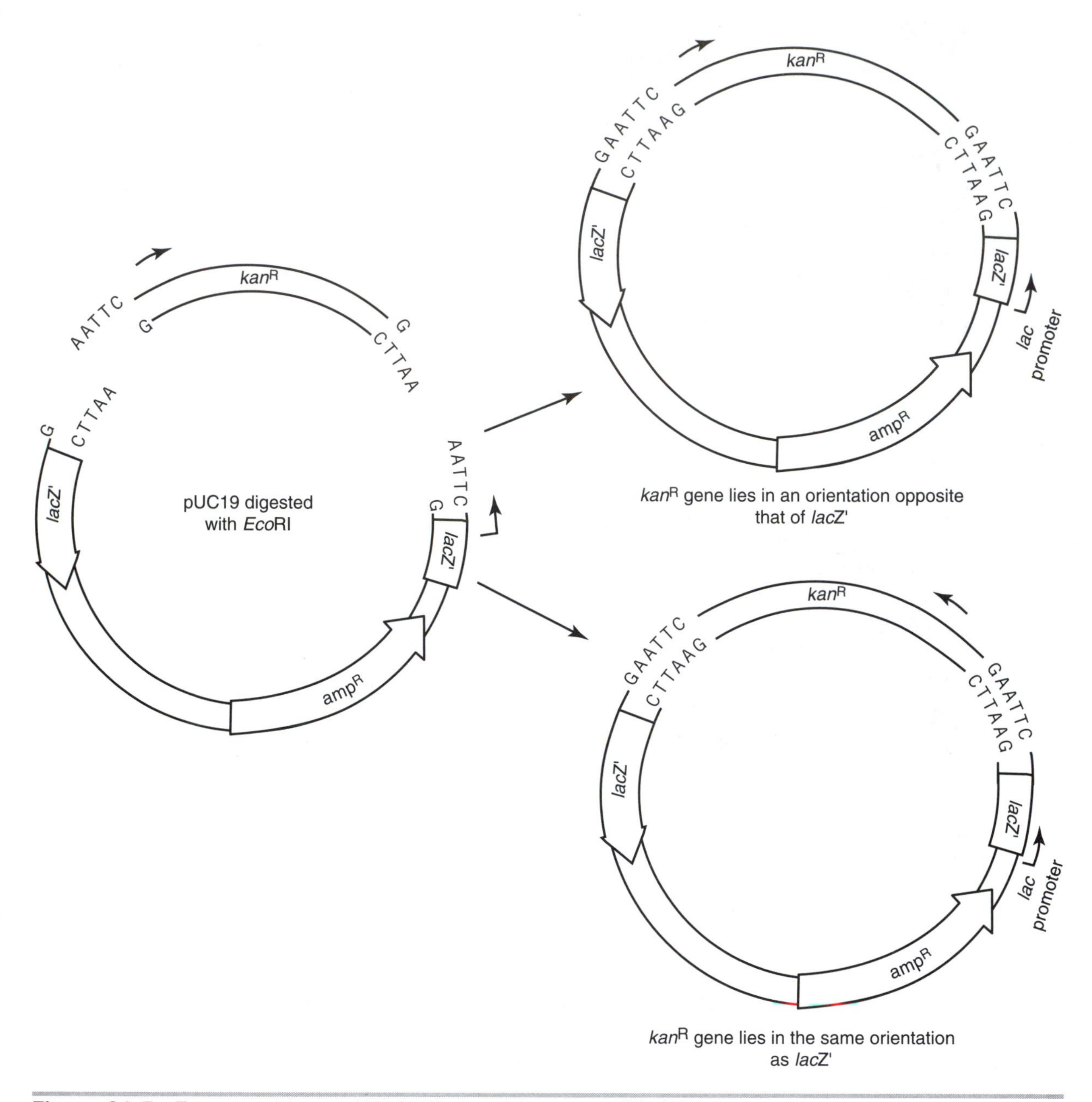

Figure 21-3 Two recombinant pUC19/4K plasmids are possible if both plasmids are digested with a single enzyme. The two resulting plasmids differ in the orientation of the kan^R gene in pUC19. The arrow (→) indicates the direction of transcription of the kan^R gene.

with compatible single-stranded DNA overhangs, the desired pUC19/4K recombinant plasmid is by no means the only plasmid that could be produced during the ligation reaction. For instance, it is possible that the *Eco*RI/*Xho*I fragment carrying the kan^R gene from pUC4K will ligate back into pUC4K. It is also possible that the *Eco*RI/*Sal*I fragment will ligate back into pUC19. The possibility of reforming either of the two parent plasmids is minimized, however, because both events would require a successful three-point ligation (see Fig. 21-4). The two-point ligation that is required to produce the desired pUC19/4K plasmid is statistically much more likely to occur.

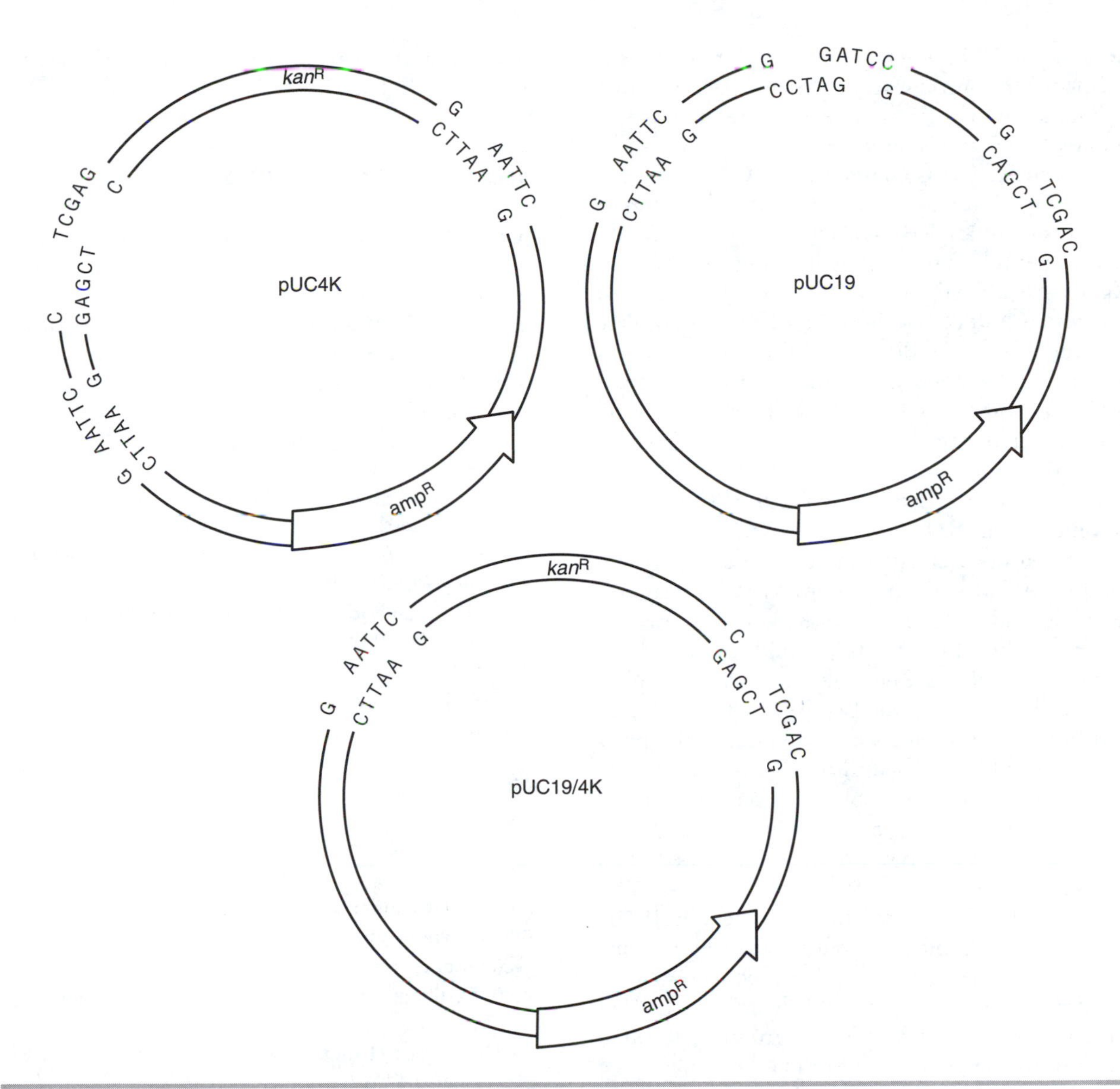

Figure 21-4 On a statistical basis, the two-point ligation required to form the recombinant pUC19/4K plasmid is much more likely than the three-point ligations required to form pUC19 or pUC4K.

How do you distinguish between cells carrying the pUC19 plasmid and the pUC19/4K recombinant plasmid? Recall that pUC19 carries the gene for amp^R, while pUC19/4K carries the genes for amp^R and kan^R. Any colony that will grow in the presence of ampicillin but not kanamycin most likely harbors pUC19. In addition to the antibiotic selection just described, pUC19 will also confer another unique phenotype to the host cell that will distinguish it from those carrying pUC19/4K. The host cell used in this experiment is *Epicurian coli* XL1-Blue. This strain contains a deletion on the chromosome of the first 450 base pairs of its *lacZ* coding sequence (the LacZα peptide). The pUC19 plasmid will be able to restore β-galactosidase activity in the host strain, since this plasmid contains an uninterrupted LacZα peptide coding sequence (the strain will be *complemented* for β-galactosidase activity). The LacZα peptide expressed from the pUC19 plasmid will be able to noncovalently interact with the C-terminal fragment of LacZ produced by the host cell, creating a functional β-galactosidase enzyme. This technique of selection is therefore referred to as *α-complementation.* If cells containing pUC19 are placed on an agar plate containing IPTG and 5-bromo-4-chloro-3-indolyl-β,D-galactopyranoside (Xgal), LacZα peptide expression will be induced, and the Xgal substrate will be converted to a blue-colored precipitate. Blue colonies on an IPTG/Xgal plate will therefore be indicative of the presence of an intact LacZα peptide sequence in the plasmid (pUC19). Since pUC19/4K will have the kan^R gene inserted into the LacZα peptide sequence, it will not be able to complement the host strain for β-galactosidase activity, and it will produce white colonies on an IPTG/Xgal plate.

How do you distinguish between cells carrying the pUC4K plasmid and the pUC19/4K recombinant plasmid? This is not as easy as the situation described above, since both plasmids contain an insertion in the LacZα peptide sequence. Both plasmids will confer kan^R and amp^R to the host strain and both will produce white colonies in the presence of IPTG and Xgal. Still, there is one difference between these two plasmids that you will be able to exploit during the selection process. Recall that the kan^R gene is *constitutively* expressed from the pUC4K plasmid, while the kan^R gene is under the control of the *lac* promoter in the pUC19/4K recombinant plasmid. Therefore, any white colony that grows well on a plate containing only kanamycin (no IPTG) most likely carries pUC4K. In contrast, any colony that grows well on a kanamycin plate containing IPTG, but poorly on a plate containing only kanamycin, most likely carries pUC19/4K. The basis for the different types of selection that will be used in this experiment are outlined in Table 21-1.

Table 21-1 Basis of Selection for Clones Carrying the Desired pUC19/4K Recombinant Plasmid

Plasmid	Kan^R?	Amp^R?	Kan^R IPTG-Inducible?	Color on IPTG/Xgal Plate
pUC19	No	Yes	—	Blue
pUC4K	Yes	Yes	No	White
pUC19/4K	Yes	Yes	Yes	White

Supplies and Reagents

pUC19 (0.25 μg/μl)—Pharmacia catalog #27-4951-01
pUC4K (0.5 μg/μl)—Pharmacia catalog #27-4958-01
Distilled water (sterile)
*Pst*I (10 units/μl) with 10X buffer—Gibco BRL catalog #15215-023
*Eco*RI (10 units/μl) with 10X buffer—Gibco BRL catalog #15202-021
*Xho*I (10 units/μl) with 10X buffer—Gibco BRL catalog #15231-020
*Sal*I (10 units/μl) with 10X buffer—Gibco BRL catalog #15217-029
*Bam*HI (10 units/μl) with 10X buffer—Gibco BRL catalog #15201-031
*Hin*dIII (10 units/μl) with 10X buffer—Gibco BRL catalog #15207-020
Water baths at 37°C, 15°C, and 42°C
Phenol (Tris-saturated)
Chloroform
Isoamyl alcohol
3 M sodium acetate (pH 5.2)
Absolute ethanol
Dry ice/ethanol bath (−70°C)
Microcentrifuge
1.5-ml plastic microcentrifuge tubes
70% vol/vol ethanol in distilled water
Vacuum microcentrifuge (Speed-Vac)
P-20, P-200, and P-1000 Pipetmen with disposable tips (sterile)
T4 DNA ligase (1 unit/μl) with 5X buffer—Gibco BRL catalog #15224-025
Epicurian coli XL1-Blue competent cells—Stratagene, catalog #200130
Yeast-tryptone (YT) broth (1% Bactotryptone, 0.5% yeast extract, 0.5% NaCl)
100 mM IPTG solution in sterile distilled water (filter-sterilized)
Ampicillin (100 mg/ml) prepared in sterile distilled water—Sigma catalog #A-9518
Kanamycin (100 mg/ml) prepared in sterile distilled water—Sigma catalog #K-4000
Agar
Culture shaker with test tube racks at 37°C
Petri plates (disposable, plastic)
Sterile glass rod (for spreading bacteria over the plate)
Sterile toothpicks
37°C incubator

Large (16 × 125 mm) glass test tubes (sterile)
1-kb DNA ladder size standard—Gibco BRL catalog #15615-016
Ethidium bromide solution in 0.5X TBE buffer (0.5 μg/ml)—*Toxic!*
Polaroid camera with film and a 256-nm light box
GTE lysis buffer (25 mM Tris, pH 8.0, 50 mM glucose, 10 mM EDTA, 10 μg/ml lysozyme)
Fresh solution of 0.2 N NaOH and 1% SDS in distilled water
Potassium acetate solution (60 ml of 5 M potassium acetate, 11.5 ml of glacial acetic acid, 28.5 ml of distilled water)
RNaseA (10 mg/ml in water)—*Boil this solution for 10 min to inactivate any DNase*
6X DNA loading buffer (0.25% wt/vol bromophenol blue, 0.25% wt/vol xylene cyanole, 30% wt/vol glycerol in distilled water)
Agarose (electrophoresis grade)
5X TBE buffer (54 g/liter Trizma base, 27.5 g/liter boric acid, 20 ml/liter of 0.5 M EDTA, pH 8.0)
Agarose-gel electrophoresis apparatus with casting box and 10-well comb
5-bromo-4-chloro-3-indolyl-β,D-galactopyranoside (Xgal), 40 mg/ml solution in dimethylformamide—*Toxic!* (Sigma catalog #B-4252)

Protocol

Day 1: Restriction Endonuclease Digestion of pUC19 and pUC4K, Ethanol Precipitation of DNA Fragments, and DNA Ligation

1. Set up the following reactions in two labeled 1.5-ml sterile microcentrifuge tubes:

Digestion of pUC19	Digestion of pUC4K
14 μl of distilled water (sterile)	14 μl of distilled water (sterile)
2 μl of 10X React Buffer 2	2 μl of 10X React Buffer 2
1 μl of pUC19 (0.25 μg/μl)	2 μl of pUC4K (0.50 μg/μl)
1 μl of *Eco*RI (10 units/μl)	1 μl of *Eco*RI (10 units/μl)
1 μl of *Bam*HI (10 units/μl)	1 μl of *Xho*I (10 units/μl)
1 μl of *Sal*I (10 units/μl)	

2. Incubate both reactions at 37°C for 1.5 hr.
3. Transfer both reactions to a single tube and add 160 μl of sterile distilled water.
4. Add 200 μl of Tris-saturated phenol:chloroform:isoamyl alcohol (25:24:1). Mix vigorously in a Vortex mixer for 1 min and centrifuge for 3 min to separate the aqueous and organic phases. This step is done to denature and remove the restriction endonucleases from the DNA solution.
5. Remove the aqueous (upper) phase and place it in a clean microcentrifuge tube. Repeat step 4 (extraction of aqueous phase) with 200 μl of chloroform:isoamyl alcohol (24:1). This step is done to remove all traces of phenol from the solution, which may denature the T4 DNA ligase during the ligation reaction. Dispose of the organic phase from this extraction in a waste beaker designated by the instructor.
6. Transfer the upper (aqueous) phase to a fresh microcentrifuge tube. *Do not transfer any of the organic phase during this process.* Dispose of the organic phase from this extraction in a waste beaker designated by the instructor.
7. Add 10 μl of 3 M sodium acetate (pH 5.2) to the aqueous phase, as well as 800 μl of absolute ethanol. Cap and invert the tube several times to mix. Submerge the closed tube in a dry ice/ethanol bath (−70°C) for 10 min. This is done to precipitate plasmid DNA.
8. Centrifuge the sample at 4°C in a microcentrifuge for 30 min. Carefully remove the supernatant from the precipitated DNA pellet. *Depending on how pure the DNA is, this pellet may be translucent and difficult to see. It may help you to place the outside hinge of the microcentrifuge tube toward the outside of the rotor during the centrifugation so that you will know where the DNA pellet is at the bottom of the tube. You may also find the use of Pellet Paint to be helpful during this procedure. This product, sold by Novagen (catalog #69049-3), is a pink chromophore that will bind the DNA and give the pellet a color that will be easier to identify at this point.*
9. Add 1 ml of ice cold 70% ethanol to the DNA pellet, centrifuge at 4°C for 5 min, and carefully remove the supernatant from the precipitated DNA pellet. This step is done to remove the salt that was added to the DNA to aid in the precipitation. *This salt, if not removed, could interfere with the DNA ligation reaction.*

10. Allow the DNA pellet to air dry (~10 min) or dry it down briefly (~2 min) in the vacuum microcentrifuge.
11. Resuspend the DNA pellet in 20 μl of sterile distilled water. Store the DNA on ice until ready to proceed with the ligation reaction.
12. Carefully mix the following together in a sterile, 1.5-ml microcentrifuge tube:

    15 μl of DNA sample (from step 11)
    4 μl of T4 DNA ligase buffer (250 mM Tris, pH 7.6, 50 mM $MgCl_2$, 5 mM ATP, 5 mM dithiothreitol, 25% wt/vol polyethylene glycol-8000)
    1 μl of T4 DNA ligase (1 unit/μl)

13. Incubate the reaction at room temperature for 1 hr and then at 15°C overnight. Store the reaction at 4°C for use on Day 2. Also, store the remainder of your DNA sample from step 11 (5 μl) at 4°C for use on Day 2.

Day 2: Transformation of *E. coli* Host Cells

1. Add 30 μl of sterile distilled water to your 20-μl ligation reaction. Add 2.5 μl of 3 M sodium acetate (pH 5.2) and 200 μl of absolute ethanol. Cap the tube and invert several times to mix. Submerge the capped tube in a dry ice/ethanol bath (−70°C) for 10 min.
2. Centrifuge the sample at 4°C for 30 min. Remove the supernatant from the precipitated DNA pellet and add 1 ml of ice cold 70% ethanol. Centrifuge the sample again at 4°C for 5 min. Remove the supernatant from the precipitated DNA pellet.
3. Air dry the DNA pellet as before or dry briefly in the vacuum microcentrifuge.
4. Resuspend the dried DNA pellet in 5 μl of sterile distilled water.
5. Obtain a microcentrifuge tube from the instructor containing 160 μl of freshly thawed *Epicurian coli* XL1-Blue competent cells (with β-mercaptoethanol added as per the manufacturer's instructions). *Keep these cells on ice at all times!*
6. Add 40 μl of these cells to four *prechilled* (on ice) round-bottomed, 15-ml polypropylene culture tubes. Again, *keep the cells on ice at all times* and label the tubes 1 to 4.
7. To tube 1, add 5 μl of the ligation mixture prepared in step 4. To tube 2, add 5 μl of a 1-ng/ml solution of untreated pUC19. To tube 3, add 5 μl of the 1X ligation buffer (no DNA). To tube 4, add 5 μl of *unligated* DNA sample left over from Day 1 (see Table 21-2 and step 13 of the Day 1 protocol).
8. Incubate the cells with these samples on ice for 15 min, swirling gently every 2 min. Quickly remove the tubes from the ice and place them in a 42°C water bath. Incubate the tubes at 42°C for *exactly* 45 sec. Quickly remove the tubes from the water bath and incubate them again on ice for 2 min. On heat shock, the competent *E. coli* cells will take up the intact plasmid DNA produced during the ligation reaction. *The competent cells are very compromised at this point, and they will die if the incubation is not performed at exactly 42°C for exactly 45 sec. The incubation on ice will prevent the cells from dying after the heat shock.*
9. Add 1 ml of YT broth (use sterile technique) and 30 μl of sterile 100 mM IPTG solution to each tube. Incubate the cells at 37°C with *gentle* shaking for 1 hr. During this time, the cells will recover from the heat shock and begin to express genes (such as kan^R in the pUC19/4K plasmid) under the control of the *lac* promoter.
10. Obtain five agar plates containing IPTG (40 μg/ml), Xgal (40 μg/ml), and ampicillin (100 μg/ml). Obtain two agar plates containing IPTG (40 μg/ml), Xgal (40 μg/ml), and kanamycin (50 μg/ml).
11. Remove 100- and 200-μl aliquots from culture tube 1 and place on two separate IPTG/Xgal/amp plates (see Table 21-3). Plate the same volumes of cells from culture tube 1 on the two separate IPTG/Xgal/kan plates.

Table 21-2 Procedure for *E. coli* Transformation

Tube	DNA Sample	Volume (μl)	Volume of *E. coli* XL1-Blue cells (μl)
1	Ligation mixture	5	40
2	pUC19 (untreated)	5	40
3	Ligation buffer (no DNA)	5	40
4	Unligated DNA	5	40

Table 21-3 Procedure for Plating and Selection of Transformed Bacteria

Transformation Tube	Volume to Be Plated (μl)	Components in Agar Plate
1	100	IPTG/Xgal/amp
1	100	IPTG/Xgal/kan
1	200	IPTG/Xgal/amp
1	200	IPTG/Xgal/kan
2	200	IPTG/Xgal/amp
3	200	IPTG/Xgal/amp
4	200	IPTG/Xgal/amp

12. Remove 200 μl from culture tube 2 and place on an IPTG/Xgal/amp plate. Remove 200 μl from culture tube 3 and place on an IPTG/Xgal/amp plate. Remove 200 μl from culture tube 4 and place on an IPTG/Xgal/amp plate. *Be sure that you label each plate so that you know what antibiotic the plate contains and what transformation mixture was placed on each plate.*
13. Using sterile technique (to be demonstrated by the instructor), use a sterile bent glass rod to spread the bacteria evenly over the surface of each plate. Allow the surface of the plates to dry for 5 min, invert the plates (agar side up), and place them in the 37°C incubator overnight. The colonies that appear on these plates will be grown in culture the next day, and plasmid DNA will be isolated from them on Day 3. The plates must be incubated at 37°C for at least 24 hr. After this incubation, the plates may be stored at 4°C for several days. *We have found that the blue color will develop much better in colonies that contain a plasmid with an intact* LacZα *peptide sequence if this is done. Incubating the plates at 4°C will make it much easier to distinguish between blue and white colonies when the plates are counted (see below).*

Off day: Growth of Transformed Cells

1. Remove your seven agar plates from the 37°C incubator. Obtain four large glass culture tubes from the instructor that each contain 3 ml of sterile YT broth supplemented with 100 μg/ml ampicillin. Label the tubes 1 to 4.
2. Analyze the two IPTG/Xgal/kan plates *containing the transformation mixture from tube 1* to find *four colonies that are white in color.* Obtain two fresh agar plates: one containing kanamycin and one containing kanamycin *and* IPTG. On the bottom surface of each plate, use a marker to divide the plate into four quadrants, labeled 1 to 4.
3. Using a sterile toothpick, stab into the middle of a white colony that is present on this IPTG/Xgal/kan plate. Remove the toothpick and stab the end that came into contact with the bacteria first on quadrant 1 of the kan plate *and then* into quadrant 1 of the IPTG/kan plate. Finally, drop the toothpick (bacteria end down) into tube 1 containing ampicillin supplemented YT broth. Repeat step 3 for each of the other quadrants on the two plates, picking a *new* white colony on the IPTG/Xgal/kan plate each time.
4. Incubate the cultures at 37°C with shaking overnight. These strains will be used on Day 3 to isolate plasmid DNA. Place the two replica plates produced in step 3, agar side up, in the 37°C incubator overnight.
5. Count the number of colonies from transformation culture tube 1 present on the two IPTG/Xgal/kan plates. How many of these colonies are blue and how many are white? Explain these results in terms of what you know about the properties of each of the plasmids that these bacterial colonies may contain.
6. Count the number of colonies from transformation culture tube 1 present on the two IPTG/Xgal/amp plates. How many of these colonies are blue and how many are white? Explain these results in terms of what you know about the properties of each of the plasmids that these bacterial colonies may contain.
7. Count the number of colonies present on the IPTG/Xgal/amp plate containing bacteria from transformation culture tube 2. Why was this control experiment performed? What color are these colonies? Explain. Describe what a low number of colonies on this plate would indicate, as well as possible causes for this result.
8. Count the number of colonies present on the IPTG/Xgal/amp plate containing bacteria from transformation culture tube 3. Why was

this control experiment performed? Describe what a large number of colonies on this plate would indicate, as well as possible causes for this result.

9. Count the number of colonies present on the IPTG/Xgal/amp plate containing bacteria from transformation culture tube 4. Why was this control experiment performed? What color are these colonies? Explain. Describe what a large number of blue or white colonies would indicate, as well as possible causes for this result.

Day 3: Isolation of Plasmid DNA

1. Remove the four culture tubes from the 37°C shaker and add 1.5 ml of each culture to four separate microcentrifuge tubes labeled 1 to 4. Cap each tube and centrifuge for 2 min at room temperature to harvest the cells. Remove the supernatant from the bacterial pellet.
2. Add the remaining 1.5 ml of each culture to the appropriate microcentrifuge tubes and repeat step 1.
3. Add 100 μl of GTE (lysis) buffer to each bacterial cell pellet. Resuspend each cell pellet thoroughly by repeated pipetting or mixing with a Vortex mixer. Incubate the tubes at room temperature for 5 min.
4. Add 200 μl of *freshly prepared* 0.2 N NaOH–1% SDS solution to each tube. Cap and invert several times to mix. The solution will turn from turbid to more translucent as the cells are completely lysed and the DNA and proteins are denatured. Incubate the tubes on ice for 5 min.
5. Add 150 μl of 3 M potassium acetate solution to each tube. Cap and invert several times to mix. The precipitate that forms contains much of the denatured proteins produced in step 4, as well as chromosomal DNA that was not able to renature correctly when the pH was quickly lowered. Incubate the tubes on ice for 10 min.
6. Centrifuge the tubes for 15 min at 4°C. Remove the *supernatant* from each tube and place in four fresh microcentrifuge tubes labeled 1 to 4. The supernatant contains the plasmid DNA, while the pellet contains denatured proteins, chromosomal DNA, membrane components, and other cell debris. The tubes containing this pellet may be discarded.
7. Add 500 μl of Tris-saturated phenol:chloroform:isoamyl alcohol (25:24:1) to each tube containing the supernatant fraction. Mix each tube with a Vortex mixer for 1 min, centrifuge for 5 min at room temperature, and transfer the aqueous (upper) phases to four fresh microcentrifuge tubes labeled 1 to 4. This step is done to remove any remaining proteins and lipid components from the plasmid DNA. Dispose of the organic phase from this extraction in a waste beaker designated by the instructor.
8. Repeat the aqueous phase extraction procedure described in step 7 with 500 μl of chloroform:isoamyl alcohol (24:1). Transfer the aqueous (upper) phases to four fresh microcentrifuge tubes. This step is done to remove all traces of phenol from the solution. Dispose of the organic phase from this extraction in a waste beaker designated by the instructor.
9. Add 1 ml of absolute ethanol to each of the four aqueous fractions. Cap the tubes and invert several times to mix. Incubate the tubes on ice for 10 min.
10. Centrifuge the tubes at 4°C for 15 min. Remove the supernatant from the precipitated DNA pellet (which may also contain significant amounts of RNA, making the pellet appear white in color). Add 1 ml of ice cold 70% ethanol to each DNA pellet, centrifuge for 5 min at 4°C, and remove the supernatant from the DNA pellet.
11. Allow the pellets to air dry for 10 min or dry briefly in the vacuum microcentrifuge. Store the four, dried DNA pellets at −20°C for use on Day 4.

Day 4: Restriction Endonuclease Digestion and Agarose-Gel Electrophoresis of Plasmid DNA

1. In this experiment, you will perform restriction digests on two of the four plasmid DNA samples that you have isolated. If it is possible, select the plasmid DNA isolated from two bacterial colonies that were able to grow well on the IPTG/kan plates, but not as well on the kanamycin plates without IPTG. This is determined by analyzing the growth of the four colonies (clones) on the kan plate and the kan/IPTG plate produced in step 3 from the

previous day. *Remember that our method of selecting between cells carrying the pUC4K and pUC19/4K plasmids is that the kan^R gene on pUC19/4K is IPTG-inducible, while the kan^R gene is expressed constitutively in cells carrying the pUC4K plasmid.* The *lac* promoter is known to be somewhat "leaky," showing some expression of genes under its control (such as the kan^R gene) even in the absence of IPTG. Therefore, you may find that cells carrying pUC19/4K may show some growth on the kan plate without IPTG. Ideally, you will want to analyze colonies (clones) that grew *well* only on the kan plate *with* IPTG. If you only have colonies that grew *equally well* on both the kan plate and the IPTG/kan plate, choose any two of the four plasmids for analysis.

2. Resuspend the two plasmid DNA sample pellets in 32 μl of sterile distilled water.
3. Set up the following reactions *for each of the two plasmid samples* in 1.5 ml microcentrifuge tubes:

***Pst*I Digests**	***Hind*III Digests**	***Sal*I Digests**
8 μl of plasmid DNA	8 μl of plasmid DNA	8 μl of plasmid DNA
1.5 μl of 10X React 2	1.5 μl of 10X React 2	1.5 μl of 10X React 10
1 μl of *Pst*I (10 units/μl)	1 μl of *Hind*III (10 units/μl)	1 μl of *Sal*I (10 units/μl)
1 μl of RNaseA	1 μl of RNAseA	1 μl of RNAseA
3.5 μl of water	3.5 μl of water	3.5 μl of water

The RNAseA is at a concentration of 10 mg/ml, and is added to completely digest all of the RNA remaining in the sample. If this is not done, a large RNA "spot" will be present on the gel following ethidium bromide staining, which will prevent you from seeing low molecular weight DNA fragments on the gel.

4. After a 1-hr incubation at 37°C, add 3 μl of 6X DNA sample buffer to each of the reaction tubes, as well as to the 8 μl of *undigested plasmid* remaining from each of the two samples (see step 2). At this point you should have eight samples for agarose-gel analysis.
5. Prepare a 1% TBE agarose gel by adding 0.5 g of electrophoresis-grade agarose to a 250-ml Erlenmeyer flask containing 50 ml of 0.5X TBE buffer (dilute the 5X TBE buffer stock 1:10 with distilled water to prepare the 0.5X working solution). *Preweigh the flask before heating and record this value.*
6. Microwave the flask for 2 to 3 min, or until the agarose has been fully dissolved (no small pieces of undissolved agarose remain). Place the flask on the balance and add distilled water to the flask until its mass is the same as that before the flask was heated. Water will evaporate from the flask as it is heated. This water must be replaced to ensure that a 1% agarose solution is maintained. Depending on the amount of water that has evaporated, the 1% agarose gel prepared in step 5 may now be greater than 1%.
7. Allow the solution to cool for 5 min, swirling the flask gently every so often to prevent the gel from hardening. When the outside of the flask is cool enough to touch, pour the solution into an agarose-gel cast fitted with a 10-well comb at one end (consult the instructor). Allow the gel to set (~40 min). Remove the comb, and place the gel in an agarose-gel electrophoresis chamber. Add 0.5X TBE buffer to the chamber *until it completely covers the gel and fills the wells.*
8. Load the samples prepared in step 4 as follows:

	Lane 1	**Lane 2**	**Lane 3**	**Lane 4**	**Lane 5**	**Lane 6**	**Lane 7**	**Lane 8**	**Lane 9**
Restriction enzyme	*	*Pst*I	*Hind*III	*Sal*I	None	*	*Pst*I	*Hind*III	*Sal*I
Plasmid	1	1	1	1	†	2	2	2	2

*This lane contains your undigested plasmid samples.

†This lane should be loaded with a solution containing 1 μg of 1-kb DNA ladder size standard, water, and DNA sample buffer (prepared by the instructor).

9. Attach the negative electrode (cathode) to the well side of the chamber and the positive electrode (anode) to the other side of the chamber. Remember that the DNA is negatively charged, and will migrate through the gel toward the positive electrode.
10. Perform the electrophoresis at 50 mA, constant current. You will see two dye fronts develop: one from the bromophenol blue dye (dark blue) and one from the xylene cyanole dye (light blue). The latter of the two dyes will migrate the same as a DNA fragment of approximately 4 kb, while the former dye will migrate the same as a DNA fragment of about 0.5 kb.
11. Continue the electrophoresis until the faster moving of the two dye fronts (bromophenol blue) migrates about three-fourths of the way through the gel.
12. Turn off the power supply, disassemble the electrophoresis apparatus, carefully remove the agarose gel, and submerge it in a small tray containing a 0.5-μg/ml solution of ethidium bromide in 0.5X TBE (enough to completely cover the gel). *Wear gloves at all times when working with ethidium bromide!* Incubate at room temperature for about 20 min. The ethidium bromide will enter the gel and intercalate between the base pairs in the DNA strands. When exposed to ultraviolet light, the ethidium bromide will fluoresce, and the DNA will appear as orange or pink bands on the gel.
13. Destain the gel for 10 min in a solution of 0.5X TBE buffer without ethidium bromide. This is done to remove ethidium bromide from all portions of the gel that do not contain DNA.
14. Place the gel in a dark box fitted with a 254 nm light source. Close the doors of the box and turn on the light source to visualize the DNA bands on the gel. Take a photograph of the gel for later analysis.
16. Prepare a plot of the number of base pairs versus the distance traveled (in centimeters) for each DNA fragment present in the 1-kb DNA ladder lane. If the 1-kb DNA size standards resolved well, you should be able to differentiate between the relative mobilities of the 0.5-, 1.0-, 1.6-, 2.0-, and 3.0-kb DNA fragments (see Fig. 21-5) for use in preparing the standard curve. *Do not attempt to determine the relative mobilities of the larger-molecular-weight DNA size standards on the gel if they did not resolve or separate well. This will only add error to the standard curve over the region where the DNA size standards did display good resolution.*

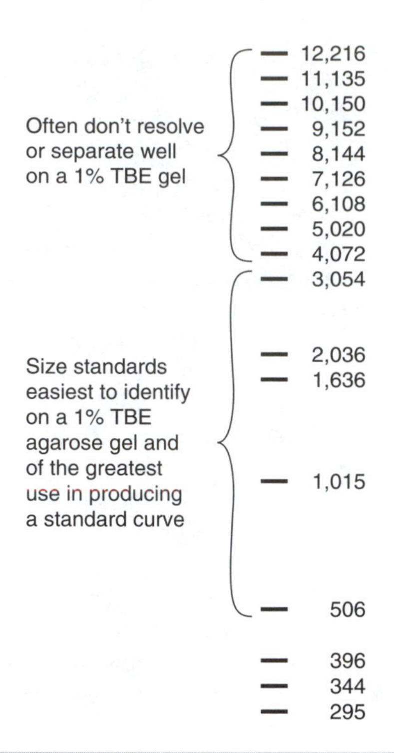

Figure 21-5 Number of base pairs in each band of the 1-kb DNA ladder.

17. Have you successfully constructed the desired pUC19/4K recombinant plasmid? Explain your answer in terms of what size DNA fragments resulted from each digest and your knowledge of the composition of the parent plasmids used in this experiment (pUC19 and pUC4K). Prepare a restriction map of the newly constructed plasmid. The map should include the total number of base pairs in the plasmid, the identity and position of all restriction sites in the plasmid, and the position of all of the different genes present in the plasmid.
18. If you do not think that you have isolated the desired pUC19/4K plasmid, prepare a map of the plasmid that you think you have isolated. Support your drawing with an explanation of how the digestion of this plasmid with the various restriction enzymes would produce the results that you obtained.

Exercises

1. Why is it important to perform a transformation control experiment using undigested plasmid? Describe two possible results that could be obtained in this control experiment, along with an explanation of what each result would indicate.
2. Why is it important to perform a transformation control experiment using no plasmid DNA? Describe two possible results that could be obtained in this control experiment, along with an explanation of what each result would indicate.
3. Why is it important to perform a transformation control experiment using unligated plasmid? Describe all the possible results that could be obtained in this control experiment, along with an explanation of what each result would indicate.
4. The pUC19 and pUC4K plasmids both confer antibiotic resistance to the *E. coli* host strains that they are transformed into. Other than drug resistance, can you think of any other gene(s) that could be carried on a plasmid that would allow you to select for cells that obtained it on transformation? For each of the genetic elements that you propose, describe the relevant features of the genotype of the host strain that you would use in the experiment, as well as how you would select for cells that obtained the plasmid following transformation.
5. Can you think of any advantages of having the *lacI* gene carried on the pUC19 plasmid?
6. You have obtained a plasmid that is able to be propagated in *Escherichia coli* but not in *Bacillus subtilis*. Why is this plasmid able to be propagated in one bacterium but not another? How would you propose to alter the plasmid so that it is able to be propagated both in *E. coli* and *B. subtilis?*
7. Where were plasmids originally isolated from? What function(s) do naturally occurring plasmids serve?

REFERENCES

Berger, S. L., and Kimmel, A. R. (1987). Guide to Molecular Cloning Techniques. *Methods Enzymol* Vol. **152**.

Garrett, R. H., and Grisham, C. M. (1995). Recombinant DNA: Cloning and Creation of Chimeric Genes. In: *Biochemistry*. Orlando, FL: Saunders.

Lehninger, A. L., Nelson, D. L., and Cox, M. M. (1993). Recombinant DNA Technology. In: *Principles of Biochemistry*, 2nd ed. New York: Worth.

Sambrook, J., Fritsch, E. F., and Maniatis, T. (1989). *Molecular Cloning*, 2nd ed. Plainview, NY: Cold Spring Harbor Laboratory Press.

Wilson, K., and Walker, J. M. (1995). Molecular Biology Techniques. In: *Principles and Techniques of Practical Biochemistry*, 4th ed. Hatfield, UK: Cambridge University Press.

In Vitro Transcription from a Plasmid Carrying a T7 RNA Polymerase–Specific Promoter

Theory

Transcription is one of two processes that allow a cell to synthesize the proteins encoded in its DNA. In the first process, the genes contained on the chromosome are *transcribed* into molecules of RNA. In the second process, these RNA messages are *translated* by the ribosomes and transfer RNAs (tRNAs) to produce proteins with specific amino acid sequences. RNA polymerase is the enzyme that synthesizes RNA. This enzyme will read a single-stranded DNA template in the 3′ to 5′ direction and construct an RNA molecule in the 5′ to 3′ direction that is complementary to it (Fig. 22-1). In some respects, RNA synthesis is similar to DNA synthesis: both molecules are synthesized in the 5′ to 3′ direction, both of the enzymes involved require a template DNA strand to direct the synthesis, and both of the enzymes involved use nucleoside triphosphates as the basic monomeric units in the synthesis reaction.

Despite these similarities, there are some very important differences between the reactions carried out by DNA polymerase and RNA polymerase. Remember that DNA is synthesized with the use of four *deoxy*ribonucleoside triphosphates (dATP, dGTP, dTTP, and dCTP). In the double-stranded DNA helix, adenine bases hydrogen-bond with thymine bases, and guanine bases hydrogen-bond with cytosine bases. RNA, however, is synthesized with the use of four ribonucleoside triphosphates (ATP, GTP, CTP, and UTP). Adenine and guanine bases in the DNA template will direct the addition of uracil and cytosine bases to the growing RNA molecule, respectively, while thymine and cytosine bases in the DNA template will direct the addition of adenine and guanine bases, respectively, to the growing RNA molecule. The net result of this is that RNA polymerase will produce an RNA molecule identical to that of the nontemplate (coding) DNA strand, with uracil in place of thymine bases.

Another difference between DNA polymerase and RNA polymerase is that the latter enzyme does not require the presence of a single-stranded nucleic acid primer to initiate the polymerization reaction. Given the correct sequence of DNA (a promoter, see below), RNA polymerase will bind the DNA template and begin transcription. This is in contrast to DNA polymerase (see Experiment 24), which requires a short nucleic acid primer to initiate synthesis of DNA. Unlike DNA polymerase, RNA polymerase does not have a 3′ to 5′ exonuclease or "proofreading" activity. As a result, an error (a single base change) will be introduced into an RNA molecule for every 10^4 to 10^6 nucleotides added to the growing macromolecule during synthesis. This error rate is acceptable, since many molecules of RNA are produced from a single DNA template, and because the population of a particular RNA molecule is constantly being regenerated or turned over with time (the half-life of mRNA is relatively short). The 3′ to 5′ proofreading activity of DNA polymerase decreases the error rate in DNA synthesis to one in 10^9 to 10^{10} bases. Because the DNA in a cell provides the *permanent* genetic information, you can understand why the accuracy of DNA synthesis during replication is more critical than that of RNA synthesis in the life cycle of a cell.

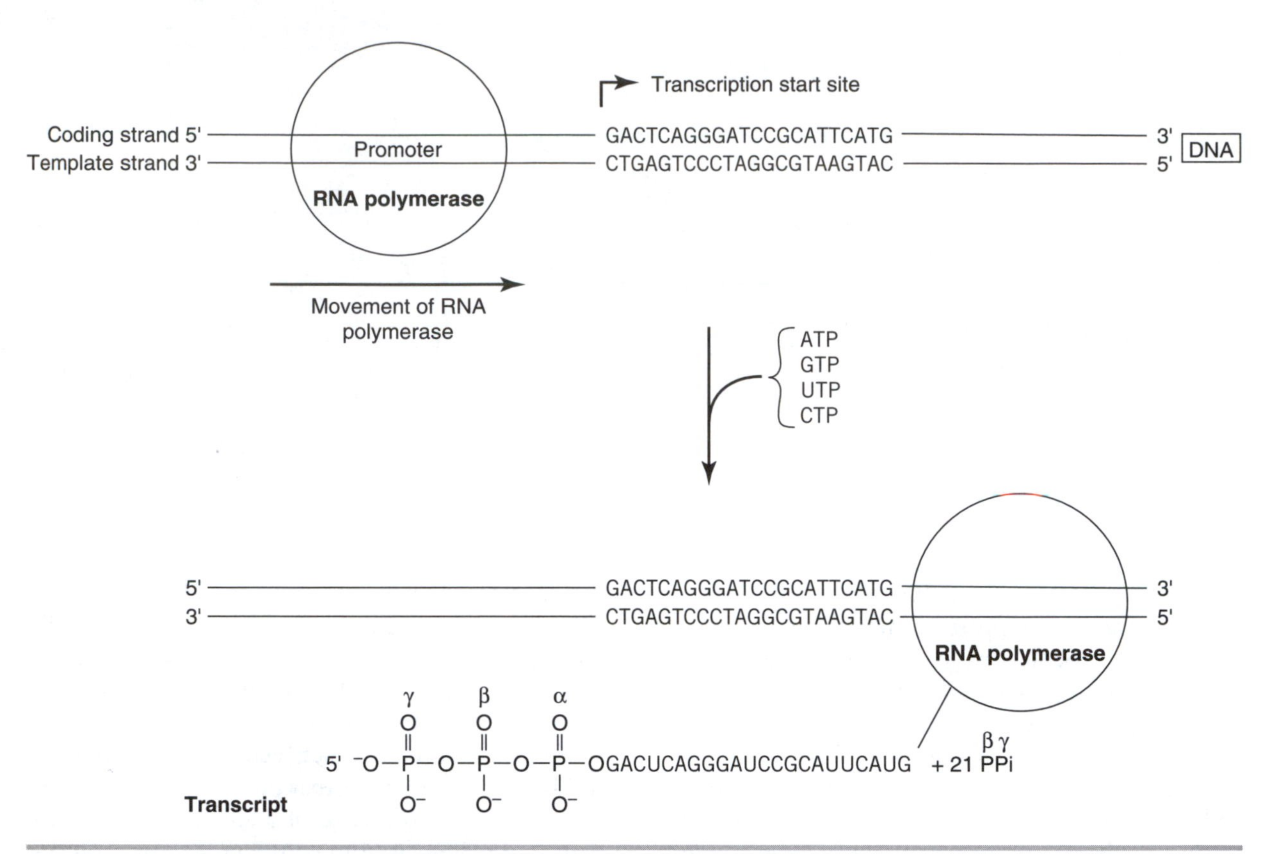

Figure 22-1 Schematic diagram of the process of transcription. RNA polymerase will read the template DNA strand 3′ to 5′ and produce a transcript in the 5′ to 3′ direction that is identical to the sequence of the coding DNA strand, with uracil (U) in place of thymine (T). A radiolabeled transcript can be produced by including an α-[^{32}P] nucleotide, which will become part of the product, or by including a γ-[^{32}P] nucleoside triphosphate that is known to be the first nucleotide in the transcript.

Where does RNA polymerase begin the process of transcription on the DNA template? The DNA element on which RNA polymerase binds and initiates transcription is termed the *promoter*. The simplest promoters are those recognized by viral RNA polymerases. Usually, these are a contiguous sequence of 15 to 30 base pairs that direct the viral RNA polymerase to the transcriptional start site (Fig. 22-2). The viral promoter sequences possess a 5′ to 3′ polarity, allowing you to determine which of the two DNA strands will act as the template for transcription. Bacterial promoters consist of two different but conserved DNA sequences. One of these elements is located about 10 base pairs on the 5′ side of the transcriptional start site, while the other element is located about 35 base pairs to the 5′ side of the transcriptional start site. Although these two elements consist of as little as six base pairs each, the sequence and spacing of these two elements are critical for allowing the bacterial RNA polymerase to bind the DNA template and initiate transcription. Promoters for eukaryotic RNA polymerases are quite variable. Still, there are some recurring elements found about 25, 40, and 100 base pairs to the 5′ side of the transcriptional start site. These sequences, as well as others that are often thousands of base pairs from the transcriptional start site, are believed to be the binding sites for transcription factors (proteins) that regulate the activity of the eukaryotic RNA polymerases. A list of different RNA polymerases, and the promoter sequences that they recognize, are listed in Table 22-1.

Where does transcription of a DNA template end? The process of transcription termination is best understood in bacteria. Rho-*independent* terminators are characterized by a self-complementary

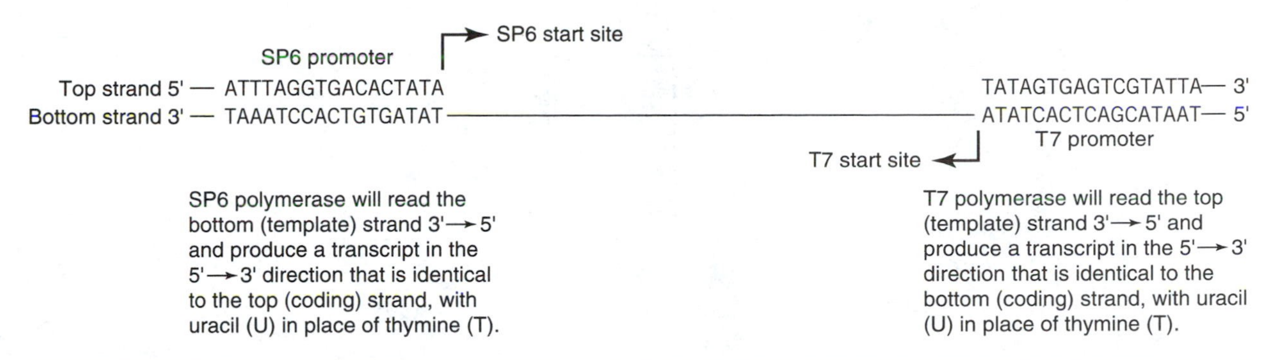

Figure 22-2 Promoters specify where RNA polymerase will bind the DNA and initiate transcription. The polarity of the promoter sequence will specify the coding and template strands of the DNA.

DNA sequence located roughly 20 bases on the 5′ side of the termination site. The RNA produced on transcription of this region is able to form a hairpin structure that causes the RNA polymerase to "stall" and eventually dissociate from the DNA template. A stretch of uridyl nucleotides at the 3′ end of the RNA hairpin is part of the rho-independent terminator sequence that is also believed to help destabilize the enzyme–DNA complex. A rho-*dependent* terminator also contains a self-complementary sequence capable of forming a hairpin structure. In an unknown mechanism, the rho protein hydrolyzes ATP and destabilizes the RNA polymerase–DNA complex as it stalls at the hairpin, eventually leading to the termination of transcription.

How were the promoter sites for the various RNA polymerases determined? If the enzyme binds a particular sequence of bases in the DNA, then that stretch of bases will be protected from chemicals and nucleases that cleave DNA. A single strand of DNA in a double helix is first labeled with a radioactive group at either the 3′ or 5′ end. The DNA duplex is then incubated with the RNA polymerase and treated with a chemical or DNase enzyme that produces a number of different-sized DNA fragments. The DNA duplex is then denatured, and the DNA fragments are resolved by polyacrylamide-gel electrophoresis. Any region of the DNA duplex protected by the polymerase during the DNase treatment will be absent on the autoradiogram of the gel, compared to the same radiolabeled DNA duplex not protected by the polymerase. The "footprint" left on the DNA by the polymerase provides insight into what sequences and what regions of the DNA are bound by the RNA polymerase (Fig. 22-3).

Table 22-1 RNA Polymerases and the Promoters That They Recognize

RNA Polymerase	Molecular Weight	Subunits	Promoter
SP6 (from SP6 phage)	~100 kDa	1	5′-ATTTAGGTGACACTATAGAACTC-3′
T7 (from T7 phage)	~100 kDa	1	5′-TAATACGACTCACTATAGGGAGA-3′
Bacterial (*E. coli*)*	~324 kDa	5	5′-TTGACA-3′ (−35 region)
			5′-TATAAT-3′ (−10 region)
Eukaryotic†	500–700 kDa	~12	5′-GGCCAATCT-3′ (−110 region)
			5′-GGGCGG-3′ (−40 region)
			5′-TATAAAA-3′ (−25 region)

*Promoter sequence recognition by bacterial RNA polymerase is governed by its sigma subunits. The sequence shown is recognized by RNA polymerase with the major subunit found in rapidly growing, well-nourished cells. Other sequences are recognized under other physiological conditions that lead to formation or activation of alternate sigma subunits.

†The promoter sequences for eukaryotic RNA polymerases I, II, and III are numerous and, in some cases, not well defined. The promoter sequence shown is the consensus sequence recognized by RNA polymerase II.

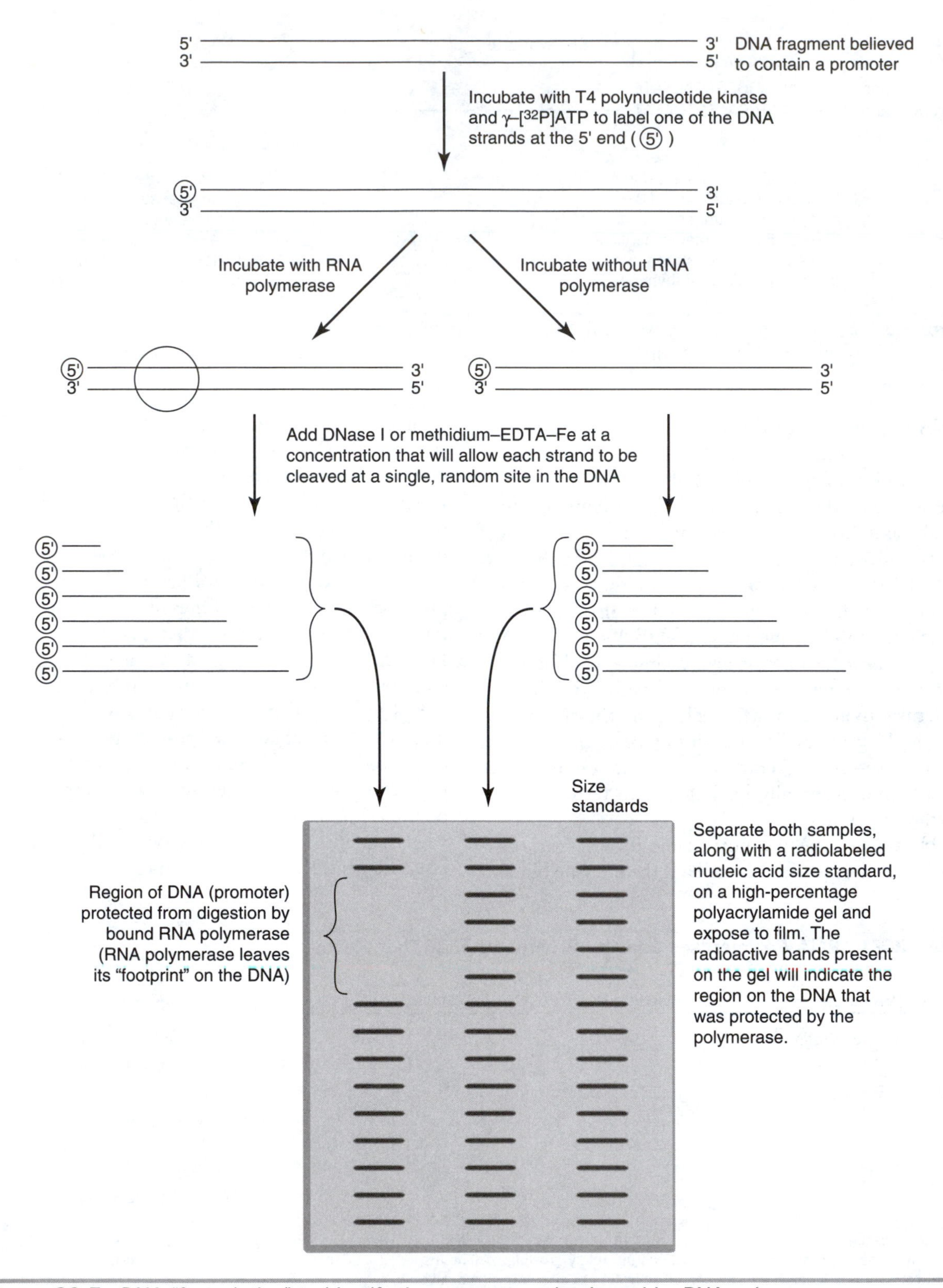

Figure 22-3 DNA "footprinting" to identify the promoter region bound by RNA polymerase.

In this experiment, you will carry out an in vitro transcription reaction from a plasmid (pSP72, Fig. 22-4) containing a T7 RNA polymerase specific promoter. A ^{32}P-labeled nucleoside triphosphate (labeled at the α position) will be included in the reaction to produce a radioactive RNA molecule. The size and sequence of the various transcripts that you will produce can be predicted by the restriction enzymes that the plasmid will be digested with prior to the beginning of the transcription reaction. Since the plasmid template for the transcription reaction will be digested, you will be preparing "run-off" transcripts from the T7 promoter. As a result, the termination of the transcription reaction will not rely on the presence of any transcriptional terminator sequence (rho-dependent or -independent) on the plasmid. The exact molecular weight or size of the various transcripts (the exact number of nucleotides that they contain) will be verified by polyacrylamide-gel electrophoresis at the end of the experiment.

Although this may appear to be a simple experiment, the same techniques may be modified to analyze the effects of different mutations in the RNA polymerase and/or promoter sequence on the process of transcription. Suppose that you wanted

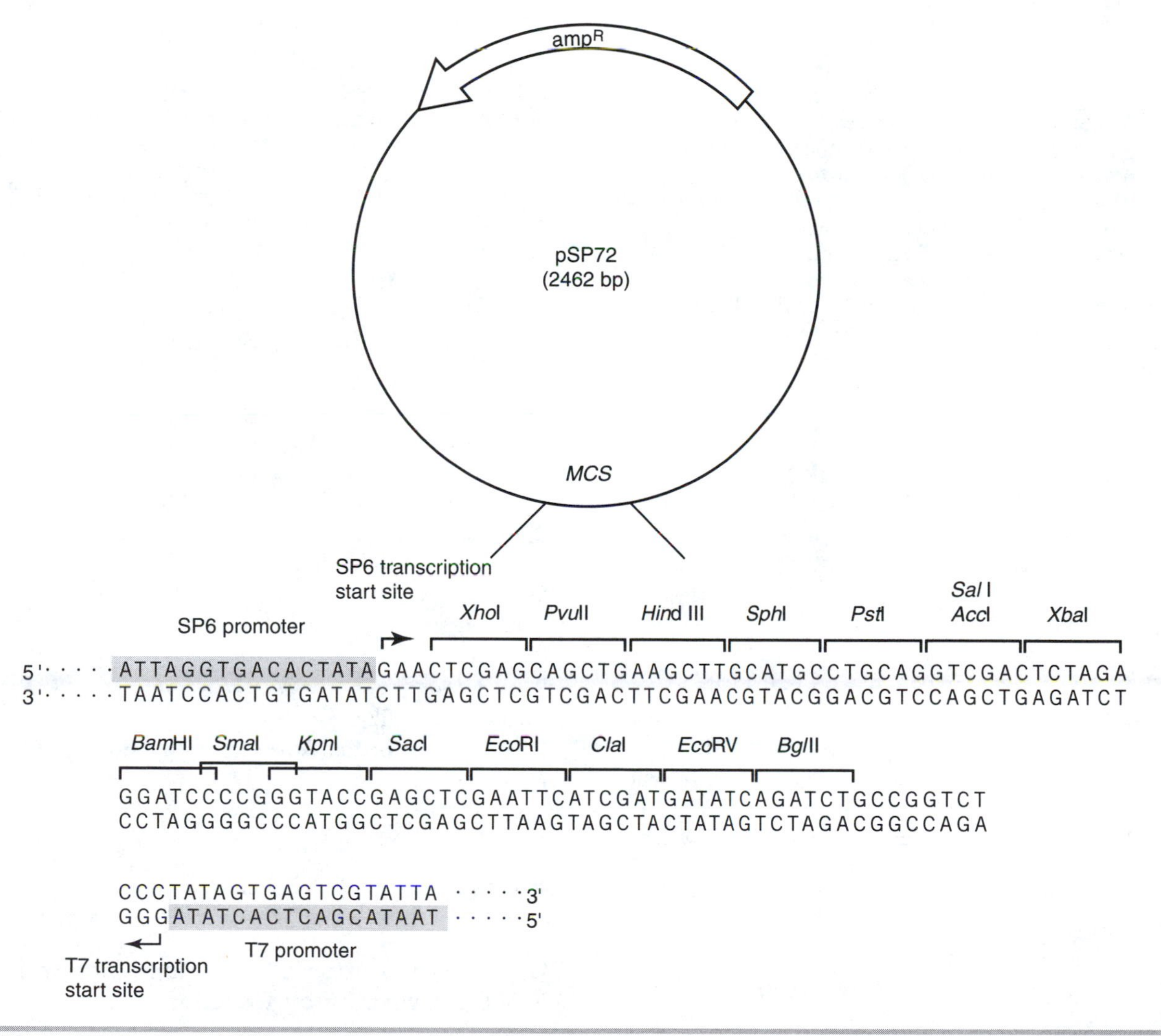

Figure 22-4 The pSP72 plasmid map (simplified).

to determine which of the bases in the T7 promoter were absolutely critical for transcription. You could produce a number of different point mutations within the T7 promoter sequence and use these as templates in the same kind of run-off transcription experiments. By comparing the results of these experiments with those that you will obtain in your experiment (using the consensus T7 promoter in the template DNA), you could make *quantitative* determinations of how each point mutation in the promoter affects transcription. In the same fashion, T7 RNA polymerase mutants could be tested against the wild-type enzyme to determine which amino acid residues are important to carry out transcription from the consensus T7 promoter.

WARNING

Caution: This experiment will require the use of significant amounts of ^{32}P-labeled nucleotide. Wear safety goggles and gloves at all times. Use Plexiglas shields when working directly with the radioisotope. Cover your work surface with absorbent paper to localize any spills that may occur. Consult the laboratory instructor for the proper disposal of the solid and liquid radioactive waste that will be produced in this experiment.

Supplies and Reagents

P-20, P-200, and P-1000 Pipetmen with sterile disposable tips
1.5-ml microcentrifuge tubes
Microcentrifuge
*Hind*III (10 units/μl) with the supplied 10X reaction buffer—Gibco BRL catalog #15207-020
*Bam*HI (10 units/μl) with the supplied 10X reaction buffer—Gibco BRL catalog #15201-031
*Eco*RI (10 units/μl) with the supplied 10X reaction buffer—Gibco BRL catalog #15202-021
pSP72 plasmid (~0.1 mg/ml in sterile distilled water)—Promega catalog #P-2191
37°C water bath
Dry ice/ethanol bath (−70°C)
Sterile distilled water
Phenol:chloroform (1:1, water saturated)
Vortex mixer
Chloroform:isoamyl alcohol (24:1)
3 M sodium acetate (pH 5.2)
Absolute ethanol
Vacuum microcentrifuge (Speed-Vac)
TE buffer (10 mM Tris, pH 7.5, 1 mM EDTA)
10X Transcription mix

- 0.4 M Tris, pH 8.0
- 0.15 M $MgCl_2$
- Bovine serum albumin (1 mg/ml)
- 0.5 mM ATP
- 10 mM CTP
- 10 mM GTP
- 10 mM UTP
- 0.1 mM dithiothreitol
- α-[^{32}P]ATP (50,000 cpm/μl)—ICN Biomedicals catalog #32007x

T7 RNA polymerase (20 units/μl)—Gibco BRL catalog #18033-100
Apparatus for polyacrylamide gel electrophoresis
Reagents for 15% polyacrylamide gel containing 7 M urea

- 40% acrylamide solution (380 g/liter acrylamide, 20 g/liter *N,N′*-methylenebisacrylamide)
- Urea
- 5X TBE buffer (see below)
- TEMED (*N,N,N′,N′*-tetramethylethylenediamine)
- 10% wt/vol ammonium persulfate in water—*prepared fresh*

5X TBE buffer (54 g/liter Tris base, 27.5 g/liter boric acid, 20 ml/liter 0.5 M EDTA, pH 8.0)
Loading buffer (7 M urea in water with 0.1% wt/vol xylene cyanole and bromophenol blue)
RNA size standards (~20, 61, and 97 nucleotides)—prepared by the instructor
Power supply
Film cassette
Plastic wrap
Razor blade
Kodak X-omat film
Film developer or Kodak developing solutions
Dark room

Protocol

Day 1: Preparation of DNA Template and in Vitro Transcription Reaction

1. Set up the following reactions in three labeled 1.5-ml microcentrifuge tubes:

Tube 1	Tube 2	Tube 3
8 μl of pSP72 plasmid	8 μl of pSP72 plasmid	8 μl of pSP72 plasmid
8 μl of 10X React 3	8 μl of 10X React 3	8 μl of 10X React 2
2 μl of *Eco*RI	2 μl of *Bam*HI	2 μl of *Hin*dIII
62 μl of water	62 μl of water	62 μl of water

Mix all the components thoroughly (do not use a Vortex mixer) and incubate in a 37°C water bath for 1.5 hr.

2. To remove the restriction endonucleases from the reaction, add 80 μl of phenol:chloroform (1:1) to each tube and mix with a Vortex mixer for 1 min. Centrifuge the samples at room temperature for 1 min, remove the aqueous phase (top layer) from each tube, and place these in three fresh microcentrifuge tubes. Add 80 μl of chloroform:isoamyl alcohol (24:1) to each of the three aqueous samples, cap, and mix with a Vortex mixer for 1 min. Centrifuge the samples at room temperature for 1 min and transfer the three aqueous phases (top layers) to three fresh microcentrifuge tubes. This last extraction is done to remove all traces of phenol from the solution. *It is critical that all of the phenol and chloroform be removed from the aqueous phases in this final extraction, since they may denature the RNA polymerase during the transcription reaction.* Dispose of the organic phases produced in these extractions in a waste container designated by the instructor.
3. Add 10 μl of 3 M sodium acetate (pH 5.2) to the aqueous phase in each of the three tubes, along with 200 μl of absolute ethanol. Cap the tubes and invert them several times to mix.
4. Incubate the tubes in a dry ice/ethanol bath (−70°C) for 10 min to precipitate the plasmid DNA. Pellet the precipitated DNA by centrifugation for 30 min at 4°C. *Place the microcentrifuge tubes in the rotor with the cap hinge facing the outside during this step. The DNA pellet will be translucent and very difficult to see, but you will know that the pellet will be on the same side of the tube as the cap hinge.* Carefully remove the supernatant from the DNA pellet. You may also find the use of Pellet Paint to be helpful during this procedure. This product, sold by Novagen (catalog #69049-3), is a pink chromophore that will bind the DNA and give the pellet a color that will be easier to identify.
5. Add 1 ml of ice cold 70% ethanol to each tube, centrifuge for 5 min at 4°C, and carefully remove the supernatant from the precipitated DNA pellet. This wash is done to remove the salts that were added to the DNA solution to aid in its precipitation, which may affect the activity of the T7 RNA polymerase.
6. Dry the DNA pellets in each of the three tubes in the vacuum microcentrifuge (~5 min), or allow the pellets to air dry for about 20 min.
7. Resuspend the DNA pellet in each of these three tubes with 8 μl of TE buffer. Label the tubes with your name and either H (*Hin*dIII), E (*Eco*RI), or B (*Bam*HI), depending on which restriction enzyme was used to treat each of the plasmid samples.
8. Add 1 μl of 10X transcription mix and 1 μl of T7 RNA polymerase to each of the three reaction tubes. Mix thoroughly but do not use a Vortex mixer. *Remember that the transcription mix is radioactive. Wear safety goggles and gloves when working with radioisotopes. Anything that touches the solution at this point will be radioactive. Be careful with disposal of radioactive pipette tips, microcentrifuge tubes, etc. Your lab instructor will provide you with radioactive waste storage containers.* Flash spin the samples to the bottom of the tubes for 1 sec and incubate all three reactions at 37°C for 1 hr.
9. After the 1-hr incubation, store the three reaction tubes at −20°C for use on Day 2.

Day 2: Polyacrylamide-Gel Electrophoresis of RNA Transcripts

1. Obtain a 10-well, 15% polyacrylamide gel containing 7 M urea from the instructor. Place the gel in the electrophoresis apparatus.
2. Add 0.5X TBE buffer to the upper and lower buffer chambers. Remove the comb from the gel. Connect the negative electrode (cathode) to the top buffer chamber and the positive

electrode (anode) to the lower buffer chamber. The RNA that is to be loaded on this gel is negatively charged and will migrate through the gel toward the positive electrode. Apply 200 V (constant voltage) to the gel for 20 min *before the samples are loaded.* This pretreatment of the gel will remove ions left over from the polymerization reaction that may affect the migration of the RNA on the gel.

3. Add 5 μl of loading buffer to each of the three samples prepared on Day 1 that have been thawed at room temperature. *Remember that these samples are radioactive. Wear safety goggles and gloves at all times, and be careful in the disposal of radioactive waste.* Also, add 5 μl of loading buffer to a 10-μl sample of RNA size markers that have been prepared by the instructor. The urea present in the loading buffer and in the gel works to disrupt any hydrogen bonding (intramolecular or intermolecular) within the RNA molecules that may cause them to run at an apparent higher or lower molecular weight than they really are.
4. Turn off the power supply to the gel and wash the wells out thoroughly with 0.5X TBE buffer to remove any precipitated urea and unpolymerized acrylamide that may be clogging the wells. Load the three samples in three consecutive wells on the gel. Next to your three samples, load the RNA size standard sample. *Since you will be using only 4 of the 10 wells, another group can skip a lane and load their four samples. Be sure that you know which of the four samples are in which lane.*
5. Reconnect the power supply and apply 200 V (constant voltage). Continue the electrophoresis until the bromophenol blue (dark blue) dye just migrates off of the bottom of the gel into the lower buffer chamber. *Any unincorporated radiolabeled nucleotide that was present in the samples will run off of the bottom of the gel with the dye. Remember that the buffer in the lower chamber is now radioactive.*
6. Turn off the power supply and remove the plates containing the gel. Separate the plates to expose the gel and put a notch in one corner of the gel with a razor blade to allow you to orient the gel. Place the gel on a sheet of Whatman 3 MM filter paper. Wrap the gel in plastic wrap and place it in the film cassette.
7. Expose the gel to film overnight (~18 hr) at −70°C. Before closing the cassette, use a felt-tipped marking pen to mark the position on the film that corresponds to the notch made on the corner of the gel. Exposing the film at low temperatures will reduce the radioactive scattering and produce more defined RNA bands on the autoradiogram. Develop the film the following day.
8. Align the autoradiogram with the gel and mark the position of the wells on the film. Identify the RNA size marker bands on the autoradiogram (three of them in one lane). The instructor will provide you with the information that you need to determine the number of nucleotides that each standard RNA marker contains.
9. Prepare a plot of number of nucleotides versus distance migrated (in centimeters) for each RNA size standard band on the autoradiogram. Using this standard curve, and the distances migrated by each of the transcripts produced in the three reactions (the other three lanes on the gel), determine the molecular weight and the number of nucleotides present in the transcription reactions containing pSP72 digested with *Eco*RI, *Bam*HI, and *Hin*dIII.
10. Using the map of the pSP72 plasmid, determine the *exact* sequence of the transcript that you would expect following run-off transcription from the T7 promoter using the pSP72 plasmid digested with the three different restriction enzymes (assume that the plasmid was digested to completion with each restriction enzyme). Do the sizes of the RNA molecules present in the three sample lanes on the autoradiogram agree well with your predicted results? Explain your answer.
11. Is there more than one radiolabeled RNA fragment present in each of the three sample lanes? If so, determine whether they are lower or higher in molecular weight than that of the predicted RNA molecule. What may have occurred to produce these RNA fragments of unexpected size?
12. Look at the autoradiogram and compare the intensities (the dark color) of each of the RNA molecules in the three sample lanes. Are all transcripts of equal radioactive intensity? Why or why not? Based on the predicted sequence of your transcripts (determined in step 10),

which of the three transcripts should be the most and the least radioactive? Be as quantitative as you can in answering this question.

Exercises

1. Determine the exact sequence of the transcripts that you would expect to be produced if the same templates used in this experiment (pSP72 digested with *Bam*HI, *Eco*RI, and *Hin*dIII) were subjected to an in vitro transcription reaction using the SP6 RNA polymerase rather than the T7 RNA polymerase.
2. In this experiment, you used α-[^{32}P]ATP as the radiolabeled nucleotide. Would your results have been any different if you had used γ-[^{32}P]GTP as the radiolabeled nucleotide? Explain your answer in terms of the mechanism of the reaction catalyzed by RNA polymerase. Could you have used another γ-labeled nucleotide and still obtained a radiolabeled RNA transcript? Explain your answer.
3. You digest 10 μl of the pSP72 plasmid (0.1 mg/ml) to completion with *Bam*HI. This digested plasmid is used as a template for in vitro transcription reactions (50 μl total reaction volume) using α-[^{32}P]CTP (1 μCi/μmol) and SP6 RNA polymerase. After the transcription reaction is complete, you isolate the RNA transcript from the unincorporated nucleotides and determine that 1 μl of the 50-μl reaction contains 20,000 dpm of ^{32}P. Calculate the number of micrograms of RNA produced in this reaction. What specific activity of γ-[^{32}P]GTP would you have had to use in this reaction to obtain the same mass of RNA transcript containing the same amount of radioactivity (the same specific activity of the RNA transcript)?
4. You are conducting a study of the mechanism of action of T7 RNA polymerase. You have purified a total of 10 point mutants in the enzyme that are unable to carry out transcription from the T7 promoter. You are not sure if these mutations affect the ability of the enzyme to recognize the promoter or if they render the enzyme unable to perform the polymerization reaction. Design a set of experiments that would allow you to distinguish between these two possibilities.
5. Often, phage genomes contain genes under the control of both viral and bacterial promoters. One gene that is commonly under the control of the bacterial promoter is that which encodes the phage RNA polymerase. In terms of the life cycle of a phage, why has the phage evolved to have genes under the control of both types of promoters? What will ultimately determine what species of bacteria a certain phage will be able to infect? (Why can a particular phage infect one species of bacteria but not another?)

REFERENCES

Chamberlin, M., McGrath, J., and Waskell, L. (1970). New RNA Polymerase Form *Escherichia coli* Infected with Bacteriophage T7. *Nature* **228**:227.

Chamberlin, M., and Ryan, T. (1982). Bacteriophage DNA-Dependent RNA Polymerases. In: Boyer, P.D., ed. *The Enzymes*, Vol. 15: Nucleic Acids, Part B, p. 87. Academic Press, New York.

Dunn, J. J., and Studier, F. W. (1983). The Complete Nucleotide Sequence of Bacteriophage T7 DNA and the Locations of T7 Genetic Elements. *J Mol Biol* **135**:917.

Lehninger, A. L., Nelson, D.L., and Cox, M. M. (1993). RNA Metabolism. In: *Principles of Biochemistry*, 2nd ed. New York: Worth.

Rosa, M. D. (1979). Four T7 RNA Polymerase Promoters Contain an Identical 23 Base Pair Sequence. *Cell* **16**:815.

Stahl, S., and Zinn, K. (1981). Nucleotide Sequence of the Cloned Gene for Bacteriophage T7 RNA Polymerase. *J Mol Biol* **148**:481.

In Vitro Translation: mRNA, tRNA, and Ribosomes

Theory

As discussed in Experiment 22, the flow of genetic information from the genes encoded by the DNA into proteins is a two-step process. During *transcription*, RNA polymerase binds at the promoter regions on the DNA and transcribes a single-stranded messenger RNA (mRNA). The RNA polymerase enzyme is instructed as to the sequence of each mRNA molecule that it produces by reading one of the two DNA strands (template strand) in the 3′ to 5′ direction and synthesizing an RNA in the 5′ to 3′ direction that is complementary to it (or identical to the sequence of the coding DNA strand, with U in place of T). Once an mRNA is constructed, it will take part in a process known as *translation*. Beginning at a *defined position* within the mRNA, the ribosome will read the base sequence of the mRNA in sets of three (a codon) and produce a protein or peptide specified by that particular messenger molecule. Translation is carried out in a four-step mechanism, which is discussed in detail below. The mechanism of translation is best understood in prokaryotes. The details of the mechanism of translation in eukaryotes remains an area of intensive research. Because of this, the four-step mechanism described below applies to the prokaryotic system. Immediately following this, the mechanism of translation in eukaryotes is described, and differences between eukaryotic and prokaryotic translation are highlighted.

Step 1: Synthesis of Aminoacyl tRNA

The first step involves activation or "charging" of the various transfer RNAs (tRNAs) in the cell with their appropriate amino acids. The general structure of the tRNA is shown in Figure 23-1. The two-dimensional "cloverleaf" structure of the molecule imparted to it by the ordered array of intramolecular hydrogen bonds between neighboring bases has two important features. First, each tRNA contains an *anticodon loop* that will allow it to base-pair with a complementary sequence (codon) on the mRNA and deliver the appropriate amino acid to the growing peptide chain. The second is the conserved 3′ end of the molecule (5′-GCCA-3′ or

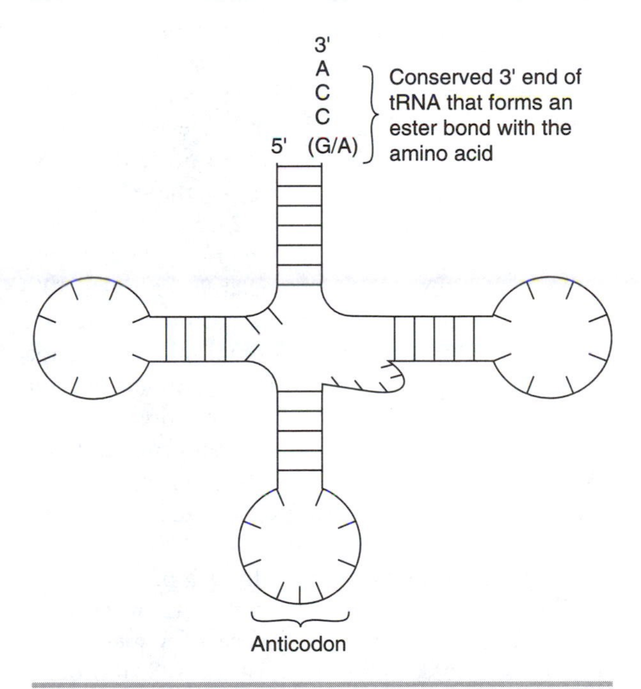

Figure 23-1 Two-dimensional structure of tRNA.

5′-ACCA-3′) that will form a covalent (ester) bond with the appropriate amino acid and deliver it to the site of translation. The sequence of the acceptor stem, which is made up of the 5′ and 3′ ends of the tRNA, is important in the recognition of a given tRNA by the correct "charging enzyme."

The enzymes that link the amino acids to the 3′ ends of the various tRNAs are called *aminoacyl tRNA synthetases.* There is at least one aminoacyl tRNA synthetase that can recognize each of the 20 amino acids. Due to the *degeneracy* of the genetic code (more than one codon that specifies a particular amino acid), a single aminoacyl tRNA synthetase may be able to charge a number of different tRNAs with the same amino acid. For example, phenylalanine is specified by the codons UUU and UUC. The aminoacyl tRNA synthetase specific for phenylalanine will therefore be able to charge a tRNA with AAA or AAG in the anticodon loop with this amino acid. The mechanism by which an amino acid is linked to the 3′ end of the tRNA is shown in Figure 23-2. The process is energetically costly, requiring the hydrolysis of two high-energy phosphate bonds.

Step 2: Initiation

Translation in prokaryotes begins with the formation of the ribosome complex at a defined position on the mRNA, termed the *Shine–Delgarno* sequence, or the ribosome binding site (RBS). In prokaryotic mRNA, this is a relatively small (4–7 nucleotides) region rich in purine nucleotides located less than 10 nucleotides to the 5′ side of the translational start site (Fig. 23-3). This Shine–Delgarno sequence is complementary to the 16S ribosomal RNA (rRNA) associated with the 30S ribosomal subunit, and directs it to bind the mRNA at that position. The 30S ribosomal subunit will not bind this region on the mRNA without the aid of an associated protein called initiation factor 3 (IF-3). The 30S ribosomal subunit will bind the mRNA in such a manner that the *peptidyl (P) site* of the complex is occupied by a specialized codon with the sequence, AUG. It is at this AUG "start" codon where translation will eventually begin.

The next step of the initiation process in prokaryotes involves the delivery of a specialized aminoacyl tRNA to the P site of the 30S ribosomal subunit. This specialized tRNA has a UAC anticodon loop sequence capable of interacting with the AUG start codon, and is charged with an unusual amino acid called *N-formylmethionine* (fMet, Fig. 23-4). The fMet-tRNA is brought to the P site through its association with a GTP binding protein called initiation factor 2 (IF-2, Fig. 23-3).

To complete the initiation process, the 50S ribosomal subunit binds to the 30S–mRNA complex, with the subsequent hydrolysis of GTP and the release of IF-2 and IF-3 from the 70S ribosomal complex (Fig. 23-3). At this point, the *aminoacyl site* (A site) of the ribosome is ready to accept the incoming aminoacyl tRNA specified by the codon that occupies this site.

Step 3: Elongation

As with initiation, elongation is a multistep process. In the first step, another GTP-binding protein, elongation factor Tu (EF-Tu), delivers the next aminoacyl tRNA to the A site on the ribosome. As this occurs, EF-Tu hydrolyzes GTP and leaves the ribosome (Fig. 23-5).

Next, fMet-tRNA will deliver the carboxyl terminus of fMet to the amino terminus of the amino acid linked to the tRNA at the A site (a *peptidyl transfer* reaction), with the subsequent formation of a peptide bond between the two amino acids. At this point, the tRNA at the A site is covalently linked to a dipeptide (Fig. 23-5).

To remove the now uncharged fMet-tRNA from the P site, the ribosomal complex will shift toward the 3′ end of the mRNA by three bases, positioning a new codon in the A site and ejecting the "spent" fMet-tRNA. At this point, the dipeptide bound by the tRNA that formerly occupied the A site now occupies the P site. This final step in translation elongation is fueled by the hydrolysis of another GTP molecule by a translocase enzyme called elongation factor G (EF-G, Fig. 23-5).

The process of elongation will continue in the 5′ to 3′ direction down the mRNA. Each time the ribosome shifts position by three bases to accept another aminoacyl tRNA, the growing peptide chain will increase in length by one amino acid. Because the carboxyl group of the peptide linked to the tRNA at the P site always forms a covalent bond with the amino group of the amino acid on the tRNA at the A site, peptides are always synthesized from the amino terminus to the carboxyl terminus.

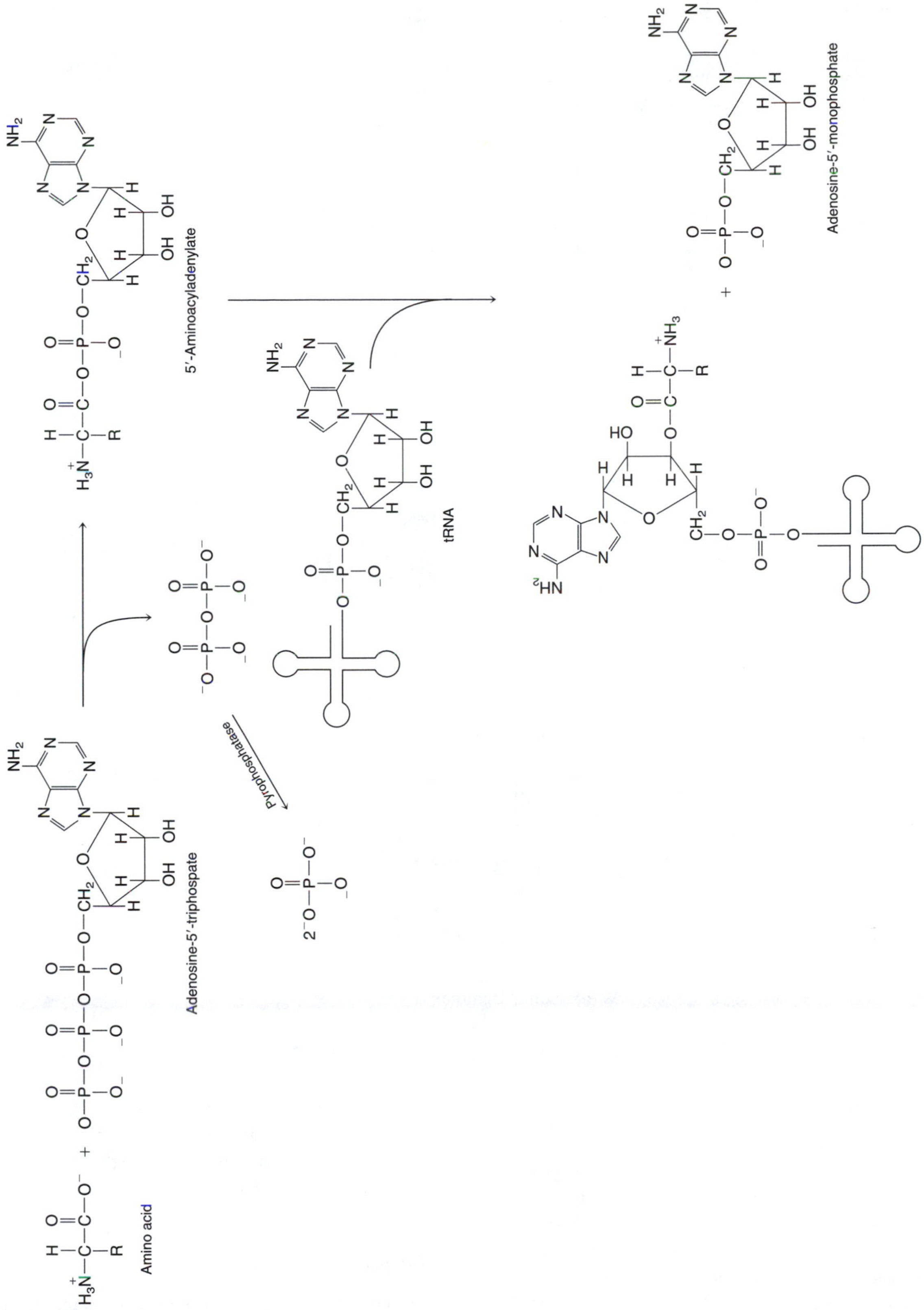

Figure 23-2 Reaction catalyzed by aminoacyl tRNA synthetase.

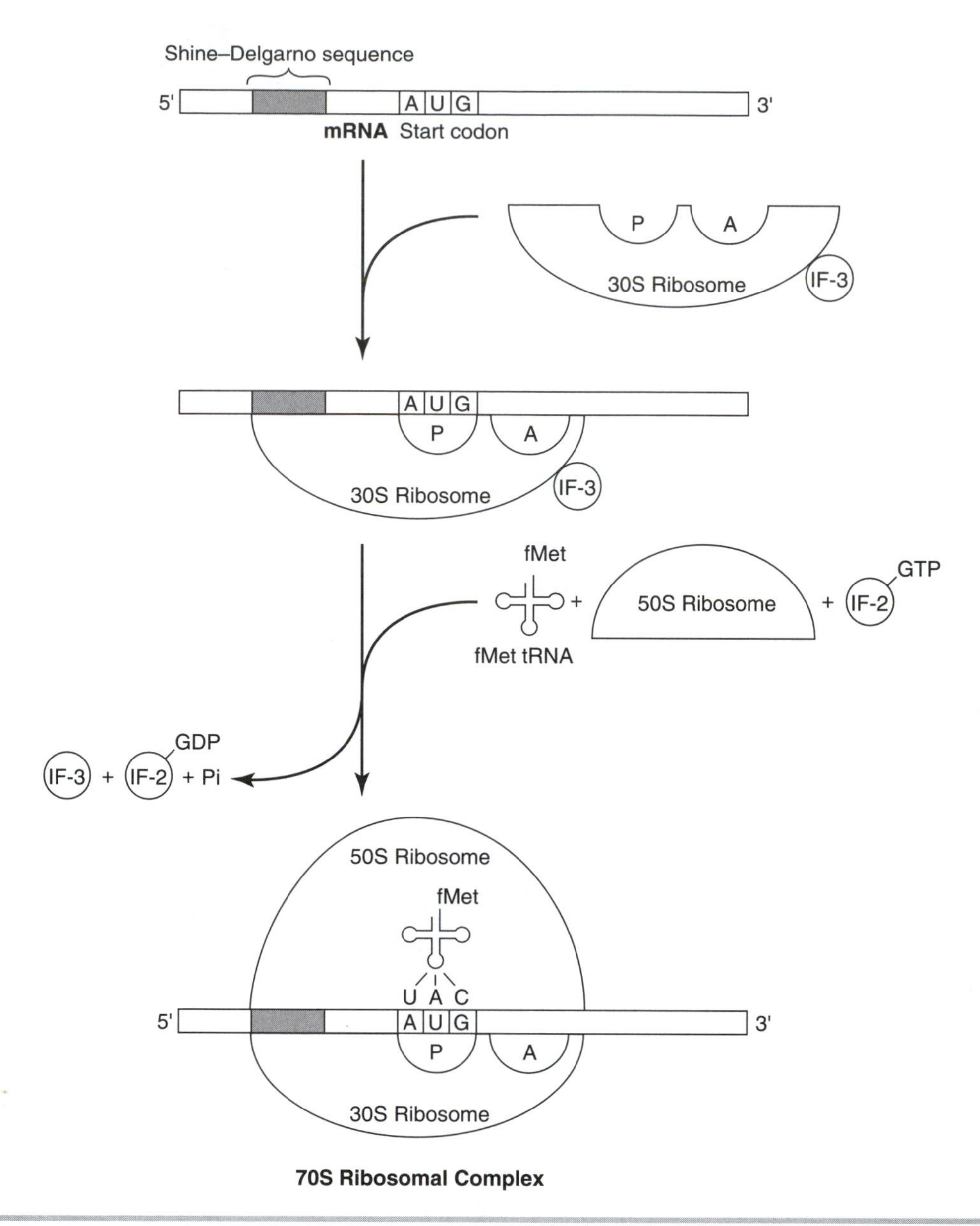

Figure 23-3 Schematic diagram of translational initiation in prokaryotes.

Step 4: Termination

Unlike translational initiation, elongation, and aminoacyl tRNA synthesis, translational termination is a spontaneous process that does not require the input of energy (GTP hydrolysis). There are three codons on the mRNA that will trigger the end of translation when they appear at the A site on the ribosome: UAA, UGA, and UAG. As the ribosome encounters these codons, one of two release factor proteins will bind at the A site and hydrolyze the ester bond linking the peptide chain to the tRNA at the P site. Following this event, the 50S and 30S ribosomal subunits will dissociate from the mRNA (Fig. 23-6).

Eukaryotic Translation

One major difference between the prokaryotic and eukaryotic translational machinery is the composition of the two ribosomal subunits. Unlike the small

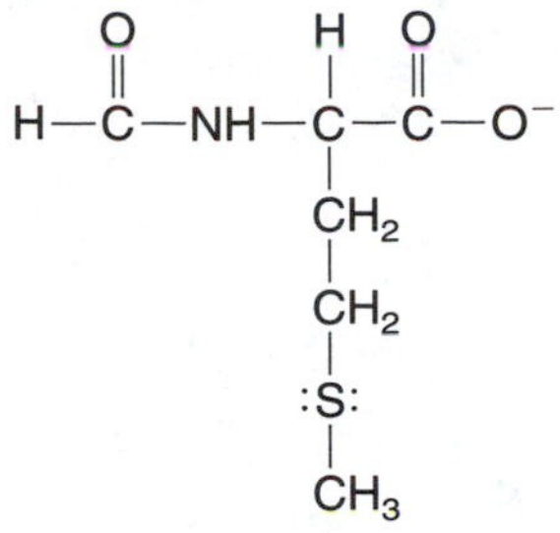

Figure 23-4 The structure of *N*-formylmethionine, which is the amino terminal residue of all newly synthesized prokaryotic proteins. In many cases, the *N*-formyl group or both the *N*-formyl group and methionine are posttranslationally cleaved from the protein.

prokaryotic ribosomal subunit (30S, with 21 proteins and 16S rRNA), the small eukaryotic ribosomal subunit (40S) consists of 33 proteins and an 18S rRNA. The large ribosomal subunit also differs between prokaryotic and eukaryotic organisms: the prokaryotic subunit (50S) is made up of 34 proteins, a 23S rRNA, and a 5S rRNA, while the eukaryotic subunit (60S) consists of about 50 proteins and three separate rRNAs (5S, 28S, and 5.8S). The two prokaryotic subunits combine to form the functional 70S ribosome, while the eukaryotic subunits combine to form the functional 80S ribosome.

Another difference between prokaryotic and eukaryotic translation is the nucleic acid sequences that recruit the small ribosomal subunit to the mRNA. As stated earlier, the Shine–Delgarno sequence interacts with the 16S rRNA of the 30S ribosomal subunit in the prokaryotic system, the critical step in translation initiation. In the eukaryotic

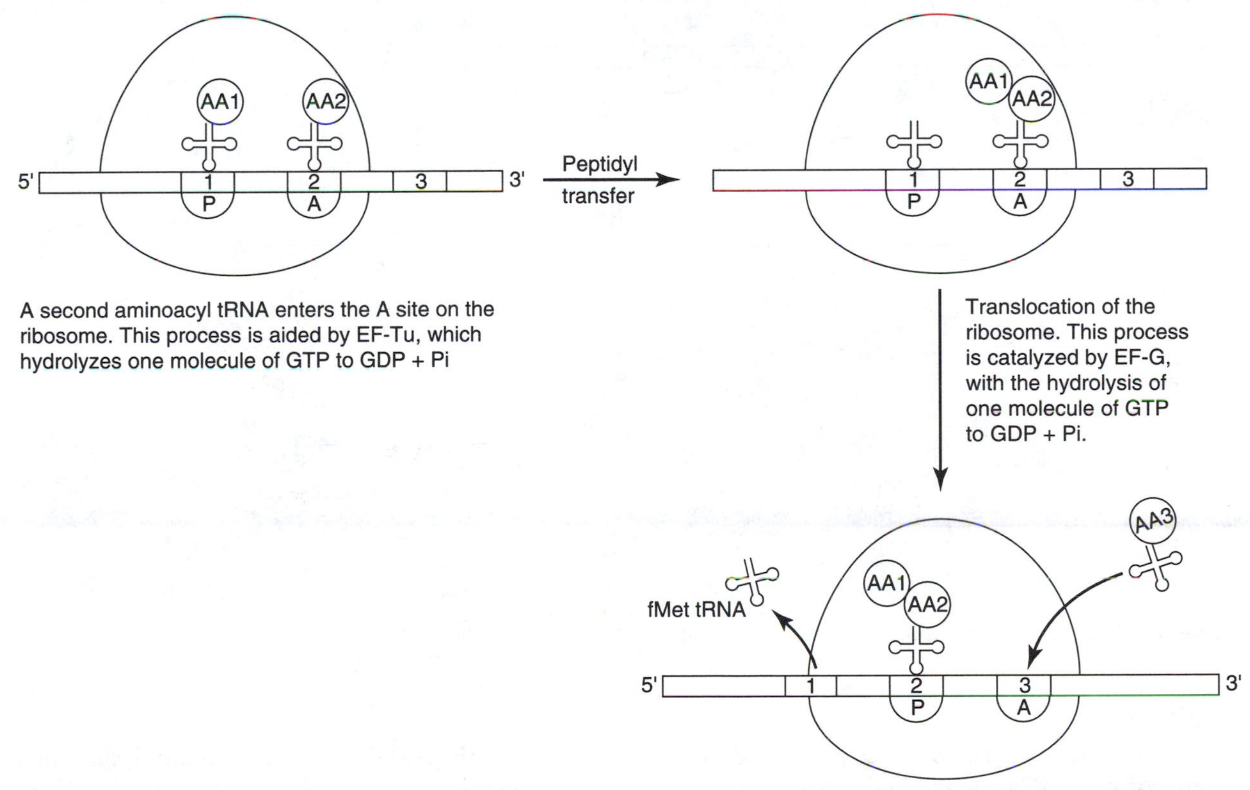

Figure 23-5 Schematic diagram of translational elongation in prokaryotes.

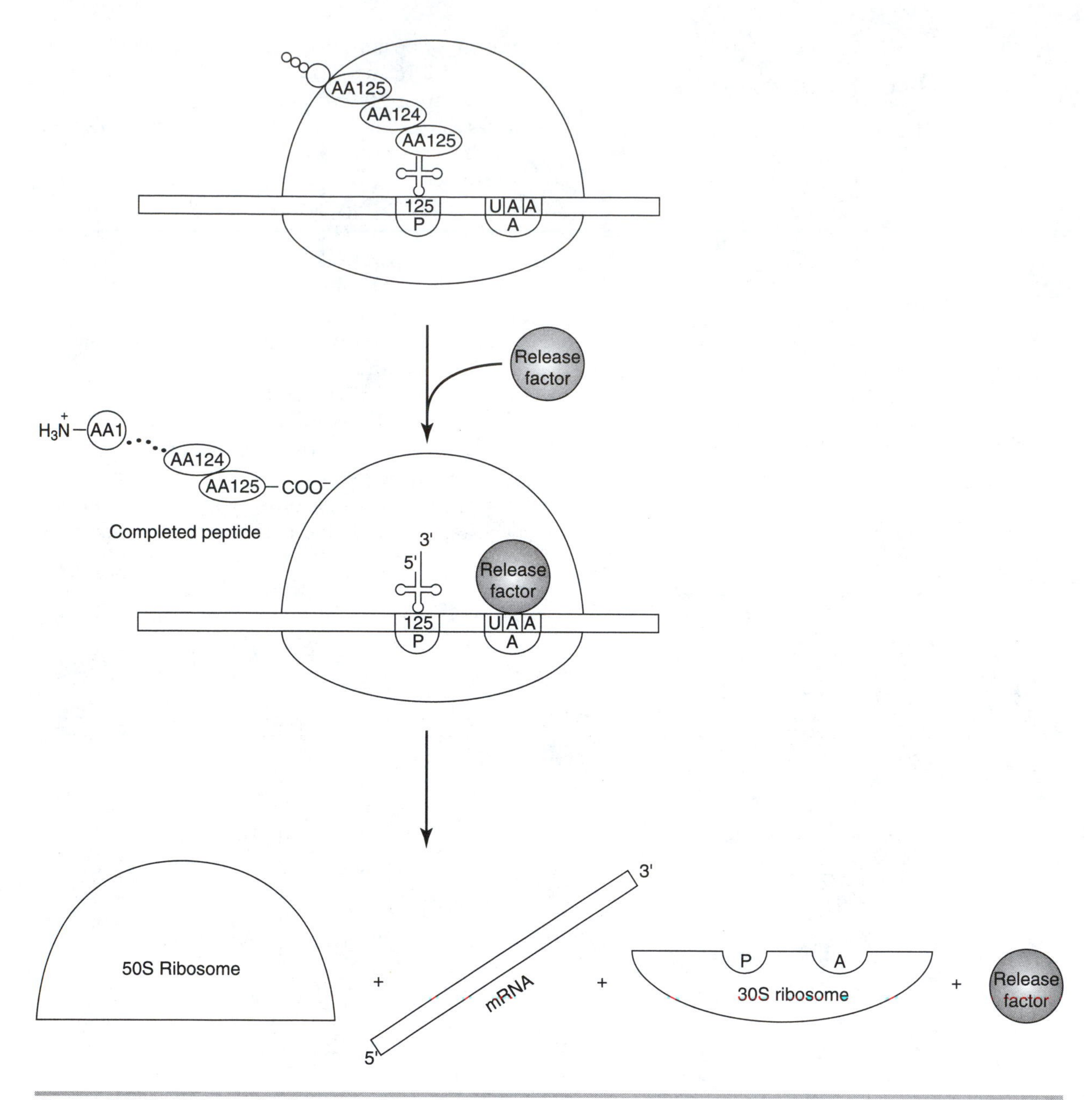

Figure 23-6 Schematic diagram of translational termination in prokaryotes.

system, a 7-methyl-guanosine "cap" is added to the 5′ end of the mRNA (Fig. 23-7). It is this 5′ cap that facilitates the binding of the 40S ribosomal subunit to the mRNA (see below).

Eukaryotic translation initiation is far more complicated than the prokaryotic system described earlier. To begin the process, two initiation factor proteins (eIF3 and eIF4C) bind to the 40S ribosomal subunit and prepare it to bind to the mRNA. Next, a cap-binding protein (CBPI) binds the 7-methyl-guanosine cap at the 5′ end of the mRNA, along with three other initiation factor proteins (eIF4A, eIF4B, and eIF4F). In an ATP-requiring process, this protein complex "scans" the mRNA until the

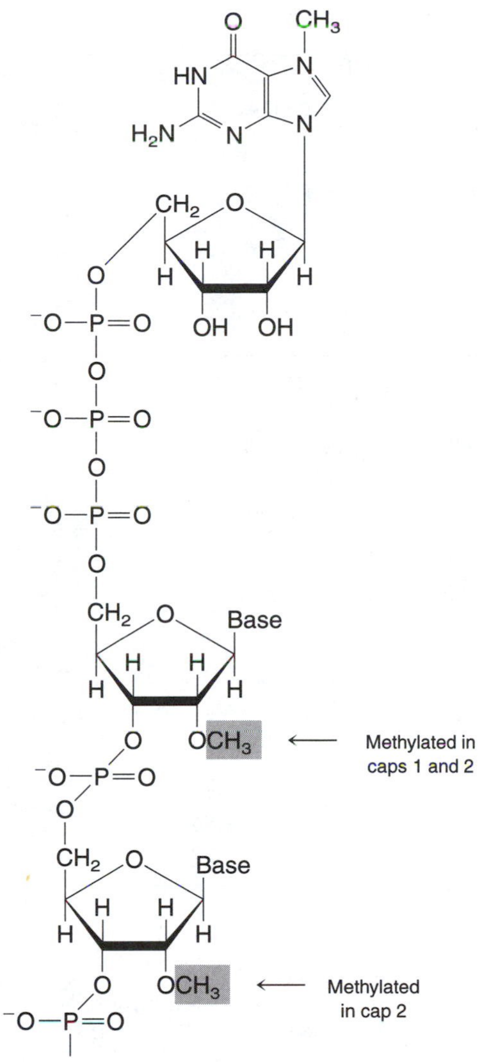

Figure 23-7 Structure of 7-methyl-guanosine capped eukaryotic mRNA. (From Stryer, L. (1988) *Biochemistry,* 3rd ed. New York: Freeman, Figure 29-27. Reprinted with permission.)

AUG start codon is identified. At this point, the 40S ribosome/eIF3/eIF4C complex binds the mRNA.

The eIF2 protein (a GTPase) delivers the first aminoacyl tRNA to the P site on the 40S ribosome. Unlike the prokaryotic system, in which translation begins with f-Met, eukaryotic peptides begin with L-methionine. Although this initiating methionine is unmodified, the tRNA that delivers it to the P site of the ribosome is different from the tRNA that delivers methionine residues (internal) to the growing polypeptide. eIF2 hydrolyzes GTP to GDP and Pi during this process. The eIF2 protein releases the GDP and recycles to the GTP form with the aid of another protein—eIF2B. Finally, the protein eIF5 binds the mRNA/Met-$tRNA^{met}$/40S ribosomal complex, causing eIF3 and eIF4C to dissociate. At this point, the 60S ribosomal subunit binds to form the functional 80S ribosome. The process of eukaryotic translation initiation is depicted in Figure 23-8.

The elongation reaction described earlier for the prokaryotic system is essentially the same as that found in the eukaryotic system. The final difference between the two systems lies in the mechanism of translational termination. Recall that the prokaryotic system utilizes two different release factor proteins. The eukaryotic system relies on a single release factor protein, eRF.

In this experiment, you will perform a number of in vitro translation reactions. The ribosomes used in this experiment were obtained from wheat germ, a eukaryotic organism. After the material is ground into a fine paste, the mixture is diluted with buffer to extract most of the proteins and other small molecules from the cells. This cellular extract is then subjected to centrifugation at $30{,}000 \times g$. The insoluble material harvested following this step contains unlysed cells, cellular debris, and intact mitochondria. The supernatant, or *S-30 fraction,* contains all of the components needed to perform in vitro translation (ribosomes, tRNA, initiation factor, elongation factor, etc.).

The mRNA that you will use in these reactions is poly(U), which will direct the ribosome to synthesize a poly-phenylalanine (poly-Phe) peptide. By including a small concentration of radiolabeled (^3H) phenylalanine (Phe) in the reaction, the relative amounts of the peptide produced in the different reactions can be determined through isolation of the peptide and scintillation counting (see below).

In vitro, the fidelity of translation is strongly influenced by the concentration of Mg^{2+} ions in the reaction. In the range from 1 to 4 mM Mg^{2+}, the ribosome will require a Shine–Delgarno sequence (prokaryotic) or a 7-methyl-guanosine cap (eukaryotic) on the mRNA before translation will initiate. For this reason, in vitro translation experiments performed with naturally occurring mRNAs are most

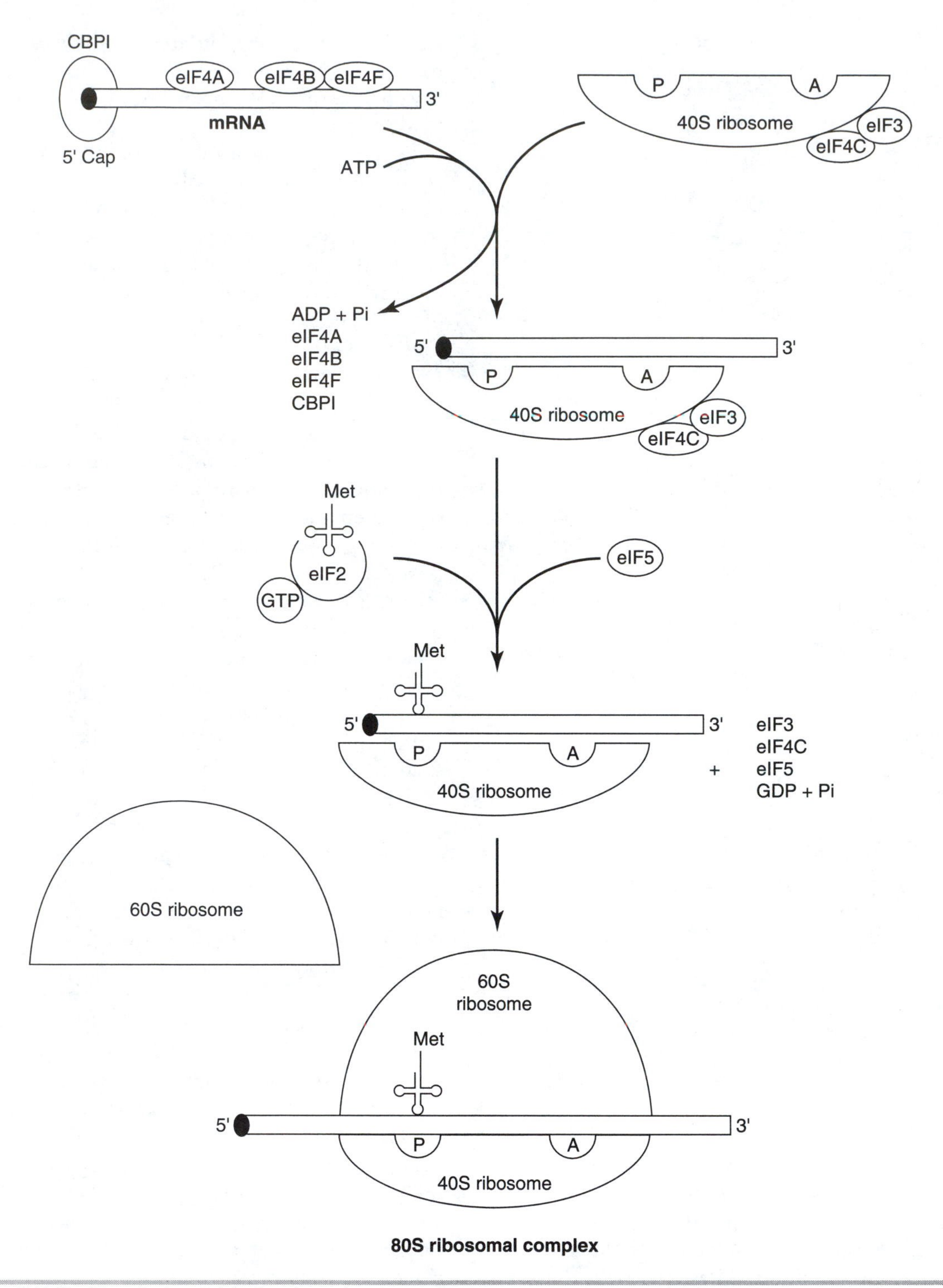

Figure 23-8 Translation initiation in eukaryotes.

often carried out at low Mg^{2+} concentrations. If the Mg^{2+} concentration is raised to the range from 8 to 12 mM, the ribosome will initiate transcription at *random* positions throughout the mRNA. Thus, by conducting your in vitro translation reactions under high Mg^{2+} concentrations, the ribosomes will efficiently translate the poly(U) mRNA into a poly-Phe peptide in the absence of any formal translational start site.

The ^{3}H-poly-Phe peptides produced in these experiments will be isolated by exploiting their ability to precipitate in a solution containing 5% trichloroacetic acid (TCA). Strong acids such as TCA are effective in inducing the precipitation of all ionic *macromolecules.* Thus, the proteins, tRNA, and mRNA present in the translation reactions will all become *insoluble* on the addition of 5% TCA. Remember that there are three individual pools of ^{3}H-Phe present in this ongoing translation reaction: the free ^{3}H-Phe, the ^{3}H-Phe covalently linked to the tRNA, and the ^{3}H-Phe that has been incorporated into the growing peptide. The free ^{3}H-Phe will remain soluble in 5% TCA due to its small size, and therefore will not be present in the precipitate. The ^{3}H-Phe covalently linked to the 3′ end of the tRNA, however, will be insoluble in 5% TCA and will be present in the precipitate (Fig. 23-9). *If this ^{3}H aminoacyl tRNA pool is not removed from the precipitate, you will overestimate the amount of ^{3}H-Phe that was incorporated into the peptide when the reactions are subjected to scintillation counting.*

How can you remove the ^{3}H-Phe tRNA from the precipitate to obtain an accurate measure of the amount of ^{3}H-poly-Phe peptide produced? There is one difference in the properties of these two molecules that you can exploit. Unlike the *peptide bonds* linking the ^{3}H-Phe monomers in the peptide, the *ester bond* linking the ^{3}H-Phe to the 3′ end of the tRNA is labile in 5% TCA at 90°C. Therefore, if the precipitate is heated at 90°C for 10 min, all of the ^{3}H-Phe associated with the tRNA will be hydrolyzed and converted to soluble (free) ^{3}H-Phe that will diffuse back into the bulk solution. *At this point, the only ^{3}H-Phe present in the solution will be that which incorporated into the peptide* (Fig. 23-9).

To quantify the relative amount of ^{3}H-poly-Phe produced from one reaction to the next, the precipitate will be filtered through a 0.45-μm membrane that will trap the insoluble material and allow the soluble ^{3}H-Phe in the reaction to pass through. *When this filter is dried and subjected to scintillation counting, the amount of ^{3}H present (in counts per minute) will be directly proportional to the amount of ^{3}H-poly-Phe produced in the reaction* (Fig. 23-9).

There are four variables in the general in vitro translation protocol that you will investigate. First, you will vary the Mg^{2+} concentration to study its effect on the fidelity of translation initiation. Second, you will investigate the specificity of the ^{3}H-Phe-tRNA and the mRNA codon with which it is able to interact. Third, you will study the effects of different ribonuclease enzymes on the various RNA components present in the reaction (see below). Finally, you will investigate the ability of the ribosome to carry out translation in the presence of three different inhibitors with different modes of action (see below).

The two RNase enzymes that will be used in this experiment are RNase A and RNase T_1. RNase A is a heat-stable nuclease that will hydrolyze the phosphodiester bonds linking the pyrimidine bases (U or C) in the molecule. RNase A will cleave both the mRNA and the tRNA present in the reaction. RNase T_1 differs from RNase A in that it is a *base-specific* nuclease. RNase T_1 will cleave the phosphodiester bonds on the 3′ side of *guanine.* Because of its specificity, RNase T_1 will act *only* on the tRNA present in the reaction (the poly(U) mRNA does not contain guanine).

The three antibiotic inhibitors of translation that will be used in this experiment are chloramphenicol, cycloheximide, and puromycin (Fig. 23-10). Chloramphenicol is specific for *prokaryotic ribosomes,* blocking the transfer of the peptide on the tRNA at the P site to the amino acid linked to the tRNA at the A site (the peptidyl transfer reaction). Since the source of the ribosomes used in this experiment is wheat germ (eukaryotic), we would predict that chloramphenicol would not have a great effect on translation. The mechanism of cycloheximide-mediated inhibition is the same as that described above for chloramphenicol, except for the fact that it is specific for the 80S *eukaryotic ribosome.* Puromycin is a more broad translational inhibitor, effective on both eukaryotic and prokaryotic ribosomes. It acts as a substrate analog of aminoacyl tRNA. When it binds at the A site of the ribosome, it induces premature termination of translation (Fig. 23-10).

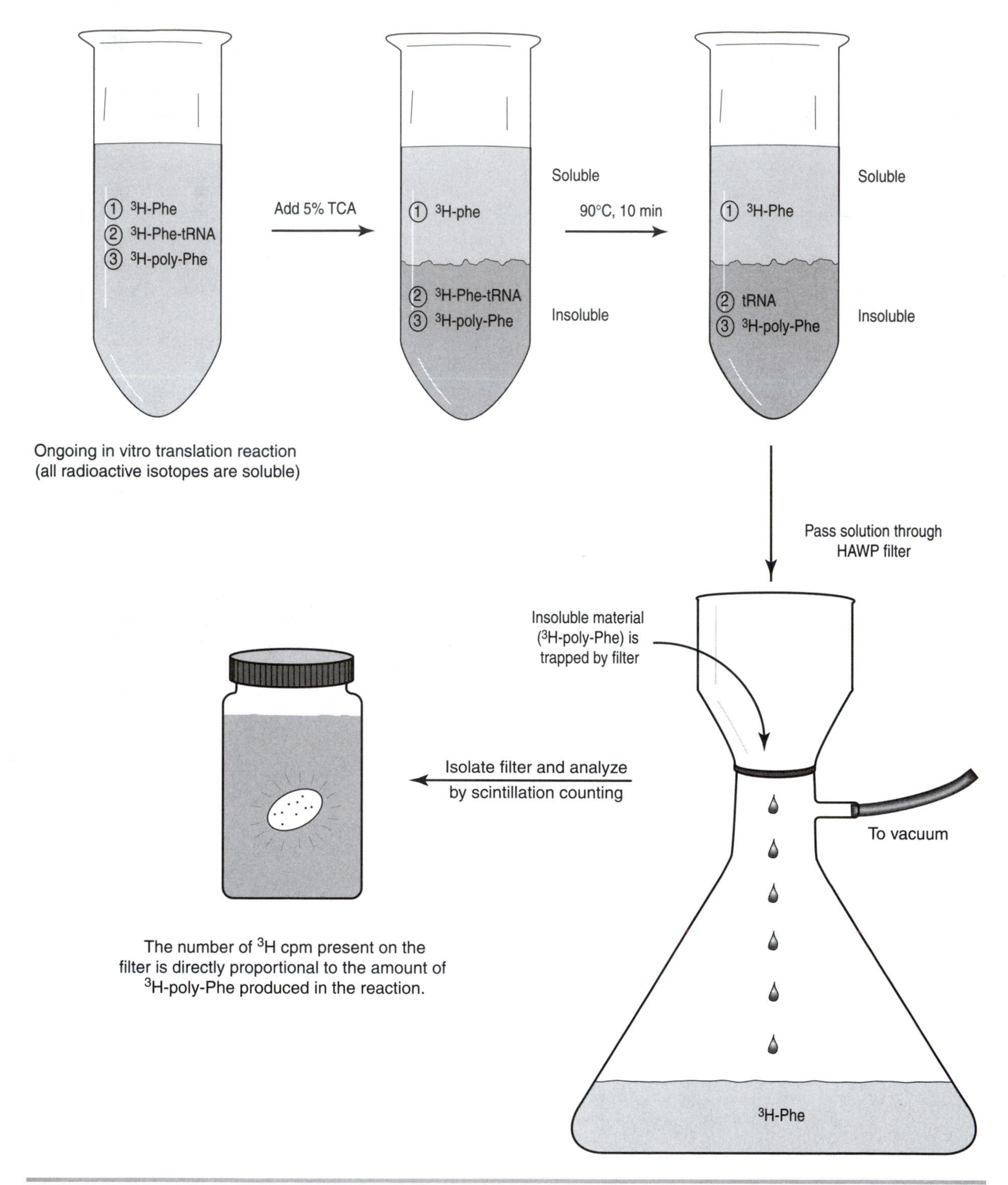

Figure 23-9 Analysis of poly-Phe peptide produced following in vitro translation.

Chloramphenicol

Cycloheximide

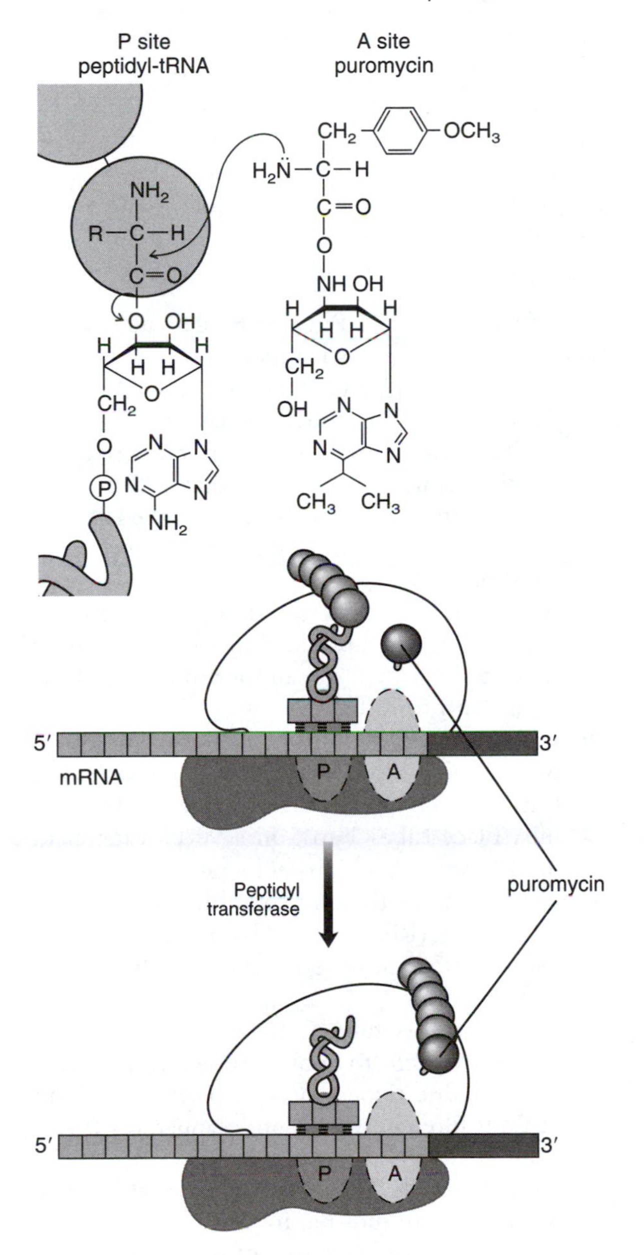

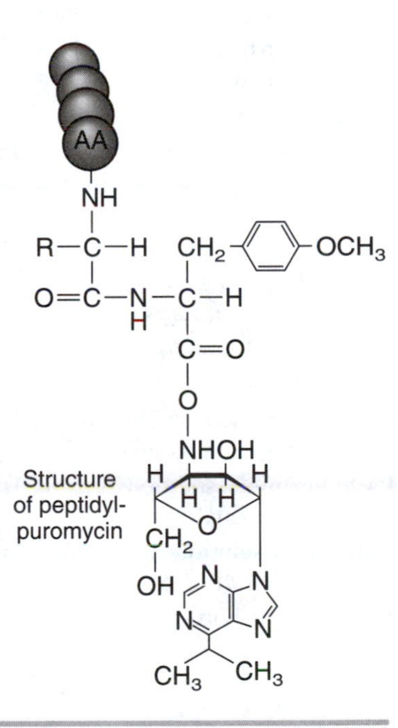

Figure 23-10 Structure of translation inhibitors, and the mechanism of action of puromycin. (From Lehninger, A.L., Nelson, D.L., and Cox, M.M. *Principles of Biochemistry,* 2nd ed. New York: Worth, Figure 26-34. Reprinted with permission.)

Supplies and Reagents

P-20, P-200, and P-1000 Pipetmen with sterile disposable tips
1.5-ml Microcentrifuge tubes
S-30 fraction from wheat germ—commercial (Fisher catalog #L-4440) or freshly prepared (see Appendix) supplemented with 125 mM ATP (potassium salt), 2.5 mM GTP (potassium salt), 100 mM creatine phosphate, and 0.5 mg/ml creatine
Reaction buffer (300 mM HEPES, 30 mM dithiothreitol [pH 7.1])
Salt solution I (1.3 M KCl, 225 mM magnesium acetate)
Salt solution II (1.3 M KCl, 25 mM magnesium acetate)
^{3}H-Phenylalanine solution (0.1 mM total concentration at an activity of 0.005 μCi/ml)
Poly(U) solution (1 mg/ml) in RNase free distilled water—Sigma catalog #P-9528
Poly(A) solution (1 mg/ml) in RNase free distilled water—Sigma catalog #P-9403
Trichloroacetic acid solution (5% wt/vol) in distilled water
Fine-tipped forceps
Millipore filters (HAWP, 0.45-μm pore size)
Filtration apparatus with vacuum pump
30°C water bath
90°C waterbath
Scintillation vials with caps
Scintillation fluid (0.5% wt/vol diphenyloxazole, 10% wt/vol Triton X-100 in toluene)
Scintillation counter
100°C oven
RNase A solution (0.5 mg/ml) in distilled water—Sigma catalog #R-5125
RNase T_1 solution (0.5 mg/ml) in distilled water—Sigma catalog #R-1003
Chloramphenicol solution (0.5 mM) in distilled water—Sigma catalog #C-0378
Cycloheximide solution (0.5 mM) in distilled water—Sigma catalog #C-7698
Puromycin solution (0.1 mM) in distilled water—Sigma catalog #P-7255
Plastic squirt bottle

WARNING
Caution: This experiment requires the use of low levels of tritium (^{3}H). Wear gloves and safety goggles at all times when working with the radioisotope. Cover your work surface with absorbent paper to localize any spills that may occur. Consult the instructor for the proper disposal of all solid and liquid radioactive waste that will be produced in the experiment.

Protocol

Day 1: Effect of Mg^{2+} and Various mRNAs on Translation

1. Set up the following reactions in five labeled microcentrifuge tubes:

	Volumes (μl)				
Component	1	2	3	4	5
Reaction buffer	2	2	2	2	2
Salt solution I	2	2	2	—	2
Salt solution II	—	—	—	2	—
1 mg/ml Poly(U)	10	10	—	10	—
1 mg/ml Poly(A)	—	—	—	—	10
S-30 fraction	20	20	20	20	20
Distilled water	6	6	16	6	6

2. Add 10 μl of ^{3}H-Phe solution to tube 1 and pipette up and down rapidly several times to mix. *Immediately* add 750 μl of 5% TCA to stop the reaction. Place tube 1 on ice for later analysis. This reaction will serve as a time zero blank for the remaining reactions that will indicate the amount of soluble ^{3}H-Phe in the reaction that will unavoidably contaminate the precipitated proteins.
3. Add 10 μl of ^{3}H-Phe solution to each of tubes 2 to 5. Pipette up and down rapidly several times to mix. Incubate all four tubes in a 30°C water bath for 30 min.
4. Following the 30-min incubation, add 750 μl of 5% TCA solution to each of tubes 2 to 5. Cap the tubes and invert them several times to mix. Place tubes 1 to 5 in a 90°C water bath for 10 min. This high-temperature incubation will hydrolyze the ester bond linking the ^{3}H-Phe to the tRNA. The ^{3}H-Phe that has been incorporated into the peptide is stable at this temperature.
5. Remove all five tubes from the 90°C water bath and allow them to cool to room temperature.
6. Using a fine-tipped forceps, place a 0.45-μm HAWP filter in the filtration apparatus (to be demonstrated by the instructor). Fill a squirt bottle with 5% TCA and apply a small volume to the filter to moisten it.
7. Apply the entire contents of tube 1 to the filter and turn on the vacuum line to the unit.

Immediately fill the now empty tube 1 with 5% TCA and pour the contents on the filter. Repeat this wash procedure on tube 1 a total of five times to harvest all of the precipitated material and transfer it to the filter.

8. Apply one last aliquot (~5 ml) of 5% TCA to the filter and allow the filter to dry under vacuum for 20 sec (allow all of the solvent to flow through the filter and continue to apply the vacuum for an additional 20 sec). This final wash step is done to remove any trace amounts of soluble ^{3}H-Phe that may be bound to the filter or to the precipitate adsorbed to the filter.
9. Place the filter in a scintillation vial and place the vial (uncapped) in a 100°C oven for *exactly* 10 min. This heat treatment will dehydrate the filter and cause it to curl up in the vial. Avoid excessive heating, which will cause the filter to turn yellow and possibly quench the ^{3}H β particles in the sample as they are being counted (*color quenching*, see introduction to Experiment 3).
10. Perform steps 6 to 9 on the contents of each of tubes 2 to 5. *You will require a clean, unused filter for each of the remaining four reaction tubes. When you are finished, you will have five scintillation vials (labeled 1 to 5), each corresponding to the appropriately numbered reaction tube.*

Reaction Time Course

11. Obtain four fresh 1.5 ml microcentrifuge tubes. Label these tubes 6 to 9, add 750 μl of 5% TCA to each, and place all four of the tubes on ice.
12. Obtain a fifth fresh microcentrifuge tube and add the following: 10 μl of reaction buffer; 10 μl of salt solution I; 50 μl of 1 mg/ml Poly(U); 100 μl of S-30 fraction; and 30 μl of distilled water.
13. Add 50 μl of ^{3}H-Phe solution to this tube, and pipette up and down rapidly several times to mix. Incubate this tube in a 30°C water bath.
14. After a total of 15 min of incubation at 30°C, remove a 50-μl aliquot of the reaction and add it to tube 6. Cap tube 6 and invert several times to mix. Place tube 6 back on ice and return the reaction tube to the 30°C water bath.
15. After a total of 30 min of incubation at 30°C, remove a 50-μl aliquot of the reaction and add it to tube 7. Cap tube 7 and invert several times to mix. Place tube 7 back on ice and return the reaction tube to the 30°C water bath.
16. After a total of 45 min of incubation at 30°C, remove a 50-μl aliquot of the reaction and add it to tube 8. Cap tube 8 and invert several times to mix. Place tube 8 back on ice and return the reaction tube to the 30°C water bath.
17. After a total of 60 min of incubation at 30°C, remove a 50-μl aliquot of the reaction and add it to tube 9. Cap tube 9 and invert several times to mix. Place tube 9 back on ice and return the reaction tube to the 30°C water bath.
18. Perform steps 4 to 9 (heating at 90°C and filtration steps) on each of the reaction timepoints in tubes 6 to 9. *At the end of the procedure, you will have a total of nine scintillation vials (labeled 1 to 9), each containing a dried filter coressponding to the appropriately numbered reaction or reaction timepoint tube.*
19. Add 5 or 10 ml of scintillation fluid to each of the nine scintillation vials and place them in a scintillation counter rack marked with your name.
20. Place the rack in the scintillation counter and count each vial for 2 min in the ^{3}H channel (consult the instructor) to obtain a ^{3}H cpm value for each filter.
21. Subtract the counts per minute (cpm) value for vial 1 (time = zero, reaction control) from all of the other cpm values obtained for vials 2 to 9. This is done to account for the fact that some *soluble* ^{3}H may not have been removed from the *insoluble* material (precipitate) during the filtration step.
22. Compare the relative level of ^{3}H-poly-Phe present in vial 2 compared to that in vial 3. Explain your results in terms of the components that were present in these two reactions.
23. Compare the relative level of ^{3}H-poly-Phe present in vial 2 compared to that in vial 4. Explain your results in terms of the components that were present in these two reactions, as well as your knowledge of the dependence of translational initiation on Mg^{2+} concentration.
24. Compare the relative level of ^{3}H-poly-Phe present in vial 2 compared to that in vial 5. Explain your results in terms of the components that were present in these two reactions, as well as your knowledge of the genetic code.
25. Using the cpm data obtained from tubes 6 to 9, prepare a plot of ^{3}H-poly-Phe cpm versus

reaction time (15, 30, 45, and 60 min). Are the kinetics of this in vitro translation reaction *linear* over 60 min (does this plot yield a straight line through the data points)? If not, for what length of time is this reaction linear with time?

26. Determine the *slope* of the line through the *linear* portion of the curve, and calculate an initial velocity (v_0) for the reaction in units of cpm of ^{3}H incorporated per min.

Day 2: Effect of Nucleases and Inhibitors on Translation

1. Obtain eight fresh 1.5-ml microcentrifuge tubes and label them 10 to 17. Set up the following reactions in these tubes:

	Volumes (μl)							
Component	10	11	12	13	14	15	16	17
Reaction buffer	2	2	2	2	2	2	2	2
Salt solution I	2	2	2	2	2	2	2	2
1 mg/ml Poly(U)	10	10	10	10	10	10	10	10
0.5 mg/ml RNase A	—	—	2	—	—	—	—	—
0.5 mg/ml RNase T_1	—	—	—	2	—	—	—	—
0.5 mM chloramphenicol	—	—	—	—	6	—	—	—
0.5 mM cycloheximide	—	—	—	—	—	6	—	—
0.1 mM puromycin	—	—	—	—	—	—	6	—
S-30 Fraction	20	20	20	20	20	20	20	20
Distilled water	6	6	4	4	—	—	—	6

2. Add 10 μl of ^{3}H-Phe solution to tube 10 and pipette up and down rapidly several times to mix. *Immediately* add 750 μl of 5% TCA to stop the reaction. Place tube 10 on ice for later analysis. This reaction will serve as a time zero blank for the remaining reactions that will indicate the amount of ^{3}H-Phe in the reaction that will unavoidably contaminate the precipitated proteins.
3. Add 10 μl of ^{3}H-Phe solution to each of tubes 11 to 17. Pipette up and down rapidly several times to mix. Incubate all seven tubes in a 30°C water bath for 30 min.
4. After the 30-min incubation, add 750 μl of 5% TCA to each of tubes 11 to 17. Cap the tubes and invert several times to mix.
5. Place tubes 10 to 16 (*but not 17, which should be kept on ice*) in a 90°C water bath for 10 min.
6. After the 10-min incubation, remove tubes 10 to 16 from the 90°C water bath and allow them to cool at room temperature.
7. Perform steps 6 to 9 from the Day 1 protocol (filtration and TCA wash steps) on the contents of tubes 10 to 17. *When you are finished, you should have eight scintillation vials (labeled 10 to 17) corresponding to each of the appropriately numbered reaction tubes.*
8. Add 5 or 10 ml of scintillation fluid to each of the nine scintillation vials and place them in a scintillation counter rack marked with your name.
9. Place the rack in the scintillation counter and count each vial for 2 min in the ^{3}H channel (consult the instructor) to obtain a ^{3}H cpm value for each filter.
10. Subtract the cpm value for vial 10 (time = zero reaction control) from all the other cpm values obtained for vials 11 to 17. This is done to account for the fact that some *soluble* ^{3}H may not have been removed from the *insoluble* material (precipitate) during the filtration step.
11. Compare the relative level of ^{3}H-poly-Phe present in vial 11 compared to that in vial 12. Explain your results in terms of the components that were present in these two reactions, and your knowledge of the mechanism of action of RNase A.
12. Compare the relative level of ^{3}H-poly-Phe present in vial 11 compared to that in vial 13.

Explain your results in terms of the components that were present in these two reactions, and your knowledge of the mechanism of action of RNase T1.

13. Compare the relative level of ^{3}H-poly-Phe present in vial 11 compared to that in vial 14. Explain your results in terms of the components that were present in these two reactions, and your knowledge of the mechanism of action of chloramphenicol.
14. Compare the relative level of ^{3}H-poly-Phe present in vial 11 compared to that in vial 15. Explain your results in terms of the components that were present in these two reactions, and your knowledge of the mechanism of action of cycloheximide.
15. Compare the relative level of ^{3}H-poly-Phe present in vial 11 compared to that in vial 16. Explain your results in terms of the components that were present in these two reactions, and your knowledge of the mechanism of action of puromycin.
16. Compare the relative level of ^{3}H-poly-Phe present in vial 11 compared to that in vial 17. Explain your results in terms how these two reactions were treated following the addition of 5% TCA.

Exercises

1. For each of the synthetic mRNAs described below, predict all the possible peptides that would be produced in an in vitro translation reaction performed under conditions of high Mg^{2+} concentration:

   a. Poly(G)
   b. Poly(A)
   c. Poly(UC)
   d. Poly(UAC)

2. You have purified protein X. You believe that it may interact with (bind) protein Y. You have not purified protein Y, but you have the gene that encodes it cloned into the pSP72 plasmid (see Experiment 22) in an orientation that places it under the control of the T7 promoter. In addition, you have a monoclonal antibody that has been raised against protein X. Describe an experiment that you could perform, involving both in vitro translation and one of the immunochemical techniques described in Section IV, that would allow you to determine quickly whether protein X interacts with protein Y. Assume that you also have a number of radiolabeled amino acids at your disposal.
3. Why is the fidelity of translation initiation in vitro dependent on the concentration of Mg^{2+}? To answer this, you must think about the interaction between the mRNA and the 30S ribosomal subunit.
4. Using your knowledge of the multiple steps involved in translation that require an input of energy, how many *total molecules* of GTP must be hydrolyzed to produce a peptide that is 10 amino acids in length?

REFERENCES

Conway, T. W., and Lipmann, F. (1964). Characterization of Ribosome-linked GTPase in *E. coli* Extracts. *Proc Natl Acad Sci USA* **52**:1462.

Gualerzi, C. O., and Pon, C. L. (1990). Initiation of mRNA Translation in Prokaryotes. *Biochemistry* **29**:5881.

Laemmli, U. K. (1970). Cleavage of Structural Proteins during the Assembly of the Head of Bacteriophage T4. *Nature* **227**:680.

Lehninger, A. L., Nelson, D. L., and Cox, M. M. (1993). Protein Metabolism. In: *Principles of Biochemistry*, 2nd ed. New York: Worth.

Moldave, K. (1985). Eukaryotic Protein Synthesis. *Annu Rev Biochem* **54**:1109.

Prescott, L. M., Harley, J. P., and Klein, D. A. (1993). Synthesis of Nucleic Acids and Proteins. In: *Microbiology*, 2nd ed. Dubuque, IA: Wm. C. Brown.

Roberts, P. E., and Paterson, B. M. (1973). Efficient Translation of TMV-RNA and Globin 9S RNA in a Cell-Free System from Commercial Wheat Germ. *Proc Natl Acad Sci USA* **70**:2330.

Amplification of a DNA Fragment Using Polymerase Chain Reaction

Theory

Polymerase chain reaction (PCR) is a technique that allows the amplification of a *specific* fragment of double-stranded DNA in a matter of hours. This technique has revolutionized the use of molecular biology in basic research, as well as in a clinical setting. PCR is carried out in a three-step process (Fig. 24-1). First, the template DNA that contains the target DNA to be amplified is heated to denature or "melt" the double-stranded DNA duplex. Second, the solution is cooled in the presence of an excess of two single-stranded oligonucleotides (primers) that are complementary to the DNA sequences flanking the target DNA. Since DNA synthesis always occurs in the 5′ to 3′ direction (reading the template strand 3′ to 5′), you must ensure that the two primers are complementary to (will *anneal* to) opposite strands of the DNA duplex that flank the region of target DNA that is to be amplified. Third, a heat-stable DNA polymerase is added, along with the four deoxyribonucleotide triphosphates (dNTPs), so that two new DNA strands that are identical to the template DNA strands can be synthesized. If this melting, annealing, and polymerization cycle is repeated, the fragment of double-stranded DNA located between the primer sequences can be amplified over a millionfold in a matter of hours.

The heat-stable DNA polymerase *(Taq)* commonly used in PCR reactions was isolated from a thermophilic bacterium, *Thermus aquaticus*. Since this enzyme is heat-stable, it can withstand the high temperatures required to denature the DNA template after each successive round of polymerization and retain its activity. Since the development of the technique, biotechnology companies have developed a number of improved and specialized polymerase enzymes for use in PCR (Table 24-1). Many of these polymerases are marketed as being more processive and/or "accurate" than the traditional *Taq* enzyme, since they display 3′ to 5′ exonuclease (proofreading) activity. The *Taq* DNA polymerase has no proofreading activity, increasing the possibility of introducing point mutations (single base pair changes) in the amplified DNA product.

In this experiment, you will amplify a fragment of pBluescript II (a plasmid), which includes the multiple cloning site (MCS) of the vector (Fig. 24-2). The pBluescript II plasmid comes in the S/K form and the K/S form. These two plasmids are identical *except for the orientation of the MCS* (see Fig. 24-2). Using restriction enzymes and agarose-gel electrophoresis, you will determine which of these two plasmids was used as a template in the PCR reaction. The sequences of the two primers that will be used in the PCR reaction are shown under "Supplies and Reagents." Primer 1 will anneal to positions 188 to 211 (5′ to 3′) on one strand of the plasmid, while Primer 2 will anneal to positions 1730 to 1707 (5′ to 3′) on the opposite strand of the plasmid (Fig. 24-3). On amplification, a 1543-base-pair fragment of DNA will be produced that includes the multiple cloning site of the plasmid. *Sst*I (an isoschizomer of *Sac*I) and *Kpn*I will then be used to determine whether the S/K or K/S form of the pBluescript II plasmid was used as a template in the amplification reaction.

Although this experiment is designed to introduce you to the basic technique of PCR, you should be aware that PCR can be used in a variety

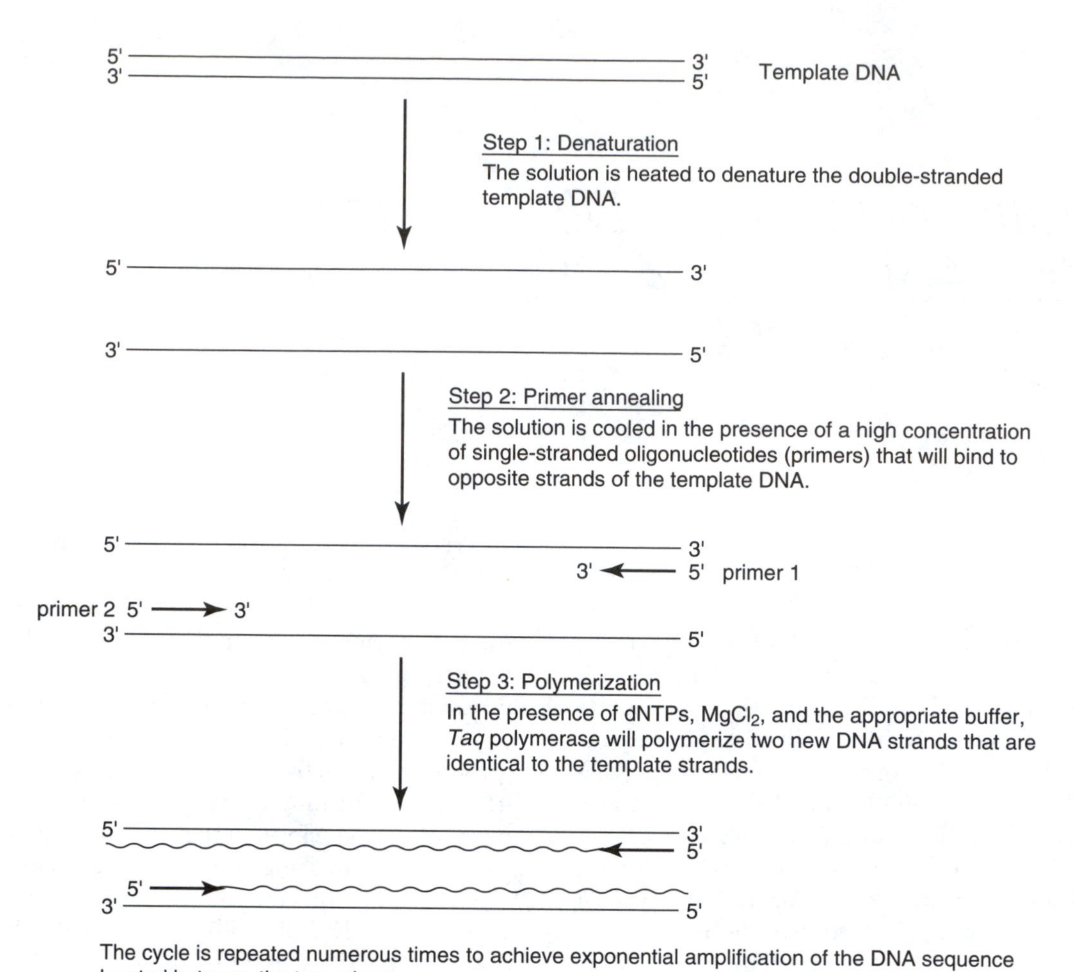

Figure 24-1 The basic principle underlying the technique of polymerase chain reaction (PCR).

Table 24-1 Some Commercially Available Polymerase Enzymes for Use with Polymerase Chain Reaction

Polymerase Enzyme	Relevant Features	Company
Pfu	Low error rate (proofreading) Produces blunt-ended PCR products	Stratagene
Taq 2000	High processivity (for long PCR products)	Stratagene
Exo^{-} *Pfu*	Specially designed for use with PCR sequencing	Stratagene
Ampli*Taq*	High polymerase temperature (minimizes false-priming)	Perkin Elmer
UlTma	Excellent proofreading activity	Perkin Elmer
rTth	High processivity (for PCR products 5–40 kb in length)	Perkin Elmer
Platinum *Taq*	Temperature activation of polymerase (minimizes false priming)	Life Technologies
Vent	Excellent proofreading activity	New England Biolabs

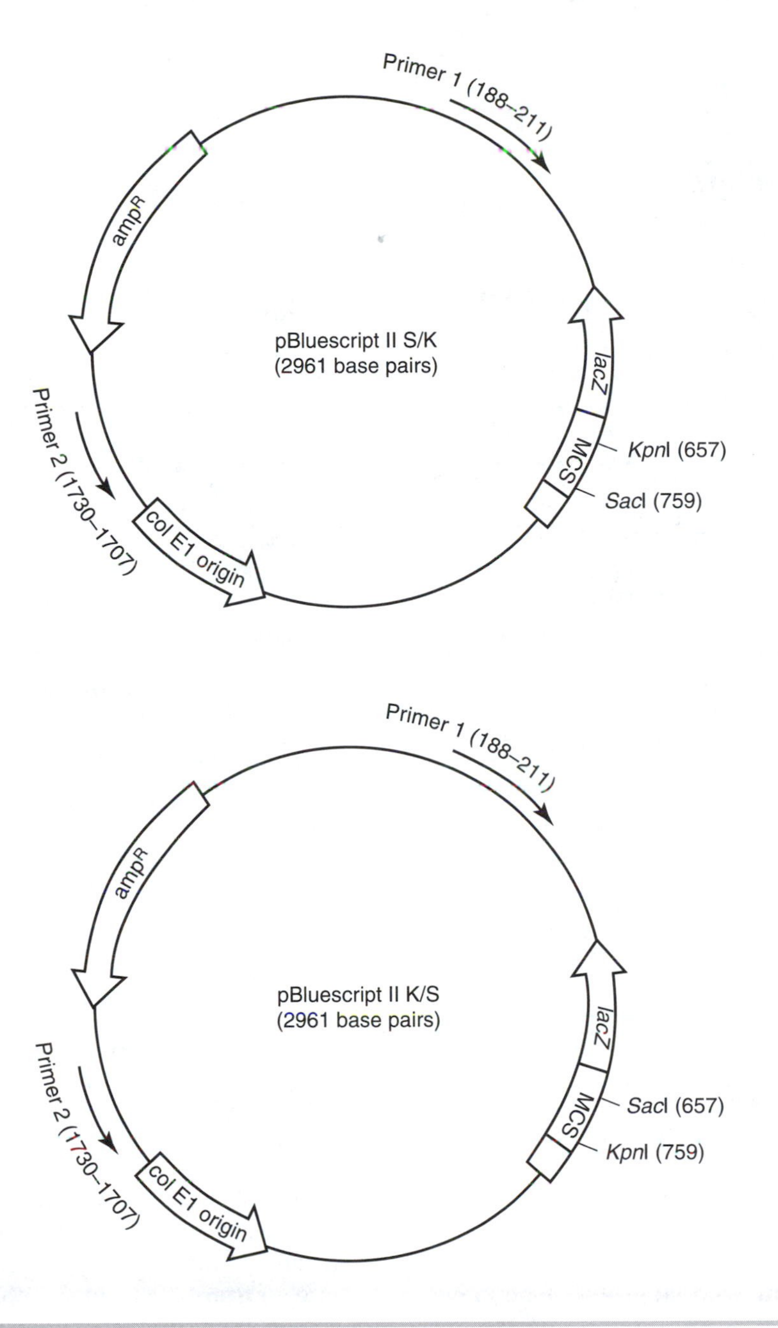

Figure 24-2 Plasmid maps of pBluescriptII (S/K) and pBluescriptII (K/S).

of different applications. One popular application for PCR is its use in the introduction of *specific* mutations in the product DNA that is amplified from the template DNA. For example, suppose you wanted to produce a PCR product that had restriction-enzyme recognition sequences at either end. If these recognition sequences are present in the single-stranded DNA primers, the target DNA will be amplified to include these sites to allow for easy cloning into a desired vector following PCR (Fig. 24-4*a*). In addition to introducing restriction recognition sequences, PCR can be used to add (Fig. 24-4*b*) or delete (Fig. 24-4*c*) small sequences from a gene of interest. Provided that the DNA primers are long enough to allow for sufficient base pairing on either side of the desired mutation, nearly any sequence can be added to, or deleted from, a gene of interest. PCR can also be used to

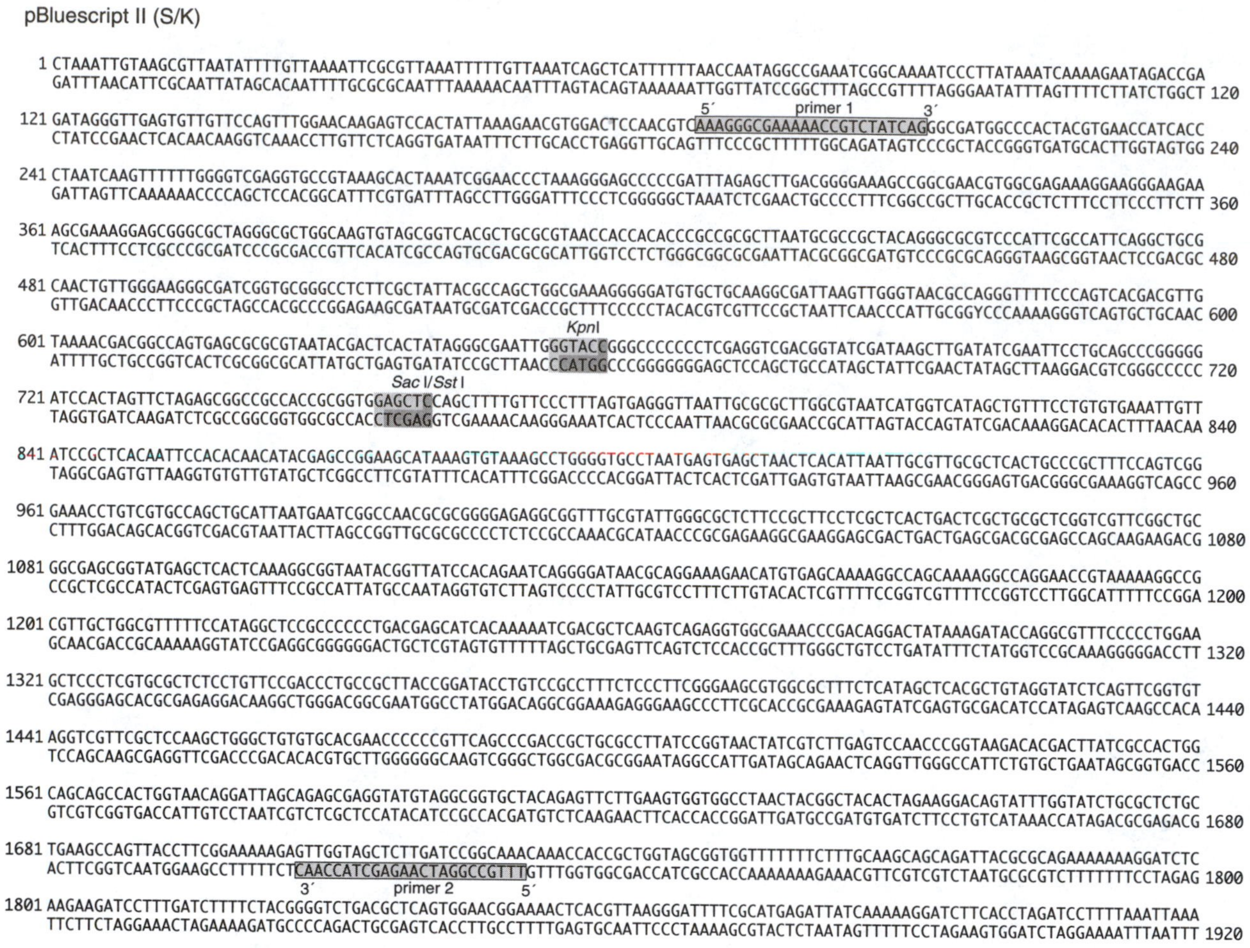

pBluescript II (S/K)

```
   1 CTAAATTGTAAGCGTTAATATTTTGTTAAAATTCGCGTTAAATTTTTGTTAAATCAGCTCATTTTTTAACCAATAGGCCGAAATCGGCAAAATCCCTTATAAATCAAAAGAATAGACCGA
     GATTTAACATTCGCAATTATAGCACAATTTTGCGCGCAATTTAAAAACAATTTAGTACAGTAAAAAATTGGTTATCCGGCTTTAGCCGTTTTAGGGAATATTTAGTTTTCTTATCTGGCT 120
                                                                                  5′      primer 1       3′
 121 GATAGGGTTGAGTGTTGTTCCAGTTTGGAACAAGAGTCCACTATTAAAGAACGTGGACTCCAACGTCAAAGGGCGAAAAACCGTCTATCAGGGCGATGGCCCACTACGTGAACCATCACC
     CTATCCGAACTCACAACAAGGTCAAACCTTGTTCTCAGGTGATAATTTCTTGCACCTGAGGTTGCAGTTTCCCGCTTTTTGGCAGATAGTCCCGCTACCGGGTGATGCACTTGGTAGTGG 240

 241 CTAATCAAGTTTTTTGGGGTCGAGGTGCCGTAAAGCACTAAATCGGAACCCTAAAGGGAGCCCCCGATTTAGAGCTTGACGGGGAAAGCCGGCGAACGTGGCGAGAAAGGAAGGGAAGAA
     GATTAGTTCAAAAAACCCCAGCTCCACGGCATTTCGTGATTTAGCCTTGGGATTTCCCTCGGGGGCTAAATCTCGAACTGCCCCTTTCGGCCGCTTGCACCGCTCTTTCCTTCCCTTCTT 360

 361 AGCGAAAGGAGCGGGCGCTAGGGCGCTGGCAAGTGTAGCGGTCACGCTGCGCGTAACCACCACACCCGCCGCGCTTAATGCGCCGCTACAGGGCGCGTCCCATTCGCCATTCAGGCTGCG
     TCACTTTCCTCGCCCGCGATCCCGCGACCGTTCACATCGCCAGTGCGACGCGCATTGGTCCTCTGGGCGGCGCGAATTACGCGGCGATGTCCCGCGCAGGGTAAGCGGTAACTCCGACGC 480

 481 CAACTGTTGGGAAGGGCGATCGGTGCGGGCCTCTTCGCTATTACGCCAGCTGGCGAAAGGGGGATGTGCTGCAAGGCGATTAAGTTGGGTAACGCCAGGGTTTTCCCAGTCACGACGTTG
     GTTGACAACCCTTCCCGCTAGCCACGCCCGGAGAAGCGATAATGCGATCGACCGCTTTCCCCCTACACGTCGTTCCGCTAATTCAACCCATTGCGGYCCCAAAAGGGTCAGTGCTGCAAC 600
                                                               KpnI
 601 TAAAACGACGGCCAGTGAGCGCGCGTAATACGACTCACTATAGGGCGAATTGGGTACCGGGCCCCCCCTCGAGGTCGACGGTATCGATAAGCTTGATATCGAATTCCTGCAGCCCGGGGG
     ATTTTGCTGCCGGTCACTCGCGCGCATTATGCTGAGTGATATCCCGCTTAACCCATGGCCCGGGGGGGAGCTCCAGCTGCCATAGCTATTCGAACTATAGCTTAAGGACGTCGGGCCCCC 720
                                  SacI/SstI
 721 ATCCACTAGTTCTAGAGCGGCCGCCACCGCGGTGGAGCTCCAGCTTTTGTTCCCTTTAGTGAGGGTTAATTGCGCGCTTGGCGTAATCATGGTCATAGCTGTTTCCTGTGTGAAATTGTT
     TAGGTGATCAAGATCTCGCCGGCGGTGGCGCCACCTCGAGGTCGAAAACAAGGGAAATCACTCCCAATTAACGCGCGAACCGCATTAGTACCAGTATCGACAAAGGACACACTTTAACAA 840

 841 ATCCGCTCACAATTCCACACAACATACGAGCCGGAAGCATAAAGTGTAAAGCCTGGGGTGCCTAATGAGTGAGCTAACTCACATTAATTGCGTTGCGCTCACTGCCCGCTTTCCAGTCGG
     TAGGCGAGTGTTAAGGTGTGTTGTATGCTCGGCCTTCGTATTTCACATTTCGGACCCCACGGATTACTCACTCGATTGAGTGTAATTAAGCGAACGGGAGTGACGGGCGAAAGGTCAGCC 960

 961 GAAACCTGTCGTGCCAGCTGCATTAATGAATCGGCCAACGCGCGGGGAGAGGCGGTTTGCGTATTGGGCGCTCTTCCGCTTCCTCGCTCACTGACTCGCTGCGCTCGGTCGTTCGGCTGC
     CTTTGGACAGCACGGTCGACGTAATTACTTAGCCGGTTGCGCGCCCCTCTCCGCCAAACGCATAACCCGCGAGAAGGCGAAGGAGCGACTGACTGAGCGACGCGAGCCAGCAAGAAGACG 1080

1081 GGCGAGCGGTATGAGCTCACTCAAAGGCGGTAATACGGTTATCCACAGAATCAGGGGATAACGCAGGAAAGAACATGTGAGCAAAAGGCCAGCAAAAGGCCAGGAACCGTAAAAAGGCCG
     CCGCTCGCCATACTCGAGTGAGTTTCCGCCATTATGCCAATAGGTGTCTTAGTCCCCTATTGCGTCCTTTCTTGTACACTCGTTTTCCGGTCGTTTTCCGGTCCTTGGCATTTTTCCGGA 1200

1201 CGTTGCTGGCGTTTTTCCATAGGCTCCGCCCCCCTGACGAGCATCACAAAAATCGACGCTCAAGTCAGAGGTGGCGAAACCCGACAGGACTATAAAGATACCAGGCGTTTCCCCCTGGAA
     GCAACGACCGCAAAAAGGTATCCGAGGCGGGGGGACTGCTCGTAGTGTTTTTAGCTGCGAGTTCAGTCTCCACCGCTTTGGGCTGTCCTGATATTTCTATGGTCCGCAAAGGGGGACCTT 1320

1321 GCTCCCTCGTGCGCTCTCCTGTTCCGACCCTGCCGCTTACCGGATACCTGTCCGCCTTTCTCCCTTCGGGAAGCGTGGCGCTTTCTCATAGCTCACGCTGTAGGTATCTCAGTTCGGTGT
     CGAGGGAGCACGCGAGAGGACAAGGCTGGGACGGCGAATGGCCTATGGACAGGCGGAAAGAGGGAAGCCCTTCGCACCGCGAAAGAGTATCGAGTGCGACATCCATAGAGTCAAGCCACA 1440

1441 AGGTCGTTCGCTCCAAGCTGGGCTGTGTGCACGAACCCCCCGTTCAGCCCGACCGCTGCGCCTTATCCGGTAACTATCGTCTTGAGTCCAACCCGGTAAGACACGACTTATCGCCACTGG
     TCCAGCAAGCGAGGTTCGACCCGACACACGTGCTTGGGGGGCAAGTCGGGCTGGCGACGCGGAATAGGCCATTGATAGCAGAACTCAGGTTGGGCCATTCTGTGCTGAATAGCGGTGACC 1560

1561 CAGCAGCCACTGGTAACAGGATTAGCAGAGCGAGGTATGTAGGCGGTGCTACAGAGTTCTTGAAGTGGTGGCCTAACTACGGCTACACTAGAAGGACAGTATTTGGTATCTGCGCTCTGC
     GTCGTCGGTGACCATTGTCCTAATCGTCTCGCTCCATACATCCGCCACGATGTCTCAAGAACTTCACCACCGGATTGATGCCGATGTGATCTTCCTGTCATAAACCATAGACGCGAGACG 1680

1681 TGAAGCCAGTTACCTTCGGAAAAAGAGTTGGTAGCTCTTGATCCGGCAAACAAACCACCGCTGGTAGCGGTGGTTTTTTTGTTTGCAAGCAGCAGATTACGCGCAGAAAAAAAGGATCTC
     ACTTCGGTCAATGGAAGCCTTTTTCTCAACCATCGAGAACTAGGCCGTTTGTTTGGTGGCGACCATCGCCACCAAAAAAACAAACGTTCGTCGTCTAATGCGCGTCTTTTTTTCCTAGAG 1800
                                3′       primer 2       5′
1801 AAGAAGATCCTTTGATCTTTTCTACGGGGTCTGACGCTCAGTGGAACGGAAAACTCACGTTAAGGGATTTTCGCATGAGATTATCAAAAAGGATCTTCACCTAGATCCTTTTAAATTAAA
     TTCTTCTAGGAAACTAGAAAAGATGCCCCAGACTGCGAGTCACCTTGCCTTTTGAGTGCAATTCCCTAAAAGCGTACTCTAATAGTTTTTCCTAGAAGTGGATCTAGGAAAATTTAATTT 1920
```

Figure 24-3 Annealing of primers to pBluescript plasmids. Note that only the portion of the plasmids between base pairs 1 and 1920 is shown.

change a single base pair in the gene of interest (Fig. 24-4*d*). Studies of protein mutants produced by these methods have proven useful in studies of protein–protein interaction, protein function, and protein structure. *The thing to keep in mind when designing mutagenic primers is that any base pair changes, deletions, or additions present in sequence of the primer, in contrast to that of the DNA template, will be present in the amplified DNA product.*

Mutation analysis is one of many applications of the technique of PCR. Variations on the basic technique described in this experiment will allow you to obtain the sequence of a DNA fragment from as little as 50 fmol of sample (PCR cycle sequencing). Once a gene has been cloned and localized to a particular region on a chromosome, PCR can be used to amplify DNA fragments on either side of the gene and "walk" down the chromosome to obtain new sequence information and identify new open reading frames (genes). It is even possible to obtain mRNA from a cell and use PCR to *reverse transcribe* the message to obtain the cDNA for a number of different genes (RT-PCR).

Since polymerase chain reaction can be used to amplify a specific DNA fragment in the presence of countless other sequences, it has proven to be a powerful tool in a number of different fields of study. One area that has benefited tremendously from this technique is the field of forensic science. *Any* biological sample recovered from a crime scene that contains DNA (a single hair, a drop of blood, skin cell, saliva, etc.) can be subjected to PCR. Primers complementary to repetitive DNA sequences found throughout the human genome are

pBluescript II (S/K)

```
   1 CTAAATTGTAAGCGTTAATATTTTGTTAAAATTCGCGTTAAATTTTTGTTAAATCAGCTCATTTTTTAACCAATAGGCCGAAATCGGCAAAATCCCTTATAAATCAAAAGAATAGACCGA
     GATTTAACATTCGCAATTATAGCACAATTTTGCGCGCAATTTAAAAACAATTTAGTACAGTAAAAAATTGGTTATCCGGCTTTAGCCGTTTTAGGGAATATTTAGTTTTCTTATCTGGCT 120
                                                                          5′       primer 1       3′
 121 GATAGGGTTGAGTGTTGTTCCAGTTTGGAACAAGAGTCCACTATTAAAGAACGTGGACTCCAACGTCAAAGGGCGAAAAACCGTCTATCAGGGCGATGGCCCACTACGTGAACCATCACC
     CTATCCGAACTCACAACAAGGTCAAACCTTGTTCTCAGGTGATAATTTCTTGCACCTGAGGTTGCAGTTTCCCGCTTTTTGGCAGATAGTCCCGCTACCGGGTGATGCACTTGGTAGTGG 240

 241 CTAATCAAGTTTTTTGGGGTCGAGGTGCCGTAAAGCACTAAATCGGAACCCTAAAGGGAGCCCCCGATTTAGAGCTTGACGGGGAAAGCCGGCGAACGTGGCGAGAAAGGAAGGGAAGAA
     GATTAGTTCAAAAAACCCCAGCTCCACGGCATTTCGTGATTTAGCCTTGGGATTTCCCTCGGGGGCTAAATCTCGAACTGCCCCTTTCGGCCGCTTGCACCGCTCTTTCCTTCCCTTCTT 360

 361 AGCGAAAGGAGCGGGCGCTAGGGCGCTGGCAAGTGTAGCGGTCACGCTGCGCGTAACCACCACACCCGCCGCGCTTAATGCGCCGCTACAGGGCGCGTCCCATTCGCCATTCAGGCTGCG
     TCACTTTCCTCGCCCGCGATCCCGCGACCGTTCACATCGCCAGTGCGACGCGCATTGGTCCTCTGGGCGGCGCGAATTACGCGGCGATGTCCCGCGCAGGGTAAGCGGTAACTCCGACGC 480

 481 CAACTGTTGGGAAGGGCGATCGGTGCGGGCCTCTTCGCTATTACGCCAGCTGGCGAAAGGGGGATGTGCTGCAAGGCGATTAAGTTGGGTAACGCCAGGGTTTTCCCAGTCACGACGTTG
     GTTGACAACCCTTCCCGCTAGCCACGCCCGGAGAAGCGATAATGCGATCGACCGCTTTCCCCCTACACGTCGTTCCGCTAATTCAACCCATTGCGGYCCCAAAAGGGTCAGTGCTGCAAC 600
                                                       KpnI
 601 TAAAACGACGGCCAGTGAGCGCGCGTAATACGACTCACTATAGGGCGAATTGGGTACCGGGCCCCCCCTCGAGGTCGACGGTATCGATAAGCTTGATATCGAATTCCTGCAGCCCGGGGG
     ATTTTGCTGCCGGTCACTCGCGGCGCATTATGCTGAGTGATATCCGCTTAACCCATGGCCCGGGGGGGAGCTCCAGCTGCCATAGCTATTCGAACTATAGCTTAAGGACGTCGGGCCCCC 720
                                 SacI/SstI
 721 ATCCACTAGTTCTAGAGCGGCCGCCACCGCGGTGGAGCTCCAGCTTTTGTTCCCTTTAGTGAGGGTTAATTGCGCGCTTGGCGTAATCATGGTCATAGCTGTTTCCTGTGTGAAATTGTT
     TAGGTGATCAAGATCTCGCCGGCGGTGGCGCCACCTCGAGGTCGAAAACAAGGGAAATCACTCCCAATTAACGCGCGAACCGCATTAGTACCAGTATCGACAAAGGACACACTTTAACAA 840

 841 ATCCGCTCACAATTCCACACAACATACGAGCCGGAAGCATAAAGTGTAAAGCCTGGGGTGCCTAATGAGTGAGCTAACTCACATTAATTGCGTTGCGCTCACTGCCCGCTTTCCAGTCGG
     TAGGCGAGTGTTAAGGTGTGTTGTATGCTCGGCCTTCGTATTTCACATTTCGGACCCCACGGATTACTCACTCGATTGAGTGTAATTAAGCGAACGGGAGTGACGGGCGAAAGGTCAGCC 960

 961 GAAACCTGTCGTGCCAGCTGCATTAATGAATCGGCCAACGCGCGGGGAGAGGCGGTTTGCGTATTGGGCGCTCTTCCGCTTCCTCGCTCACTGACTCGCTGCGCTCGGTCGTTCGGCTGC
     CTTTGGACAGCACGGTCGACGTAATTACTTAGCCGGTTGCGCGCCCCTCTCCGCCAAACGCATAACCCGCGAGAAGGCGAAGGAGCGACTGACTGAGCGACGCGAGCCAGCAAGAAGACG 1080

1081 GGCGAGCGGTATGAGCTCACTCAAAGGCGGTAATACGGTTATCCACAGAATCAGGGGATAACGCAGGAAAGAACATGTGAGCAAAAGGCCAGCAAAAGGCCAGGAACCGTAAAAAGGCCG
     CCGCTCGCCATACTCGAGTGAGTTTCCGCCATTATGCCAATAGGTGTCTTAGTCCCCTATTGCGTCCTTTCTTGTACACTCGTTTTCCGGTCGTTTTCCGGTCCTTGGCATTTTTCCGGA 1200

1201 CGTTGCTGGCGTTTTTCCATAGGCTCCGCCCCCCTGACGAGCATCACAAAAATCGACGCTCAAGTCAGAGGTGGCGAAACCCGACAGGACTATAAAGATACCAGGCGTTTCCCCCTGGAA
     GCAACGACCGCAAAAAGGTATCCGAGGCGGGGGGACTGCTCGTAGTGTTTTTAGCTGCGAGTTCAGTCTCCACCGCTTTGGGCTGTCCTGATATTTCTATGGTCCGCAAAGGGGGACCTT 1320

1321 GCTCCCTCGTGCGCTCTCCTGTTCCGACCCTGCCGCTTACCGGATACCTGTCCGCCTTTCTCCCTTCGGGAAGCGTGGCGCTTTCTCATAGCTCACGCTGTAGGTATCTCAGTTCGGTGT
     CGAGGGAGCACGCGAGAGGACAAGGCTGGGACGGCGAATGGCCTATGGACAGGCGGAAAGAGGGAAGCCCTTCGCACCGCGAAAGAGTATCGAGTGCGACATCCATAGAGTCAAGCCACA 1440

1441 AGGTCGTTCGCTCCAAGCTGGGCTGTGTGCACGAACCCCCCGTTCAGCCCGACCGCTGCGCCTTATCCGGTAACTATCGTCTTGAGTCCAACCCGGTAAGACACGACTTATCGCCACTGG
     TCCAGCAAGCGAGGTTCGACCCGACACACGTGCTTGGGGGGCAAGTCGGGCTGGCGACGCGGAATAGGCCATTGATAGCAGAACTCAGGTTGGGCCATTCTGTGCTGAATAGCGGTGACC 1560

1561 CAGCAGCCACTGGTAACAGGATTAGCAGAGCGAGGTATGTAGGCGGTGCTACAGAGTTCTTGAAGTGGTGGCCTAACTACGGCTACACTAGAAGGACAGTATTTGGTATCTGCGCTCTGC
     GTCGTCGGTGACCATTGTCCTAATCGTCTCGCTCCATACATCCGCCACGATGTCTCAAGAACTTCACCACCGGATTGATGCCGATGTGATCTTCCTGTCATAAACCATAGACGCGAGACG 1680

1681 TGAAGCCAGTTACCTTCGGAAAAAGAGTTGGTAGCTCTTGATCCGGCAAACAAACCACCGCTGGTAGCGGTGGTTTTTTTGTTTGCAAGCAGCAGATTACGCGCAGAAAAAAAGGATCTC
     ACTTCGGTCAATGGAAGCCTTTTTCTCAACCATCGAGAACTAGGCCGTTTGTTTGGTGGCGACCATCGCCACCAAAAAAAGAAACGTTCGTCGTCTAATGCGCGTCTTTTTTTCCTAGAG 1800
                               3′        primer 2        5′
1801 AAGAAGATCCTTTGATCTTTTCTACGGGGTCTGACGCTCAGTGGAACGGAAAACTCACGTTAAGGGATTTTCGCATGAGATTATCAAAAAGGATCTTCACCTAGATCCTTTTAAATTAAA
     TTCTTCTAGGAAACTAGAAAAGATGCCCCAGACTGCGAGTCACCTTGCCTTTTGAGTGCAATTCCCTAAAAGCGTACTCTAATAGTTTTTCCTAGAAGTGGATCTAGGAAAATTTAATTT 1920
```

Figure 24-3 (*continued*)

used to produce a set of amplified DNA products. If these PCR fragments are subjected to Southern blotting (see introduction to Section V) after being digested with a number of different restriction enzymes, the size and pattern of restriction fragments produced provide a DNA "fingerprint" (Fig. 24-5). Traditionally, forensic science has relied only on blood chemistry and restriction mapping of whole chromosomal DNA, which often require larger biological samples and are less discriminating.

Polymerase chain reaction has also proven to be useful in the fields of archaeology and evolution. Ancient biological samples recovered from digs and expeditions can be subjected to PCR to gain insight into the genetic composition of extinct species, including some of our earliest ancestors. The information obtained from these analyses can provide a basis for a detailed study on the process of evolution. Considering the enormous impact that the technique of PCR has had over a wide range of fields in the past 10 years, it is not surprising that the Nobel Prize was awarded for this discovery.

Supplies and Reagents

0.5-ml thin-walled PCR tubes

P-20, P-200, and P-1000 Pipetmen with sterile disposable tips

1.5-ml plastic microcentrifuge tubes

pBluescript II (S/K) plasmid in TE buffer (1 $\mu g/\mu l$)—Stratagene catalog #212205

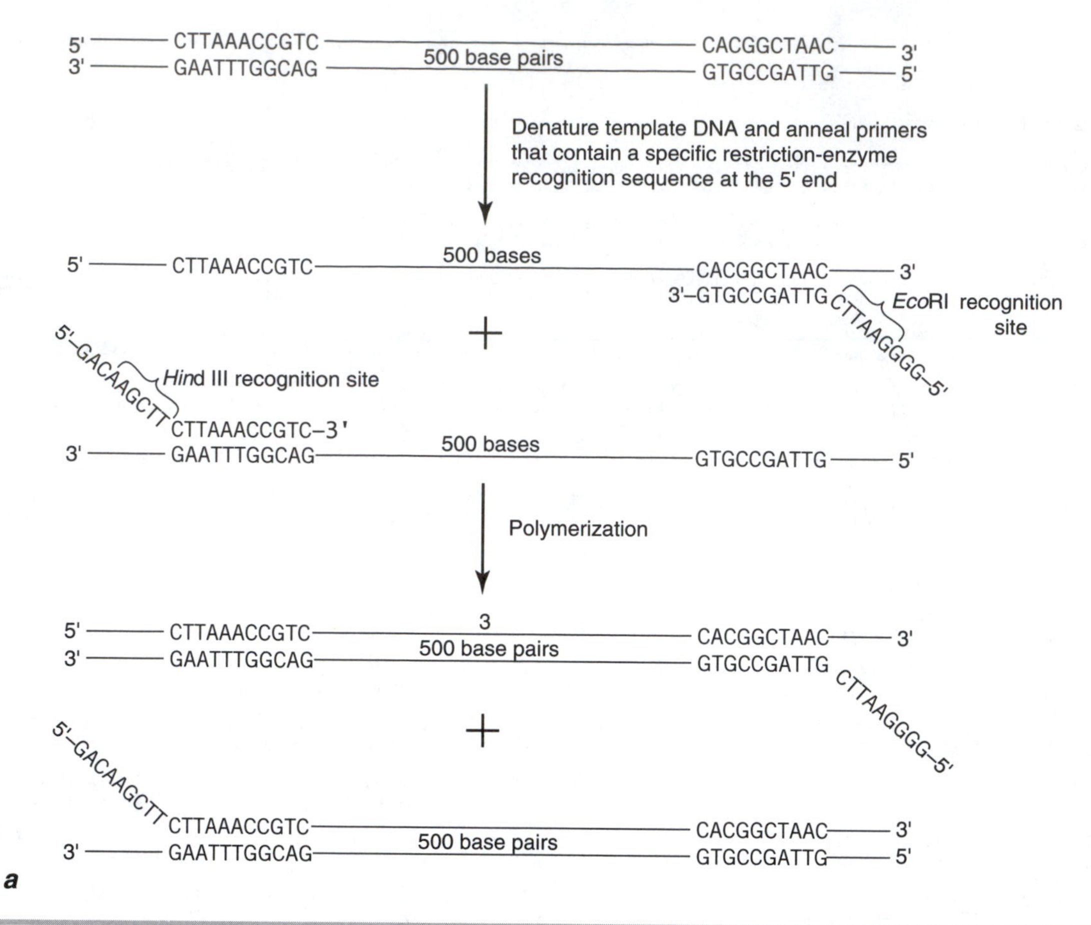

Figure 24-4 (*a*) Using PCR to introduce restriction sites on either end of the amplified DNA product. The amplified DNA in subsequent cycles will include the *Eco*RI site and *Hin*dIII site at either end for easy cloning into a desired vector. (*b*) Using PCR to add base pairs within a DNA fragment of interest. The amplified DNA in subsequent cycles will include the five additional base pairs specified by the mutagenic primer. (*c*) Using PCR to delete base pairs within a DNA fragment of interest. The amplified DNA in subsequent cycles will be missing the five base pairs not present in the mutagenic primer. (*d*) Using PCR to change a single base pair in the template DNA. The amplified DNA in subsequent cycles will include the single-base-pair change encoded by the mutagenic primer.

pBluescript II (K/S) plasmid in TE buffer (1 μg/μl)—Stratagene catalog #212207

Sterile distilled water

TE buffer (10 mM Tris, pH 7.5, 1 mM EDTA)

12.5 mM $MgCl_2$ in distilled water

Deoxyribonucleotide (dNTP) solution—1.25 mM each of dATP, dCTP, dGTP, and dTTP in water

Primer 1—10 μM in distilled water (5′-AAAGGGC-GAAAAACCGTCTATCAG-3′)

Primer 2—10 μM in distilled water (5′-TTTGCCG-GATCAAGAGCTACCAAC-3′)

Taq polymerase (1 unit/μl)—Gibco BRL catalog #18038-042

10X *Taq* Polymerase buffer (supplied with enzyme)—Gibco BRL

PCR thermocycler

Agarose (electrophoresis grade)

5X TBE buffer (54 g/liter Tris base, 27.5 g/liter boric acid, 20 ml/liter of 0.5 M EDTA, pH 8.0)

6X agarose gel DNA sample buffer (0.25% (wt/vol) bromophenol blue, 0.25% (wt/vol) xylene cyanole, 30% (wt/vol) glycerol in water)

1 kb-DNA ladder size marker—Gibco BRL catalog #15615-016

Casting box and apparatus for agarose-gel electrophoresis

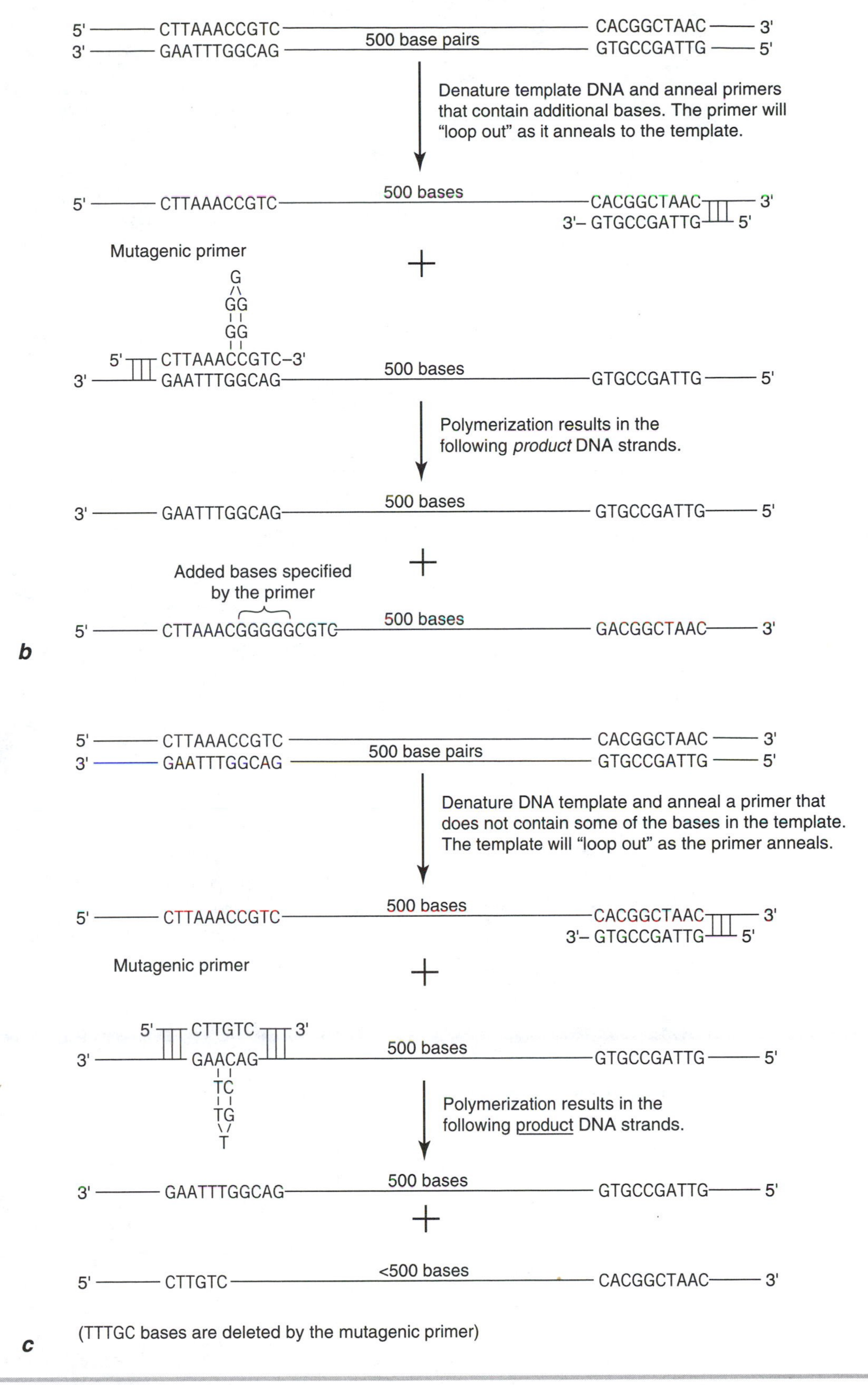

Figure 24-4 (*continued*)

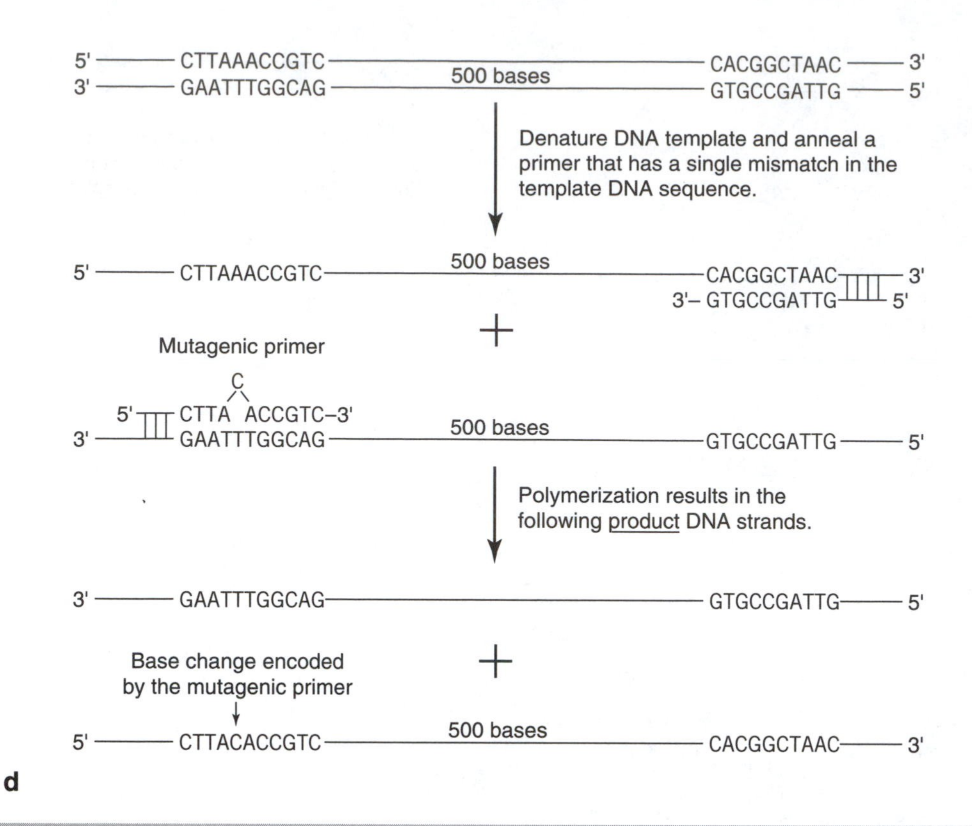

Figure 24-4 *(continued)*

Power supply

*Kpn*I (10 units/μl) with 10X reaction buffer—Gibco BRL catalog #15232-036

*Sst*I (10 units/μl) with 10X reaction buffer—Gibco BRL catalog #15222-037

0.5 μg/ml solution of ethidium bromide in 0.5X TBE buffer

Light box with a 256-nm light source

Polaroid camera with film

Protocol

Day 1: Standard PCR Amplification Reaction

1. Obtain four thin-walled PCR tubes containing the following:

Tube	Volume of pBluescript II DNA	Concentration of pBluescript II	Vol. of Water
A	—	—	10 μl
B	10 μl	1 fg/μl	—
C	10 μl	1 pg/μl	—
D	10 μl	1 ng/μl	—

Remember the metric units: f = femto = 10^{-15}, p = pico = 10^{-12}, n = nano = 10^{-9}. Store these tubes on ice while you prepare the Master Mix for the PCR reaction.

2. Prepare the following Master Mix in a sterile, 1.5-ml microcentrifuge tube (mix all components thoroughly by repeated pipetting up and down):

Component	Volume of Stock Solution	Stock Solution
$MgCl_2$	68 μl	12.5 mM
dNTPs	68 μl	1.25 mM (of each dNTP)
Primer 1	42.5 μl	10 μM
Primer 2	42.5 μl	10 μM
Taq buffer	42.5 μl	10X
Sterile water	110.5 μl	—
Taq polymerase	8.5 μl	1 unit/μl

3. Add 90 μl of this Master Mix to each of the four PCR reaction tubes prepared in step 1 (A–D) and mix thoroughly.

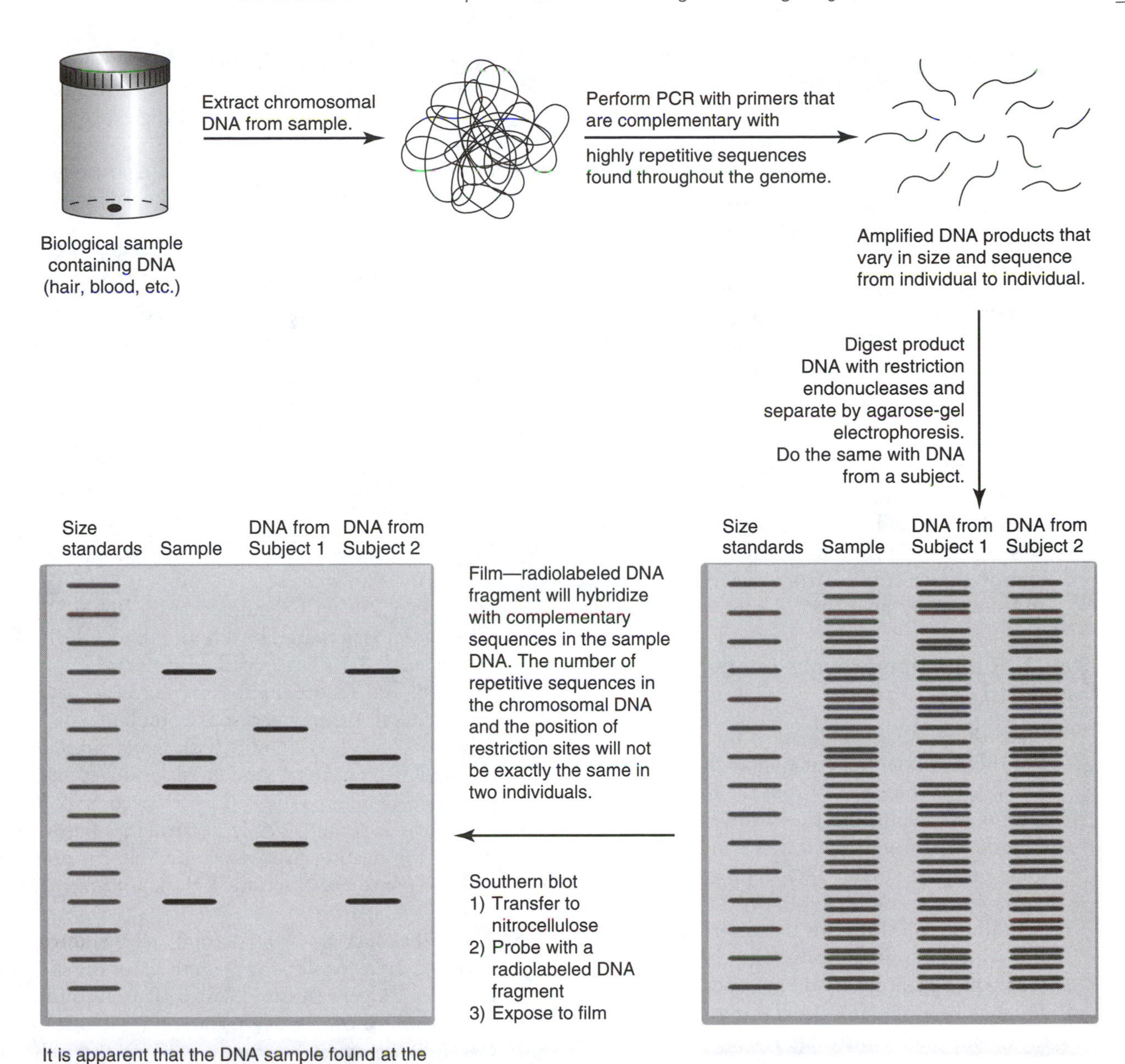

Figure 24-5 Use of PCR in forensic science.

4. Overlay each sample in tubes A to D with 40 μl of mineral oil. *Some thermocyclers are equipped with heated lids that make the mineral oil unnecessary. The mineral oil will keep the reaction volume at the bottom of the tube as the solution is repeatedly heated and cooled during the reaction. If this were not done, the solution would begin to condense on the walls of the tube and possibly alter the concentrations of the different components in the reaction.*

5. Label the four tubes with your name and place them in the thermocycler set to room temperature.
6. Perform 30 PCR cycles using the following parameters:

   a. Ramp from room temperature to 94°C in 50 sec and hold at 94°C for 1 min (denaturation).
   b. Ramp down to 55°C in 40 sec and hold at 55°C for 1 min (primer annealing).
   c. Ramp up to 72°C in 10 sec and hold at 72°C for 2 min (polymerization).
   d. Ramp from 72 to 94°C in 15 sec (denaturation).

   The 30-cycle program will take approximately 2.5 hr. After completion of the 30 cycles, a "soak file" should be included to hold the samples at 4°C until they are ready to be removed.
7. Remove the samples from the thermocycler and store them at −20°C for use on Day 2.

Day 2: Restriction-Enzyme Analysis of Amplified DNA Product

1. Place the samples obtained at the end of Day 1 at room temperature and allow them to thaw. Remove the mineral oil from the four PCR reactions by poking the tip of the P-200 Pipetman down through the mineral oil to the bottom of each tube. Draw up the bottom (aqueous) phase into the tip up to the level of the mineral oil. Remove the tip from the tube and wipe the mineral oil off of the outside of the tip using laboratory tissue paper. Expel each aqueous PCR reaction (A–D) into a new, sterile 1.5-ml microcentrifuge tube.
2. Prepare a sample of the PCR reaction *in tube D* for restriction digest in two separate 1.5-ml microcentrifuge tubes as follows:

*Kpn*I Digest	*Sst*I Digest
5 μl of PCR reaction from tube D	5 μl of PCR reaction from tube D
16.5 μl of sterile water	16.5 μl of sterile water
2.5 μl of 10X *Kpn*I buffer	2.5 μl of 10X *Sst*I buffer
1 μl of *Kpn*I (10 units/μl)	1 μl of *Sst*I (10 units/μl)

   Incubate both reactions at 37°C for 1 hr.

3. Add 6 μl of 6X DNA sample buffer to each of these two reactions following the 1-hr incubation. Label these tubes 7 (*Kpn*I digest) and 8 (*Sst*I digest).
4. Set up six additional samples for agarose-gel electrophoresis, each containing the following:

Tube	Volume of DNA Sample (Undigested)	Volume of 6X DNA Sample Buffer
1	5 μl (PCR reaction A)	1 μl
2	5 μl (PCR reaction B)	1 μl
3	20 μl (PCR reaction B)	4 μl
4	5 μl (PCR reaction C)	1 μl
5	5 μl (PCR reaction D)	1 μl
6	5 μl of 0.2 μg/μl 1-kb DNA ladder	1 μl

5. Prepare a 10-well, 1% TBE agarose gel by adding 0.5 g of agarose to 50 ml of 0.5X TBE buffer in a 250 ml Erlenmeyer flask (the 0.5X TBE buffer is prepared by diluting the 5X TBE buffer stock 1:10 with distilled water). Weigh the flask and record its mass. Microwave the flask until all of the agarose is completely dissolved (~2 min). Weigh the flask again and add distilled water until the flask reaches its preweighed mass. This is done to account for the fact that some water is lost to evaporation during the heating process. If this water is not added back to the flask, the gel may not actually be 1% agarose.
6. Allow the flask to cool at room temperature (swirling occasionally) for ~5 min. Pour the solution into a gel cast fitted with a 10-well comb at one end. Allow the agarose to set (~40 min).
7. Carefully remove the comb, remove the gel from the cast, and place it in the agarose-gel electrophoresis unit. Add 0.5X TBE buffer to the chamber until the buffer covers the top of the gel and fills all of the wells.
8. Load the DNA samples in order (1–8) into separate wells of the gel. Connect the negative electrode (cathode) to the well side of the unit and the positive electrode (anode) to the other side of the unit (the side opposite the wells in the gel). Remember that the DNA is negatively charged and will migrate toward the positive electrode.

9. Apply 50 mA (constant current) and continue the electrophoresis until the bromophenol blue (dark blue) tracking dye has migrated about three-fourths of the way to the end of the gel.
10. Turn off the power supply, disconnect the electrodes, and place the gel in a solution of 0.5X TBE containing 0.5 μg/ml ethidium bromide. *Wear gloves at all times when handling ethidium bromide.* Incubate the gel at room temperature for 20 min. During this time, the ethidium bromide will enter the gel and intercalate between the base pairs in the DNA. The DNA will appear as pink or orange bands on the gel when exposed to ultraviolet light, as the ethidium bromide–DNA complex fluoresces.
11. Destain the gel by placing it in a solution of 0.5X TBE buffer without ethidium bromide. Incubate at room temperature for 10 min. This is done to remove any ethidium bromide from areas of the gel that do not contain DNA.
12. Place the gel in a light box fitted with a camera, appropriate filter, and a 254-nm light source. Take a photograph of your gel for later analysis.
13. Prepare a plot of number of base pairs versus distance traveled (in centimeters) for as many DNA fragments present in the 1-kb DNA ladder lane as possible. If the 1-kb DNA ladder size standards resolved well, you should be able to differentiate accurately between the relative mobilities of the 0.5-, 1.0-, 1.6-, 2.0-, and 3.0-kb fragments for use in preparing the standard curve (*do not try to calculate the relative mobilities of the higher-molecular-weight DNA size standard fragments if they did not resolve well* [see Experiment 21, Day 4, step 16]). Also, refer to Figure 21-5 for the sizes of the DNA fragments present in the 1-kb DNA ladder.
14. Using the standard curve prepared in step 13, calculate the number of base pairs present in all of the other sample lanes on the gel. What is the size of the DNA fragment produced in PCR reactions A to D? Is there any DNA present in the lane containing a sample of PCR reaction A? Explain. Which of the PCR reactions (B, C, or D) produced the most and the least amount of product? Explain.
15. Based on the *intensities* of the DNA bands present in the lanes containing samples of PCR reactions B to D, *estimate* the total amount of product produced in each PCR reaction. *Hint: the 1636 bp DNA band in the 1-kb DNA ladder lane contains approximately 100 ng of DNA. Therefore, if the PCR product band is of the same intensity as this marker, then the PCR product band also contains about 100 ng of DNA.*
16. Based on the size of the DNA fragments produced from the *Kpn*I and *Sst*I digests, determine whether you amplified a portion of the pBluescript II (S/K) plasmid or the pBluescript II (K/S) plasmid. Justify your answer in terms of what size DNA fragments were produced in each digest.
17. Determine whether there were any "unexpected" DNA bands in the lanes containing samples of PCR reactions B to D. If there were, explain what you think may have caused these, as well as how you might alter the conditions of the PCR reaction to eliminate them.
18. Determine the number of molecules (N_0) of the 2961-base-pair pBluescript II plasmid that were present in each of PCR reactions B to D. Based on your estimation of the mass of DNA product present in each of these PCR reactions (step 15) and the molecular weight of the product (320 Da/nucleotide, or 640 Da per base pair), calculate the number of molecules of DNA product (N) that were produced in each PCR reaction. From these two pieces of information, estimate the amplification efficiency (E) for each PCR reaction using the following equation:

$$N = N_0 (1 + E)^n$$

where n = the number of amplification cycles (30). The amplification efficiency (E) can have a value of between 0 and 1.0. Zero represents no amplification and 1.0 represents 100% amplification efficiency. Which of the three PCR reactions (B, C, or D) showed the highest and the lowest amplification efficiency? Explain why you think the amplification efficiency might vary with respect to template DNA concentration.
19. Calculate the mass of DNA product that would be present in PCR reaction tubes B to D if the amplification efficiency were 100% (E = 1.0).

Exercises

1. The maximum number of molecules of product produced (assuming E = 1.0) is often not possible due to the fact that the amount of primers and/or dNTPs may be limiting. Based on the number of primers present in reactions B to D, calculate the maximum number of micrograms of product that could be produced. Based on the number of dNTPs present in reactions B to D, calculate the maximum number of micrograms of product that could be produced. Which of the two components, dNTPs or primers, would ultimately limit the amount of product that can be produced in these reactions?
2. Calculate the melting temperature (T_m) of the 1543-base-pair product and the 2961-base-pair pBluescript II plasmid used in this experiment with the following formula:

$$T_m = 81.5°C + 16.6 \log [Na^+] + (0.41)(\% \text{ GC}) - (675/\text{number of base pairs}) - (\% \text{ formamide}) - (\text{number of mismatched base pairs})$$

   The $[Na^+]$ in these reactions is 0.05 M, the % GC of the product is 56.2, and the % GC of the plasmid is 50.4.
3. You wish to amplify a portion of a bacterial gene that has homologs both in yeast and mice. This gene has already been cloned from both of these organisms, and it has been found that two amino acid sequences at the amino and carboxyl termini are conserved in both proteins:

   mouse: NH_2 — LKVAPWYVDGSE — (105 amino acids) —LFGLCTANDHKVQ — COOH

   yeast: NH_2 — PRYAPWYVDGTC — (105 amino acids) — GRILCTANDHGRN — COOH

   Design two degenerate primers (21 nucleotides in length) that could be used to try to amplify the homologous gene from the bacterial chromosome. You will need a chart of the genetic code to design this primer pair. The degeneracy of the primer arises due to the fact that there is "wobble" in the genetic code at the third position for many amino acids. Calculate the degeneracy of each primer that you have designed. This can be done simply by multiplying the number of possible codons for each amino acid over the length of the amino acid sequence that the primer spans. For example, glycine has four possible codons and phenylalanine has two possible codons. Therefore, the degeneracy of a primer spanning these two amino acids is 8. Also, calculate the molecular weight and the number of base pairs that you may expect if the PCR reaction is successful.

REFERENCES

Arnheim, M., and Erlich, H. (1992). Polymerase Chain Reaction Strategy. *Annu Rev Biochem* **61**:131.

Erlich, H. A., Gelfand, D., and Sninsky, J. J. (1991). Recent Advances in Polymerase Chain Reaction. *Science* **252**:1643.

Lehninger, A. L., Nelson, D. L., and Cox, M. M. (1993). Recombinant DNA Technology. In: *Principles of Biochemistry*, 2nd ed. New York: Worth.

Saiki, R. K., Gelfand, D. H., Stoeffel, B., Scharf, S. J., Higuchi, R., Horn, G. T., Mullis, K. B., and Erlich, H. A. (1988). Primer Directed Amplification of DNA with a Thermostable DNA Polymerase. *Science* **239**:487.

Sambrook, J., Fritsch, E. F., and Maniatis, T. (1989). *Molecular Cloning*, Plainview, NY: Cold Spring Harbor Laboratory Press.

SECTION VI

Information Science

Experiment 25 **Obtaining and Analyzing Genetic and Protein Sequence Information via the World Wide Web, Lasergene, and RasMol**

SECTION VI

Information Science

Introduction

In the past several decades, research scientists have cloned and sequenced hundreds of thousands of genes from a variety of organisms. More recently, the scientific community has undertaken the massive project of sequencing entire genomes from a selected set of model organisms (including humans) that are most often used in studies of biochemistry, molecular biology, and genetics. The overwhelming amount of DNA and protein sequence information generated by these projects necessitated the development of biological databases that could organize, store, and make the information accessible to research scientists around the world. There has also been a striking increase in the numbers of high-resolution structures of biological macromolecules determined by x-ray diffraction studies or by nuclear magnetic resonance (NMR) spectroscopy. This rich database of important biological information must also be made readily available to the research community. This need has led to the growth of a new area of research commonly referred to as *information science* or *bioinformatics*. In a cooperative effort, research scientists and computer scientists have established a large number of biological databases, most of which are accessible via the World Wide Web.

How do you find the different biological databases on the Internet? In an effort to publicize the numerous sites that are being maintained, several Internet sites have been created to act as directories. Two of these database directories that we have found to be quite useful are the *Biology Workbench* (http://biology.ncsa.uiuc.edu) and *Pedro's BioMolecular Research Tools* (http://www1.iastate.edu/~pedro/research_tools.html).

The Biology Workbench is a site that has been developed by the Computational Biology Group at the National Center for Supercomputing Applications (NCSA) at the University of Illinois. Pedro's BioMolecular Research Tools is a similar directory maintained by a group at Iowa State. Both of these sites serve as search engines for a number of general and specialized DNA and protein databases, some of which are described below. Many of these databases are directly accessible through the Biology Workbench. A more recent directory site has been developed by Christopher M. Smith of the San Diego Supercomputer Center. The CMS Molecular Biology Resources site (http://www.sdsc.edu/ResTools/cmshp.html) lists nearly 2000 biological Web sites. Unlike the other two directories, this one lists sites according to the desired application. For instance, if you wish to analyze the coding region(s) within a fragment of DNA, it will list sites that will aid you. If you wish to perform a phylogenetic analysis on a set of related genes, it will list sites that will aid you.

GenBank

GenBank is the National Institutes of Health sequence database. It is the oldest, largest, and most complete general database. Together, GenBank and the EMBL database (see below) comprise the *International Nucleotide Sequence Database Collaboration*. Newly discovered genes must be submitted to either GenBank or EMBL before they can be published in a research journal. Gene sequences can be located on GenBank either by submitting the name of the gene or the GenBank accession number that is assigned to the gene. Often, a search performed on the basis of the gene name will identify several homologs of the gene that have been identified in different organisms. GenBank is most easily accessed from an Internet site maintained by the *National Center for Biotechnology Information* (http://www.ncbi.nlm.nih.gov/).

European Molecular Biology Laboratory (EMBL)

The EMBL database is maintained by the Hinxton Outstation (Great Britain) of the European Molecular Biology Laboratory, which specializes in the area of bioinformatics. Since EMBL is constantly exchanging information with GenBank, it too is quite large and current. The EMBL database has numerous specialized databases that draw information from it (see Table VI-1).

ExPASy and ISREC

ExPASy is a database maintained by the *Swiss Institute of Bioinformatics* (SIB) that is dedicated to the organization of protein sequences. ISREC is a similar database maintained by the *Bioinformatics Group*

Table VI-1 An Incomplete Listing of Biological Databases and Other Useful Sites on the World Wide Web

Database	Source	Area of Specialization
3D-ALI	EMBL	Protein structure based on amino acid sequence
AA Analysis	EMBL PIR	Protein identification based on amino acid sequence
AA CompIdent	SwissProt ExPASy	Protein identification based on amino acid sequence
AbCheck	Kabat	Antibody sequences
ALIGN	EERIE MIPS	Sequence alignments
AllAll	SwissProt	Protein sequence alignments
ASC	EMBL	Protein analytical surface calculations
ATLAS	MIPS	General DNA and protein searches
BERLIN	CAOS/ CAMM	RNA databank (5S rRNA sequences)
BLAST	GenBank EERIE	Alignment of sequences
BLOCKS	EMBL	Conserved sequences in proteins
BioMagResBank	BIMAS	Protein structure determined by NMR
BMCD	CARB	Crystallization of macromolecules
Coils	ISREC	Protein structure (coiled coil regions)
DrugBank	NIH	Structure of different drugs
DSSP	EMBL	Protein secondary structure
EMP Pathways	NIH	Metabolic pathways
ENZYME	ExPASy	Enzymes
EPD	EMBL	Eukaryotic promoters
FlyBase	Harvard	Sequences identified in *Drosophila*
FSSP	EMBL	Groups proteins into families based on structure
Gene Finder	Baylor	Introns, exons, and RNA splice sites
GuessProt	SwissProt ExPASy	Identify proteins by isoelectric point and molecular weight
HOVERGEN	CAOS/ CAMM	Homologous genes identified in different vertebrates
Kabat	Kabat	Proteins related to immunology

LIGAND	GenomeNet	Ligands (chemicals) for different enzymes
MassSearch	EMBL SwissProt	Identifies proteins on the basis of mass
MGD	Jackson Laboratory	Mouse genome sequences
MHCPEP	WEHI	Binding peptides for major histocompatibility complex
MPBD	EMBL	Molecular probes
NRSub	National Institute of Genetics (Japan)	Sequences from the genome of *Bacillus subtilis*
NuclPepSearch	SwissProt	General DNA searches
NYC-MASS	Rockefeller	Protein mass spectrometry
PHD	EMBL	Protein secondary structure
Phospepsort	EERIE	Phosphorylated peptides
pI/Mw	ExPASy	Calculation of isoelectric point and molecular weight
PMD	EMBL GenomeNet	Mutant proteins
ProDom	EMBL	Protein domains
ProfileScan	ExPASy	Searches based on sequence profile
PROPSEARCH	EMBL	Protein homolgy based on amino acid composition
PROSITE	ExPASy	Protein sites and patterns
PSORT	GenomeNet	Prediction of protein sorting signals based on amino acid composition
RDP	University of Illinois	Ribosomal proteins
ReBase	EMBL	Restriction enzymes
REPRO	EMBL	Repeated sequences in proteins
SAPS	ISREC	Protein sequence statistical analysis
SSPRED	EMBL	Predict protein secondary structure
Swiss-2DPage	ExPASy	Two-dimensional PAGE
TFD	NIH	Transcription factors
TFSITE	EMBL	Transcription factors
TMAP	EMBL	Transmembrane regions in proteins
TMpred	ISREC	Transmembrane regions in proteins
VecBase	CAOS/ CAMM	Sequence of cloning vectors

at the Swiss Institute for Experimental Cancer Research. As shown in Table VI-1, a number of specialized databases draw information from these two databases.

SwissProt

SwissProt is a computational biology database specializing in protein sequence analysis maintained by the *Swiss Federal Institute of Technology* (ETH) in Zurich, Switzerland. Like the other general databases described above, a number of more specialized biological databases draw information from this source.

Specialized Biological Databases

As science has become increasingly specialized over the past two decades, so too has the area of bioinformatics. There are numerous specialized databases that deal with different aspects of biology. Some of these specialize in the area of proteins, while others specialize in the area of DNA. In general, these sites differ in two respects: the general database(s) from which they draw information, and the parameters that they require the user to define to conduct the search. For instance, the *AA Analysis* database searches the more general SwissProt database to identify or group proteins on the basis of their amino acid composition. If you were to isolate a protein and determine its amino acid composition, you could enter the data into the AA Analysis database in an attempt to identify it. Keep in mind that new databases are constantly being developed. Table VI-I presents an entensive, but incomplete, list of specialized databases.

Computer Software Programs for Analyzing Sequences

Although biological databases contain a great deal of sequence information, they are very limited in their ability to analyze DNA and protein sequences. Fortunately, computer software is available that will allow a researcher to import a sequence of interest from a database and subject it to a variety of different analytical applications. Of the many software programs currently on the market, we find Lasergene by DNASTAR, Inc. (Madison, WI) to be the most comprehensive and easy to use. This software is available in both PC and Macintosh formats. Lasergene is under copyright by DNASTAR, Inc. *It is illegal to install the software on a new computer without purchasing it.* If you would like to use Lasergene *strictly for teaching purposes at no charge to your institution*, you may contact DNASTAR (1–608–258–7420) to establish an educational license agreement. DNASTAR is very helpful in the installation of the software, and they provide a wealth of literature for students to help them understand the operation of the various applications contained in the software. A brief description of the various applications in Lasergene is found below. These descriptions are intended to inform you of the capabilities of the Lasergene software. A more detailed description of the software is available in the "User's Guide" supplied by DNASTAR. A less detailed but adequate description is provided in the "Installing, Updating, and Getting Started" manual also provided by DNASTAR.

DNASTAR is one of many companies that supplies computer software useful in the field of bioinformatics. Other popular computer software packages are offered by DNAStrider, the Genetic Computer Group, and GeneJockey II. If you have experience with these software packages, you may contact the company to establish an educational license agreement.

EDITSEQ

The EDITSEQ application allows you to manually enter DNA or protein sequence information into your computer. This application has several features that make it useful to the research scientist. First, it can identify open reading frames (possible gene sequences) within a DNA sequence. Second, it can provide the percent base composition (A,G,C,T), the percent GC, the percent AT, and the melting temperature of the entire sequence or a small subset of that sequence. Third, EDITSEQ can translate a nucleotide sequence into a protein sequence. Finally, the application is capable of translating or reverse translating a nucleotide sequence of interest using codes other than the standard genetic code.

GENEMAN

The GENEMAN application is a tool that allows you to access and search for DNA and protein sequences located in six different biological databases. The search for a sequence of interest can be made as broad or restrictive as desired, since there are 12 different "fields" (definition, reference, source, accession number, etc.) to choose from when the search is performed. In addition to performing database searches to find sequences of interest, GENEMAN allows you to search the database for sequences that share homology with the sequence of interest, or for entries that contain a particular conserved sequence. Any number of different DNA or protein sequences found in these databases can be isolated and stored as a sequence file for later analysis.

MAPDRAW

The MAPDRAW application provides a detailed restriction map of the DNA sequence of interest, whether it has been entered manually or imported from a database. Since MAPDRAW is able to identify 478 different restriction endonuclease recognition sequences, you may wish to simplify your restriction map by applying selective *filters*. These filters will identify restriction endonuclease sites specifically on the basis of name, the 5′ or 3′ single-stranded overhangs that they produce, the frequency (number) of sites contained within the sequence, and/or the restriction endonuclease class (type I or type II, see Section V). In addition, a detailed description of any restriction endonuclease can be obtained with this application simply by "clicking" on the enzyme of interest. MAPDRAW will display the amino acid sequence of the double-stranded DNA in all six potential open reading frames (three for the top strand, and three for the bottom strand). This feature enhances the researcher's ability to identify the correct reading frame within a large DNA sequence.

MEGALIGN

The MEGALIGN application is a comprehensive tool for establishing relationships between different DNA and protein sequences. If multiple sequences are to be compared, the *Clustal* and the *Jotun Hein* algorithms are at your disposal. If only two sequences are to be compared, the *Wilbur–Lipman*, *Martinez–Needleman–Wunsch*, and *Dotplot* algorithms can be applied. The results of these alignments can be viewed or displayed in a variety of formats with MEGALIGN. First, the alignment can be displayed as a consensus sequence that the "majority" (a parameter that can be defined by the operator) of the sequences in the alignment contain. Second, you may wish to view the alignment in a tabular format showing the percent similarity and the percent divergence among the sequences in the set. Finally, you have the option of viewing the alignment in the form of a phylogenetic tree, indicating how closely related two sequences are. You can also examine DNA or protein sequences that may have evolved from a common ancestor.

PRIMERSELECT

The PRIMERSELECT module is an extremely valuable tool for designing oligonucleotides for molecular biology applications, including polymerase chain reaction (PCR), DNA sequencing, and Southern blotting. The application assists in identifying suitable regions within a DNA template sequence to which an oligonucleotide with a complementary sequence will hybridize with a high degree of specificity. The PRIMERSELECT application searches for primer pairs on the template based on the PCR product length, the upper primer range, the lower primer range, or combinations of the three. In addition to these parameters, you can restrict the search for primers by increasing the stringency of eight different criteria to which the search will adhere (primer length, number of bases in the primers capable of forming inter- or intramolecular hydrogen bonds with neighboring bases, etc.). As the sets of primer pairs are selected, they are presented along with thermodynamic and statistical data that you may find useful. In addition to altering the parameters for primer selection, PRIMERSELECT also allows you to alter various conditions for the proposed PCR reaction, including primer concentration and ionic strength.

PROTEAN

The PROTEAN application provides a great deal of general information about protein sequences, entered manually or imported into the program from a database. A single report provides the protein's molecular weight, amino acid composition, extinction coefficient, isoelectric point, and theoretical titration curve. The application also provides a large number of different secondary structure predictions for the protein. The *Garnier–Robson* and *Chou–Fasman* algorithms predict alpha helices, beta sheets, and turn regions based on the linear amino acid sequence of the protein. The algorithm of *Kyte and Doolittle* identifies hydrophobic stretches of amino acids, indicating possible transmembrane regions in the protein. An antigenic index of the protein calculated by the *Jameson–Wolf* method indicates possible antigenic peptides that could be used to raise antibodies specific for that protein. The *Emini* algorithm predicts amino acid residues that are likely to reside on the surface of the native protein, and so forth.

Two Lasergene applications that are not explored in Experiment 25 are SEQMAN II and GENEQUEST. SEQMAN II is capable of aligning over 64,000 individual DNA sequences into a single, continuous sequence (a *contig*). Since it is capable of accepting information generated in EDITSEQ or derived from selected automated DNA sequencers, it is suited to both large- and small-scale sequencing projects. GENEQUEST is designed to greatly simplify the identification of specific genetic elements within a DNA sequence, such as open reading frames, transcription start sites, transcription stop sites, translation start sites, translation stop sites, binding sites for transcription factors, and so forth.

The Protein Data Bank (PDB) and RasMol

The Protein Data Bank is a unique database in that it specializes in the three-dimensional structures of proteins and other biomolecules. It is maintained by the *Brookhaven National Laboratory* (http://www.pdb.bnl.gov/). (Note that this database will move to Rutgers University in the future.) This database allows you to visualize a protein in three dimensions, provided that atomic coordinate information for it is available from crystallography or NMR studies. Once you have accessed PDB at the above address, you will select *Software and Related Information* on the home page. Here, you will find a free molecular visualization program called *RasMol* (http://www.umass.edu/microbio/rasmol/rasquick.htm). After you have accessed the RasMol home page, you will find a detailed description of how to install the RasMol software on your computer (select *Getting and Installing RasMol* on the home page). At this point, you are ready to search the PBD for a three-dimensional view of a protein or other molecule of interest.

Once the coordinates of a molecule have been imported into RasMol, there are a number of analytical tools that can be used to visualize the many different interactions taking place within it. For instance, RasMol allows you to determine how many peptide chains there are in the protein, the positions of particular atoms within the protein, the positions of different ions and hydrophobic amino acids, alpha helices and beta sheets, the distance between two particular atoms or residues in the protein, points of contact between the protein and its ligands, the amino terminal and carboxy terminal amino acids, inter- and intramolecular disulfide bonds within the protein, and hydrogen bonds present between different atoms in the protein.

In addition, RasMol has the capability to rotate the molecule on its X, Y, and Z axes, giving a true three-dimensional view of the molecule. If you are interested in a quick tutorial session, you can go to http://www.umass.edu/microbio/rasmol/rastut.htm. If you are interested in how to investigate the different interactions listed above, you can go to http://www.umass.edu/microbio/rasmol/raswhat.htm.

REFERENCES

An, J., Nakama, T., Kubota, Y., and Sarai, A. (1998). 3DinSight: An Integrated Relational Database and Search Tool for the Structure, Function and Properties of Biomolecules. *Bioinformatics* **14**:188.

Barlow, D. J., and Perkins, T. D. (1990). Applications of Interactive Computer Graphics in Analysis of Biomolecular Structures. *Nat Prod Rep* **7**:311.

Brazma, A., Jonassen, I., Eidhammer, I., and Gilbert, D. (1998). Approaches to the Automatic Discovery of Patterns in Biosequences. *J Comput Biol* **5**:279.

Clewley, J. P. (1995). Macintosh Sequence Analysis Software. DNAStar's LaserGene. *Mol Biotechnol* **3**:221.

Doolittle, R. F. (1996). Computer Methods for Macromolecular Sequence Analysis. *Methods Enzymology*, **266**.

Froimowitz, M. (1993). HyperChem: A Software Package for Computational Chemistry and Molecular Modeling. *Biotechniques* **14**:1010.

Gasterland, T. (1998). Structural Genomics: Bioinformatics in the Driver's Seat. *Nat Biotechnol* **16**:625.

Hellinga, H. W. (1998). Computational Protein Engineering. *Nat Struct Biol* **5**:525.

Kanehisa, M. (1998). Grand Challenges in Bioinformatics. *Bioinformatics* **14**:309.

Moszer, I. (1998). The Complete Genome of *Bacillus subtilis:* From Sequence Annotation to Data Arrangement and Analysis. *FEBS Lett* **430**:28.

Plasterer, T. N. (1997). MAPDRAW: Restriction Mapping and Analysis. *Methods Mol Biol* **70**:241.

Plasterer, T. N. (1997). PRIMERSELECT: Primer and Probe Design. *Methods Mol Biol* **70**:291.

Plasterer, T. N. (1997). PROTEAN: Protein Sequence Analysis and Prediction. *Methods Mol Biol* **70**:227.

Sanchez-Ferrer, A., Nunez-Delicado, E., and Bru, R. (1995). Software for Retrieving Biomolecules in Three Dimensions on the Internet. *Trends Biochem Sci* **20**:286.

Sayle, R. A., and Milner-White, E. J. (1995). RASMOL. Biomolecular Graphics for All. *Trends Biochem Sci* **20**:374.

Smith, T. F. (1998). Functional Genomic—Bioinformatics Is Ready for the Challenge. *Trends Genet* **7**:291.

Obtaining and Analyzing Genetic and Protein Sequence Information via the World Wide Web, Lasergene, and RasMol

This computer exercise is designed to introduce you to the wealth of DNA and protein sequence information available on the World Wide Web. Using the Biology Workbench and Lasergene software (DNASTAR, Madison, WI), you will search GenBank for a particular sequence of interest and study it in depth. We use glutathione-*S*-transferase, MAP kinase, *Eco*RI, and *Che*Y for this exercise. We encourage you to experiment with different genes for this exercise, perhaps one that is of interest to you in your own research. The only requirement is that you must use a gene that encodes a protein for which the atomic coordinates are known (consult the instructor before the beginning of the experiment if you are not sure).

Supplies and Reagents

Lasergene software (contact DNASTAR, Inc. in Madison, WI)
200-Mhz computer with access to the Internet
Computer printer

Protocol

> **WARNING**
> **The following protocol explains the application of the Lasergene and RasMol software in the Macintosh format. If you are using IBM computers, this protocol must be modified for use in the IBM format. A Lasergene "User's Guide" can be obtained from DNASTAR, Inc. RasMol software compatible for IBM computers can be installed after accessing the RasMol home page.**

1. Using the Biology Workbench (http://biology.ncsa.uiuc.edu) or the GENEMAN application of Lasergene, search GenBank for a gene of interest (enter the name of the gene or protein for the query).
2. What are some of the "fields" that you can use to restrict your search of the GenBank database? How many entries did you obtain in your search? Explain why you may have obtained several entries in response to your search, as well as what they represent.
3. Select *one* of these entries and save it as a Sequence file in your computer for later analysis.
4. Using the EDITSEQ application of Lasergene, obtain some statistical information about the DNA sequence that you have selected, search for open reading frames (ORFs) within the sequence, and translate the ORFs into protein sequences.

   a. Open your sequence file (saved in step 3) by selecting **Open** from the **File** menu.
   b. Use **Command-A** to select the entire DNA sequence and select **Find ORF** from the **Search** menu. If there is more than one ORF present in your DNA sequence, the **Find ORF** function can be repeated several times to identify all of them.
   c. Click on the largest ORF in your DNA sequence to highlight it, then select **Translate DNA** from the **Goodies** menu. At this point, you should see a new window containing the amino acid sequence encoded by the DNA.
   d. To obtain some statistical information about your DNA sequence, use **Command-A** to

highlight the entire sequence and select **DNA statistics** from the **Goodies** menu. What is the %A, %G, %T, and %C in your sequence? What is the %A-T and %G-C composition of your DNA sequence? What is the melting temperature of this DNA sequence as determined by the Davis–Botstein–Roth method?

e. Print out the report of the DNA statistics and select **Quit** from the EDITSEQ **File** menu.

5. Using the MAPDRAW application of Lasergene, generate a detailed restriction map of your DNA sequence.

   a. To open your DNA sequence, select **Open** from the **File** menu. At this point, your DNA sequence (double-stranded) will be showing with the positions of all of the potential 478 restriction endonuclease recognition sequences that it may contain. Below this, you will see the amino acid sequence specified by your DNA sequence in all six reading frames (three for the top strand and three for the bottom strand).
   b. Select **Unique sites** from the **Map** menu to show the position of all of the restriction endonuclease sites that appear *only once* within your DNA sequence. Print out this report, which you will need for a later exercise.
   c. Select **Absent sites** from the **Map** menu to obtain a list of restriction endonucleases *that will not cleave* within your DNA sequence (the recognition sequence for these enzymes are not present within your DNA sequence). Print out this report, which you will need for a later exercise.
   d. Experiment with some of the different restriction enzyme "filters" available in the MAPDRAW application (apply the **New Filter** option under the **Enzyme** menu). When you feel that your restriction map is complete and easy to read, print out the report displayed in the window for use in a later exercise.
   e. To view all of the potential ORFs in your DNA sequence, select **ORF map** from the **Map** menu. You will see the ORFs encoded by the top strand displayed in red and the ORFs encoded by the bottom strand displayed in green. You will likely see a "cluster" of ORFs specified by either the top or the bottom strand that all *end* in the same place, but have *different* start sites. These "staggered" start sites represent all *potential* ATG (met) start codons in the interior of the full-length gene. Single-click with the mouse on the *largest* ORF specified by the top strand. In the upper left portion of the screen, you will see the range of base pairs specified by this ORF. *Record the range of this ORF (where it begins and ends) in your notebook.* You will need this information later when you design a set of primers that could be used to amplify the gene by polymerase chain reaction (PCR).
   f. Select **Quit** from the MAPDRAW **File** menu.

6. Using the information obtained in step 5 and your knowledge of molecular biology techniques (see Experiment 21 and Section V), design a protocol that will allow you to clone your gene into the multiple cloning site of a vector of your choice (pUC18/19, pBluescript S/K or K/S, etc.). Your instructor will provide you with a list of restriction-enzyme recognition sequences, where these enzymes cleave within this sequence, and restriction maps of a number of different cloning vectors for this exercise. If you cannot find restriction sites within your sequence that will allow you to easily clone the entire gene, you may be forced to design a protocol that will allow you to clone a portion of the gene. Alternatively, your instructor may have previously introduced "phantom" restriction endonuclease recognition sites at the 5′ and/or 3′ end of the gene to make this gene sequence easier to work with. Your detailed protocol should include the following:

   a. The restriction endonucleases with which you will digest your DNA sequence.
   b. The restriction endonucleases with which you will digest your cloning vector.
   c. A complete map of the recombinant plasmid that would result if your ligation reaction were successful.
   d. The host strain into which you would transform your ligation mixture.

e. The method that you would use to select both for transformed cells and for those cells that may contain the desired recombinant plasmid.

f. A set of restriction endonuclease digestions and agarose-gel electrophoresis experiments that would allow you to verify the structure (sequence) of the recombinant plasmid after you have re-isolated it from the host cell. (What enzymes would you digest the plasmid with and what size DNA fragments would you expect to be produced in these digests?)

7. Using the PRIMERSELECT application of Lasergene, determine a set of primers that could be used to amplify your DNA sequence using PCR.

 a. Select **Initial conditions** from the **Conditions** menu to view the salt concentration and primer concentration at which the proposed PCR experiment will be performed. These are fairly standard conditions, so you will not need to change them for purposes of this exercise. Record these values in your notebook and close this window.

 b. Select **Primer characteristics** from the **Conditions** menu. This window will allow you to increase or decrease the stringency of your search for suitable primer pairs. We will experiment later with the primer length minimum, primer length maximum, the maximum dimer duplexing, and the maximum hairpin values, *so make a note of their locations in this window.*

 c. Open your DNA sequence by selecting **Enter sequence** from the **File** menu. Next, select **Primer locations** from the **Conditions** menu and select **upper and lower primer ranges** from the pull-down menu. Using the upper and lower ranges for your gene obtained with the MAPDRAW application of Lasergene (step 5E), enter the appropriate ranges *that will encompass the entire coding region of the gene.*

 d. Select **PCR primer pairs** from the **Locate** menu to obtain a set of primers that fit the criteria specified by the default values (step 7B). If your parameters were too stringent, you may not have received any entries. Experiment with the different parameters described in step 7B to obtain *5 to 10 primer pairs that are 17 to 24 bases in length.* The located primer pairs will be listed in order of decreasing "score."

 e. Double-click on the primer pair with the *highest score* to obtain a schematic view of the DNA product that would be produced in a PCR reaction performed with this set of primers.

 f. Select **Amplification summary** from the **Report** menu and print a copy of this report. What is the 5′ to 3′ sequence of your two primers? To which strands of the DNA template (top or bottom) will each primer anneal? What are the predicted melting temperatures (T_m) of these two primers? What is the predicted length of the DNA product that would result in a PCR reaction performed with this set of primers? What annealing temperature would you use in this PCR reaction?

 g. Select **Composition summary** from the **Report** menu and print a copy of this report. What is the molecular weight of your two primers? What is the conversion factor (nM/A_{260} and $\mu g/A_{260}$) for each primer? What is the %G, %C, %A, and %T composition of your two primers?

 h. Select **Quit** from the PRIMERSELECT **File** menu.

8. Using the PROTEAN application of Lasergene, predict some of the secondary structural features of your protein based on its amino acid (primary) sequence.

 a. Open your protein sequence file by selecting **Open** from the **File** menu.

 b. Select **Composition** from the **Analysis** menu and print a copy of this report. What is the molecular weight of your protein? How many amino acids does it contain? What is the molar extinction coefficient (at 280 nm) for your protein? What is the isoelectric point for your protein? What is the percent (by weight and by frequency) of the charged, acidic, basic, hydrophobic, and polar amino acids in your protein?

 c. Select **Titration curve** from the **Analysis** menu to obtain a theoretical titration curve for your protein and print out a copy of this report. What is the net charge on your protein at pH 5.0, 7.0, 8.0, and 9.0?

d. Return to the original PROTEAN window to display a list of secondary structure predictions for your protein. Print out a copy of this report. Using the Chou–Fasman and Garnier–Robson algorithms, locate the predicted alpha helices, beta sheets, and turn regions in your protein. What stretches of your protein are predicted to be quite hydrophobic? What residues in your protein are likely to be found on the surface? What peptide contained within your protein could you use to raise an antibody against your protein (the sequence of the peptide)?

9. Using the RasMol program, visualize your protein in three dimensions and compare its structure to the secondary structural predictions for your protein generated by the PROTEAN application of Lasergene.

   a. Open your sequence file and the RasMac program by selecting **Open** from the respective **File** Menus. Be sure that the viewing window and command line window are visible.
   b. You can rotate your protein by clicking and dragging within the viewing window. Experiment with the RasMol viewing tools until you have obtained a good view of every surface of your protein.
   c. Experiment with different options in both the **Display** and **Colours** menu to obtain several different views or models of your protein (ribbon diagram, space-filling model, atomic backbone, etc.). Which of these models do you find to be most useful and why? If your protein is an enzyme, can you locate some of the residues that may comprise the active site?
   d. Experiment with the different **Select** and **Label** commands to highlight some different residues in your protein.
   e. Using the **Cartoons** command in the **Display** menu and the **Structure** command in the **Colours** menu, "select" and "label" both the alpha helices and beta sheets in your protein. Compare these experimental results with the secondary structural predictions generated by the Chou–Fasma and Garnier–Robson algorithms in the PROTEAN application of Lasergene. Which of the two algorithms performed better in predicting the structural characteristics of your protein? Which algorithm did a better job at predicting the α-helical regions within your protein sequence? Which algorithm did a better job at predicting the β-sheet regions within your protein sequence? Is the same true for other proteins that members of your class have analyzed?

APPENDIX

Reagents and Materials Suggestions for Preparation and Teaching Staff

Appendix: Reagents and Materials Suggestions for Preparation and Teaching Staff

The following appendix lists details for the preparation of reagents and equipment for 100 students working in pairs (50 groups). It allows for student waste of reagents. Smaller classes will need a fraction of these quantities. Lesser quantities of reagents will often suffice if reagent usage is carefully controlled by an instructor. At times, specific commercial sources of reagents are listed. These choices are mostly for convenience; alternative sources of these reagents may be equally satisfactory.

Experiment 1: Photometry

Spectrophotometer with cuvettes—At least 1 required

Examination of an Absorption Spectrum

Adenosine solution (5×10^{-5} M)—Dissolve 2.67 mg of adenosine (free base) in 200 ml of water. Store at 4°C.

Riboflavin solution (5×10^{-5} M)—Dissolve 3.76 mg of riboflavin in 200 ml of water. Store at 4°C.

Folin–Ciocalteau Assay

Lysozyme solution (1 mg/ml)—Dissolve 200 mg of lysozyme in 200 ml of water. Store at 4°C.

200 ml of lysozyme at "unknown" concentrations (store at 4°C). Each group requires 2.5 ml of *one* of these "unknown" solutions:

- 2 mg/ml lysozyme—Dissolve 150 mg of lysozyme to 75 ml of water.
- 1.5 mg/ml lysozyme—Add 18.7 ml of 2 mg/ml lysozyme to 6.3 ml of water.
- 1 mg/ml lysozyme—Add 12.5 ml of 2 mg/ml lysozyme to 12.5 ml of water.
- 0.5 mg/ml lysozyme—Add 12.5 ml of 1 mg/ml lysozyme to 12.5 ml of water.
- 0.25 mg/ml lysozyme—Add 12.5 ml of 0.5 mg/ml lysozyme to 12.5 ml of water.

1% (wt/vol) $CuSO_4 \cdot 5H_2O$—Dissolve 2 g of $CuSO_4 \cdot 5H_2O$ in 200 ml of distilled water.

2% (wt/vol) Sodium tartrate—Dissolve 4 g of sodium tartrate $\cdot$ $2H_2O$ in 200 ml of distilled water.

2% (wt/vol) Na_2CO_3 in 0.1 N NaOH—Dissolve 20.0 g of NaOH and 100 g of Na_2CO_3 in 5 liters of distilled water.

Folin–Ciocalteau Phenol Reagent (2 N)—200 ml required. (Sigma catalog #F-9252).

Large (16 × 125 mm) glass test tubes—500 required.

Experiment 2: Chromatography

Chromatography columns—25 pieces; 5- to 8-mm inside diameter clean glass tubing approximately 15 in. long, or 50 equivalent chromatography columns.

CM-Sephadex (G-50) in 1.0 M potassium acetate, pH 6.0—Stir 150 g of fresh or used CM-Sephadex (G-50, medium grain size) into 2.5 liters of distilled water and vacuum filter (Büchner funnel) off the excess fluid. Rinse the accumulated filter cake on the filter with 2 volumes of 95% ethanol, followed by 2 volumes of distilled water. Remove and resuspend the filter cake in 4 volumes of 0.1 N HCl (415 ml concentrated HCl into 4.6 liters of distilled water), then vacuum filter and rinse with 2 volumes of distilled water as before. Remove and resuspend the filter cake in 2 volumes of 1.0 M potassium acetate, pH 6.0 (i.e., dissolve 490 g of potassium acetate in 4.75 liters of distilled water and titrate to pH 6.0 with 15 to 25 ml of glacial acetic acid), and vacuum filter again. Repeat this suspension and filtration step in 2 volumes of 1.0 M potassium acetate, pH 6.0, a second time. Finally, resuspend the filter cake in 2 volumes of 1.0 M potassium acetate, pH 6.0. At this point, you will have approximately 1 liter of the swollen resin.

CM-Sephadex (G-50) in 0.1 M potassium acetate, pH 6.0—Perform the ethanol, acid, and 1.0 M potassium acetate, pH 6.0 washings of CM-Sephadex (G-50) as described above. After the second wash, resuspend the filter cake in 4 volumes of 0.1 M potassium acetate, pH 6.0 (i.e., 1 part 1.0 M potassium acetate, pH 6.0, + 9 parts distilled water). Remove the excess fluid with vacuum filtration and resuspend the filter cake in 2 volumes of 0.1 M potassium acetate, pH 6.0. At this point, you will have approximately 1 liter of swollen resin.

1.0 M potassium acetate, pH 6.0—Dissolve 490 g potassium acetate in 4.75 liters of distilled water and titrate to pH 6.0 with 15 to 25 ml of glacial acetic acid. Dilute to 5.0 liters.

0.1 M potassium acetate, pH 6.0—Mix 0.5 liters of 1.0 M potassium acetate, pH 6.0 with 4.5 liters of distilled water.

Glass wool—10 in^3 required.

Chromatography sample—dissolve 25 mg of Blue Dextran, 50 mg of cytochrome *c*, and 25 mg of DNP-glycine together in 25 ml of distilled water.

Tape—1 roll required.

Pasteur pipettes with dropper bulbs—50 required.

Experiment 3: Radioisotope Techniques

^3H and ^{14}C unknowns—*In a fume hood,* mix sufficient quantities of ^3H-toluene, ^{14}C-toluene and unlabeled toluene to obtain 2 to 10 nCi of ^3H-toluene and 1 to 2 nCi of ^{14}C-toluene per 0.1 ml of toluene solution. You will need 10 ml of this solution for 50 groups to perform the experiment.

^{14}C-toluene standard—*In a fume hood,* mix sufficient ^{14}C-toluene and unlabeled toluene to yield a known quantity of ^{14}C-toluene (e.g., 1 to 2 nCi of ^{14}C-toluene) per 0.1 ml of toluene solution. Label the sample with the ^{14}C-toluene/ml with known disintegrations per minute (dpm). You will need 250 ml of this solution. Save 50 ml of this reagent for the quenched ^{14}C unknown. Dispense the other 200 ml to the students.

Volumetric dispensing bottle—Provide one bottle or apparatus capable of dispensing 5 or 10 ml of scintillation fluid, depending on the size of the scintillation vials to be used.

Scintillation fluid—Dissolve 25 g of 2,5-diphenyloxazole in 5 liters of reagent-grade toluene.

Chloroform—50 ml required.

Quenched ^{14}C unknown—In a hood, mix 37.5 to 47.5 ml of the ^{14}C-toluene standard with 2.5 to 12.5 ml of chloroform to obtain 50 ml.

Scintillation vials—400 required.

Scintillation counter—1 required.

Marking pen—preferably 1 per group.

Experiment 4: Electrophoresis

NOTE: The quantities of materials to be used and details of preparation of the PAGE gels for electrophoresis will depend on the number of students who share a single electrophoresis gel and the characteristics of the electrophoresis apparatus to be used. Appropriate adjustments must be made to fit the specific circumstances of your laboratory. The volumes given below are convenient volumes for preparation, but are generally well in excess of the amount required for 50 student pairs.

1.5 M Tris chloride, pH 8.8—Dissolve 91 g of Tris (tris-hydroxymethyl-aminomethane) free base in about 375 ml of distilled water. Titrate the solution to pH 8.8 with concentrated HCl and dilute to 500 ml with distilled water.

1.0 M Tris chloride, pH 6.8—Dissolve 30.5 g Tris (tris-hydroxymethyl-aminomethane) free base in about 175 ml of distilled water. Titrate the solution to pH 6.8 with concentrated HCl and dilute to 250 ml with distilled water.

30% acrylamide (292 g/liter acrylamide, 8 g/liter *N,N'*-methylenebisacrylamide)—*Caution: Acrylamide is toxic!* Dissolve 146 g of acrylamide and 4 g of *N,N'*-methylene-bis-acrylamide in 500 ml of distilled water. Filter through a 0.45-μm-pore-size membrane (to remove undissolved particulate matter) and store in a dark bottle labeled with the above warning at 4°C.

10% (wt/vol) SDS—Dissolve (*gently to avoid foaming*) 5 g of sodium dodecyl sulfate in 50 ml of distilled water.

TEMED (*N,N,N',N'*-tetraethylethylenediamine)—Use the commercial liquid neat. (*Caution: unpleasant odor.*) Approximately 500 μl will be required.

10% (wt/vol) ammonium persulfate in water—*This solution must be freshly prepared by the students.* Provide ammonium persulfate as a dry powder. Students should prepare the solution needed by dissolving 20 mg in 200 μl of distilled water immediately before use.

SDS–PAGE electrophoresis buffer—Dissolve 30 g of Tris free base, 188 g of glycine, and 10 g of SDS *(gently to avoid foaming)* in 9 liters of distilled water. Adjust the pH to 8.3 with concentrated HCl, and dilute to 10 liters with distilled water.

4X SDS–PAGE sample buffer—Mix together 17.5 ml of distilled water, 12.5 ml of 1.0 M Tris · HCl (pH 7.0), 15 ml of glycerol, and 5 ml of 2-mercaptoethanol. *(Caution: unpleasant odor.)* Dissolve 4 g of SDS and 2.0 mg of bromophenol blue in this mixture *carefully, to avoid foaming.* The stock 1.0 M Tris chloride, pH 7.0, is prepared by dissolving 60.5 g of Tris free base in about 400 ml of distilled water, titrating to pH 7.0 with concentrated HCl, and diluting to 500 ml with distilled water.

Coomassie Blue staining solution—Dissolve 1.25 g of Coomassie Brilliant Blue R-250 in 500 ml of destaining solution (see next entry).

Destaining solution—Mix 2 liters of methanol and 2.5 liters of distilled water, then carefully add 500 ml of glacial acetic acid.

Power supplies and electrophoresis apparatus—8 required.

Protein molecular weight standards—Suggested standards are phosphorylase (97,400 Da), bovine serum albumin (66,200 Da), ovalbumin (45,000 Da), carbonic anhydrase (31,000 Da), soybean trypsin inhibitor (21,500 Da) and lysozyme (14,400 Da) at 1 mg/ml each in a single mixture in 0.01 M Tris chloride buffer, pH 7.0 (prepared by dilution of the stock 1 M buffer above). Approximately 2 ml of this solution will be required.

Unknown protein solutions (approximately 1 mg/ml in 0.01 M Tris chloride buffer, pH 7.0)—It is convenient to prepare the unknowns in larger volumes (100 to 500 μl) and to dispense them to students as 20-μl samples in microcentrifuge tubes. Any of the proteins used as standards or other readily available proteins with molecular weights within the range of the standards may be used.

Experiment 5: Acid–Base Properties of Amino Acids

Amino acid unknowns—5 g each; any of the common amino acids can be used (aspartic acid, leucine, isoleucine, arginine, and cystine are less suitable for titrations than the others); also suitable are salts of weak acids (e.g., sodium formate, sodium acetate) and bases (e.g., tris-hydroxymethylaminomethane). Take care to record the precise salt form (sodium salt, hydrochloride) and hydration of the salt used. Each group will require 800 mg of their unknown.

pH meters—50 required.

2 N HCl—Slowly add 415 ml of concentrated HCl to 2.085 liters of distilled water. Keep tightly stoppered. The precise concentration should be determined by titration with commercial standard base.

2 N NaOH—Dissolve 200 g of NaOH in 500 ml of distilled water and dilute to 2.5 liters. Protect from CO_2 with a soda-lime tube. The precise concentration should be determined by titration with standard acid.

Buffers for standardizing pH meters—Use commercial preparations; 4.0, 7.0, and 10.0 are suitable. We perform a 2-point calibration for the acid titration with pH 4.0 and 7.0. For the base titration, we recalibrate the pH meter with solutions at 7.0 and 10.0. Approximately 500 ml of each standard will be required.

10-ml Burettes and holders—50 required.

Magnetic stirring motors with stirring bars—50 required.

NOTE: Instruction in the operation and principles of operation of pH meters may be necessary. Provide burette brushes and solvent ($CHCl_3$) for cleaning burettes and removing stopcock grease, unless burettes with Teflon stopcocks are used.

Experiment 6: Sequence Determination of a Dipeptide

1-fluoro-2,4-dinitrobenzene (FDNB) solution—Add 1.25 ml of FDNB to 23.75 ml of absolute ethanol.

Ninhydrin spray—Dissolve 5 g of ninhydrin in 2.5 liters of *n*-butanol. Add 2.5 ml of 2,4-collidine *immediately before use.*

Dipeptides (gly-leu, ala-phe, leu-gly, ala-val)—Prepare 25 glass test tubes (13 × 100 mm) and 25 microcentrifuge tubes each containing 2 mg of one of the dipeptides. Number each of the tubes in matching sets. Keep a record of the students' names and their unknown numbers. Any dipeptide can be used, provided that cysteine, tryptophan, arginine, glutamine, and asparagine are not present.

DNP-amino acid standards (for thin-layer chromatography)—Dissolve 1 mg of each standard in 1.5 ml of acetone. DNP-gly, DNP-ala, DNP-leu, DNP-ser, and DNP-phe are useful for the four dipeptides listed above. If the dipeptides that you choose contain other N-terminal amino acids, DNP-amino acid standards of them must be prepared.

Amino acid standards (for paper chromatography)—Dissolve 2 mg of each amino acid in 2 ml of distilled water. Gly, leu, ala, phe, val, gln, lys, and ser are useful. Store at 4°C.

Peroxide-free diethyl ether—Place two 500-ml canisters of this reagent in the fume hood.

Whatman 3 MM chromatography paper—Each group requires an 8.5-in.-by-8.5-in. piece for paper chromatography.

4-ml glass (hydrolysis) vials with Teflon-lined screw caps—100 required.

6 N HCl—Add 50 ml of concentrated HCl to 50 ml of distilled water.

4.2% $NaHCO_3$—Dissolve 10.5 g of $NaHCO_3$ in 250 ml of water.

Silica-gel thin-layer chromatography plates (Fisher catalog #05-713-317)—Heat at 100°C for 10 min just before use. (Store dessicated at room temperature.) Twenty-five will be required; we try to fit two groups on one plate.

pH paper (1–11)—One roll for every 2 to 4 groups is required.

Heating block—Each group will require two compartments in the heating block that will hold the 4-ml glass vials.

Pasteur pipettes—500 to 1000 required.

Large (16 × 125 mm) and small (13 × 100 mm) glass test tubes—Approximately 500 of each required.

Vacuum centrifuge—1 required.

Mobile phase for thin-layer chromatography—Mix 1.5 liters of chloroform, 600 ml of t-amyl alcohol, and 60 ml of glacial acetic acid. Store tightly covered in a dark bottle at room temperature.

Mobile phase for paper chromatography—Mix 1.8 liters of *n*-butanol, 450 ml of glacial acetic acid, and 750 ml of water. Store tightly covered in a dark bottle at room temperature.

Wash reagent for PTC-coupling of amino acids—Mix 6 ml of absolute ethanol, 6 ml of triethylamine, and 3 ml of water. Store tightly covered at 4°C.

PTC coupling reagent—Mix 21 ml of absolute ethanol, 3 ml of triethylamine, 3 ml of phenylisothiocyanate (Sigma catalog #P-1034), and 3 ml of water. Store tightly covered at 4°C.

HPLC unit—1 required.

UV/Vis Spectrophotometer and chart recorder for HPLC—1 of each required.

Alltech Adsorbosphere HS C18 reverse-phase HPLC column (250 × 4.6 mm)—1 required (Alltech catalog #28935).

PicoTag Eluant A (Waters catalog #88108)—2 liters required.

140 mM potassium acetate, pH 5.3 (optional)—Dissolve 27.48 g of potassium acetate in 1.9 liters of distilled water. Adjust to pH 5.3 with 2 M glacial acetic acid. Bring the final volume of the solution to 2 liters with distilled water. Filter sterilize and de-gas the solution before use in HPLC.

60% (vol/vol) Acetonitrile—Mix 1.2 liters of HPLC-grade acetonitrile with 800 ml of HPLC-grade water.

Amino Acid Standard H (Sigma catalog #AA-S-18)—Approximately 1 ml required.

Pressurized N_2 tank—1 required.

Chromatography chambers for thin layer chromatography—for 25 plates.

Chromatography jars for paper chromatography—50 required.

Experiment 7: Study of the Properties of β-Galactosidase

0.08 M sodium phosphate buffer, pH 7.7—Dissolve 191 g of Na_2HPO_4 (anhydrous) and 18 g of $NaH_2PO_4 \cdot H_2O$ in 14 liters of distilled water. Bring the pH to 7.7 with 10 M HCl or 10 M NaOH. Bring the final volume to 15 liters with distilled water. Store at room temperature.

β-galactosidase stock solution—Dissolve 5000 units of lyophilized β-galactosidase (Sigma catalog #G-6008) in 10 ml of 0.08 M sodium phosphate buffer, pH 7.7. Make dilutions of this stock solution in the same buffer containing 1 mg/ml bovine serum albumin (1:10 to 1:500). Using these dilute solutions, determine what dilution of the stock enzyme solution will give a ΔA_{420}/min of about 0.10, using the same protocol as described in Day 1 of the protocol. For the class, prepare a stock β-galactosidase solution that is *five* times as concentrated as the dilution that gave a ΔA_{420}/min of about 0.10. Prepare 250 ml of the appropriate dilution and divide this into 1.0-ml aliquots. Each group will require about 4 ml of this enzyme solution to perform the experiment: 1 ml on Day 1, 2 ml on Day 2, and 1 ml on Day 3.

0.08 M sodium phosphate buffers, pH 6.4, 6.8, 7.2, and 8.0—Dissolve 11.04 g of $NaH_2PO_4 \cdot H_2O$ in 900 ml of distilled water. Bring the solutions to the appropriate pH with 10 M NaOH. Bring the final volume of each solution to 1.0 liter with distilled water. Store at room temperature.

2.5 mM *o*-nitrophenyl-β,D-galactopyranoside (ONPG)—Dissolve 1.507 g of ONPG in 2.0 liters of distilled water. Store at 4°C in small aliquots.

10 mM *o*-nitrophenyl-β,D-galactopyranoside (ONPG)—Dissolve 0.6 g of ONPG in 200 ml of distilled water. Store at 4°C.

750 mM methyl-β,D-galactopyranoside (MGP)—Dissolve 5.826 g of MGP in 40 ml of distilled water. Store at 4°C.

150 mM methyl-β,D-thiogalactoside (MTG)—Dissolve 1.261 g of MTG in 40 ml of distilled water. Store at 4°C.

200 mM isopropylthio-β,D-galactopyranoside (IPTG)—Dissolve 1.430 g of IPTG in 30 ml of distilled water. *Filter sterilize* the solution and store at 4°C.

100 mM adenosine 3′,5′ cyclic monophosphate (cAMP)—Dissolve 527 mg of cAMP (sodium salt) in 15 ml of distilled water. *Filter sterilize* the solution and store at 4°C.

20% (wt/vol) glucose—Dissolve 10 g of D-glucose in 50 ml of distilled water. *Filter sterilize* the solution and store at 4°C.

1.0 M Na_2CO_3—Dissolve 132.5 g of Na_2CO_3 in 1 liter of distilled water. Bring the final volume to 1.25 liters with distilled water, and store at room temperature.

0.1% sodium dodecyl sulfate (SDS)—Dissolve 50 mg of SDS in 50 ml of distilled water. Store at room temperature.

YT broth—Dissolve 50 g of bactotryptone, 25 g of yeast extract, and 25 g of NaCl in 4 liters of distilled water. Bring the final volume to 5 liters with distilled water, and autoclave in 1-liter aliquots. Store at 4°C; warm to room temperature before use.

Chloroform—About 50 ml required.

Small (13 × 100 mm) glass test tubes—3000 required.

Large glass test tubes (16 × 125 mm, sterile) with caps—1200 required.

Colorimeter tubes or cuvettes for spectrophotometer—50 required.

Heated water baths—5 required.

1.5-ml Microcentrifuge tubes—Approximately 500 required.

Spectrophotometer—1 required.

E. coli strains EM1, EM1327, EM1328, EM1329—The equivalent strains can be obtained from the E. coli Genetic Stock Center, Department of Biology, Yale University, New Haven, CN (website address: http://cgsc.biology.yale.edu). These are EM1 = CA8000, EM1327 = CA8445-1 *(crp⁻)*, EM1328 = CA8306 *(cya⁻)*, and EM1329 = 3.300 *(lacI⁻)*.

Experiment 8: Purification of Glutamate–Oxaloacetate Transaminase from Pig Heart

0.05 M potassium maleate buffer (pH 6.0)—Dissolve 29 g of maleic acid and 9.3 g of disodium EDTA · 2 H_2O in 4 liters of distilled water. Adjust pH to 6.0 with KOH. Bring the total volume of the solution to 5 liters with distilled water. Store at 4°C. If the buffer is to be stored for more than a week, one drop of toluene or chloroform may be added to prevent bacterial growth.

0.04 M α-ketoglutarate—*Prepare this reagent just before use and store at 4°C.* Dissolve 17.5 g of α-ketoglutaric acid in 2.6 liters of distilled water. Adjust the pH to 6.0 with NaOH. Bring the total volume of the solution to 3.0 liters with distilled water.

0.03 M sodium acetate buffer—Dilute 17.2 ml of glacial acetic acid to 9 liters with distilled water. Adjust the pH to 5.4 with NaOH. Bring the total volume of the solution to 10 liters with distilled water. Store at room temperature. If the buffer is to be stored for more than a week, one drop of toluene or chloroform may be added to prevent bacterial growth.

Saturated ammonium sulfate solution—Dissolve 7 kg of ammonium sulfate (reagent grade) in 10 liters of distilled water. Store at room temperature.

1% $CuSO_4 \cdot 5H_2O$—Dissolve 5 g of $CuSO_4 \cdot 5H_2O$ in 500 ml of water. Store at room temperature.

2% sodium tartrate solution—Dissolve 10 g of sodium tartrate · $2H_2O$ in 500 ml of distilled water. Store at room temperature.

2% Na_2CO_3 in 0.1 N NaOH—Dissolve 20 g of NaOH and 100 g of Na_2CO_3 in 5 liters of distilled water. Store at room temperature.

Folin–Ciocalteau Phenol Reagent (2 N)—Approximately 1 liter will be required. (Sigma catalog #F-9252).

1 mg/ml bovine serum albumin (or lysozyme) standard—Dissolve 600 mg of bovine serum albumin (fraction V) or lysozyme in 600 ml of distilled water. Store at 4°C.

0.10 M potassium phosphate buffer (pH 7.5)—Dissolve 14.1 g of anhydrous K_2HPO_4 and 2.18 g of anhydrous KH_2PO_4 in 800 ml of distilled water. Adjust to pH 7.5. Bring the total volume of the solution to 1.0 liter with distilled water. Store at 4°C. If the buffer is to be stored for more than a week, one drop of toluene or chloroform may be added to prevent bacterial growth.

0.10 M potassium phosphate/0.12 M aspartate buffer—Dissolve 261 g of anhydrous K_2HPO_4 and 240 g of L-aspartic acid in 12 liters of distilled water. Adjust to pH 7.5 with dilute KOH. Bring the total volume of the solution to 15 liters with distilled water. Store at 4°C. If the buffer is to be stored for more than a week, one drop of toluene or chloroform may be added to prevent bacterial growth.

0.10 M α-ketoglutarate (pH 7.5)—*Prepare this reagent just before use and store at 4°C.* Dissolve 8.76 g of α-ketoglutaric acid in 480 ml of distilled water. Adjust the pH to 7.5 with NaOH. Bring the total volume of the solution to 600 ml.

250-ml plastic centrifuge bottles—50 required.

50-ml plastic centrifuge tubes—50 required.

Bunsen burners—50 required.

Pasteur pipettes—200 required.

Spectrophotometer with cuvettes—At least 1 required.

Small (13 × 100 mm) glass test tubes—500 required.

Large (16 × 125 mm) glass test tubes—500 required.

Cheesecloth—1 large roll required.

Glass wool

Chromatography columns (20 ml capacity) with tubing—50 required.

Meat grinder—1 required.

Blender—1 required.

3 mM NADH—*Prepare this reagent just before use and store at 4°C.* Dissolve 68.7 mg of $Na_2 \cdot NADH \cdot 3H_2O$ (if other salt and hydration forms are supplied, adjust weight accordingly) in 30 ml of 0.10 M potassium phosphate buffer, pH 7.5. *Because this reagent is expensive and unstable, only the quantity needed for each day should be prepared at the start of each laboratory period. This reagent is stable for 2 to 4 hr.*

Malic dehydrogenase—*Prepare this reagent just before use and store at 4°C.* Dilute commercially available enzyme in 0.10 M potassium phosphate buffer, pH 7.5, in which 1 mg/ml bovine serum albumin has been dissolved. The final concentration of malic dehydrogenase should be 200 units/ml. Prepare a total of 20 ml of this reagent.

0.03 M sodium acetate buffer, pH 5.0—Dilute 34.4 ml of glacial acetic acid to 19 liters with distilled water. Adjust the pH to 5.0 with NaOH. *Do not overtitrate or back titrate with acid. It is essential that the buffer does not contain excess salt.* Bring the total volume of the solution to 20 liters with distilled water. Store at 4°C. If the buffer is to be stored for more than a week, one drop of toluene or chloroform may be added to prevent bacterial growth.

0.08 M sodium acetate buffer, pH 5.0—Dilute 4.6 ml of glacial acetic acid to 900 ml with distilled water. Adjust the pH to 5.0 with NaOH. Bring the total volume of the solution to 1 liter with distilled water. Store at 4°C. If the buffer is to be stored for more than a week, one drop of toluene or chloroform may be added to prevent bacterial growth.

Carboxymethylcellulose slurry (equivalent of 75 g of dry material)—Slowly wet 75 g of dry resin in distilled water. Draw off the supernatant (after the resin has settled) with vacuum filtration. Resuspend the filtered resin cake in 1.5 L of 0.5 M NaOH (30 g of NaOH dissolved in 1.5 L of distilled water). Allow the resin to settle, draw off the supernatant, and wash the resin cake twice, as before, in 2.5 L volumes of distilled water. Resuspend the resin cake in 2.5 L of 0.5 M HCl (100 ml of concentrated HCl in 2.4 L of distilled water). Allow the resin to settle, draw off the supernatant, and wash the resin cake twice, as before, in 2.5 L volumes of distilled water. Repeat the wash procedure described above with 1.5 L volumes of 0.03 M sodium acetate buffer, pH 5.0, until the pH and ionic strength of the drawn-off supernatant is the same as that of the sodium acetate buffer (pH and conductivity meter). Resuspend the resin cake in 1.5 L 0.03 M sodium acetate buffer, pH 5.0. Add 2 g/liter sodium azide for storage to prevent bacterial growth. Remove the azide by washing again in 0.03 M sodium acetate buffer, pH 5.0 when it is ready to be used by the class.

Fresh pig hearts—10 required.

Dialysis tubing (8000–12000 molecular weight cutoff)—Heat tubing at 100°C for 15 to 20 min in a solution containing 1 g of $NaHCO_3$ and 0.1 g of disodium EDTA per 100 ml. Rinse thoroughly in distilled water after the tubing has cooled and store wet and covered at 4°C. Pierce Chemical Co. offers SnakeSkin dialysis tubing in molecular weight cutoffs from 3500 to 10,000 Da. It is supplied in a dry roll ready to use (no boiling or hydration required as with conventional dialysis tubing). It is very convenient for use with large laboratory groups.

Supplies and reagents for SDS–PAGE—See Experiment 4. We normally use a 12% polyacrylamide gel to analyze the purity of GOT in this experiment.

Experiment 9: Kinetic and Regulatory Properties of Aspartate Transcarbamylase

Escherichia coli strain EK1104 containing plasmid pEK2. Obtain strain No. 700695 from American Type Culture Collection, 10801 University Blvd., Manassas, VA 20110.

Materials for Growth Medium and Culturing Bacteria

The growth and harvesting of the bacterial cells for this experiment is usually performed by a teaching assistant or a technician in advance of the experiment. The cells and cell extracts can be stored frozen at −20 or −70°C before the class will use them. The amount of growth medium required will depend on the amount of cells required. Generally, 1 liter of bacterial culture will provide sufficient enzyme for a class of 50 students working in pairs.

Prepare the growth medium as follows (the amounts given are for *1 liter* of growth medium): Dissolve the following in 870 ml of distilled water: 6 g of Na_2HPO_4, 3 g of KH_2PO_4, 1 g of NH_4Cl, 5 g of Casamino acids, and 0.5 g of NaCl. Autoclave in a capped 2- or 3-liter culture flask. Separately autoclave (in separate flasks or bottles): 4 g of glucose in 50 ml of distilled water and 12 mg of uracil in 25 ml of distilled water. Prepare (by *filter sterilization*) solutions of 40 mg of L-tryptophan in 50 ml of distilled water and 50 mg of ampicillin in 2.0 ml of distilled water. *When the sterilized solutions have cooled, add the small-volume components to the large flask (using sterile technique), holding back 0.5 ml of the uracil solution to add to the inoculation medium (see next paragraph).* Separately, prepare autoclaved stock solutions of the following components: 1 M $MgSO_4 \cdot 7H_2O$ (24.65 g/100 ml H_2O); 0.1 M $CaCl_2 \cdot 2H_2O$ (1.47 g/100 ml H_2O), 0.5% thiamine (50 mg/10 ml), 1 mM $FeSO_4 \cdot 7H_2O$ (27.8 mg/100 ml), and 1 mM $ZnSO_4 \cdot 7H_2O$ (28.7 mg/100 ml). Using sterile technique, add 1 ml of each of these solutions to the large flask.

Prepare the inoculation medium as follows: Using sterile technique, withdraw 10 ml from the growth medium described above and place it in a sterile growth tube. Add 0.5 ml of sterile uracil from the previous paragraph to bring the inoculation medium to 30 μg/ml. *Add 150 μg ampicillin per ml of the medium by adding 30 μl of a 50-mg/ml solution of ampicillin in H_2O, which has been filter-sterilized.* Grow and harvest the bacteria as described in the protocol for Experiment 9.

Shaking incubator at 37°C.

0.1 M Tris chloride buffer, pH 9.2—Dissolve 6.05 g Tris free base in distilled water, titrate to pH 9.2 with concentrated HCl and dilute to 500 ml with distilled water. Store at 1 to 4°C.

Sonicator—1 required.

30 and 60°C Water baths—One of each required.

0.08 M Sodium phosphate buffer, pH 7.0—Dissolve 21.25 g of $NaH_2PO_4 \cdot H_2O$ and 35.0 g of Na_2HPO_4 (anhydrous) in distilled water to a final volume of 5 liters. The pH should be 7.0.

0.125 M Sodium L-aspartate—Dissolve 8.3 g of L-aspartic acid in 350 ml of distilled water, bring the pH to 7.0 with NaOH (2 N or more concentrated; addition of NaOH may be necessary to dissolve all of the aspartic acid), and dilute to 500 ml with distilled water.

0.036 M Carbamyl phosphate—Dissolve 690 mg of dilithium carbamyl phosphate in sufficient distilled water (use weight appropriate to hydration state indicated on label) to yield 125 ml of solution. If the solution is cloudy, warm it briefly to clarify. *Prepare just before the experiment and store at 1 to 3°C.*

1.0 mM Carbamyl aspartate standard—Dissolve 17.5 mg of carbamyl aspartate in distilled water, titrating with dilute NaOH to bring the solid into solution (do not go above pH 7.0). Dilute to 100 ml with distilled water. *Prepare just before experiment and store at 1 to 3°C.*

Antipyrine-H_2SO_4 reagent—*Carefully* add 1.25 liters of concentrated H_2SO_4 to 1.25 liters of distilled water and allow to cool. Dissolve 12.5 g of antipyrine in the H_2SO_4 solution. Label with warning: *strong acid.*

Diacetylmonoxime reagent—Prepare in advance 1.5 liters of 5% acetic acid (75 ml glacial acetic acid diluted to 1.5 liters with distilled water). *Just before the experiment,* dissolve 12 g of diacetylmonoxime (butanedione monoxime) in the 1.5 liters of 5% acetic acid. Keep refrigerated in a foil-covered bottle.

2% Perchloric acid—Add 30 ml 70% $HClO_4$ to 1 liter of distilled water.

40 mM ATP—Dissolve 585 mg of $Na_2ATP \cdot 2H_2O$ (use weight appropriate to salt and hydration form indicated on the label) in 20 ml of distilled water, titrate to pH 6.0 to 8.0 with 1 N NaOH, and dilute to 25 ml with distilled water. *Store frozen; thaw just before use and keep at 1 to 3°C.*

10.0 mM CTP—Dissolve 138 mg of $Na_2CTP \cdot H_2O$ (use weight appropriate to salt and hydration form indicated on the label) in 20 ml of distilled water, titrate to pH 6.0 to 8.0 with 1 N NaOH, and dilute to 25.0 ml with distilled water. *Store frozen; thaw just before use and keep at 1 to 3°C.*

5 mM *p*-Mercuribenzoate—Dissolve 95 mg of sodium *p*-chloromercuribenzoate in 50 ml of 5 mM KOH. *Prepare on the day of use. Store in a foil-wrapped con-*

tainer to protect from light. The 5 mM KOH is prepared by dissolving 14 mg of KOH in a total of 50 ml of distilled water.

Experiment 10: Affinity Purification of Glutathione-S-Transferase

pGEX-2T/*E. coli* culture stock—We purchase pGEX-2T from Pharmacia (catalog #27-4801-01). This 25-μg package size of DNA can be resuspended in 25 μl of TE buffer (see Experiment 24). To prepare a culture stock, transform *E. coli* strain TG-1 (Stratagene catalog #200123) with 10 μl of a 1 ng/ml solution of the plasmid prepared in sterile distilled water. If you prefer, *E. coli* strain DH5α (Gibco-BRL catalog #18265-017) or BL21 (Pharmacia catalog #27-1542-01) would perform well. After obtaining colonies on an agar plate supplemented with 100 μg/ml ampicillin, grow a 2-ml culture of this stock (from a single colony) in YT broth containing 100 μg/ml ampicillin at 37°C for several hours. When the 2-ml solution achieves an OD_{600} of about 0.3, add 140 μl of dimethylsulfoxide (DMSO). Allow the culture to incubate at 37°C for about 30 min with gentle agitation, and freeze the culture in a sterile vial at −70°C. Each semester, a sterile inoculation loop can be used to streak out the culture on a fresh agar YT-agar plate containing 100 μg/ml ampicillin. *NOTE:* If you are unfamiliar with bacterial transformation and preparation of competent *E. coli cells*, refer to Experiment 21 and Table V-6.

YT broth—Dissolve 20 g of bactotryptone, 10 g of yeast extract, and 10 g of NaCl in 1 liter of distilled water. Bring the final volume to 2 liters with distilled water, and autoclave in a single, 4-liter culture flask.

Sonicator—1 required.

PBS buffer—Dissolve 13.15 g of NaCl, 3.4 g of Na_2HPO_4, and 0.72 g of NaH_2PO_4 in 1.2 liters of distilled water. Adjust pH to 7.3 with HCl or NaOH if required. Bring the total volume of the solution to 1.5 liters with distilled water.

Glutathione Sepharose 4B—We obtain this from Pharmacia (catalog #17-0756-01). Prepare a 50% slurry of this resin in PBS buffer according to the manufacturer's instructions. *NOTE:* The resin can be regenerated and used in following semesters by washing the resin in 3 M NaCl. The Sepharose 4B matrix is chemically stable in the pH range of 4 to 9.

10 mM reduced glutathione solution—Dissolve 0.3 g of Tris free base in 40 ml of distilled water. Adjust the pH to 8.0 with HCl. Add 0.15 g of reduced glutathione to the solution, adjust the pH to 8.0 again (if necessary), and bring the final volume of the solution to 50 ml with distilled water. Prepare this reagent just before use.

IPTG solution for strain induction—Dissolve 0.24 g of isopropylthio-β,D-galactopyranoside in 2 ml of distilled water. Filter-sterilize this solution. Add this entire solution to a growing 2-liter culture of *E. coli* strain TG1 harboring plasmid pGEX-2T to induce expression of glutathione-*S*-transferase.

Preparative centrifuge with a rotor for 250-ml centrifuge bottles—You will need one centrifuge and eight 250-ml polypropylene centrifuge bottles rated for up to 17,000 to 20,000 × *g*.

Reagents and supplies to perform SDS–PAGE—See Experiment 4. We suggest using a 12 or 15% polyacrylamide gel for this experiment.

Purified GST standard (prepared by instructor)—Prepare a 1-liter culture of the GST expression strain, harvest the cells, and purify the protein according to the protocol. Use 1 ml of the immobilized glutathione Sepharose 4B per 1 liter of culture. Adjust the concentration of the purified protein to 1 mg/ml, and store at −20°C in 1-ml aliquots. One preparation of GST will give enough protein to act as a standard for many semesters.

Experiment 11: Microanalysis of Carbohydrate Mixtures by Isotopic, Enzymatic, and Colorimetric Methods

Unknown Carbohydrate Solutions

Prepare separate 30 mg/ml stock solutions of D-glucose, D-ribose, and D-glucosamine by dissolving 300 mg of each carbohydrate in 10 ml of distilled water. Prepare "unknown" solutions containing between 5 and 10 mg/ml of each carbohydrate by mixing different proportions of these stock solutions with distilled water. Each group will require about 1 ml of the "unknown" carbohydrate solution.

Reduction of Carbohydrates

[^{14}C]glucose—Dilute 2 μl of a 200- to 300-mCi/mmol solution of [^{14}C]glucose in 1 ml of distilled water. Each group requires 5 μl of this solution. Dilute commercial ^{14}C-glucose (sp act ≥ 200 mCi/mmol) to yield a solution with about 2,000 cpm/μl in the $^{14}C/^{3}H$ window. Store cold or frozen, tightly stoppered. From this solution, *accurately* withdraw a 5 or 10 μl sample and transfer quantitatively to a 1 × 1/2 inch segment of Whatman No. 1 paper, and determine, using the same counting conditions as the class, (1) the cpm in the $^{14}C/^{3}H$ channel per μl glucose solution, and (2) the cpm in a full ^{3}H channel per μl glucose solution. (1) *plus* (2) should equal the total ^{14}C cpm (which you can also determine by counting the full ^{14}C channel). From these data, calculate and post for class use in analyzing data:

$$\frac{\text{Fraction of total } ^{14}\text{C}}{\text{in } ^{14}\text{C}/^{3}\text{H channel}} = \frac{(1)}{(1) + (2)},$$

and

$$\frac{\text{Fraction of total } ^{14}\text{C}}{\text{in } ^{3}\text{H channel}} = \frac{(2)}{(1) + (2)}$$

NaB_3H_4—Dissolve 5 mg of 15- to 150-mCi/mmol NaB_3H_4 (15–25 mCi/mmol is most convenient and economical; New England Nuclear Corp. or Amersham/Searle can supply ^{3}H-$NaBH_4$) in 250 μl 0.1 N NaOH. Store frozen; stable about 1 month. Determine the specific activity accurately as follows: Mix 5 μl of a glucose solution of known concentration (~20 μmoles/ml is convenient) with 5 μl ^{14}C-glucose (same as used by class) and 5 μl ^{3}H-$NaBH_4$ solution just prepared. Proceed as described in Experiment 14. From the ^{3}H-cpm in the ^{3}H-glucitol on the chromatogram, determine the specific activity of the ^{3}H-$NaBH_4$ (the value to be given to the class), expressing it as ^{3}H-cpm per mole of monosaccharide reduced. This value will depend on the efficiency of counting, so it should be determined under conditions identical to those used by the class. Each group requires 5 μl of this solution.

0.75 N H_2SO_4—Add 21 μl of concentrated HCl to 979 ml of distilled water. Each group requires 5 μl of this solution.

Mobile-phase solvent for paper chromatography—Mix together 45.4 ml of 88% formic acid, 154.6 ml of distilled water, 120 ml of glacial acetic acid, and 720 ml of ethyl acetate. Store tightly stoppered in a dark glass bottle at room temperature.

Scintillation fluid—Dissolve 30 g of diphenyloxazole in 6 liters of toluene. Store tightly stoppered in a dark glass bottle at room temperature.

Scintillation vials—2000 required.

Scintillation counter—1 required.

Heat blocks—5 to 10 required.

Descending chromatography tank—1 to 5 required, depending on its size.

0.5-ml microcentrifuge tubes—50 required.

1.5-ml microcentrifuge tubes—200 required.

Fume hood—1 required (it is very important for the students' safety that this is provided).

Whatman 3 MM chromatography paper—50 strips (1 × 22 in. each).

4-ml Glass (hydrolysis) vials with Teflon-lined screw caps—400 required.

Enzymatic Determination of Glucose

Glucose hexokinase reagent—Dissolve the contents of two Glucose (HK) reagent vials (Sigma catalog #16-50) each in 50 ml of distilled water. Store at 4°C. Each group requires 2 ml of this reagent.

Elson–Morgan Determination of Glucosamine

1 M Na_2CO_3—Dissolve 21.2 g of Na_2CO_3 in a total volume of 200 ml of distilled water.

Glucosamine standard solution (0.5 μmol/ml)—Add 98.8 μl of the 30-mg/ml D-glucosamine stock solution to 27.4 ml of distilled water.

Acetylacetone reagent—Add 6 ml of acetylacetone to 294 ml of 1 M Na_2CO_3. *This reagent must be prepared on the day that the experiment is performed.*

95% ethanol—Add 50 ml of distilled water to 950 ml of absolute ethanol. Store tightly capped at 4°C.

p-Dimethylaminobenzaldehyde reagent—Dissolve 7.83 g of *p*-dimethylaminobenzaldehyde in 300 ml of a 1:1 mixture of 95% ethanol and concentrated HCl. Store tightly covered at 4°C. *Prepare on the day of the experiment.*

Spectrophotometer with cuvettes—At least 1 required.

Experiment 12: Glucose-1-Phosphate: Enzymatic Formation from Starch and Chemical Characterization

NOTE: The measurements in Experiment 12 will be seriously interfered with if residues of phosphate-containing laboratory detergents are present in the glassware. Students should be made aware of the potential problem, and a supply of phosphate-free detergent, such as sodium dodecyl sulfate, should be made available for washing glassware during this experiment.

Phenylmercuric nitrate—2.5 g required.

0.8 M Potassium phosphate buffer, pH 6.7—Dissolve 950 g of $K_2HPO_4 \cdot 3\ H_2O$ and 545 g of KH_2PO_4 in approximately 9 liters of distilled water and adjust to pH 6.7 with HCl or KOH, as necessary, before making up to 10 liters.

Soluble starch—500 g required.

Filter aid (Celite 535 or similar material)—125 g required.

14% NH_4OH—Mix 500 ml of concentrated NH_4OH with 500 ml of distilled water.

2.0 N HCl—Mix 165 ml of concentrated HCl with 850 ml of distilled water. (It is important that both this reagent and the 2.0 N NaOH below be exactly 2.0 N, or that equal volumes of each reagent yield a neutral [pH 6 to 8] solution. A teaching assistant should check this by titration.)

2.0 N NaOH—Dissolve 80 g of NaOH in 1 liter of distilled water. (See note under 2.0 N HCl above.)

1.0 N HCl—Mix 41.5 ml of concentrated HCl with 460 ml of distilled water. (It is important that equal volumes of this reagent and the 1.0 N NaOH described in the next entry neutralize each other [pH 6 to 8]. This should be checked by titration.)

1.0 N NaOH—Dissolve 20 g of NaOH in a total volume of 500 ml of distilled water.

Potatoes—Stored 24 hr at 1 to 5°C and tested for phosphorylase activity. Do not freeze. Approximately 30 potatoes will be required.

Blenders—1 or 2 required.

Cheesecloth—1 or 2 rolls required.

$Mg(Acetate)_2 \cdot 4\ H_2O$—1.5 kg required

Dowex 50 (H^+ form)—Mix 1 kg of Dowex 50, X-8, 20 to 50 mesh, with 20 liters of dilute HCl (3 liters of concentrated HCl + 17 liters of distilled water), stir for 5 min, fill with additional H_2O, and decant. Wash by decantation or suspension and vacuum filtration until the eluate no longer yields any titratable acidity (pH 3 to 4). Drain the resin until a moist paste is obtained. Each pair of students will need about 350 ml of moist Dowex 50. The resin should be collected at the end of the experiment *(Do not allow accidental mixing with IR-45)* and promptly regenerated by the above procedure. Store tightly covered under a water layer. (See note following IR-45 preparation.)

IR-45 (OH^- form)—Add 5 liters of wet IR-45 to a battery jar. Wash with 10 liters of 4% Na_2CO_3 and decant. Wash with 10 liters of 4% NH_4OH and decant. Wash with 10 liters of 4% NaOH and decant. Rinse with deionized water with stirring and decanting until the pH of wash water is 9 or lower. Store under water. Each pair of students will need about 250 ml of moist resin. The resin should be collected at the end of the experiment (*Do not allow accidental mixing with Dowex 50.*) and promptly regenerated as follows: Wash the resin with 5% KOH (50 g/liter, about 10 liters for the above 2.5-liter quantity of resin), decant, and wash thoroughly with distilled water until the pH is 9.5 to 10.5. Store tightly covered under water. A few days before use of the regenerated IR-45 resin, remove yellow organic material that bleeds from the beads by washing thoroughly with methanol, and then again with distilled water. The same procedure can be used with Dowex-50 if yellow material accumulates in the water in which it is stored.

Absolute methanol—Freshly opened reagent-grade solvent should be used. Approximately 30 liters will be required.

Glass tubes—Approximately 1 in. in diameter and 2 ft in length, or equivalent columns for chromatography. The group will require 50 of these tubes.

Glass wool.

Reducing agent—Dissolve 75 g of $NaHSO_3$ and 25 g of *p*-methylaminophenol in 2.5 liters of distilled water. Store in a brown bottle. Stable for 10 days.

Acid molybdate reagent—*Cautiously* add 680 ml concentrated H_2SO_4 to 1.8 liters of distilled water and allow to cool. Dissolve 125 g $(NH_4)_6Mo_7O_{24} \cdot 4\ H_2O$ in 2.5 liters of distilled water and add to the above sulfuric acid solution. Make up to 5 liters with distilled water.

1.0 mM Phosphate standard—Dissolve 680 mg of KH_2PO_4 in 5 liters of distilled water.

5% KOH—Dissolve 500 g KOH in 10 liters of distilled water.

0.01 M KI, 0.01 M I_2—Dissolve 450 mg of KI in 250 ml of distilled water. Add 325 mg of I_2 and mix until dissolved. Stable about 1 month.

Charcoal (finely powdered, e.g., Norit)—100 g required.

Vacuum desiccator containing fresh P_2O_5—1 required.

High-speed centrifuges—1 or 2 required.

300-ml Centrifuge bottles or cans—50 to 100 required.

Nelson's A reagent—Dissolve 62.5 g of Na_2CO_3 (anhydrous), 62.5 g of potassium sodium tartrate, 50 g $NaHCO_3$, and 500 g Na_2SO_4 (anhydrous) in 1.75 liters of distilled water and dilute to 2.5 liters with distilled water.

Nelson's B reagent—Dissolve 37.5 g $CuSO_4 \cdot 5\ H_2O$ in

250 ml H_2O, and add 5 drops of concentrated H_2SO_4.

Arsenomolybdate reagent—Dissolve 125 g $(NH_4)_6Mo_7O_{24} \cdot 4\,H_2O$ in 2.25 liters of distilled water, and add 105 ml of concentrated H_2SO_4. Dissolve 15.0 g of $Na_2HAsO_4 \cdot 7\,H_2O$ in 125 ml of distilled water, and add to the above acid molybdate solution. Store in a brown bottle for 24 hr at 37°C. This reagent should be yellow with no green tint.

Glucose standard (1.0 μmol/ml in 1.0 M NaCl)—Dissolve 36 mg (weighed to 0.1 mg) and 11.2 g NaCl in distilled water and dilute to 200 ml with distilled water.

Large sheets Whatman No. 1 chromatography paper—20 sheets required.

1% Glucose-1-phosphate standard—Dissolve 100 mg glucose-1-phosphate dipotassium salt in 10 ml of distilled water; store at 0 to 5°C.

1% Glucose-6-phosphate standard—Dissolve 100 mg glucose-6-phosphate Na or K salt in 10 ml of distilled water; store at 0 to 5°C.

1% Glucose standard in 1 N HCl—Dissolve 100 mg glucose in 10 ml of 1 N HCl (concentrated HCl:H_2O; 1:11); store at 0 to 5°C.

1% Inorganic phosphate standard—Dissolve 100 mg of KH_2PO_4 in 10 ml of distilled water.

10% $Mg(NO_3)_2$ in ethanol—Dissolve 5 g of $Mg(NO_3)_2 \cdot 6\,H_2O$ in 10 ml of absolute ethanol.

Methanol—2.5 liters required.

88% Formic acid—500 ml required.

Aniline–acid–oxalate spray—Dissolve 4.5 g oxalic acid in 1 liter of distilled water, add 9 ml of aniline. Store in a brown bottle.

Modified Hanes–Ischerwood spray reagent—Mix the following: 250 ml of 4% ammonium molybdate (10 g $(NH_4)_6Mo_7O_{24} \cdot 4\,H_2O$ dissolved in H_2O to yield 250 ml of solution), 100 ml of 1.0 N HCl (see instructions earlier in Experiment 12 list), 50 ml of 70% perchloric acid, and 600 ml of distilled water. *Prepare this reagent on the day of use.* Use phosphate (detergent)-free glassware for mixing, storing, and spraying.

Acid-$FeCl_3$ in acetone—Mix 750 mg of $FeCl_3 \cdot 6H_2O$, 15 ml of 0.3 M HCl (12.5 ml concentrated HCl + 487.5 ml of H_2O), and 485 ml of acetone until dissolved. Store in a tightly stoppered bottle.

1.25% Sulfosalicylic acid in acetone—Dissolve 6.25 g of sulfosalicylic acid in 500 ml of acetone. *Prepare on the day of the experiment.* Store in a tightly stoppered bottle.

UV lamp—1 or 2 required.

100°C oven—1 required.

Experiment 13: Isolation and Characterization of Erythrocyte Membranes

Fresh pig blood—Approximately 4 liters required.

Heparin sulfate/EDTA solution—Dissolve 80 mg of heparin sulfate in 40 ml of 0.5 M EDTA. *Stir this solution into 4 liters of pig blood immediately after it is collected to prevent clotting.*

Isotonic phosphate buffer (310 mosM, pH 6.8)—Dissolve 85.6 g of NaH_2PO_4 (monobasic monohydrate) and 88.0 g of Na_2HPO_4 (dibasic anhydrous) in 9.6 liters of distilled water. Chill the solution, adjust the pH to 6.8 with HCl, and bring the final volume to 10 liters with distilled water.

Hypotonic phosphate buffer (20 mosM, pH 6.8)—Add 600 ml of isotonic phosphate buffer to 8.7 liters of distilled water. Adjust the pH to 6.8 if necessary.

Plastic centrifuge bottles (250–300 ml)—At least six required.

Plastic centrifuge tubes (30–50 ml)—50 required.

15-ml Plastic conical tubes with screw caps—100 required.

1.5-ml Microcentrifuge tubes—500 required.

Disposable Pasteur pipettes—Approximately 500 required.

4-ml Glass vials with Teflon-lined screw caps—Approximately 400 required.

100°C Oven—One required.

Silica-gel thin-layer chromatography plates—We purchase these from Fisher (catalog #05-713-317). The plates are prepared (activated) by heating in a 100°C oven for 5 to 10 min to remove any moisture from the matrix. Remove from the oven and store desiccated at room temperature until they are ready to be used. Approximately 25 plates will be required (1 plate/2 groups).

Large (16 × 125 mm) glass test tubes—500 required.

Thin-layer chromatography tanks—5 to 10 required, depending on their size.

I_2 Chamber—Fill the bottom of a rectangular glass chamber with I_2 (solid) and cover tightly. Allow 6 to 8 hr for the chamber to saturate with I_2 vapors before the thin-layer chromatography plates are developed.

Pressurized N_2 tank—1 required.

Perchloric acid (70%)—Aldrich catalog #31, 142-1 or equivalent. 200 ml required.

Cholesterol assay reagent, pH 6.5 (See Supplies and Reagents list for the formulation)—We purchase this reagent in dry, mixed form from Sigma (catalog #352-50) and reconstitute each vial according to the manufacturer's instructions. Each group will require 6 ml.

Cholesterol standard solutions (1 mg/ml, 2 mg/ml, and 4 mg/ml)—We purchase prepared sets of these standards from Sigma (catalog #C-0534). These standards are supplied as 100, 200, and 400 mg/dl solutions, which are equivalent to 1, 2, and 4-mg/ml solutions, respectively, in an aqueous/isopropanol solvent. They are ready to use in the cholesterol quantitation assay.

Phospholipid standard solutions in $CHCl_3$—We purchase sphingomyelin from Sigma (catalog #S-1131). Dissolve 25 mg of sphingomyelin in 0.5 ml of $CHCl_3$. We purchase L-α-phosphatidylcholine from Sigma (catalog #P-6638). It is supplied as a 10 mg/ml solution in $CHCl_3$ and is ready to use as a standard. We purchase L-α-phosphatidylethanolamine from Sigma (catalog #P-9137). It is supplied as a 10 mg/ml solution in $CHCl_3$ and is ready to use as a standard. We purchase L-α-phosphatidyl-L-serine from Sigma (catalog #P-7769). Dissolve 25 mg of L-α-phosphatidyl-L-serine in 0.5 ml of $CHCl_3$. We purchase L-α-phosphatidylinositol from Sigma (catalog #P-2517). It is supplied as a 10 mg/ml solution in $CHCl_3$ and is ready to use as a standard. We purchase cholesterol from Sigma (catalog #8667). Dissolve 500 mg of cholesterol in 10 ml of $CHCl_3$. Store all of these solutions *tightly capped* at −20°C.

Heating block—5 to 10 required, depending on the capacity of the blocks for the 4-ml vials.

Preparative centrifuge—At least one required.

Clinical tabletop centrifuge (low speed)—At least one required.

Spectrophotometer to read absorbance at 500 and 660 nm with cuvettes—At least one required.

Chloroform:methanol (1:2)—Mix 150 ml of chloroform with 300 ml of methanol. Store tightly covered in a dark bottle at room temperature.

0.1 M HCl—Add 500 μl of concentrated HCl to 60 ml of distilled water.

Mobile phase for thin-layer chromatography—Mix 1.25 liters of chloroform, 750 ml of methanol, 200 ml of glacial acetic acid, and 50 ml of distilled water. Store tightly covered in a dark bottle at room temperature.

1% Copper sulfate solution—Dissolve 1 g of $CuSO_4 \cdot 5H_2O$ in 100 ml of distilled water.

2% Sodium tartrate solution—Dissolve 2 g of sodium tartrate in 100 ml of distilled water.

2% Sodium carbonate solution—Dissolve 40 g of Na_2CO_3 and 8 g of NaOH in 2 liters of distilled water.

Folin–Ciocalteau reagent (2 N)—200 ml required.

Lysozyme solution (1 mg/ml)—Dissolve 100 mg of lysozyme in 100 ml of distilled water. Store at 4°C.

Phosphate standard solution—Dissolve 13.8 mg of $NaH_2PO_4 \cdot H_2O$ in 100 ml of distilled water.

5% Ammonium molybdate solution—Dissolve 5 g of ammonium molybdate in 100 ml of distilled water.

Amidol reagent—Dissolve, *in order*, 25 g of sodium bisulfite and 1 g of 2,4-diaminophenol in 100 ml of distilled water. *Make this reagent fresh on the day that the assay is performed.*

SDS cleaning solution—Dissolve 10 g of sodium dodecyl sulfate in 2 liters of distilled water.

Experiment 14: Electron Transport

Oxygen electrode system with recorder—For example, Yellow Spring Instruments Model 53 oxygen electrode (Yellow Springs, OH), or equivalent, attached to a linear 10-mV recorder. Five of these units will be required.

Oxygen electrode electrolyte—Fluid should be that specified by the manufacturer of the oxygen electrode. Approximately 250 ml of this solution will be required.

Oxygen electrode membranes—As specified by the manufacturer of the oxygen electrode—5 required.

Magnetic stirrers and stirring bars—Standard magnetic stirrers with stirring bars (or cut up paper clips) small enough to fit in sample chamber of the oxygen electrodes—5 required.

Sample injection syringe with tubing end (5 required)—Equip a calibrated 1-ml syringe with a short needle and slip a piece of polyethylene tubing 1 to 2 in. in length over the needle. The tubing diameter and length should be the minimum necessary to reach to the bottom of the sample chamber of the oxygen electrode.

Fresh pig heart—5 required.

Meat grinder—1 or 2 required.

1 mM Ethylenediaminetetraacetate (EDTA), pH 7.4—Dissolve 7.44 g of disodium EDTA in 20 liters of distilled water and titrate to pH 7.4 with NaOH. (You will use 5 to 10 liters of this to prepare other reagents.)

Blender—1 or 2 required.

0.02 M Potassium phosphate, 1 mM EDTA (pH 7.4)—Dissolve 13.95 g of K_2HPO_4 and 2.7 g of KH_2PO_4 in 5 liters of 1 mM EDTA, pH 7.4.

High-speed centrifuge—This centrifuge must be capable of centrifuging 300 to 600 ml of fluid at at least 5000 × *g*. A Sorvall RC2B centrifuge is satisfactory.

0.25 M Potassium phosphate, 1 mM EDTA (pH 7.4)—Dissolve 35.0 g of K_2HPO_4 and 6.7 g of KH_2PO_4 in 1 liter of 1 mM EDTA, and adjust to pH 7.4.

1.0 M Acetic acid—Dilute 29.5 ml of glacial acetic acid with 470 ml of distilled water.

1.0 mM NAD^+—Dissolve 40 mg of $Na_2NAD^+ \cdot 4\,H_2O$ in 50 ml of distilled water. Refrigerate until use. Stable for 1 week.

0.3 M Potassium phosphate (pH 7.4)—Dissolve 210 g of K_2HPO_4 and 40.5 g of KH_2PO_4 in 5 liters of distilled water; adjust pH to 7.4 if necessary with 1.0 M KOH or HCl.

0.05 M Sodium succinate—Dilute 0.5 M sodium succinate, prepared as below (5 ml of 0.5 M sodium succinate + 45 ml of H_2O). Refrigerate. Stable for 1 week.

0.5 M Sodium succinate—Dissolve 33.9 g of sodium succinate in 250 ml of distilled water. Refrigerate. Stable for 1 week.

0.25 M Malate (Na salt)—Dissolve 1.675 g of malic acid and 500 mg of NaOH in approximately 40 ml of distilled water; adjust pH to 6 to 8 with 1 N NaOH, and dilute to 50 ml with distilled water. Refrigerate. Stable for 1 week.

0.25 M β-Hydroxybutyrate—Dissolve 1.3 g of β-hydroxybutyric acid in 25 ml of distilled water by gradual addition of 12.5 ml of 1 N NaOH (final pH should be in the range from 6 to 8); dilute to a final volume of 50 ml with distilled water. Alternatively, use the sodium salt of β-hydroxybutyrate (1.575 g) and adjust the pH with HCl. Refrigerate. Stable for 1 week.

6×10^{-4} M Cytochrome *c*—Quantities used depend on purity of starting material. The molecular weight is 13,000. Prepare 50 ml of this solution. Refrigerate. Stable for 1 week.

Ascorbic acid or sodium ascorbate—Solutions are prepared by the students just before use (see Experiment 14). Approximately 25 g will be required.

1 N NaOH—Dissolve 2.0 g of NaOH in 50 ml of distilled water. (Not needed if sodium ascorbate is provided.)

0.5 M Malonate—Suspend 2.6 g of malonic acid in 6 ml of distilled water. Adjust to pH 6 to 8 with 4.0 N NaOH (80 g NaOH/500 ml) and dilute to 50 ml with distilled water.

0.1 M KCN—Dissolve 325 mg of KCN in 50 ml of distilled water on the day of use. Keep in a tightly stoppered bottle marked *Poison*. Provide a propipette for dispensing.

0.01 M Methylene blue—Dissolve 935 mg of methylene blue in 250 ml of distilled water.

Antimycin A (180 μg/ml 95% ethanol)—Dissolve 9.0 mg of Antimycin A in 50 ml 95% (vol/vol) ethanol.

95% (vol/vol) Ethanol—Mix 47.5 ml of absolute ethanol with 2.5 ml of distilled water.

6 M KOH—Dissolve 17 g of KOH in 50 ml of distilled water.

DPIP solution—Dissolve 67.5 mg of 2,6-dichlorophenol-indophenol (DPIP) solution in 250 ml of distilled water.

Experiment 15: Study of the Phosphoryl Group Transfer Cascade Governing Glucose Metabolism: Allosteric and Covalent Regulation of Enzyme Activity

1 M Tris, pH 8.0—Dissolve 88.8 g of Tris · HCl and 53 g of Tris free base in about 700 ml of distilled water. The pH of this solution will be approximately 8.0. Adjust the pH if necessary with dilute NaOH or HCl. Bring the final volume of the solution to 1 liter with distilled water.

1 M Glycerophosphate, pH 8.0—Dissolve 108 g of glycerophosphate (disodium hydrate, Sigma catalog #G-6501) in about 350 ml of distilled water. Adjust the pH to 8.0 with HCl. Bring the final volume of the solution to 500 ml with distilled water. Filter-sterilize the solution and store at 4°C.

60 mM $Mg(acetate)_2$, pH 7.3—Dissolve 0.64 g of magnesium acetate tetrahydrate in 40 ml of distilled water. Adjust the pH to 7.3 with dilute NaOH. Bring the final volume of the solution to 50 ml with distilled water. Filter-sterilize the solution and store at 4°C.

Concentrated enzyme storage buffer—Add 0.5 ml of 1 M Tris, pH 8.0 (see above), 0.5 ml of 1 M glycerophosphate, pH 8.0 (see above), 10 μl of β-mercaptoethanol, and 40 μl of 0.5 M EDTA, pH 8.0, to 6 ml of distilled water. The 0.5 M EDTA, pH 8.0, is prepared as follows: dissolve 186 g of disodium EDTA dihydrate in 800 ml of distilled water. Adjust the pH to 8.0 with concentrated NaOH. Bring the final volume of the solution to 1 liter with distilled water. Add 4 ml of glycerol to this solution and mix thoroughly.

Reaction buffer—Dilute 2.5 ml of 1 M Tris, pH 8.0, and 2.5 ml of 1 M glycerophosphate, pH 8.0, in 45 ml of distilled water.

2 mM ATP solution—Dissolve 11 mg of adenosine 5′-triphosphate (disodium salt) in 10 ml of 60 mM Mg(acetate)$_2$, pH 7.3. Divide into 20- to 50-μl aliquots and store at −70°C. Thaw on ice just before use.

Spectrophotometer with cuvettes or colorimeter tubes—At least 1 required.

Glycogen phosphorylase *b* stock solution—Dissolve 25 mg (~750 units) of glycogen phosphorylase *b* (Sigma catalog #P-6635) in approximately 750 μl of concentrated enzyme storage buffer (resuspend the lyophilized protein gently, do not use a vortex mixer). Divide into 20 μl aliquots and store at −20°C. Dilute this ~1000-unit/ml stock solution to the required concentration (see protocol) just before use. Store the dilute enzyme on ice until it is ready to be used.

Glycogen phosphorylase kinase stock solution—Dissolve a 2500-unit package size of glycogen phosphorylase kinase (Sigma catalog #P-2014) in 1 ml of concentrated enzyme storage buffer (gently, do not use a vortex mixer). Divide into 20-μl aliquots and store at −20°C. Dilute this 2500-unit/ml stock solution to the appropriate concentration (see protocol) just before use. Store the dilute enzyme on ice until it is ready to be used.

4X SDS/EDTA buffer—Dissolve 3.51 g of Tris · HCl and 0.335 g of Tris free base in 50 ml of 0.5 M EDTA, pH 8.0. Adjust the pH to 7.0 with dilute NaOH or HCl if necessary. Add 30 ml of glycerol, 10 ml of β-mercaptoethanol, and 10 ml of distilled water. Dissolve (avoid foaming) 8 g of SDS into this solution, as well as 4 mg of bromophenol blue. Store the solution tightly capped at room temperature.

γ-[^{32}P]ATP (4000 Ci/mmol)—We purchase this from ICN Biochemicals (catalog #35001x). We dilute this solution 1:80 with distilled water for use in the in vitro phosphorylation assay. At this specific activity, a 12- to 18-hr exposure time is usually required. If you choose to do a greater dilution of this reagent, longer exposure times will be required. For safety purposes, we keep this reagent shielded and in the hands of the teaching assistant. The students bring their prepared reactions to the teaching assistant, who adds the reagent for them.

Prestained high-molecular-weight protein markers—We purchase these from Gibco BRL (catalog #26041-020) and prepare them according to the manufacturer's instructions.

Plastic wrap—1 large roll required.

Whatman 3 MM filter paper—Approximately 8 to 10 sheets required.

Film cassette for autoradiography—2 to 5 required, depending on the size of the gels and cassettes.

Dark room with solutions for film development.

Kodak X-omat film—2 to 5 large pieces required.

1.5-ml Microcentrifuge tubes—1000 required.

Glass test tubes (13 × 100 mm)—1000 required.

10-ml Pipettes with bulbs—50 required.

30 mM cysteine, pH 7.0—Dissolve 527 mg of L-cysteine hydrochloride monohydrate in 80 ml of distilled water. Adjust the pH to 7.0 by slowly adding solid sodium bicarbonate. Bring the final volume of the solution to 100 ml with distilled water. Prepare this reagent on the day that it is to be used.

80 mM cysteine, pH 6.8—Dissolve 1.4 g of L-cysteine hydrochloride monohydrate in 80 ml of distilled water. Adjust the pH to 6.8 by slowly adding solid sodium bicarbonate. Bring the final volume of the solution to 100 ml with distilled water. Prepare this reagent on the day that it is to be used.

250 mM Tris, pH 7.7—Dissolve 28.6 g of Tris · HCl and 8.3 g of Tris free base in 800 ml of distilled water. The pH of the solution should be at or near pH 7.7. Adjust the pH to 7.7 with dilute HCl or NaOH if necessary. Bring the final volume of the solution to 1 liter with distilled water.

250 mM glycerophosphate in 250 mM Tris, pH 7.7—Dissolve 54 g of glycerophosphate (disodium salt) in 700 ml of 250 mM Tris, pH 7.7. Adjust the pH to 7.7 with dilute NaOH or HCl, if necessary, and bring the final volume of the solution to 1 liter with 250 mM Tris, pH 7.7.

40 mM glycerophosphate in 80 mM cysteine, pH 6.8—Dissolve 864 mg of glycerophosphate (disodium salt) in 70 ml of 80 mM cysteine, pH 6.8. Bring the final volume of the solution to 100 ml with 80 mM cysteine, pH 6.8.

Glycogen phosphorylase kinase solution—Dilute 40 μl of the 2500-unit/ml glycogen phosphorylase kinase stock solution in 4.96 ml of 30 mM cysteine, pH 7.0.

30 mM ATP solution—Dissolve 165 mg of adenosine 5′-triphosphate (disodium salt) in 8 ml of distilled water. Adjust the pH to 7.7 with very dilute NaOH and bring the final volume of the solution to 10 ml with distilled water. Divide the solution into 20- to 50-μl aliquots and store at −70°C. Thaw on ice just before use.

100 mM Mg(acetate)$_2$, pH 7.7—Dissolve 21.45 g of magnesium acetate terahydrate in 900 ml of distilled water. Adjust the pH to 7.7 with NaOH, and bring the final volume of the solution to 1 liter with distilled water.

500 mM potassium phosphate, pH 6.8—Dissolve 19.05 g of monobasic potassium phosphate (anhydrous)

and 62.71 g of dibasic potassium phosphate (anhydrous) in 800 ml of distilled water. The pH of the solution should be near 7.2. Adjust the pH of the solution to 6.8 with HCl, and bring the final volume of the solution to 1 liter with distilled water.

4% (wt/vol) glycogen solution—Dissolve 10 g of rabbit liver glycogen (Sigma catalog #G-8876) in distilled water to a total volume of 250 ml. Filter sterilize the solution and store at 4°C.

300 mM $MgCl_2$—Dissolve 30.5 g of $MgCl_2 \cdot 6\ H_2O$ in distilled water to a final volume of 500 ml.

100 mM EDTA, pH 8.0—Dissolve 18.61 g of disodium EDTA dihydrate in 400 ml of distilled water. Adjust the pH to 8.0 with NaOH and bring the final volume of the solution to 500 ml with distilled water.

6.5 mM nicotinamide adenine dinucleotide phosphate solution (NADP)—We purchase this reagent in preweighed vials from Sigma (catalog #240-310). To prepare the 6.5 mM reagent, dissolve 10-mg vial of NADP in 2.0 ml of distilled water. Prepare this reagent immediately before use.

0.1% (wt/vol) α-D-glucose-1,6-diphosphate solution—Dissolve 0.1 g of α-D-glucose-1,6-diphosphate (potassium salt, hydrate) in 100 ml of distilled water. Filter-sterilize the solution and store in small aliquots at −20°C.

Phosphoglucomutase solution (10 units/ml)—We purchase this enzyme from Sigma (catalog #P-3397) as a crystalline suspension in 2.5 M ammonium sulfate. Consult the specification sheet for the product (lot-specific, usually 100–300 units/mg of protein) to determine a dilution that will yield a 10-unit/ml solution. The dilution of this enzyme is done in distilled water immediately before use. Store the remainder of the crystalline suspension at 4°C for use in future semesters.

Glucose-6-phosphate dehydrogenase solution (10 units/ml)—We purchase this enzyme from Sigma (catalog #G-6378), which is derived from baker's yeast. Reconstitute the enzyme in distilled water as per the manufacturer's instructions. Consult the specification sheet (lot-specific, usually 200–400 units/mg of protein) to determine a dilution of the enzyme stock that will yield a 10-unit/ml solution (dilution is performed in distilled water). Store the remainder of the reconstituted stock enzyme solution in small aliquots at −20°C for use in future semesters.

Glycogen phosphorylase *a* enzyme solution (2 units/ml)—We purchase this enzyme from Sigma (catalog #P-1261). Dissolve 25 mg (~625 units) of glycogen phosphorylase *a* (gently, do not use a Vortex mixer) in 625 μl of concentrated enzyme storage buffer. Divide into 20-μl aliquots and store at −20°C. Dilute this 1000-units/ml stock enzyme solution 1:500 in 40 mM glycerophosphate, pH 6.8, shortly before use. Store on ice.

Glycogen phosphorylase *b* control solution (2 units/ml)—Dilute the 1000-units/ml enzyme stock solution (see above) 1:500 in 40 mM glycerophosphate, pH 6.8.

Glycogen phosphorylase *b* control solution (2 units/ml in 60 mM AMP)—Dilute the 1000-units/ml enzyme stock solution (see above) 1:500 in 40 mM glycerophosphate containing 60 mM AMP. This solution is prepared by dissolving 2.08 g of adenosine 5′-monophosphate (monosodium salt) in 100 ml of 40 mM glycerophosphate, pH 6.8.

Reagents and supplies for SDS–PAGE—See Experiment 4.

Experiment 16: Experiments in Clinical Biochemistry and Metabolism

Fresh pig blood—Approximately 100 to 200 ml required.

Heparin sulfate/EDTA solution—Dissolve 4 mg of heparin sulfate in 2 ml of 0.5 M EDTA. *Stir this solution into 200 ml of pig blood immediately after it is collected to prevent clotting.*

5-ml and 10-ml Pipettes—50 of each required.

Small (13 × 100 mm) glass test tubes—1000 required.

Spectrophotometer with cuvettes—At least 1 required.

Glucose standard stock solution (44.48 mM)*—Dissolve 80.1 mg of D-glucose in 10 ml of distilled water.

3.00 mM glucose standard solution*—Add 67 μl of 44.48 mM glucose standard stock solution to 933 μl of distilled water.

11.10 mM glucose standard solution*—Add 250 μl of 44.48 mM glucose standard stock solution to 750 μl of distilled water.

16.65 mM glucose standard solution*—Add 374 μl of 44.48 mM glucose standard stock solution to 626 μl of distilled water.

4.44 mM lactate standard—Dissolve 4.0 mg of D-lactic acid (free acid, crystalline) in 10 ml of distilled water.

*Sigma offers a combined glucose/urea standard set (catalog #16-11) that can be used for these assays. These standards contain 800 mg/dL (44.48 mM glucose, 133.3 mM urea), 300 mg/dl (16.65 mM glucose, 50 mM urea), and 100 mg/dl (5.55 mM glucose, 16.66 mM urea) of each compound. These standards can be diluted appropriately to prepare the standard solutions required for this assay.

Urea standard stock solution (35.70 mM)*—Dissolve 21.4 mg of urea in 10 ml of distilled water.

3.57 mM urea standard*—Add 100 μl of 35.70 mM urea standard stock solution to 900 μl of distilled water.

8.92 mM urea standard*—Add 250 μl of 35.70 mM urea standard stock solution to 750 μl of distilled water.

10.71 mM urea standard*—Add 300 μl of 35.70 mM urea standard stock solution to 700 μl of distilled water.

Serum lactate reagent—Sigma catalog #735-10. Prepare by adding distilled water to the dry mixture as per the instructions. Each group will require 6 ml.

Serum urea nitrogen reagent—Sigma catalog #66-20. Prepare by adding distilled water to the dry mixture as per the instructions. Each group will require 12 ml.

Serum glucose reagent—Sigma catalog #16-50. Prepare by adding distilled water to the dry mixture as per the instructions. Each group will require 10 ml.

Serum creatine kinase reagent—Sigma catalog #DG-147-K. Prepare Reagents A and B by adding distilled water to the dry mixtures as per the supplier's instructions. Mix Reagents A and B at the recommended ratios. Each group will require 5 ml of this working solution.

Experiment 17: Partial Purification of a Polyclonal Antibody, Determination of Titer, and Quantitation of an Antigen Using the ELISA

10X Phosphate-buffered saline (10X PBS)—Dissolve 5.12 g of $NaH_2PO_4 \cdot H_2O$, 44.96 g of $Na_2HPO_4 \cdot 7H_2O$, and 87.6 g NaCl in 1.5 liter of distilled water. Adjust the pH to 7.3 with HCl or NaOH as needed. Bring the final volume of the solution to 2 liters with distilled water. Store short term at room temperature or long term at 4°C. Dilute this solution 1:10 with distilled water to prepare the PBS solution (1X) used in the blocking and wash solutions described below.

Normal or pre-immune rabbit serum—Each group will require 180 μl. Store at −20°C. We have also purchased normal rabbit serum from Sigma (catalog #R-4505).

*Sigma offers a combined glucose/urea standard set (catalog #16-11) that can be used for these assays. These standards contain 800 mg/dL (44.48 mM glucose, 133.3 mM urea), 300 mg/dl (16.65 mM glucose, 50 mM urea), and 100 mg/dl (5.55 mM glucose, 16.66 mM urea) of each compound. These standards can be diluted appropriately to prepare the standard solutions required for this assay.

Anti-β-galactosidase rabbit serum*—Each group will require 180 μl. Store at −20°C

Polyoxyethylenesorbitan monolaurate (Tween 20, Sigma catalog #P-5927)—about 10 ml required.

Immulon-2 96-well flat-bottomed microtiter (ELISA) plate (50 required)—We purchase these from Fisher (catalog #1424561), but any standard ELISA plate would suffice.

Blocking solution—Dissolve 50 g of bovine serum albumin (BSA) and 2.5 ml of polyoxyethylenesorbitan monolaurate (Tween 20) in 5 liters of 1X PBS.

Wash solution—Dissolve 5 ml of Tween 20 in 10 liters of 1X PBS.

β-Galactosidase stock solution—Dissolve 5 mg of β-galactosidase (Sigma catalog #G-6008) in 50 ml of 1X PBS. Store at −20°C.

β-Galactosidase solution (10 μg/ml)—Add 20 ml of the β-galactosidase stock solution to 180 ml of 1X PBS. Store at −20°C.

β-Galactosidase solution (40 μg/ml)—Add 8 ml of β-galactosidase stock solution to 12 ml of 1X PBS. Store at −20°C.

β-Galactosidase solutions ("unknown concentrations")—Prepare about 5 ml of each unknown concentration by making the appropriate dilution of the β-galactosidase stock solution in 1X PBS. We have found that unknown concentrations in the range from 10 μg/ml to 100 μg/ml work well in the competitive ELISA. Store each unknown solution at −20°C.

4 N H_2SO_4—Dilute 666 ml of 12 N H_2SO_4 to 2 liters with distilled water. Store covered at room temperature. Label with the warning *Caution: Strong Acid.*

HRP substrate—Mix equal volumes of each of the 3,3′,5,5′-tetramethylbenzidene and H_2O_2 solutions (Kirkegaard & Perry catalog #50-7600) to prepare the substrate. *This should be done right before the students are ready to develop their microtiter plates.* About 750 ml of this substrate is required for the full three-day experiment.

* As stated in the text, the dilutions for the primary antibodies used in this experiment are approximate. We have found that the titer of the J1 monoclonal antibody varies somewhat from lot to lot. Obviously, the titer of the rabbit polyclonal antibody will also vary from animal to animal. Because of this, *we suggest that the experiment be performed in advance of the class in case some small adjustments of the dilutions stated in the text are necessary.* The secondary antibody dilutions described in the text are much less variable from semester to semester. We raise the rabbit-anti-β-galactosidase polyclonal antibody by shipping the antigen (Sigma catalog #G-6008) to CoCalico Biologicals, Inc. in Reamstown, PA. This company will raise the antibody and send you batches of the serum. The company will test the serum antibodies in an ELISA to determine if they are adequate for your application.

Mouse-anti-β-galactosidase monoclonal antibody (J1)*—Keep the stock antibody (Sigma catalog #G-8021) frozen at −80°C. We have found that the titer drops dramatically after short-term storage at −20°C. Dilute the antibody immediately before use in blocking solution (about a 1:1000 dilution). Each group will require about 1.2 ml of the diluted antibody to perform the three-day experiment.

HRP-conjugated-goat-antirabbit IgG*—Each group will prepare 2.5 ml of this solution by making serial dilutions of the stock antibody (Kirkegaard & Perry catalog #074-1506) in blocking solution. The final dilution should be about 1:9000. Prepare immediately before use.

HRP-conjugated-goat-antimouse IgG*—Each group will prepare 0.8 ml of this solution by making serial dilutions of the stock antibody (Kirkegaard & Perry catalog #074-1802) in blocking solution. The final dilution should be about 1:10,000. Prepare immediately before use.

Saturated ammonium sulfate solution—Dissolve 350 g of ammonium sulfate (reagent grade) in 0.5 liter of distilled water. Store at room temperature.

Spectrophotometer with cuvettes—At least 1 required.

10-ml Pasteur pipettes—50 required.

1.5-ml microcentrifuge tubes—500 required.

96-Well microplate spectrophotometer—1 required, with a printer.

Experiment 18: Western Blot to Identify an Antigen

Reagents and apparatus for SDS–PAGE—See Experiment 4.

Power supplies—Enough to power all available SDS–PAGE and Western blot transfer units.

Prestained high-molecular weight protein markers—We purchase these from Gibco-BRL (catalog #26041-020) and prepare them according to the manufacturer's instructions.

Coomassie Blue staining solution—Dissolve 400 ml of methanol, 100 ml of glacial acetic acid, and 2.5 g of Coomassie Brilliant Blue R-250 in 500 ml of distilled water. Store tightly covered at room temperature.

Coomassie Blue destaining solution—Dissolve 400 ml of methanol and 100 ml of glacial acetic acid in 500 ml of distilled water. Store tightly covered at room temperature.

Methanol—200 ml required to prepare 1 liter of transfer buffer.

Transfer buffer—*We suggest that this solution be prepared fresh on the day of the experiment.* Dissolve 11.6 g of Tris base, 5.86 g of glycine, and 0.038 g of sodium dodecyl sulfate in 700 ml of distilled water. Adjust pH to 9.2 with HCl or NaOH as needed. Add 200 ml of methanol. Bring the final volume of the solution to 1 liter with distilled water. Store this solution tightly covered at room temperature.

Immobilon-P PVDF membrane (0.45-μm pore size)—We purchase this from Millipore (catalog #IPVH 000 10). One large roll required.

Whatman 3 MM filter paper—Approximately 50 sheets required, depending on the size of the gels and the type of transfer apparatus used.

Western blot blocking solution—*We suggest that this solution be prepared fresh on the day of the experiment.* Dissolve 2.42 g of Tris base, 14.62 g of NaCl, and 86.22 g of nonfat dry (powdered) milk to 700 ml of distilled water. Bring the final volume of the solution to 1 liter with distilled water.

Tween 20 (polyoxyethylenesorbitan monolaurate)—We purchase this from Sigma (catalog #P-5927). Approximately 5 ml will be required.

Western blot wash solution—*We suggest that this solution be prepared fresh on the day of the experiment.* The amount given is sufficient for *one group of students.* Dissolve 2.42 g of Tris base, 14.62 g of NaCl, and 86.22 g of nonfat dry (powdered) milk to 700 ml of distilled water. Add 0.5 ml of polyoxyethylenesorbitan monolaurate (Tween 20). Bring the final volume of the solution to 1 liter with distilled water.

Mouse-antiglutathione-*S*-transferase monoclonal antibody—We purchase this from Sigma (catalog #G-1160). This antibody is supplied as a sterile liquid. Dilute as specified in the protocol in blocking solution. Prepare about 5 liters (~100 ml per group).

HRP-conjugated goat-antimouse IgG—We purchase this reagent from Pierce Chemical Co. (catalog #31430). This reagent is supplied as a lyophilized powder. Reconstitute and store as specified by the

* As stated in the text, the dilutions for the primary antibodies used in this experiment are approximate. We have found that the titer of the J1 monoclonal antibody varies somewhat from lot to lot. Obviously, the titer of the rabbit polyclonal antibody will also vary from animal to animal. Because of this, *we suggest that the experiment be performed in advance of the class in case some small adjustments of the dilutions stated in the text are necessary.* The secondary antibody dilutions described in the text are much less variable from semester to semester. We raise the rabbit-anti-β-galactosidase polyclonal antibody by shipping the antigen (Sigma catalog #G-6008) to CoCalico Biologicals, Inc. in Reamstown, PA. For a very fair price, this company will raise the antibody and send you batches of the serum. The company will even test the serum antibodies in an ELISA to determine if they are adequate for your application.

manufacturer's instructions. Dilute as specified in the protocol in blocking solution. Prepare about 5 liters (~100 ml per group).

Razor blades—10 to 20 required (students can share these).

Aluminum foil—1 large roll required.

Small metal or plastic trays for membrane incubations—50 required.

SuperSignal West Pico Chemiluminescent Substrate—Pierce catalog #34080. Prepare the working solution immediately before use by mixing equal volumes of the SuperSignal luminol/enhancer solution and the stable peroxide solution. Each group will require 5 to 10 ml, depending on the size of the membrane.

Plastic wrap—1 large roll required.

Kodak X-omat film—5 to 7 large pieces required, depending on the size of the membranes and film cassettes.

Film cassettes—5 to 7 required, depending on the size of the membranes.

Darkroom—1 required, preferably fitted with a "safe light."

Transfer apparatus—We use the BIORAD Transblot SD model. With mini-gels, we can fit at least three blots per unit.

Reagents for Colorimetric Detection of HRP (optional)

10 mM Tris-buffered saline (TBS), pH 7.4—Dissolve 1.8 g of Tris free base and 13 g of NaCl to a total volume of 1.4 liters with distilled water. Adjust the pH to 7.4 with 1 M HCl. Bring the final volume of the solution to 1.5 liters with distilled water.

4-Chloro-1-naphthol—This reagent can be purchased from Pierce Chemical Co. (catalog #34010). See the protocol for the preparation of this substrate solution.

Absolute ethanol—200 ml required.

3% (wt/wt) hydrogen peroxide in distilled water—Prepare this reagent immediately before use by adding 10 ml of a 30% (wt/wt) hydrogen peroxide stock solution (Sigma catalog #H-1009) to 90 ml of distilled water.

Experiment 19: Isolation of Bacterial DNA

For Growth of Wild-Type (Trp^+) *Bacillus subtilis*

Usually a teaching assistant or technician performs this portion of the experiment. This section is designed to yield cells for 10 students.

Stock culture of wild-type *B. subtilis* (Bacillus Genetic Stock Center strain 1A2)—1 required.

Inoculating needle—1 required.

Luria broth—Luria broth: Prepare five 100-ml and five 10-liter solutions containing 10 g/liter Bacto-Tryptone, 5 g/liter Bacto-Yeast extract, 5 g/liter NaCl, 0.5 g/liter antifoam agent (e.g., Polyglycol P-2000), and 1 g/liter glucose (autoclave separately). Autoclave the main solution and separate solution of concentrated glucose for 20 min at 105 to 110°C and cool; use sterile conditions to add the glucose solution to the main solution and use within 24 hr.

37°C Incubator-shaker—1 or 2 required.

Fermenter capable of handling 50 liters of growth medium.

Centrifuges—For harvesting the 10-liter culture, a continuous-flow centrifuge such as a Sharples is most convenient. For washing the cells, a refrigerated high-speed laboratory centrifuge (e.g., Sorvall RC-2B) is necessary.

Sterile saline solution—Dissolve 4.25 g of NaCl in 500 ml of distilled water and autoclave (20 min at 105–110°C).

Parafilm or wax paper—1 or 2 rolls required.

Bacterial cell paste—If performing Experiment 20 also, grow wild-type *Bacillus subtilis* (Bacillus Genetic Stock Center strain 1A2, Columbus, OH) as described for Experiment 20 or buy wild-type *B. subtilis* from Grain Processing Corp. Muscatine, IA, or General Biochemicals, Chagrin Falls, OH. If not performing Experiment 20, grow or buy any bacterial species (late log phase). Provide frozen cells for students. Approximately 50 g will be required.

0.15 M NaCl, 0.1 M Na_2EDTA—Dissolve 21.9 g of NaCl and 93 g of disodium EDTA · $2H_2O$ in 2.5 liters of distilled water.

Lysozyme solution, 10 mg/ml—Dissolve 1.25 g of crystalline egg white lysozyme in 125 ml of distilled water. Refrigerate until use. Stable for 24 to 48 hr.

25% sodium dodecyl sulfate solution—Dissolve 125 g of sodium dodecyl sulfate in 400 ml of distilled water and dilute to 500 ml with distilled water.

5.0 M $NaClO_4$—Dissolve 1.53 kg of $NaClO_4$ (anhydrous) in 2 liters of distilled water and dilute to 2.5 liters with distilled water.

Chloroform-isoamyl alcohol (24:1)—Mix 4.8 liters of chloroform and 200 ml of isoamyl alcohol. Store tightly capped in a dark bottle at room temperature.

95% Ethanol—Add 250 ml of distilled water to 4.75 liters of absolute ethanol. Store tightly capped.

Dilute (1/10 X) saline-citrate (0.015 M NaCl, 1.5 mM

Na_3citrate)—Dilute 25 ml of concentrated (10X) saline-citrate solution (see below) with 2.475 liters of distilled water. Stable for 1 month.

Standard (1X) saline-citrate (0.15 M NaCl, 15 mM Na_3citrate)—Dilute 250 ml of concentrated (10X) saline-citrate solution (see below) with 2.25 liters of distilled water. Stable for 1 month.

Concentrated (10X) saline-citrate (1.5 M NaCl, 0.15 M Na_3citrate)—Dissolve 43.85 g of NaCl and 22 g of Na_3citrate · $2H_2O$ in 450 ml of distilled water and dilute to 500 ml with distilled water. Use 275 ml of this solution to make up the dilutions of saline-citrate above and dispense the rest to the students. Stable for 1 month.

3.0 M Sodium acetate, 1 mM EDTA, pH 7.0—Dissolve 61.5 g of sodium acetate (anhydrous) and 95 mg of disodium EDTA · $2H_2O$ in 175 ml of distilled water; titrate to pH 7.0 with a few drops of 1.0 M HCl (concentrated HCl:H_2O = 1:11) and dilute to 250 ml with distilled water.

Isopropanol—1 liter required.

Spectrophotometer with quartz cuvettes—At least 1 required.

Chloroform—50 ml required.

Experiment 20: Transformation of a Genetic Character with Bacterial DNA

For Growth of Competent Recipient *Bacillus subtilis* Strain 168 (Trp^-)

Usually a teaching assistant or technician performs this portion of the experiment. This section is designed to yield cells for 30 to 40 pairs of students.

Stock culture of *B. subtilis* strain 168 (Bacillus Genetic Stock Center strain 1A1, Columbus, OH).

Inoculating needle—1 required.

0.8% Nutrient broth, 0.5% glucose, 50 μg/ml tryptophan—Dissolve 0.24 g of nutrient broth in 10 ml of distilled water, 0.15 g of glucose in a second 10 ml of distilled water, and 1.5 mg of tryptophan in a third 10 ml of distilled water. Autoclave (20 min at 105 to 110°C) all samples separately, cool, and aseptically mix in a sterile 100-ml flask.

37°C incubator-shaker—1 required.

Competence growth medium I (SPI)—First prepare 750 ml of SPI medium; 500 ml of this will be used to prepare the SPII medium (see next entry). The SPI medium is composed of the following four separately autoclaved solutions—*Salts solution:* Dissolve 10.5 g of K_2HPO_4, 4.5 g of KH_2PO_4, 1.5 g of $(NH_4)_2SO_4$, 750 mg of Na_3citrate · $2H_2O$, 375 mg of $MgSO_4$ · $7H_2O$, and 750 mg of yeast extract in 689 ml of distilled water. *Glucose solution:* Dissolve 3.75 g of glucose in 10 ml of distilled water. *Casamino acids solution:* Dissolve 50 mg of Casamino acids in 1 ml of distilled water. *Tryptophan solution:* Dissolve 37.5 mg of L-tryptophan in 50 ml of distilled water. Autoclave (20 min at 105 to 110°C) all samples separately, cool, and aseptically mix in a sterile 1-liter flask. Transfer 250 ml to another sterile 1-liter flask, and retain the rest for preparation of SPII medium (next entry).

Competence growth medium II (SPII)—Prepare stock solutions of 1.0 M $CaCl_2$ (1.47 g of $CaCl_2$ · $2H_2O$ in 10 ml of H_2O) and 1.0 M $MgCl_2$ (2.03 g of $MgCl_2$ · $6H_2O$ in 10 ml of H_2O). Sterilize these solutions by autoclaving (20 min at 105 to 110°C), allow them to cool, and store at room temperature. On the day of use for preparation of competent cells, prepare 500 ml of SPII medium by adding 225 μl of sterile 1.0-M $CaCl_2$ and 1.125 ml of sterile 1.0-M $MgCl_2$ to 499 ml of sterile SPI medium from the preceding entry.

Minimal medium agar Petri plate (see section on transformation below)—1 required.

Minimal medium agar Petri plate supplemented with 50 μg/ml tryptophan (see section on transformation below)—1 required.

Centrifuge—1 required.

Sterile centrifuge bottles—Autoclave a sufficient number of bottles to centrifuge 500 ml of culture medium.

Sterile glass rod—Place a fire-polished glass rod in a capped test tube and autoclave (20 min at 105 to 110°C).

Sterile synthetic medium—Prepare a 10-ml and two 200-ml solutions containing: 2 g/liter $(NH_4)_2SO_4$, 14 g/liter K_2HPO_4, 6 g/liter KH_2PO_4, 1.9 g/liter Na_3citrate · $2H_2O$, 0.72 g/liter $MgSO_4$, and 50 mg/liter of each of the following amino acids: L-histidine, L-tryptophan, L-arginine · HCl, L-valine, L-lysine · HCl, L-threonine, L-aspartic acid, and L-methionine. Glucose (5 g/liter final medium) should be prepared as a 50-fold concentrated solution and autoclaved separately. Autoclave the main solution and the concentrated glucose solution for 20 min at 105 to 110°C and cool; use sterile conditions to add glucose to the main solution, and use the medium within 24 hr.

Sterile 10 ml pipettes—Autoclave in a pipette can as for sterile pipettes in the section on transformation. At least 2 required.

Sterile 1.5-ml microcentrifuge tubes (for aliquots of competent cells if these are to be frozen)—Approximately 80 of these will be required.

Sterile 100% glycerol (if competent cells are to be stored frozen)—Autoclave for 20 min at 105 to 110°C. Approximately 5 ml will be required.

For Transformation

Sterile test tubes with caps—Autoclave (20 min at 105 to 110°C) test tubes (preferably 13 × 100 mm tubes) with sterile caps loosely set over tops. Approximately 150 of these will be required.

Sterile 0.1- and 1.0-ml pipette tips: Autoclave (20 min at 105–110°C) in covered racks. You will require about 400 of the 0.1-ml tips and about 25 of the 1.0 ml tips.

Sterile minimal medium—Prepare and autoclave (20 min at 105–110°C) 25 capped 18 × 150 mm test tubes, each containing 10 ml of the following medium: 2 g/liter $(NH_4)_2SO_4$, 14 g/liter K_2HPO_4, 6 g/liter KH_2PO_4, 0.2 g/liter $MgSO_4$, 1 g/liter $Na_3citrate \cdot 2H_2O$, 1 mg/liter yeast extract, 5 g/liter monosodium glutamate (autoclave separately), and 5 g/liter glucose (autoclave separately). *Prepare 5 liters of minimal medium, because it is also needed for preparation of the Petri plates (see below).*

DNA from wild-type *B. subtilis*—As isolated and dissolved in sterile 1X saline-citrate (a *minimum* of 0.6 ml per pair of students) as described in Experiment 19. Prepare about 15 ml of this reagent.

Standard (1X) saline-citrate (0.15 M NaCl, 0.015 M Na_3-citrate) (sterile)—Prepare as in the Appendix to Experiment 19. Autoclave (20 min at 105–110°C). Prepare about 250 ml of this reagent.

Pancreatic DNase (1 mg/ml in sterile 1X saline-citrate)—Dissolve 1 mg of pancreatic DNase (Worthington Biochemical Corp., Freehold, N.J.) in 5 ml of sterile 1X saline-citrate. Refrigerate or freeze until use. If possible, pass through a sterile bacterial filter (e.g., Millipore, type HAW) into a sterile test tube. Stable for 48 hr.

Recipient *B. subtilis* strain 168 cells in sterile saline—These cells are conveniently provided as thawed 0.6-ml aliquots of frozen competent cells prepared as described in Experiment 20. You will require approximately 250 ml of these cells.

Sterile 0.1 M EGTA, pH 7.9—Dissolve 1.9 g of EGTA (ethylene glycol-bis(β-aminoethyl ether) N, N, N', N'-tetraacetic acid) in 35 ml of distilled water, titrate to pH 7.9 with 1 N NaOH, and dilute to 50 ml final volume with distilled water. Autoclave in a capped test tube (20 min at 105–110°C) and cool.

40 to 45°C water bath—1 required.

37°C incubator-shaker—1 required.

95% Ethanol—Add 125 ml of distilled water to 2.375 liters of absolute ethanol. Store in a tightly capped bottle.

Bent glass rods—Bend Pyrex glass rods so they fit in a 150-ml beaker. Fire polish (O_2 torch) the ends. See Figure 20-1 for details. Prepare 25 of these.

Minimal medium agar Petri plates—Add 33.75 g of agar to sufficient salts and yeast extract to prepare 2.25 liters of the minimal medium *described above* (medium lacking the monosodium glutamate and glucose samples, which are autoclaved separately). Warm (~50°C) this mixture in sufficient water to dissolve (leaving room for later additions of sterile Na-glutamate and glucose solutions). Autoclave (20 min at 105–110°C) this solution and separate Na-glutamate and glucose solutions. Just after autoclaving (while still hot), aseptically mix the three solutions to create minimal medium agar solution; pour 10-ml aliquots into each of 225 sterile Petri plates. Cover the plates, allow medium to solidify, invert the plates, and store (refrigerated) until use. Stable for 1 week.

Minimal medium agar Petri plates supplemented with 50 μg/ml tryptophan—Add 18.75 g of agar and 62.5 mg of L-tryptophan to sufficient salts and yeast extract to prepare 1.25 liters of the minimal medium *described above* (medium lacking the monosodium glutamate and glucose samples, which are autoclaved separately). Autoclave this solution, and separate solutions of Na-glutamate and glucose for 20 min at 105–110°C. Just after autoclaving (while still hot), aseptically mix the three solutions and then pour 125 sterile agar medium-containing plates as above. Refrigerate; stable for 1 week.

Tubes containing 9.9 ml of sterile saline—Dissolve 12.75 g of NaCl in 1.5 liters of distilled water. Dispense 9.9-ml aliquots of this solution into each of 135 test tubes (18 × 150 mm) with loose-fitting sterile caps. Autoclave (20 min at 105–110°C), cool, and store capped until use. Save remaining saline solution for dilution tubes (see following entry).

Tubes containing 0.9 ml of sterile saline—Dispense 0.9-ml aliquots of above saline solution (12.75 g NaCl/1.5 liters H_2O) into each of 60 test tubes (18 × 150 mm) with loose-fitting caps. Autoclave (20 min at 105–110°C), cool, and store capped until use.

Masking tape (for taping stacks of Petri plates together during incubation)—1 or 2 rolls required.

Experiment 21: Constructing and Characterizing a Recombinant DNA Plasmid

TE buffer—Prepare a 1 M stock solution of Tris by dissolving 121.1 g of Tris free base in 700 ml of distilled water. Adjust the pH to 7.5 with concentrated HCl. Prepare a 0.5 M EDTA stock solution by dissolving 93 g of $Na_2EDTA \cdot 2H_2O$ in 400 ml of distilled water. Adjust the pH to 8.0 by slowly dissolving in NaOH pellets. Bring the final volume of the solution to 500 ml with distilled water. To prepare the TE buffer, add 1 ml of the 0.5 M EDTA stock solution and 10 ml of the 1 M Tris stock solution to 989 ml of distilled water.

pUC19 (0.25 μg/μl)—Pharmacia catalog #27-4951-01. This plasmid is supplied in 25-μg quantities (0.5 μg/μl). Dilute the contents of the tube with an equal volume of TE buffer. Store in 10-μl aliquots at −20°C.

pUC4K (0.5 μg/μl)—Pharmacia catalog #27-4958-01. Use as supplied.

*Hin*dIII (10 units/μl) with 10X reaction buffer supplied—We obtain this from Gibco BRL (catalog #15207-020), but any other supplier of quality restriction endonucleases will do.

*Bam*HI (10 units/μl) with 10X reaction buffer supplied—We obtain this from Gibco BRL (catalog #15201-031), but any other supplier of quality restriction endonucleases will do.

*Eco*RI (10 units/μl) with 10X reaction buffer supplied—We obtain this from Gibco BRL (catalog #15202-021), but any other supplier of quality restriction endonucleases will do.

*Xho*I (10 units/μl) with 10X reaction buffer supplied—We obtain this from Gibco BRL (catalog #15231-020), but any other supplier of quality restriction endonucleases will do.

*Sal*I (10 units/μl) with 10X reaction buffer supplied—We obtain this from Gibco BRL (catalog #15217-029), but any other supplier of quality restriction endonucleases will do.

*Pst*I (10 units/μl) with 10X reaction buffer supplied—We obtain this from Gibco BRL (catalog #15215-023), but any other supplier of quality restriction endonucleases will do.

Heated water baths—One at 37°C, one at 15°C (place in cold room and adjust the temperature), and one at 42°C.

3 M sodium acetate, pH 5.2—Dissolve 40.8 g of sodium acetate $\cdot$ $3H_2O$ in 70 ml of distilled water. Adjust the pH to 5.2 with glacial acetic acid. Bring the final volume of the solution to 100 ml with distilled water. Autoclave or filter sterilize the solution and store at room temperature.

Phenol (Tris-saturated)—Add 500 ml of liquefied phenol to 500 ml of 0.5 M Tris $\cdot$ HCl, pH 8.0. Gently but thoroughly mix the solution for 15 min. Allow the two phases to separate, and aspirate off as much of the aqueous (upper) phase as possible. Repeat this extraction two more times with (500 ml per extraction) with 100 mM Tris $\cdot$ HCl, pH 8.0. After the final extraction, leave a sufficient layer of Tris $\cdot$ HCl, pH 8.0, over the equilibrated phenol. Store this solution in a tightly capped dark bottle at 4°C (stable for several weeks).

Phenol:chloroform:isoamyl alcohol (25:24:1)—Mix together 100 ml of Tris-saturated phenol, 96 ml of chloroform, and 4 ml of isoamyl alcohol. Store tightly capped in a dark bottle at room temperature.

Chloroform:isoamyl alcohol (24:1)—Mix together 192 ml of chloroform and 8 ml of isoamyl alcohol. Store tightly capped in a dark bottle at room temperature.

Absolute ethanol—Approximately 500 ml required.

Dry ice/ethanol bath—Grind or crush enough dry ice to fill a large tray about half full. Slowly add ethanol or acetone to the dry ice, stirring constantly with a spatula, until a bubbling gel or slurry is achieved. Allow this slurry to stand for 10 min before the precipitations are performed.

1.5-ml microcentrifuge tubes—2000 required.

70% (vol/vol ethanol)—Mix 300 ml of distilled water with 700 ml of absolute ethanol. Store tightly capped at −20°C and place on ice for the students to use.

Vacuum microcentrifuge—1 required.

Incubating culture shaker with test tube racks—1 required.

Sterile glass rods for spreading cultures over petri plates—50 required.

Sterile toothpicks (autoclaved)—Wash wooden toothpicks thoroughly with distilled water (to remove any preservatives that may inhibit bacterial growth). Place in a shallow container with a cover and autoclave for 20 min.

37°C Incubator—1 required.

T4 DNA ligase (1 unit/μl) with 5X reaction buffer—We obtain this from Gibco BRL (catalog #15224-025), but any other supplier of quality T4 ligase will do.

100 mg/ml ampicillin solution—Dissolve 1 g of ampicillin (sodium salt, Sigma catalog #A-9518) in a total of 10 ml of distilled water. Filter-sterilize the solution and store at −20°C.

50 mg/ml kanamycin sulfate solution—Dissolve 0.5 g of kanamycin sulfate (Sigma catalog #K-4000) in

10 ml of distilled water. Filter-sterilize the solution and store at −20°C.

40 mg/ml Isopropylthio-β,D-galactopyranoside (IPTG) solution—Dissolve 0.4 g of IPTG (Sigma catalog #I-6758) in 10 ml of distilled water. Filter-sterilize the solution and store at −20°C.

40 mg/ml 5-bromo-4-chloro-3-indolyl-β,D-galactopyranoside (Xgal) solution—Dissolve 0.4 g of Xgal (Sigma catalog #B-4252) in 10 ml of dimethylformamide (DMF). *DMF is very toxic and can be absorbed through the skin. Avoid contact with this reagent and do not breathe in its vapors. Work in the fume hood.* Filter-sterilize the solution and store at −20°C.

100 mM IPTG solution—Dissolve 0.715 g of IPTG in 30 ml of distilled water. Filter-sterilize the solution and store at 4°C.

Epicurian coli XL1-Blue subcloning grade competent cells—Stratagene catalog #200130.

Disposable plastic petri plates—500 required.

Large (16 × 125 mm) glass culture tubes with caps—250 required. Sterilize by autoclaving for 15 min prior to use.

YT broth—Dissolve 10 g of Bactotryptone (Fisher catalog #DFO 123), 5 g of yeast extract (Fisher catalog #DFO 127), and 5 g of NaCl in 800 ml of distilled water. Bring the final volume of the solution to 1 liter with distilled water. Autoclave for 25 min to sterilize. You will need 600 ml of this media for the growth of the students cultures on the "off day" of the experiment. To prepare the YT media agar plates, add 15 to 20 g/liter agar (Fisher catalog #DFO 140) to each 1 liter batch of YT media before autoclaving. You will need to make ~6.25 liters of this YT agar media for the IPTG/Xgal/amp plates. As each 1-liter batch cools following the autoclave step, add 1 ml of 40-mg/ml IPTG solution, 1 ml of 40-mg/ml Xgal solution, and 1 ml of 100-mg/ml ampicillin solution to the YT agar media. After the addition of these reagents and thorough mixing, pour about 25 ml of the YT/agar media per plate. You will also need to make ~2.5 liters of this YT agar media for the IPTG/Xgal/kan plates. As each 1-liter batch cools following the autoclave step, add 1 ml of 40-mg/ml IPTG solution, 1 ml of 40-mg/ml Xgal solution, and 1 ml of 50-mg/ml kanamycin solution to the YT agar media. After the addition of these reagents and thorough mixing, pour about 25 ml of the YT/agar media per plate. After the plates have cooled and hardened, store them at 4°C in plastic bags.

5X TBE agarose-gel electrophoresis buffer—Refer to Experiment 4 ("Electrophoresis") for this recipe. The stock solution will be diluted 1:10 to produce a 0.5X solution for use in this experiment. You will need 1 to 2 liters of the 5X TBE stock solution, depending on the size of your electrophoresis chambers.

GTE lysis buffer—Dissolve 1.8 g of D-glucose, 0.606 g of Tris free base, and 0.744 g of disodium EDTA · $2H_2O$ in 150 ml of distilled water. Adjust the pH to 8.0 with *dilute* HCl. Bring the final volume of the solution to 200 ml with distilled water. Autoclave the solution and store at 4°C. If you wish, you may add lysozyme to this solution to a final concentration of 10 μg/ml. We find that this is not necessary.

Solution of 0.2 N NaOH and 1% SDS—Prepare a 10 N NaOH stock solution by dissolving 200 g of NaOH in 300 ml of water. Bring the final volume of the solution to 500 ml with distilled water and store in a plastic (not glass) bottle at room temperature. Prepare a separate 10% (wt/vol) SDS solution by dissolving 10 g of sodium dodecyl sulfate in 70 ml of water. You may have to heat the solution slightly to allow all of the SDS to dissolve. Bring the final volume of the solution to 100 ml with distilled water and store at room temperature. To prepare the 0.2 N NaOH/1% SDS solution immediately before the experiment, mix 2 ml of the 10 N NaOH stock solution and 10 ml of the 10% (wt/vol) SDS stock solution with 88 ml of distilled water.

5 M potassium acetate solution—Dissolve 29.4 g of potassium acetate in distilled water to a final volume of 60 ml. Add 11.5 ml of glacial acetic acid and 28.5 ml of distilled water. Autoclave the solution and store at 4°C.

6X DNA sample buffer—Dissolve 0.25 g of xylene cyanole FF and 0.25 g of bromophenol blue in 70 ml of distilled water. Add and thoroughly mix in 30 ml of glycerol. Store the solution at room temperature.

1-kb DNA ladder size standard—We purchase this from Gibco BRL (catalog #15615-016). Add 50μl of 1-kb DNA ladder to 775 μl of distilled water. Add 175 μl of 6X DNA sample buffer. Mix thoroughly, but gently, and store at 4°C. Each group will load 20 μl of this solution.

Ethidium bromide (0.5 μg/ml) in 0.5X TBE buffer—Dissolve 0.5 mg of ethidium bromide in 1 liter of 0.5X TBE buffer. Place the entire 1-liter solution in a shallow tray for students to incubate their gels in. A more concentrated solution of ethidium bromide (1–2 μg/ml) can be prepared if you wish to decrease the required staining time. **Caution: ethidium bromide is a suspected carcinogen.**

Polaroid camera fitted to a 256-nm light box—1 required.

Film for camera—50 exposures required.

RNaseA—We purchase this enzyme from Sigma (catalog #R-5125). Dissolve 100 mg gently in 10 ml of distilled water. Boil this solution for 5 min to inactivate any DNase that may be present (RNaseA is heat stable while DNase is not). Allow the solution to cool to room temperature and store at 4°C.

Agarose (electrophoresis grade)—25 g required.

Agarose-gel electrophoresis apparatus with casting molds and 10 well combs—50 required.

Power supplies—Enough to power all required electrophoresis chambers.

Experiment 22: In Vitro Transcription from a Plasmid Carrying a T7 RNA Polymerase Specific Promoter

*Hin*dIII (10 units/μl) with 10X reaction buffer supplied—We obtain this from Gibco BRL (catalog #15207-020), but any other supplier of high-quality restriction endonucleases will do.

*Bam*HI (10 units/μl) with 10X reaction buffer supplied—We obtain this from Gibco BRL (catalog #15201-031), but any other supplier of high-quality restriction endonucleases will do.

*Eco*RI (10 units/μl) with 10X reaction buffer supplied—We obtain this from Gibco BRL (catalog #15202-021), but any other supplier of high-quality restriction endonucleases will do.

pSP72 plasmid (100 μg/ml)—Promega catalog #P-2191. This plasmid is supplied as a 1 μg/μl solution in TE buffer. Dilute this solution 10-fold with TE buffer. Each group will require 24 μl of the plasmid to perform the experiment. Store at −20°C.

37°C Water bath—1 required.

Dry ice/ethanol bath—See Experiment 21.

Phenol:chloroform (1:1)—Mix 250 ml of chloroform with 250 ml of Tris-saturated phenol (see Experiment 21). Add about 75 ml of water to the solution and allow the phases to separate. Leave a layer of water over the organic phase and store in a dark bottle tightly capped at room temperature.

Chloroform:isoamyl alcohol (24:1)—See Experiment 21.

3 M sodium acetate, pH 5.2—See Experiment 21.

0.4 M Tris, pH 8.0—Dissolve 3.55 g of Tris · HCl and 2.12 g of Tris free base in 100 ml of distilled water. The pH should be at or near 8.0. If necessary, adjust the pH to 8.0 with dilute HCl or NaOH. Store at room temperature.

100 mM nucleotide solutions (UTP, ATP, CTP, GTP)—We buy these as a set from Boehringer Mannheim (catalog #1277057). Each nucleotide is supplied separately in the set at a concentration of 100 mM.

0.15 M $MgCl_2$—Dissolve 3.05 g of $MgCl_2 \cdot 6H_2O$ in 100 ml of distilled water. Filter-sterilize or autoclave the solution and store at room temperature.

20 mg/ml Bovine serum albumin—Dissolve 2 g of bovine serum albumin (RNase-free) in 100 ml of distilled water. Filter-sterilize the solution and store at −20°C.

TE buffer—See Experiment 21.

0.1 M dithiothreitol solution—Dissolve 1.54 g of DL-dithiothreitol in 100 ml of distilled water. Filter-sterilize (do not autoclave) and store at −20°C in 1.0-ml aliquots.

α-[^{32}P]ATP—Obtain 250 μCi of a solution at a specific activity of 3,000 Ci/mmol. We use ICN catalog #32007x.

T7 RNA polymerase—We obtain this in 2500-unit (50 units/μl) aliquots from Gibco BRL (catalog #18033-100). The enzyme is diluted immediately before use to 20 units/μl in sterile distilled water.

10X Transcription mix—Mix the following reagents together to prepare 100 μl of this solution. *Prepare this reagent immediately before use:*

40 μl of 0.4 M Tris-HCl, pH 8.0
15 μl of 0.15 M $MgCl_2$
5 μl of 20 mg/ml bovine serum albumin
0.5 μl of 100 mM ATP
10 μl of 100 mM CTP
10 μl of 100 mM GTP
10 μl of 100 mM UTP
1 μl of 0.1 M dithiothreitol (DTT)
6 μl of sterile distilled water
2.5 μl of a 1:100 dilution of the α-[^{32}P]ATP stock in sterile distilled water

Loading buffer—Dissolve 42 g of urea, 0.25 g of xylene cyanole FF, and 0.25 g of bromophenol blue in distilled water to a final volume of 100 ml. Store at room temperature.

Absolute ethanol—Approximately 100 ml required.

15% Polyacrylamide gel—Refer to Experiment 4.

0.5X TBE buffer—See Experiment 21.

Film cassettes—5 to 10 required, depending on the size of the gels.

Plastic wrap—1 roll required.

Razor blades—5 to 10 required (students can share these).

Kodak X-omat film—5 to 10 pieces required, depending on the size of the gels and film cassettes.

Darkroom and solutions for film development.

Electrophoresis chambers and power supplies—25 required.

RNA size markers—Add 20 μl of the pSP72 plasmid to three microcentrifuge tubes. One of these plasmid samples will be digested with *Xho*I, the other with *Xba*I, and the third with *Eco*RV. Add 6 μl of the appropriate 10X restriction-enzyme buffers to each of the three tubes. Add 29 μl of sterile distilled water to each tube, as well as 5 μl of the appropriate restriction enzyme (10 units/μl). Incubate these tubes at 37°C for 2 hr. Ethanol-precipitate the digested plasmid in each of these tubes by adding 8 μl of 3 M sodium acetate, 200 μl of absolute ethanol, and incubating in a dry ice/ethanol bath for 10 min. Centrifuge at 4°C for 30 min to harvest the DNA and discard the supernatant. After a 70% ethanol wash, dissolve the DNA pellets in 6 μl of sterile distilled water. Add the following to each tube:

4 μl of 5X T3/T7 RNA polymerase buffer (supplied with enzyme)

2 μl of 0.1 M dithiothreitol

4 μl of a 2.5 mM solution of GTP, CTP, and UTP (see below)

1 μl of a 1 mM ATP solution (see below)

1 μl of RNasin (20–40 units/μl) (Promega catalog #N2511)

2 μl of the undiluted 10mCi/ml solution of α-[^{32}P]ATP

2 μl of T7 RNA polymerase (20 units/μl)

The 2.5 mM CTP, UTP, and GTP solution is prepared by adding 1 μl of *each* 100 mM nucleotide stock solution to 47 μl of distilled water. The 1 mM ATP solution is prepared by adding 1 μl of the 100 mM ATP stock to 99 μl of distilled water. Incubate these three transcription tubes at 37°C for 1 hr. Store at −20°C. To prepare the markers, the reactions will be mixed together in a defined ratio. You will have to experiment with an exact dilution and ratio of these transcripts to produce RNA size markers that will be visible as defined bands on the film after an 18-hr exposure. The *Xho*I transcript will be 97 nucleotides in length, the *Xba*I transcript will be 61 nucleotides in length, and the *Eco*RV transcript will be 20 nucleotides in length. Remember that the *Eco*RV- and *Xba*I-digested transcripts will have less radioactivity per mole than the *Xho*I-digested transcript.

Experiment 23: In Vitro Translation: mRNA, tRNA, and Ribosomes

S-30 Fraction from wheat germ—This can be obtained commercially (Fisher catalog #L-4440) in a form that is ready to use in the in vitro reaction. If you wish to prepare it yourself, perform the following protocol. Place 6.75 g of wheat germ and 6.75 g of acid-washed sand in a cold mortar. Vigorously grind the dry wheat germ and sand with a cold pestle for 1 to 2 min. Add 5 ml of cold grinding buffer (20 mM HEPES, 100 mM KCl, 1 mM magnesium acetate, 2 mM $CaCl_2$, 6 mM dithiothreitol, pH 7.6) and continue grinding until you obtain an even mush (~1 min more). Continue grinding and adding 5-ml aliquots of fresh grinding buffer until a total of 25 ml have been added. Centrifuge the homogenate (30,000 × *g*) for 10 min at 4°C and collect the S-30 supernatant, being careful to avoid the yellow fatty surface layer above the cell debris and mitochondria. Pass 15 ml of the supernatant over a 50 × 2 cm Sepadex G-25 column pre-equilibrated with at least 600 ml of buffer (20 mM HEPES, 120 mM KCl, 5 mM magnesium acetate, 6 mM dithiothreitol, pH 7.6). Collect the first 10 fractions (1.5 ml each) as they elute from the column. Using a Pasteur pipette, *slowly* drip the fractions into a pool of liquid nitrogen. Allow each drop to freeze and sink to the bottom of the vessel. When all of the material has been applied to the liquid nitrogen, frozen, and sunk to the bottom of the vessel, pour off the N_2 and place the frozen S-30 fraction in small, airtight vials to be frozen at −80°C.

1.5-ml Microcentrifuge tubes—2,000 required.

Fine-tipped forceps—50 required.

Millipore filters (HAWP, 0.45-μm pore size)—2000 required.

Filtration apparatus with vacuum pump (or house vacuum)—50 required.

Water baths at 30°C and 90°C—One of each required.

100°C Oven—1 required.

Scintillation vials—2000 required.

Scintillation counter—1 required.

Plastic squirt bottles—50 required.

Reaction buffer—Dissolve 7.15 g of *N*-[2-hydroxyethyl]piperazine-*N*′-[2-ethanesulfonic acid] (Sigma catalog #H-7523) and 0.46 g of dithiothreitol (Sigma catalog #D-5545) in 80 ml of distilled water. Adjust the pH of the solution to 7.1 with dilute NaOH. Bring the final volume of the solution to 100 ml with distilled water and store at 4°C.

Salt solution I—Dissolve 9.7 g of KCl and 4.83 g of magnesium acetate tetrahydrate in 80 ml of distilled water. Bring the final volume of the solution to 100 ml with distilled water and store at 4°C.

Salt solution II—Dissolve 9.7 g of KCl and 0.483 g of magnesium acetate tetrahydrate in 80 ml of distilled water. Bring the final volume of the solution to 100 ml with distilled water and store at 4°C.

^{3}H-Phenylalanine solution—Dissolve 1.65 mg of L-phenylalanine (Sigma catalog #P-8324) in 100 ml of distilled water. Add an amount of high specific activity (greater than 1000 Ci/mmol) [^{3}H]-phenylalanine that contains a total of 500 μCi. Store at 4°C.

Poly(U) solution—Dissolve 10 mg of Sigma catalog #P-9528 in 10 ml of distilled water. Store at −20°C.

Poly(A) solution—Dissolve 25 mg of Sigma catalog #P-9403 in 25 ml of distilled water. Store at −20°C.

Trichloroacetic acid solution—Dissolve 250 g of trichloroacetic acid (Sigma catalog #T-9159) in 5 liters of distilled water. Store at 4°C.

Scintillation fluid—Dissolve 50 g of diphenyloxazole and 100 ml of Triton X-100 in 9 liters of toluene. Store tightly covered in a dark bottle at room temperature.

RNase A solution—Dissolve 5 mg of Sigma catalog #R-5125 in 10 ml of distilled water. Store at 4°C.

RNase T_1 solution—We use product #R-1003 from Sigma, which is supplied as a crystalline suspension in 3.2 M ammonium sulfate (300,000–600,000 units/mg of protein). Transfer a volume equivalent to 1 mg of protein (consult the specification sheet) to a microcentrifuge tube and centrifuge for 5 min to collect the crystals. Draw off the supernatant and resuspend the solid crystals in 2 ml of distilled water.

Chloramphenicol solution—Dissolve 16.2 mg of Sigma catalog #C-0378 in 100 ml of distilled water. Store at −20°C in small aliquots.

Cycloheximide solution—Dissolve 14.1 mg of Sigma catalog #C-7698 in 100 ml of distilled water. Store at −20°C in small aliquots.

Puromycin solution—Dissolve 5.4 mg of Sigma catalog #P-7255 in 100 ml of distilled water. Store at −20°C in small aliquots.

Experiment 24: Amplification of a DNA Fragment Using Polymerase Chain Reaction

TE buffer—See Experiment 21.

0.5-ml Thin-walled PCR tubes—250 required.

1.5-ml Microcentrifuge tubes—500 required.

Primers—We order these from Integrated DNA Technologies, Inc., 1710 Commercial Park, Coralville, IA 52441. This company can be reached at 1-800-328-2661, or on the World Wide Web (http://www.idtdna.com). The sequence of the primers needed for this experiment is shown in the Supplies and Reagents list for this exercise. We usually order 100 nmol of each primer. Consult the specification sheet to determine exactly how much of each primer is present when you receive it. Add 0.5 to 1.5 ml of sterile distilled water to each tube to bring the final concentration of each primer to 100 pmol/μl. Each group will need 42.5 μl of each primer at a concentration of 10 pmol/μl (10 μM). Dilute each of the primer stock solutions 10-fold in sterile distilled water to achieve this concentration. Store the remainder of the primer stocks at −20°C for use in future semesters (100 nmol of primers is enough for approximately two semesters).

Template DNA—The pBluescript II (S/K) and (K/S) plasmids can be purchased from Stratagene (catalog #212205 and 212207, respectively). The 20-μg package size for each plasmid is enough to perform the experiment for approximately 40 semesters. Using *serial dilutions*, prepare 1 ng/μl, 1 pg/μl, and 1 fg/μl stock solutions.

dNTPs—We purchase 100 mM stock solutions of dATP, dTTP, dGTP, and dCTP from Gibco-BRL (catalog #10297-018). Add 62.5 μl of each 100 mM stock dNTP solution to 4.750 ml of sterile distilled water to prepare the 1.25 mM dNTP stock solution. Each group will need 68 μl of the dNTP solution to perform the experiment. Store the remainder of the 100 mM dNTP stock solutions (in small aliquots) at −20°C for use in future semesters.

Taq polymerase—We purchase this from Gibco-BRL (catalog #18038-042). The enzyme is supplied with 10X Taq polymerase dilution buffer and a vial of 50 mM $MgCl_2$. Each group will require 68 μl of 12.5 mM $MgCl_2$. To prepare, add 1 ml of 50 mM $MgCl_2$ stock to 3 ml of sterile distilled water. To adjust the 5-units/μl Taq polymerase stock to 1 unit/μl for use in the experiment, add 100 μl of 5-units/μl Taq polymerase stock solution to 400 μl of 1X Taq polymerase buffer. This 1X Taq dilution buffer is prepared by adding 100 μl of the 10X Taq polymerase buffer stock to 900 μl of distilled water. *The dilution of the Taq polymerase to 1 unit/μl should be performed immediately before the start of the experiment to preserve the activity of the enzyme.* Each group will require 8.5 μl of the 1-unit/μl Taq polymerase solution to perform the experiment.

1-kb DNA ladder—We purchase these from Gibco-BRL (catalog #15615-016). To prepare the 0.2 μg/μl solution, add 50 μl of the 1 μg/μl stock solution to 200 μl of TE buffer.

5X TBE agarose-gel electrophoresis buffer—See Experiment 21.

*Kpn*I (10 units/μl) with 10X reaction buffer supplied—We obtain this from Gibco-BRL (catalog #15232-036), but any supplier of high-quality restriction endonucleases will do.

*Sst*I (10 units/μl) with 10X reaction buffer supplied—We obtain this from Gibco-BRL (catalog #15222-037), but any supplier of high-quality restriction endonucleases will do.

Agarose-gel electrophoresis chambers with casting molds and 10 well combs—25 required.

Power supplies—Enough to power 25 agarose-gel electrophoresis chambers.

6X DNA sample buffer—See Experiment 21.

Agarose (electrophoresis grade)—Approximately 12.5 g required.

Ethidium bromide solution (0.5 μg/ml) in 0.5X TBE buffer—See Experiment 21.

Polaroid camera fitted to a 256-nm light box—1 required.

Film for camera—At least 25 exposures will be required.

Index